AF361753

Power System Dynamics and Renewable Energy Integration—Volume II

Editor

Juri Belikov

Basel • Beijing • Wuhan • Barcelona • Belgrade • Novi Sad • Cluj • Manchester

Editor
Juri Belikov
Tallinn University of Technology
Estonia

Editorial Office
MDPI
St. Alban-Anlage 66
4052 Basel, Switzerland

This is a reprint of articles from the Special Issue published online in the open access journal *Energies* (ISSN 1996-1073) (available at: https://www.mdpi.com/journal/energies/special_issues/Power_System_Dynamics_Renewable_Energy_Integration).

For citation purposes, cite each article independently as indicated on the article page online and as indicated below:

Lastname, A.A.; Lastname, B.B. Article Title. *Journal Name* **Year**, *Volume Number*, Page Range.

Volume II
ISBN 978-3-0365-8206-1 (Hbk)
ISBN 978-3-0365-8207-8 (PDF)
doi.org/10.3390/books978-3-0365-8207-8

Set
ISBN 978-3-0365-6966-6 (Hbk)
ISBN 978-3-0365-6967-3 (PDF)

Contents

About the Editor

Juri Belikov

Juri Belikov, Ph.D., is an Assistant Professor in the Department of Software Science, Tallinn University of Technology (since 2018). His research interests lie on the edge of nonlinear control theory, power systems, and computer science. He received a BSc (cum laude) in Mathematics from Tallinn University, and MSc and PhD degrees in Computer and Systems Engineering from Tallinn University of Technology in 2006, 2008, and 2012, respectively. In 2015–2017, he was a Postdoctoral Fellow in the Faculty of Mechanical Engineering and The Andrew & Erna Viterbi Faculty of Electrical Engineering, Technion—Israel Institute of Technology.

Preface

As environmental concerns continue to grow and the push for carbon neutrality intensifies, the question is no longer whether renewable energy is a viable solution, but rather how to successfully transition towards it. This raises an intriguing discussion regarding the inherent limitations of traditional power systems. Specifically, there is a clear gap between today's emerging generation technologies and our current understanding of the dynamics of complex power systems—a gap that is partly covered by the research results in this book.

Juri Belikov
Editor

Article

Green Solution for Insulation System of a Medium Frequency High Voltage Transformer for an Offshore Wind Farm

Mohammad Kharezy [1,2,*], Hassan Reza Mirzaei [3], Torbjörn Thiringer [2] and Yuriy V. Serdyuk [2]

1 RISE Research Institutes of Sweden, 50462 Borås, Sweden
2 Department of Electrical Engineering, Chalmers University of Technology, 41296 Göteborg, Sweden; torbjorn.thiringer@chalmers.se (T.T.); yuriy.serdyuk@chalmers.se (Y.V.S.)
3 Department of Electrical Engineering, University of Zanjan, Zanjan 45371-38791, Iran; hr.mirzaei@znu.ac.ir
* Correspondence: mohammad.kharezy@ri.se; Tel.: +46-010-528-5228

Abstract: High Voltage Direct Current (HVDC) transmission represents the most efficient way for transporting produced electrical energy from remotely located offshore wind farms to the shore. Such systems are implemented today using very expensive and large power transformers and converter stations placed on dedicated platforms. The present study aims at elaborating a compact solution for an energy collections system. The solution allows for a minimum of total transformer weight in the wind turbine nacelle reducing or even eliminating the need for a sea-based platform(s). The heart of the project is a Medium Frequency Transformer (MFT) that has a high DC voltage insulation towards ground. The transformer is employed in a DC/DC converter that delivers the energy into a serial array without additional conversion units. The insulation design methodology of an environmentally friendly HV insulation system for an MFT, based on pressboard and biodegradable oil, is introduced. The measurement method and results of the measurements of electrical conductivities of the transformer oil and Oil Impregnated Pressboard (OIP) are reported. The measurements show that the biodegradable ester oil/OIP conductivities are generally higher than the mineral oil/OIP conductivities. Numerical simulations reveal that the performance of the insulation system is slightly better when ester oil is used. Additionally, a lower temperature dependency for ester oil/OIP conductivities is observed, with the result that the transformer filled with ester oil is less sensitive to temperature variations.

Keywords: DC-DC power converters; design methodology; HVDC transmission; insulation design; power transformer; wind energy

Citation: Kharezy, M.; Mirzaei, H.R.; Thiringer, T.; Serdyuk, Y.V. Green Solution for Insulation System of a Medium Frequency High Voltage Transformer for an Offshore Wind Farm. *Energies* **2022**, *15*, 1998. https://doi.org/10.3390/en15061998

Academic Editor: Juri Belikov

Received: 6 February 2022
Accepted: 7 March 2022
Published: 9 March 2022

Publisher's Note: MDPI stays neutral with regard to jurisdictional claims in published maps and institutional affiliations.

1. Introduction

The potential to harness energy from wind is enormous and offshore wind power generation is one of the most rapidly developing methods to utilize this potential. This type of electricity generation contributes to the reduction of CO_2 emissions from electricity generation worldwide [1]. However, it remains of great importance to reduce the cost of these energy sources throughout their life cycle. Thus, the energy generated by the offshore windfarms is transferred to the shore by means of submarine high voltage cables. Since AC transmission becomes inefficient at distances longer than approximately 60 km [2], it is necessary to switch to High Voltage Direct Current (HVDC) transmission for sea-based wind farms. Today's HVDC concept of offshore wind farms require a transformer and a converter station platform, which serves as the hub for the collection network and as a connection point of the HVDC cable to land. This solution is economically inefficient, and therefore, a reduction in costs of platforms is highly desirable.

The present work contributes to a green solution that partly reduces the need for offshore platforms as well as costs and weight in the wind turbines themselves. The suggested concept foresees replacing heavy 50 Hz transformers with lighter and smaller

Medium Frequency Transformers (MFT) in container-mounted DC/DC converters that delivers the electrical energy to series coupled converters to achieve a high DC voltage sufficient for direct transfer of the generated energy to land [3] (see Figure 1). The design of the MFT, which is the key component of such a system, presents a very challenging task as its AC insulation level is much lower compared with a conventional converter transformer and its volume can be considerably decreased by utilizing higher operation frequency.

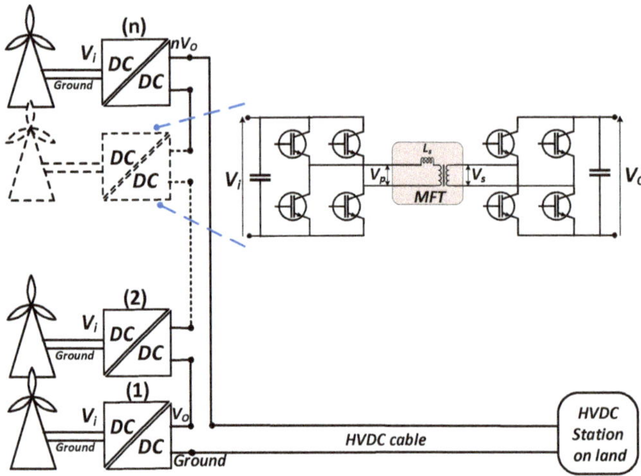

Figure 1. The series DC concept where the DC outputs of the wind turbines are directly delivering the high voltage to be transferred to land.

Another advantage of the series concept is the ability to deliver as much energy as possible in a low wind situation. This can be achieved by minimizing the amount of magnetic and Ohmic losses in the power supply circuit using a DC/DC converter and an MFT.

A significant potential for the series DC concept was identified in [4], through considerable cost savings from a Life Cycle Cost (LCC) perspective, compared to today's technology with internal 'parallel-coupled' AC radials. The main cost reduction was due to the replacement of AC transformers with significantly smaller DC/DC converters, which eliminates the need for a dedicated offshore platform [4]. The results of that work were expected to be used by offshore windfarm integration-network designers and manufacturers of equipment for wind farms, as well as manufacturers of MFTs for insulated DC/DC converters. Furthermore, during the previous phases of the project, design and construction processes of prototype transformers that can withstand a high DC offset voltages were reported [5,6] and the challenges in designing such transformers were reviewed [7]. In the present paper, the next step of the project is reported, which focuses on implementation of an environmentally friendly solution of the insulation system of MFT, based on biodegradable transformer oil and Oil Impregnated Pressboard (OIP).

Design principles of the insulation of MFTs operating at very high DC voltages to ground have not been widely presented in the literature [8]. Most of the publications deal with insulation systems of conventional HVDC converter station transformers, which contain large amounts of mineral oil with relatively long insolation distances inside the transformer tank. Due to much higher operation frequency, an MFT is much more compact compared with a 50 Hz transformer. When a biodegradable oil is used, as this article reveals, the performance of the transformer insulation is different, especially during the transient operations of energizing or loading.

Among the transformer oils, mineral oil bears the most well-known characteristics for AC and HVDC insulated large converter transformers [9]. Still, there is an increasing trend and interest to use biodegradable oils specially the ester-based oils for transformer insulation systems [10]. To secure proper functioning of the insulation system, its design should be based on well-established properties of the constituting materials. In [11] the various aspects of using the synthetic ester oil in the high voltage equipment including its breakdown performance under different electrical stress conditions are reviewed. The design of an HVDC insulation system strongly depends on the electrical conductivity values of oil and OIP under very high DC voltages. However, the electrical conductivity values given in the literature are typically obtained at low test voltages and therefore, are hardly applicable for designing an HVDC insulation [12]. Information regarding conductivity of biodegradable synthetic ester oil under HVDC stress is rather limited [13,14] and even less data is available for ester OIP [15].

The purpose of the present study is to contribute to the development of an MFT bearing a high DC insulation strength and at the same time a much more compact design compared with conventional power transformers. The new design uses biodegradable insulation materials instead of mineral oil, which is undesirable from an environmental point of view, especially for offshore applications. The electrical conductivities of mineral oil and OIP insulation materials were measured previously by the authors and reported in [16]. In this work, the properties of the biodegradable dielectric oil and OIP are measured under an extensive effort and presented, and the results are used for the insulation design of an MFT introduced in the paper. Additionally, a 10 MVA HVDC MFT with an insulation system using biodegradable oil and OIP is designed based on the method developed previously and verified by experimental testing [5]. Consequently, the performance of the ester-based insulation system is analyzed and compared to the case when mineral oil is used. Furthermore, in addition to previously considered safety factors of the insulating oil gaps and creepage paths, a novel criterion based on the combined path is suggested.

First in Section 2, an overview of the design, manufacturing, and verification tests of a mineral-oil-based high voltage DC medium frequency transformer prototype is presented. In Section 2.1 the implemented design process is reviewed, and the performed verification tests are reported. Later in Section 2.2, a real scale 10 MW 200 KV DC transformer suitable for an offshore wind farm based on the DC series system is introduced. The insulation system behavior under DC stress is explained later under Section 2.3. Section 3 describes the details of the method used for the measurement of temperature and stress dependent conductivities of a biodegradable oil/OIP. The results are compared with the previously performed measurement results of a sample mineral oil/OIP. In Section 4, FEM simulations are applied to present a reliable design of a high-power medium frequency transformer filled with an environment-friendly oil which is to be subjected to a very high offset DC voltage. Finally, a discussion on the results is presented in Section 5.

2. Overview of the Design of the Mineral-Oil-Based HVDC Medium Frequency Transformer

2.1. Design Method Development and Verification

After successfully manufacturing two downscaled prototypes [6], a set of adopted methods for designing an oil insulated transformer having a very high DC isolation to ground was developed. The methods were verified by manufacturing a 50 kW, 0.42/4.5 kV AC 5 kHz and a 125 kV DC prototype transformer [5] (see Figure 2).

(a)

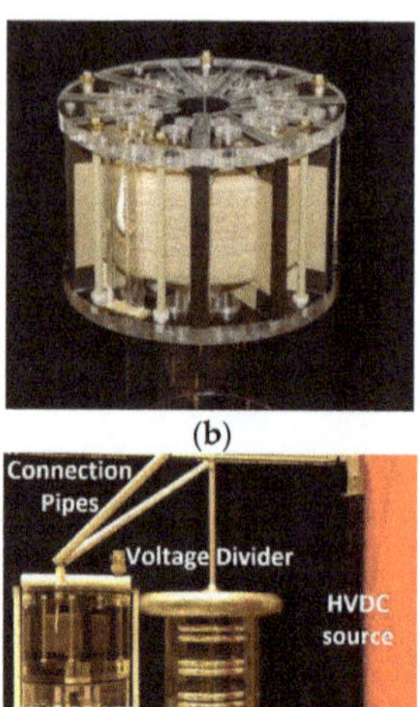

(b)

(c)

Figure 2. Manufactured prototype: (**a**) Complete transformer; (**b**) Active part; (**c**) DC insulation test setup.

As seen from Figure 2, the core consists of ten rectangularly shaped ferrite core sets and the windings made of stranded Litz conductors are placed inside the core window. The structure is fixed between PMMA plates using plastic tighteners. The transformer is immersed in oil and main insulation is provided by paper and pressboard elements. The background for such a design and the procedure of manufacturing the insulation system is described in [5].

The equation relating the number of turns and the core cross section for a given flux is:

$$\varphi_{\mathrm{m}} = \frac{V_{\mathrm{P}} D^{\mathrm{T}}/2}{N} = \frac{V_{\mathrm{P}}}{4\,N\,f} \tag{1}$$

where T is the time period of the wave and here it is assumed that the duty cycle of the applied rectangular wave D is 0.5. Therefore, the needed core area is:

$$A_C = \frac{\varphi_{\mathrm{m}}}{B_{\mathrm{m}}} = \frac{V_{\mathrm{P}}}{4\,N f B_{\mathrm{m}}} \tag{2}$$

where V_{P} is the rms of the DC voltage on the primary winding, A_{c} is the effective cross section of the transformer core, φ_{m} is the maximum flux in the core, B_{m} is the maximum flux density of the core, N is the number of turns for the primary winding and f is the switching frequency [17].

For specific voltage conditions and to achieve a certain power, the optimal switching frequency of a Dual Active Bridge (DAB) converter for high power applications is practically limited to the range of 6–20 kHz [18]. However, the rise time of the voltage pulses are very short which means harmonic components of much higher frequencies. This dictates special criteria for the selection of core material as well as the winding wire types and construction [5]. In the case of a core, the main material requirements are low loss, high saturation flux density, and high continuous operating temperature [19]. In case of a suitable winding conductor to minimize the extra losses caused by high frequency current harmonics, stranded Litz conductors are used to minimize the proximity and skin effects.

In this article, the focus centers on the DC insulation design of the transformer, which is mostly based on temperature dependent high voltage conductivity values of the insulation materials. The losses in transformer components like the core and windings, increase the temperature of the insulation materials. Therefore, it is crucial to characterize the main loss components of the transformer which are in the core and windings.

The core loss is highly dependent on the duty cycle and the rise time of the excitation wave shape. Increasing the rise time, results in higher core loss and increasing the duty cycle reduces the losses. A set of adopted versions of that equation used by the designers is introduced in [19].

At high frequencies, the winding losses increase considerably because of the skin effect in the conductors, and the proximity effect of the adjacent conductors. The winding losses can be calculated by a numerical method proposed by the authors, using (Finite Element Method) FEM simulations among the other methods [20].

To achieve a soft switching possibility in the DAB and to consequently increase the efficiency of the DC/DC converter, it is of great importance to set a maximum limit to the leakage inductance of the MFT. That is why the precise calculation of the inductance is crucial for the optimized operation of a DAB converter. In [21] a relation for calculation of the leakage inductance required for a DAB is presented. For an MFT, the leakage inductance can be calculated using the total magnetic energy stored inside the core window [22]. By increasing the isolation distance between the windings, the amount of stored energy in this region will also increase [22,23]. Consequently, the leakage inductance will increase and may pass the upper limit set by the soft switching requirement. Simultaneously, a high insulation distance between the windings is required to withstand the high applied DC voltage between the windings. Therefore, special care must be paid to these two contradicting requirements for the gap between the windings. The authors have presented an effective analytical method, for calculating the leakage inductance of shell-type transformers with circular windings in [24].

2.2. A Real Scale 10 MW HVDC MFT

In [4] a life cycle cost and the energy efficiency of three different offshore windfarms of AC/AC, AC/DC and DC/DC were determined and compared using an example of a real 1 GW wind park containing 100 wind turbines of 10 MW capacity and 200 KV DC output voltage. For the DC series system, a series connection of several 1.8/18 kV DC/DC converters is proposed employing 5 kHz transformers inside, each bearing 200 kV DC to ground insulation strength (see Figure 1). For such an operating voltage, the DC isolation should be designed for the level of at least 250 kV (HV winding to ground) to ensure a safety margin.

In the developed MFT prototype, the number of turns, N_1 and N_2 are 12 and 120, respectively. A ferrite magnetic core having 2240 mm^2 as its physical core area including 10 core stacks is selected. A Litz wire bearing 9000 strands, each with 0.2 mm in diameter and a physical dimension of 26.6 × 17.5 mm^2 is used. The primary winding is the layer type with 4 layers, each consisting of 3 turns. Each turn is made from 8 parallel Litz conductors. A 20 mm oil gap between the layers are considered for cooling purposes. The secondary winding is a disc type winding, each disc consisting of 6 turns. The HV winding contains twenty discs with a 5.5 mm oil gap between two adjacent discs.

Compared with the previously manufactured prototype MFT [5], the isolation distances are increased, and a higher number of barriers and oil gaps and higher oil gap width, and a greater OIP barriers thickness are used. The insulation design aspects are thoroughly discussed in Section 4. The dimensions of the described real scale 10 MVA HVDC MFT are presented in Figure 3.

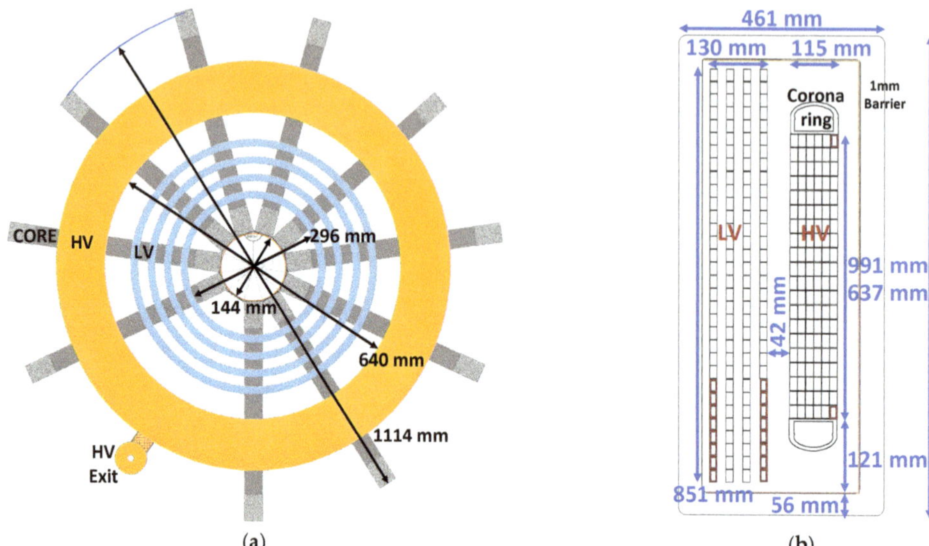

Figure 3. Dimensions of the real scale Ferrite 10 MW, 1.8/18 kV, 0.32 T, 5555/555 A, 12/120 Turns, 5 kHz MFT: (**a**) Top view; (**b**) Side view.

2.3. Insulation System Behaviour under DC Stress

The insulation design of AC transformers has been performed for several decades and well-known methods are developed for this purpose [25]. The AC stress distribution in these transformers is based on dielectrics' permittivities, and as the model is purely capacitive, it is time invariant. Similarly, when applying a step DC voltage to the transformer windings, the initial stress distribution is determined by the insulations' permittivities. However, when the transient is attenuated, conduction currents are developed inside insulations which are determined by their electric conductivities as well as the electric field intensity. Since these currents are not the same in the series insulations, time varying space charges will be accumulated at their interfaces. The resulting electric field stress distribution stabilizes when the space charge is established and come into a steady state condition. At this time, the stress distribution is determined by materials' conductivities. This procedure is explained in the Linear Maxwell–Wagner (LMW) model. The differential equation of this model is explained in [5]. The LMW model is widely used for the DC dielectric design for oil–paper converter transformers [26,27]. According to this model, directly after applying a DC voltage, the ratio of the initial electric fields in the materials is proportional to the inverse of their dielectric constants' ratios, which are 2.2/4.4 = 0.50 for mineral oil/OIP, and 3.5/4.6 = 0.76 for ester oil/OIP. Therefore, the electric field intensity in the oil is higher than the paper. But at the final steady state, in which the conductivity of materials determines the stress; a material with greater conductivity will take less field intensity, and vice versa [8]. For example, since the ratio of the mineral oil/OIP conductivities may vary from 2 to 1000 [16,28], the paper will be more stressed than the oil at steady state. Therefore, the conductivity plays a crucial role in an insulation system which is stressed by a DC voltage.

For each dielectric, a time constant is defined as the ratio of its electrical conductivity to its permittivity. In a system formed by layered dielectric materials with different time constants, time dependent electric field transients appear. While the electric potential distribution in dielectric materials depends on their capacitances at the instant of the HVDC application, it is determined by their resistances at the final steady state. The stabilization time may last up to 10,000 s or longer in the case of using mineral oil [27,29,30].

It is known that the electrical conductivity of dielectrics depends on temperature and electric field intensity [16,30–32]. Therefore, the LMW model is not fully suitable for determining the electrical stress distribution in the transformer's dielectric structure under very high DC voltage application. In [5], the authors investigated and demonstrated the non-linear dependence of the mineral oil/OIP conductivities to the variation of the applied voltage as well as the temperature. Additionally, a Non-Linear Maxwell–Wagner (NLMW) model is developed by implementing FEM to consider the nonlinear behavior of the insulation materials with respect to the applied electric field.

In the next section, the temperature and stress dependent conductivities of ester oil/OIP are measured and compared with the conductivities of mineral oil/OIP. Consequently, in Section 4, the measured values are implemented using the NLMW method for dielectric design of the real scale MFT, comparing ester and mineral oils/OIPs application.

3. Conductivity Measurement

The electrical conductivity values given in the literature are typically obtained at low test voltages and therefore are hardly applicable for designing HVDC insulation. Moreover, insulation materials behave nonlinearly under high DC voltage stresses. In addition, additives such as inhibitors, pour-point depressants, antioxidants or antistatic agents that are added to the transformer oils, introduce large variations of the real conductivity values [33]. Even specific features of production facilities or the season of production and packaging can make an effect. The conductivity of OIP is also dependent on the cellulose fiber density and properties of the oil used for the impregnation [34]. Studies show that the standard production and treatment methods do not guarantee reproducible and comparable measurements results under an HVDC test or service conditions [16]. Furthermore, parameters such as temperature, field strength, pre-stressing and duration of stress bear a direct effect on the conductivity values under high DC stress [35]. Hence, for a successful insulation design, actual conductivity values are needed and should be measured within the actual ranges of electric field strength and temperature variations that are expected to appear under real working conditions.

The method of conductivity measurements including the procedures for preparation of the test samples, the design and setup of the measuring equipment, and methods of treating the measurement data are presented by the authors in their previously published works [5,16]. To achieve reliable results, the following activities should be carefully planned and implemented: the test cell design and temperature chamber preparation, the preparation of oil and solid insulation samples as well as the high voltage supply system and its safety circuit design, the current measuring system and its protection circuit design and the mitigation of noises. In addition, the very low current values recorded under very high voltage stress and under noisy environmental conditions should be treated using a proper method [16,36].

As the MFT introduced here is intended for installation in the sea environment, the use of a biodegradable insulation oil is considered as the substitute to the traditional mineral oil and its electrical properties are investigated in the present work. Previously, a complete set of conductivity measurements was performed by the authors for mineral oil and OIP and the results were comparable with the presented values in literature [16]. Here, the same method is applied to measure the conductivities of ester oil/OIP under a similar range of high voltage stresses and temperatures. MIDEL 7131 synthetic ester oil is used.

As the preliminary tests showed unstable results for the polarization current of ester oil, it was thus decided to perform conditioning for the ester oil and to increase polarization time from 1 h (which was used for mineral oil) to 3 h [15]. Ester OIP is tested similarly to the mineral OIP for 3 h. Figure 4 demonstrates the conditioning process and the test duration for ester oil (a) and ester OIP (b), respectively.

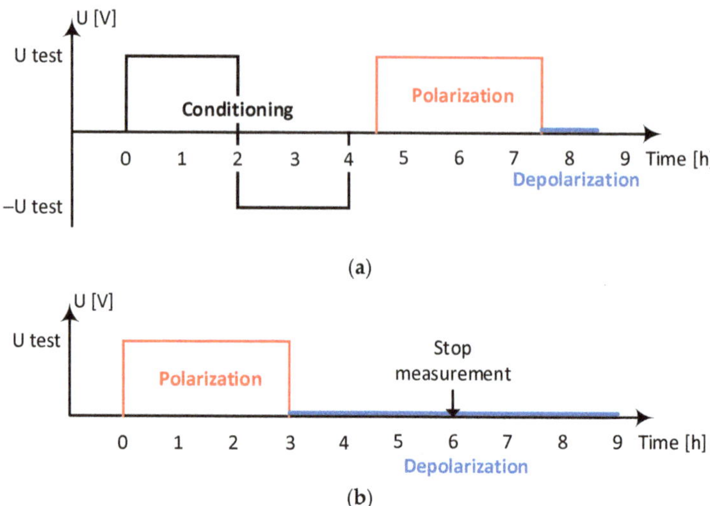

Figure 4. Conditioning and measurement cycle for: (**a**) ester oil; (**b**) ester OIP.

Tables 1 and 2 summarize the conductivity measurements results for ester and mineral oils/OIPs whereas Figure 5 presents them graphically.

Table 1. Conductivity measurement results for mineral oil/OIP and their conductivities ratio (CR).

		1 kV/mm	3 kV/mm	6 kV/mm	12 kV/mm
Mineral Oil	30 °C	5.0×10^{-14}	3.9×10^{-14}	5.8×10^{-14}	-
	50 °C	1.2×10^{-13}	9.5×10^{-14}	1.1×10^{-13}	-
	90 °C	5.2×10^{-13}	4.6×10^{-13}	3.9×10^{-13}	-
Mineral OIP	30 °C	1.7×10^{-16}	2.2×10^{-16}	2.7×10^{-16}	4.0×10^{-16}
	50 °C	2.6×10^{-15}	3.3×10^{-15}	4.0×10^{-15}	5.8×10^{-15}
	90 °C	2.6×10^{-13}	3.0×10^{-13}	3.8×10^{-13}	5.1×10^{-13}
CR	30 °C	294	177	215	-
	50 °C	46	29	27	-
	90 °C	2	1.5	1	-

Table 2. Conductivity measurement results for ester oil/OIP and their conductivities ratio (CR).

		1 kV/mm	3 kV/mm	6 kV/mm	12 kV/mm
Ester Oil	30 °C	1.4×10^{-11}	1.0×10^{-11}	0.8×10^{-11}	-
	50 °C	5.0×10^{-11}	3.6×10^{-11}	3.2×10^{-11}	-
	90 °C	2.9×10^{-10}	2.3×10^{-10}	2.3×10^{-10}	-
Ester OIP	30 °C	6.1×10^{-13}	6.9×10^{-13}	4.9×10^{-13}	3.4×10^{-13}
	50 °C	2.0×10^{-12}	2.2×10^{-12}	1.6×10^{-12}	1.2×10^{-12}
	90 °C	1.4×10^{-11}	1.5×10^{-11}	1.2×10^{-11}	1.1×10^{-11}
CR	30 °C	23	15	16	-
	50 °C	25	16	20	-
	90 °C	21	15	19	-

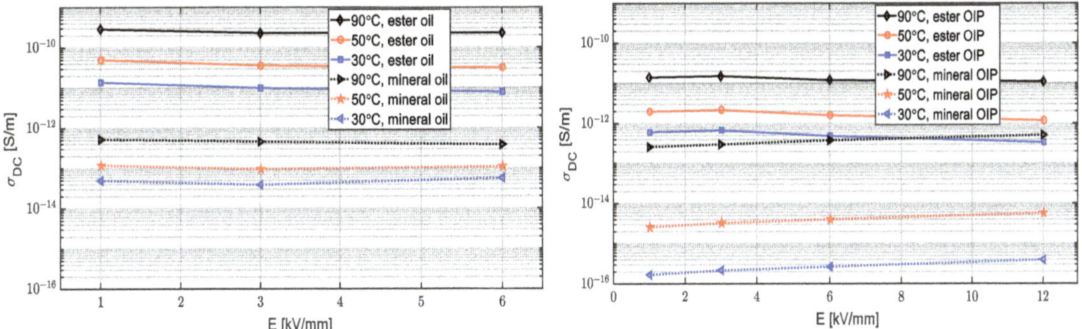

Figure 5. The measurement results on the ester and mineral oils/OIPs.

It is noticeable that the ester oil/OIP conductivities are generally higher than the mineral oil/OIP conductivities. Quantitatively, the differences are 100–130 times at 30 °C, 320–430 times at 50 °C, and 450–480 times at 90 °C. This causes a lower time constant and reduces the dynamic stress duration of the insulation system, which consequently results in a shorter time to reach the steady state condition in cases when ester oil is employed. The Conductivity Ratio (CR) values of mineral oil/OIP indicate stronger temperature dependence compared to the ester oil/OIP. This renders the transformers filled with the mineral oil very sensitive to the temperature variation during energizing or loading as well as operational environmental temperature. Since the CR determines the stress ratio of the oil to OIP at steady sate condition, a high and unstable variation in CR for mineral oil, results in the unstable stress shift from oil to OIP or vice versa. In the case of ester oil/OIP, the CR variation is very low, and the stress distribution condition is more stable. Further comparisons are presented in the next sub-section.

4. FEM Simulation

It can be clearly noticed from Figure 1 that the applied HVDC in the last unit of the DC/DC converters is at maximum. Therefore, the insulation design of the real scale MFT, which is placed inside the last converter unit, is selected to be investigated here as the worst case. The dielectrics of the MFT will be under time varying electric field stresses, which change from the initial permittivity-based distribution to the final conductivity-based distribution, including the transient phase in-between. To find exact values of the stresses on dielectric materials, computer simulations were conducted utilizing time dependent electric current physics in COMSOL Multiphysics software. Owing to the vertical symmetry at the core window section, only the upper part of the design is considered. A DC voltage is applied to the HV winding while both the LV winding and the core are supposed to be at ground potential (Figure 6). For the combined AC and DC voltages on the HV winding of the MFT, the magnitude of the DC voltage (250 kV) is much higher than the AC voltage (18 kV). In this way, applying just a step HVDC to this winding can be considered enough to simulate the operational condition. The logarithmic time steps from t = 1 s (the instant of HVDC application) to t = 10,000 s (when the steady state condition supposed to be reached) are considered to find the solutions. The conductivities of the ester oil/OIP and mineral oil/OIP materials depend on the temperature and the instantaneous electric field intensity. Therefore, as a basis for a NLMW solution, three separate sets of simulations were performed for 30, 50 and 90 °C temperatures for ester oil/OIP, and similarly, for mineral oil/OIP. In each set, the measured results from Tables 1 and 2 were used as two input local tables of stress dependent conductivities in COMSOL. The electric field distribution was investigated during three distinguished time periods: initial, final and transient stages. Subsequently, a smart algorithm for dielectric stress evaluation is applied.

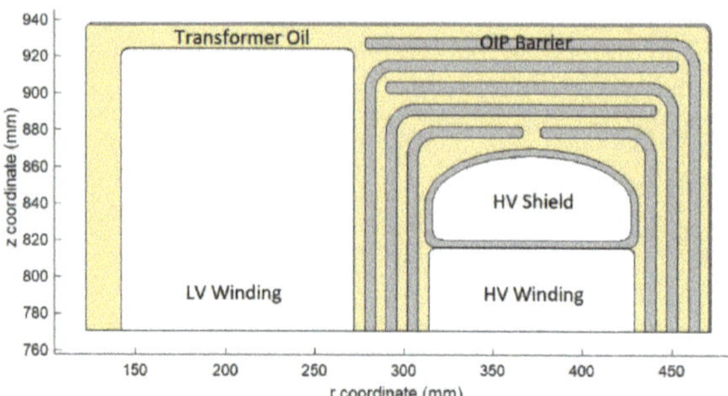

Figure 6. The dielectric structure of the real scale MFT.

4.1. Initial Stress Distribution

Figure 7 demonstrates the initial distribution of the electric stress at the insulation system immediately after the application of 250 kV DC which is based on the Permittivity Ratio (PR) of the oil to OIP. The dielectric permittivity does not depend on temperature, and therefore, the initial distribution of the stresses is temperature independent. Since the permittivities of the oils are lower than the OIPs in both ester and mineral oils, the stresses in the oil channels are higher than in the OIP barriers. However, since the PR of the ester oil to ester OIP (3.5 to 4.6) is lower than the PR of the mineral oil to mineral OIP (2.2 to 4.4), for the case of ester oil, the initial stresses in OIPs is higher than in the case when mineral oil is used. As shown in Figure 7, the distribution of the streamlines is nearly the same for ester and mineral oils and the highest stresses in both cases are at the sides of the HV shields and in the oil. Moreover, as can be seen from the color bars, the maximum stress in the mineral oil is slightly higher than in ester oil.

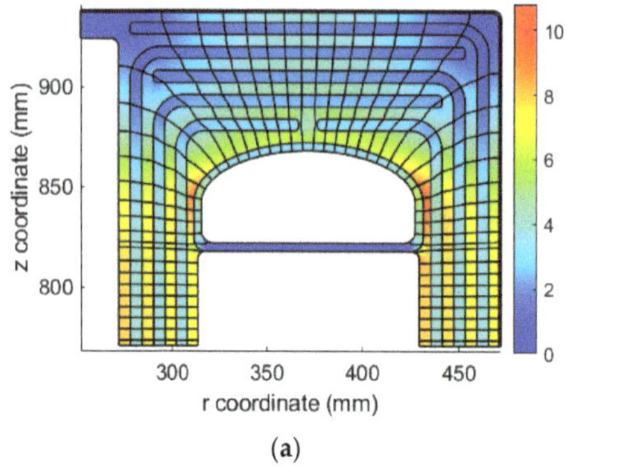

(a)

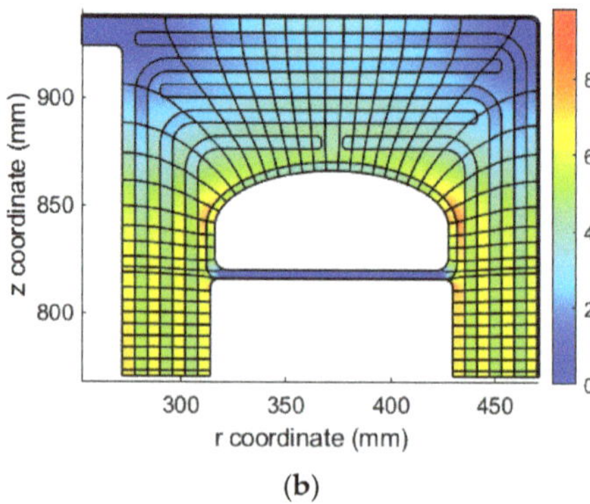

(b)

Figure 7. The initial permittivity-based stress distribution in kV/mm and the electric field streamlines in the dielectric structure of the MFT using: (**a**) mineral oil; (**b**) ester oil.

4.2. Final Stress Distribution

Figure 8 illustrates the final steady state stress distribution in the MFT insulation system by using both ester and mineral oils. It can be easily seen that they are temperature dependent due to the temperature dependence of the conductivity ratio (CR) of the oil to OIP. However, the temperature dependency of CR of the ester oil is weaker than of the mineral oil, and hence, higher consistency of stress distribution regarding the temperature variation of the operation condition can be observed in the case of ester oil usage.

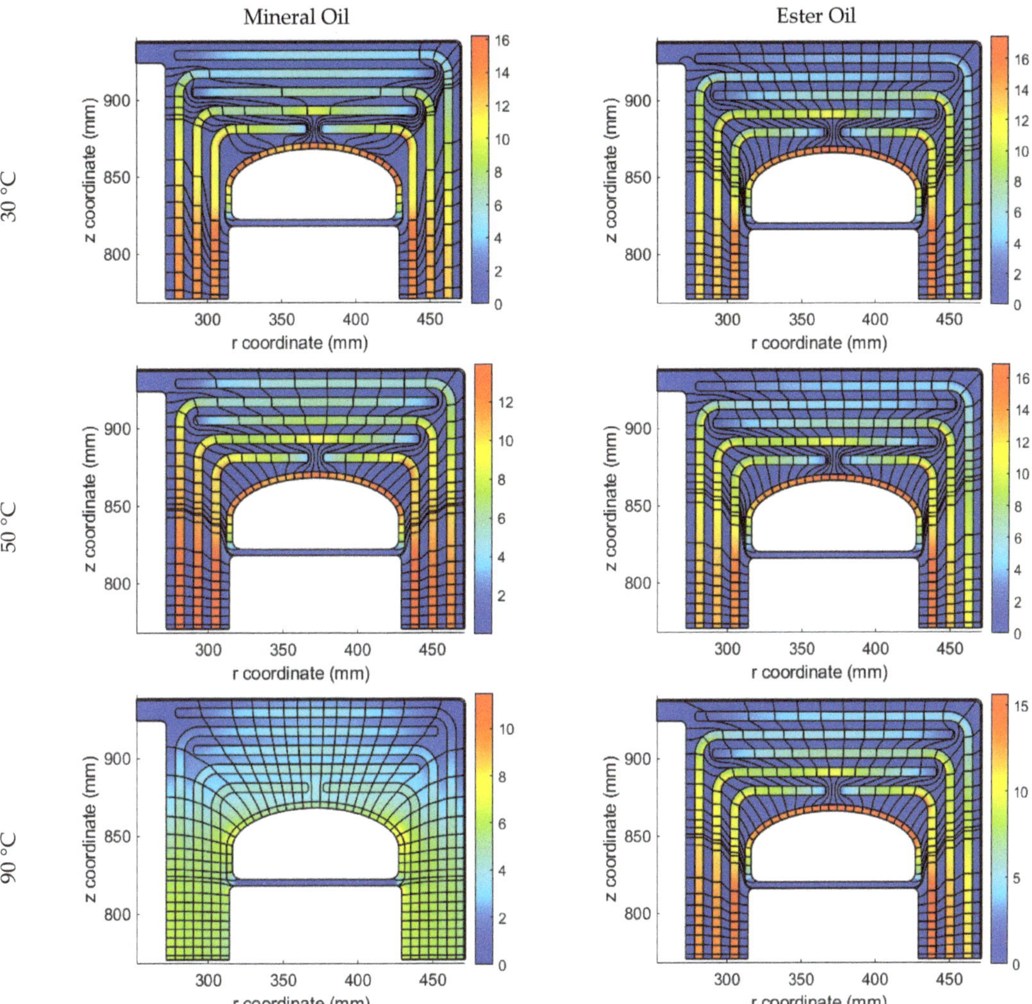

Figure 8. The final steady state stress distribution in kV/mm and the electric field streamlines in the dielectric structure of the MFT at different temperatures using: (**left**) mineral oil; (**right**) ester oil.

4.3. Transient Stress Distribution

The transient phase, which occurs between the initial and final states, may last for some hours. To investigate the time-dependent behavior of the stress in the dielectric structure of the MFT, seven various typical points were selected in this structure as shown in Figure 9. The points P1, P3, P4, P6 and P8 indicate the locations inside OIPs and the other remaining points are in the oil channels. The curves presenting time dependent stresses at these points are demonstrated for each temperature (30, 50 and 90 °C) in Figure 10. The time axis is plotted up to 160 min for mineral oil and 2 min for the ester oil. It can be noticed that the major differences caused by using ester and mineral oils are in the duration of the transient phase, the location of the maximum stresses, the final steady state distribution, and the effect of the temperature. As mentioned previously, the initial stress distribution immediately after the HVDC application depends on the PR of the oil to OIP, and therefore, using ester oil causes higher stresses in OIPs with respects to the mineral oil application. Moreover, using ester oil results in a shorter transient phase and faster transition to the steady state condition with respect to mineral oil. This is because of the lower time constants (the ratio of electrical conductivity to the permittivity of the insulation materials) when using the ester oil, which originates from its much higher conductivity with respect to the mineral oil. Additionally, an even faster transition can be observed at higher temperatures in both oils for the same reason.

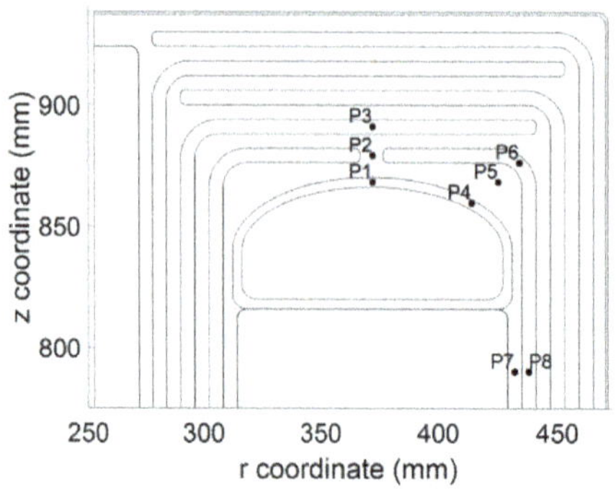

Figure 9. The position of some typical points in the dielectric structure of the MFT to investigate their stresses in time domain.

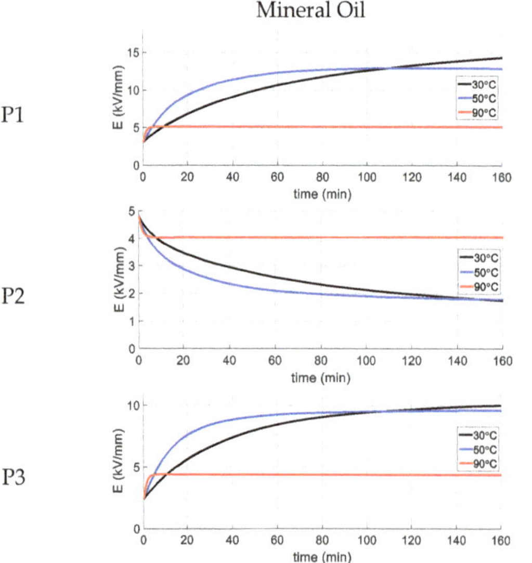

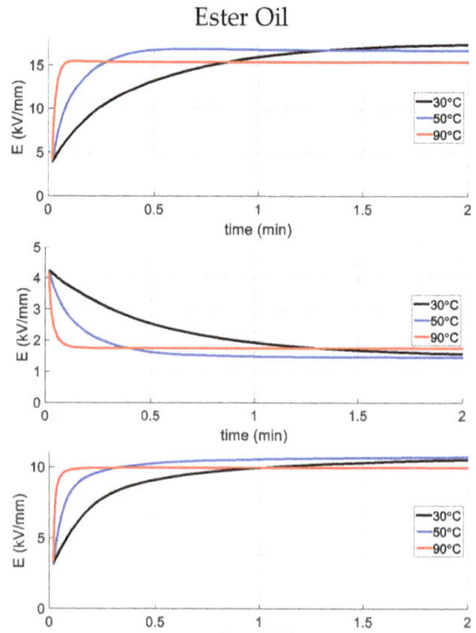

Figure 10. *Cont.*

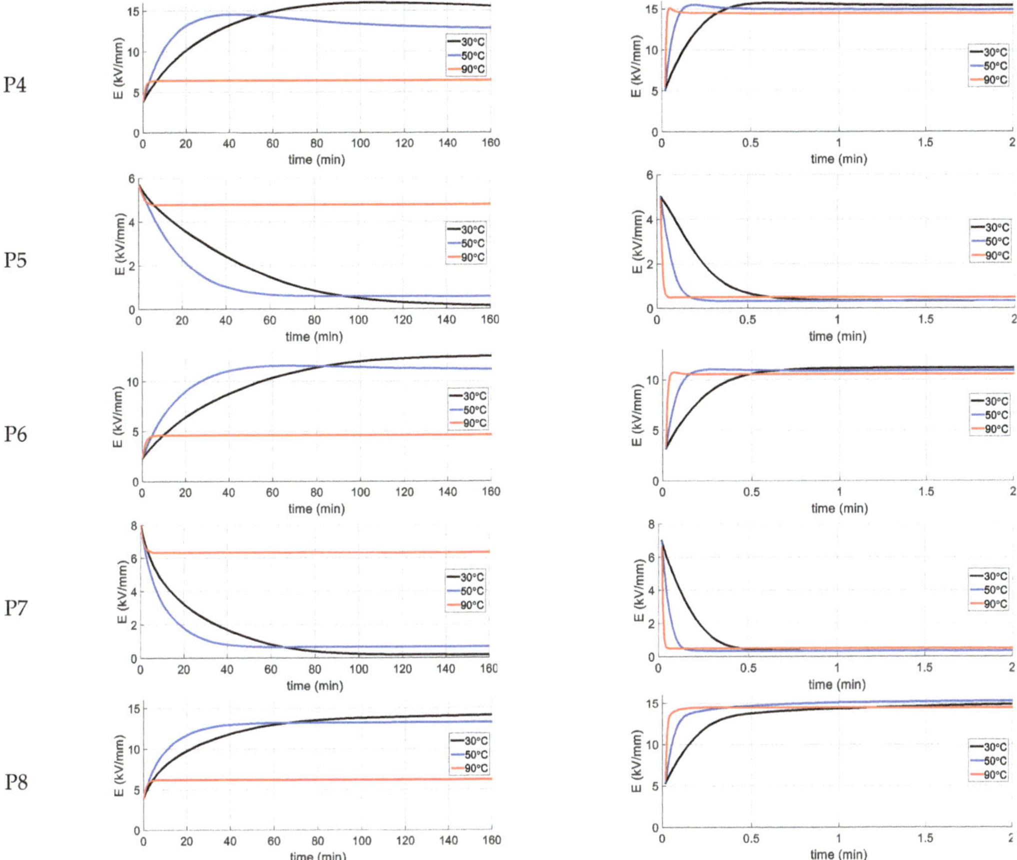

Figure 10. The time variation of stress curves of P1 to P8 using: (**left**) mineral oil; (**right**) ester oil.

For both oils, the stress in OIPs at the steady state is greater than their values at the initial state and the opposite is observed for stresses at oil gaps. Additionally, it can be noticed that the final steady state values for mineral oil at 90 °C differ considerably with respect to 30 and 50 °C, while this is not the case when using ester oil. This is caused by the fact that the CR of the oil to OIP has greater divergence with respect to the temperature in mineral oil than for ester oil, see Tables 1 and 2. Moreover, at points P1, P4 and P6, it can be observed that that the local maximum stress can be higher than its initial and final values. This fact emphasizes that considering just the initial and final stress distribution is insufficient to obtain an accurate dielectric design evaluation, since the worst cases (in terms of stress in the dielectrics) may occur during the transient phase.

4.4. Dielctirc Design Evaluation

Here, the dielectric design of the real scale MFT with application of ester oil/OIP is evaluated and investigated from the dielectric withstand viewpoint, and then the design is compared with mineral oil/OIP application. To evaluate the dielectric design automatically, COMSOL Live Link with MATLAB was utilized to identify the minimum safety factors in the oil gaps and creepage surfaces and maximum stress in the OIPs at various temperatures and for all the time steps. At every single time step of each time dependent FEM simulation, the maximum stress in OIPs as well as the minimum safety factors in oil gap paths, creepage paths and combined paths were investigated.

4.4.1. The Maximum Stress in OIP

The OIPs bear a definite maximum permissible electric stress, below which they can work safely. This maximum stress is usually determined by experiments and is typically about 20 kVrms/mm according to [25]. Therefore, the maximum stress in the OIPs must be checked to be less than this value. For each streamline path of a FEM simulation and at each time step, the OIP path sections are extracted. The maximum electric field at all points of these sections are evaluated as a measure of the severity of the stress in OIPs. This procedure is illustrated in Figure 11.

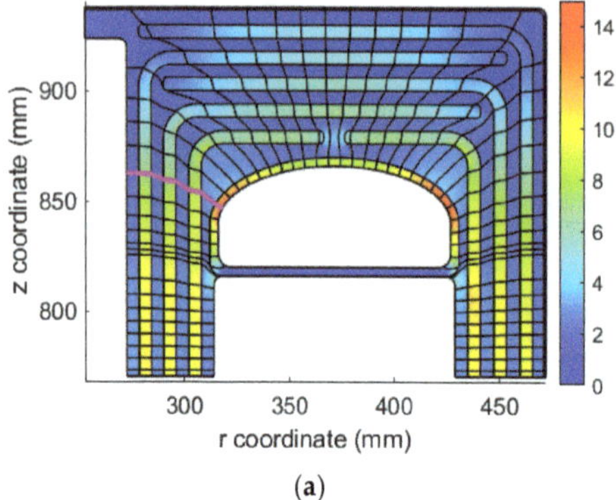

(a)

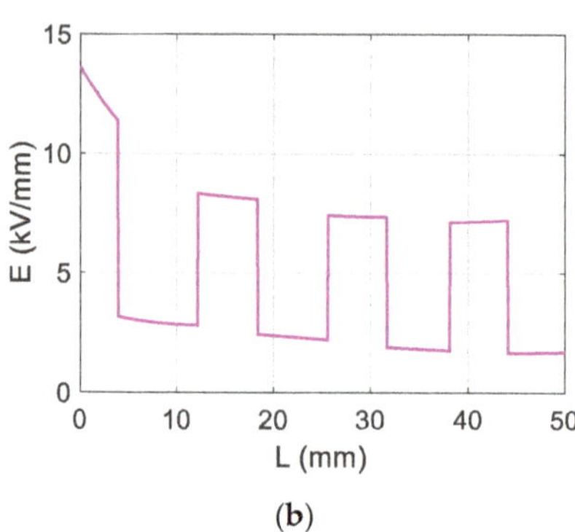

(b)

Figure 11. The streamlines by using ester oil at 10 s and 30 °C: (**a**) A sample streamline (thick pink line); (**b**) The electric field curve along this streamline starting from the rightmost point.

4.4.2. The Safety Factors in Oil Gaps

Similarly, for each streamline path of the FEM simulation at each time step, the oil gap path sections are extracted (Figure 11), and then, the minimum safety factor (SF) for each path is calculated using the method described in [5,25]. To obtain the SF for each oil gap path, the electric field curve along this path is extracted and rearranged in a descending order. Then, the cumulative stress curve, E_{av}, is obtained as follows [5,25]:

$$E_{\mathrm{av}} = \frac{1}{z} \int_0^z E(z').dz' \tag{3}$$

where E is the rearranged electric field stress curve in kV/mm in the oil gap and z is the length of the oil gap in mm. Then the safety factor curve of the oil gap is calculated by dividing the cumulative stress curve to the breakdown strength curve of the oil (E_{bd}) [5,25]:

$$SF = \frac{E_{\mathrm{bd}}}{E_{\mathrm{av}}} \tag{4}$$

$$E_{\mathrm{bd}} = E_0 d^{-0.37} \tag{5}$$

where E_0 is 17.5 or 21.5 kVrms/mm for bare and covered electrodes, respectively, for mineral oil application, and d is distance in mm. The safety factor of an oil gap is considered as the minimum value of the related safety factor curve. It is worth mentioning that the partial discharge inception voltage and breakdown voltages of synthetic ester oil is slightly higher than mineral oil [37], and therefore, (5) is assumed to be valid for ester oil with a safer margin.

4.4.3. The Safety Factors in Creepage Surfaces

The common surfaces of the oils and OIPs are disposed to creepage discharges created by the tangential field. According to [5,25], the withstand level of such creepage surfaces can be considered as 0.7 of the oil gaps withstand level. To investigate the creepage strength of the design, the tangential electric field on the creepage surface of each barrier is extracted. Then, its related curve along the surface path is divided into multiple sections at places where its value falls below 0.2 kV/mm [26] or its direction changes (Figure 12). Afterwards, the SF of each creepage path sections is calculated using a procedure like the one discussed for oil gaps; however, the factor 0.7 is introduced to account for the weakening of the strength at the surface due to possible creepage:

$$\mathrm{SF} = \frac{0.7 E_{\mathrm{bd}}}{E_{\mathrm{av}}} \tag{6}$$

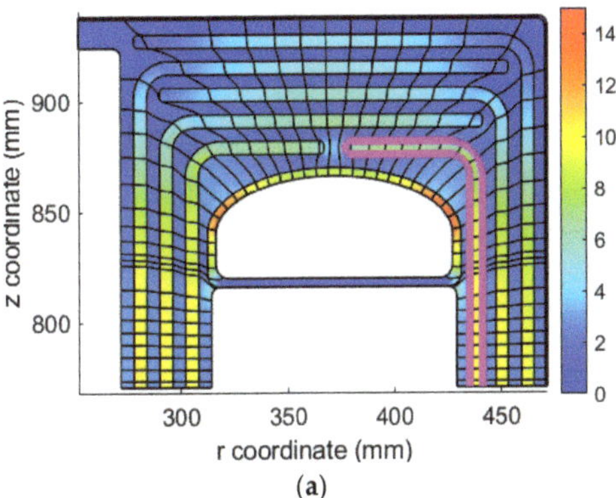

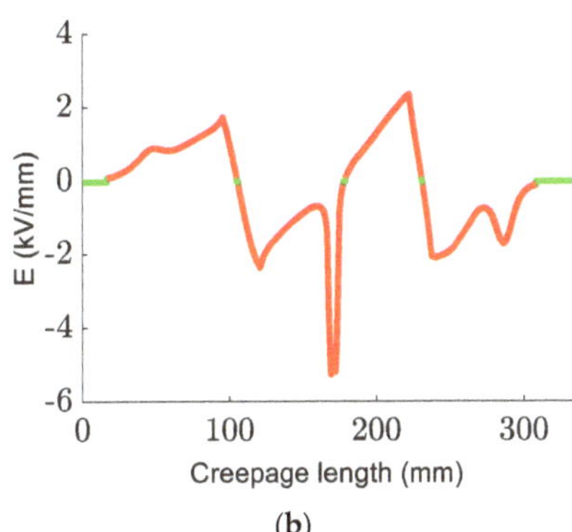

(a) (b)

Figure 12. The streamlines by using ester oil at 10 s and 30 °C: (**a**) A sample creepage surface (thick pink line); (**b**) The electric field curve along this creepage surface (accounted from the right bottom).

As a replacement to (6), one can consider the same breakdown voltage for both oil gap and creepage paths, and instead, the E_{av} stress curve shall be divided by 0.7, giving:

$$\mathrm{SF} = \frac{E_{\mathrm{bd}}}{E_{\mathrm{av}}/0.7} \tag{7}$$

4.4.4. The Safety Factors in the Combined Oil Gaps and Creepage Surfaces

It is known that electrical discharges can be initiated in oil gaps and then continue their path on the creepage surfaces. Additionally, the breakdown in the oil gaps and creepage surfaces along the oil gaps may be considered to have a similar mechanism. Therefore, the oil gap strength and creepage strength cannot be assumed to be completely independent. Thus, it is reasonable to calculate the safety factors in combined oil gaps and creepage paths. To explain the case, consider a section of a streamline in an oil gap that is confined between two barriers as shown in Figure 13. This streamline can be extended from two ends on the barriers, in the same direction of the tangential electric field on both surfaces. These protractions would be extended along the barriers until the tangential electric field on the surface falls below 0.2 kV/mm or its direction changes. The stress curve on this path is obtained by the stress curve for an oil gap path combined with the tangential stress curves at both creepage paths. As shown in (7), the stress curves of creepage paths shall be divided by factor 0.7 before combination with the stress curve of the oil gap.

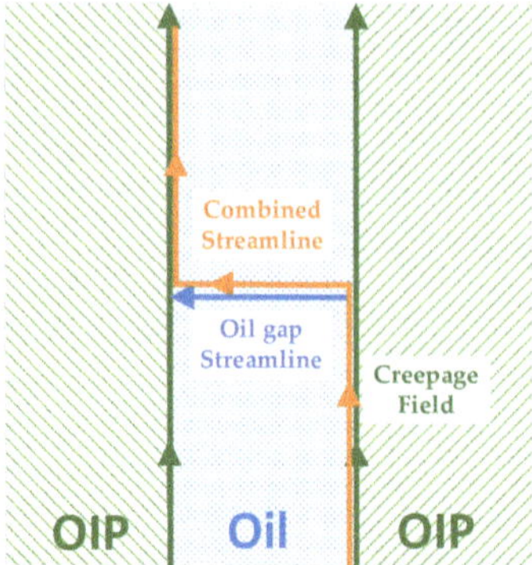

Figure 13. Extracting a combined path from an oil gap path and the linked creepage paths.

Considering this procedure, one can perceive that the minimum SF of a combined path must be equal (in a good dielectric design) or less than to the minimum SF of its related oil gap path and the minimum SFs of the related two creepage paths. One can consider the combined path, which is illustrated in Figure 14a by a pink line, which is composed by two creepage paths connected to the two ends of an oil gap path. The stress curves along this combined path as well as on its forming sections in the oil gap and on creepage surface are shown in Figure 14b. The sorted electric field stress curves (E curve from (3)) of the mentioned paths are depicted in Figure 14c. Since the E curves of oil path and creepage paths are the rearranged form of their stress curves in descending order, the maximum of stresses in each of these paths could be adhered together in the E curve of the combined path. Therefore, it can be clearly seen that the E curve of the combined path is greater than the E curves of the oil gap and creepage paths. Consequently, the E_{av} curve of the combined path will be higher compared to that of the oil gap and creepage paths. Consequently, according to (4), this can lead to a lower minimum SF for combined path with respect to the SFs of its forming oil gap and creepage paths (Figure 14d).

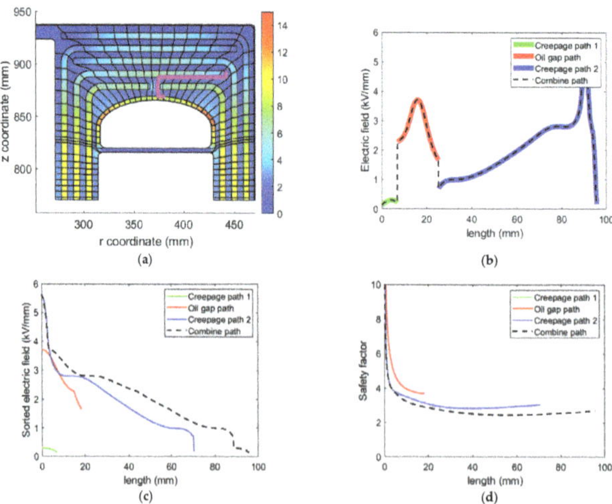

Figure 14. The streamlines by using ester oil at 10 s and 30 °C: (**a**) A sample combined path in an oil gap and two creepage paths (thick pink line, creepage path 1 is on the insulation of the HV shield and creepage path 2 is on the oil duct barrier); (**b**) The electric field curve along this combined path; (**c**) the sorted electric field curves; (**d**) the safety factor curves (the safety factor of the creepage path 1 is higher than 40 and that is why the green line is not shown).

4.4.5. The Automatic Dielectric Evaluation

The above-mentioned procedure for the evaluation of the dielectric strength is implemented to be run automatically in MATLAB using COMSOL Live Link and is clarified in the flowchart in Figure 15. The final results from this automatic evaluation for ester and mineral oils at all time steps and all temperatures (30, 50 and 90 °C) are summarized in Table 3.

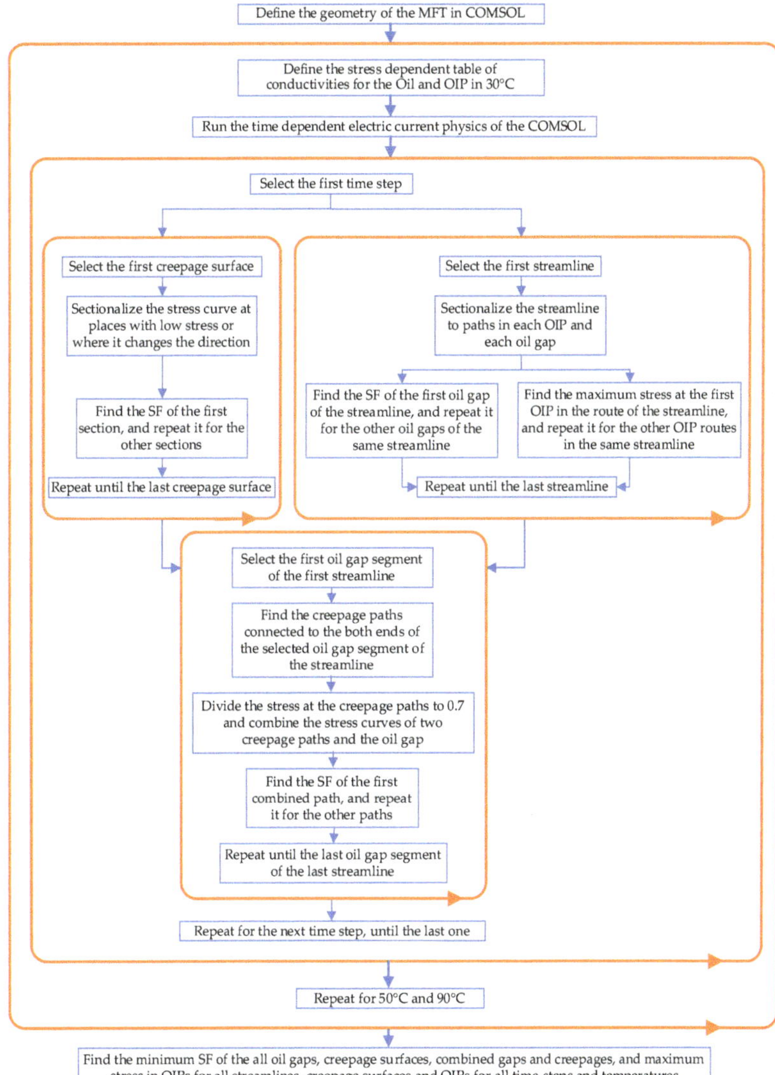

Figure 15. The flowchart of automatic insulation design evaluation applied in COMSOL Live Link with MATLAB.

Table 3. The maximum stresses in OIPs, the minimum *SF*s in the oil gaps, on the interface creepage paths and combined paths and the related time of occurrences.

	Temp. (°C)	Max. Stress in OIPs (kV/mm)	Time (s)	Min. *SF* in Oil Gaps	Time (s)	Min. *SF* at Creepage Surfaces	Time (s)	Min. *SF* at Combined Paths	Time (s)
mineral oil	30	17.5	4467	1.5	1	2.4	2239	1.5	1
	50	16.2	1778	1.5	1	2.4	891	1.5	1
	90	8.3	398	1.5	1	2.5	158	1.5	1
ester oil	30	17.7	28	1.7	1	2.3	12.6	1.6	1
	50	17.7	8	1.7	1	2.4	4	1.6	1
	90	17.3	2	1.8	1	2.5	1.4	1.7	1

It can be seen in Table 3 that the maximum stresses in OIPs and minimum SFs in creepage paths occur not in the initial state nor in the final state, but rather in the transient phase. This fact again emphasizes the importance of the time dependent analysis of stresses in the dielectrics. The minimum SFs in the oil gaps happen right after the HVDC switching. The same is valid for the combined paths.

As was mentioned previously, the minimum SF of the combined paths will be equal or less than the minimum of the SFs of the oil gap paths and the SFs of a creepage paths. In Table 3, by comparing the minimum SFs of the oil gap paths, the creepage paths and combined paths and the time of their occurrences, it can be concluded that the SFs of the combined paths are generally determined by the oil gaps in the dielectric design of the real scale MFT. However, just in one of the simulations, using ester oil at 90 °C, the minimum SF of the combined path (1.7) is slightly lower than the SF of the oil gap paths (1.8).

Regarding the maximum stress in OIPs, it can be seen that by using ester oil instead of mineral oil, a slightly higher stress at 30 and 50 °C occurs. But at 90 °C, the maximum stress in mineral OIP is not much lower than that of ester OIP, even much lower than mineral OIP at 30 and 50 °C. Therefore, this issue cannot be considered as a superiority for mineral oil application.

As the final deduction from Table 3, it can be indicated that the SFs of all oil gaps, creepage paths as well as the combined paths at all temperatures and time steps are far higher than 1 and the maximum stresses in OIPs are less than 20 kV/mm. Therefore, the merit of the dielectric design of the provided MFT using the biodegradable ester oil can be verified. Additionally, the minimum SFs of all oil gaps, creepage paths and nearly all combined paths by using ester oil are higher than the case when the mineral oil is used. In addition, the maximum stress at all temperatures and all time-steps in the case of ester oil (17.7 kV/mm) is only slightly higher than the mineral oil application (17.5 kV/mm). Therefore, from the viewpoint of the dielectric withstand level, there is no restriction for using ester oil instead of mineral oil.

5. Discussion

The results of the measurements and the analysis of the electric stresses in the insulation system reveal:

- The measured conductivity of the oil/OIP insulation materials are temperature and electric stress dependent. This fact is considered in the FEM simulations adopted for finding the insulation withstand level of the transformer design.
- The conductivity values of the ester oil/OIP are generally higher than the mineral oil/OIP, which causes lower time constants for ester oil/OIP and consequently faster convergence to the steady state condition (shorter transient phase) by using ester oil.
- For both ester and mineral oils, the lower the operational temperature, the lower the conductivity values. As a result, the transient state is longer at low temperatures.
- The temperature dependency of ester oil/OIP conductivities are lower, which causes the transformer to be less sensitive to temperature variations during energization or

variable loading conditions. As a result, the transformer filled with mineral oil behaves completely different at 90 °C compared to lower temperatures.

- For a successful insulation design, it is insufficient to check the stress distribution only during the initial and steady state conditions, but also during the transient state when instantaneous maximums in the field strength may occur.
- By using ester oil, the stress in the OIP at steady state condition is at the same level as for a similar transformer filled with mineral oil. Similarly, the minimum SFs at the oil gaps, creepage paths as well as the combined paths are almost at the same order for both ester and mineral oils applications.
- The introduced combined method for safety factor calculation can be used effectively in a transformer insulation design evaluation. The SF value of an insulation design discovered by this method is equal or lower than the minimum of the SF values found by conventional methods on the independent oil gaps and creepage paths. Therefore, the new proposed method can be considered as a conservative method for insulation design evaluation.

6. Conclusions

The design aspects regarding a combined DC/AC transformer for a cost-effective integration of wind farms to land without the need for a huge and expensive power transformer and converter stations have been presented. The insulation design process of the real size medium-frequency biodegradable oil insulated transformer was introduced based on a previous study [5], where the verified method for characterizing high voltage materials and a high voltage DC insulation, manufacturing and testing a prototype transformer were reported.

The extensive measurements are performed on a biodegradable transformer oil and OIP. The article reviews the method employed by the authors for characterization measurements and reveals the measurement results of the biodegradable solid and liquid insulation materials.

The key result of the present study is the insulation design methodology for high power Medium Frequency Transformers considered for a series DC connection, filled with an environment-friendly oil. These kinds of transformers are to be subjected to a very high offset DC voltage.

An innovative solution for identifying the minimum safety factor in the insulation system is proposed. The solution is based on evaluations of the safety factor along combined paths, instead of the conventional approach when the oil gaps and creepage paths are considered separately. It is suggested that the combined path evaluation should be considered during the transformer insulation design process, in addition to the evaluation of the oil gaps and creepage paths. The developed combined design method, presented in the paper shall be verified by experimental investigations on simple insulation systems as well as more complex insulation system models.

As another future work, thermal simulations shall be constructed to provide a more exact picture of the temperature gradient in the different parts of the transformer during energization, in the transient phase, and the steady state conditions. Afterwards, more precise temperature-dependent insulation design FEM simulations can be performed.

Author Contributions: Conceptualization, T.T. and M.K.; methodology, M.K. and H.R.M.; software, M.K. and H.R.M.; validation, M.K., H.R.M. and Y.V.S.; formal analysis, M.K. and H.R.M.; investigation, M.K. and H.R.M.; resources, T.T. and M.K.; data curation, M.K. and H.R.M.; writing—original draft preparation, M.K. and H.R.M.; writing—review and editing, All authors; visualization, M.K. and H.R.M.; supervision, T.T. and Y.V.S.; project administration, T.T. and M.K.; funding acquisition, M.K. and T.T. All authors have read and agreed to the published version of the manuscript.

Funding: This research was funded by Swedish Energy Agency (Energimyndigheten, Project No. 43048-1) and Rise Research Institutes of Sweden.

Acknowledgments: Funding from the Swedish Energy Agency and financial support Rise Research Institutes of Sweden is gratefully acknowledged, as are helpful research assistance from Morteza Eslamian and Oriol Guillén Sentís and conductivity measurements assistance from Joakim Rastamo and Marcus Svensson. Any remaining errors reside solely with the authors.

Conflicts of Interest: The authors declare no conflict of interest. The funders had no role in the design of the study; in the collection, analyses, or interpretation of data; in the writing of the manuscript, or in the decision to publish the results.

Abbreviations

HVDC	High Voltage Direct Current
MFT	Medium Frequency Transformer
OIP	Oil Impregnated Pressboard
DAB	Dual Active Bridge
LCC	Life Cycle Cost
LMW	Linear Maxwell–Wagner
NLMW	Non-Linear Maxwell–Wagner
FEA	Finite Element Analysis
FEM	Finite Element Method
CR	Conductivity Ratio

References

1. IEA. Offshore Wind Outlook 2019. Available online: https://www.iea.org/reports/offshore-wind-outlook-2019 (accessed on 20 November 2021).
2. U.S. Department of Energy. Independent Statistics & Analysis. Assessing HVDC Transmission for Impacts of Non-Dispatchable Generation. June 2018. Available online: https://www.eia.gov/analysis/studies/electricity/hvdctransmission/pdf/transmission.pdf (accessed on 20 November 2021).
3. Lundberg, S. Wind Farm Configuration and Energy Efficiency Studies—Series DC versus AC Layouts. Ph.D. Thesis, Chalmers University of Technology, Göteborg, Sweden, 2006.
4. Guillén Sentís, O. Feasibility of Sea Based Wind Park. Master's Thesis, Chalmers University of Technology, Göteborg, Sweden, 2020. Available online: https://hdl.handle.net/20.500.12380/301473 (accessed on 20 November 2021).
5. Kharezy, M.; Mirzaei, H.R.; Serdyuk, Y.; Thiringer, T.; Eslamian, M. A Novel Oil-Immersed Medium Frequency Transformer for Offshore HVDC Wind Farms. *IEEE Trans. Power Deliv.* **2021**, *36*, 3185–3195. [CrossRef]
6. Bahmani, M.A.; Thiringer, T.; Kharezy, M. Optimization and experimental validation of medium-frequency high power transformers in solid-state transformer applications. In Proceedings of the IEEE Applied Power Electronics Conference and Exposition APEC '16, Long Beach, CA, USA, 20–24 March 2016; pp. 3043–3050.
7. Kharezy, M.; Thiringer, T. Challenges with the design of cost-effective series DC collection network for offshore windfarm. In Proceedings of the 17th Wind Integration Workshop on Large Scale Integration of Wind Power into Power Systems as well as on Transmission Networks for Offshore Wind Power Plants, Stockholm, Sweden, 17 October 2018.
8. Kharezy, M.; Eslamian, M.; Thiringer, T. Insulation Design of a Medium Frequency Power Transformer for a Cost-Effective Series High Voltage DC Collection Network of an Offshore Wind Farm. In Proceedings of the 21st International Symposium on High Voltage Engineering ISH '19, Budapest, Hungary, 26–30 August 2019; pp. 1406–1417.
9. Fritsche, K.L.R.; Trautmann, F.; Hammer, T. Transformers for High Voltage Direct Current Transmission—Challenge, Technology and Development. 2016. Available online: https://www.researchgate.net/publication/315657537 (accessed on 8 March 2022). (In German).
10. Mohan Rao, U.; Fofana, I.; Jaya, T.; Rodriguez-Celis, E.M.; Jalbert, J.; Picher, P. Alternative Dielectric Fluids for Transformer Insulation System: Progress, Challenges, and Future Prospects. *IEEE Access* **2019**, *7*, 184552–184571. [CrossRef]
11. Rozga, P.; Beroual, A.; Przybylek, P.; Jaroszewski, M.; Strzelecki, K. A Review on Synthetic Ester Liquids for Transformer Applications. *Energies* **2020**, *13*, 6429. [CrossRef]
12. Cigré Technical Brochure 646, HVDC Transformer Insulation: Oil Conductivity. Electra ELT_285_3. 2016. Available online: https://e-cigre.org/publication/ELT_285_3-hvdc-transformer-insulation-oil-conductivity (accessed on 17 December 2021).
13. Aakre, T.G.; Ve, T.A.; Hestad, Ø.L. Conductivity and Permittivity of Midel 7131: Effect of Temperature, Moisture Content, Hydrostatic Pressure and Electric Field. *IEEE Trans. Dielectr. Electr. Insul.* **2016**, *23*, 2957–2964. [CrossRef]
14. Rumpelt, P.; Jenau, F. Investigation on DC conductivity of alternative insulating oils as an application for HVDC converter transformers. In Proceedings of the IEEE 19th International Conference on Dielectric Liquids ICDL '17, Manchester, UK, 25–29 June 2017; pp. P.1–P.10.

15. Rumpelt, P.; Jenau, F. Oil Impregnated Pressboard Barrier Systems Based on Ester Fluids for an Application in HVDC Insulation Systems. *Energies* **2017**, *10*, 2147. Available online: https://www.mdpi.com/1996-1073/10/12/2147#cite (accessed on 17 December 2021). [CrossRef]

16. Kharezy, M.; Mirzaei, H.R.; Rastamo, J.; Svensson, M.; Serdyuk, Y.; Thiringer, T. Performance of Insulation of DC/DC Converter Transformer for Offshore Wind Power Applications. In Proceedings of the Conference on Electrical Insulation and Dielectric Phenomena CEIDP '20, Virtual, East Rutherford, NJ, USA, 18–30 October 2020; pp. 382–385.

17. Bahmani, M.A.; Thiringer, T.; Kharezy, M. Design Methodology and Optimization of a Medium-Frequency Transformer for High-Power DC-DC Applications. *IEEE Trans. Ind. Appl.* **2016**, *52*, 4225–4233. [CrossRef]

18. Mogorovic, M.; Dujic, D. Sensitivity Analysis of Medium-Frequency Transformer Designs for Solid-State Transformers. *IEEE Trans. Power Electron.* **2019**, *34*, 8356–8367. [CrossRef]

19. Agheb, E.; Bahmani, M.A.; Høidalen, H.K.; Thiringer, T. Core loss behavior in high frequency high power transformers—II: Arbitrary excitation. *J. Renew. Sustain. Energy* **2012**, *4*, 033113. [CrossRef]

20. Kharezy, M.; Eslamian, M.; Thiringer, T. Estimation of the winding losses of Medium Frequency Transformers with Litz wire using an equivalent permeability and conductivity method. In Proceedings of the 22nd European Conference on Power Electronics and Applications EPE '20 ECCE Europe, Virtual, Lyon, France, 7–11 September 2020; pp. P.1–P.7.

21. Bahmani, M.A.; Thiringer, T. Accurate Evaluation of Leakage Inductance in High Frequency Transformers Using an Improved Frequency-Dependent Expression. *IEEE Trans. Power Electron.* **2015**, *30*, 5738–5745. [CrossRef]

22. Eslamian, M.; Kharezy, M.; Thiringer, T. Calculation of the Leakage Inductance of Medium Frequency Transformers with Rectangular-Shaped Windings using an Accurate Analytical Method. In Proceedings of the 21st European Conference on Power Electronics and Applications EPE '19 ECCE Europe, Genova, Italy, 2–5 September 2019; pp. P.1–P.10.

23. Kulkarni, S.V.; Khaparde, S.A. *Transformer Engineering Design, Technology, and Diagnostics*; CRC Press, Taylor & Francis Group: Boca Raton, FL, USA, 2017.

24. Eslamian, M.; Kharezy, M.; Thiringer, T. An Accurate Method for Leakage Inductance Calculation of Shell-Type Multi Core-Segment Transformers with Circular Windings. *IEEE Access* **2021**, *9*, 111417–111431. [CrossRef]

25. Moser, H.P. *Transformerboard*; Weidmann AG: Rapperswil, Switzerland, 1979.

26. Yea, M.; Han, K.J.; Park, J.; Lee, S.; Choi, J. Design optimization for the insulation of HVDC converter transformers under composite electric stresses. *IEEE Trans. Dielectr. Electr. Insul.* **2018**, *25*, 253–262. [CrossRef]

27. Okubo, H. HVDC electrical insulation performance in oil/pressboard composite insulation system based on Kerr electro-optic field measurement and electric field analysis. *IEEE Trans. Dielectr. Electr. Insul.* **2018**, *25*, 1785–1797. [CrossRef]

28. Okubo, H.; Sakai, T.; Furuyashiki, T.; Takabayashi, K.; Kato, K. HVDC electric field control by pressboard arrangement in oil-pressboard composite electrical insulation systems. In Proceedings of the IEEE Conference on Electrical Insulation and Dielectric Phenomena CEIDP '16, Toronto, ON, Canada, 16–19 October 2016.

29. Hao, M.; Zhou, Y.; Chen, G.; Wilson, G.; Jarman, P. Space charge behavior in oil gap and impregnated pressboard combined system under HVDC stresses. *IEEE Trans. Dielectr. Electr. Insul.* **2016**, *23*, 848–858. [CrossRef]

30. Schober, F.; Küchler, A.; Liebschner, M.; Berger, F.; Exner, W.; Krause, C.; Fritsche, R.; Wimmer, R. Dielectric Behavior of HVDC Insulating Materials, Polarization and Conduction Processes in Mineral Oil and in Pressboard Impregnated with Different Fluids. In Proceedings of the 19th International Symposium on High Voltage Engineering ISH '15, Pilsen, Czech Republic, 23–28 August 2015.

31. Vahidi, F.; Tenbohlen, S.; Rösner, M.; Perrier, C.; Fink, H. The investigation of the temperature and electric field dependency of mineral oil electrical conductivity. In Proceedings of the ETG Power Engineering Society Symposium, Dresden, Germany, 12 November 2013.

32. Vahidi, F.; Tenbohlen, S.; Rösner, M.; Perrier, C.; Fink, H. Influence of electrode material on conductivity measurements under DC stresses. In Proceedings of the IEEE International Conference on Dielectric Liquids ICDL '14, Bled, Slovenia, 29 June–3 July 2014.

33. Jeroense, M.J.P.; Kreuger, F.H. Electrical conduction in HVDC mass-impregnated paper cable. *IEEE Trans. Dielectr. Electr. Insul.* **1995**, *2*, 718–723. [CrossRef]

34. Schober, F.; Harrer, S.; Küchler, A.; Berger, F.; Exner, W.; Krause, C. Transient and steady-state dc behavior of oil-impregnated pressboard. *IEEE Electr. Insul. Mag.* **2016**, *32*, 8–14. [CrossRef]

35. Kato, K.; Okubo, H.; Endo, F.; Yamagishi, A.; Miyagi, K. Investigation of charge behavior in low viscosity silicone liquid by Kerr electro-optic field measurement. *IEEE Trans. Dielectr. Electr. Insul.* **2010**, *17*, 1214–1220. [CrossRef]

36. Küchler, A. Condition Assessment of Aged transformer Bushing Insulation. *Sess. Pap. Cigré Study Com. A2* **2006**, *A2-104*, P.1–P.10.

37. Dolata, B.; Borsi, H.; Gokenbach, E. Comparison of Electric and Dielectric properties of Ester Fluids with Mineral based Transformer oil. In Proceedings of the 15th International Symposium on High Voltage Engineering (ISH), Ljubljana, Slovenia, 27–31 August 2007.

 *energies*

Article

A Group-Based Droop Control Strategy Considering Pitch Angle Protection to Deloaded Wind Farms

Hui Liu [1,2], Peng Wang [1,2], Teyang Zhao [1,2,*], Zhenggang Fan [1,2] and Houlin Pan [1,2]

1   School of Electrical Engineering, Guangxi University, Nanning 530004, China; hughlh@gxu.edu.cn (H.L.);
     1912301051@st.gxu.edu.cn (P.W.); 1912301006@st.gxu.edu.cn (Z.F.); 1912391051@st.gxu.edu.cn (H.P.)
2   Guangxi Key Laboratory of Power System Optimization and Energy Technology, Guangxi University,
     Nanning 530004, China
*   Correspondence: 2112401004@st.gxu.edu.cn; Tel.: +86-157-7712-6023

Abstract: To promote the frequency stability of a system with high penetration of wind power integrated into it, this paper presents a systematic frequency regulation strategy for wind farms (WFs). As preparation for frequency response, a coordinated deloading control (CDC) scheme combining the over-speed control (OSC) and the pitch angle control (PAC) methods is proposed for wind turbine generators (WTGs) to preserve power reserve. The novelty lies in the consideration of high wind speed situations and pitch angle protection. Then, a group-based droop control (GBDC) scheme is proposed for a WF consisting of WTGs with the CDC. In this scheme, WTGs are divided into two groups for different controls. To improve the frequency response performance and ensure stable operation, the droop coefficients of the WF, groups, and all WTGs are determined according to their frequency regulation capabilities (FRCs). Moreover, pitch angle protection during the frequency response process is considered in this scheme. The effectiveness of the GBDC scheme is verified by comparing it with several existing droop control schemes in various situations.

Keywords: frequency stability; deloading control; frequency regulation capability; droop control; exponential membership function

Citation: Liu, H.; Wang, P.; Zhao, T.; Fan, Z.; Pan, H. A Group-Based Droop Control Strategy Considering Pitch Angle Protection to Deloaded Wind Farms. *Energies* **2022**, *15*, 2722. https://doi.org/10.3390/en15082722

Academic Editor: Juri Belikov

Received: 2 March 2022
Accepted: 31 March 2022
Published: 8 April 2022

Publisher's Note: MDPI stays neutral with regard to jurisdictional claims in published maps and institutional affiliations.

1. Introduction

Owing to the properties of cleanness, renewability, and abundance, the exploitation of wind energy for power generation has been identified with vital significance by many countries in recent decades [1]. As wind penetration increases, the intermittency of wind power and the de-inertia behavior of wind turbine generators (WTGs) bring unprecedented challenges to the frequency stability of the power system [2]. In this situation, the ancillary service of frequency regulation provided by WTGs becomes indispensable.

To obtain the optimal wind energy capture efficiency, WTGs usually operate in the maximum power point tracking (MPPT) state [3,4]. Once frequency fluctuations are detected, WTGs will temporarily deviate from the MPPT state to provide extra active power by releasing the rotational kinetic energy in their rotors to support frequency stability. There are different kinds of control methods for WTGs in the MPPT state to participate in frequency regulation, including droop control, virtual inertia control [5–10], fixed trajectory control [11–15], and combinations of the above control methods [16,17]. However, after the early power-increasing stage, the rotor speeds need to be restored and returned to their optimal value, which may lead to a secondary dip in the system frequency [6].

To address this problem, WTGs can preserve a part of the output reserve for frequency regulation through a so-called deloading control. Generally, there are two kinds of deloading controls: pitch angle control (PAC) and over-speed control (OSC). With the PAC method in [18,19], power reserves are produced by increasing the pitch angles of WTGs to reduce the wind power coefficients before frequency events. Although the PAC method can complete the deloading task at various ranges of wind speeds, the frequent action on the blade pitch angle can easily cause wear on the mechanical devices of WTGs and, consequently, shorten their service lives [20]. With the OSC method, the rotor speed is adjusted to exceed the optimal value to obtain a sub-optimal wind power coefficient so that an active power headroom can be produced for frequency regulation. Considering convenience and the harmlessness property, references [17,21] suggest that WTGs can accomplish a deloading task through an independent OSC approach. However, the optimal rotor speed gradually approaches its limit with increasing wind speed, which indicates the attenuation of the exceeding room. Given the wide range of operating environments, the OSC method may not fulfill the deloading requirements, especially when an order of large deloading extent coincides with a high wind speed situation. Considering the harmful property of the PAC and the narrow control range of the OSC [22], a sophisticated deloading scheme combining the PAC and the OSC is demanded as preparation for frequency regulation.

The frequency regulation controls of deloading WTGs are similar to those of WTGs in the MPPT state. Affected by the wake effect [23–25], the frequency regulation capability (FRC) of each WTG in a wind farm (WF) can be different because of the various arriving wind speeds. Thus, setting the same control parameters for different WTGs will not only waste resources but even endanger operating stability [26]. In order to solve this problem, some works adjust the control parameters of WTGs in the MPPT state according to the kinetic energy stored in their rotors [7,9,27–29]. For WTGs in the deloading state, references [30,31] proposed reserve-based control schemes where the droop loop gains are generally proportional to the power reserve margins. However, large droop coefficients for WTGs at high wind speed will inevitably lead to large-scale movements of the blade pitch angle, which will have a negative effect on the WTGs' long-term service. Moreover, the FRCs of WTGs are underestimated by reserve-based strategies. Therefore, a novel frequency regulation strategy from the perspective of a WF is urgently demanded.

This paper proposes a systematic control strategy to enable WFs to participate in system frequency regulation. The main contributions of this paper are as follows: First, this paper presents a coordinated deloading scheme (CDS) combining the OSC and the PAC methods. Compared with the traditional deloading scheme, the CDS applies to the situation of high wind speed where the OSC cannot accomplish the deloading target alone. The problem of frequent pitch angle actions caused by the PAC is also addressed through the CDS. Second, the FRCs of deloading WTGs and WFs are evaluated based on the mechanism of power point tracking control, which is more precise than the prevalent reserve-based evaluation. Third, a group-based droop control (GBDC) scheme considering the diversity of WTGs is proposed for deloading WFs. In this scheme, WTGs are divided into two groups for different controls. The performance of frequency regulation is improved since the droop coefficients are determined according to the FRCs. On the premise of improving frequency response guarantee, pitch angle protection is also considered in this scheme. Moreover, since the control methods proposed for over-frequency and under-frequency events are symmetrical, this paper is presented from an under-frequency perspective for concision and comprehensibility.

2. Cooperative Deloading Scheme

2.1. MPPT Control Principle

Doubly fed induction generators (DFIGs) and direct-drive permanent magnet synchronous generators (PMSGs) are currently the most prevalent wind generators. These two types of WTGs are both composed of a wind turbine, a generator, and a couple of back-to-back power electronic converters. As these WTGs adopt analogous power control frameworks, the research in this paper applies to both DFIGs and PMSGs. In this paper, the simplified model of WTGs in [32] is applied and modified for simulation efficiency. Since the research is mainly focused on the mechanism between the system and the frequency response power with different schemes, parametric uncertainty is not considered in this paper.

Ignoring the loss caused by mechanical friction, the mechanical power, P_m, extracted by the wind turbine can be calculated by

$$P_\mathrm{m} = \frac{1}{2}\rho\pi R^2 C_\mathrm{p}(\lambda,\beta)v^3 \tag{1}$$

where ρ, R, v, and β are the air density, blade radius, wind speed, and blade pitch angle, respectively. The power coefficient, C_p, is defined as

$$C_\mathrm{p} = 0.5176\left(\frac{116}{\lambda_\mathrm{i}} - 0.4\beta - 5\right)\cdot e^{-\frac{21}{\lambda_\mathrm{i}}} + 0.0068\lambda \tag{2}$$

where

$$\frac{1}{\lambda_\mathrm{i}} = \frac{1}{\lambda + 0.08\beta} - \frac{0.035}{\beta + 1} \tag{3}$$

The λ among the above equations represents the tip–speed ratio, which can be obtained by

$$\lambda = \frac{\omega_\mathrm{r} R}{v} \tag{4}$$

where ω_r is the angular speed of the WT. In this paper, a one-mass drive train model is adopted, and the rotating equation of the WTG is described as

$$P_\mathrm{e} - P_\mathrm{m} = -J\omega_\mathrm{r}\frac{\mathrm{d}\omega_\mathrm{r}}{\mathrm{d}t} \tag{5}$$

The mechanical power and the power coefficient curves of wind turbines are presented in Figures 1 and 2, respectively. For each definite wind speed, there is a so-called optimal tip–speed ratio, λ_opt, and a corresponding rotor speed, $\omega_{\mathrm{r\,opt}}$, enabling maximum power extraction efficiency. The black dashed curve in Figure 1 is the well-known MPPT curve, consisting of an ordinary operation zone, Zone2 (Segment BC); two constant speed zones, Zone1 and Zone3 (Segments AB and CD); and a constant power zone, Zone4 (Ray DE). WTGs in Zone1, 3, and 4 are under protection controls, which are not suitable for participating in frequency regulation. Thus, the studies in this paper are mainly focused on the WTGs in Zone2.

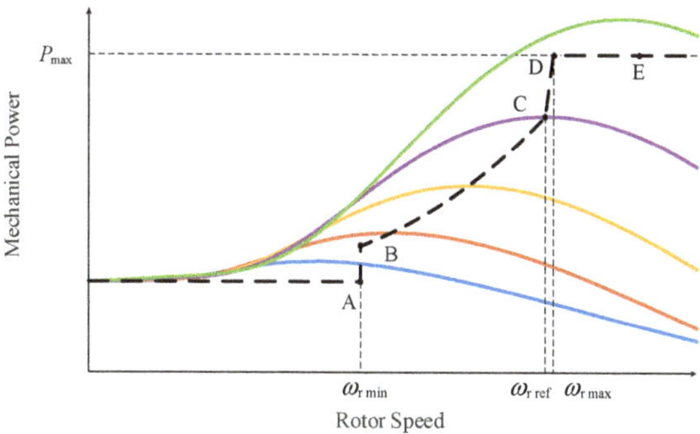

Figure 1. Mechanical power and the MPPT curves.

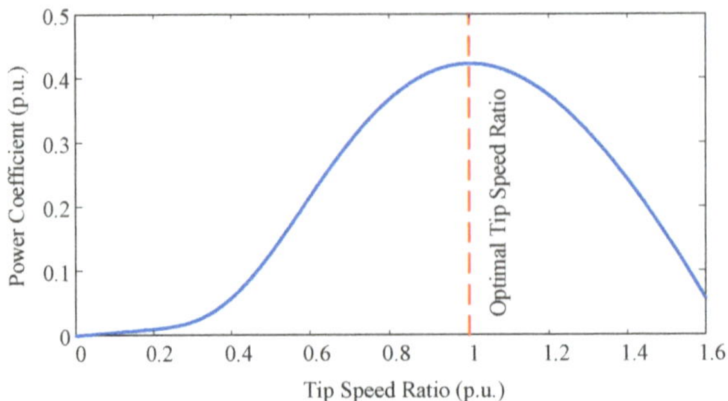

Figure 2. Power coefficient curve.

The optimal active power reference of a WTG in Zone2 is given by

$$P_{\mathrm{MPPT}} = K_{\mathrm{opt}}\omega_r^3 \tag{6}$$

where K_{opt} is the MPPT coefficient, whose expression is

$$K_{\mathrm{opt}} = \frac{1}{2}\rho\pi\frac{R^5}{\lambda_{\mathrm{opt}}^3}C_{\mathrm{p}}(\lambda_{\mathrm{opt}}, 0) \tag{7}$$

As explained above, WTGs in Zone2 are in safe operation; thus, β is set to zero in order to acquire the maximum wind energy harvest.

2.2. Coordinated Deloading Scheme

To address the secondary-frequency-dip issue, WTGs are generally released from the MPPT state to preserve active power reserve margins for frequency regulation. The deloading ratio is generally determined according to the system structure and the operating status of the WF. With an appropriate frequency regulation strategy, 10% deloading work can provide adequate headroom in a system with wind penetration up to 50% [21]. For a given deloading ratio, $d\%$, the deloading power reference is

$$P_{\mathrm{del}} = (1 - d\%)P_{\mathrm{MPPT}} \tag{8}$$

As illustrated in Figure 3, WTGs can reach the assigned deloading points by adjusting the power point tracking curve. The brown dashed curve is the prevalent over-speed curve (deloading without the PAC). When the wind speed is 9.5 m/s (a low wind speed case), there is adequate margin between the MPPT point, B, and the reference rotor speed for speed protection, $\omega_{r\,\mathrm{ref}}$. Thus, WTGs can switch to the deloading point, B1, through the traditional OSC method. The active power reference under this control is given by

$$P_{\mathrm{del}} = K_{\mathrm{del}}\omega_r^3 \tag{9}$$

The K_{del} in (9) is the deloading power point tracking (DPPT) coefficient, which is calculated by

$$K_{\mathrm{del}} = \frac{1}{2}\rho\pi\frac{R^5}{\lambda_{\mathrm{del}}^3}C_{\mathrm{p}}(\lambda_{\mathrm{del}}, 0) \tag{10}$$

where λ_{del} is the corresponding deloading tip–speed ratio, which can be obtained from (2–4) or Figure 2. Since the WTG is at low wind speed and only the OSC is used for deloading, the pitch angle, β, in (10) is equal to 0.

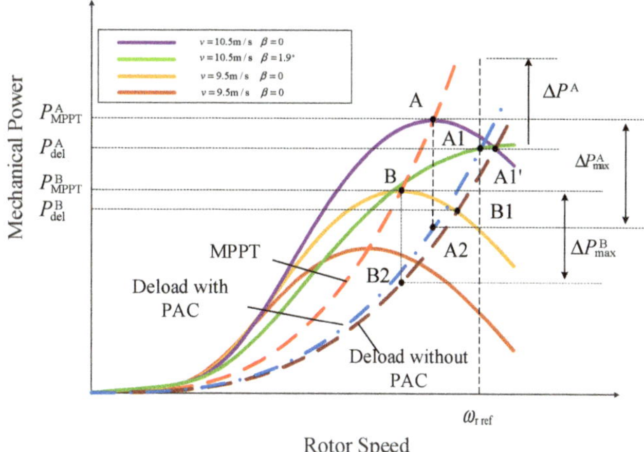

Figure 3. Principle of the cooperative deloading control.

When the wind speed is 10.5 m/s (a high wind speed case), the optimal rotor speed, $\omega_{\mathrm{r\,opt}}$, at the MPPT point, A, is close to $\omega_{\mathrm{r\,ref}}$. The rotor speed at the OSC-deloading point, A1', exceeds the reference rotor speed, which means the deloading requirement is beyond the OSC-deloading potential. In such a situation, the PAC should be applied to help accomplish the required deloading ratio.

To minimize the action of the pitch angle and maximize the stored kinetic energy, the deloading power point, A1, is located at the maximum speed line. Then, we can find a deloading pitch angle β_0 ($\beta_0 = 1.9$ when $d\% = 10\%$), with which the mechanical power curve at 10.5 m/s wind speed intersects the $\omega_{\mathrm{r\,ref}}$ line at A1. The blue dashed line is the DPPT line when $v = 10.5$ m/s and $d\% = 10\%$. As the deloading rotor speed is at the reference value, the DPPT coefficient should be further adjusted to

$$K_{\mathrm{del}} = \frac{P_{\mathrm{del}}}{\omega^3_{\mathrm{r\,ref}}} = \frac{(1 - d\%)K_{\mathrm{opt}}\omega^3_{\mathrm{r\,opt}}}{\omega^3_{\mathrm{r\,ref}}} \tag{11}$$

As $\omega_{\mathrm{r\,opt}}$ varies with wind speed, K_{del} can be considered as a function of v and $d\%$. The deloading pitch angle can be obtained through a lookup table method. Table 1 shows part of the β_0 table.

Table 1. Lookup table for deloading pitch angle.

Wind Speed (m/s)	10% Deloading	20% Deloading
	Deloading Pitch Angle β_0 (Degree)	
10.4	1.947	2.899
10.7	1.806	2.753
11.0	1.649	2.553
11.3	1.493	2.308
11.6	1.347	2.042
11.9	1.216	1.788

The boundary between high wind speed and low wind speed is given by

$$v_{\mathrm{b}} = \frac{\omega_{\mathrm{r\,ref}}R}{\lambda_{\mathrm{del}}} \tag{12}$$

The definition of v_b is the maximum wind speed where WTGs can fulfill the deloading requirement only through the OSC. Because of the correlation between λ_{del} and $d\%$ depicted in Figure 2, v_b also varies with $d\%$. The DPPT coefficient is rewritten as

$$K_{del} = \begin{cases} \frac{1}{2}\rho\pi\frac{R^5}{\lambda_{del}^3}C_p(\lambda_{del},0)\,, & v_0 \leq v \leq v_b \\ \frac{1}{2}\rho\pi\frac{R^5}{\lambda_{opt}^3}C_p(\lambda_{opt},0)\cdot\frac{(1-d\%)\omega_{r\,opt}^3}{\omega_{r\,ref}^3}\,, & v_b \leq v \leq v_1 \end{cases} \tag{13}$$

where v_0 and v_1 are the cut-in wind speed and the minimum wind speed in the constant speed interval (Zone3), respectively.

The procedure for each control cycle of the CDS scheme is described in detail in Figure 4. When receiving deloading signals from the system operator, the control center broadcasts the deloading order, $d\%$, and wind speed boundary, v_b, to all the WTGs. Then, WTGs in low wind speed adjust the DPPT coefficients, K_{del}, according to (13), and they accomplish the deloading task through the OSC. For the WTGs in high wind speed, the PAC and the OSC are both applied to deload. The corresponding K_{del} can be calculated from (13), and the deloading pitch angle is obtained through the β_0 table.

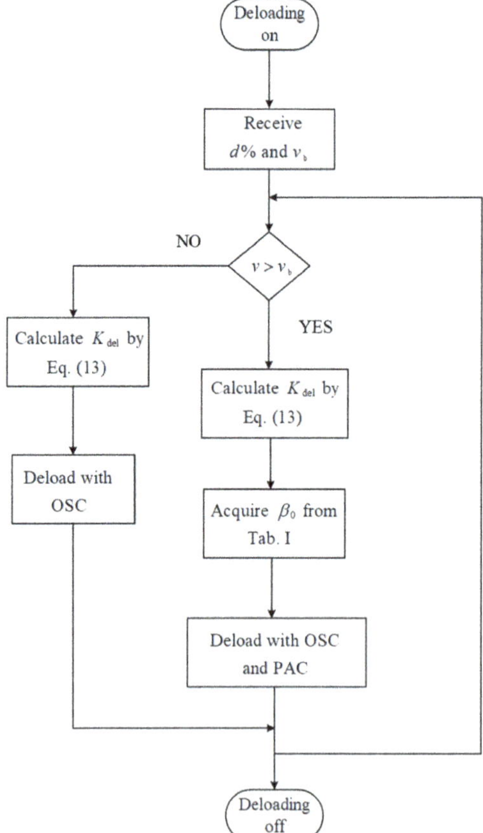

Figure 4. Algorithm of the CDS.

2.3. The Developed Pitch Angle Control

The frequency response of deloading WTGs is realized by extracting the reserve power stored by the OSC and the PAC. Figure 5 presents the active power reference of a WTG, which can be calculated by

$$
\begin{aligned}
P_{\text{ref}}^{*} &= P_{\text{ref}} + \Delta P_{\text{ref}} \\
&= K_{\text{del}}\omega_{\text{r}}^{3} - K_{\text{p}}\Delta f - K_{\text{d}}\frac{df}{dt}
\end{aligned}
\tag{14}
$$

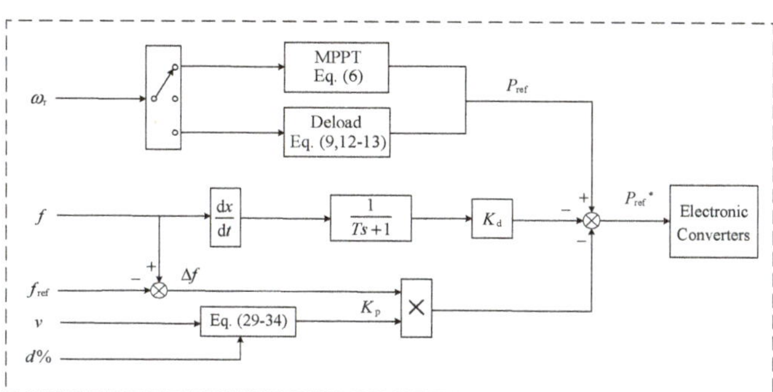

Figure 5. Configuration of active power reference.

For the WTGs with the OSC, an increase in active power reference, ΔP_{ref}, produces an imbalance between electrical and mechanical power, causing a reduction in the rotor speed, ω_{r}, according to (5). The mechanical power thus increases with ω_{r} approaching the optimal speed, $\omega_{\text{r opt}}$. In this process, the reserve power stored by the OSC can be naturally extracted. However, the reserve power provided by the PAC cannot be released spontaneously because of the lack of frequency response from the pitch angle.

To overcome this problem, a developed pitch angle control (DPAC) scheme, as shown in Figure 6, is introduced in this paper. β_f in Figure 6 is the frequency response component of the pitch angle. To release the power reserve provided by the PAC, the pitch angle decreases with the rotor speed in this scheme. As is shown in Figure 7a, β is proportional to ω_{r} within the interval $[\omega_{\text{r opt}}, \omega_{\text{r ref}}]$. Thus, the expression of β_f is

$$
\beta_f = \begin{cases} -\beta_0 \,, & \omega_{\text{r}} \leq \omega_{\text{r opt}} \\ K_\beta \omega_{\text{r}} + B_\beta \,, & \omega_{\text{r opt}} < \omega_{\text{r}} \leq \omega_{\text{r ref}} \\ 0 \,, & \omega_{\text{r}} > \omega_{\text{r ref}} \end{cases}
\tag{15}
$$

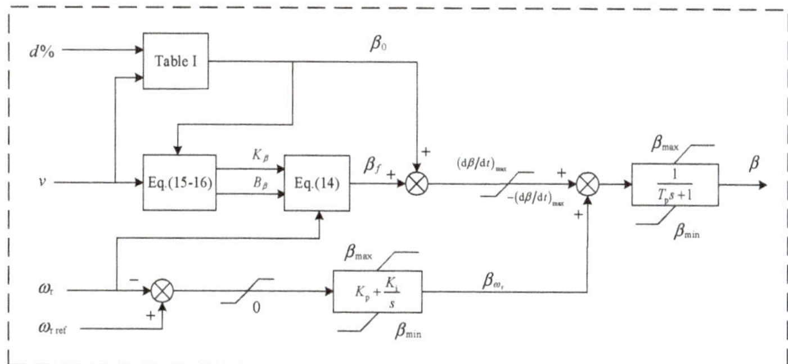

Figure 6. Configuration of DPAC.

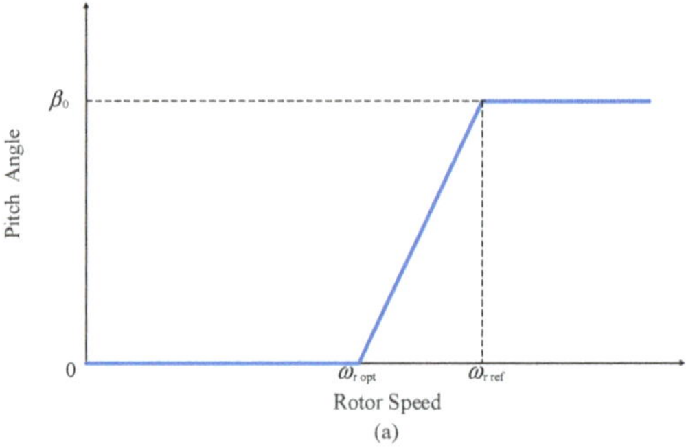

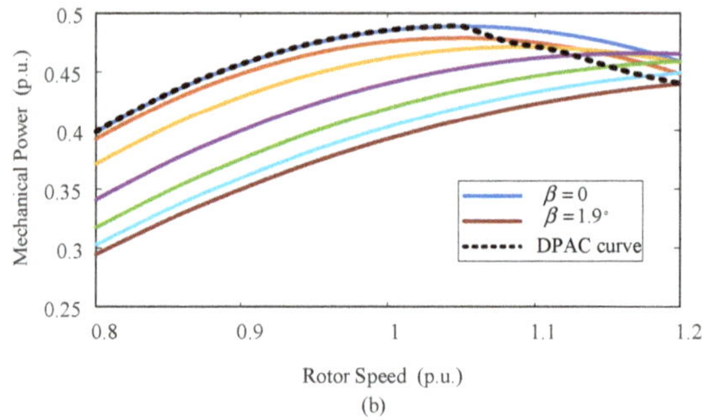

Figure 7. Schematic of DPAC: (**a**) turning the pitch angle depending on rotor speed; (**b**) mechanical power with the DPAC scheme.

The coefficients K_β and B_β are given by

$$K_\beta = \frac{\beta_0}{\omega_{\text{r ref}} - \omega_{\text{r opt}}} \tag{16}$$

$$B_\beta = \frac{\beta_0 \cdot \omega_{\text{r opt}}}{\omega_{\text{r ref}} - \omega_{\text{r opt}}} \tag{17}$$

As the value of $\beta_0 + \beta_f$ decreases to 0 at the optimal rotor speed, $\omega_{\text{r opt}}$, the synchronization between the extractions of the power reserve stored by the OSC and the PAC can be realized. The above conclusion can also be verified in Figure 7b. With the DPAC scheme, the mechanical power curve reaches the maximum value through an optimal route. Additionally, the proposed PAC scheme retains the speed protection function. When the reference rotor speed is exceeded, the pitch angle will increase to keep the rotor speed within a safe range. β_{ω_r} in Figure 6 is the speed protection component of the pitch angle.

3. Evaluation of Frequency Regulation Capability

As calculated in (14), the active power reference is adjusted according to the frequency deviation, Δf, and the rate of change of frequency (ROCOF), df/dt. To avoid operating on the left of the MPPT curve, which causes instability of small disturbances, ΔP_{ref} in (12) cannot transcend a certain limit. Inspired by reference [17,33], the limit is obtained by

$$\begin{aligned} \Delta P_{\text{max}} &= P_{\text{MPPT}} - K_{\text{del}}\omega_{\text{r opt}}^3 \\ &= K_{\text{opt}}\omega_{\text{r opt}}^3 - K_{\text{del}}\omega_{\text{r opt}}^3 \end{aligned} \tag{18}$$

We can readily acquire the limits at wind speeds of 10.5 m/s and 9.5 m/s from Figure 3, which are Segment AA2 and Segment BB2, respectively. Naturally, the power increase limit of the WF is the sum of those of the WTGs, which is

$$\Delta P_{\mathrm{max}}^{\mathrm{WF}} = \sum_{i=1}^{n} \Delta P_{\mathrm{max}}^{\mathrm{WTG}\,i} \tag{19}$$

where $\Delta P_{\mathrm{max}}^{\mathrm{WTG}\,i}$ is the limit of the i-th WTG, and n is the number of WTGs in the WF.

In the frequency regulation process, the active power increases with the drop in the system frequency. The FRC in this paper is defined as the difference between the power increase and the limit. The FRC of the i-th deloading WTG can be calculated by

$$P_{\mathrm{cap}}^{\mathrm{WTG}\,i} = \Delta P_{\mathrm{max}}^{\mathrm{WTG}\,i} - \Delta P_{\mathrm{ref}}^{\mathrm{WTG}\,i} \tag{20}$$

The total FRC of the WF can be estimated by

$$P_{\mathrm{cap}}^{\mathrm{WF}} = \sum_{i=1}^{n} P_{\mathrm{cap}}^{\mathrm{WTG}\,i} \tag{21}$$

The limits $\Delta P_{\mathrm{max}}^{\mathrm{WTG}\,i}$ and $\Delta P_{\mathrm{max}}^{\mathrm{WF}}$ are the FRCs prior to disturbances, which are also defined as the maximum FRCs. For the long-term service of WTGs, frequent actions on pitch angle should be inhibited during frequency regulation. Therefore, WTGs in the WF are divided into two groups according to whether the PAC is applied in the deloading process. The two groups, G_1 and G_2, are defined by

$$G_1 = \{i \,|\, v_0 \le v_i \le v_b\} \tag{22}$$

$$G_2 = \{i \,|\, v_b \le v_i \le v_1\} \tag{23}$$

The WTGs of G_1 are at low wind speed, and the deloading tasks of them are accomplished only with the OSC. Inversely, the WTGs of G_2 are at high wind speed and are deloaded with the OSC and the PAC.

The power increase limit of these two groups can be calculated by

$$\Delta P_{\mathrm{max}}^{G_1} = \sum_{i \in G_1} \Delta P_{\mathrm{max}}^{\mathrm{WTG}\,i} \tag{24}$$

$$\Delta P_{\mathrm{max}}^{G_2} = \sum_{i \in G_2} \Delta P_{\mathrm{max}}^{\mathrm{WTG}\,i} \tag{25}$$

Similarly, the FRCs of these two groups are calculated by

$$P_{\mathrm{cap}}^{G_1} = \sum_{i \in G_1} P_{\mathrm{cap}}^{\mathrm{WTG}\,i} \tag{26}$$

$$P_{\mathrm{cap}}^{G_2} = \sum_{i \in G_2} P_{\mathrm{cap}}^{\mathrm{WTG}\,i} \tag{27}$$

4. The Group-Based Droop Control

As a commonly applied method of frequency regulation, droop control is favorable for both transient and steady-state characteristics of the system frequency [10,35]. Based on the evaluation of FRC in Section 3, a novel droop scheme for WFs in CDS-deloading operation is presented in this section. Considering the sensitivity to measurement errors [10,34], the virtual inertia loop gains are universally set to 0.

4.1. Total Droop Coefficient of the WF

In the frequency regulation process, the control center determines the total droop coefficient of the WF. The total droop coefficient is defined as the sum of all droop gains of WTGs:

$$
\begin{aligned}
K_p^{WF} &= K_p^{G_1} + K_p^{G_2} \\
&= \textstyle\sum_{i \in G_1} K_p^{WTG\ i} + \sum_{i \in G_2} K_p^{WTG\ i} \\
&= \sum_{i=1}^{n} K_p^{WTG\ i}
\end{aligned}
\tag{28}
$$

where $K_p^{G_1}$ and $K_p^{G_2}$ are the total coefficients of G_1 and G_2, respectively.

To improve the frequency response of the WF, K_p^{WF} should be determined depending on the FRC of the WF. When the FRC is large, a large droop gain can be set to make full use of the resources; when the FRC is small, the droop gain should be sufficiently small in case the headroom of the power increase is exceeded. The droop gain of WFs is determined as follows:

$$
K_p^{WF} = \frac{\Delta P_{max}^{WF}}{|\Delta f|_{max}}
\tag{29}
$$

where $|\Delta f|_{max}$ is the threshold of low-frequency load shedding. As such, K_p^{WF} is proportional to the maximum FRC, and the frequency response power can be calculated by

$$
\Delta P_{ref} = \frac{\Delta P_{max}^{WF}}{|\Delta f|_{max}} \cdot \Delta f
\tag{30}
$$

when the frequency deviation touches the threshold, $|\Delta f|_{max}$, ΔP_{ref} is equal to the maximum FRC of the WF.

4.2. Total Droop Coefficients of the Groups

As mentioned in Section 3, WTGs in G_2 deload with the OSC and the PAC methods. The frequency response of these WTGs is probably likely to cause the action of pitch angles. Thus, the total droop coefficients (TDCs) of G_1 and G_2 should be determined reasonably in different disturbances so that the pressure of the mechanical devices of WTGs in G_2 can be alleviated. When the system frequency fluctuates gently, the response of WTGs in G_2 should be minimized, and the frequency regulation task is mainly completed by G_1. However, when the fluctuation is intense, the response of WTGs in G_1 alone can no longer meet the demand of frequency regulation in the WF. In this case, the resources in G_2 should be effectively extracted to support frequency stability.

Therefore, a pair of exponential membership functions are introduced to allocate the frequency regulation tasks of G_1 and G_2. The variable of the functions is the ratio of the FRC of G_1 and the power increase limit of the WF, calculated by

$$
x = \frac{P_{cap}^{G_1}}{\Delta P_{max}^{WF}}
\tag{31}
$$

The participation degrees of these two groups are then obtained by

$$
\frac{K_p^{G_1}}{K_p^{WF}} = \frac{\Delta P_{max}^{G_1}}{\Delta P_{max}^{WF}} \left(1 - \frac{1}{1 + e^{-a(x-0.1)}} \right) \left(\frac{1}{1 + e^{-a(x+0.1)}} \right)
\tag{32}
$$

$$
\frac{K_p^{G_2}}{K_p^{WF}} = 1 - \frac{K_p^{G_1}}{K_p^{WF}}
\tag{33}
$$

The constant, a, is the conversion factor. A larger a means a higher participation degree of the frequency response from G_1 and more protection for WTGs in G_2. However, a too-large value of a may also weaken the frequency regulation effect, as it limits the participation of G_2 to a large extent. Considering the effect from both aspects, a is set to 800 in this paper.

The dynamic properties of the participating extent of the two groups $K_p^{G_1}/K_p^{WF}$ and $K_p^{G_2}/K_p^{WF}$ are depicted in Figure 8. When $P_{cap}^{G_1}/\Delta P_{max}^{WF}$ is greater than 0.1, it means that there are still sufficient frequency regulation resources in G_1. Thus, $K_p^{G_1}/K_p^{WF}$ at this time equals 1. WTGs in G_1 take over the whole droop control task of the WF, while WTGs in G_2 only add a small amount of active power through the virtual inertia control. Inversely, when $P_{cap}^{G_1}/\Delta P_{max}^{WF}$ is less than 0.1, the resources of G_1 are close to running out. Thus, $K_p^{G_1}/K_p^{WF}$ decreases gradually with $P_{cap}^{G_1}/\Delta P_{max}^{WF}$. The task undertaken by G_1 gradually decreases while that of G_2 increases. To prevent an excessive response to the frequency deviation, which endangers the stable operation of the WTGs of G_1, the slope of the curves around 0.1 is sufficiently large. As the $P_{cap}^{G_1}/\Delta P_{max}^{WF}$ ratio decreases further, which means the resources of both G_1 and G_2 are running out, $K_p^{G_1}/K_p^{WF}$ and $K_p^{G_2}/K_p^{WF}$ converge to $\Delta P_{max}^{G_1}/\Delta P_{max}^{WF}$ and $\Delta P_{max}^{G_2}/\Delta P_{max}^{WF}$, respectively. At this time, the frequency regulation tasks of G_1 and G_2 are assigned according to their own resources.

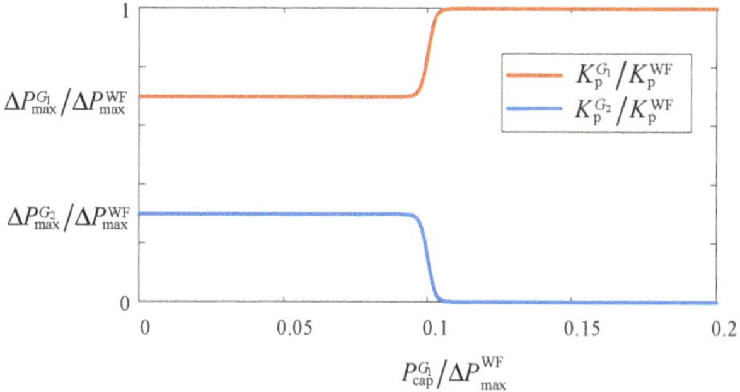

Figure 8. Exponential membership functions.

4.3. Droop Coefficients of WTGs

Due to the wake effect, the arriving wind speed of a WF possesses a declining trend while traversing each row of WTGs. The universal parameter setting of the conventional droop control is not suitable for WTGs with different FRCs. Moreover, FRCs also vary because of wind fluctuation and power variation in the frequency response process. In this situation, the droop coefficients of WTGs in each group are proportional to their real-time FRCs, i.e.,

$$K_p^{WTG\,i} \propto P_{cap}^{WTG\,i} \tag{34}$$

Considering the constraint of (19), $K_p^{WTG\,i}$ for WTGs in the two groups can be calculated by

$$K_p^{WTG\,i} = \begin{cases} K_p^{G_1} \cdot \dfrac{P_{cap}^{WTG\,i}}{P_{cap}^{G_1}}, & i \in G_1 \\[2ex] K_p^{G_2} \cdot \dfrac{P_{cap}^{WTG\,i}}{P_{cap}^{G_2}}, & i \in G_2 \end{cases} \tag{35}$$

WTGs with greater capabilities provide more output in frequency regulation while those with less capability give a small output increase to guarantee stable operation. Therefore, the resources of frequency regulation can be efficiently utilized with the proposed coefficients.

5. Case Studies

The simulations in this paper are implemented using MATLAB/Simulink. A developed low-order power system model [32] (Figure 9) is adopted to verify the proposed GBDC scheme of a deloading WF. The voltage control involving the dynamics of the system's reactive power is not considered. The system contains a 20 MW small hydropower station, a 120 MW thermal power plant with combined-cycle gas turbines, and a wind farm. All of the conventional power plants are equipped with droop control to participate in frequency regulation, and the gains of the droop loops, K_p, are set to 20. H_e in Figure 9 is the equivalent inertia time constant of the system, which can be calculated by [35,36]

$$H_e = r_W H_W + r_T H_T \tag{36}$$

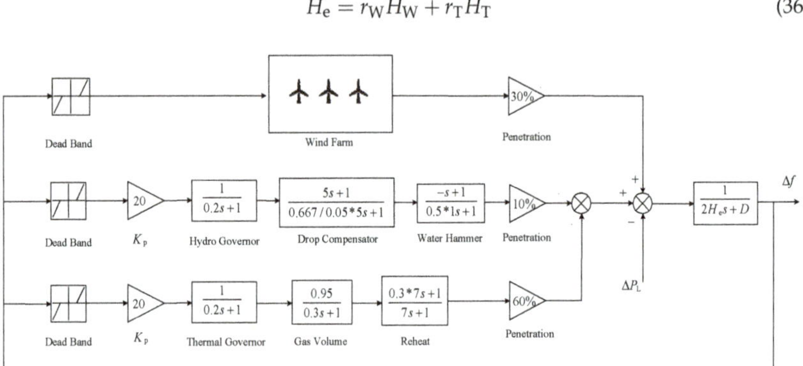

Figure 9. Primary frequency control of the low-order system.

H_W and H_T are the inertia time constants of the hydropower station and the thermal power plant, and the values of them are 3.6 s and 6.5 s, respectively. r_W and r_T are the hydro and thermal penetrations, which are 10% and 50%, respectively. The load damping, D, is 0.5 in each unit.

The wind farm is composed of twenty 3 MW DFIGs, arranged in four rows (as shown in Figure 10). The wind power penetration of the system is 30%. The GBDC scheme is compared with three different droop control schemes, namely, (1) traditional fixed droop control (FDC), (2) adaptive droop control (ADC), and (3) no droop control (NDC), for wind farms. In the FDC scheme, every WTG responds to the frequency with a fixed droop coefficient. The droop coefficients of the FDC scheme are obtained by the trial and error method to ensure that all WTGs operate stably in the frequency response process. The ADC scheme has no function of pitch angle protection; thus, the droop coefficients of all WTGs are proportional to their maximum FRCs, which are calculated by

$$K_p^{WTG\,i} = \frac{\Delta P_{max}^{WTG\,i}}{|\Delta f|_{max}} \tag{37}$$

The WTGs in all these schemes operate with a 10% deloading ratio through the CDS strategy. The gains of the virtual inertial control are all set to 0. Additionally, all the generators in this system respond to the frequency deviation through a ±0.01 Hz dead band.

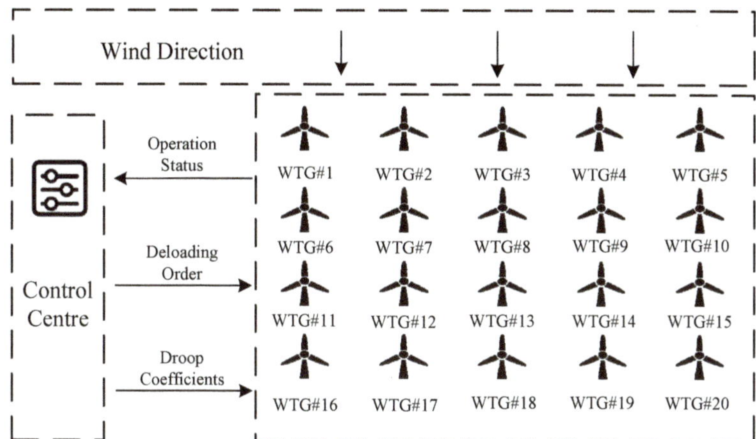

Figure 10. Structure of the wind farm.

5.1. Case 1: Small Load Disturbance

The free wind speed in this case is 11.00 m/s and remains unchanged. The wind direction is shown in Figure 10. Due to the wake effect, the actual wind speed received by each row of WTGs gradually decreases. In this paper, the arriving wind speed of each row is calculated by the Jansen wake model [24]. The wind speeds from front to back are set as 11.00 m/s, 9.57 m/s, 9.41 m/s, and 9.36 m/s. The WTGs in the first row are under high wind speed, while the others are under low wind speed. At $t = 30$ s, the system load suddenly increases from 120 MW to 126 MW.

The results of Case 1 are presented in Figure 11. Since both the GBDC and the ADC schemes determine the total droop coefficient according to the FRC of the wind farm, the total active power increase in the wind farm is greater than in the case of the FDC scheme. With the GBDC scheme, the frequency nadir, f_{nadir}, is 49.83 Hz, which is almost the same as that with the ADC scheme. f_{nadir} with the GBDC and the ADC schemes is higher than that with the FDC and the NDC schemes by 0.06 Hz and 0.15 Hz, respectively, and the steady-state frequency is closer to the standard frequency, 50 Hz. It needs to be pointed out that the frequency curves of the proposed scheme and the ADC scheme are slightly different. The reason for this is that the different droop coefficient distributions in these two schemes cause different rotor speeds of the WTGs; thus, the sum of P_{ref} in (14) is also different.

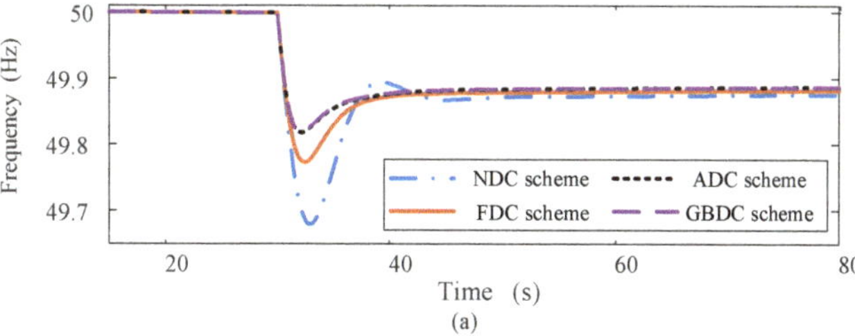

Figure 11. *Cont.*

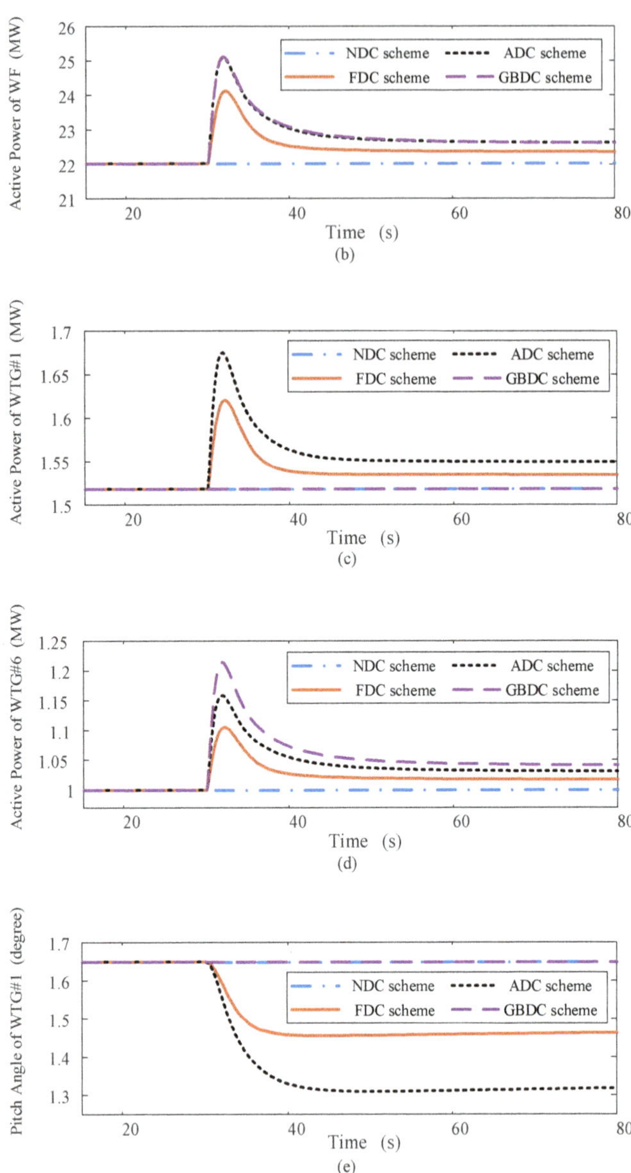

Figure 11. Simulation results of Case 1: (**a**) system frequency; (**b**) active power of WF; (**c**) active power of WTG#1; (**d**) active power of WTG#6; (**e**) pitch angle of WTG#1.

The dynamics of WTGs are also presented in Figure 11 to prove the advancement of the GBDC scheme. We can observe in Figure 11c,d that, when the disturbance is small, the resources of frequency regulation in G_1 (represented by WTG#6) are adequate; thus, there is no need for WTGs in G_2 (represented by WTG#1) to participate in the droop control of the WF. When the proposed scheme is adopted, the power increase in WTG#1 is negligible, which is the same as that using the NDC scheme and much smaller than those with the FDC and ADC schemes. The power increase in WTG#6 with the proposed droop scheme is larger than those with the other schemes, which suggests that the power deficit is supplemented by the WTGs in G_1. As the WTGs all deload with the CDS, the pitch angles of WTG#1 are all $1.649°$ in the initial stage of the simulation. The pitch angle curve of the proposed scheme is the same as that of the NDC scheme and almost remains unchanged. With the FDC scheme, the pitch angle is gradually reduced to $1.460°$. Due to the larger power increase, the change in the pitch angle of the ADC scheme is greater, which is reduced to $1.312°$.

5.2. Case 2: Large Load Disturbance

The wind speed of the wind farm is the same as that in Case 1. At $t = 30$ s, the load increases from 120 MW to 136 MW.

The results of Case 2 are presented in Figure 12. Because of the FRC-based droop coefficients, the power increases in the WF with the GBDC and the ADC schemes are very close, and they are both higher than those with the other two schemes. f_{nadir} with the GBDC scheme is 49.71 Hz, which is the same as that with the ADC scheme and higher than that with the FDC and NDC schemes by 0.30 Hz and 0.55 Hz, respectively. The steady-state frequency with the GBDC and the ADC schemes is 49.71 Hz, which is higher than those with the FDC and NDC schemes by 0.01 Hz and 0.03 Hz, respectively.

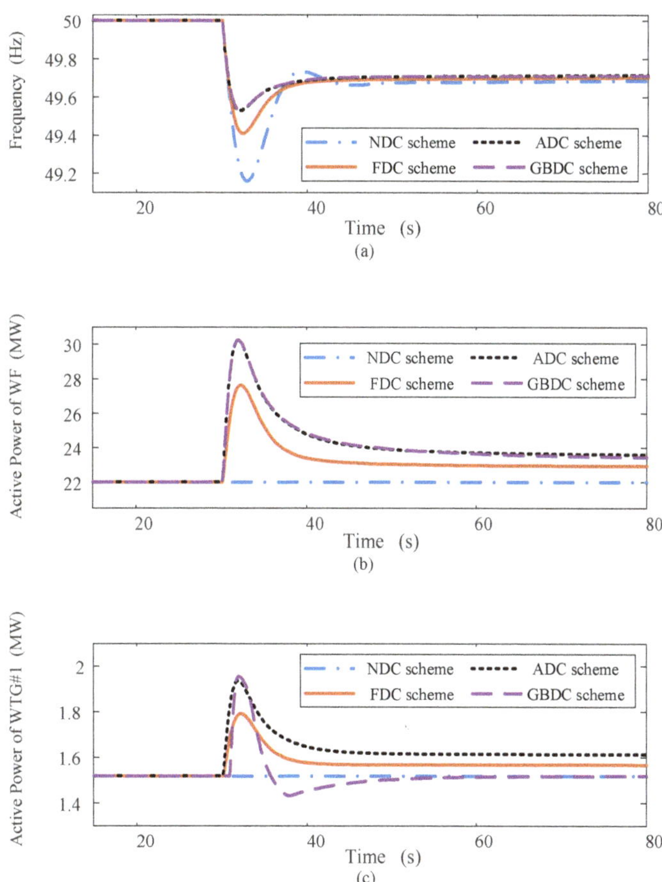

Figure 12. *Cont.*

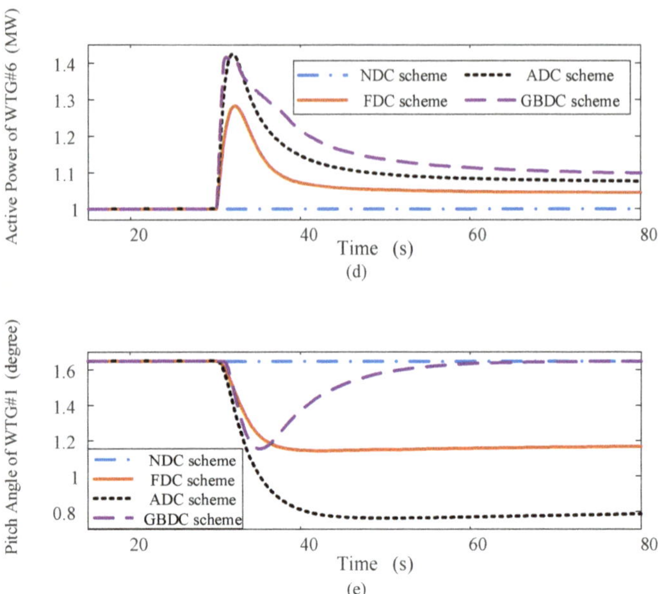

Figure 12. Simulation results of Case 2: (**a**) system frequency; (**b**) active power of WF; (**c**) active power of WTG#1; (**d**) active power of WTG#6; (**e**) pitch angle of WTG#1.

The power and pitch angle curves of WTG#1 and WTG#6 are also presented in Figure 12. In the GBDC scheme, G_1 and G_2 are both appointed to participate in frequency regulation because of the large frequency deviation. The power increases in WTG#1 and WTG#6 are larger than those in the FDC scheme. In the ADC scheme, the peak powers of WTG#1 and WTG#6 are almost the same as those in the GBDC scheme. As the deviation decreases in the frequency regulation process, the GBDC scheme reduces the participating degree of G_2. This is why the power of WTG#1 decreases faster than that in the ADC scheme. In the ADC scheme, the pitch angle of WTG#1 is reduced from $1.649°$ to $0.770°$ to capture more mechanical power. In the FDC scheme, the settled value is $1.160°$. In the GBDC scheme, the pitch angle is reduced to $1.158°$ at $t = 34.9$ s and finally returns to $1.649°$.

Compared with the traditional FDC scheme, the GBDC and the ADC schemes can both improve the frequency response performance. For the same extent of frequency improvement, the GBDC scheme makes the pitch angle move less than the ADC scheme.

5.3. Case 3: Wind Disturbance

In this case, the system load remains unchanged at 120 MW. Initially, the wind speed of the WF is 11.00 m/s, and it drops to 10.30 m/s at $t = 30$ s. To simplify the simulation, the time delay of the wind speed drops of each row of WTGs is ignored. Therefore, the arriving wind speeds of each row are reduced from 11.50 m/s, 10.09 m/s, 9.93 m/s, and 9.88 m/s to 10.50 m/s, 9.05 m/s, 8.89 m/s, and 8.83 m/s at $t = 30$ s, respectively.

The results of Case 3 are presented in Figure 13. It can be seen in Figure 13b that, due to the decrease in wind speed, the power of the WF decreases at $t = 30$ s, producing a disturbance in the system frequency. When detecting the frequency deviation, the WF begins to increase its output to participate in frequency regulation. The power increases in the GBDC and the ADC schemes are larger than that in the FDC scheme because of the larger droop coefficients. Therefore, the frequency curves of these two schemes are higher than those of the FDC and the NDC schemes in the whole simulation. The f_{nadir} of the NDC scheme is 49.82 Hz and appears at 34.33 s. At the same time, the frequencies of the proposed scheme and the ADC scheme are both 49.89 Hz, while the frequency value of the FDC scheme is 49.87 Hz.

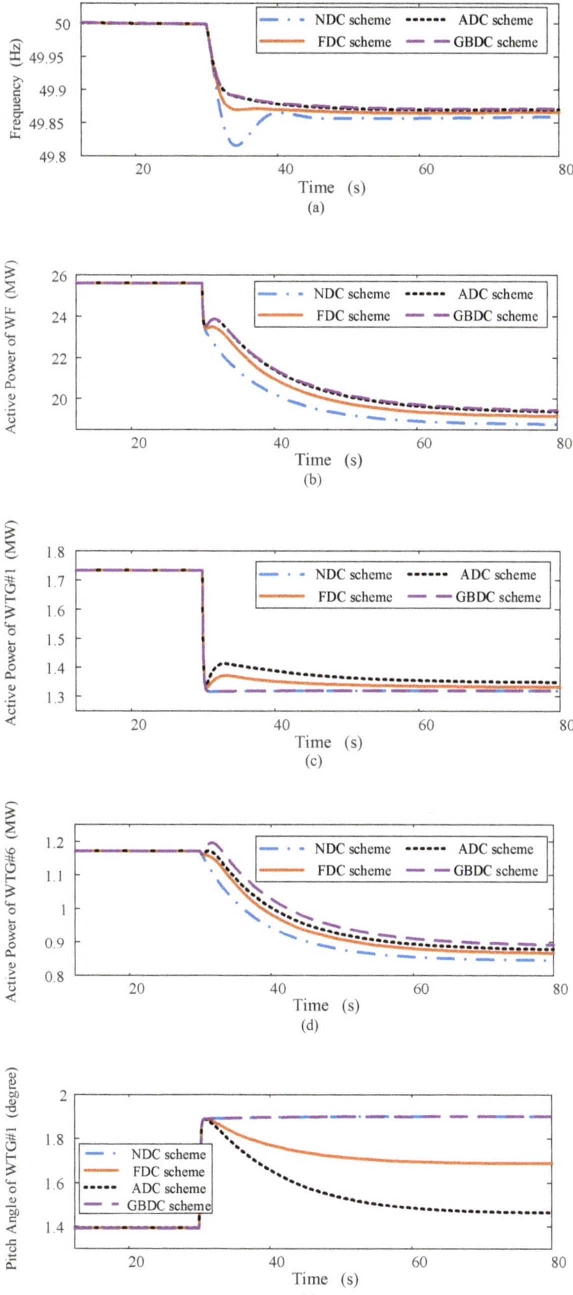

Figure 13. Simulation results of Case 3: (**a**) system frequency; (**b**) active power of WF; (**c**) active power of WTG#1; (**d**) active power of WTG#6; (**e**) pitch angle of WTG#1.

As the deloading pitch angle, β_0, varies with the wind speed, the pitch angle of the WF increases to a new value at $t = 30$ s. After that, the pitch angle of the FDC and the ADC schemes is reduced so that the power reserve stored by the PAC can be extracted. The active power curves of WTG#1 and WTG#6 are presented in Figure 13c,d, which represent the WTGs of G_2 and G_1, respectively. With the proposed scheme, the power increase in WTG#1 is the same as that with the NDC scheme, which is negligible. The power increase in the ADC scheme is the largest, as the first row of WTGs possess the largest FRC. As for the power increase in WTG#6, the schemes, in descending order, are the proposed scheme, the ADC scheme, the FDC scheme, and the NDC scheme. With the proposed scheme, the pitch angle of WTG#1 remains at the new deloading value, $1.900°$, the same as that with the NDC scheme. The curves of the ADC and the FDC schemes decrease to $1.469°$ and $1.691°$, respectively.

5.4. Case 4: Random Wind Conditions

To verify the practicality of the proposed scheme, a simulation under continuous wind and load disturbances is carried out in this section. The wind speed of each row is shown in Figure 14a, while the load disturbance is presented in Figure 14b. Moreover, the uncertainty of the frequency measurement is also included as a disturbance to test the robustness of the schemes. The measurement error is shown in Figure 14c.

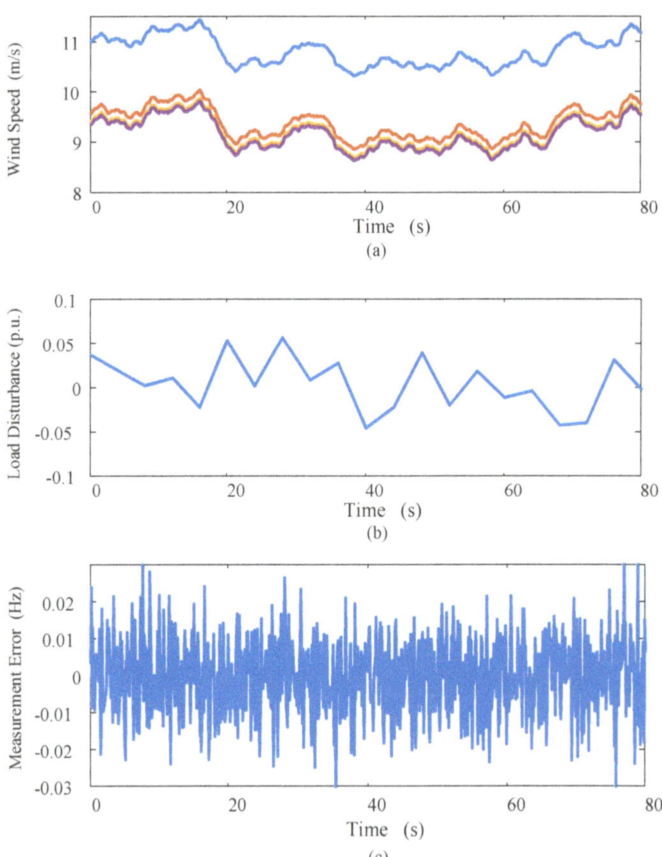

Figure 14. Disturbances: (**a**) wind disturbance; (**b**) load disturbance; (**c**) frequency measurement error.

The simulation results are presented in Figure 15. Compared with the NDC scheme, where WTGs do not participate in frequency regulation, the other schemes can all mitigate the frequency fluctuations. The frequency curves of the GBDC and ADC schemes are coincident and fluctuate more gently than that of the FDC scheme most of the time. Since there is no frequency regulation control, the pitch angle in the NDC scheme only changes with the wind speed. Therefore, the pitch angle of the NDC scheme can be applied as a reference to evaluate the fluctuation of the other schemes. Compared with the FDC and ADC schemes, the pitch angle curve of the GBDC scheme is closer to that of the NDC scheme, which means that the pitch angle changes more gently.

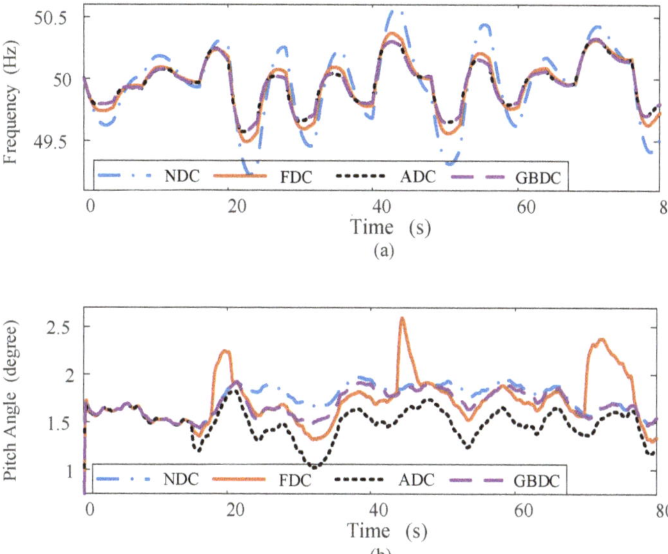

Figure 15. Simulation results of Case 4: (**a**) frequency; (**b**) pitch angle of WTG#1.

To further verify the superiority of the GBDC scheme, four indices are applied to evaluate the performances of the schemes. As introduced in [37], the expressions of the indices are as follows:

$$\begin{cases} \text{ISE}_{\Delta f} = \int_0^{t_s} [\Delta f(t)]^2 dt \\ \text{ITSE}_{\Delta f} = \int_0^{t_s} t[\Delta f(t)]^2 dt \\ \text{IAE}_{\Delta f} = \int_0^{t_s} |\Delta f(t)| dt \\ \text{ITAE}_{\Delta f} = \int_0^{t_s} t|\Delta f(t)| dt \end{cases} \tag{38}$$

$$\begin{cases} \text{ISE}_{\Delta \beta} = \int_0^{t_s} [\Delta \beta(t)]^2 dt \\ \text{ITSE}_{\Delta \beta} = \int_0^{t_s} t[\Delta \beta(t)]^2 dt \\ \text{IAE}_{\Delta \beta} = \int_0^{t_s} |\Delta \beta(t)| dt \\ \text{ITAE}_{\Delta \beta} = \int_0^{t_s} t|\Delta \beta(t)| dt \end{cases} \tag{39}$$

where t_s is the simulation time, and $\Delta \beta(t)$ is the difference between the pitch angles of the evaluated and the NDC schemes.

The curves of the indices are presented in Figures 16 and 17. The comparison proves that the ADC and the GBDC schemes can both improve the frequency to a satisfactory level.

As for the pitch angle indices, the accumulative error of the GBDC is closest to the reference (curves of the NDC scheme), which means that the fluctuation is the most gentle.

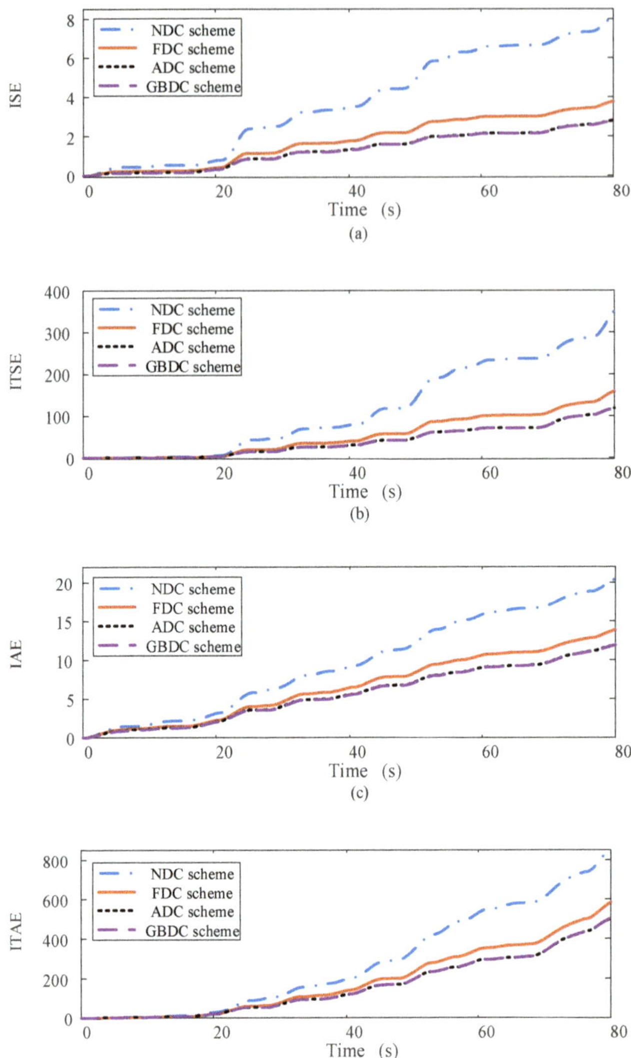

Figure 16. Indices of frequency deviation: (**a**) ISE; (**b**) ITSE; (**c**) IAE; (**d**) ITAE.

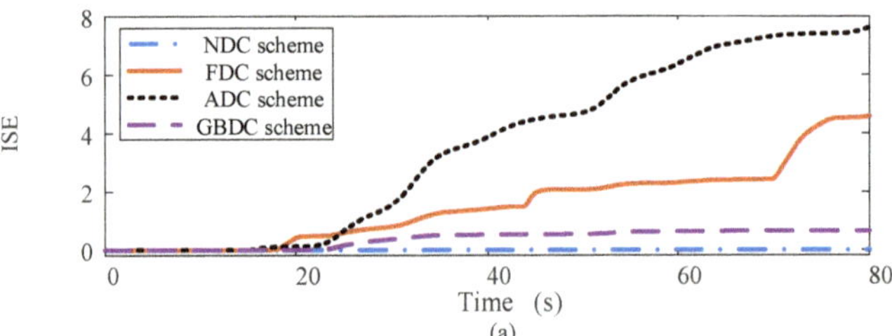

Figure 17. *Cont.*

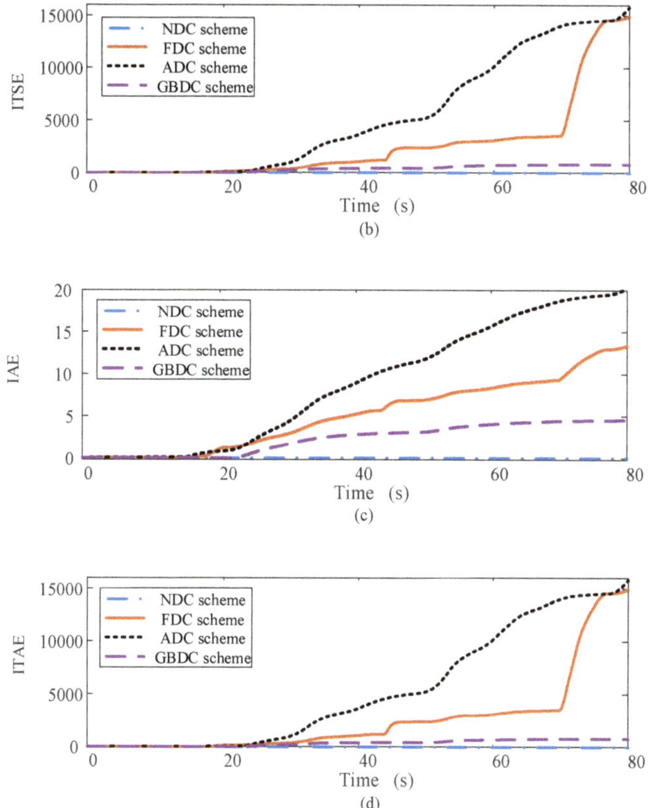

Figure 17. Indices of pitch angle deviation: (**a**) ISE; (**b**) ITSE; (**c**) IAE; (**d**) ITAE.

6. Conclusions

This paper proposes a sophisticated frequency regulation strategy for WFs. To prepare an adequate active power reserve for frequency response, a CDC scheme combining the OSC and the PAC is presented. In this scheme, WTGs at different ranges of wind speed deload using different methods. For WTGs at low wind speed, only the OSC is adopted so that the frequent pitch angle action caused by the PAC is avoided. For WTGs at high wind speed, the deloading work is accomplished by the coordination of the OSC and the PAC. In this study, the PAC is only adopted as a supplementary control method. Compared with the OSC scheme, the coordination enables the WTGs at high wind speed to possess a large deloading ratio.

For the frequency response of WFs consisting of WTGs with the CDC scheme, a GBDC scheme is proposed in this paper. In this scheme, the droop coefficients of the WF and the WTGs are determined according to the precisely evaluated FRCs. To improve the frequency response and ensure stable operation, the droop coefficients are proportional to the FRCs in the whole process. Moreover, to reduce the pitch angle action in the frequency response process, the WTGs are divided into two groups according to whether the PAC is adopted in the deloading process. Then, a pair of exponential functions are adopted to determine their participation degrees of frequency response. The simulations demonstrate that the GBDC scheme can improve the system frequency to the same extent as the ADC scheme, while the pitch angle action of the GBDC scheme is smaller. In some small disturbances, the GBDC scheme can even make the pitch angle fluctuate more gently than the conventional FDC scheme.

One consideration for future work is the influence of parametric uncertainty. In this paper, the interest is mainly focused on the mechanism between the system frequency and the frequency response power of WFs with different schemes. The WTG model is simplified to improve simulation efficiency. Considering various problems that may be encountered in practice, including the difficulty of parameter measurement, the frequency regulation methods based on parametric uncertainty and their influence on the frequency stability are of vital significance. Moreover, wind power fluctuation is a main disturbance source to the frequency. An accurate wind power prediction is helpful to develop the frequency regulation strategy. Therefore, a novel strategy that contains deloading and frequency response schemes based on wind power prediction is under ongoing investigation.

Author Contributions: Conceptualization, H.L.; methodology, H.L. and P.W.; software, P.W.; validation, H.L., T.Z. and Z.F.; formal analysis, T.Z.; investigation, P.W.; resources, P.W. and T.Z.; data curation, P.W.; writing—original draft preparation, P.W. and T.Z.; writing—review and editing, T.Z.; visualization, H.L. and H.P.; supervision, H.L. and Z.F. All authors have read and agreed to the published version of the manuscript.

Funding: This research was funded in part by the Natural Science Foundation for Distinguished Young Scholars of Guangxi (Grant No.2018GXNSFFA281006).

Institutional Review Board Statement: Not applicable.

Informed Consent Statement: Not applicable.

Data Availability Statement: Not applicable.

Conflicts of Interest: The authors declare no conflict of interest.

References

1. Vargas, S.A.; Esteves, G.R.T.; Maçaira, P.M.; Bastos, B.Q.; Cyrino Oliveira, F.L.; Souza, R.C. Wind Power Generation: A Review and a Research Agenda. *J. Clean. Prod.* **2019**, *218*, 850–870. [CrossRef]
2. Ye, L.; Zhang, C.; Xue, H.; Li, J.; Lu, P.; Zhao, Y. Study of Assessment on Capability of Wind Power Accommodation in Regional Power Grids. *Renew. Energy* **2019**, *133*, 647–662. [CrossRef]
3. Kumar, D.; Chatterjee, K. A Review of Conventional and Advanced MPPT Algorithms for Wind Energy Systems. *Renew. Sustain. Energy Rev.* **2016**, *55*, 957–970. [CrossRef]
4. Hu, L.; Xue, F.; Qin, Z.; Shi, J.; Qiao, W.; Yang, W.; Yang, T. Sliding Mode Extremum Seeking Control Based on Improved Invasive Weed Optimization for MPPT in Wind Energy Conversion System. *Appl. Energy* **2019**, *248*, 567–575. [CrossRef]
5. Bonfiglio, A.; Invernizzi, M.; Labella, A.; Procopio, R. Design and Implementation of a Variable Synthetic Inertia Controller for Wind Turbine Generators. *IEEE Trans. Power Syst.* **2019**, *34*, 754–764. [CrossRef]
6. Cheng, Y.; Azizipanah-Abarghooee, R.; Azizi, S.; Ding, L.; Terzija, V. Smart Frequency Control in Low Inertia Energy Systems Based on Frequency Response Techniques: A Review. *Appl. Energy* **2020**, *279*, 115798. [CrossRef]
7. Yang, D.; Jin, Z.; Zheng, T.; Jin, E. An Adaptive Droop Control Strategy with Smooth Rotor Speed Recovery Capability for Type III Wind Turbine Generators. *Int. J. Electr. Power Energy Syst.* **2022**, *135*, 107532. [CrossRef]
8. Boyle, J.; Littler, T.; Muyeen, S.M.; Foley, A.M. An Alternative Frequency-droop Scheme for Wind Turbines That Provide Primary Frequency Regulation via Rotor Speed Control. *Int. J. Electr. Power. Energy Syst.* **2021**, *133*, 107219. [CrossRef]
9. Wu, Y.; Yang, W.; Hu, Y.; Dzung, P.Q. Frequency Regulation at a Wind Farm Using Time-varying Inertia and Droop Controls. *IEEE Trans. Ind. Appl.* **2019**, *55*, 213–224. [CrossRef]
10. Van de Vyver, J.; De Kooning, J.D.M.; Meersman, B.; Vandevelde, L.; Vandoorn, T.L. Droop Control as an Alternative Inertial Response Strategy for the Synthetic Inertia on Wind Turbines. *IEEE Trans. Power Syst.* **2016**, *31*, 1129–1138. [CrossRef]
11. Prakash, V.; Kushwaha, P.; Sharma, K.C.; Bhakar, R. Frequency Response Support Assessment from Uncertain Wind Generation. *Int. J. Electr. Power Energy Syst.* **2022**, *134*, 107465. [CrossRef]
12. Hafiz, F.; Abdennour, A. Optimal Use of Kinetic Energy for the Inertial Support from Variable Speed Wind Turbines. *Renew. Energy* **2015**, *80*, 629–643. [CrossRef]
13. Wang, H.; Chen, Z.; Jiang, Q. Optimal Control Method for Wind Farm to Support Temporary Primary Frequency Control with Minimised Wind Energy Cost. *IET Renew. Power Gener.* **2015**, *9*, 350–359. [CrossRef]
14. Yang, D.; Kim, J.; Kang, Y.C.; Muljadi, E.; Zhang, N.; Hong, J.; Song, S.; Zheng, T. Temporary Frequency Support of a DFIG for High Wind Power Penetration. *IEEE Trans. Power Syst.* **2018**, *33*, 3428–3437. [CrossRef]
15. Altin, M.; Hansen, A.D.; Barlas, T.K.; Das, K.; Sakamuri, J.N. Optimization of Short-term Overproduction Response of Variable Speed Wind Turbines. *IEEE Trans. Sustain. Energy* **2018**, *9*, 1732–1739. [CrossRef]

16. Ruttledge, L.; Flynn, D. Emulated Inertial Response from Wind Turbines: Gain Scheduling and Resource Coordination. *IEEE Trans. Power Syst.* **2016**, *31*, 3747–3755. [CrossRef]
17. Wang, S.; Tomsovic, K. A Novel Active Power Control Framework for Wind Turbine Generators to Improve Frequency Response. *IEEE Trans. Power Syst.* **2018**, *33*, 6579–6589. [CrossRef]
18. Wilches-Bernal, F.; Chow, J.H.; Sanchez-Gasca, J.J. A Fundamental Study of Applying Wind Turbines for Power System Frequency Control. *IEEE Trans. Power Syst.* **2016**, *31*, 1496–1505. [CrossRef]
19. Liu, Z.; He, H.; Jiang, S.; Yu, H.; Xiao, S. The Effects of Wind Turbine and Energy Storage Participating in Frequency Regulation on System Frequency Response. In Proceedings of the 2021 IEEE 5th Advanced Information Technology, Electronic and Automation Control Conference (IAEAC), Chongqing, China, 12–14 March 2021; Volume 5, pp. 283–288.
20. Lyu, X.; Zhao, J.; Jia, Y.; Xu, Z.; Wong, K.P. Coordinated Control Strategies of PMSG-based Wind Turbine for Smoothing Power Fluctuations. *IEEE Trans. Power Syst.* **2019**, *34*, 391–401. [CrossRef]
21. Ye, H.; Pei, W.; Qi, Z. Analytical Modeling of Inertial and Droop Responses from a Wind Farm for Short-term Frequency Regulation in Power Systems. *IEEE Trans. Power Syst.* **2016**, *31*, 3414–3423. [CrossRef]
22. Fernández-Bustamante, P.; Barambones, O.; Calvo, I.; Napole, C. Provision of Frequency Response from Wind Farms: A Review. *Energies* **2021**, *14*, 6689. [CrossRef]
23. Abraham, A.; Hong, J. Dynamic Wake Modulation Induced by Utility-scale Wind Turbine Operation. *Appl. Energy* **2020**, *257*, 114003. [CrossRef]
24. Gao, X.; Wang, T.; Li, B.; Sun, H.; Yang, H.; Han, Z.; Wang, Y.; Zhao, F. Investigation of Wind Turbine Performance Coupling Wake and Topography Effects Based on LiDAR Measurements and SCADA Data. *Appl. Energy* **2019**, *255*, 113816. [CrossRef]
25. Brogna, R.; Feng, J.; Sørensen, J.N.; Shen, W.; Porté-Agel, F. A New Wake Model and Comparison of Eight Algorithms for Layout Optimization of Wind Farms in Complex Terrain. *Appl. Energy* **2019**, *259*, 114189. [CrossRef]
26. Aziz, A.; Oo, A.M.T.; Stojcevski, A. Frequency regulation capabilities in wind power plant. *Sustain. Energy Technol. Assess.* **2018**, *26*, 47–76. [CrossRef]
27. Choi, S.; Kang, Y.C.; Kim, K.; Lee, Y.; Terzija, V. A Frequency-responsive Power-smoothing Scheme of a Doubly-fed Induction Generator for Enhancing the Energy-absorbing Capability. *Int. J. Electr. Power Energy Syst.* **2021**, *131*, 107053. [CrossRef]
28. Li, Y.; Xu, Z.; Zhang, J.; Wong, K.P. Variable Gain Control Scheme of DFIG-based Wind Farm for Over-frequency Support. *Renew. Energy* **2018**, *120*, 379–391. [CrossRef]
29. Lee, J.; Muljadi, E.; Srensen, P.; Kang, Y.C. Releasable Kinetic Energy-based Inertial Control of a DFIG Wind Power Plant. *IEEE Trans. Sustain. Energy* **2016**, *7*, 279–288. [CrossRef]
30. Vidyanandan, K.V.; Senroy, N. Primary Frequency Regulation by Deloaded Wind Turbines Using Variable Droop. *IEEE Trans. Power Syst.* **2013**, *28*, 837–846. [CrossRef]
31. Mahish, P.; Pradhan, A.K. Distributed Synchronized Control in Grid Integrated Wind Farms to Improve Primary Frequency Regulation. *IEEE Trans. Power Syst.* **2020**, *35*, 362–373. [CrossRef]
32. Ochoa, D.; Martinez, S. Frequency Dependent Strategy for Mitigating Wind Power Fluctuations of a Doubly-fed Induction Generator Wind Turbine Based on Virtual Inertia Control and Blade Pitch Angle Regulation. *Renew. Energy* **2018**, *128*, 108–124. [CrossRef]
33. Ding, L.; Yin, S.; Wang, T.; Jiang, J.; Cheng, F.; Si, J. Integrated Frequency Control Strategy of DFIGs Based on Virtual Inertia and Over-speed Control. *Power Syst. Technol.* **2015**, *35*, 2385–2391.
34. Datta, U.; Kalam, A.; Shi, J. Frequency Performance Analysis of Multi-gain Droop Controlled DFIG in an Isolated Microgrid Using Real-time Digital Simulaton. *Eng. Sci. Technol. Int. J.* **2020**, *23*, 1028–1041.
35. Hwang, M.; Muljadi, E.; Park, J.; Sørensen, P.; Kang, Y.C. Dynamic Droop–based Inertial Control of a Doubly-fed Induction Generator. *IEEE Trans. Sustain. Energy* **2016**, *7*, 924–933. [CrossRef]
36. Xu, B.; Zhang, L.; Yao, Y.; Yu, X.; Yang, Y.; Li, D. Virtual Inertia Coordinated Allocation Method Considering Inertia Demand and Wind Turbine Inertia Response Capability. *Energies* **2021**, *14*, 5002. [CrossRef]
37. Zafran, M.; Khan, L.; Khan, Q.; Ullah, S.; Sami, I. Finite-Time Fast Dynamic Terminal Sliding Mode Maximum Power Point Tracking Control Paradigm for Permanent Magnet Synchronous Generator-Based Wind Energy Conversion System. *Appl. Sci.* **2020**, *10*, 6361. [CrossRef]

 *energies*

Article

Arc Voltage Distortion as a Source of Higher Harmonics Generated by Electric Arc Furnaces

Zbigniew Olczykowski

Faculty of Transport, Electrical Engineering and Computer Science, Kazimierz Pulaski University of Technology and Humanities, Malczewskiego 29, 26-600 Radom, Poland; z.olczykowski@uthrad.pl

Abstract: Due to high unit capacities, electric arc furnaces are among the receivers that significantly affect the power system from which they are supplied. Arc furnaces generate a number of disturbances to the power grid, including fast-changing voltage fluctuations causing the phenomenon of flickering light, asymmetry, and deformation of the voltage curve. The main issues discussed in the article are problems related to the distortion of current and voltage waveforms, resulting from the operation of electric arc furnaces. An analysis of the indices characterizing the voltage distortion recorded in the supply network of the arc furnaces is presented. The changes in the range of current and voltage waveform deformation in individual smelting phases in the arc furnace are also presented. Furthermore, the changes in the degree of deformation of the current and voltage waveforms in the individual smelting phases in an arc furnace are presented. A multi-voltage electric arc model used in computer simulations is proposed.

Keywords: arc furnaces; deformation of the arc voltage; multi-voltage electric arc model power quality; melting power

Citation: Olczykowski, Z. Arc Voltage Distortion as a Source of Higher Harmonics Generated by Electric Arc Furnaces. *Energies* **2022**, *15*, 3628. https://doi.org/10.3390/en15103628

Academic Editor: José Matas

Received: 18 April 2022
Accepted: 13 May 2022
Published: 16 May 2022

Publisher's Note: MDPI stays neutral with regard to jurisdictional claims in published maps and institutional affiliations.

1. Introduction

Arc furnaces are among the receivers whose load changes in both dynamic and stochastic ways. These changes are caused by the technological process carried out in the furnace, which uses an electric arc to melt the charge. The mathematical description of an electric arc is very complex, as it includes electromagnetic and gasodynamic phenomena, as well as chemical reactions. The knowledge of the arc model is necessary to determine the optimal operating point of the arc device, i.e., ensuring the maximization of the arc power during scrap melting [1,2]. The stabilization of the operating point is achieved through the use of electrode shift control systems, which use algorithms dependent on the adopted arc model [3,4].

The nature and value of the arc device current are influenced by many factors. They can be divided into several groups. The first includes factors related to the parameters of the supply line. These are mainly the supply voltage level, resistance, and reactance of the supply network, which determine the amount of short-circuit power at the connection point of the arc furnaces. The second group depends on the parameters of the arc furnace: the furnace power in relation to the short-circuit power of the network, the construction of the high-current circuit, and the measures used to limit the impact of the furnace on the power system. Another group consists of factors related to the method of smelting, including the chemical composition of the medium in which the arc takes place, its pressure and temperature, the distance between the electrodes, their material, scrap quality, and the amount of injected oxygen [5,6].

The complexity and variability of the physicochemical processes occurring at various phases of the technological process, in the arc itself, as well as in the environment of the electrodes, and the conditions of supplying the arc from the power system make the analysis of the operation of a circuit with an electric arc a complicated task. The result has been the publication of a great variety of arch models. One of the first models of the arc was proposed by Cassie, in which he assumed that the arc has a cylindrical column with a uniform temperature and current density [7,8]. Mayer proposed an improved model in which he assumed that the arc had a constant diameter with varying temperature and conductivity. He considered that the power losses were due to heat conduction at low currents. This means that the conductance is strongly dependent on the temperature and independent of the arc cross-sectional area [9]. The models proposed by Cassie and Mayr were analyzed, among others, in [10–12].

The literature on the subject presents many different arc models, including nonlinear resistance models [13,14], models using current–voltage characteristics [15–19], models represented by voltage sources [20,21], models based on the relationship between arc length and voltage [22,23], and models using neural networks and fuzzy logic [24–27]. In many publications, the influence of arc furnaces on voltage distortion in the supply network was presented on the basis of modeling with the use of simulation programs [28–30].

The present article is a continuation of the issues concerning the impact of electric arc furnaces on the power system, described by the author in [31–34]. In [32,33], voltage fluctuations generated by arc furnaces were characterized in detail. The results of measurements of indicators characterizing energy quality (including light flicker coefficients) were presented. The results of model tests were proposed in which the electric arc of the electric arc furnace was replaced with a voltage whose value depends on the arc length. The phenomenon of current and voltage asymmetry in the lines supplying arc furnaces was presented in [34].

The aim of the research presented in the article was to evaluate the influence of electric arc furnaces on the deformation of the voltage supplying the arc devices.

The presented results are an extension of the information presented in [31]. An analysis of changes in the indices characterizing distorted voltage waveforms (*THD*, U_h) recorded in the lines supplying the arc furnaces was performed.

The measurements were carried out for various supply conditions of the steelworks (at different short-circuit powers of the supply networks) and for arc furnaces with different power of furnace transformers.

In the model tests, the electric arc model of the electric arc furnace developed by the author was used. The presented model studies were aimed at estimating how the deformation of the electric arc voltage affects the power system from which the arc furnaces are supplied. The main objective of the model tests was to determine the influence of the supply conditions (short-circuit power of supply lines) in metallurgical plants (steel mills). The influence of the arc voltage deformation on the melting power (active power consumed by the arc furnace) was also determined.

The article does not deal with the issues related to limiting the distortion of currents and voltages generated by arc furnaces.

2. Deformation of Voltages and Currents Arising during Steel Smelting in an Electric Arc Furnace

The source of higher harmonics in the supply networks of arc furnaces is the nonlinear nature of the arc. The arc voltage changes its shape during the smelting from close to rectangular in the initial stage of smelting the scrap, and from triangular to sinusoidal in the final stage of smelting.

Arc furnaces generate the greatest disturbances in the initial phase of scrap melting. This is caused by dynamic changes in the electric arc voltage. The arc furnace operates from the short-circuit of the electrodes with the charge to the idle state (electric arc break). After ignition of the electric arc between the charge and the electrodes, the arc voltage curve changes very dynamically (Figure 1a).

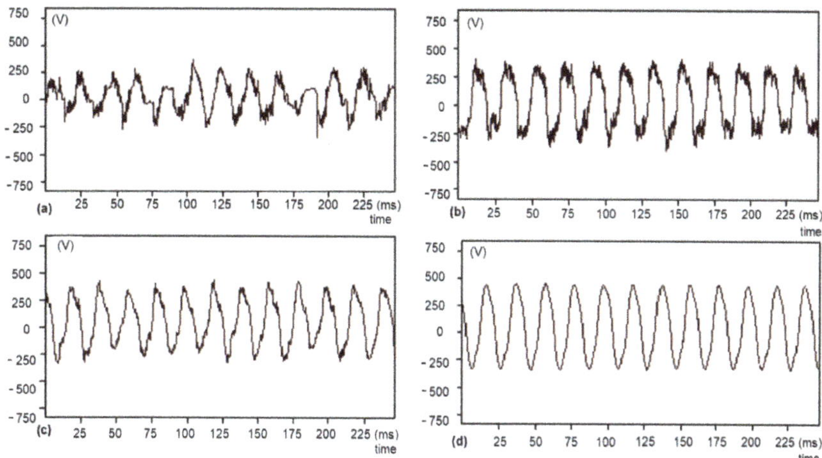

Figure 1. The shape of the arc voltage curve: (**a**) arc ignition start; (**b**) smelting scrap; (**c**) the main phase of melting the scrap; (**d**) final stage of smelting [31].

The electric arc of the electric arc furnace changes its length, which causes a change in the voltage value between the charge and the electrodes. As the scrap melts, the arc voltage at this stage is closer to the square wave (Figure 1b), which can be used to extract the arc voltage frequency. In the next step of smelting the scrap, the arc voltage is triangular in shape (Figure 1c). In the final stage of smelting the scrap, the arc voltage is sinusoidal (Figure 1d). The electric arc is covered with foamed slag, which causes thermal stabilization of the furnace operation [31].

Changes in the arc voltage waveform distort the arc furnace current waveform. The result of current distortion in the mains supplying the arc devices is a distorted voltage drop. As a result, the supply voltage of the steelworks has a deformed course.

Figure 2 presents the changes in the value of the harmonic distortion of the THD_I current and the THD_U voltage, recorded during one smelting in the arc furnace. There is a significant correlation between the THD_I and THD_U waveforms. The correlation coefficient is $r_{THDI\text{-}U} = 0.71$.

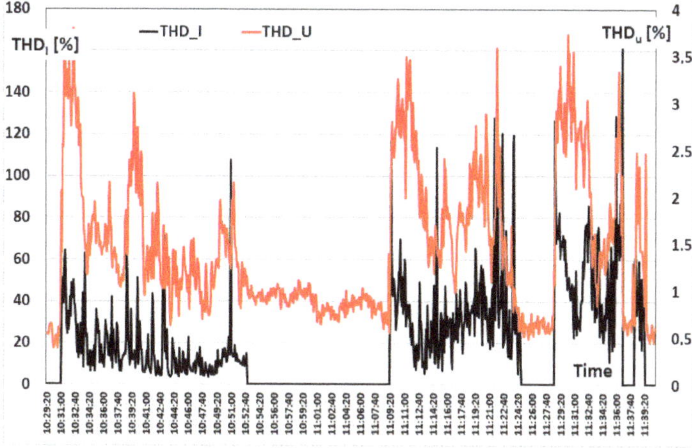

Figure 2. Changes in THD_I and THD_U coefficients in the different smelting phases in the electric arc furnace.

Analytical determination of the indices characterizing the deformation of voltages and currents in the circuits of arc furnaces is very difficult due to the fact that the values of these indices are random variables. Additionally, the shape of the current (similar to the arc voltage—Figure 1) changes with the time of smelting. The harmonic content of the voltage in the electric arc furnace supply network increases with the increase in the furnace power and decreases with the increase in the short-circuit power of the network.

The main cause of the arc furnace current distortion is the arc voltage distortion between the electrodes and the charge. The distorted current is the cause of the distorted voltage drop on the impedance of the arc supply line. As a result, the supply voltage of the steel plant does not have a perfect sinusoidal wave.

Figure 3 shows the current waveform recorded in the initial stage of smelting in the arc furnace supply network with the power of a 20 MVA furnace transformer. In this phase, there are changes in which the frequency of the current cannot be clearly determined because it is not periodic.

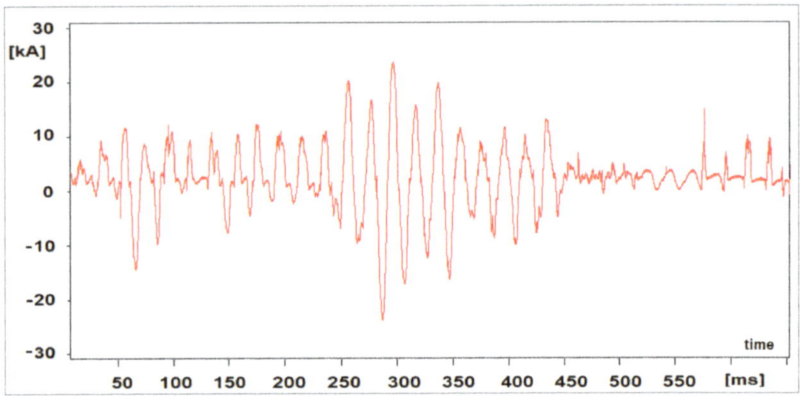

Figure 3. The oscillogram of the arc current in the initial phase of melting the scrap.

In the next part of the smelting, powerless periods occur sporadically. The current, however, shows dynamic changes in value, which in turn causes voltage fluctuations in the supply network (Figure 4).

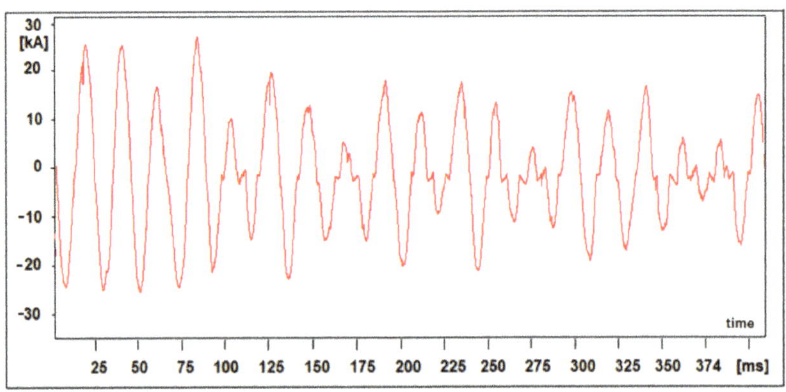

Figure 4. The oscillogram of the arc furnace current recorded in the next phase of scrap melting.

In the last stage of the smelting, with the melt and the smelting with foamed slag, it is possible to speak of a furnace current whose shape is similar to a sinusoidal course (Figure 5).

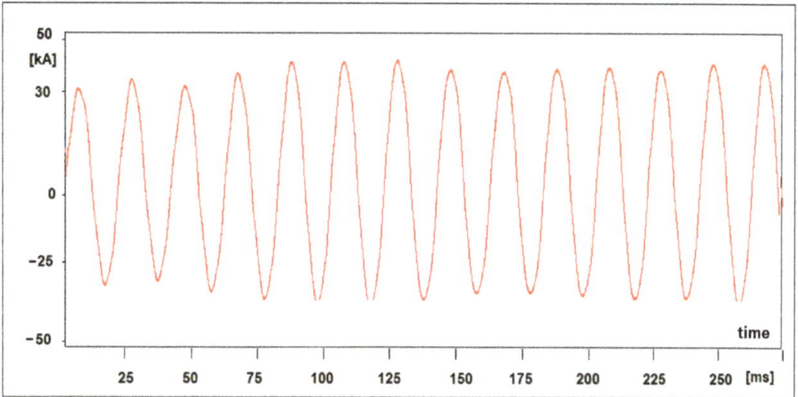

Figure 5. Oscillogram of the current with complete melting of the charge in the electric arc furnace.

In order to assess the voltage distortion in the networks supplying arc devices, measurements were made, and then an analysis of changes in the indices characterizing distorted waveforms (*THD*, U_h) was carried out. Figure 6 shows the changes in the THD_U harmonic distortion coefficient recorded in the steelwork supply network (110 kV) and in the arc furnace supply line (30 kV).

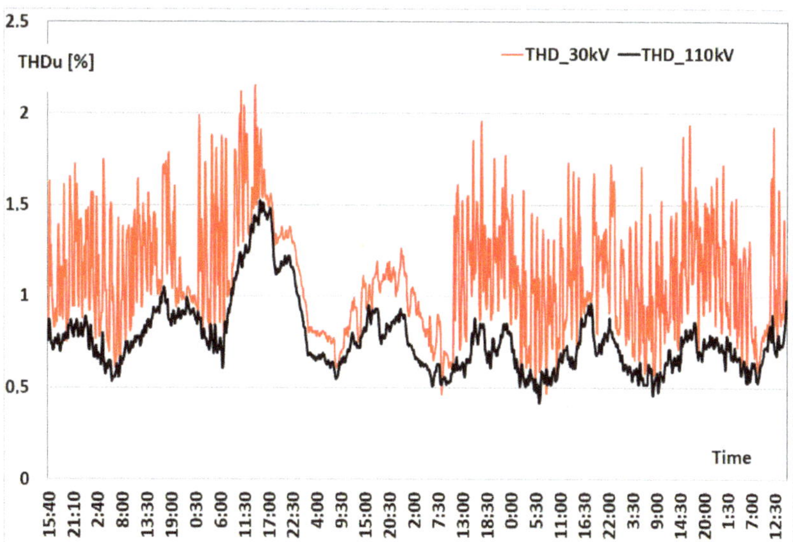

Figure 6. Harmonic distortion changes THD_U recorded during the measurement week in the steel plant supply network and the arc furnace supply line.

For the presented power supply system of the steelworks (rated voltage of the supply line $U_N = 110$ kV), the ratio of the short-circuit power of the network to the power of the arc furnace is $S_{cc}/S_{trc} = 134$. It was estimated that the influence of the electric arc furnace on voltage distortion at the level of 110 kV is small. Changes of the THD_U_110 kV coefficient are typical for the waveforms corresponding to the daily changes of the power system load. In the case of the furnace transformer supply line (rated line voltage $U_N = 30$ kV), the influence of the arc furnace on voltage distortion is much greater. There is a visible increase in THD_U_30 kV during the operation of the arc furnace. For the analyzed line, the ratio of the short-circuit power of the network to the power of the electric arc furnace is $S_{cc}/S_{trc} = 20$.

During the individual stages of steel smelting in the electric arc furnace, the values of the parameters characterizing the energy quality change. The reason for this is the dynamically changing current of the arc furnace. The indices determining the deformation of the supply voltage, i.e., THD_U and U_h, also change during the smelting. Figure 7 shows the changes in the total harmonic distortion of the THD voltage recorded during one smelting.

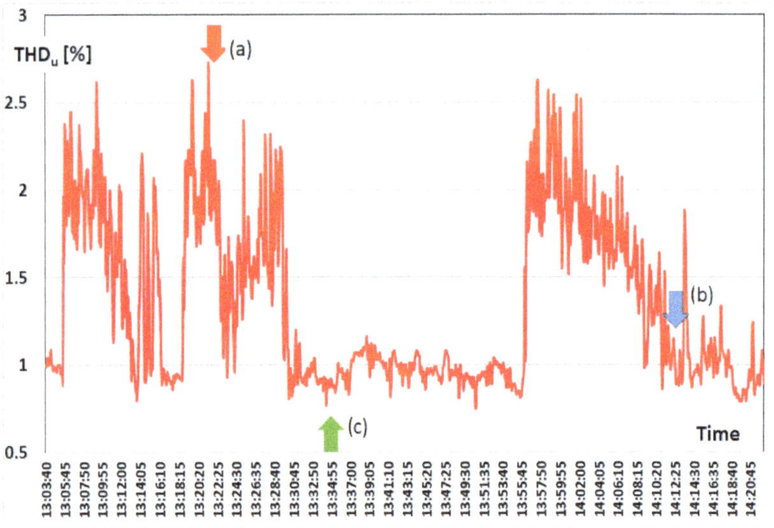

Figure 7. Changes in THD coefficient in the different smelting phases in the electric arc furnace (a) start of scrap melting, (b) final smelting phase—molten scrap, (c) arc furnace off.

After starting steel smelting in the electric arc furnace, when the arc voltage is the most deformed (the shape of the voltage curve is close to a square wave), the greatest deformation of the furnace current also occurs. This results in the greatest distortion of the voltage supplied to the arc furnaces (Figure 7, point a). Then, there are odd harmonics ($U_{(3)} = 1.238\%$, $U_{(5)} = 1.913\%$, $U_{(7)} = 0.428\%$, and $U_{(9)} = 0.151\%$) and even harmonics ($U_{(2)} = 0.372\%$, $U_{(4)} = 0.252\%$, $U_{(6)} = 0.267\%$, $U_{(8)} = 0.191\%$, and $U_{(10)} = 0.151\%$), as shown in Figure 8a When the scrap is melted, the voltage distortion and the content of individual harmonics are reduced, e.g., $U_{(2)} = 0.111\%$, $U_{(4)} = 0.116\%$, $U_{(6)} = 0.03\%$, $U_{(8)} = 0.015\%$, $U_{(10)} = 0.035\%$ and odd $U_{(3)} = 0.186\%$, $U_{(5)} = 0.982\%$, $U_{(7)} = 0.554\%$, and $U_{(9)} = 0.015\%$ (Figure 8b). Figure 8c shows the spectrum of voltage harmonics recorded with the arc furnace off, where $U_{(5)} = 0.644\%$ of the fundamental harmonic, while $U_{(3)} = 0.367\%$ and $U_{(7)} = 0.267\%$.

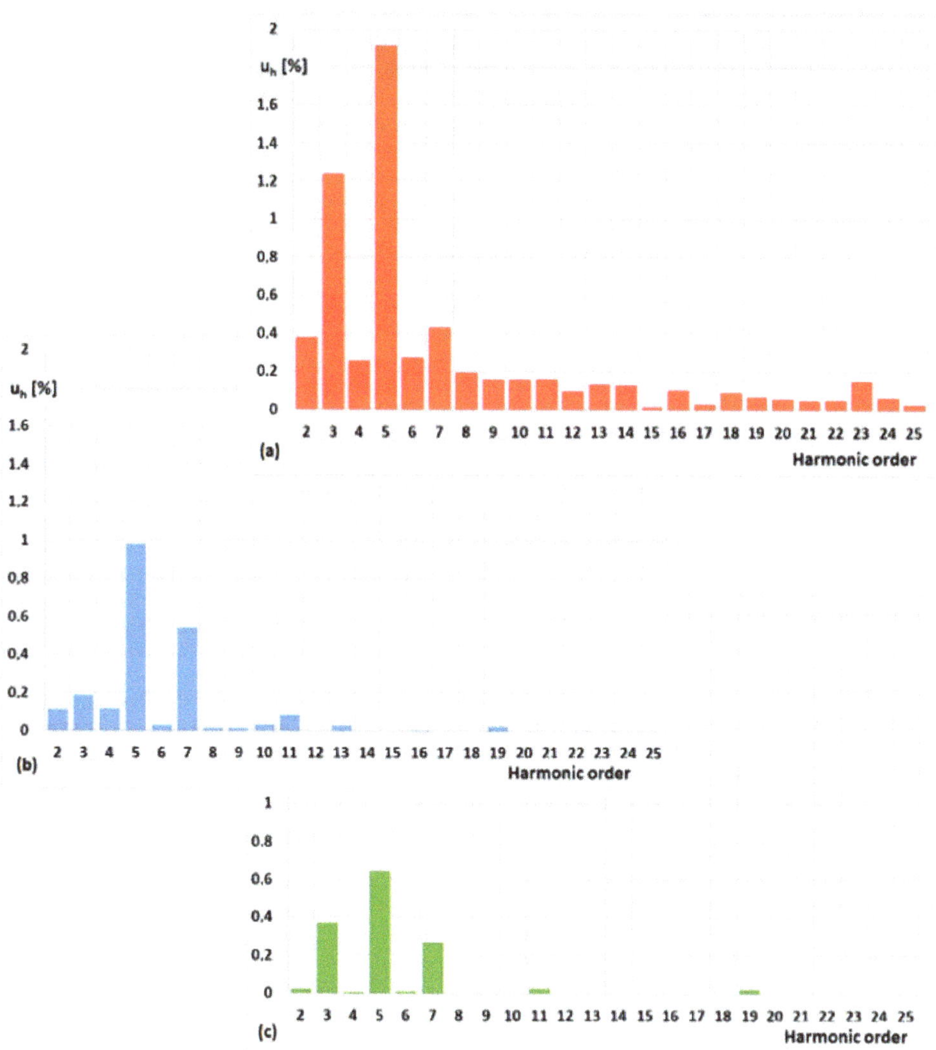

Figure 8. Spectrum of voltage harmonics in different melting phases in the arc furnace: (**a**) commencement of melting the charge; (**b**) melted charge; (**c**) when the furnace is turned off (loading the scrap into the furnace).

The changing spectrum of harmonics, in terms of both the value of individual components and their order (odd and even harmonics appearing—Figures 9 and 10), may cause disturbances in the operation of the furnace system equipment.

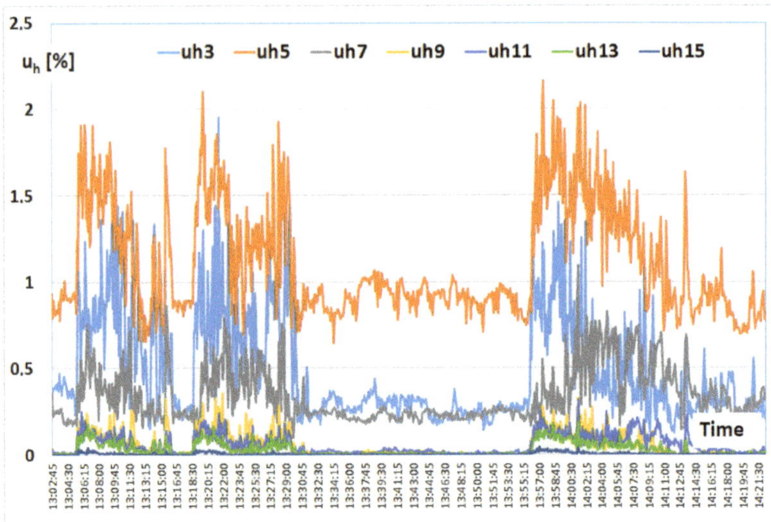

Figure 9. Changes of the amplitude in odd harmonics of the voltage during the smelting process in the arc furnace.

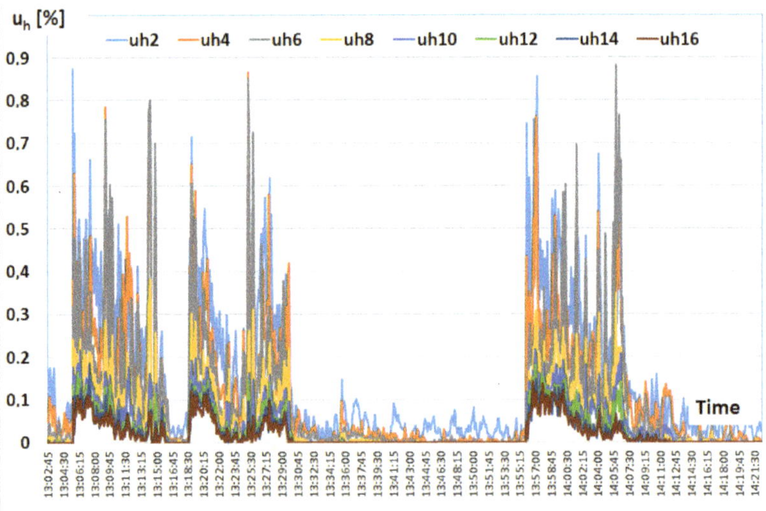

Figure 10. Even harmonic amplitude changes of voltage during smelting in an arc furnace.

Capacitor batteries used to compensate for the reactive power of the arc furnace are particularly at risk. A voltage surge may occur at the capacitor terminals due to resonance conditions. Figure 11 shows a damaged capacitor battery used to compensate for reactive power of the arc furnace. The cause of the capacitor failure was resonance phenomena causing an increase in the voltage supplying the capacitor bank. The next section of the article presents model studies presenting the influence of power supply conditions for arc furnaces (short-circuit power of the network) on transient phenomena resulting from switching on the capacitor bank.

Figure 11. Damaged capacitor battery used to compensate for the reactive power of the arc furnace.

3. Multi-Voltage Model of the Arc Device

Voltage distortion in arc supply networks depends both on disturbances caused by arc furnaces and on the influence of other nonlinear loads supplied from the power system. This is confirmed in Figure 6, showing changes in the THD_U coefficient recorded over a week in the power line supplying arc furnaces (at the voltage level of 110 kV or 30 kV). The value of the THD_U factor also depends on the short-circuit power of the network from which the steel plant is supplied and changes with the phases of smelting in the arc furnace. Additionally, during smelting, the spectrum of harmonics of the current and arc voltage changes. These factors make modeling of voltage distortions caused by arc devices extremely difficult. Due to the very high rated powers of furnace transformers and the dynamically changing load, arc furnaces are complex research facilities. Therefore, a certain compromise should be taken into account between the degree of complexity of the model and the accuracy (approximation) in reproducing the real conditions and its practical use.

Due to the fact that, in individual phases, there is a very large correlation between the changes in the total distortion factor voltage harmonics, by analyzing the voltage distortion caused by arc furnaces, the article considers the phase with the greatest deformation (the highest THD_U value), which results from a very large correlation between changes in the THD_U value in individual phases, e.g., for the steelworks supply line (Figure 12), i.e., $r_{THD12} = 0.995$, $r_{THD23} = 0.995$, and $r_{THD31} = 0.995$.

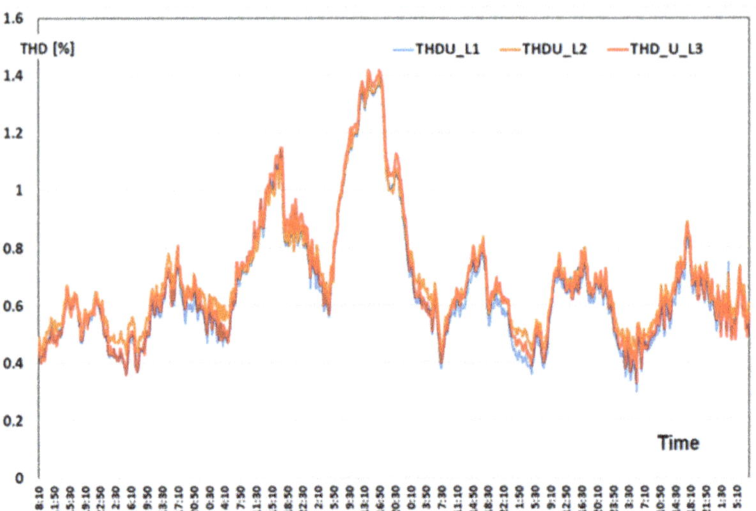

Figure 12. Changes in the total harmonic distortion THD_U recorded in three phases of the steel supply network during the week of measurements.

Analyzing the THD_U changes recorded in the individual phases of the steel supply line (Figure 13), during one melt (at 5 s measurement intervals), a significant correlation was also found, i.e., $r_{THD12} = 0.905$, $r_{THD23} = 0.880$, and $r_{THD31} = 0.897$.

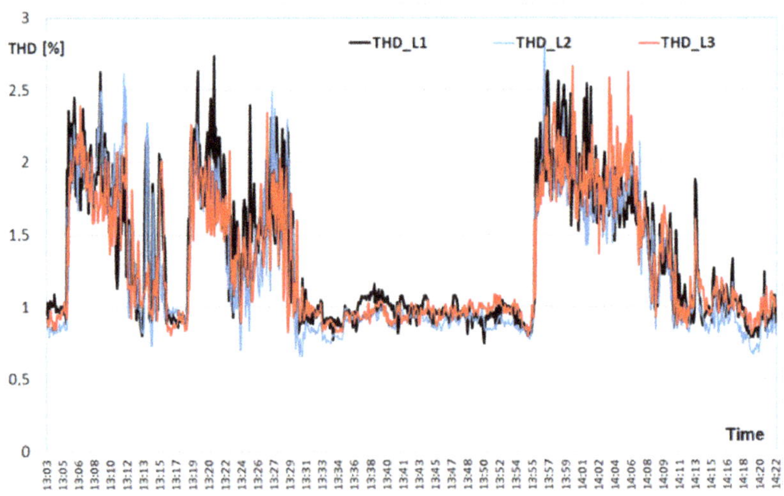

Figure 13. Changes in the total harmonic distortion THD recorded in three phases of the steelwork supply network during one melting in the arc furnace.

Taking into account the significant correlation of THD changes in individual phases of the supply line to the arc furnaces, a single-phase equivalent diagram of the arc device with a multi-voltage electric arc model is proposed in Figure 14 [31].

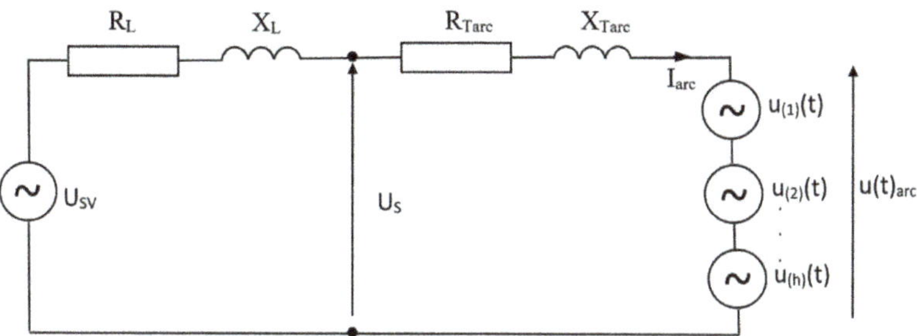

Figure 14. Single-phase diagram of equivalent power supply for an arc furnace with a multi-voltage electric arc model [31].

The following designations are adopted in Figure 14: U_{SV}—rated voltage, U_S—steelwork supply voltage, R_L, X_L—power line resistance and reactance, R_{Tarc}, X_{Larc}—resistance and reactance high-current path with the furnace transformer, I_{arc}—arc furnace current, $U_{(1)}(t)$, $U_{(2)}(t), \ldots , U_{(h)}(t)$—voltage harmonics, and $U(t)_{arc}$—arc voltage.

The article proposes a model based on harmonic voltage (harmonic voltage source model—HVSM). In this model, the arc furnace is represented as a series connection of the resistance R_T and the reactance X_T representing the high-current path and the arc voltage $U_{arc}(t)$ (Figure 15).

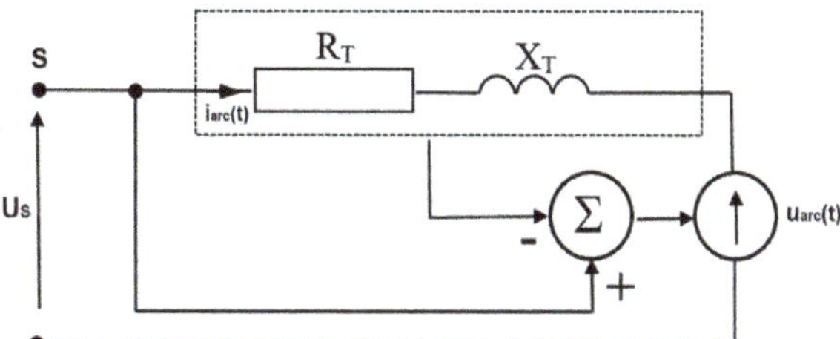

Figure 15. The arc furnace model based on the harmonics of the electric arc voltage (HVSM).

The arc voltage is calculated as the difference between the measured voltage $U_S(t)$,

$$U_S(t) = \sum_{h=1}^{n} U_h \sin(h\omega t + \varphi_h),\tag{1}$$

and that determined on the basis of the measured current, $i_{arc}(t)$, with the voltage $U_S(t)$ on the impedance of the high-current circuit.

$$U_S(t) = R_T i_{arc}(t) + L_T \frac{di_{arc}(t)}{d(t)}.\tag{2}$$

Hence,

$$U_{arc}(t) = U_S(t) - U_{SV}(t).\tag{3}$$

The presented two models based on the measurements of real voltages and currents in networks supplying arc devices can be classified as dynamic models of the arc device.

Parallel to the measurements of indicators characterizing the quality of electric energy (e.g., THD_U, U_h) in the lines supplying the arc furnaces, the voltage of the electric arc was recorded. The recorded waveforms of the electric arc voltage were the basis for the development of a multi-voltage model of the arc device. The proposed model takes into account changes in the arc voltage shape during the smelting process. This model consists of series connected voltage sources of successive harmonics representing the changing arc voltage. The change in both the value and the shape of the arc voltage is taken into account. Figure 16a shows an example of the recorded arc voltage. The spectrum of the true arc voltage is determined using FFT. Knowing the spectrum of voltage harmonics, the arc device was modeled using the proposed model (Figure 16b) [31].

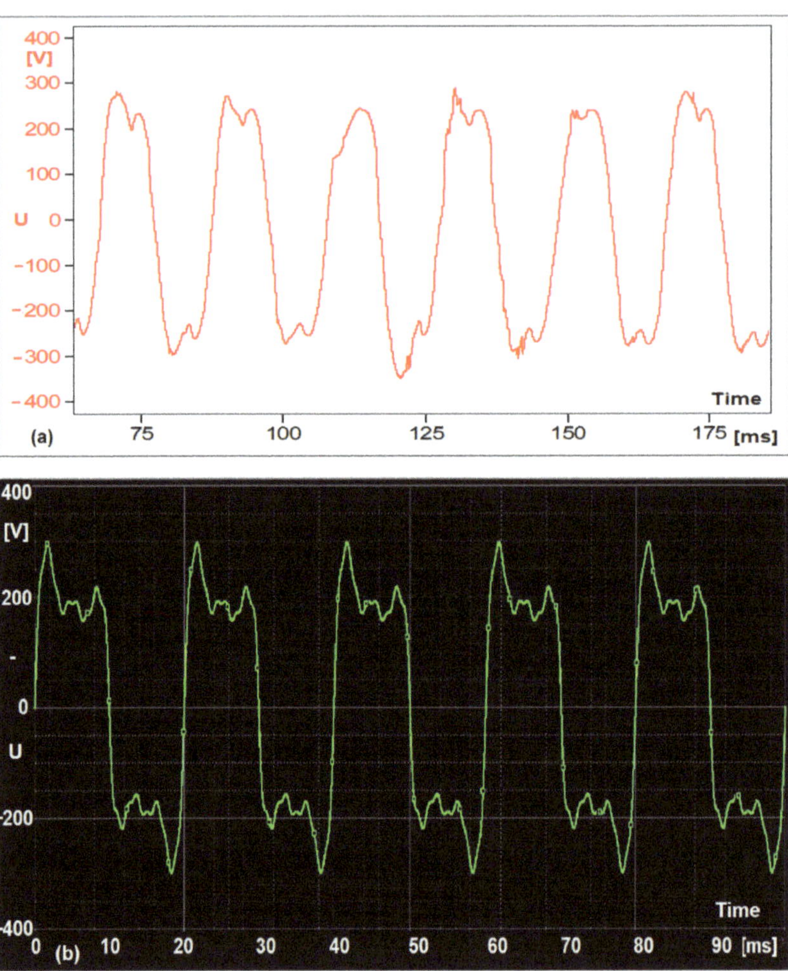

Figure 16. Electric arc voltage oscillograms recorded in real conditions (**a**) and determined using the proposed model (**b**)—initial stage of melting [31].

Taking into account the change in the arc voltage shape and the change in the content of higher harmonics of this voltage, the operation of the arc device in the phase after melting the scrap in the arc furnace was modeled (Figure 17).

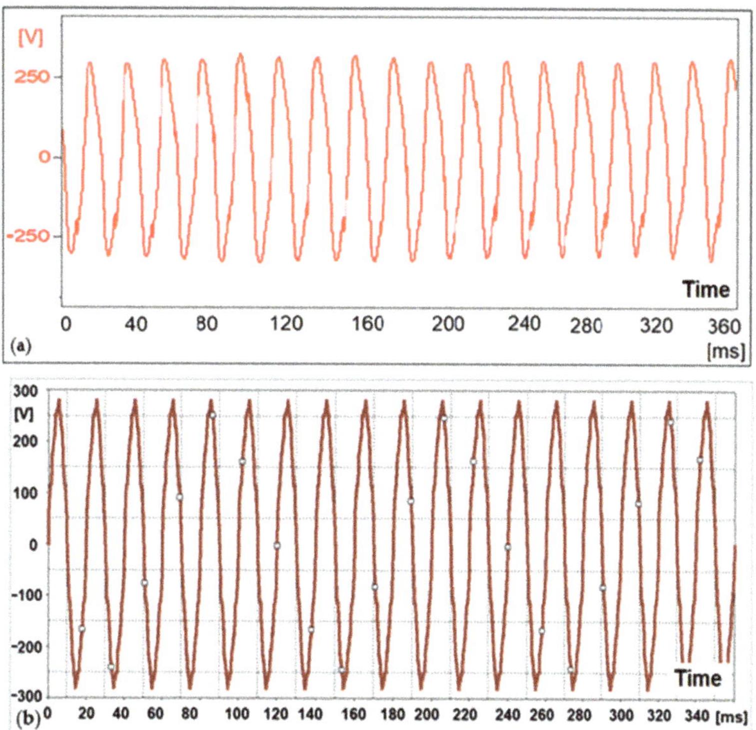

Figure 17. Electric arc voltage oscillograms recorded in real conditions (**a**) and determined using the proposed model (**b**)—after melting the charge [31].

The power supply conditions of the steelworks have a decisive influence on the deformation of the voltage supplying arc furnaces. These conditions are determined by the ratio of the short-circuit power of the network supplying the arc devices to the power of the arc furnace at the short-circuit of the electrodes with the charge—S_{cc}/S_{ts}. Assuming the actual supply conditions of the steel mill, two different ratios of the short-circuit power of the network to the power of the furnace at the short-circuit of the electrodes with the charge were assumed: S_{cc}/S_{ts} = 134—Figure 18a and S_{cc}/S_{ts} = 20—Figure 18b.

The article proposes a single-phase equivalent diagram of the arc device. A certain simplification of the presented electric arc furnace based on the multi-voltage model of the electric arc is the two-voltage model of the arc device—Figure 19.

Figure 20 shows the spectrum of voltage harmonics for an ideal square wave. In the case of a square wave, the harmonic distortion factor THD_{Urec} is about 0.483, assuming the amplitude of the first harmonic as 1.

The harmonic distortion factor THD for a rectangular arc voltage can also be determined from Equation (4).

$$THD_{Urec} = \frac{\sqrt{\sum_{v=2}^{\infty} U_{recv}^2}}{U_{rec1}} = \sqrt{\frac{U_{rec0}^2 - U_{rec1}^2}{U_{rec1}^2}} = \sqrt{\frac{U_{rec0}^2}{U_{rec1}^2} - 1} = \sqrt{\frac{\pi^2}{8} - 1} = 0.4834 \quad (4)$$

Figure 21 shows the changes in the value of the arc voltage distortion coefficient THD_{Uarc} as a function of parameter b. For the parameter $b = 0$, the arc voltage waveform is rectangular (THD_{Uarc} = 0.483); when $b = 1$, the arc voltage shape is sinusoidal (THD_{Usin} = 0).

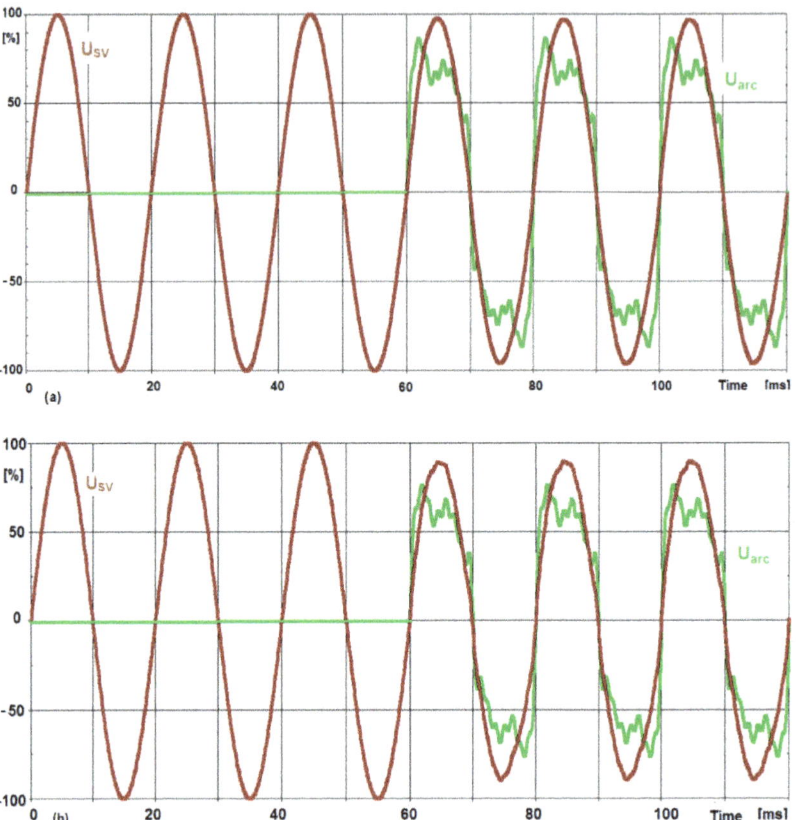

Figure 18. Oscillograms of the voltage in the supply line $U_{SV}(t)$ and the arc voltage $U_{arc}(t)$ for $S_{cc}/S_{ts} = 134$ (**a**) and $S_{cc}/S_{ts} = 20$ (**b**) [31].

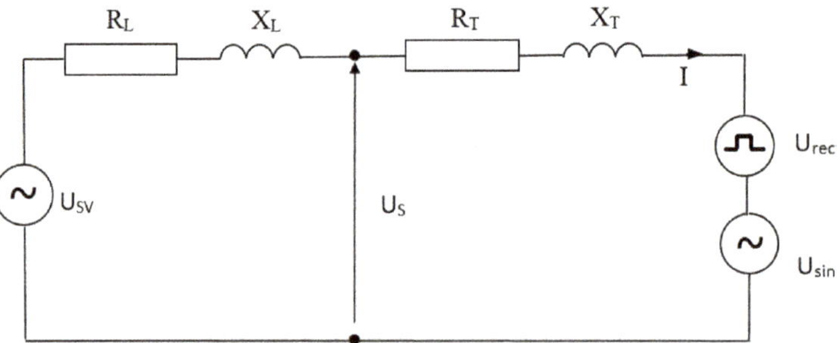

Figure 19. Single-phase equivalent circuit with a two-voltage electric arc model.

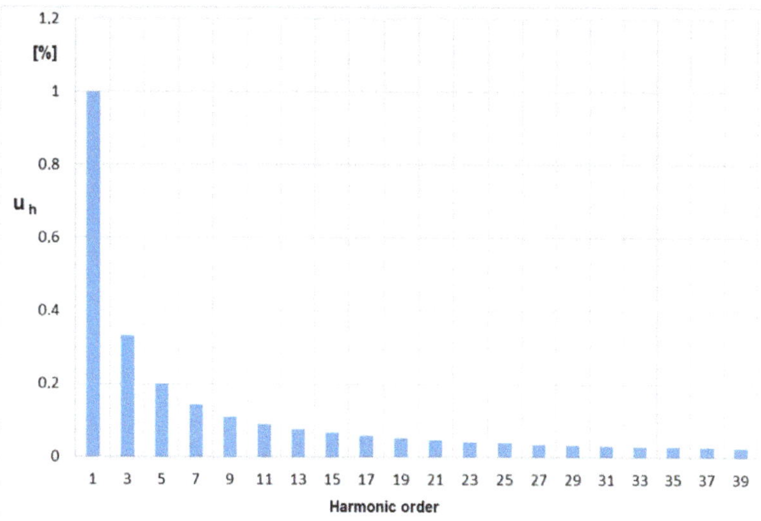

Figure 20. The spectrum of higher harmonics for the rectangular waveform of the arc voltage.

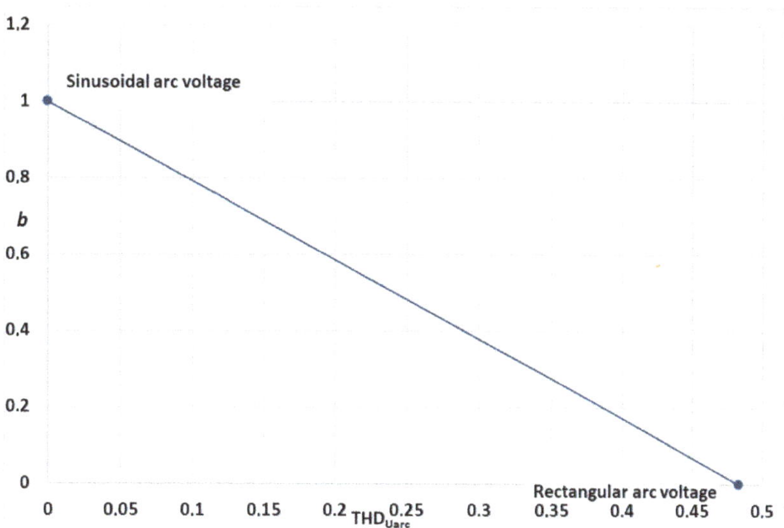

Figure 21. Dependence of changes in the value of the arc voltage distortion coefficient on the value of parameter *b*.

The THD_{Uarc} factor for a given arc voltage waveform distortion is described by Equation (5).

$$THD_{Uarc} = (1 - b)\sqrt{\frac{\pi^2}{8} - 1}. \tag{5}$$

In the PSpice program, model tests were carried out with the use of a three-voltage model of the electric arc furnace. The purpose of the model tests was, among others, to estimate the distortion of the supply voltage depending on the degree of deformation of the arc voltage and the supply conditions of the arc furnaces. Figure 21 shows the spectrum of harmonics and the waveform of the voltage supplied to the steel mill U_S. A rectangular shape of the arc voltage U_{arc} was assumed, and the ratio of the short-circuit power of the S_{cc} network to the short-circuit power of the electrodes with the S_{trc} charge was $S_{cc}/S_{trc} = 134$—Figure 22.

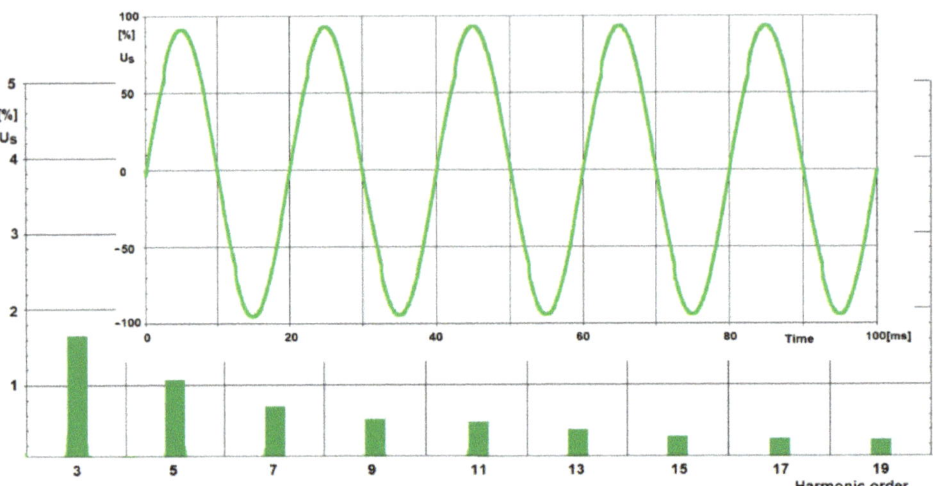

Figure 22. Oscillogram and harmonics spectrum of the supply voltage of the steel plant for $S_{cc}/S_{trc} = 134$.

Figure 23 shows the spectrum of harmonics and the voltage waveform of the U_S steelwork supply, at the rectangular arc voltage U_{arc}, for the ratio of the short-circuit power of the S_{cc} network to the short-circuit power of the electrodes with the charge S_{trc} equal to $S_{cc}/S_{trc} = 20$. This corresponds to the actual supply conditions for the metallurgical plant and the steel plant from which the arc furnaces are powered.

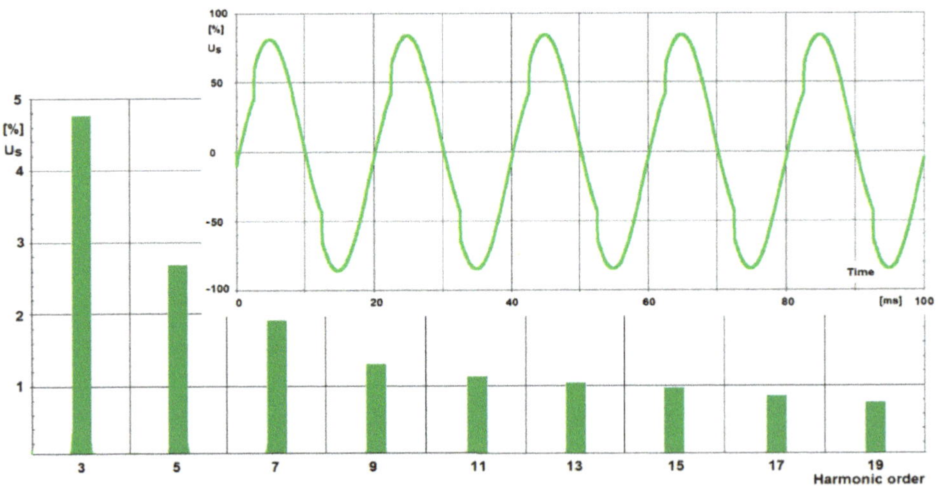

Figure 23. Oscillogram and harmonics spectrum of the supply voltage of the steel plant for $S_{cc}/S_{trc} = 20$.

Arc furnaces are characterized by a high consumption of inductive reactive power. This is due to the high inductance of the high-current circuit supplying the arc. Therefore, it becomes necessary to use reactive power compensation, mainly with the use of capacitors.

When the arc furnace is turned on (the arc burns without interruption, and the arc voltage has a rectangular waveform $U_{arc}(t)$), the voltage drops on the impedance of the supply network, which causes a decrease in the value and voltage distortion on the rails of the $U_S(t)$ steel plant (Figure 24).

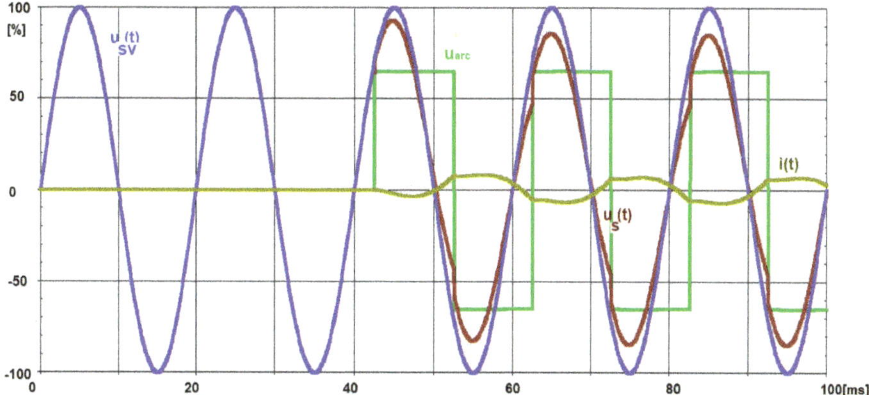

Figure 24. Oscillograms of the supply (source) voltage $U_{SV}(t)$, the arc furnace supply voltage $U_S(t)$, the arc voltage $U_{arc}(t)$, and the arc furnace current $i(t)$—switching on the arc furnace.

Using the proposed model of the electric arc, the influence of the connected capacitor bank on the improvement of the power supply conditions was analyzed.

Figure 25 shows the course of the supply voltage $U_S(t)$ at the moment of switching on the capacitor bank. A sharp increase in the U_S voltage is visible. The arc voltage waveform $U_{arc}(t)$ has a rectangular shape [31].

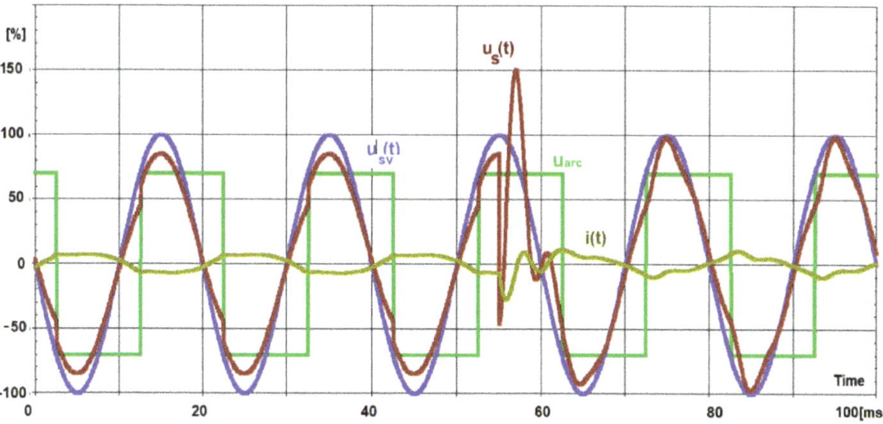

Figure 25. Oscillograms of the supply (source) voltage $U_{SV}(t)$, the arc furnace supply voltage $U_S(t)$, the arc voltage $U_{arc}(t)$, and the arc furnace current $i(t)$—switching on the capacitor bank [31].

Depending on the supply conditions of arc furnaces (supply line parameters: resistance R_L and reactance X_L), there may be transients of different duration. The waveforms for the supply voltage $U_{SV}(t)$, the voltage steelworks $U_S(t)$, and the current $i(t)$ during the capacitors' connection are shown in Figure 26. The supply line is characterized by a low short-circuit power in relation to the power of arc furnaces.

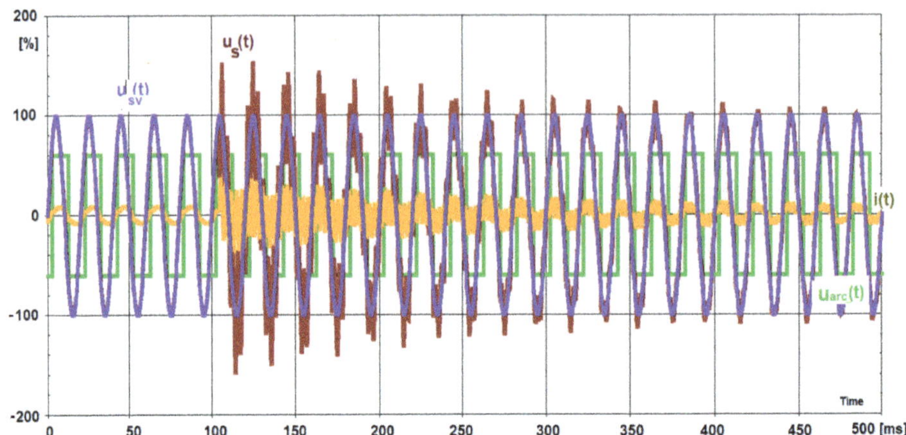

Figure 26. Oscillograms of the supply (source) voltage $U_{SV}(t)$, the arc furnace supply voltage $U_S(t)$, the arc voltage $U_{arc}(t)$, and the arc furnace current $i(t)$—switching on the capacitor bank, with high short-circuit power of the network [31].

In the case of worse supply conditions for the arc furnace (lower short-circuit power), the furnace has a greater influence on the value and deformation of the voltage supplying the steel plant. Sudden voltage increases due to transients occurring during the capacitor bank switching on can be limited by the use of suppressors.

4. Influence of the Arc Voltage Distortion on the Melting Power of the Electric Arc Furnace

On the basis of the proposed model of the arc device, an analysis was made of the influence of the electric arc voltage deformation on the scrap melting power (maximum electric arc power). The adoption of the single-phase circuit of the arc device was based on, inter alia, on analyzing the results of measurements of indicators recorded in the networks supplying arc devices and symmetry of the high-current track structure with elements of the furnace installation, conducting the steel smelting process under operational symmetry, i.e., with the same currents and arc voltages in each phase. Figure 27 shows the single-phase equivalent power supply diagram of the arc device. The impedance of the supply line to the smelter was omitted (as opposed to the diagram shown in Figures 14 and 19). The voltage *Us* is in this case the supply voltage of the steelworks.

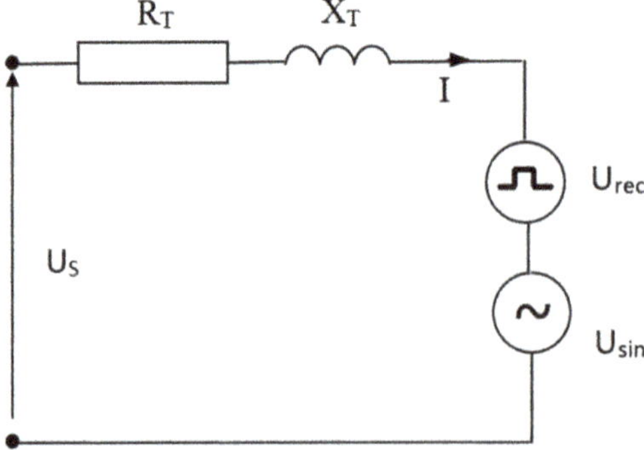

Figure 27. Single-phase equivalent power supply diagram of the arc device.

This circuit includes the arc supply system, consisting of an equivalent, ideal voltage source, with the phase voltage $U_S(t)$,

$$U_S(t) = U_S \sin(\omega t + \psi), \tag{6}$$

and equivalent resistance R_T and inductance L_T or, for fundamental harmonic, reactance $X_T = \omega L_T$.

The electric arc of the electric arc furnace is replaced by a system of ideal voltage sources connected in series, one with a sinusoidal waveform $U_{sin}(t)$ and the other with a square waveform $U_{rec}(t)$,

$$U_{arc}(t) = U_{sin}(t) + U_{rec}(t) = U_{sin}\,\sin \omega t + U_{rec}sign(i) = bU_{arc1}\sin \omega t + \frac{(1-b)\pi}{4}U_{arc1}sign(i), \tag{7}$$

where U_{arc1} is the amplitude of the fundamental harmonic of the arc voltage.

The waveform $i(t)$, in the circuit given in Figure 27, is determined by the following differential equation:

$$U_S(t) = R_T i(t) + L_T \frac{di}{dt} + U_{sin}(t) + U_{rec}(t). \tag{8}$$

The instantaneous value of the current in the circuit (in steady state) resulting from this equation is the sum of three waveforms,

$$i(t) = i_1(t) + i_{sin}(t) + i_{rec}(t). \tag{9}$$

The value of $i_1(t)$ is forced by the voltage of the supply source Formula (10) and is equal to

$$i_1(t) = \frac{U_S}{Z_T}\sin(\omega t + \psi - \varphi), \tag{10}$$

where

$$Z_T = \sqrt{R_T^2 + (\omega L_T)^2}, \tag{11}$$

$$\varphi = arctg(\omega L_T / R_T). \tag{12}$$

The $i_{sin}(t)$ waveform is forced by the sinusoidal component of the arc voltage $U_{sin}(t)$, i.e., the first of the expressions in

$$i_{sin}(t) = \frac{U_{sin}}{Z_T}\sin(\omega t - \varphi), \tag{13}$$

and the wave of $i_{rec}(t)$, forced by the rectangular component of the arc voltage $U_{rec}(t)$, i.e., the second of the expressions in

$$i_{rec}(t) = A - \left(A + \frac{2U_{rec}}{3R}\right)(1 - \exp(-\omega t / \tau)), \tag{14}$$

where the auxiliary quantities are

$$\tau = \frac{L_T}{\omega R_T}, \tag{15}$$

$$A = \frac{2U_{rec}}{3R}B, \tag{16}$$

$$B = C(4 - 4C + C^2)/(1 + (1 - C)^3), \tag{17}$$

$$C = 1 - \exp\left(-\frac{\pi}{6}/\tau\right). \tag{18}$$

The instantaneous value of the arc power is given by the formula

$$p_{arc}(t) = u_{arc}(t)i = (U_{sin}(t) + U_{rec}(t))(i_1(t) + i_{sin}(t) + i_{rec}(t)). \tag{19}$$

The calculations for an exemplary arc device are presented below. The calculations were performed in relative units, taking the following as the basic quantities:

Supply phase amplitude,

$$U_S = 1 p.u.(100\%); \tag{20}$$

arc device rated current,

$$I_N = 1 p.u.(100\%). \tag{21}$$

The reactance of the supply circuit was determined assuming that the theoretical short-circuit current of the electrodes with the charge is twice as high as the rated one.

$$I_{tsc} = U_S / X_T = 2 p.u.(200\%). \tag{22}$$

The assumed reactance value is

$$X_T = U_S / I_{tsc} = 0.5 p.u.(50\%). \tag{23}$$

The circuit resistance was assumed to be five times lower than the reactance, i.e.,

$$R_T = X_T / 5 = 0.1 p.u.(10\%). \tag{24}$$

Similar assumptions were made when analyzing the voltage fluctuations generated by the arc furnaces presented in [31,32]. The operating characteristics of the power released in the arcs as a function of the current consumed by the arc device were determined (Figure 28).

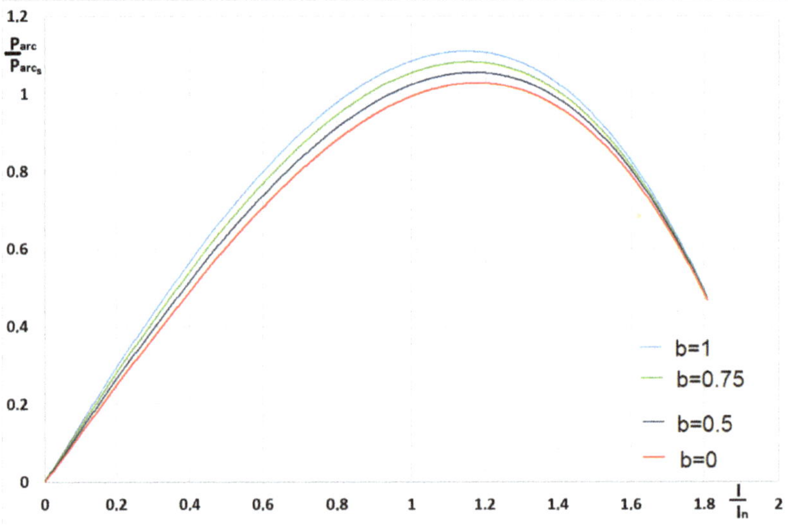

Figure 28. Arc force as a function of the arc current for different values of the parameter *b*.

This power is expressed as a percentage of the maximum value occurring at the sinusoidal current (assumption $b = 1$, i.e., $U_{rec} = 0$).

Maximum power occurs at the following current:

$$I_M = I_N \sqrt{2} \sqrt{1 - \cos \varphi}. \tag{25}$$

The value of the maximum arc power decreases with the increase in the nonlinearity of the arc, i.e., with the decrease in the parameter *b* value from unity $U_{rec} = 0$ to zero I $U_{sin} = 0$.

The relative reduction of the maximum arc power as a function of parameter b is shown in Figure 29.

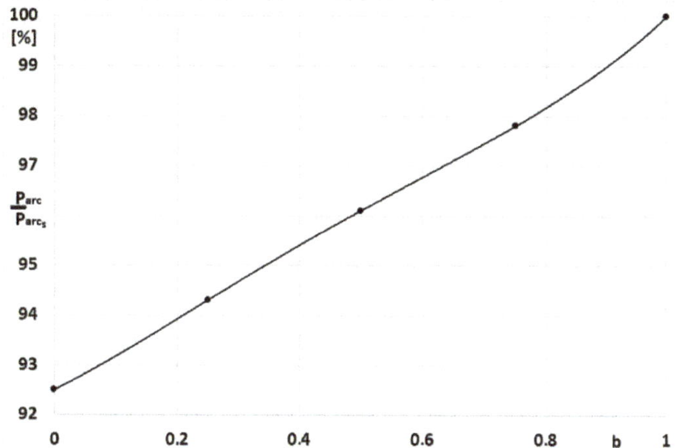

Figure 29. Relative decrease in maximum arc power as a function of parameter b.

The arc power value decreases gradually from the value $p = 100\%$ for $b = 1$ to $p = 97.8\%$ for $b = 0.75$, through $p = 96.1\%$ for $b = 0.5$, $p = 94.3\%$ for $b = 0.25$, and $p = 92.5\%$ for $b = 0$.

5. Summary

In the case of complex research objects such as arc devices, a certain compromise should be taken into account between the degree of model complexity and the accuracy (approximation) in reproducing the real conditions and its practical use.

The presented measurements show that the voltage distortion in the networks supplying the arc devices is influenced by both the arc furnaces and other nonlinear loads supplied from the same part of the power system. This is evidenced by the waveforms of the total coefficient of distortion of the *THD* voltage, the changes of which recorded during the week of measurements are typical for daily changes in the power system load. The *THD* value also depends on the short-circuit power of the network from which the steel plant is supplied. Additionally, during smelting, the spectrum of harmonics of the current and arc voltage changes. All these factors make modeling of voltage distortions caused by arc devices extremely difficult.

The multi-voltage model of the arc furnace presented in the article is a compromise between the accuracy of the determined deformation of the supply voltage and its practical application. In this model, the electric arc voltage waveforms registered in industrial conditions are used.

On the basis of the analysis of the results of the measurements of arc voltage and furnace current recorded herein, several conclusions can be drawn. The changing shape of the arc voltage and its length cause distortion of the furnace current and changes in its value, which are the reason for the generation of higher harmonics and voltage fluctuations by arc furnaces. The waveforms of the arc voltage and furnace current, especially in the initial stage of smelting, are not symmetrical with respect to the time axis and are also not periodic. During the smelting process in the arc furnace, the shape and value of the arc voltage change, starting from an almost rectangular waveform, through a triangular curve, ending with a waveform close to a sinusoidal one. The time when the arc furnace causes the greatest disturbances in the power system (the beginning of smelting), to a large extent, depends on the quality of the scrap, which affects the speed of the formation of the melt wells under the electrodes. Loading subsequent baskets with scrap causes a renewed increase in disturbances generated by arc furnaces. The greatest disturbances occur when loading the first basket (commencement of smelting).

The conclusion from the conducted theoretical considerations and model calculations is that the maximum power generated in the arcs occurs at a very similar value of the arc current (with an accuracy of less than 0.5%), and the value of this maximum decreases with an increase in the nonlinearity of the arc (a decrease in the value of parameter b).

Funding: Stakeholder of the Faculty of Transport, Electrical Engineering and Computer Science: Zakład Usług Technicznych, Energoaudyt sp. z o.o., Radom, Poland.

Institutional Review Board Statement: Not applicable.

Informed Consent Statement: Not applicable.

Conflicts of Interest: The author declares no conflict of interest.

Abbreviations

The following nomenclature is used in this manuscript:

I_{arc}	arc furnace current
I_N	rated arc furnace current
I_W	scrap melting current with maximum power
R_L	power line resistance
X_L	power line reactance
R_T	resistance high-current path with the furnace transformer
X_T	reactance high-current path with the furnace transformer
S_{cc}	short-circuit power
S_{tr}	power of the arc furnace transformer
S_{trc}	power at short-circuit of electrodes with charge
THD_U	voltage total harmonic distortion
THD_I	current total harmonic distortion
U_h	percentage of voltage higher harmonic in relation to the fundamental harmonic
U_{SV}	supply voltage
U_S	steelworks supply voltage
U_{arc}	arc voltage
$U_{sin}(t)$	sinusoidal arc voltage
$U_{rec}(t)$	rectangular arc voltage
r	correlation coefficient
P_{arc}	melting power for scrap metal
P_{arcmax}	maximum melting power for scrap metal

References

1. Wąsowski, A. The impact of the actual operating conditions of a three-phase arc furnace on the criterion of maximum efficiency and overall efficiency. (In Polish: Wpływ rzeczywistych warunków eksploatacyjnych trójfazowego pieca łukowego na kryterium maksymalnej wydajności oraz na sprawność ogólną). *Jakość I Użytkowanie Energii Elektr.* **2000**, *6*, 71–76.
2. Wąsowski, A. *Adaptation of the Three-Phase Arc Device to the Power System (Dopasowanie Trójfazowego Urządzenia Łukowego do Systemu Elektroenergetycznego, (In Polish))*; Kazimierz Pulaski University of Technology and Humanities: Radom, Poland, 2009.
3. Andriy Lozynskyy, A.; Kozyra, J.; Łukasik, Z.; Aldona Kuśmińska-Fijałkowska, A.; Andriy Kutsyk, A.; Paranchuk, Y.; Kasha, L. A Mathematical Model of Electrical Arc Furnaces for Analysis of Electrical Mode Parameters and Synthesis of Controlling Influences. *Energies* **2022**, *15*, 1623. [CrossRef]
4. Lozynskyy, A.; Perzyński, T.; Kozyra, J.; Biletskyi, Y.; Kasha, L. The Interconnection and Damping Assignment Passivity-Based Control Synthesis via the Optimal Control Method for Electric Vehicle Subsystems. *Energies* **2021**, *14*, 3711. [CrossRef]
5. Karbowniczek, M.; Kamiński, P. Wpływ palników gazowo-tlenowych na roztapianie wsadu w elektrycznym piecu łukowym. *Hut. Wiadomości Hut.* **2003**, *70*, 142–149.
6. Karbowniczek, M.; Sadowski, A.; Hryniewicz, M. Możliwości zastosowania odpadów poszlifierskich z produkcji łożysk jako wsadu w piecu łukowy. *Hut. Wiadomości Hut.* **2005**, *72*, 371–377.
7. Cassie, A.M. *Theorie Nouvelle des Arcs de Rupture et de la Rigidité des Circuits*; CIGRE: Paris, France, 1939; Volume 102, pp. 588–608.
8. Cassie, A.M. *Arc Rupture and Circuit Severity/Theorie Nouvelle Des Arcs De Rupture Et De La Rigidité Des Circuits*; CIGRE: Paris, France, 1939; pp. 588–608.

9. Mayr, O. Beitrage zur Theorie des Statischen und des Dynamischen Lichthogens. *Arch. Für Elektrotechnik* **1943**, *37*, 588–608. [CrossRef]

10. Bhonsle, D.C.; Kelkar, R.B. Analyzing power quality issues in electric arc furnace by modeling. *Energy* **2016**, *115*, 830–839. [CrossRef]

11. Bhonsle, D.C.; Kelkar, R.B. Simulation of Electric Arc Furnace Characteristics for Voltage Flicker study using MATLAB. In Proceedings of the International Conference on Recent Advancements in Electrical, Electronics and Control Engineering, Sivakasi, India, 15–17 December 2011.

12. Samet, H.; Farjah, E.; Zahra Sharifi, Z. A dynamic, nonlinear and time-varying model for electric arc furnace. *Int. Trans. Electr. Energy Syst.* **2015**, *25*, 2165–2180. [CrossRef]

13. Alonso, M.A.P.; Donsion, M.P. An improved time domain arc furnace model for harmonic analysis. *IEEE Trans. Power Deliv.* **2004**, *19*, 367–373. [CrossRef]

14. Varadan, S.; Makram, E.B.; Girgis, A.A. A new time domain voltage source model for an arc furnace using EMTP. *IEEE Trans. Power Deliv.* **1996**, *11*, 1685. [CrossRef]

15. Cano Plata, E.A.; Tacca, H.E. Arc Furnace Modeling in ATP-EMTP. In Proceedings of the 6th International Conference on Power Systems Transients, Montreal, QC, Canada, 20–23 June 2005; pp. 19–23.

16. Höke, W.; Brethauer, K. Die Schwankungen des Spannungs-bedarfs des Hochstrolichtbogens in einem Lichtbogrnofen infolge der Bogenbewegung. In *Elektrowarme Inrernätional*; Vulkan Verlag GmbH: Essen, Gemany, 1981; Volume 39, pp. 274–282.

17. Lozynskyi, O.Y.; Paranchuk, Y.S.; Paranchuk, R.Y.; Matico, F.D. Development of methods and means of computer simulation for studying arc furnace electric modes. *Electr. Eng. Electromechan.* **2018**, *3*, 29–36. [CrossRef]

18. Odenthal, H.J.; Kemminger, A.; Krause, F.; Sankowski, L.; Uebber, N.; Vogl, N. Review on Modeling and Simulation of the Electric Arc Furnace (EAF). *Steel Res. Int.* **2018**, *89*, 1700098. [CrossRef]

19. Sadeghian, A.; Lavers, J.D. Dynamic reconstruction of nonlinear v–i characteristic in electric arc furnaces using adaptive neuro-fuzzy rule-based networks. *Appl. Soft Comput.* **2011**, *11*, 1448. [CrossRef]

20. Mokhtari, H.; Hejri, M. A new three phase time-domain model for electric arc furnace using MATLAB. In Proceedings of the IEEE Asia-Pacific Transmission and Distribution Conference and Exhibition, Yokohama, Japan, 6–10 October 2002; Volume 3, pp. 2078–2083.

21. Tang, L.; Kolluri, S.; McGranaghan, M.F. Voltage flicker prediction for two simultaneously operated AC arc furnaces. *IEEE Trans. Power Deliv.* **1997**, *12*, 985–992. [CrossRef]

22. Panoiu, M.; Panoiu, C.; Ghiormez, L. Modeling of the Electric Arc Behavior of the Electric Arc Furnace. In *Advances in Intelligent Systems and Computing*; Springer: Berlin/Heidelberg, Germany, 2013; pp. 195–261.

23. Panoiu, M.; Panoiu, C.; Iordan, A.; Ghiormez, L. Artificial neural networks in predicting current in electric arc furnaces. In *IOP Conf. Series: Materials Science and Engineering, Proceedings of the International Conference on Applied Sciences (ICAS2013), Seoul, Korea, 29–31 August 2014*; IOP: Bristol, UK, 2020; p. 57.

24. Chang, G.W.; Cheng, C.I.; Liu, Y.J. A neural-network-based method of modeling electric arc furnace load for power engineering study. *IEEE Trans. Power Syst.* **2009**, *25*, 138. [CrossRef]

25. Chang, G.W.; Liu, Y.J.; Chen, C.I. Modeling voltage-current characteristics of an electric arc furnace based on actual recorded data: A comparison of classic and advanced models. In Proceedings of the IEEE Power and Energy Society General Meeting—Conversion and Delivery of Electrical Energy in the 21st Century, Pittsburgh, PA, USA, 20–24 July 2008; pp. 1–6.

26. Wang, F.; Jin, Z.; Zhu, Z. Modeling and Prediction of Electric Arc Furnace Based on Neural Network and Chaos Theory, Advances in Neural Networks. In *Lecture Notes in Computer Science*; Springer: Berlin/Heidelberg, Germany, 2005; Volume 3498, p. 819.

27. Garcia-Segura, R.; Vázquez Castillo, J.; Martell-Chavez, F.; Longoria-Gandara, O.; Ortegón Aguilar, J. Electric Arc FurnaceModeling with Artificial Neural Networks and Arc Length with Variable Voltage Gradient. *Energies* **2017**, *10*, 1424. [CrossRef]

28. Acha, E.; Madrigal, M. *Power Systems Harmonics: Computer Modeling and Analysis*; John Wiley & Sons: Hoboken, NJ, USA, 2001.

29. Acha, E.; Semlyen, A.; Rajakovic, N. A harmonic domain computational package for nonlinear problems and its application to electric arcs. *IEEE Trans. Power Deliv.* **1990**, *5*, 1390–1397. [CrossRef]

30. Vinayaka, K.U.; Puttaswamy, P.S. Review on characteristic modeling of electric arc furnace and its effects. In Proceedings of the 2017 International Conference on Intelligent Computing, Instrumentation and Control Technologies (ICICICT), Kerala, India, 6–7 July 2017.

31. Łukasik, Z.; Olczykowski, Z. Estimating the impact of arc furnaces on the quality of power in supply systems. *Energies* **2020**, *13*, 1462. [CrossRef]

32. Olczykowski, Z. Modeling of Voltage Fluctuations Generated by Arc Furnaces. *Appl. Sci.* **2021**, *11*, 3056. [CrossRef]

33. Olczykowski, Z.; Łukasik, Z. Evaluation of Flicker of Light Generated by Arc Furnaces. *Energies* **2021**, *14*, 3901. [CrossRef]

34. Olczykowski, Z. Electric Arc Furnaces as a Cause of Current and Voltage Asymmetry. *Energies* **2021**, *14*, 5058. [CrossRef]

 *energies*

Review

The Architecture Optimization and Energy Management Technology of Aircraft Power Systems: A Review and Future Trends

Tao Lei [1,2,*], Zhihao Min [2], Qinxiang Gao [1,2], Lina Song [1,2], Xingyu Zhang [2] and Xiaobin Zhang [1,2]

[1] Department of Electrical Engineering, School of Automation, Northwestern Polytechnical University, Xi'an 710129, China; gaoqinxiang@mail.nwpu.edu.cn (Q.G.); s18182668697@mail.nwpu.edu.cn (L.S.); dg1907@126.com (X.Z.)

[2] Key Laboratory of Aircraft Electric Propulsion Technology, Ministry of Industry and Information Technology of China, Xi'an 710072, China; minzhihao@outlook.com (Z.M.); jensonzhangbuaa@126.com (X.Z.)

* Correspondence: lttiger@nwpu.edu.cn

Citation: Lei, T.; Min, Z.; Gao, Q.; Song, L.; Zhang, X.; Zhang, X. The Architecture Optimization and Energy Management Technology of Aircraft Power Systems: A Review and Future Trends. *Energies* 2022, 15, 4109. https://doi.org/10.3390/en15114109

Academic Editor: Silvio Simani

Received: 17 April 2022
Accepted: 24 May 2022
Published: 2 June 2022

Publisher's Note: MDPI stays neutral with regard to jurisdictional claims in published maps and institutional affiliations.

Abstract: With the development of More/All-Electric Aircraft, especially the progress of hybrid electrical propulsion or electrical propulsion aircraft, the problem of optimizing the energy system design and operation of the aircraft must be solved regarding the increasing electrical power demand-limited thermal sink capability. The paper overviews the state of the art in architecture optimization and an energy management system for the aircraft power system. The basic design method for power system architecture optimization in aircraft is reviewed from the multi-energy form in this paper. Renewable energy, such as the photo-voltaic battery and the fuel cell, is integrated into the electrical power system onboard which can also make the problem of optimal energy distribution in the aircraft complex because of the uncertainty and power response speed. The basic idea and research progress for the optimization, evaluation technology, and dynamic management control methods of the aircraft power system are analyzed and presented in this paper. The trend in optimization methods of engineering design for the energy system architecture in aircraft was summarized and derived from the multiple objective optimizations within the constraint conditions, such as weight, reliability, safety, efficiency, and characteristics of renewable energy. The cost function, based on the energy efficiency and power quality, was commented on and discussed according to different power flow relationships in the aircraft. The dynamic control strategies of different microgrid architectures in aircraft are compared with other methods in the review paper. Some integrated energy management optimization strategies or methods for electrical propulsion aircraft and more electric aircraft were reviewed. The mathematical consideration and expression of the energy optimization technologies of aircraft were analyzed and compared with some features and solution methods. The thermal and electric energy coupling relationship research field is discussed with the power quality and stability of the aircraft power system with some reference papers. Finally, the future energy interaction optimization problem between the airport microgrid and electric propulsion aircraft power system was also discussed and predicted in this review paper. Based on the state of the art technology development for EMS and architecture optimization, this paper intends to present the industry's common sense and future trends on aircraft power system electrification and proposes an EMS+TMS+PHM to follow in the electrified aircraft propulsion system architecture selection

Keywords: More/All Electric Aircraft (MEA/AEA); energy optimization and evaluation methods; energy management of power system; electric propulsion aircraft (EPA); renewable energy uncertainty; load power stochastic model; stability analysis; physical healthy management; electric thermal coupling

1. Introduction

Traditionally, the energy or power system in aircraft can be classified as the primary power system and secondary power system. The primary power is thrust mainly provided by the petroleum oil engines, the secondary power can be divided into hydraulic, pneumatic, and electrical power, which can extract mechanical energy from the engine for control and maneuverability. There is a growing trend toward electrification of the aircraft power system for various aircraft segments. The major motivation for this includes increased efficiency, reduced CO_2 emissions, and lower operating costs [1]. In the electrified aircraft concept, the duct or propeller fan is driven by an electric motor, whereas, in a conventional aircraft, a gas turbine engine drives the duct fan. Secondary power systems allow for aircraft safe operation and ensure passengers' comfort. For conventional aircraft, secondary power systems combined pneumatic, hydraulic, mechanical, and electric power, and their energy consumption represents approximately 5% of the total fuel burned during the flight stage [2].

With the advent of the More Electric Aircraft (MEA) and All-Electric Aircraft (AEA) initiatives, electric power systems are progressively taking the place of pneumatic, hydraulic, and mechanical power systems [3]. Electric power is the only energy form on board. In recent years, with the development of green and clean aviation, electric aircraft is a path to zero-emission air travel [4]. In particular, the aims of this research are focused on the onboard power systems of new All-Electric Aircraft and electrical propulsion aircraft, where a crucial design point is related to the electrical energy optimization management and power control [5–7].

In the "all-electric" concept, where pneumatic and hydraulic power systems are eliminated to improve aviation costs and environmental impact, the dynamics of electrical power balance are to be characterized and managed to avoid excessive peaks with respect to generators' limited capabilities. Especially for electric propulsion aircraft (EPA), the thrust force can be partly or completely provided by an electric power system [8]. With the development of power electronics, the HVDC system is becoming a trend for MEA/AEA and EPA [9]. The energy storage system such as the Lithium-ion battery is often integrated into the electrical power system to improve the performance of the power system. The aircraft's electrical power system is a "flying microgrid" to realize the different flight functions.

Renewable energy, such as the fuel cell and the photovoltaic battery, has been integrated into the aircraft electrical power system as the auxiliary power unit [10], emergency power [11], or main propulsion hybrid energy source [12]; the output power characteristics of renewable energy further degraded the power quality of the electrical power system on board. The necessity of energy and power management is obvious to the engineering designer. In recent years, the aircraft electric propulsion has become an important research topic, this trend gives a higher requirement for the energy system, especially for the electric power system in the aircraft. Due to their performance variation with power and energy requirements and the size of the components, a mix of chemical, electrical and mechanical components can be integrated in aircraft electric propulsion systems [13,14]. To evaluate the mass and performance index of these systems, one method is to use a constant specific power density, or power-to-mass ratio, and efficiency for each device in the aircraft power system for economic issues [15]. The stability and power quality of aircraft power systems must be considered with the optimization of energy management, while the thermal energies are generated and transmitted onboard due to power loss from the typical energy component.

The optimization of energy onboard the aircraft can be classified into two main fields: (1) Static architecture and configuration optimization, evaluation for power system; (2) Dynamic energy and power planning or management methods to fulfill the real-time requirement of efficiency, stability, reliability, and safety for the power system. It is not similar to the terrestrial energy system which is mainly focused on the efficiency and cost, and reliability of system operation without considering the weight, volume and power loss, and security of the energy system [16,17]. The power system in aircraft must have very high reliability and safety; there is a lot of research regarding the fault tolerance, safety, and reliability of power systems in aircraft [18–23]. Compared with vehicle and ship power systems, the safety requirement for aircraft is more stringent. So the energy optimization for aircraft must include health management and fault analysis, considering the different voltage levels, load characteristics, and architecture, compared with ground vehicles and ships [24].

For the static optimization and assessment of the energy and power system of aircraft, there are component levels [25,26], subsystem levels [27–31], and system levels [32–35] in the design optimization and evaluation. The number of main designing parameters of typical energy conversion devices, such as the generator and motor, and power converter, is very high and the coupling relationship of different energy is very complex [36]. So the methods of evaluation and optimization in the energy system in the MEA/AEA or EPA must be built based on the mathematical multi-parameter programming [37] and power flow analysis [38]. In some situations, the designing variable and optimization variable parameters include the integer number, for example, some of the optimization Boolean variables are 0 or 1, such as the switching operating status decision, and the integer number of the battery's cell or generator in a power system [39,40]. Therefore, these optimization problems can also be changed into multi-parametric mixed-integer linear programming (MILP). To formulate and solve an optimization problem in energy systems, there are numerous elements that need to be defined, including the system's parameter boundaries, optimization criteria, decision variables, and objective functions according to different energy architectures of aircraft. Assessment tools are developed for the overall MEA power system [41].

In addition, for dynamic energy and the power management system onboard, traditional MEA power demands are not changed greatly because of the small percentage of electrical power out of the overall power from the engine [2], so the energy management framework and control methods are simple. However, for all-electric and electrical propulsion aircraft, the energy demand from the EPA is greatly variable during the flight stage; according to the aerodynamic control requirements, the power balance must be maintained in every time instant [42], so the real-time dynamic energy and power management strategies must be considered. Although the dynamic energy management methods are very conservative in traditional aircraft power systems, the regenerated energy is not allowed to feed into the aircraft electric power system. Furthermore, the fixed priority of the electrical load and the active energy management strategies are still beneficial to the aircraft, while considering the demand of EPA [43–45]. Many energy and power management system architectures are proposed from the power source, power distribution network, and power load side to coordinate these components [46–53]. The energy and power management control methods are elaborated upon to realize the optimization objective such as fuel consumption minimization, flight cost minimization, system reliability maximum, quick power response, stability enhancement, and power quality improvement. The energy and power optimization is applied to flight control subsystem [47], environmental control system [46], electrical power generation and distribution system [49–51,53], and load management system [48,52].

This article aims to present a comprehensive review of electrified aircraft electric power and propulsion systems. The key technology progress and trends such as architecture optimization and energy management control were discussed here. The major topic items can be listed as follows:

(1)   Architecture evaluation and optimization of the aircraft power systems;
(2)   Energy power source analysis for power systems with different sizes of aircraft;
(3)   Power load characteristic and requirements analysis for aircraft power system;
(4)   Energy management strategies and control architecture for aircraft power systems;
(5)   Electrical thermal coupling and integrated control methods for aircraft power system;

(6) The energy or power interaction analysis and control between the electric aircraft power system and airport microgrid can improve the performance of both systems;

In this paper, the architecture research and design methods of the aircraft system are reviewed in Section 2 The characteristics of energy power for aircraft, especially with renewable energy integrated, has been analyzed, and the challenging problem for energy source and power management strategies of the electric power system was presented, regarding the uncertainty from the power source and load power demand, in Sections 3 and 4. The optimization problem of energy efficiency and power distribution systems in aircraft was discussed in Section 5. Research progress of different energy and power management architecture and strategies are also presented in Section 5. Section 6 gives a review of thermal and electrical coupling issues in aircraft power systems. The developing trend for the connection of the airport microgrid and EPA is provided in Section 7, the energy management control research issues can be derived from their interaction. Finally, the conclusion of the state art EMS was presented, to predict the arriving era of EPA, in Section 8.

2. The Review of Power Architecture Research of Aircraft Power Systems

In this section, the definition of the power system consists of a system in which a group of components works together as a system and delivers electric power and thrust power to the aircraft. The aircraft power system is sometimes split in to two different categories: the electric propulsion system responsible for providing the electric thrust power and the electric power system response for supplying electric power for avionic equipment or an electric actuator.

2.1. The Selection of the Architecture of Aircraft Power Systems

The typical power system architecture of aircraft is shown in Figure 1a. The power flow of multiple energy domains at the system level, for a generic aircraft power architecture, illustrates the decomposition that will be utilized to cope with the complexity of this multi-physics domain model. Traditionally, the power subsystem design is independently carried out by different single energy forms, leading to overly conservative system sizing results [54]. Except for generating the thrust force for aircraft, the engine model also provides mechanical, hydraulic, and pneumatic power from shaft power, while also acting as a sink for waste heat via bypass duct heat exchangers. The electrical system converts mechanical power to electrical power with power loss or waste heat as a byproduct. The thermal management system uses electrical and pneumatic power to move and reject thermal energy around the aircraft. With the development of renewable energy, the PV and fuel cell, or hydrogen energy, can be used for the aircraft electrical power system [12,13]. Compared with engine-driven generators, the energy efficiency of green power energy is a big advantage. However, their energy density is limited, so these kinds of energy are often used with the oil-engine-driven generator. For future aircraft power systems, renewable energy such as photovoltaic (PV) systems and fuel cells can be integrated into the power system of aircraft, while the energy extraction from the engine can be reduced to improve the overall efficiency of aircraft power systems. The architecture of more electrical aircraft power systems is shown in Figure 1b. In this architecture, the bleed air system can be eliminated, and the ECS and ice protection system can get the electrical energy from the electrical power systems [2]. All-electric and electrical propulsion aircraft are presented in Figure 1c. This is the turbo electrical power system architecture for medium- and large-size aircraft [42]. A typical power system architecture of unmanned air vehicles (UAV), and small size or light aircraft integrated with renewable energy power sources, is given in Figure 1d [55,56]. In some situations, the different configurations of the renewable power source, such as PV or fuel cell, can be realized depending on the size of the aircraft and the power demand from the load [57].

As power systems become more integrated, traditional design silos are being broken down in favor of cooperative design between thermal, electrical, and mechanical engineers [8]. The optimization design of aircraft power system structure is basically a complex optimization problem, which is to find the optimal solution in a huge parameter space. The selection of architecture for power systems must be regarded with volum e, weight, reliability, and efficiency. According to the above constraints, the evaluation and optimization methods can be applied to be an iterative process to find the optimal structure of the aircraft power system.

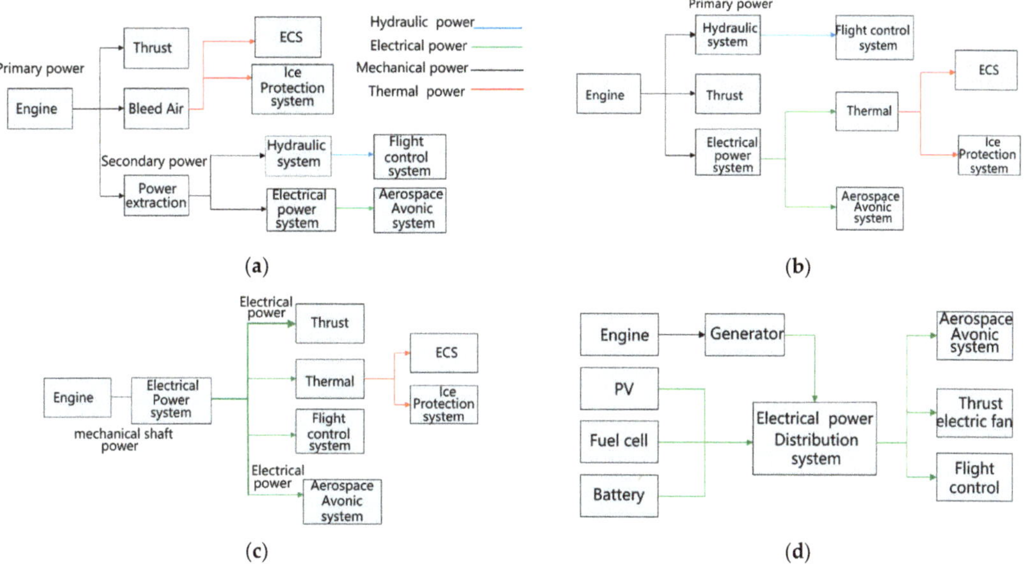

Figure 1. (**a**) The traditional architecture of aircraft power systems; (**b**) The more electric architecture of aircraft power systems; (**c**) The all-electric and electrical propulsion architecture of aircraft power systems; (**d**) The unmanned air vehicle or small size aircraft integrated with renewable energy power source.

2.2. The Optimization and Evaluation of Architecture for Aircraft Power Systems

In the aircraft power system architecture design, the modeling of typical components is very important, where power equipment/subsystems coming from different physical energy domains (thermodynamic, aerodynamic, hydraulic, electrical, etc.) work and interact with each other [58]. It is clearly concerned with finding the initial equilibrium condition and defining solver settings, such as the numerical method and the chosen time step [59], while regarding the integration of complex power models describing various physical phenomena with respect to quite different frequency contents.

The basis of the analysis tool for different architectures of AEPS is a library of component models covering the main power subsystems; these component models can be integrated to form user-defined power system architectures, which can be quantitatively assessed about some performance index, such as both mass and energy or exergy efficiency, throughout various operating modes, or flight cycles, of the aircraft power system [58]. According to the second law of the thermodynamic system, the energy is the available energy of the working medium, and it is used to determine the portion of given energy that is likely to be useful in a given state of a thermodynamic system.

The traditional small-signal modeling methods are often used for power system stability analysis and control loop design [16]. However, these methods are not fit for the multi-source microgrid application; some methods are given in some papers to overcome this obstacle to meet the demands from the AEA and EPA [59]. Although the multi-generator VF electric power system, based on the dynamic phasor model, is provided [60], this is very difficult to integrate with thermal energy flow analysis.

In [54,58], posing the problem of safe, robust, and efficient design and control, it needs a new method or idea for a basic multi-layered modeling framework that may lend itself to ripping potential benefits from the electrification of terrestrial power systems, ships, or aircraft. The modeling methods, based on exergy analysis with the second law of thermodynamics, were presented for the power system. For the multi-physical systems, such as terrestrial energy systems, aircraft, and ship power systems, the interconnected systems are modeled as dynamically interacting power modules with different time scales. This approach is shown to be particularly well-suited for the scalable optimization of large-scale complex power systems. The proposed multi-layered modeling of system dynamics in the energy and exergy space offers a promising basic method for modeling and controlling inter-dependencies across multi-physics subsystems, for ensuring both feasible and near-optimal operation.

From an operational research point of view, architecture optimization of the power system is the process of maximizing or minimizing a linear or nonlinear analytical objective function, regarding various equality or inequality constraints, for a number of design variables or parameters; for each of these, a range of value exists. Set more simply and practically, optimization design problem involves finding the best possible configuration and variables solution for a given problem with reasonable constraints.

Optimal conditions are generally strongly dependent on the chosen objective function, which can involve the figure of merits (FOM) for the aircraft power system. However, several aspects of the performance index are often important in practical applications for the aircraft power system. In aircraft thermal and energy systems design, efficiency (energy and/or exergy), power production rate, reliability, power quality, and heat transfer rate are common quantities or FOM that are to be maximized, while cost, weight, fuel consumption, environmental impact, and power loss are quantities to be minimized. Any of these can be chosen as the objective function for an optimization problem, but it is usually more meaningful and useful to consider more than one objective function. In a common situation, several optimization objects maybe contradictory to each other, while one objective function is optimal and the other is not optimal under the same operating points. The minimum and maximum of a single variable function are able to be determined by users of simple linear optimization, and first or second derivative techniques to find the optimal value of a given function can be utilized. At the advanced level, an optimum value of multivariable nonlinear functions can be found by users of optimization. In addition, multivariable optimization problems with nonlinear constraints can be solved. A constrained optimization problem is an important research subject in scientific practice since most real-world problems contain constraints [61].

Multi-objective optimization has been extensively used and studied for aerospace power systems [12,14,21,28,33]. There exist many algorithms and application case studies involving the multi-objective optimization of aircraft power systems [30,35,37].

One of the common approaches for dealing with multiple objective functions is to combine them with some weighting factor into a single objective function that is then minimized or maximized. For example, in the optimal design of heat exchangers and cooling systems for electronic equipment in aircraft power systems, it is desirable to minimize the exergy destruction rate and maximize the heat transfer rate [62]. However, this often comes at the price of increased fluid flow rates and corresponding frictional pressure losses for cooling systems. A multi-objective optimization problem has objective functions that are either minimized or maximized [54]. As with single-objective optimization, multi-objective optimization involves several linear or nonlinear constraints that any feasible solution, including the optimal solution, must satisfy. In addition, electric, hydraulic, and pneumatic energy are integrated into aircraft power systems [63]. The heterogeneous energy optimization problem often makes the objective function very complex and parameter normalization is not very easy to tackle. The electrification trend of the aircraft power system can mitigate the complexity.

The power management optimization problem is formulated as Equation (1).

Minimize/maximize

$$f_n(x) \ \ n = 1, 2, \ldots N \tag{1}$$

Subject to

$$g_j(x) > 0, \ \ j = 1, 2, \ldots J \tag{2}$$

$$h_k(x) = 0, \ \ k = 1, 2, \ldots K \tag{3}$$

$$x_i^{(L)} < x_i < x_i^{(u)}, \ \ i = 1, 2, \ldots n \tag{4}$$

In the above equation, f is the objective function, g the inequality constraints, and h the equality constraints. A solution to this problem is x, which is a vector of n decision variables or design parameters. The last set of constraints in Equation (4) is called the variable bounds, which restrict the boundary of searching space. Any solution of the decision variables should be within limits with a lower bound $x_i^{(L)}$ and upper bound $x_i^{(u)}$. To illustrate this problem, consider this situation: the multi-objective optimization model has two objective functions, f_1 and f_2 [61]. It is assumed that these two functions are to be minimized (although maximization can be similarly handled since it is equivalent to the minimization of the negative of the function). The values for the two objective functions at five different design points are shown in Figure 2A. Design point 2 in this figure is clearly found to be preferable to design 4 because both objective functions f_1 and f_2 are smaller for design 2 compared to design 4. Similarly, objective functions in design 3 are optimal for design 5. In addition, designs 1, 2, and 3 are not predominated, by any other parameter design variation.

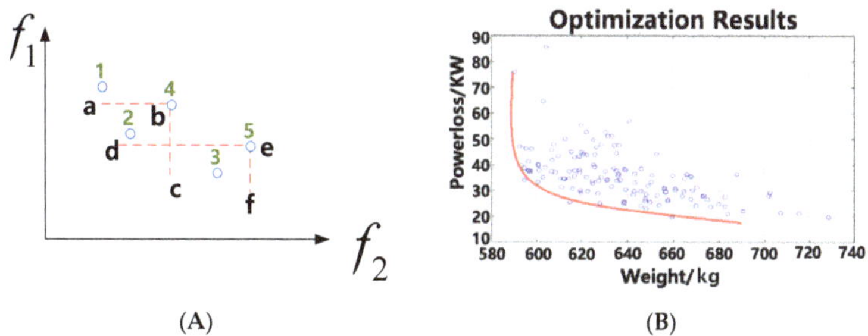

Figure 2. Multi-objective optimization problem. (**A**) Shows values for the two objective functions at five design points. (**B**) The set of GA optimization designs is introduced as the Pareto frontier curve and represents the best collection of design parameter points for aircraft power systems.

The aggregation of non-dominated designs is formed as the Pareto frontier curve, representing the best collection of design parameter points of the aircraft power system. Figure 2B shows this curve. Note that an optimal design condition can be selected from any point on the Pareto frontier. The choice of a specific design from the set of points forming the Pareto frontier curve is at the experience and judgment of the optimal decision maker. The aircraft power system maybe has many optimization objectives from the energy or power field such as weight, reliability, efficiency, volume and cost for fuel consumption, which can be expressed in Equation (1). The power flow balance relationship is shown in Equation (3). Some variation rates of power output for the generator or battery can be formatted in inequality constraints (2). The decision variable or design variable in the optimization model is sometimes stochastic, such as the load power demand or some power source like PV, so the stochastic optimization and robust optimization methods must be provided.

For the aircraft electrical power system, the energy efficiency function of power generators can be written in the following equation [41]:

$$E_{G_i}(t) = \alpha_{G_i} P_{G_i}^2(t) + \beta_{G_i} P_{G_i}(t) + \gamma_{G_i} \tag{5}$$

where $\alpha_{G_i}, \beta_{G_i}, \gamma_{G_i}$ are the coefficients for the efficiency of the power generator G_i from the polynomial fitting according to the data from the experiments and simulations. The different typical efficiency curves of power converter components in aircraft power systems are shown in Figure 3a–d. By using this method, the model of the auto-transformer rectifier unit (ATRU), dual active bridge (DAB) converter, some DC–DC converters, and a voltage source inverter (VSI) can be built to achieve the power system level efficiency optimization [53]. From Equation (5), the efficiency of the power converter is nearly the convex function, and this will reduce the difficulty of the optimization problem.

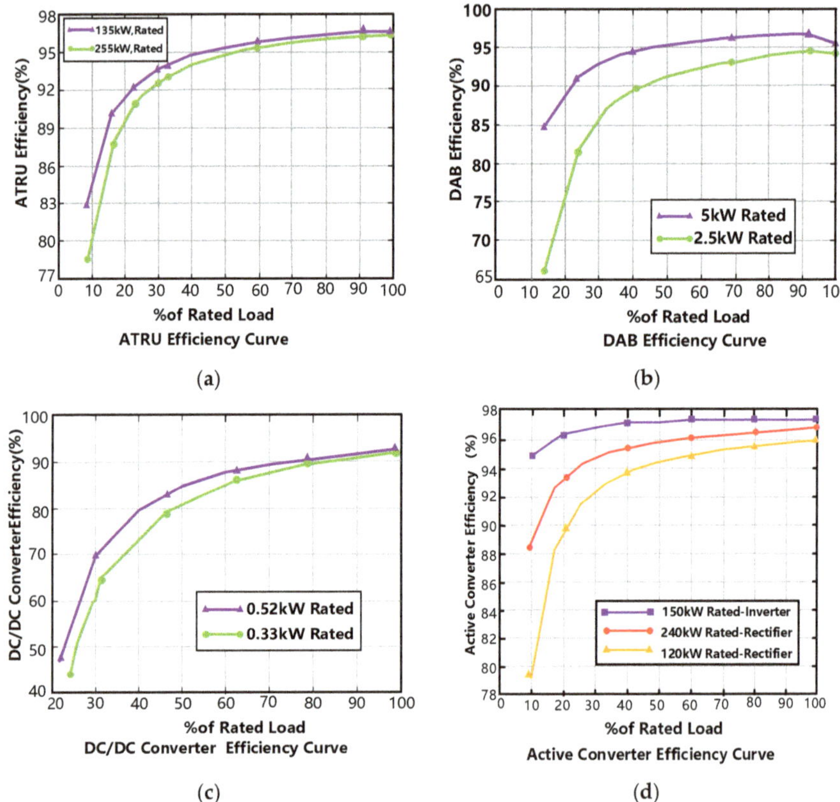

Figure 3. The efficiency curve of typical electrical power converter components in aircraft. (**a**) ATRU efficiency curve. (**b**) DAB efficiency curve. (**c**) DC/DC converter efficiency curve. (**d**) Active converter efficiency curve.

Optimization problem reformulation of the architecture of the aircraft power system can be expressed by the following, with regard to the system's power efficiency:

Problem: PMS with multiple objective optimization for the aircraft power system

Minimize

$$\varepsilon = \left\{ E_{G_i} | G_i \in g \right\} \cup \left\{ E_{C_i} | C_i \in C \right\} \tag{6}$$

where the E_{G_i} is the efficiency function of power generators, $G_i \in g$, the g is the set of power generators in EPS; $C_i \in C$, the C is the set of power converters in EPS. E_{Ci} is the efficiency function of power converters for an aircraft power system.

In an aircraft power system, the energy static optimization problem formation must be regarded with many constraints such as volume, weight, heat dissipation, and reliability. The optimal architecture and configuration for aircraft power systems can be derived based on optimal parameters and optimal distribution from the different power source. The power optimal scheduling problems to achieve a good balance on multiple objectives include ensuring optimal operating ranges for generators, obtaining a higher efficiency area of generators and power converters by utilizing the electric storage system according to variation in the power demand, and also maintaining the power priority in the connections of generators, power buses, and electric loads at the optimal performance index. There are several equipment examples in AEPS to explain these constraints [64].

1. Power Bus and Load Priority Constraints: Two kinds of priority constraints in the traditional AEPS can be regarded: the electric load priority constrants in the matter of critical and noncritical electric loads, and bus priority constraints in the matter of preset priority lists or the connections between each generator and primary power buses, and between each primary bus and secondary distributed power buses. The priority constraints are realized by adding a sequence of penalty factors in an objective function for different power buses and loads—the higher the priority, the larger the penalty factors [64].

2. Generator and Bus Power Capacity Constraints: The power generated by each generator must be held within its limits of power rating change and capacity, while the power transferred through each bus should not exceed the upper bound. In some situations, this constraint is formed as the combined optimization problem [65].

3. Power Balance Constraints With Consideration of Power Efficiency: The required power to the main power buses may only be supplied by one main generator. Similar to the secondary bus power allocation, a redundant or emergency bus should be considered in case of a failure of the first allocated generator. In addition, the power balance equation is normally a quadratic function; this can reduce the convexity of the objective function [66].

4. Bus Connection Constraints: At each time instant, the main bus should only be connected to one generator. According to the optimization model of the aircraft power system, the objective function, decision variables, and constraints have been presented based on the above consideration. The distributed solid-state power distribution system can control the load by connecting or disconnecting the secondary bus by a discrete switching signal. The optimization objective function is a nonlinear function and is non-differential in parameter space [67].

In general, classical optimization techniques are useful to search for the optimum solution or unconstrained maximum or minimum of continuous and differential functions. It is very easy to find the optimal value. Some specifications for numerical optimization can be selected based on this understanding, as described below briefly:

- Linear programming (LP). The optimization objective function is a linear function for decision variation, while the constraints for variables are linear. Basically, this optimization model is a convex optimization. The simplex methods and interior points methods can solve this problem. For AEPS's simple architecture or small-size dc power system, the optimization of energy efficiency can be realized as the LP problem [12,45].

- Quadratic Programming (QP): Allows the objective function to have quadratic terms, while set A must be specified with linear equalities and inequalities. Although some optimization models are not convex optimization, the relaxation and approximation of non-convex optimization can achieve the optimal solution accurately. For the AC power system of large-sized aircraft, the power scheduling optimization can be regarded as the QP problem [30,64].

- Nonlinear programming: Applies to the general case in which the objective function, or the constraints, or both, contain nonlinear parts. Regarding aircraft multi-energy power systems including electric power systems, hydraulic systems, and thermal management systems, the energy optimization problem is a nonlinear programming problem [27,37].
- Stochastic programming: Applies to the case in which some of the constraints depend on random variables. With the renewable energy power source such as PV and fuel cell installed onboard, this uncertainty from the power source and electric load can change the optimization problem into SP [29,62].
- Combinatorial optimization: Concerns problems where the set of feasible solutions is discrete or can be reduced to a discrete one. For example, some power loads with a different priority can be connected to a secondary power bus with limited power capacity, the optimal configuration of the electric load must be found with the different flight stages. In this situation, the aircraft power system optimization problem is a mixed integer linear programming problem (MILP) [31,48], which can be solved by some mature solver such as Gurobi, CPLEX, GAMS, and the special power system planning software.
- Evolutionary algorithm: Involves numerical methods based on a random search. Other heuristic-based methods such as particle swarm optimization (PSO), fuzzy logic-optimization, simulated annealing methods (SA), and the genetic algorithm (GA) can be used to find the optimal value space in aircraft power system architecture design [68]. While the optimization function of aircraft power systems is nonlinear and non-convex, heuristic methods can be applied to solve this problem. The solution to this optimization maybe not a global optimal solution, but a sub-optimal solution. The static optimization of the architecture of the power system is not enough, while the system's configuration is optimal in terms of the size of the overall system. Therefore, the dynamic optimization of energy management for aircraft power systems is introduced in part 5.

3. The Energy Power Source Onboard the Aircraft (Uncertainty Analysis)

The different kinds of power sources can be configured into an integrated power system according to the flight mission and the type of aircraft. The load power profile is changed with different flight stages shown in Figure 4a. For EPA, this power profile is shown in Figure 4b, for different power architectures such as CFAC or HVDC. From this figure, it can be seen that the power demand is very large at the climbing and descending stages, and the power is very stable during the cruising periods; therefore, the reasonable power supply configuration and the combination become very important for aircraft.

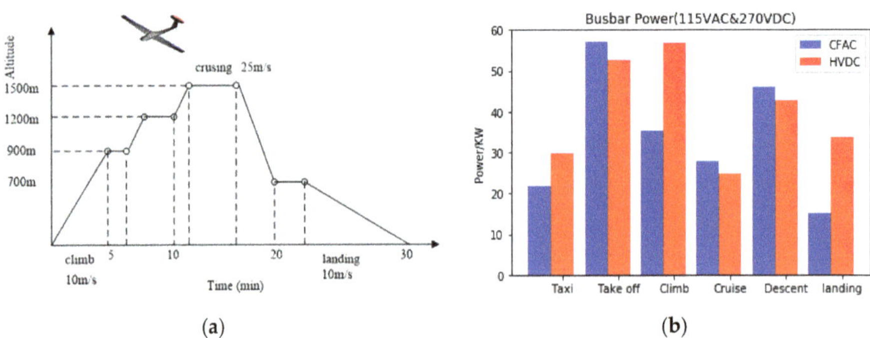

Figure 4. The typical flight profile (**a**) and power demand (**b**) for the small size aircraft.

1. Small aviation aircraft and UAV

The power demand is comparably low for small size aircraft and unmanned air vehicles (UAVs), so the new renewable energy, such as PV and the fuel cell, can be applied to this type of aircraft because of limited power density and power response speed. When the aircraft is at a high altitude, there are many environmental factors that affect the operating characteristics of a PV cell and its power generation capability [12]. The two main environmental parameters are solar irradiance G, measured in W/m^2, and ambient temperature T, measured in degrees Celsius (°C). The relationship between these two factors and the operating electric characteristics for PV can be modeled mathematically [69]. The power–voltage curve can be calculated based on the I–V curve of a PV cell or panel. Figure 5a presents the power–voltage curve for the P–V curve under different ambient temperature, and Figure 5b is given in different solar irradiances, where the MPP is the maximum power point labeled with a circle in this curve. For the solar aircraft or UAV, both the solar irradiance, angle, and temperature will change with the aircraft's altitude, sometimes gradually (minutes to hours) and sometimes quickly (seconds), for example, passing clouds and temperature variation affect the aircraft's roll angle. Considering the I–V curves as the characteristics for just an instant of time, the PV system must be integrated with the other storage energy system to meet the requirements of aircraft power systems [12]. Two different design methods must be considered to enhance the performance of a solar-powered aircraft: (1) Collection of more solar energy for a given wing area and application of gravitational potential energy; (2) Better utilization of collected solar power. The collection of solar radiation can, thus, be increased if the solar cells placed on the wing of the aircraft are tilted toward the sun. This is realized by light control and attitude adjustment. However, this will add a penalty in the form of air drag, resulting in a higher power requirement to fly. This penalty can be minimized or traded off if the optimal value of the PV bank angle is considered dynamic and the optimal flight route or trajectory is determined. This is still an open research topic. Solar energy is not fit for the medium- and large-type aircraft because of the limited energy density and power density of traditional energy storage devices.

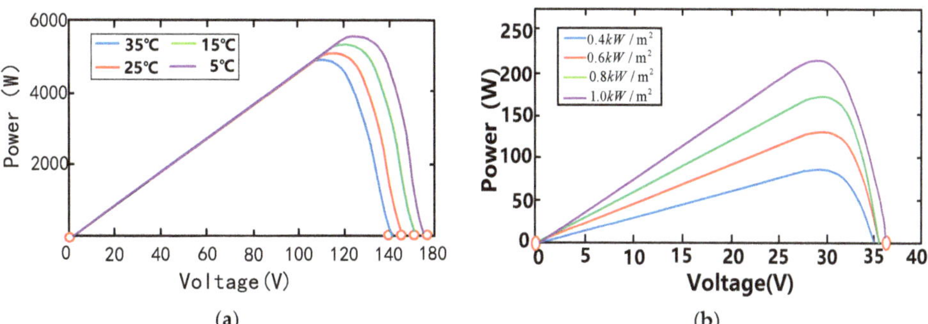

Figure 5. The power characteristics of PV modules under different ambient temperatures (**a**) and solar irradiance (**b**).

For fuel cell UAV, an unmanned aerial vehicle (UAV) in a mission that may be flying at a high altitude, this will cause a change in environmental temperature and air pressure. The temperature, for each rise of 1000 m above sea level, will reduce by approximately 6 degrees Celsius, and for every rise of 900 m above sea level, the air pressure will reduce by approximately 10 kpa (0.1 bar). Therefore, in the study of unmanned aerial vehicle (UAV) hybrid power supply, EMS will not be able to ignore environmental factors due to altitude change and the effects on the output power characteristics of PEMFC. The characteristics of a fuel cell are shown in Figure 6 [69]. In order to maintain the optimal operation or high efficiency area of a fuel cell, It is important to design the proper energy management strategies and power control methods to tackle the variation in the environment during the flight.

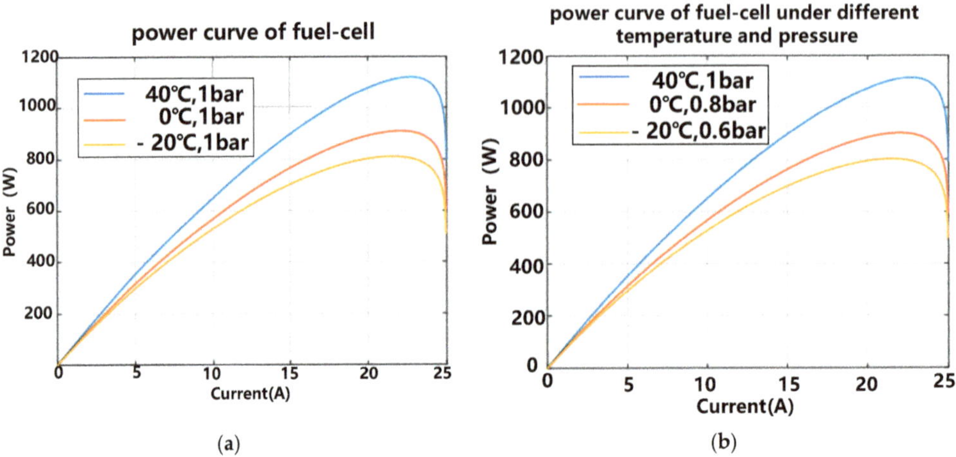

Figure 6. The output power characteristics of fuel cell with different temperatures (**a**) and air pressure (**b**).

2.　Regional aircraft refers to short-haul aircraft flights

Regional aircraft refers to short-haul flights with up to 100 passengers. In the traditional MEA and future mid-size or large electrical propulsion aircraft, the main electrical energy source is the engine-driven generators. The power source is mainly for a generator which can provide the large proportional electric energy for the power load. The battery packs are often used as an energy storage system to regulate the operation points of the engine. The fuel cell provides the electric energy for the auxiliary power unit (APU) and will act as an emergency power source in the future because of the green energy demand [70–72]. Due to the slower internal electro-chemical and mechanical dynamics of fuel cells, the response to the fast electrical load transients is slow. Load transient variation or power disturbance during the flight of an aircraft produces a harmful low-reactant condition inside the fuel cells and shortens their life cycle. The difference between the time constants of the fuel cell and electrical load calls for an electric energy storage unit that would supplement the peak power demand for the fuel cell during transient states for an aircraft such as takeoff, climbing and yaw control. An auxiliary electric energy source, such as a battery or super-capacitor, has the following functions: (1) Compensating for the slow power dynamics of a main power source like fuel cells; (2) Reducing the response time to the fast-changing electrical load during transient state; and (3) Supplying power to the load until the output power of the fuel cell is adjusted to match the new steady-state average power demand.

Although fuel cells have a higher energy density compared with batteries, they are very sensitive to low-frequency ripple currents [73]. While providing the low-frequency alternating current to the AC electric load from a fuel-cell-based power source, a second harmonic component of the AC current may appear at the fuel cell stack [13]. The low-frequency ripple current reaching the fuel cell may move the operating point from the region of ohmic polarization to the region of concentration polarization, thus leading to the unstable operation of the fuel cell system. This may result in the malfunction of the fuel cell power unit, and hence, the system may shut down and be damaged. Therefore, this low-frequency current should be absorbed and eliminated by the power electronic converter. The high power density DC–DC interleaved boost converter can be used to realize this aim to be interfaced with the DC bus. In some situation, the multi-cell or multiple parallel power converter maybe a better choice to solve this problem.

3. Narrowbody and widebody or two-aisle aircraft

Aircraft with up to 295 passengers are referred to as narrowbody or single-aisle aircraft, while widebody or two-aisle aircraft refers to aircraft that can carry between 250 and 600 passengers. Lifting electrical power consumption from a hundred kilowatts in MEA toward MW-level AEA EPS design is extraordinary work. This variation translates to tackling several technological challenges: high voltage transmission and distribution [74], superconductivity [75], thermal management and cooling [76], and large power generation. The large power, high temperature, super-conductive generation system is the only viable path to reduce the high losses in MW-level architectures and increase power density of the whole aircraft EPS [77]. However, there are still several drawbacks and challenges for superconducting architectures. For example, to achieve superconductive effect, a cryogenic cooling system is needed to affect size or volume, weight, efficiency, and the specific power of the material or electric components. Due to the uncertainty of the power source, there is no clarity when it comes to the operation of such devices at high altitude for the aircraft power system. In order to reduce the emissions of CO_2, as an alternative to the direct burning of hydrogen, a turbo electric propulsion system (TEPS) can be utilized [74]. The turbo electric distributed propulsion can be applied to aircraft, which has the potential to be the next disruptive technological breakthrough for aircraft. The turbo electric propulsion solution can minimize the overall fuel or gas consumption by letting the hydrogen turbine operate at its optimum efficiency point during the whole flight stage (improved gas turbine cycle). Combined with the fuel cell, the TEPS based on LH_2 can take better advantage of the high level of synergy of LH_2 as a fuel and as a cryogenic cooling medium for superconducting power conversion [13]. It also can be integrated with a fuel cell as the emergency power source or APU.

The uncertainty of a renewable energy source integrated with the aircraft power system must be considered; the safety and reliability of the power source is the most important performance index. Onboard electrical power source systems (EPSs) undergo significant changes in order to provide substantially increased power demands while meeting extremely strict requirements for weight and volume, safety and reliability, electric power quality, availability, etc. In the future, the aircraft electric power system must meet the requirements of green aviation and environmental protection.

Regardless of the EPS architecture, it should provide the loads of electric power with better power quality according to the established aerospace standards. One should note that for new power platforms, updated standardization documents are required since many requirements of MIL-STD-704F [73] are of a legacy nature (distortion harmonic spectrums, electric-magnetic emissions, voltage modulation envelopes, etc.) and certain aspects of future power architectures, such as higher voltage levels or grid frequency range, are not covered.

4. The Power Characteristic of Load in Aircraft (Load Stochastic Analysis)

The loads on an air vehicle power system occur with a variety of time scales:

- Continuous-steady (e.g., avionics equipment)
- Occasional-steady (e.g., landing gear retract and the braking system engages)
- Impulsive (e.g., radar, electronic warfare, DEW)
- Continuous-variable (e.g., flight controls, fuel pump)

The basic characteristic power load of traditional aircraft is basically fixed in priority, and changes with the flight profile. The loads have changing priority (landing gear, de-icing system), and the load is commonly linear. When the aircraft is in the different flight stages such as taxing, take off, and cruising, the load characteristic is the summation of different duty cycle loads. This mixed load adds to the design sizing challenges for power and thermal management systems. This is displayed in Figure 7a. Therefore, how to distribute the power energy onboard optimally to the load is a very important problem of optimization. As shown in Figure 7b, some optimal management load methods can alleviate the peak and valley value of power demand.

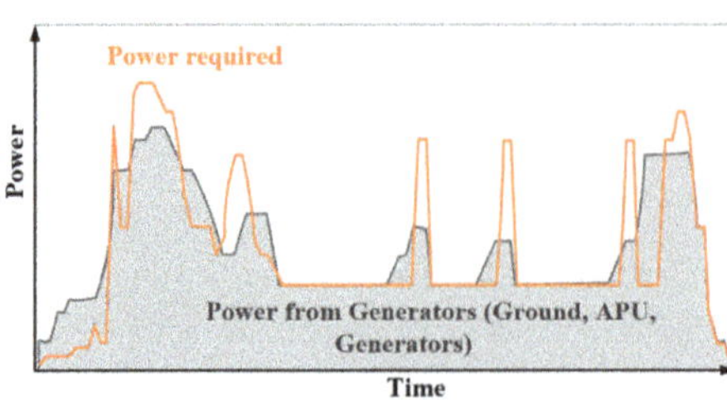

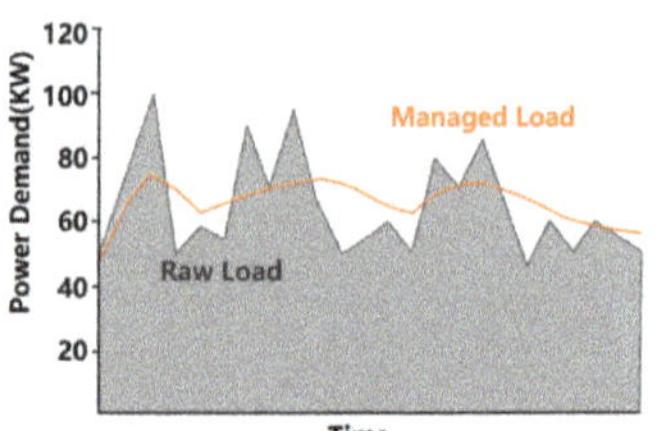

(**a**) Mixed power load for required generator   (**b**) Raw load and managed load

Figure 7. The electrical load profile for the aircraft power system.

Two typical loads are constant power load (CPL) and pulse power load (PPL), whose dynamics cause critical issues in the stable and reliable operation of aircraft DC microgrids.

1. CPL

Power electronic converter and power electric drive loads, when tightly controlled, behave as CPLs. CPLs are extensively integrated into a DC microgrid. Typical examples of CPLs are a DC/DC converter feeding resistive loads and DC/AC inverter-driven electric motors. The flight control electric actuator is a typical CPL load. Figure 8 demonstrates the latter example as a demonstration of the characteristics of CPL. When driving an electric motor with a rotating load, the DC/AC inverter has a one-to-one torque speed characteristic.

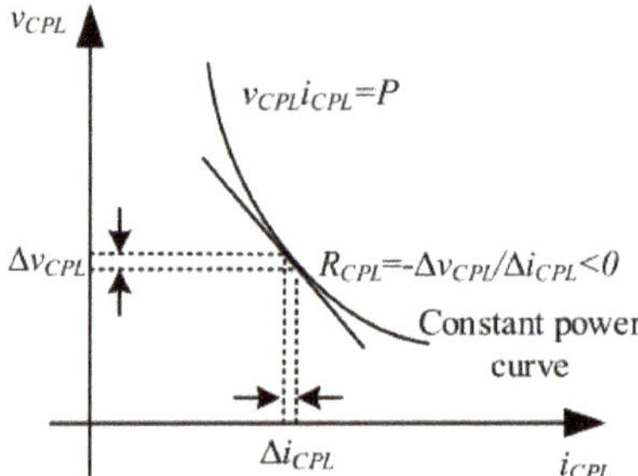

Figure 8. Voltage–current characteristics of CPLs.

In the distributed electrical propulsion aircraft, the thrust power load is variable with the flight control requirements. The load power consumption and waste heat profile are in the stochastic distribution. The probability analysis methods for the power demand loads are very necessary to save the design space for the power system. The load power flow onboard is followed by the normal, or Gauss distribution density function [29]. So the load power peak demand and average value can be reduced greatly. In addition, if the coordinated energy flight scheduling of the aircraft is a highly nonlinear complicated multi-objective programming problem, which cannot be directly solved, stochastic characteristics of the load power must be considered [38]. The flight modeling and propulsion load modeling must be given for the energy optimization management system.

2. PPL

In some situations, especially for modern military aircraft, the largest pulsed power loads vary from new weapon technologies to advanced avionics and other electrical equipment [62,78]. Pulsing power loads emulate a pulse width modulated signal, which has non-linear destabilizing effects on the electrical system. The power and DC bus voltage characteristics of PPL can be shown in Figure 9. The different duty cycles of PPL can also affect the stability of aircraft electric power systems.

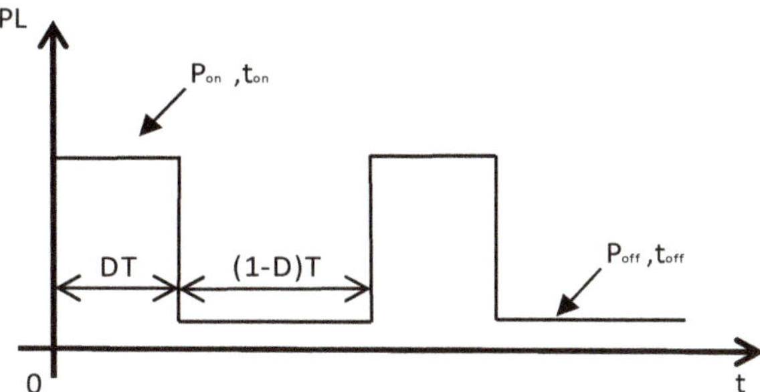

Figure 9. The power and bus voltage characteristics of PPL under different pulse periods.

Next, the mean of the load flow can be found using a steady-state analysis of the mean inputs. With the mean and variance of the load flow in hand, all that remains is to find the mean and variance of the absolute value of the load flow. The stochastic modeling methods can describe the accurate changes for the load compared with traditional power load modeling, such as the ZIP model in power system. With the development of AI and machine learning technology, the power demand model of EPA can be represented by a long-time probability distribution function for the data set coming from the MCMC sampling methods to approximate the real flight power demand.

3.   Electro-thermal load

Additionally, these large power PPL have thermal properties that can induce electrical power stability issues at low and high temperatures and various pulsing load conditions according to the system cooling requirements. These research topics are investigated in [62,63,78,79]. Regions of complete stability, metastability, marginal metastability, and instability are determined by bus voltage transient tolerances. Analyzing the marginally metastable boundary layer, thermal analysis is performed at different points of equivalent average power and varying pulse energy for an aircraft power system. This topic is discussed in part VII of this paper.

5. Energy and Power Management System and Strategies for AEA/MEA and EPA Power Systems

5.1. Energy Management Optimization and Power Control Question Formation

Increasingly stringent demands have been placed on aircraft power systems. Larger power volume, higher efficiency, and a sufficient cooling capability are required, with constraints such as weight, physical volume, and power density. Coordinated control between thermal and electrical power system is very necessary for managing the generation units, distribution units, and consumption of power onboard the aircraft. Coordinated power control and energy management can benefit from reducing the gap between performance demands and capabilities of current generation aircraft, or contributing to improve the capability for the design and sizing of next generation aircraft. Furthermore, an aircraft is a system of systems with various energy flow being converted and consumed between multiple energy systems, as shown in Figure 10. The power flow in an aircraft coupled with thermal energy is very complex [62,63], and dynamic time scales from the sub-millisecond electrical voltage regulation to the minute level control for fuel tanks and passenger cabins are shown in Figure 11. Therefore, it is essential to design an optimization model and controller structures that can cope with the temporal and spatial disparity that exist within these complex power systems. The modeling of the power system in aircraft is a challenging task because of the disparity. Except for the volume and weight constraints, the optimization model for aircraft energy and power systems must be derived regarding many constraints such as the generator power limits, thermal capability, battery's SOC, and PV power source power limitations [80]. Therefore, the power management architecture and strategies are also very important to realize energy generation and distributed optimization in aircraft. The problem of energy optimization and power control is basically a multi-spatiotemporal scale optimization problem with different constraints.

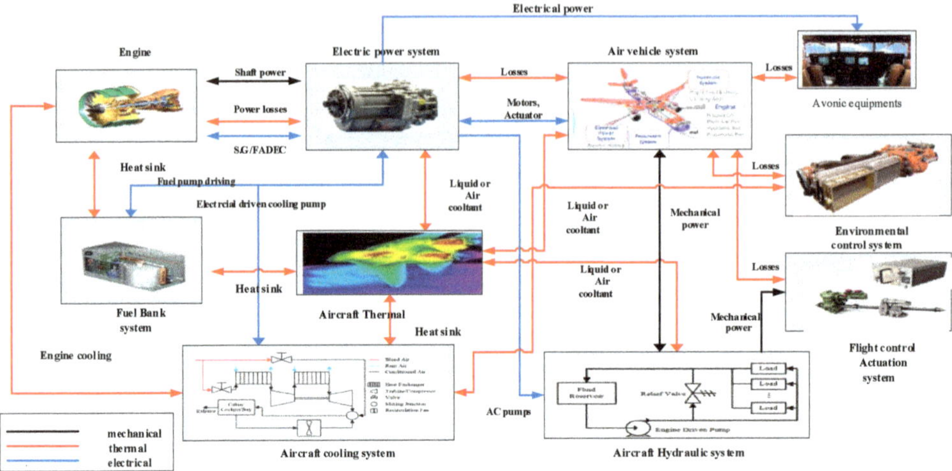

Figure 10. The multi-energy domain interactions in an aircraft power system.

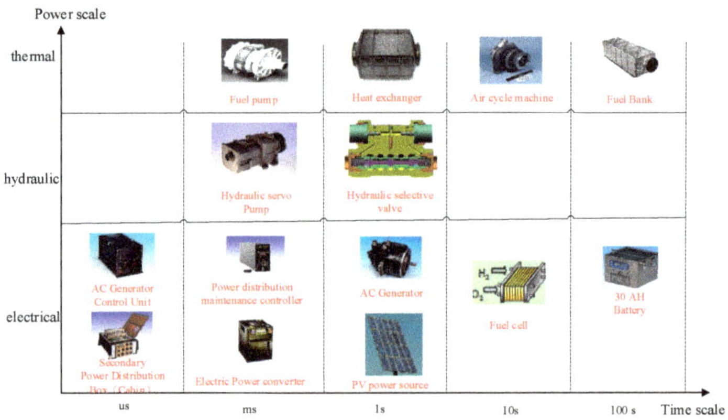

Figure 11. Temporal scale separation of major thermal, hydraulic and electrical system devices in aircraft.

The energy dynamic management and control needs some special technology to be realized; the optimal control theory is often used with PMP and dynamic programming (DP). PMP is a common optimal method; it overcomes the defect that the variational method cannot find the extremum of constrained control variables and objective functions [81]. Compared with the dynamic programming algorithm, its computation is greatly reduced, but it is still only used in offline situations. In the process of energy management strategy research of electric power systems, the performance index of the energy management system, with the constraints described above, is transformed into a Hamilton function minimization problem [82,83], so as to obtain the optimal trajectory of control quantity. DP applies to the case in which the optimization strategy is based on dividing the problem into smaller sub-problems. DP strategy has been used by many scholars to develop a hybrid energy flow management strategy, and is recognized as a relatively ideal hybrid energy management method [84]. The results are often used in the offline optimization management calculation of fixed working conditions, and are also used to evaluate the merits of other control algorithms for energy management.

Table 1 presents the energy optimization index of the aircraft power system. The optimization index or multiple optimization objectives can be combined as the optimization objective function, and the multi-objective optimization model of the electric propulsion power system can be formed by combining the constraints of working characteristics and environmental constraints of different electric power systems.

Table 1. Optimization index of different electric power architecture.

Energy and Power Architectural Form	Performance Index of Optimization Function
All types of electric power system architecture	The highest overall efficiency of the power system Strong robustness, reliability, and fault tolerance Minimal emissions of pollutants such as carbon dioxide
Lithium battery, fuel cell hybrid electric propulsion architecture	The longest service life of lithium batteries and fuel cells
Fuel cell, oil gas, battery hybrid-electric propulsion architecture	Minimization of fuel or hydrogen consumption
Solar and fuel cell electric propulsion architecture	Longest flight range and flight time

5.2. Energy and Power Management System Structure and Architecture

The energy and power management problems in aircraft are often realized by the electrical power system because of the trend of More Electric and All-Electric Aircraft, especially for the electrical propulsion aircraft (EPA). So the architecture of the power system can become microgrids, similar to the terrestrial microgrid system. In order to describe the architecture of onboard microgrids in aircraft, the different kinds of the architecture of microgrids in aircraft are presented in Figure 12. There are a variety of architectures for the aircraft onboard microgrid, such as star architecture, multiple-star architecture, ring architecture, and hybrid architecture. In order to get the best architecture for a reliable electric power supply, evaluation methods for the aircraft onboard microgrid are needed to evaluate the performance of different architectures according to certain criteria. The criteria include energy system efficiency, reliability, and fuel consumption economy. The evaluation and selection of the most appropriate aircraft power system architecture is an interactive process, made of multiple trade-offs, detailed analyses using requirements and constraints as the input variables, and detailed analysis to feed a decision matrix that is used to compare the different architectures against the functional objectives. In considering the electrical propulsion needs, the load demand must be considered with the electrical propeller load characteristics. The energy management of the power system is referred to as the long time interval energy optimization problem, which can include the above criteria. The power management of the power system is relative to the short time scale and transient time power response optimization problem, which can include the system stability, transient power response, power quality, and fault protection. Moreover, aircraft microgrids have a variety of power supply types, operating and protective modes, control topologies, and power distribution network structures, which can strongly impact their dynamic characteristics, so it is very challenging to develop an accurate, complete energy, and power management model and strategies. The hierarchical microgrid energy and power management system contains three layers to achieve different control objectives: Level 1, for generation processes based on internal control, response speed, and for system stability based on the primary control; Level 2, for high power quality and power sharing based on the secondary control; Level 3, for power economic dispatching management based on the tertiary control, system reliability and economic planning based on policy control. This situation can be shown in Figure 13a.

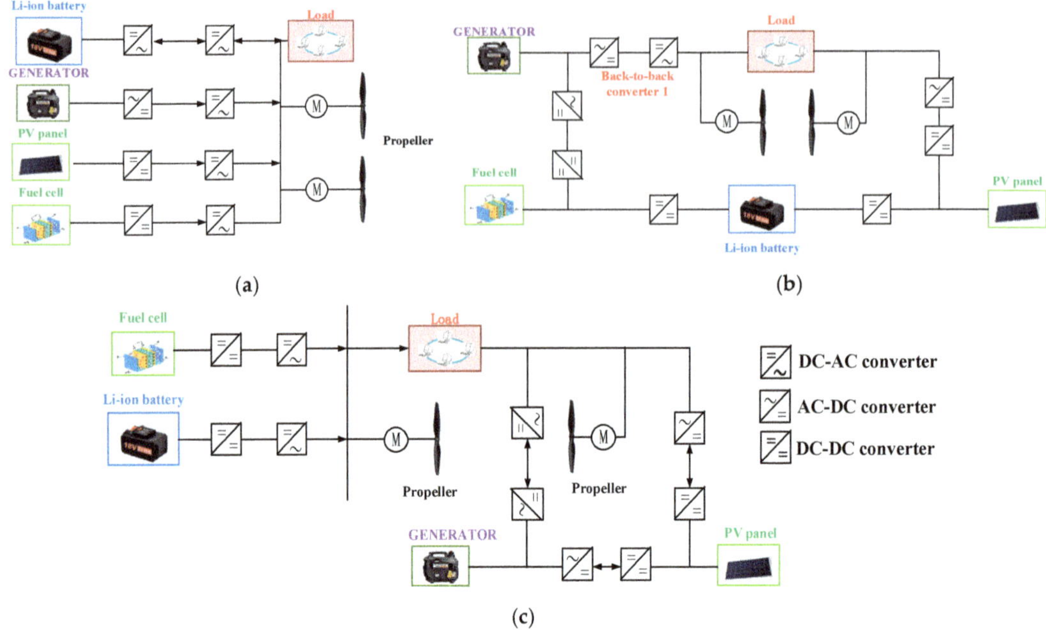

Figure 12. The different architecture of electrical power system in aircraft microgrid. (**a**) The star architecture of aircraft microgrid structure; (**b**) The ring architecture of aircraft microgrid structure; (**c**) The hybrid architecture of aircraft microgrid structure.

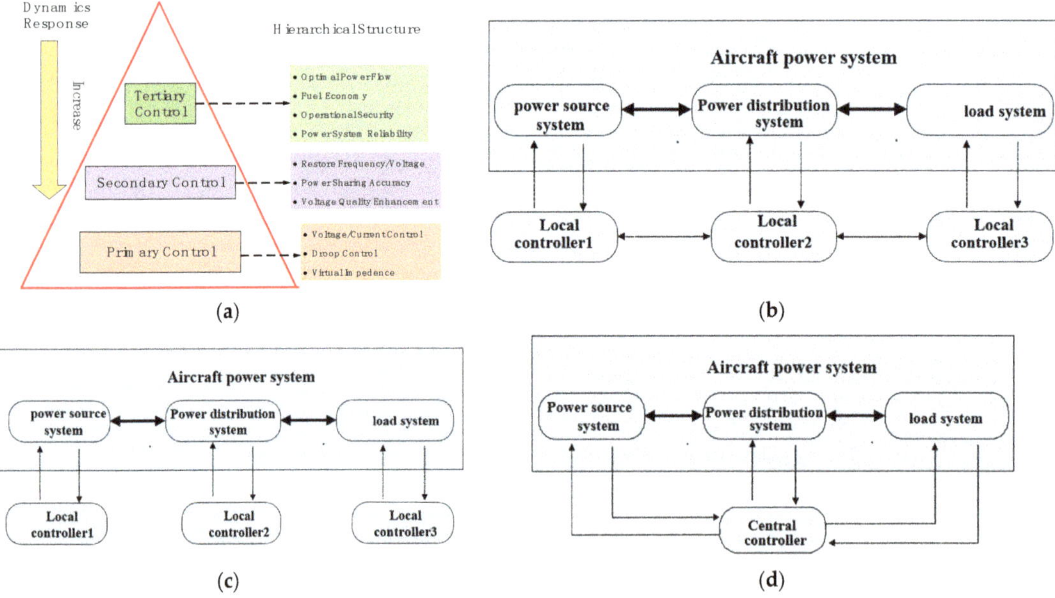

Figure 13. Different controller architecture for the aircraft power system [31]. (**a**) The hierarchical structure of control method. (**b**) The distribution controller architecture. (**c**) The decentralized controller architecture. (**d**) The centralized controller architecture.

Based on the optimal energy management and power system architecture, the dynamic power management strategies were very important for the operation of the aircraft power system. It is proposed to apply distributed control strategies to the power systems of aircraft, which allow information exchange among local controllers by establishing a communication network topology among them. In fact, distributed control strategies can be considered as a trade-off between centralized control and decentralized control by combining their advantages [85]. In aircraft, local subsystems are assumed to be developed independently by competing entities and companies, requiring that any coordination must consider the privacy of the controllers and account for differences in their update rates [86]. It is assumed that the subsystems are developed and manufactured by subcontractors who are potentially in competition with each other across various developing projects. Thus, the design details of the subsystems and their controllers may be regarded as intellectual property (IP) and their designers would be unwilling to using model-based coordination that could risk exposing it. Therefore, distributed control is better than centralized control methods just to exchange some interactive information or variables. For decentralized control architecture, the optimal performance of energy management is not as good as distributed control because of the absence of exchange information between subsystems [84]. Therefore, it is necessary to choose the proper controller architecture according to the energy system's dynamic state trajectory and control input decision variations.

In Figure 13a, The tertiary control is mainly for the onboard microgrid energy optimal distribution among different power sources, with the power consumption minimization, and fuel economic optimal dispatch, with an additional object such as power stability, safety, and fault tolerance. The real and reactive power sharing accuracy is deteriorated when the ratio of line-resistance-to-line-reactance is high in an aircraft power system. The secondary control is proposed and applied to solve such problems. Its main objective is to restore the frequency and voltage to their nominal values in the aircraft ac power system. Additionally, appropriate control methods are also proposed to enhance the voltage quality including compensating for the voltage unbalance and harmonic distortion in the secondary control.

In aircraft power systems, the different power source subsystems such as an electrical generator, fuel cell, storage energy power system such as a lithium battery, and super-capacitor have different time and spatial scale dynamic characteristics. The centralized controller architecture is prone to single point fault from the different subsystem controller. The controlling real-time characteristic is very important for microgrid of aircraft. The typical management and controller architecture of power systems in aircraft can be shown in Figure 13b–d. Figure 13b shows the distributed controller architecture. Figure 13c shows the decentralized controller architecture. Figure 13d shows the centralized controller architecture. With more penetration of electrification in the aircraft power system, the linkage between different power electric subsystems is greater, and they are more interrelated with each other. The controller architecture for each subsystem must be chosen by considering different factors with the dynamic characteristic, communication capability, time scale, and exchanged information. The distributed energy and power management architecture is more promising as a multi-agent system for future controller architecture because of the reliability and safety benefit for power systems. Also, for the aircraft electric load system, during some typical operating scenarios such as the actuation of flight surfaces during takeoff, and performing evasive high-thrust turns while using DEW shots for a military aircraft or a hybrid-propulsion craft accelerating through both a jet engine and electric motors, all these operating modes can cause a large power transient and voltage bus stability issues. To mitigate the power disturbance from fast variation of power demand, controller's architecture, or control or energy management strategies can be used.

5.3. The Strategies of Energy and Power Management

Traditionally, energy and power management strategies for power systems can be classified into two types: rule-based heuristic methods and optimization based methods. The dynamic energy optimization and power management is also very important for the micro grid in an aircraft [87]. As discussed above, renewable energy source integrated with the traditional aircraft EPS can improve the energy efficiency and reduce the oversize of electrical generators. However, the intermittent characteristic and electrical–chemical response from renewable energy make the power optimal distribution more complex. The load electrical power demand is very closely coupled with the flight control system, which often varies abruptly in different flight stages such as takeoff or landing. The stochastic changes in aircraft electric propulsion load makes the optimal power distribution control very difficult.

With the development of electrical propulsion and hybrid electrical propulsion aircraft, the aircraft power system can be viewed as a "multi-energy mobile microgrid." It is distinct from the land-based microgrid and has multiple energy transformations, which are subject to complicated flight condition constraints, and its flight durability fully depends on its energy utilization efficiency [83]. While most of the existing research works focus on the hybrid-electric propulsion aircraft (HEPA)'s real-time power balancing control, rare research works have been reported on its energy optimization problem [88], i.e., how to optimally schedule its energy consumption to achieve the best techno-economic performance. This is also an advanced research topic for the aircraft power system. The energy management strategies must be designed with multi-disciplinary optimization contents. The following table lists the main energy and power management methods onboard with different control and optimization objects according to different platforms of the aircraft. The summarized energy and power management methods for different aircraft platforms are presented in Table 2.

Table 2. The summarized energy and power management methods for different aircraft platforms.

Energy and Power Management Strategies	Aircraft Platform	Control and Optimization Objective	Features	Reference Paper
Rule-based	UAV, Aircraft APU system; More electric engine	Fuel or hydrogen consumption minimization	State machine; Power distribution based on expert experience	[50,51,71,89–91]
Fuzzy logic	UAV, Aircraft APU system, Aircraft HVDC	Hydrogen consumption minimization, voltage stability	FLC to specific flight profile or APU load demand	[53,71,89,92]
	Hybrid electrical propulsion UAV	Fuel Consumption minimization	Equivalent consumption minimization strategy plus FLC	[93]
	Fuel cell UAV	Hydrogen consumption minimization, voltage stability	PSO+FLC	[81,87]
Meta-heuristic	Aircraft emergency power system	Less hydrogen consumed; Optimal life time of electrical sources;	Artificial bee colony algorithm; Grey wolf optimization algorithm	[94]
MPC-based	Hybrid-electrical propulsion aircraft	System efficiency Fuel-consumption Voltage stability	Conventional MPC	[66,95]
	Aircraft APU system	System efficiency	Traditional MPC	[89]
	Aircraft distributed power system	Minimum switching of generator and power load shedding; power quality, minimization of THD	Stochastic MPC; INA-SQP MPC	[47,96]
	Aircraft energy storage system	Reduce the DC bus voltage transient; Minimization of current draw from the storage system	SQP solver	[78]
	MEA power system with PV	Stability of system; Steady and transient state Current control	FCS MPC + stability constraining dichotomy solution	[97]
	MEA Electrical power system and Fuel thermal management system	Fuel consumption minimization; Increased capability of thermal management	Hierarchy MPC control with thermal coupling	[98,99]
	Engines and electrical generation system in MEA	Fuel efficiency and emissions	Distributed MPC with ADMM algorithm	[86]
DP	Small-size aircraft, UAV	Fuel consumption minimization	Lookup-table	[65]
	Parallel hybrid-electric aircraft	Total mission fuel burn minimization over the flight envelope	Offline optimization	[100]
	MEA power distribution system	Power loss minimization Amount of switching is minimized	Optimal reliability configuration	[48]

Table 2. *Cont.*

Energy and Power Management Strategies	Aircraft Platform	Control and Optimization Objective	Features	Reference Paper
Adaptive energy management (ADP or RL)	more electrical aircraft with pulse load	Lowest possible rate of change of the main source of power or offers a fixed minimum rate of change in power	Integrated variable rate-limit of power	[101]
	Electrical power system in MEA	Optimization problem for power scheduling and allocation; minimize the fluctuation in power generation system; bus voltage regulation	Optimal adaptive control with MIQP, off-policy integral reinforcement learning	[82]
	Electrical emergency power system in MEA	Fuel consumption and overall efficiency optimization under the false data injection attack and denial of service attack from the critical measurements	Adaptive neuro-fuzzy inference system and specific fuzzy deep belief network	[102]
	Fuel cell electric UAV	Hydrogen consumption and battery lifetime	ADP and RL	[103]
Power decoupling methods based on filter frequency	Aircraft Fuel cell and battery hybrid emergency power system	Voltage stability; Minimization of power loss; Extend the lifetime of power source;	Slow power response for fuel cell; Fast power response for battery	[11]
Optimal control and PMP	All electrical Propulsion Aircraft	Combination of time-related and battery charge costs	Pontryagin's Minimum Principle	[83]
	Hybrid electrical aircraft	Minimization of fuel consumption with thermal management for the battery pack	Pontryagin's Minimum Principle	[104]
Combinatorial optimization	MEA aircraft ECS system	Electrical consumption minimization	MILP	[46]
	MEA Power distribution	Three-phase load balanced; Increase the lifetime of power converter	Nonlinear optimization; Convex optimization	[51,105]
	UAV	Flight mission planning and recharging optimization	SDP and QP	[106]
	UAV	Minimize the fuel consumption (FC) and polluted gas emission	Bender decomposition-based method; (MIQP)	[107]
	MEA power generation and distribution system	Optimal power allocation; Optimal generator sizing	MILP	[108]
	MEA low voltage distribution system	Load allocation on the EPS	Knapsack problem	[67]
Droop control methods	Aircraft HVDC system	DC bus stability	Active stabilization methods, load sharing	[70,109,110]
	Aircraft APU	Dynamic response for power allocation optimization	Virtual impedance Droop control	[72]

From the above Table 2, energy and power management strategies are presented for a power generation system, power distribution system, and load-side control onboard. The load-side control is similar to the demand-side management (DSM) in terrestrial microgrid. Rule-based, fuzzy logic and MPC-based energy management strategies are often used for UAV, MEA, and EPA. The dynamic transient power management strategies are mainly relative to droop control methods. A new dynamic optimization strategy for energy management of More Electric Aircraft based on hybrid systems theory was examined in the paper [66]. An expressive MPC framework is developed to address the pertinent issues for energy management in aircraft systems, creation of optimization metrics, implementation of necessary computational tools, and initial verification through simulation. A hierarchical model predictive control (MPC) framework was presented in the paper [95] for hybrid power or electrical propulsion systems that can be used in future energy-optimized aerospace systems. This framework can cover a wide bandwidth of model fidelity and enforce all necessary physical laws in the models using both MPCs together. State variable constraints are explicitly included in the controller formulation by using the equality constraints (i.e., discrete time model or system state transfer equations) to transform the state constraints into the control constraints. The hierarchical MPC framework allows operational constraints to be assigned to different levels of MPC schemes. Simulation results show that the hierarchical control system operates well in optimizing for both energy flow and dynamic current/voltage regulations under the dynamic conditions of the aircraft power system.

In the optimization of energy management for the aircraft power system, the complex system's state must be aggregated to overcome the problem of "curse of dimension" of DP. In addition, The traditional EMS, including MPC, is prone to the uncertainty or variation of the power system's model, This can further deteriorate the performance of EMS. The model-free and AI-based EMS can effectively solve this problem. Therefore, reinforcement learning or approximated dynamic programming (ADP) can be applied to the optimization problem of the aircraft power system. The optimization value function can be approximated by different methods such as TD, NN methods to fetch. Some probability functions can be used to describe the stochastic characteristic of the power source and load variation for EPA. Deep reinforcement learning (DRL) is presented in [111] to develop EMSs for a series of HEVs. Due to DRL's advantages of requiring no future driving information in derivation, good generalization in solving energy management problems can be formulated as a Markov decision process (MDP). As shown in Figure 14, The overall schematic of the proposed energy management method also can be realized for electric propulsion UAV. Generally, the implementation of this method is divided into three stages: This period is the training of EMSs. For the EPA, the EMS here is represented by a neural network, and the DDPG algorithm is adopted to update parameters of the EMS in a simulation environment. The update is based on data (state, action, reward) generated by iterative interactions among the hybrid electric UAV model, its history flighting cumulative trajectory information, flighting environment and EMS. When the training result achieves convergence, parameters and the structure of the neural network are saved as the trained EMS. The downloaded EMS can be directly used for online applications by simply mapping state-to-actions for energy optimization objects of the aircraft power system. The uncertainty and stochastic characteristic from the renewable energy and power load in the aircraft must be considered while designing the energy management strategies. The energy management strategies based on stochastic ADP methods were presented in tethered microgrid and a vehicle powertrain. These methods also can be extended to the aircraft microgrids [103].

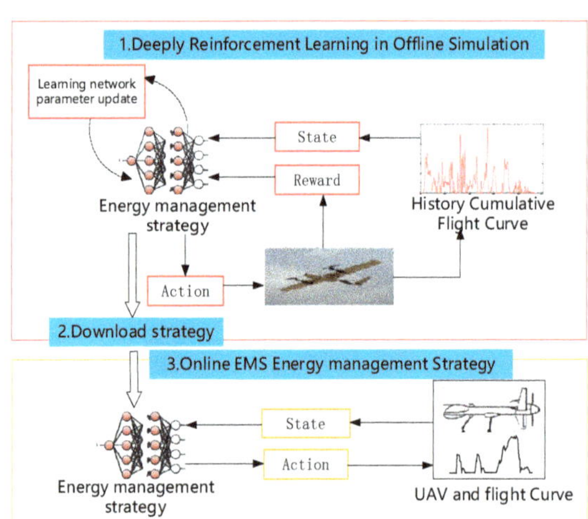

Figure 14. The overall schematic of DRL-based EMS for the UAV model.

For the electrical propulsion UAV or aircraft, the flight load power demand is stochastic in nature because of the external disturbance from wind turbulence or wind shear. In Figure 15, The flight dynamic power variation can be analyzed by the air dynamic flight equation of UAV which can describe the thrust power. The total thrust power includes the steady state thrust power and dynamic thrust power, in which the air density, area of airfoil, lift-to-drag ratio, climb angle, and roll (bank) angle can be integrated into the thrust force power equation. The experimental flight data such as the power demand of UAV can also be measured to verify the data from the digital simulation model. The transition between flight power demand can be regarded as the Markov decision process. Then, the probability model for the thrust power for UAV can be built with the nearest neighborhood methods; the transition probability matrix for the electrical propulsion power can be given for a three-dimensional dataset. The stochastic optimization model for the EMS of UAV can be derived according to Figure 15 to realize the optimal distribution of the energy of the powertrain with the Q-learning methods in reinforcement learning strategies for EPA. For large-and medium-size civil EPA, the power rating will increase greatly, so the thermal constraints caused by the power loss must be integrated with the EMS for the aircraft power system. This challenging topic must be tackled in EMS design.

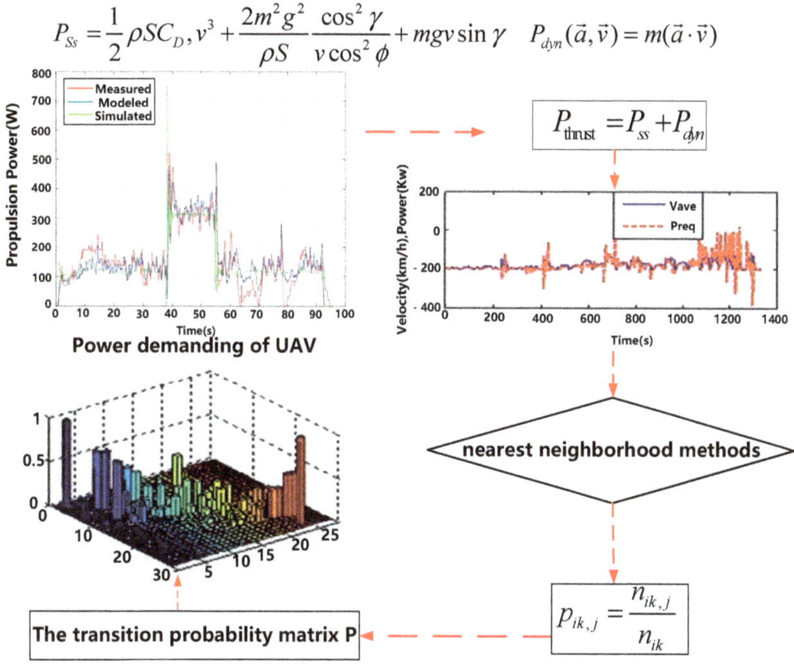

Figure 15. The probability matrix of load power demand for EMS strategy of the electrical propulsion UAV.

6. The Electrical–Thermal Coupling Control and Energy Management of AEPS

Moving toward the MEA/AEA and EPA involves increasing the volume of electrical power generation, heat dissipation potential, and distribution capability of the aircraft to supply most of the loads. This electrification trends rests on the development of power electronics (PE) and thermal management. Enabling technology for power electronics can contribute to high-efficiency improvements in the aircraft power system, based on distinctive features such as high power capability and controllable system [112]. For the PE converter, the power loss from switching cannot be avoided, so thermoelectric coupling is a very important factor which can affect the stability and reliability of the power system. In another aspect, the large electrical power demand can also cause thermal stress in traditional electrical power generation systems such as PMSG and SRMG, due to the constraints of volume and weight in the aircraft. The cryogenic temperature for power electronics is a beneficial method. Some research has already shown that some power semiconductor devices have improved performance at low temperatures, such as lower on-resistance and a faster switching speed, which means that making power electronics systems work at cryogenic temperatures can contribute to lower power dissipation and a smaller volume and weight [76]. The phase change coolant from the waste heat also can be reused to form a re-circular energy source to provide the electrical energy for onboard equipment.

PE technology is laying a foundation for the more electric engine, more electric loads, and electric propulsion loads in the aircraft. One key shortcoming of PE-driven loads is that they are prone to instability as the aircraft electrical network becomes larger and more complex, and the multitude of PE-based loads such as CPL can challenge the stability of the electrical power system (EPS) [113]. The stability assessment is thus crucial in the design of PE systems. The system stability has to be analyzed both at the small- and large-signal level. Small-signal analysis investigates the stability of an EPS when it is subject to small disturbances such as input voltage modulation or frequency modulation. The analysis is performed on a linearized system model regarding a certain operating point. In contrast, large-signal stability analysis investigates the system's behavior under large disturbances, including sudden large changes in loads. For small-signal analysis methods, the stability of EPS is generally assessed by using traditional stability analysis techniques. These include the eigenvalue root trajectory method and impedance methods based on the Nyquist stability criterion. The large signal stability is based on the mixed potential function, Lyapunov function theory, and a nonlinear dynamic system. An EPS can be viewed as a cascade of its source and load components. Plenty of papers were focused on this topic to analyze the stability and power quality issue of AEPS [79,114–118], but the thermal effect for the stability of AEPS was not fully studied in detail with the consideration of a high power pulse load and thermal storage couple in the aircraft power system.

The thermal question can cause damage to the distribution power system by the fault current onboard [119], and power loss and an overload situation may accelerate insulation aging of the aerospace motor [120]. This is a challenging issue for the stability of aircraft electrical power systems because of the limited volume and weight constraints in aircraft.

The energy optimization technology in aircraft must be focused on the exergy analysis according to the integration of the first and second laws of thermodynamics [121]. It can effectively reveal the irreversibility and inefficiencies of the thermodynamic processes, which provides a convenient approach to designing and optimizing the components and power systems in aircraft. Some thermal and electrical stability and power quality problems in local and subsystems of AEA and EPA were presented in [98,122–124]. In [123], thermal energy inherent in the cabin air and aircraft fuel, as a dynamic management solution to offset stochastic load power in the MEA power system, was introduced. The research focuses on a power electronic-controlled environmental control system (ECS), which can provide dynamic thermal inertia and act as an effective electric swing bus to mitigate power variability. The thermal and electrical coordinate control methods are proposed in [82,124]; this method can improve the stability area margin of the power system in aircraft for a high power pulse radar load. The hierarchical optimal control problem of aircraft electro-thermal systems was discussed in [82]. The coordination of these electrical and thermal systems is performed by using a hierarchical MPC control approach that decomposes the multi-energy domain and constrained optimization problems into smaller, more computationally efficient problems to cover the different timescale issues.

The larger pulsed power payloads were developed for the modern military aircraft varying from new DEW weapon technologies to advanced avionics such as radar and other electrical actuator equipment. A pulse width modulated power signal was emulated for pulsing power loads which have non-linear impairing effects on the stability of the electrical system. Additionally, these onboard devices have thermal flux properties during a very short time that can induce electrical stability issues at low and high temperatures and various pulsing load conditions due to the limiting capability of the heat sink of the aircraft. These non-linear electrical stability issues also transfer to the mechanical and thermal systems of the aircraft and can damage or degrade electronic components. The EMT model demonstrates the destabilizing effects caused by both the thermal coupling of the pulsing load and the large signal analysis of the PWM signal in [122]. The boundary conditions of stability for aircraft power systems were derived based on the Monte Carlo simulation method regarding the duty ratio and frequency. In the electric vehicle, the thermal–electrical component model such as ECS, and a battery can be built to analyze the energy relationship between the thermal and electrical energy [125–127]. This work can also be extended to the aircraft platform. It is very important to consider the coordinated energy management strategies or dynamic power control methods between electric and thermal energy to improve the stability and thermal ejective problem. A typical example of an aircraft power system based on an electric–thermal coupling control issue can be given in the next paragraph.

For the architecture of an aircraft networked microgrid in a single channel, shown at the top of Figure 16, minimizing the energy contribution from ESS and system-wide exergy destruction are reasonable optimization objects, in line with minimizing heat signature and maximizing flight endurance. The aircraft microgrid is comprised of power generator, power distribution system, and some electric load such as electrical actuator, avionic equipments, radar and pulse power load. The power consumption of each electric component generates heat loss, which is cooled or rejected by the onboard thermal management system as shown in Figure 16. In [121], there were three electrical power buses in the aircraft, the mission bus (MB), flight control (FC) bus and high power (HP) bus, operating at 270 V. Each bus has resistive storage elements R_{MB}, R_{FC}, and R_{HP} and capacitive storage components C_{MB}, C_{FC}, and C_{HP}. Two 480 V generators, V_{HPS} and V_{LPS}, were considered, located on the high pressure spool (HPS) and the low pressure spool (HPS) of a gas turbine, respectively. Through the buck power converters with duty ratio control inputs of D_{HPS} and D_{LPS}, generators are connected to the MB and HP power buses. The buck converters had been developed with resistive R_{HPS} and R_{LPS} and inductive energy storage elements L_{HPS} and L_{LPS}. The MB and HP bus were connected with each other through buck-boost bidirectional DC–DC converters, where control of the duty cycles was provided by D_{MBFC} and D_{HPFC}. The resistive and inductive elements of the connecting bidirectional DC–DC converters were R_{MBFC}, R_{HPFC}, and L_{MBFC}, L_{HPFC}, respectively.

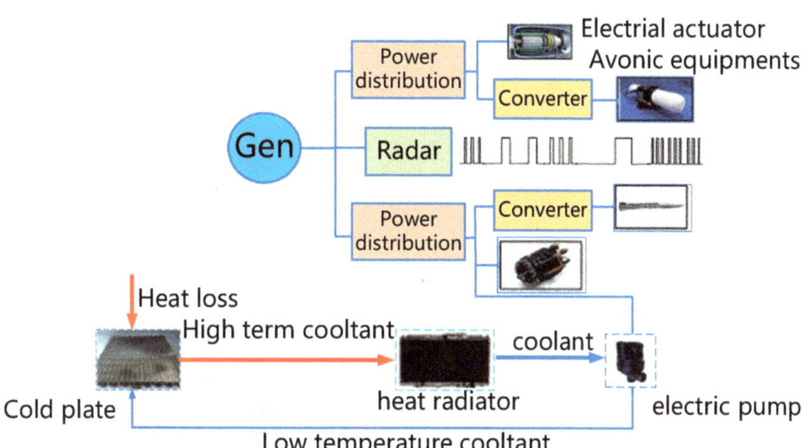

Figure 16. The aircraft electrical power system architecture for large power pulse load with cooling function (single channel).

An exergy optimal objective function based on ohmic losses within the power converters is given in Equation (7) as applied to the aircraft energy optimization problem. When minimizing object function $f(x, u)$, the optimal solution must satisfy the power flow equations constraints given in Equations (8) and (9), where the dynamics of the power system have been considered a steady state and the storage contributions set to zero.

$$\min_{x \in A} f(x, u) = \min \left[\frac{1}{2} \left(i_{HPS}^2 R_{HPS} + i_{MBFC}^2 R_{MBFC} + i_{LPS}^2 R_{LPS} + i_{HPFC}^2 R_{HPFC} \right) \right] \quad (7)$$

$$h_1(x_1, u_1) = 0 : \begin{cases} -R_{HPS} i_{HPS} - v_{MB} + D_{HPS}(v_{HPS}) = 0 \\ -R_{LPS} i_{LPS} - v_{HP} + D_{LPS}(v_{LPS}) = 0 \\ -R_{MBFC} i_{MBFC} + (1 - D_{MBFC}) v_{MB} - D_{MBFC} v_{FC} = 0 \\ -R_{HPFC} i_{HPFC} + (1 - D_{HPFC}) v_{HP} - D_{HPFC} v_{FC} = 0 \end{cases} \quad (8)$$

$$h_2(x_2, u_2) = 0 : \begin{cases} (1 - D_{MBFC}) i_{MBFC} - \frac{v_{MB}}{R_{MB}} - i_m = 0 \\ (1 - D_{HPFC}) i_{HPFC} - \frac{v_{HP}}{R_{HP}} - i_{Pl} = 0 \\ D_{MBFC} i_{MBFC} + D_{HPFC} i_{HPFC} - \frac{v_{FC}}{R_{FC}} = 0 \end{cases} \quad (9)$$

where the i_{HPS}, i_{LPS}, i_{MBFC}, i_{HPFC} is the current for the MB and HP bus, FC bus, i_m is the electrical current draw from the cooling system. V_{MB}, V_{FC} and V_{HP} are the voltage for the MB and HP bus, FC bus, respectively.

The optimization problem of the electrical and thermal management problem for aircraft can be solved based on the model of Equations (8) and (9). A block diagram of the DNN deep Q-learning and optimization strategy are shown in Figure 17. Using this energy management strategy and controlling methods, the electric and thermal system model can be neglected. The critic, actor and environment can be built with the deep Q-learning methods, and the rewards function can be updated by the interactive learning scheme. This method enhances the robustness of the system in terms of uncertainty, which comes from the change of power load or power source and power conversion.

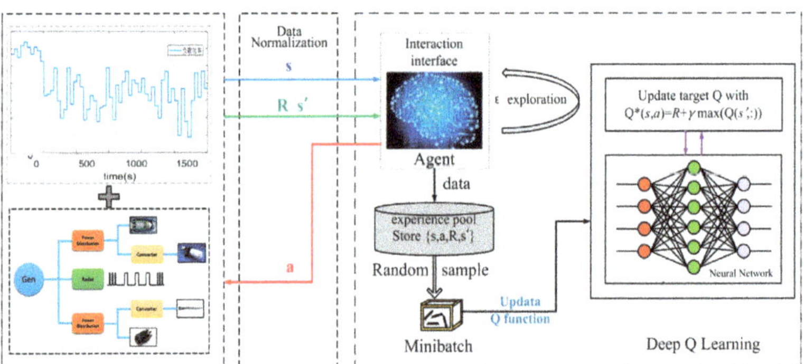

Figure 17. The control loop for the exergy based optimal energy management system for large power pulse load of aircraft.

Based on the above example, The electric–thermal coordinated management or control research topics are carried out from the industrial and academic fields. Honeywell has developed aircraft power and a thermal management system (PTMS) for application in the future. This system combines the functions of an auxiliary power unit (APU), emergency power unit (EPU), environmental control system (ECS), and thermal management system (TMS) in one integrated system [6]. The power control and energy management between the electric and thermal subsystems is still an open topic. So the coupling relationship between the various energy sources must be considered with some optimal control or management methods in next-generation aircraft power systems.

7. The Developing Trends for EMS of EPA in the Future

Under the pressure of green aviation transportation and energy consumption, and environmental protection, the airport microgrid and an energy system with renewable energy such as PV, wind, or a fuel cell can be possible in the future. As for the aircraft, an electrical propulsion aircraft (EPA) was developed because of flexibility and efficiency of energy generation and distribution. For each EPA, the electrical power system onboard was regarded as a "flying microgrid", in which the architecture of the power system is shown in Figure 18. This is the architecture of the power system for a hybrid electric propulsion aircraft, and the turbine-driven generator, hybrid energy storage, fuel cell emergency system, and HVDC power system can be combined together. The power allocation and energy optimization for the airport and EPA are necessary, similar to the terrestrial microgrid and smart grid [128]. The EPA powertrains include generators, power distribution subsystems, energy storage, and several distributed propulsion motors. Except for the EMS of the EPA on board, similar to the seaport and electrical propulsion ships [129,130], the electrification of airports and aircraft are both irreversible trends, which will greatly change the operating patterns of air transportation systems, i.e.,

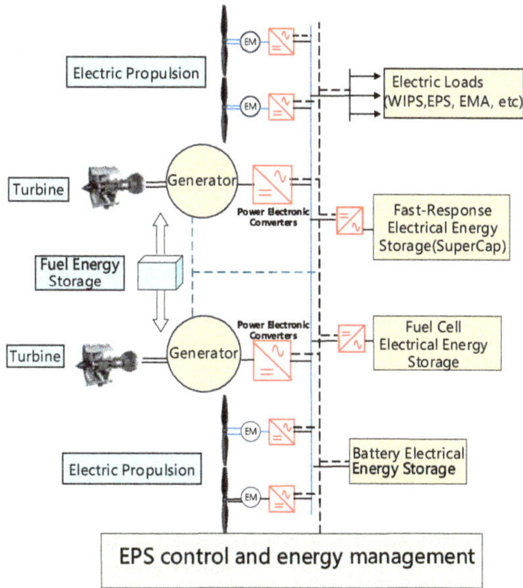

Figure 18. The power system architecture of hybrid electric propulsion aircraft.

The interaction between airport and EPA aircraft is no longer limited to passengers and logistics, but expands to the electric-side and energy-side, which makes future air transportation management a complex transportation power multi-microgrid coordination problem.

This multiple microgrid coordination problem generally has multiple objectives, i.e., the airport and EPA aircraft always have different operators and administrators, and involve great numbers of new subsystems and operation scenarios or modes that have not been ever considered in terrestrial power systems, such as the airport flight schedule, lounge bridge allocation, and aircraft route planning. The architecture of the airport microgrid is shown in Figure 19. The airport is connected with the utility grid, and different renewable energies such as PV, and fuel cell are integrated. When the aircraft arrives at the airport, the airport will provide the ground power supply and allocate the lounge bridge service to the aircraft. The airport power control central will give both power and logistic control signals to each subsystem and aircraft in the airport. In some situations, when the vacant location on the lounge bridge is not enough, the aircraft must be docked in the far parking yard. The passengers must be moved from the plane to the terminal by shuttle bus. Logistics tracks and shuttle buses can also be driven by electricity. The energy power supply must be provided by the airport energy system. The power demands of the airport will frequently change since the electric aircraft are continuously taking off and landing according to airport traffic flow. The system's power control loop can be seen in Figure 20. The airport microgrid EMS and the electrical aircraft EMS can be controlled and managed in three-level-control architecture to assure the energy balance between the airport microgrid and electric aircraft, and power response characteristics for the onboard load.

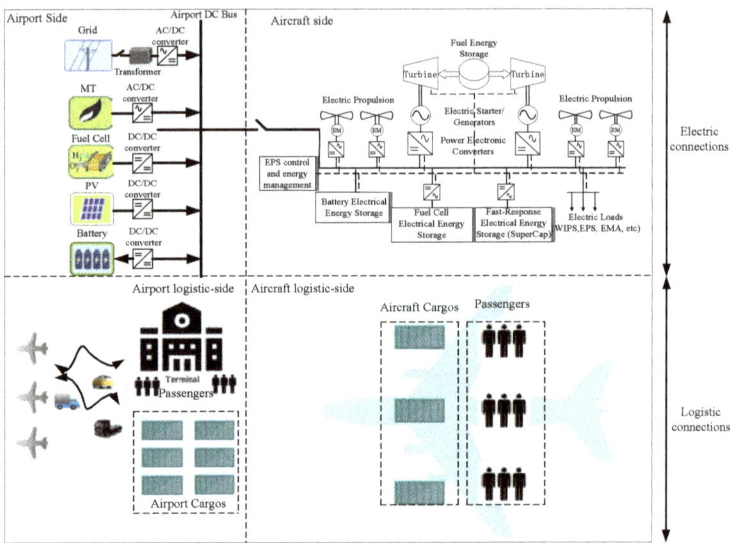

Figure 19. The energy and power interaction between the airport microgrid and electric aircraft.

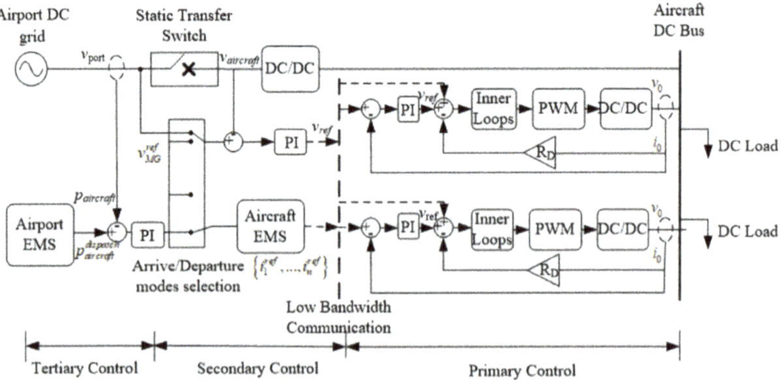

Figure 20. The power control loop architecture between airport and electrical propulsion aircraft.

In addition to the above analysis, deeper market penetration of greener and more silent electric ground and air vehicles will result in notable enhancements with regard to noise and pollution. The VTOL (vertical takeoff and landing) air vehicle will realize rapid, reliable transportation between suburbs and cities and, ultimately, within cities. The development of infrastructure to support the UAM (urban air mobility) will likely have significant cost advantages over charging stations, which can get electric energy from the microgrid in the airport. VTOL aircraft will make use of electric propulsion so they have zero operational emissions and will likely be quiet enough to operate in cities without disturbing the neighbors. The energy management and path optimization planning for electrical aircraft must be coordinated in some situations.

The verification of energy management of the airport and electrical aircraft can be carried out on HIL platforms. This situation is shown in Figure 21. Because of the size and complexity of the overall system, the algorithm of energy management must be tested in different operational scenarios, in which the digital simulation's efficiency is very low [121]. Therefore, the real-time platform can be applied to realize the system's simulation. In Figure 21, the airport microgrid virtual model can be built and dispatched in a typhoon-604 HIL terminal, the electrical aircraft also can be integrated into typhoon-HIL, and the EMS of airport microgrid can be formulated and solved in an RT-LAB 5700 platform. Based on this system configuration, the algorithm of EMS such as heuristic methods and optimization methods can be de applied to the overall energy system to improve the simulation speed and efficiency. The digital and analog signals can be transmitted between the typhoon HIL and the RT-LAB platform.

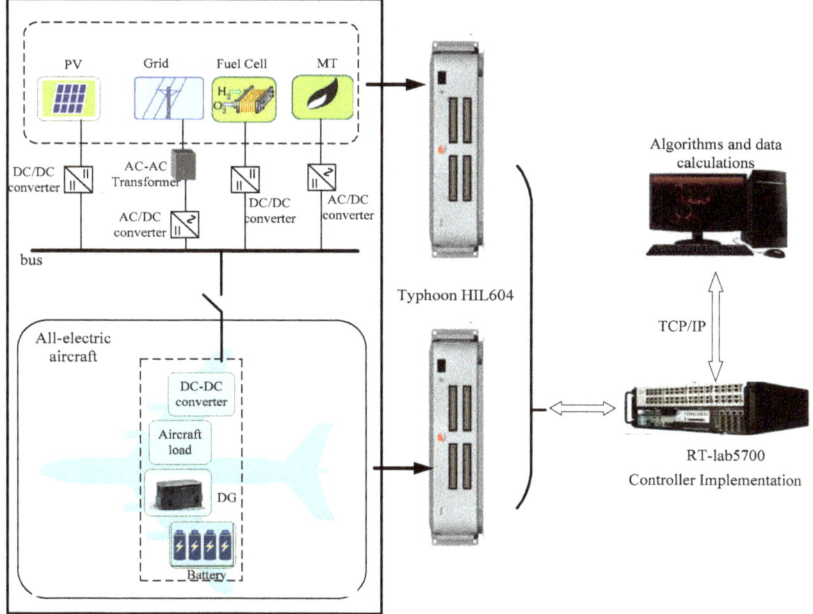

Figure 21. The hard-in-loop test bench for the airport microgrid management system.

The typical operational mode of future fully electric airports and EPA must be summarized. The hierarchical control or energy management framework is needed to be proposed for future airports. The airport is actually a hybrid AC/DC microgrid power system. The power flow calculation and optimization [131,132], analysis and distributed control methods, adaptive energy management strategy, and multiple functional optimizations of the electric energy storage system are several promising research topics in this field.

8. Conclusions

In this review paper, the energy optimization and energy management problem for MEA/AEA and EPA aircraft power systems was discussed. The technology architecture for the aircraft power system was presented with the consideration of MEA/AEA and

EPA technology trends. The optimization and evaluation of aircraft power systems also are analyzed with the state of art technology in this field. The architecture and strategies of the energy system are reviewed with the electrification trend for the aircraft power system. The uncertainty and stochastic changes from the renewable energy applied to aircraft power systems must be tackled with some useful energy management methods, the power demand resulting from the electrical propulsion of aircraft is basically stochastic because of the air turbulence movement for the flight control system in the aircraft. Thermal and electrical energy must be controlled coordinately in aircraft to achieve optimal energy application and stability. The EMS architecture must be considered with the power structure of the aircraft power system. The load-side or demand-side management for the aircraft's EPS is also important to reduce the power generation mass. With the requirement of safety and energy efficiency of the electrical propulsion aircraft, the EMS must be integrated with the PHM and TMS of the power system on board. The PHM refers to the sense of the state of power system, predicting the fault or lifetime of the device or subsystem in an aircraft power system. This trend can form new conceptions such as "EMS+PHM+TMS" for an aircraft power system. This conception also brought new and significant technology challenges and architectural complexity for aircraft. The EMS strategies have also evolved from the traditional rule-based methods or analytic optimization into dynamic time horizon rolling optimization (MPC), artificial intelligence (AI) such as deep reinforcement learning (DRL), and stochastic optimization methods. The AI-based EMS can overcome the shortcomings of traditional EMS such as the accuracy of the model, uncertainty of parameters, and robustness against disturbance. It can be forecasted that AI-based EMS must be integrated with multi-physical domains such as an electric, thermal, hydraulic, and mechanic systems to solve the complexity of modeling these large-size complex power systems.

With the development of aircraft electrical propulsion technology in the future, the large-scale, multi-disciplinary, multi-temporal and spatial-scale energy management strategies for aircraft power systems is becoming the hottest research topic in this field. This trend will drive the advancement of energy management and optimization for More and All-Electric Aircraft technology. Furthermore, the energy management interaction relationship between the airport microgrid and electric propulsion aircraft must be considered to improve renewable energy penetration and enhance the grid storage capacity by selling the electricity to the market through the main grid. The EMS and coordinated control strategies of different microgrids are the research topics regarding the future technology roadmap.

Author Contributions: T.L. and X.Z. (Xiaobin Zhang) proposed the research for the Architecture Optimization and Energy Management Technology of Aircraft power systems and present the analysis of state of art for power system; Z.M. and L.S. collect the research papers about the EMS strategies for power system of aircraft and give the evaluation for different architectures of aircraft power system; Z.M., T.L. and X.Z. (Xingyu Zhang) presented the HIL experiments for aircraft hybrid power system of EPA aircraft for airport microgrid; Z.M., Q.G. and T.L. analyzed the merits of different EMS strategies for aircraft power system and electric thermal coupling of aircraft power system; T.L., Z.M. and L.S. wrote the paper. All authors have read and agreed to the published version of the manuscript.

Funding: This research work is supported by the National Natural Science Foundation of China Grant No: 51877178.

Conflicts of Interest: There is no conflict of interest.

Abbreviations

Th ADP	Adaptive Dynamic Programming
ADMM	Alternating Direction Method of Multipliers
AEA	All-Electric Aircraft
AEPS	Aircraft Electrical Power System
AI	Artificial Intelligence

APU	Auxiliary Power Unit
ATRU	Auto Transformer Rectifier Unit
BPCU	Bus Power Control Unit
CPL	Constant Power Load
CFAC	Constant Frequency Alternating Current
DAB	Dual Active Bridge
DEW	Direct Energy Weapon
DDPG	Deep Deterministic Policy Gradient
DP	Dynamic Programming
DRL	Deep Reinforcement Learning
ECS	Environmental Control System
EPA	Electrical Propulsion Aircraft
EPS	Electrical Power System
EPU	Emergency Power Unit
EMA	Electrical Mechanical Actuator
EMS	Energy Management System
EMT	Electricity, machinery and heat
FC	Fuel Cell
GA	Genetic Algorithm
HEP	Hybrid Electric Propulsion Aircraft
HEV	Hybrid Electric Vehicle
HPS	High Pressure Spool
HSPMSM	High Speed Permanent Magnet Synchronous Machine
HSSPFC	Hamiltonian Surface Shaping Power Flow Control
HVAC	High Voltage Alternating Current
HVDC	High Voltage Direct Current
LPS	Low Pressure Spool
MB	Mission Bus
MCMC	Markov Chain–Monte Carlo
MDP	Markov Decision Process
MILP	Mixed Integer Linear Programming
MIQP	Mixed Integer Quadratic Programming
MPC	Model Predictive Control
PE	Power Electronics
PHM	Prognostics and Health Management
PMP	Pontryagin's Minimum Principle
PMSG	Permanent Magnetic Synchronous Generator
PSD	Power Spectrum Density
PSO	Particle Swarm Optimization
PTMS	Power and Thermal Management System
PV	Photo Voltaic
PWM	Pulse Width Modulation
QP	Quadratic Programming
RL	Reinforcement Learning
RSS	Root Square Summation
SCDS	Stability Constraining Dichotomy Solution
SDP	Stochastic Dynamic Programming
SHEV	Series Hybrid Electric Vehicle
SRMG	Switched Reluctance Magnetic Generator
TMS	Thermal Management System
TD	Time Difference
NN	Neural Network
UAM	Urban Air Mobility
UAV	Unmanned Air Vehicle
VF	Variable Frequency
VSI	Voltage Source Inverter

VTOL	Vertical Takeoff and Landing
WIPS	Wing Ice Protector System
ZIP	Constant Impedance (Z), Constant Current (I), and Constant Power Loads (P)

Glossary

f	the objective function
g	the inequality constraints
h	the equality constraints.
x	a solution of this problem, which is a vector of n decision variable(s) or design parameters
$x_i^{(L)}$	a lower bound
$x_i^{(u)}$	upper bound
G_i	power generator
$\alpha_{G_i}, \beta_{G_i}, \gamma_{G_i}$	the coefficients for the efficiency of power generator G_i from the polynomialfitting
E_{G_i}	the efficiency function of power generators
$G_i \in g$	the g is the set of power generators in EPS
$C_i \in C$	the C is the set of power converters in EPS.
E_{ci}	the efficiency function of power converters for aircraft power system
$R_{MB}, R_{FC}, R_{HP}, R_{HPS}$ and R_{LPS}	resistive storage elements
$C_{MB}, C_{FC},$ and C_{HP}	capacitive storage components
V_{HPS} and V_{LPS}	two HVDC generators which are considered located on the high pressure spool (HPS) and the low pressure spool (HPS) of a gas turbine respectively
$D_{HPS}, D_{LPS}, D_{MBFC}$ and D_{HPFC}	duty ratio control inputs
L_{HPS} and L_{LPS}	inductive energy storage elements
$f(x, u)$	the minimizing object function for aircraft electric-thermal system
h_1 and h_2	the equality constraints
$i_{HPS}, i_{LPS}, i_{MBFC}, i_{HPFC}$	the current for the MB and HP bus, FC bus
i_m	the electrical current draw from the cooling system.
V_{MB}, V_{FC} and V_{HP}	the voltage for the MB and HP bus, FC bus.
ρ	the air density
S	area of airfoil
γ	climb angle
ϕ	roll (bank) angle
v	velocity
$\vec{v}$	velocity vector
P_{ss}	the steady state thrust power
P_{dyn}	dynamic thrust power
$\vec{a}$	aircraft acceleration vector
C_D	zero-lift drag coefficients
$P_{ik,j}$	the transition probability
m	aircraft mass
g	acceleration of gravity
x_1, x_2	state variable of aircraft power system
$x_1 = [R_{HPS}, R_{LPS}, R_{MBFC}, R_{HPFC}, i_{HPS}, i_{LPS}, i_{MBFC}, i_{HPFC}, V_{MB}, V_{HP}, V_{HPS}, V_{LPS}, V_{FC}]^T$	
$x_2 = \left[R_{HP}, R_{FC}, R_{MB}, l_{MBFC}, i_{MBFC}, i_{HPFC}, i_m, i_{pl}, V_{MB}, V_{HP}, V_{FC}\right]^T$	
u_1, u_2	input control variable
$u_1 = [D_{HPS}, D_{LPS}, D_{MBFC}, D_{HPFC}]^T$	
$u_2 = [D_{MBFC}, D_{HPFC}]^T$	

References

1. Duffy, M.; Sevier, A.; Hupp, R.; Perdomo, E.; Wakayama, S. Propulsion Scaling Methods in the Era of Electric Flight. In Proceedings of the 2018 AIAA/IEEE Electric Aircraft Technologies Symposium, AIAA Propulsion and Energy Forum, Cincinnati, OH, USA, 9–11 July 2018; pp. 1–23.
2. Said, W. Powering Commercial Aircraft: The Next Logical Step in Vehicle Electrification. *IEEE Electrif. Mag.* **2017**, *5*, 4–8. [CrossRef]
3. Emadi, A.; Ehsani, M.; Miller, J.M. *Vehicular Electric Power Systems Land, Sea, Air and Space Vehicles*; CRC Press: Boca Raton, FL, USA, 2003.
4. Ansell, P.J.; Haran, K.S. Electrified Airplanes: A path to Zero-emission air travel. *IEEE Electrif. Mag.* **2020**, *8*, 18–26. [CrossRef]
5. O'Connell, T.; Russell, G.; McCarthy, K.; Lucus, E.; Zumberge, J.; Wolff, M. Energy Management of an Aircraft Electrical System. In Proceedings of the 46th AIAA/ASME/SAE/ASEE Joint Propulsion Conference & Exhibit, Nashville, TN, USA, 25–28 July 2010.
6. Ganev, E.; Koerner, M. Power and Thermal Management for Future Aircraft. In Proceedings of the SAE 2013 Aerotech Congress and Exhibition, Montreal, QC, Canada, 24–26 September 2013.
7. Ganev, E.D. Electric Drives for Electric Green Taxiing Systems Examining and Evaluating the Electric Drive System. *IEEE Electrif. Mag.* **2017**, *5*, 10–24. [CrossRef]
8. Buticchi, G.; Bozhko, S.; Liserre, M.; Wheeler, P.; Al-Haddad, K. On-Board Microgrids for the More Electric Aircraft-Technology Review. *IEEE Trans. Ind. Electron.* **2018**, *66*, 5588–5599. [CrossRef]
9. Benzaquen, J.; Fateh, F.; Shadmand, M.B.; Mirafzal, B. Performance Comparison of Active Rectifier Control Schemes in More Electric Aircraft Applications. *IEEE Trans. Transp. Electrif.* **2019**, *5*, 1470–1479. [CrossRef]
10. Fernandes, M.D.; Andrade, S.D.P.; Bistritzki, V.N.; Fonseca, R.M.; Zacarias, L.G.; Gonçalves, H.N.C. SOFC-APU systems for aircraft: A review. *Int. J. Hydrogen Energy* **2018**, *43*, 16311–16333. [CrossRef]
11. Turpin, C.; Morin, B.; Bru, E.; Rallières, O.; Roboam, X.; Sareni, B.; Arregui, M.G.; Roux, N. Power for Aircraft Emergencies: A hybrid proton-exchange membrane H_2/O_2 fuel cell and ultracapacitor system. *IEEE Electrif. Mag.* **2017**, *5*, 72–85. [CrossRef]
12. Dantsker, O.D.; Caccamo, M.; Imtiaz, S. Electric Propulsion System Optimization for Long-Endurance and Solar-Powered Unmanned Aircraft. In Proceedings of the AIAA Propulsion and Energy 2019 Forum, Indianapolis, IN, USA, 19–22 August 2019.
13. Nøland, J.K. Hydrogen Electric Airplanes: A disruptive technological path to clean up the aviation sector. *IEEE Electrif. Mag.* **2021**, *9*, 92–102. [CrossRef]
14. Riboldi, C.E.D. An optimal approach to the preliminary design of small hybrid-electric aircraft. *Aerosp. Sci. Technol.* **2018**, *81*, 14–31. [CrossRef]
15. Goldberg, C.; Nalianda, D.; Sethi, V.; Pilidis, P.; Singh, R.; Kyprianidis, K. Assessment of an energy-efficient aircraft concept from a techno-economic perspective. *Appl. Energy* **2018**, *221*, 229–238. [CrossRef]
16. Wu, Y.; Wu, Y.; Guerrero, J.M.; Vasquez, J.C.; Li, J. Ac Microgrid Small-signal Modelling-Hierarchical control structure challenges and solutions. *IEEE Electrif. Mag.* **2019**, *7*, 81–88. [CrossRef]
17. Sahoo, S.K.; Sinha, A.K.; Kishore, N.K. Control Techniques in AC, DC, and Hybrid AC-DC Microgrid: A Review. *IEEE J. Emerg. Sel. Top. Power Electron.* **2018**, *6*, 738–759. [CrossRef]
18. Sebastian, R.K.; Perinpinayagam, S.; Choudhary, R. Health Management Design Considerations for an All Electric Aircraft. In Proceedings of the 5th International Conference on Through-Life Engineering Services (TESConf 2016), Cranfield, UK, 1–2 November 2016; pp. 102–110.
19. Flynn, M.; Sztykiel, M.; Jones, C.E.; Husband, M. Protection and Fault Management Strategy Maps for Future Electrical Propulsion Aircraft. *IEEE Trans. Transp. Electrif.* **2019**, *5*, 1458–1469. [CrossRef]
20. Xiong, Q.; Feng, X.; Gattozzi, A.L. Series Arc Fault Detection and Localization in DC Distribution System. *IEEE Trans. Instrum. Meas.* **2020**, *69*, 122–134. [CrossRef]
21. Wang, Y.S.; Lei, H.; Hackett, R.; Beeby, M. Safety Assessment Process Optimization for Integrated Modular Avionics. *IEEE AE Syst. Mag.* **2019**, *34*, 58–67. [CrossRef]
22. Xing, L.; Wu, X.; Lents, C.E. Arc Detection for High Voltage Aircraft Distribution Systems. In Proceedings of the AIAA Propulsion and Energy Forum, Indianapolis, IN, USA, 19–22 August 2019.
23. Kulkarni, C.; Corbetta, M. Health Management and Prognostics for Electric Aircraft Powertrain. In Proceedings of the AIAA Propulsion and Energy Forum, Indianapolis, IN, USA, 19–22 August 2019.
24. Alexander, R.; Meyer, D.; Wang, J. A Comparison of Electric Vehicle Power Systems to Predict Architectures, Voltage Levels, Power Requirements, and Load Characteristics of the Future All-Electric Aircraft. In Proceedings of the 2018 IEEE Transportation Electrification Conference and Expo (ITEC), Long Beach, CA, USA, 13–15 June 2018.
25. Ismagilov, F.R.; Kiselev, M.A. Design Algorithm of an Aircraft Power Generation System. *IEEE Trans. Aerosp. Electron. Syst.* **2019**, *55*, 2899–2910. [CrossRef]
26. Vratny, P.C.; Hornung, M. Sizing Considerations of an Electric Ducted Fan for Hybrid Energy Aircraft. *Transp. Res. Procedia* **2018**, *29*, 410–426. [CrossRef]
27. Recalde, A.A.; Bozhko, S.; Atkin, J. Design of More Electric Aircraft DC Power Distribution Architectures considering Reliability Performance. In Proceedings of the AIAA Propulsion and Energy Forum, Indianapolis, IN, USA, 19–22 August 2019.

28. Dunker, C.; Bornholdt, R.; Thielecke, F.; Behr, R. Architecture and Parameter Optimization for Aircraft Electro-Hydraulic Power Generation and Distribution Systems. In Proceedings of the SAE 2015 AeroTech Congress & Exhibition, Seattle, WA, USA, 22–24 September 2015. SAE Technical Paper 2015-01-2414. [CrossRef]
29. Schley, W. Electric versus Hydraulic Flight Controls: Assessing Power Consumption and Waste Heat Using Stochastic System Methods. *SAE Int. J. Aerosp.* **2017**, *10*, 32. [CrossRef]
30. Annighöfer, B.; Kleemann, E. Large-Scale Model-Based Avionics Architecture Optimization Methods and Case Study. *IEEE Trans. Aerosp. Electron. Syst.* **2019**, *55*, 3424–3441. [CrossRef]
31. Saenger, P.; Devillers, N.; Deschinkel, K.; Péra, M.C.; Couturier, R.; Gustin, F. Optimization of Electrical Energy Storage System Sizing for an Accurate Energy Management in an Aircraft. *IEEE Trans. Veh. Technol.* **2017**, *66*, 5572–5583. [CrossRef]
32. Chen, J.; Wang, C.; Chen, J. Investigation on the Selection of Electric Power System Architecture for Future More Electric Aircraft. *IEEE Trans. Transp. Electrif.* **2018**, *4*, 563–576. [CrossRef]
33. Lawhorn, D.; Rallabandi, V.; Ionel, D.M. Scalable Graph Theory Approach for Electric Aircraft Power System Optimization. In Proceedings of the AIAA Propulsion and Energy 2019 Forum, Indianapolis, IN, USA, 19–22 August 2019.
34. Kuhn, M.R.; Otter, M.; Raulin, L. A Multi Level Approach for Aircraft Electrical Systems Design. In Proceedings of the 6th International Modelica Conference, Bielefeld, Germany, 3–4 March 2008.
35. Ounis, H.; Sareni, B.; Roboam, X.; de Andrade, A. Multi-level integrated optimal design for power systems of more electric aircraft. *Math. Comput. Simul.* **2016**, *130*, 223–235. [CrossRef]
36. Shang, Y.; Li, X.; Wu, S.; Jiao, Z. A Novel Electro Hydrostatic Actuator System with Energy Recovery Module for More Electric Aircraft. *IEEE Trans. Ind. Electron.* **2020**, *67*, 2991–2999. [CrossRef]
37. Dowdle, A.P.; Hall, D.K.; Lang, J.H. Electric Propulsion Architecture Assessment via Signomial Programming. In Proceedings of the 2018 AIAA/IEEE Electric Aircraft Technologies Symposium, Cincinnati, OH, USA, 9–11 July 2018; pp. 1–22.
38. Hendricks, E.S.; Chapman, J.W.; Aretskin-Hariton, E.D. Load Flow Analysis with Analytic Derivatives for Electric Aircraft Design Optimization. In Proceedings of the AIAA Propulsion and Energy 2019 Forum, Indianapolis, IN, USA, 19–22 August 2019.
39. Misra, A. Energy Storage for Electrified Aircraft: The Need for Better Batteries, Fuel Cells, and Super-capacitors. *IEEE Electrif. Mag.* **2018**, *6*, 54–61. [CrossRef]
40. Shibata, K.; Maedomari, T.; Rinoie, K.; Morioka, N.; Oyori, H. Aircraft Secondary Power System Integration into Conceptual Design and Its Application to More Electric System. In Proceedings of the SAE 2014 Aerospace Systems and Technology Conference, Cincinnati, OH, USA, 23–25 September 2014. SAE Technical Paper 2014-01-2199. [CrossRef]
41. Telford, R.; Jones, C.; Norman, P.; Burt, G. Analysis Tool for Initial High Level Assessment of Candidate MEA Architectures. In Proceedings of the SAE 2016 Aerospace Systems and Technology Conference, Hartford, CT, USA, 27–29 September 2016. SAE Technical Paper 2016-01-2015. [CrossRef]
42. Thalin, P. *Electric Aircraft Series—Fundamentals of Electric Aircraft*; SAE International: Warrendale, PA, USA, 2019.
43. Maldonado, M.A.; Shah, N.M.; Cleek, K.J.; Korba, G.J. Power Management and Distribution System for a More-Electric Aircraft (MADMEL)- Program Status. In Proceedings of the 34th Intersociety Energy Conversion Engineering Conference Vancouver, British, Columbia, 2–5 August 1999. SAE Technical Paper 1999-01-2547.
44. Schlabe, D.; Lienig, J. Energy Management of Aircraft Electrical Systems –State of the Art and Further Directions. In Proceedings of the 2012 Electrical Systems for Aircraft, Railway and Ship Propulsion, Bologna, Italy, 16–18 October 2012; pp. 1–6.
45. Yang, Y.; Gao, Z. Power optimization of the environmental control system for the civil more electric aircraft. *Energy* **2019**, *172*, 196–206. [CrossRef]
46. Bornholdt, R.; Thielecke, F. Optimization of the Power Allocation for Flight Control Systems. In Proceedings of the SAE 2014 Aerospace Systems and Technology Conference, Cincinnati, OH, USA, 23–25 September 2014; SAE Technical Paper 2014-01-2188. [CrossRef]
47. Shahsavari, B.; Massoumy, M.; Sangiovanni, A.; Horowitz, R. Stochastic Model Predictive Control Design for Load Management System of Aircraft Electrical Power Distribution. In Proceedings of the American Control Conference (AACC), Palmer House Hilton, Chicago, IL, USA, 1–3 July 2015; pp. 3649–3665.
48. Liu, Y.; Yang, S.; Wang, L. Dynamic Programming Algorithm for Management of Aircraft Power Supply System. In Proceedings of the International Conference on Electrical Systems for Aircraft, Railway, Ship Propulsion and Road Vehicles & International Transportation Electrification Conference, IEEE ESARS_ITEC, Nottingham, UK, 7–9 November 2018; pp. 1–6.
49. Rubino, L.; Iannuzzi, D.; Rubino, G.; Coppola, M.; Marino, P. Concept of Energy Management for Advanced Smart-Grid Power Distribution System in Aeronautical Application. In Proceedings of the 2016 International Conference on Electrical Systems for Aircraft, Railway, Ship Propulsion and Road Vehicles & International Transportation Electrification Conference (ESARS-ITEC), Toulouse, France, 2–4 November 2016; pp. 1–6.
50. Terorde, M.; Wattar, H.; Schulz, D. Phase Balancing for Aircraft Electrical Distribution Systems. *IEEE Trans. Aerosp. Electron. Syst.* **2015**, *51*, 1781–1792. [CrossRef]
51. Terorde, M.; Schulz, D. New Real-Time Heuristics for Electrical Load Rebalancing in Aircraft. *IEEE Trans. Aerosp. Electron. Syst.* **2016**, *52*, 1120–1131. [CrossRef]
52. Zhang, H.; Mollet, F.; Breban, S.; Saudemont, C.; Robyns, B. Power flow management strategies for a Local DC distribution system of More Electric Aircraft. In Proceedings of the 2010 IEEE Vehicle Power and Propulsion Conference, Lille, France, 1–3 September 2010; pp. 1–6.

53. Donateo, T.; De Giorgi, M.G.; Ficarella, A.; Argentieri, E.; Rizzo, E. A General Platform for the Modeling and Optimization of Conventional and More Electric Aircrafts. In Proceedings of the SAE 2014 Aerospace Systems and Technology Conference, Cincinnati, OH, USA, 23–25 September 2014. SAE Technical Paper 2014-01-2187. [CrossRef]

54. Lić, M.D.; Jaddivada, R. Multi-layered interactive energy space modeling for near-optimal electrification of terrestrial, shipboard and aircraft systems. *Annu. Rev. Control.* **2018**, *45*, 52–75.

55. Rouser, K.; Lucido, N.; Durkee, M.; Bellcock, A.; Zimbelman, T. Development of Turboelectric Propulsion and Power for Small Unmanned Aircraft. In Proceedings of the AIAA Propulsion and Energy Forum, Cincinnati, OH, USA, 9–11 July 2018.

56. Sliwinski, J.; Gardi, A.; Marino, M. Hybrid-electric propulsion integration in unmanned aircraft. *Energy* **2017**, *140*, 1407–1416. [CrossRef]

57. Frosina, E.; Caputo, C.; Marinaro, G. Modelling of a Hybrid-Electric Light Aircraft. In Proceedings of the 72nd Conference of the Italian Thermal Machines Engineering Association, ATI2017, Lecce, Italy, 6–8 September 2017.

58. Dong, Z.; Li, D.; Wang, Z.; Sun, M. A review on exergy analysis of aerospace power systems. *Acta Astronaut.* **2018**, *152*, 468–495. [CrossRef]

59. Cao, Y.; Williams, M.A.; Kearbey, B.J.; Smith, A.T.; Krein, P.T.; Alleyne, A.G. 20x-Real Time Modeling and Simulation of More Electric Aircraft Thermally Integrated Electrical Power Systems. In Proceedings of the 2016 International Conference on Electrical Systems for Aircraft, Railway, Ship Propulsion and Road Vehicles & International Transportation Electrification Conference(ESARS-ITEC), Toulouse, France, 2–4 November 2016; pp. 1–6.

60. Yang, T.; Bozhko, S.; Le-Peuvedic, J.M.; Asher, G.; Hill, C.I. Dynamic Phasor Modeling of Multi-Generator Variable Frequency Electrical Power Systems. *IEEE Trans. Power Syst.* **2016**, *31*, 563–571. [CrossRef]

61. Dincer, I.; Rosen, M.A.; Ahmadi, P. *Optimization of Energy Systems*; John Wiley & Sons Ltd.: Chichester, UK, 2017.

62. Williams, M.A. A Framework for the Control of Electro-Thermal Aircraft Power Systems. Ph.D. Thesis, University of Illinois, Urbana-Champaign, IL, USA, 2017.

63. Williams, M.; Sridharan, S.; Banerjee, S.; Mak, C.; Pauga, C. PowerFlow: A Toolbox for Modeling and Simulation of Aircraft Systems. In Proceedings of the SAE AeroTech Congress and Exhibition, AEROTECH 2015, Seattle, WA, USA, 22–24 September 2015; SAE Technical Paper 2015-01-2417. [CrossRef]

64. Zhang, Y.; Peng, G.O.H.; Banda, J.K.; Dasgupta, S.; Husband, M.; Su, R.; Wen, C. An Energy Efficient Power Management Solution for a Fault-Tolerant More Electric Engine/Aircraft. *IEEE Trans. Ind. Electron.* **2019**, *66*, 5663–5675. [CrossRef]

65. Bongermino, E.; Mastrorocco, F.; Tomaselli, M.; Monopoli, V.G.; Naso, D. Model and energy management system for a parallel hybrid electric unmanned aerial vehicle. In Proceedings of the 2017 IEEE 26th International Symposium on Industrial Electronics (ISIE), Edinburgh, UK, 19–21 June 2017; pp. 1868–1873.

66. Yasar, M.; Kwatny, H.; Bajpai, G. Aircraft Energy Management: Finite-time Optimal Control with Dynamic Constraints. In Proceedings of the AIAA Guidance, Navigation, and Control Conference, Kissimmee, FL, USA, 5–9 January 2015.

67. Spagnolo, C.; Madonna, V.; Bozhko, S. Optimized low voltage loads allocation for MEA electrical power systems. In Proceedings of the AIAA Propulsion and Energy Forum, Indianapolis, IN, USA, 19–22 August 2019.

68. Wang, R.; Liu, Y.; Lei, T.; Chang, J. A evaluation and optimization methods for electrical power system of large civil aircraft. In Proceedings of the CSAA/IET International Conference on Aircraft Utility Systems (AUS 2020), Online, 18–21 September 2020.

69. Blaabjerg, F.; Ionel, D.M. *Renewable Energy Devices and Systems with Simulations in Matlab and ANSYS*; CRC Press: Boca Raton, FL, USA; Taylor & Francis Group: Boca Raton, FL, USA, 2017.

70. Magne, P.; Nahid-Mobarakeh, B.; Pierfederici, S. Active stabilization of DC microgrids without remote sensors for more electric aircraft. *IEEE Trans. Ind. Appl.* **2013**, *49*, 2352–2360. [CrossRef]

71. Motapon, S.N.; Dessaint, L.A.; Al-Haddad, K. A comparative study of energy management schemes for a fuel-cell hybrid emergency power system of more-electric aircraft. *IEEE Trans. Ind. Electron.* **2014**, *61*, 1320–1334. [CrossRef]

72. Chen, J.; Song, Q. A Decentralized Energy Management Strategy for a Fuel Cell/Supercapacitor-Based Auxiliary Power Unit of a More Electric Aircraft. *IEEE Trans. Ind. Electron.* **2019**, *66*, 5736–5747. [CrossRef]

73. Department of Defense. *MIL-STD-704F. Aircraft Electric Power Characteristics*; Department of Defense Interface Standard: Lakehurst, NJ, USA, 2004; 38p.

74. Cano, T.C.; Castro, I.; Rodríguez, A.; Lamar, D.G.; Khalil, Y.F.; Albiol-Tendillo, L.; Kshirsagar, P. Future of Electrical Aircraft Energy Power Systems: An Architecture Review. *IEEE Trans. Transp. Electrif.* **2021**, *7*, 1915–1929. [CrossRef]

75. Alrashed, M.; Nikolaidis, T.; Pilidis, P.; Jafari, S. Utilisation of turboelectric distribution propulsion in commercial aviation: A review on NASA's TeDP concept. *Chin. J. Aeronaut.* **2021**, *34*, 48–65. [CrossRef]

76. Gui, H.; Chen, R.; Niu, J.; Zhang, Z.; Tolbert, L.M.; Wang, F.; Blalock, B.J.; Costinett, D.; Choi, B.B. Review of Power Electronics Components at Cryogenic Temperatures. *IEEE Trans. Power Electron.* **2020**, *35*, 5144–5156. [CrossRef]

77. Golovanov, D.; Gerada, D.; Sala, G. 4MW Class High Power Density Generator for Future Hybrid-Electric Aircraft. *IEEE Trans. Transp. Electrif.* **2021**, *7*, 2952–2964. [CrossRef]

78. Herrera, L.; Tsao, B. Analysis and Control of Energy Storage in Aircraft Power Systems with Pulsed Power Loads. *SAE Int. J. Aerosp.* **2016**, *9*, 8. [CrossRef]

79. Wen, B.; Boroyevich, D.; Burgos, R.; Mattavelli, P.; Shen, Z. Small-Signal Stability Analysis of Three-Phase AC Systems in the Presence of Constant Power Loads Based on Measured d-q Frame Impedances. *IEEE Trans. Power Electron.* **2015**, *30*, 5952–5963. [CrossRef]

80. Schettini, F.; Denti, E.; di Rito, G. Development of a simulation platform of all-electric aircraft on-board systems for energy management studies. *Aeronaut. J.* **2017**, *121*, 710–719. [CrossRef]

81. Zhou, Y.; Tao, L. The Testing Platform of Hybrid Electric Power System for a Fuel Cell Unmanned Aerial Vehicle. In Proceedings of the 2018 IEEE International Conference on Electrical Systems for Aircraft, Railway, Ship Propulsion and Road Vehicles & International Transportation Electrification Conference (ESARS-ITEC), Nottingham, UK, 7–9 November 2018.

82. Zhang, Y.; Yu, Y.; Su, R.; Chen, J. Power Scheduling in More Electric Aircraft based on An Optimal Adaptive Control Strategy. *IEEE Trans. Ind. Electron.* **2020**, *67*, 10911–10921. [CrossRef]

83. Wang, M.; Mesbahi, M. Energy Management for Electric Aircraft via Optimal Control: Cruise Phase. In Proceedings of the Aiaa Propulsion and Energy Forum, Virtual, 24–28 August 2020.

84. Boukoberine, M.N.; Zhou, Z.; Benbouzid, M. A critical review on unmanned aerial vehicles power supply and energy management: Solutions, strategies, and prospects. *Appl. Energy* **2019**, *255*, 113823. [CrossRef]

85. Guo, F.; Wen, C.; Song, Y. *Distributed Control and Optimization Technologies in Smart Grid Systems*; CRC Press: Boca Raton, FL, USA; Taylor & Francis Group: Boca Raton, FL, USA, 2017.

86. Dunham, W.; Hencey, B.; Girard, A.R.; Kolmanovsky, I. Distributed Model Predictive Control for More Electric Aircraft Subsystems Operating at Multiple Time Scales. *IEEE Trans. Control. Syst. Technol.* **2020**, *28*, 2177–2190. [CrossRef]

87. Tao, L.E.I.; Zhou, Y.A.N.G.; Zicun, L.I.N.; Zhang, X. The state of art on energy management strategy for hybrid-powered unmanned aerial vehicle. *Chin. J. Aeronaut.* **2019**, *32*, 1488–1503.

88. Xie, Y.; Savvarisal, A.; Tsourdos, A.; Zhang, D.; Gu, J. Review of hybrid electric powered aircraft, its conceptual design and energy management methodologies. *Chin. J. Aeronaut.* **2021**, *34*, 432–450. [CrossRef]

89. Motapon, S.N.; Dessaint, L.A.; Al-Haddad, K. A robust H2 consumption-minimization-based energy management strategy for a fuel cell hybrid emergency power system of more electric aircraft. *IEEE Trans. Ind. Electron.* **2014**, *61*, 6148–6156. [CrossRef]

90. Lee, B.; Kwon, S.; Park, P.; Kim, K. Active Power Management System for an Unmanned Aerial Vehicle Powered by Solar Cells, a Fuel Cell, and Batteries. *IEEE Trans. Aerosp. Electron. Syst.* **2014**, *50*, 3167–3177. [CrossRef]

91. Kratz, J.L.; Culley, D.E.; Thomas, G.L. A Control Strategy for Turbine Electrified Energy Management. In Proceedings of the AIAA Propulsion and Energy 2019 Forum, Indianapolis, IN, USA, 19–22 August 2019.

92. Zhang, H.; Mollet, F.; Saudemont, C.; Robyns, B. Experimental validation of energy storage system management strategies for a local DC distribution system of more electric aircraft. *IEEE Trans. Ind. Electron.* **2010**, *57*, 3905–3916. [CrossRef]

93. Xie, Y.; Savvaris, A.; Tsourdos, A. Fuzzy logic based equivalent consumption optimization of a hybrid electric propulsion system for unmanned aerial vehicles. *Aerosp. Sci. Technol.* **2019**, *85*, 13–23. [CrossRef]

94. Zhao, J.; Ramadan, H.S.; Becherif, M. Metaheuristic-based energy management strategies for fuel cell emergency power unit in electrical aircraft. *Int. J. Hydrogen Energy* **2019**, *44*, 2390–2406. [CrossRef]

95. Jiang, Z.; Raziei, S.A. Hierarchical Model Predictive Control for Real-Time Energy-Optimized Operation of Aerospace Systems. In Proceedings of the AIAA Propulsion and Energy 2019 Forum, Indianapolis, IN, USA, 19–22 August 2019.

96. Nademi, H.; Burgos, R.; Soghomonian, Z. Power Quality Characteristics of a Multilevel Current Source with Optimal Predictive Scheme From More-Electric-Aircraft Perspective. *IEEE Trans. Veh. Technol.* **2018**, *67*, 160–170. [CrossRef]

97. Ma, Z.; Zhang, X.; Huang, J.; Zhao, B. Stability-Constraining-Dichotomy-Solution-Based Model Predictive Control to Improve the Stability of Power Conversion System in the MEA. *IEEE Trans. Ind. Electron.* **2019**, *66*, 5696–5706. [CrossRef]

98. Koeln, J.P. Hierarchical Power Management in Vehicle Systems. Ph.D. Thesis, University of Illinois at Urbana-Champaign, Urbana-Champaign, IL, USA, 2016.

99. Koeln, J.P.; Pangborn, H.C.; Williams, M.A. Hierarchical Control of Aircraft Electro-Thermal Systems. *IEEE Trans. Control. Syst. Technol.* **2020**, *28*, 1218–1232. [CrossRef]

100. Trawick, D.; Milios, K.; Gladin, J.C. A Method for Determining Optimal Power Management Schedules for Hybrid Electric Airplanes. In Proceedings of the AIAA Propulsion and Energy Forum, Indianapolis, IN, USA, 19–22 August 2019.

101. Wu, D.; Todd, R.; Forsyth, A.J. Adaptive Rate-Limit Control for Energy Storage Systems. *IEEE Trans. Ind. Electron.* **2015**, *62*, 4231–4240. [CrossRef]

102. Kamal, M.B. Intelligent Control of Emergency APU for More Electric Aircraft. Master's Thesis, University of Akron, Akron, OH, USA, 2018.

103. Gao, Q.; Lei, T.; Deng, F.; Min, Z.; Yao, W.; Zhang, X. A Deep Reinforcement Learning based energy management strategy for fuel-cell electric UAV. In Proceedings of the 2022 International Conference on Power Energy Systems and Applications (ICoPESA), Virtual, 25–27 February 2022.

104. Misley, A.; D'Arpino, M.; Ramesh, P.; Canova, M. A Real-Time Energy Management Strategy for Hybrid Electric Aircraft Propulsion Systems. In Proceedings of the AIAA Propulsion and Energy Forum, Virtual, 9–11 August 2021.

105. Raveendran, V.; Andresen, M.; Liserre, M. Improving Onboard Converter Reliability for More Electric Aircraft with Lifetime-Based Control. *IEEE Trans. Ind. Electron.* **2019**, *66*, 5787–5796. [CrossRef]

106. Tseng, C.-H.; Chau, C.-H.; Elbassioni, K.; Khonji, M. Autonomous Recharging and Flight Mission Planning for Battery-operated Autonomous Drones. *arXiv* **2017**, arXiv:1703.10049.

107. Fang, S.; Xu, Y. Multiobjective Coordinated Scheduling of Energy and Flight for Hybrid Electric Unmanned Aircraft Microgrids. *IEEE Trans. Ind. Electron.* **2019**, *66*, 5686–5695. [CrossRef]

108. Wang, X.; Atkin, J.; Hill, C.; Bozhko, S. Power Allocation and Generator Sizing Optimization of More-Electric Aircraft Onboard Electrical Power during Different Flight Stages. In Proceedings of the AIAA Propulsion and Energy Forum, Indianapolis, IN, USA, 19–22 August 2019.

109. Karanayil, B.; Ciobotaru, M.; Agelidis, V.G. Power flow management of isolated multiport converter for more electric aircraft. *IEEE Trans Power Electron.* **2017**, *32*, 5850–5861. [CrossRef]

110. Gao, F.; Bozhko, S.; Asher, G.; Wheeler, P.; Patel, C. An Improved Voltage Compensation Approach in a Droop-Controlled DC Power System for the More Electric Aircraft. *IEEE Trans. Power Electron.* **2016**, *31*, 7369–7383. [CrossRef]

111. Li, Y.; He, H.; Peng, J.; Wang, H. Deep Reinforcement Learning-Based Energy Management for a Series Hybrid Electric Vehicle Enabled by History Cumulative Trip Information. *IEEE Trans. Veh. Technol.* **2019**, *68*, 5663–5675. [CrossRef]

112. Sumsurooah, S.; Odavic, M.; Bozhko, S.; Boroyevic, D. Toward Robust Stability of Aircraft Electrical Power Systems—Using a μ-based structural singular value to analyze and ensure network stability. *IEEE Electrif. Mag.* **2017**, *5*, 62–71. [CrossRef]

113. Areerak, K.N.; Bozhko, S.V.; Asher, G.M.; Lillo, L.D.; Thomas, D.W.P. Stability study for a hybrid ac-dc more-electric aircraft power system. *IEEE Trans. Aerosp. Electron. Syst.* **2012**, *48*, 329–347. [CrossRef]

114. Mallik, A.; Khaligh, A. Intermediate DC-Link Capacitor Reduction in a Two-Stage Cascaded AC/DC Converter for More Electric Aircrafts. *IEEE Trans. Veh. Technol.* **2018**, *67*, 935–947. [CrossRef]

115. Zadeh, M.K.; Gavagsaz-Ghoachani, R.; Nahid-Mobarakeh, B.; Pierfederici, S.; Molinas, M. Stability analysis of hybrid AC/DC power systems for More Electric Aircraft. In Proceedings of the 2016 IEEE Applied Power Electronics Conference and Exposition (APEC), Long Beach, CA, USA, 20–24 March 2016.

116. Wen, B.; Burgos, R.; Boroyevich, D.; Mattavelli, P.; Shen, Z. AC Stability Analysis and dq Frame Impedance Specifications in Power-Electronics-Based Distributed Power Systems. *IEEE J. Emerg. Sel. Top. Power Electron.* **2017**, *5*, 1455–1465. [CrossRef]

117. Gavagsaz-Ghoachani, R.; Saublet, L.M.; Martin, J.P.; Nahid-Mobarakeh, B.; Pierfederici, S. Stability Analysis and Active Stabilization of DC Power Systems for Electrified Transportation Systems, Taking into Account the Load Dynamics. *IEEE Trans. Transp. Electrif.* **2016**, *3*, 3–12. [CrossRef]

118. Rygg, A.; Molinas, M. Apparent Impedance Analysis: A Small-Signal Method for Stability Analysis of Power Electronic-Based Systems. *IEEE J. Emerg. Sel. Top. Power Electron.* **2017**, *5*, 1474–1486. [CrossRef]

119. Jones, C.E.; Norman, P.J.; Sztykiel, M.; Alzola, R.P.; Burt, G.M.; Galloway, S.J. Electrical and Thermal Effects of Fault Currents in Aircraft Electrical Power Systems with Composite Aerostructures. *IEEE Trans. Transp. Electrif.* **2018**, *4*, 660–670. [CrossRef]

120. Madonna, V.; Giangrande, P.; Gerada, C.; Galea, M. Thermal Overload and Insulation Aging of Short Duty Cycle, Aerospace Motors. *IEEE Trans. Ind. Electron.* **2020**, *67*, 2618–2629. [CrossRef]

121. Trinklein, E.H.; Cook, M.D.; Parker, G.G.; Weaver, W.W. Exergy optimal multi-physics aircraft microgrid control architecture. *Int. J. Electr. Power Energy Syst.* **2020**, *114*, 105403. [CrossRef]

122. Ilic, M.D.; Jaddivada, R. Exergy/energy dynamics-based integrative modeling and control for difficult hybrid aircraft missions. In Proceedings of the AIAA Propulsion and Energy Forum, Indianapolis, IN, USA, 19–22 August 2019.

123. Cao, Y.; Williams, M.A.; Krein, P.T. Mitigating Power Systems Variability in More Electric Aircraft Utilizing Power Electronics Implemented Dynamic Thermal Storage. In Proceedings of the Applied Power Electronics Conference and Exposition (APEC), Tampa, FL, USA, 26–30 March 2017; pp. 1412–1418.

124. Dillon, J.A. *Electro-Mechanical-Thermal Modeling and Stability of Pulsed Power Loads on a DC Network*; Michigan Technological University: Houghton, MI, USA, 2018.

125. Peng, Q.; Du, Q. Progress in Heat Pump Air Conditioning Systems for Electric Vehicles—A Review. *Energies* **2016**, *9*, 240. [CrossRef]

126. Un-Noor, F.; Padmanaban, S. A Comprehensive Study of Key Electric Vehicle (EV) Components, Technologies, Challenges, Impacts, and Future Direction of Development. *Energies* **2017**, *10*, 1217. [CrossRef]

127. Alhanouti, M.; Gießler, M.; Blank, T. New Electro-Thermal Battery Pack Model of an Electric Vehicle. *Energies* **2016**, *9*, 563. [CrossRef]

128. Karapetyan, A.; Khonji, M.; Chau, C.K.; Elbassioni, K.; Zeineldin, H.H. Efficient algorithm for scalable event-based demand response management in microgrids. *IEEE Trans. Smart Grid* **2018**, *9*, 2714–2725. [CrossRef]

129. Fang, S.; Wang, Y.; Gou, B.; Yu, Y. Toward Future Green Maritime Transportation: An overview of Seaport Microgrids and All-Electric Ships. *IEEE Trans. Veh. Technol.* **2020**, *69*, 207–219. [CrossRef]

130. Li, Z.; Xu, Y.; Fang, S.; Wang, Y.; Zheng, X. Multiobjective Coordinated Energy Dispatch and Voyage Scheduling for a Multienergy Ship Microgrid. *IEEE Trans. Ind. Appl.* **2020**, *56*, 989–999. [CrossRef]

131. Taheri, S.; Kekatos, V. Power Flow Solvers for Direct Current Networks. *IEEE Trans. Smart Grid* **2020**, *11*, 634–643. [CrossRef]

132. Molzahn, D.K.; Hiskens, I.A. A survey of relaxations and approximations of the power flow equations. *Found. Trends® Electr. Energy Syst.* **2019**, *4*, 1–221. [CrossRef]

Article

A Modified SOGI-PLL with Adjustable Refiltering for Improved Stability and Reduced Response Time

Gilberto A. Herrejón-Pintor [1,2], Enrique Melgoza-Vázquez [1,*] and Jose de Jesús Chávez [1,3]

1   Tecnológico Nacional de México/Instituto Tecnológico de Morelia, Morelia 58120, Mexico;
    d94120125@morelia.tecnm.mx or alejandro.hp@zamora.tecnm.mx (G.A.H.-P.);
    jose.cm@morelia.tecnm.mx or j.j.chavezmuro@tudelft.nl (J.d.J.C.)
2   Tecnológico Nacional de México/Instituto Tecnológico de Estudios Superiores de Zamora,
    Zamora de Hidalgo 59720, Mexico
3   Faculty of Electrical Engineering, Mathematics and Computer Science, Delft University of Technology,
    2628 CD Delft, The Netherlands
*   Correspondence: enrique.mv@morelia.tecnm.mx

Abstract: The controls of most power electronic inverters connected to an electrical power system (EPS) rely on the precise determination of the voltage magnitude, frequency, and phase angle at the point of common coupling. One of the most widely used approaches for measuring these quantities is the phase-locked loop (PLL); however, the precision of this measurement is affected during transients in the EPS and is a function of the type of event and the architecture of the PLL. PLLs based on the second-order generalized integrator (SOGI) are widely used in power converter synchronization, offering an adaptive or fixed-parameter prefilter with low-pass and band-pass characteristics. This article proposes a variant of the SOGI-PLL that offers improved stability and a faster response time. This is accomplished by decoupling the effect of the SOGI's gains and adding feedback. The modification is carried out in the state space model of the SOGI. Manipulating the attenuation moves the poles of the SOGI to improve the stability. The performance of the proposed PLL is verified and validated under the processor-in-the-loop (PIL) approach.

Keywords: synchronization; phase-locked loop; renewable energy resources; processor-in-the-loop

Citation: Herrejón-Pintor, G.A.; Melgoza-Vázquez, E.; Chávez, J.d.J. A Modified SOGI-PLL with Adjustable Refiltering for Improved Stability and Reduced Response Time. *Energies* **2022**, *15*, 4253. https://doi.org/10.3390/en15124253

Academic Editor: Juri Belikov

Received: 11 May 2022
Accepted: 6 June 2022
Published: 9 June 2022

Publisher's Note: MDPI stays neutral with regard to jurisdictional claims in published maps and institutional affiliations.

1. Introduction

In recent years, the integration of distributed generation (DG) units to the electrical power system (EPS) has increased. A fundamental aspect of a correct interconnection is the DG inverter synchronization algorithm. For a stable and reliable operation, this algorithm must ensure that the inverter voltage waveform synchronizes with the main voltages, even under disturbances within the EPS [1–5]. The most widely used synchronization approach is the phase-locked loop (PLL). However, the operation of the PLL is compromised when the network voltage and current contain harmonics and when the network is subject to faults and disturbances [2,6–9]. No matter how reliable the transmission and distribution system is, unbalanced voltages and harmonic distortions at the PLL input are unavoidable [2,9].

It must be taken into account that the grid frequency can show considerable fluctuations during transients and faults in power systems with high DG penetration [10]. This implies that the synchronization system must be insensitive to network frequency variations [11]. On the other hand, large voltage sags and other transient events can cause incorrect transient frequency measurements in the PLL [12–16]. PLLs that include non-adaptive strategies, as well as notch filters, present problems when the EPS frequency deviates from its nominal value. The problem can become serious in the presence of large deviations from the nominal frequency of the EPS, particularly under severe asymmetrical voltage sags or faults [5,12].

Phase-to-ground faults constitute up to 95% of the faults that occur in the EPS [17]. This type of fault causes voltage sags in the distribution and transmission systems. When this and other types of faults are cleared, transients appear that range from nominal frequency levels to levels the tens of kHz [18]. On the other hand, in general, a harmonic of order h present in the input signal becomes a harmonic of order $(h-1)$ (if it is a positive-sequence harmonic) or $(h+1)$ (if it is a negative-sequence harmonic) in the dq frame [19]. Therefore, the fundamental negative-sequence component causing the asymmetry in the grid voltage becomes the second harmonic ripple in the dq frame, and the displacement dc becomes the fundamental component. Consequently, the loop filter must be able to attenuate the second harmonic and the fundamental component. Eliminating the fundamental negative-sequence component without compromising the dynamic performance remains a demanding task in PLL design [20]. To achieve this, the bandwidth of the PLL must be drastically reduced [2,12,21]. However, with low bandwidth, the transient response becomes slow [15,22–24]. The main challenge associated with PLLs is to achieve a fast dynamic response without compromising the harmonic rejection capability while ensuring that disturbances in the electrical network do not drive the PLL to instability and the loss of synchronism [24–27]. The filter bandwidth should be a trade-off between filtering performance, time response, and stability [2,9,22].

One of the most frequently used synchronization algorithms is the second-order generalized integrator phase-locked loop (SOGI-PLL) due to its simplicity, low computational cost, and independence from network frequency, and because it avoids filtering delays [28–30]. In this paper, a modification of the SOGI-PLL architecture is proposed; the modification improves the stability of the quadrature signal generator (QSG) stage and increases the bandwidth of the PLL structure, achieving a fast response of the PLL as a whole, without sacrificing the harmonic rejection capability of the traditional SOGI-PLL. This is achieved by decoupling the effect of the single SOGI-QSG gain, allowing a second QSG gain to be introduced, and adding feedback from the band-pass filter output signal to the QSG input. The new QSG gain permits the manipulation of the extra attenuation while moving the QSG poles to improve stability. The decoupling of the SOGI gain is carried out by converting the transfer function of the SOGI-QSG [6,28] block diagram to a model in the state space in the controllable canonical form, since in the state space there is direct access to the sections of the QSG algorithm where gains appear. A functional structure of the ARF-SOGI-QSG is reported in the state space and also in block form. The proposed structure is a PLL with adjustable re-filtering based on a second-order generalized integrator, abbreviated to ARF-SOGI-PLL from now on.

The paper is organized as follows: Section 2.1 describes the PLL performance evaluation criteria, Section 2.2 summarizes the main characteristics of the SOGI-PLL and develops the structure of the modified SOGI-QSG, whose characteristics are further discussed in Sections 2.3 and 2.4. In Section 2.5, the proposed ARF-SOGI-PLL structure is presented; its performance is assessed for different disturbances in the electrical system in Section 3. The computational resources and execution times of the tested PLLs are also compared. The conclusions are presented in Section 4.

2. Development of the Improved PLL

In this section, an improved SOGI-PLL is proposed. The desirable improvements in performance are listed in the following section. The key characteristics of the traditional SOGI-PLL are reviewed in order to motivate the proposed modifications.

2.1. The PLL Performance Evaluation Criteria

The total harmonic distortion (THD) of the unit vectors and the synchronization time will be used as the criteria for assessing the performance of the PLL. A THD of less than 1% will be deemed as acceptable. The unit vectors are the evaluation of the sine and cosine functions of the angle calculated by the PLL. These vectors must be in synchrony with the main voltages but without harmonic contamination, since the reference signals for the control of the inverter are obtained from the angle calculated by the PLL.

Very short synchronization times generally imply a degradation in the quality of the unit vectors. On the other hand, high-frequency deviations lead to longer synchronization times or a failed re-synchronization. In both cases, the extraction of the frequency from the network will be incorrect during the PLL transient. Therefore, there is a need to improve the PLL response in two regards: increase the speed with which the correct value of the frequency is reported (without degrading the quality of the unit vectors), and reduce the error in the reported frequency during the PLL transient [27].

The disconnection times of the converter in case of voltage variations are established in the IEEE 1547, IEC61727, and VDE0126-1-1 standards [12,31,32]. Similarly, the rules applied in the case of frequency deviations and the allowable ranges are discussed in [12,31,32]. The IEEE 1547 standard states that for disturbances for which the system frequency remains between 58.8 Hz and 61.2 Hz, the converter must remain operational and must continue to provide active power. The low-frequency ride-through period is 299 s in the range $57.0 \leq f \leq 58.8$ Hz, while the high-frequency ride-through range is $61.2 < f \leq 61.8$ Hz. Within these ranges and periods, the converter must not trip and must continue to provide active power. For a frequency deviation greater than 3.5 Hz, the converter is disconnected after a period of 0.16 s. Therefore, a PLL must avoid erroneously reporting a deviation of this magnitude for more than 0.16 s.

2.2. Proposed ARF-SOGI-QSG Structure

The PLL proposed in this article is based on the modification of the SOGI-PLL, whose structure is shown in Figure 1. The SOGI-PLL is based on a QSG and generally consists of three stages: a phase detector, a controller (also called loop filter), and a frequency and phase-angle generator. The phase detector is constructed with a QSG and a Park transformation stage. The QSG is built around a SOGI, which helps in providing adaptive filter characteristics to the QSG. Therefore the complete structure of the SOGI-PLL contains, in addition to a loop filter, an adaptive pre-filtering stage, contributing to the improvement of the harmonic rejection and to the SOGI-PLL's tolerance to frequency changes. The main characteristics of the SOGI-PLL can be summarized as follows:

1. The value of the single gain k of the adaptive prefilter affects all the characteristics of the SOGI-QSG;
2. The magnitude of the line voltage is calculated by the Park transformation and reported as V_d, as shown in Figure 1;
3. The frequency calculated by the PLL is fed back to the SOGI block;
4. The angle is calculated by the integrator or VCO and fed back to the Park transformation block.

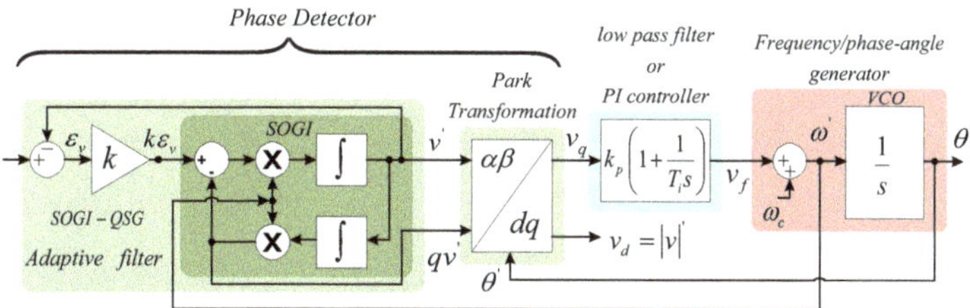

Figure 1. Structure of the traditional SOGI-PLL.

The transfer functions that characterize the SOGI-QSG and the prefiltering process are given by:

$$SOGI(s) = \frac{v'}{k\varepsilon_v}(s) = \frac{\omega' s}{s^2 + \omega'^2} \tag{1}$$

$$D(s) = \frac{v'}{v}(s) = \frac{k\omega's}{s^2 + k\omega's + \omega'^2} \tag{2}$$

$$Q(s) = \frac{qv'}{v}(s) = \frac{k\omega'^2}{s^2 + k\omega's + \omega'^2}. \tag{3}$$

The transfer Functions (2) and (3) are obtained from the transfer Function (1) of the SOGI. These transfer functions show that the bandwidth of the adaptive filter based on SOGI is not a function of the center frequency and that it only depends on the gain k [12]. Moreover, it is clear that the single gain k directly impacts three parts of the algorithm. This suggests a modification of the SOGI-QSG structure, consisting in distributing the influence of the gain over each prefilter stage. To do this, the model is represented in the state space. The controllable canonical form of Equation (2), which corresponds to the band-pass filter, is given by:

$$\underbrace{\begin{bmatrix} \dot{x}_1 \\ \dot{x}_2 \end{bmatrix}}_{\dot{x}} = \underbrace{\begin{bmatrix} -k\omega' & -\omega'^2 \\ 1 & 0 \end{bmatrix}}_{A} \underbrace{\begin{bmatrix} x_1 \\ x_2 \end{bmatrix}}_{x} + \underbrace{\begin{bmatrix} 1 \\ 0 \end{bmatrix}}_{b} v \tag{4}$$

$$y = \underbrace{\begin{bmatrix} k\omega' & 0 \end{bmatrix}}_{c} \begin{bmatrix} x_1 \\ x_2 \end{bmatrix}, \tag{5}$$

while the corresponding expressions for the low-pass filter (Equation (3)) are:

$$\begin{bmatrix} \dot{x}_1 \\ \dot{x}_2 \end{bmatrix} = \begin{bmatrix} -k\omega' & -\omega'^2 \\ 1 & 0 \end{bmatrix} \begin{bmatrix} x_1 \\ x_2 \end{bmatrix} + \begin{bmatrix} 1 \\ 0 \end{bmatrix} v \tag{6}$$

$$y = \begin{bmatrix} 0 & k\omega'^2 \end{bmatrix} \begin{bmatrix} x_1 \\ x_2 \end{bmatrix}. \tag{7}$$

The proposed SOGI-QSG form is shown in Figure 2. The new structure is based on Equations (4) to (7). Figure 2 shows, in the inferior part, the multiplication of matrix **A** by the vector of states $\dot{x}$, and in the upper part (which contains the output signals), the multiplication of only the state of interest with the part of the vector **c** that corresponds to both the band-pass filter and the output low-pass filter (y). The use of this structure makes it possible to modify the value of the gains independently.

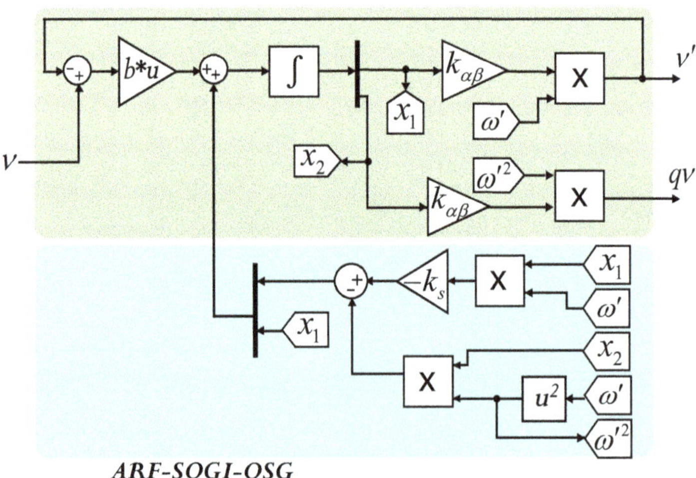

ARF-SOGI-QSG

Figure 2. State space architecture of the ARF-SOGI-QSG.

It is possible to add a negative feedback loop from the v' output of the SOGI band-pass filter to its input. This extra closed loop provides a higher attenuation in the outputs v' and qv', in contrast to the traditional adaptative filter.

The model in the state space allows access to the gains in the different sections in the QSG, and with this, the performance of the ARF-SOGI-QSG can be improved. With the proposed method arises the possibility to use three different gains. However, finding the values of the gains is not a trivial task. The next remarks must be taken into account:

1. Since the gain k_s in Figure 2 can be adjusted in value without unbalancing the magnitude of the responses v' and qv', the change in the value of this gain does not affect the orthogonality of the two signals;
2. Using different gain values at the outputs v' and qv' of the filtering could only be possible with very slight differences (of the order of 10^{-3}). Greater differences cause a different attenuation in the magnitude of signals v' and qv', which can produce imbalances and, therefore, oscillations at the PLL output (at twice the fundamental frequency of the input signal). Hence, it is recommended that these two gains from the ARF-SOGI-QSG output are equal, and from now on, the single gain is called $k_{\alpha\beta}$, as seen in Figure 2;
3. If $k_s = k_{\alpha\beta}$, the attenuation suffered by the 30 Hz and 120 Hz signals is approximately 3.9 dB; this is an advantage over the traditional SOGI, but the cost is an attenuation of approximately 6 dB in the fundamental frequency signal. However, adjustment of the gain k_s corrects this attenuation, as discussed in Section 3.5;
4. The smaller the value of k_s, the closer the ARF-SOGI-QSG is to the traditional SOGI-QSG;
5. The greater the k_s, the greater the attenuation suffered by the fundamental frequency of the voltage signal.

2.3. ARF-SOGI-QSG Transfer Functions

The form of the proposed architecture in the frequency domain is shown in Figure 3.

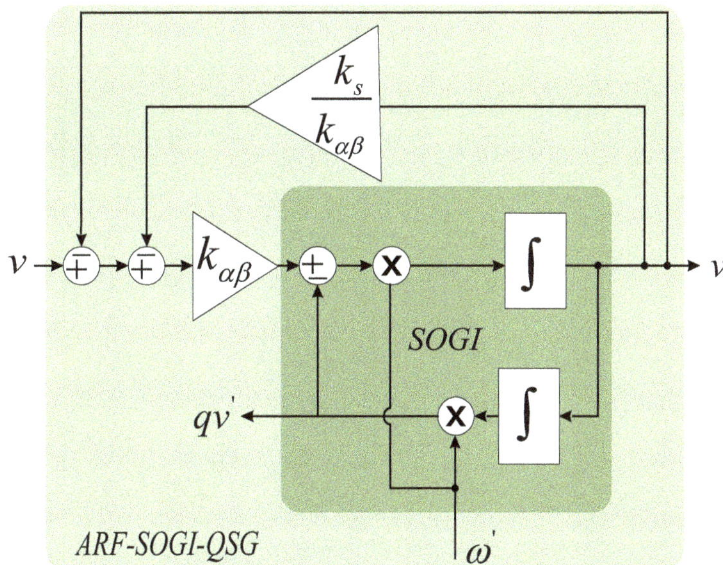

Figure 3. Block diagram representation of the ARF-SOGI-QSG.

The transfer functions of the ARF-SOGI-QSG can be obtained from the diagram in Figure 3, and is given by:

$$SOGI_m(s) = \frac{v'}{\varepsilon_v} = \frac{k_{\alpha\beta}\omega's}{s^2 + \omega'^2} \tag{8}$$

The transfer function of the signal v' band-pass characteristic is given by:

$$D_m(s) = \frac{v'}{v}(s) = \frac{k_{\alpha\beta}\omega's}{s^2 + \omega's(k_s + k_{\alpha\beta}) + \omega'^2} \tag{9}$$

The transfer function of the quadrature signal qv' low-pass characteristic is given by:

$$Q_m(s) = \frac{qv'}{v}(s) = \frac{k_{\alpha\beta}\omega'^2}{s^2 + \omega's(k_s + k_{\alpha\beta}) + \omega'^2} \tag{10}$$

It is clearly seen, from the transfer functions (9) and (10), that the smaller the k_s, the closer the transfer functions are to the original transfer functions (Equations (2) and (3)) from the SOGI-QSG.

2.4. Comparative Analysis of SOGI-QSG and ARF-SOGI-QSG

Figures 4 and 5 compare the Bode diagrams corresponding to the transfer functions (2) and (3) and the proposed ARF-SOGI-QSG. The diagrams confirm the remarks of the previous section: the types of filtering implicit in the SOGI are preserved in the proposed modification, but a greater attenuation is seen within a range of approximately $\left[\frac{1}{2}f, 2f\right]$. Such extra attenuation is more significant at the fundamental frequency. It is also observed that the attenuation in this range can be controlled by varying the gain k_s. Likewise, the orthogonality characteristic is preserved, and it is seen that the extra attenuation reported by the ARF-SOGI-QSG comes from the modification made in the denominator of the transfer functions.

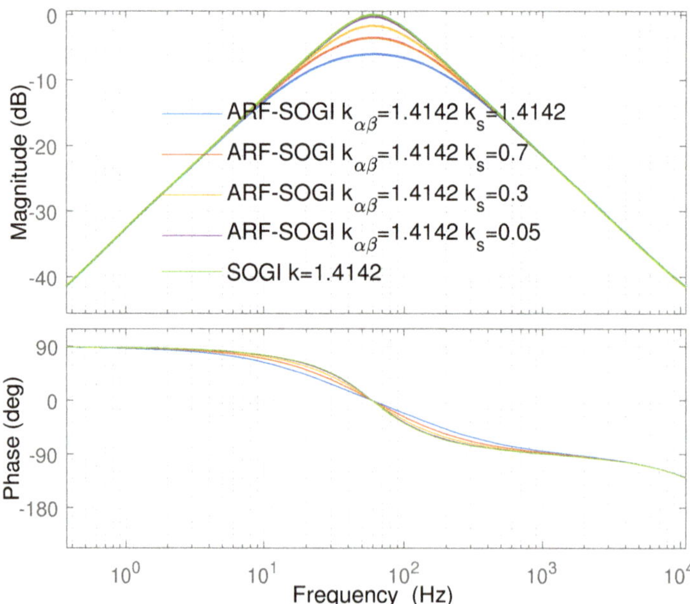

Figure 4. Bode diagrams of the SOGI-QSG versions for v'.

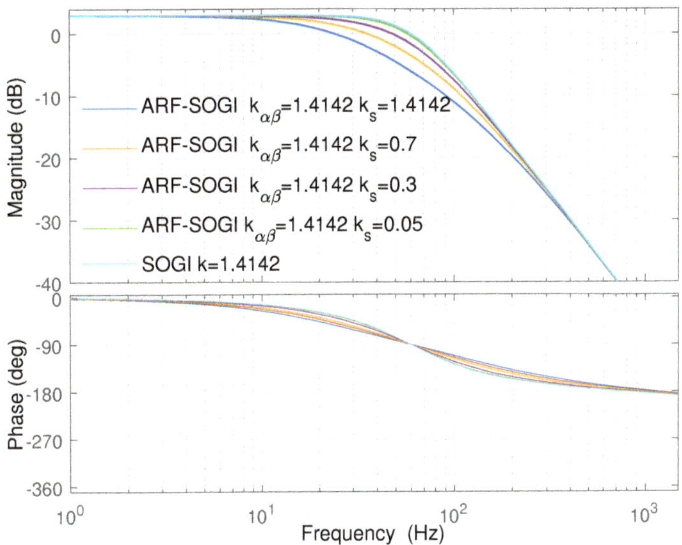

Figure 5. Bode diagrams of the SOGI-QSG versions for qv'.

Regarding the stability of the SOGI-QSG and the ARF-SOGI-QSG, Figures 6 and 7 show the effect of different values of k, k_s, and $k_{\alpha\beta}$. When the bandwidth is reduced, the systems approximate the stability limit. Figures 6 and 7 show one of the advantages of the ARF-SOGI-QSG, namely, the feasibility of tuning the filter gains (k_s, $k_{\alpha\beta}$) independently. If $k_{\alpha\beta}$ is set higher than k_s, the position of the poles is moved to the left, compared to the poles of the SOGI, moving away from the instability region with the option to have the bandwidth reduced only slightly.

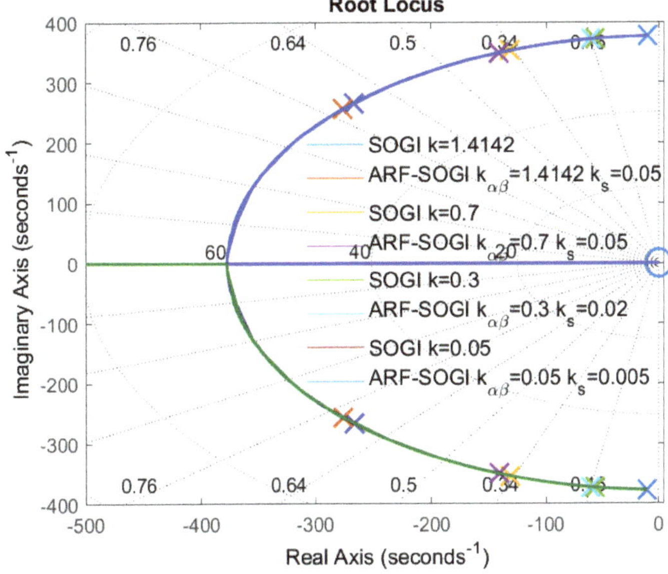

Figure 6. SOGI-QSG and ARF-SOGI-QSG band-pass filter v' poles.

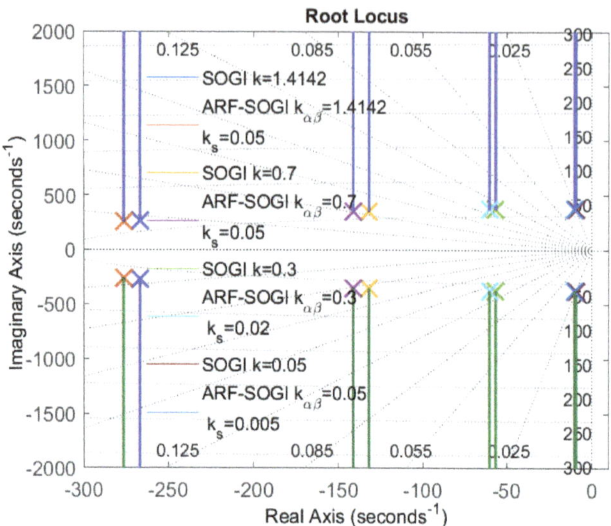

Figure 7. SOGI-QSG and ARF-SOGI-QSG low-pass filter qv' poles.

It is worth noticing that as stability is improved, the frequencies of interest are attenuated while the bandwidth is reduced; however, it is possible to find a combination that is acceptable, where k_s is not so small with respect to $k_{\alpha\beta}$. It is recommended to select k_s and $k_{\alpha\beta}$ values so that, in the unit-step test, the ARF-SOGI-QSG filters do not overshoot, in a similar manner to that of a critically damped system or to an over-damped system. This prevents large transient overshoots during phase jump disturbances and frequency deviations. As seen in Figures 8 and 9, when the value of k_s tends to the value of $k_{\alpha\beta}$, the ARF-SOGI-QSG poles are real and lie in the left half-plane, but one of the poles is approaching the imaginary axis. This implies that the system is over-damped (for instance, look at the poles when $k_s = k_{\alpha\beta} = 1.4142$).

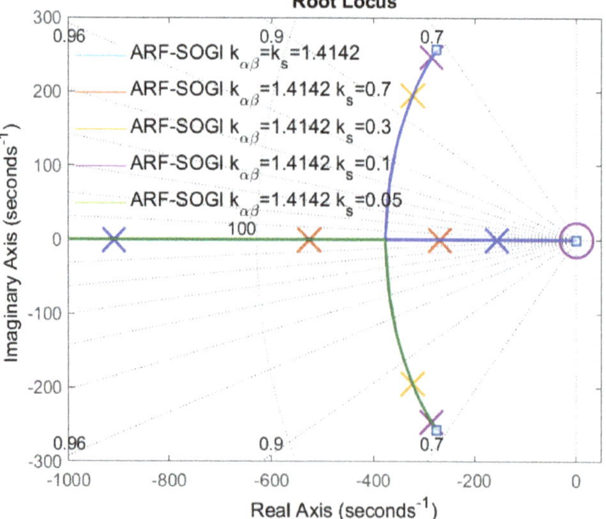

Figure 8. v' ARF-SOGI-QSG poles.

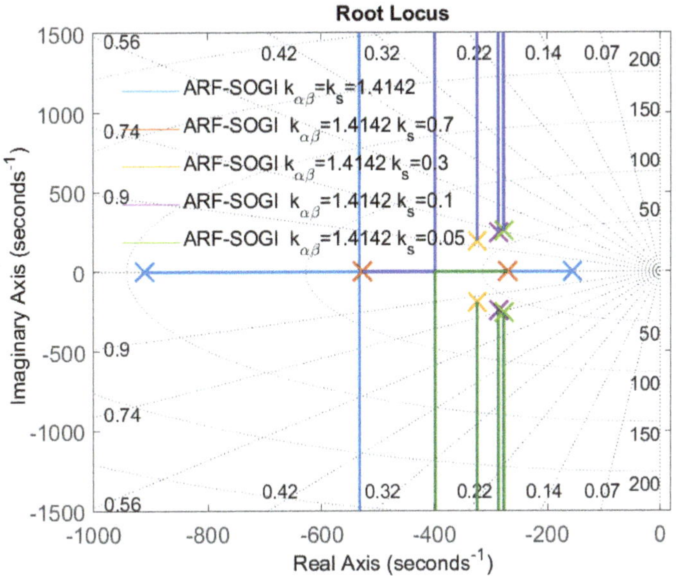

Figure 9. qv' ARF-SOGI-QSG poles.

An improved unit-step response is another advantage of the ARF-SOGI-QSG. In Figure 10, it can be seen that the unit-step response with $k = 1.4142$ in the traditional SOGI exhibits an overshoot before stabilizing. The same figure (Figure 10) shows how, in the ARF-SOGI, it is possible to eliminate that overshoot by tuning $k_{\alpha\beta} = 1.4142$, $k_s = 0.3$, achieving the steady state earlier than the traditional SOGI-QSG. As mentioned before, the closer k_s is to $k_{\alpha\beta}$, the more the attenuation of the fundamental frequency is increased, which is undesirable. However, in Figure 10, it can be seen how k_s, in this case, can be adjusted to a value lower than $k_s = 0.3$ without a significant overshoot. This improves the stability of the system, reduces the level of attenuation in the fundamental, and speeds up the response. A similar behavior occurs in the output corresponding to the band-pass effect of the ARF-SOGI-QSG.

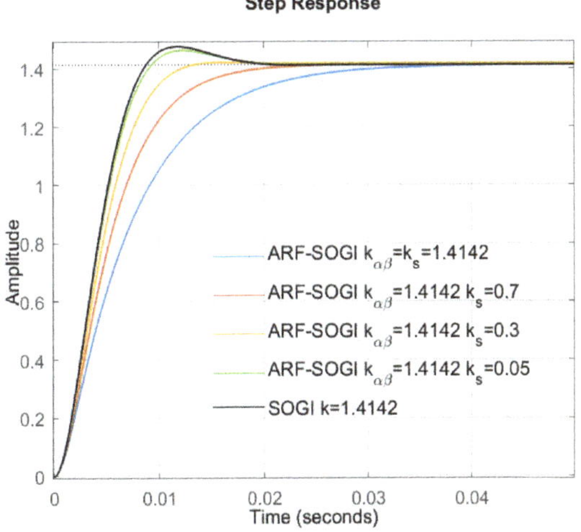

Figure 10. qv' low-pass filter-step response of the SOGI-QSG versions.

2.5. The ARF-SOGI-PLL

To implement a PLL based on ARF-SOGI-QSG, the architecture of an SRF-PLL is added to the ARF-SOGI-QSG, as shown in Figure 11.

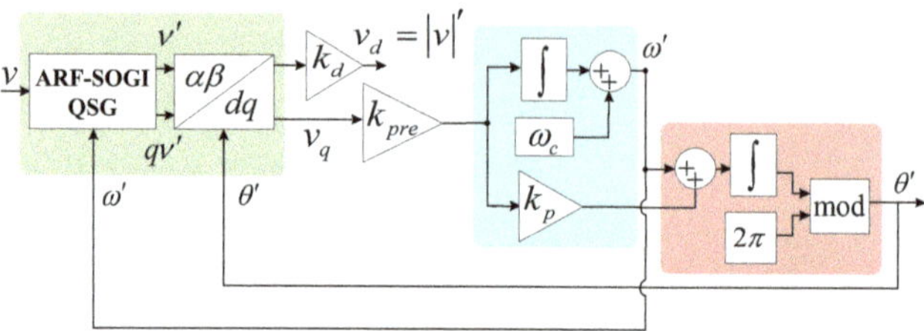

Figure 11. Structure of the ARF-SOGI-PLL.

Since the ARF-SOGI-QSG is more robust than the traditional SOGI-QSG, it is possible to increase the bandwidth of the loop filter. This is accomplished by increasing the value of the k_p and k_i gains in the PLL, or by selecting a large value for k_i under a standard design method for the selection of parameters [21]. In the proposed PLL, the multiplication of k_{pre} by the loop PI filter gains increases the bandwidth of the PLL. The PLL based on the ARF-SOGI is composed of the following stages:

1. The modified SOGI-QSG block (ARF-SOGI-QSG);
2. The k_{pre} gain, which compensates for the extra attenuation caused by the ARF-SOGI and increases the loop bandwidth of the SRF-PLL embedded in the PLL;
3. An SRF-PLL with frequency extraction using an integral controller.

To assess the performance of the proposed PLL, its response is compared with that of the traditional SOGI-PLL and the SOGI-PLL with an estimated frequency extracted from the PI controller integrator output [5] (referred to as SOGI-PLL-EFI). The comparison is made under the following conditions:

- Frequency deviations of -6 Hz and -14 Hz;
- Harmonic content: 0.04 pu of the fifth harmonic and 0.0295 pu of the seventh harmonic, for a THD = 4.99%;
- Phase jump of 75^{0};
- Voltage sag of 80%;
- Phase-to-ground short-circuit;
- Gain $k = 0.5$, to reduce the bandwidth of the SOGIs and better deal with the harmonic content.

The simulations are carried-out using a processor-in-the-loop approach based on the F28335 DSC, so that the real implementation and discretization issues are adequately taken into account. In the co-simulations of this and the following sections, different sets of gains are used to highlight the fact that the performance of the ARF-QSG-PLL is inherent to its architecture regardless of the assigned bandwidths. Three sets of tuning parameters have been used, as detailed in Tables 1–3, corresponding to different bandwidths for the QSG and the loop filter.

Table 1. Gains for tuning set S1 (small SOGI bandwidth, typical PLL bandwidth).

	ARF-SOGI				PLL
$k_{\alpha\beta}$	k_s		k_{pre}	k_p	k_i
0.5	0.5		1.4	184.7	8479.16
	SOGI				PLL
	k			k_p	k_i
	0.5			184.7	8479.16

Table 2. Gains for tuning set S2 (small SOGI bandwidth, large PLL bandwidth).

	ARF-SOGI			PLL
$k_{\alpha\beta}$	k_s	k_{pre}	k_p	k_i
0.5	0.5	1.4	563.67	50,116.247
	SOGI			PLL
	k		k_p	k_i
	0.5		563.67	50,116.247

Table 3. Gains for tuning set S3 (typical SOGI bandwidth, typical PLL bandwidth).

	ARF-SOGI			PLL
$k_{\alpha\beta}$	k_s	k_{pre}	k_p	k_i
1.4142	0.05	1.4	184.7	8479.16
	SOGI			PLL
	k		k_p	k_i
	1.4142		184.7	8479.16

3. Results

In this section, several verification tests are carried out in order to verify the performance of the proposed ARF-SOGI-PLL. The tests are made with co-simulations implemented under the processor-in-the-loop (PIL) [33,34] approach in MATLAB/Simulink with Code Composer Studio (CCS) and the digital signal controller TMS320F28335 from Texas Instruments.

3.1. Response to Frequency Deviation

For this test, the SOGI and ARF-SOGI gains are tuning sets S1 and S2, as specified in Tables 1 and 2. The response of the schemes to a frequency deviation is shown in Figure 12. It can be seen that the ARF-SOGI-PLL reduces the overshoot in the frequency to lower levels than those reported by both the SOGI-PLL and the SOGI-PLL-EFI. In addition, it also achieves the correct measurement and the steady state in synchronization. For the large bandwidth tuning, this occurs 2.8 cycles before the SOGI-PLL-EFI, as shown in Figure 13.

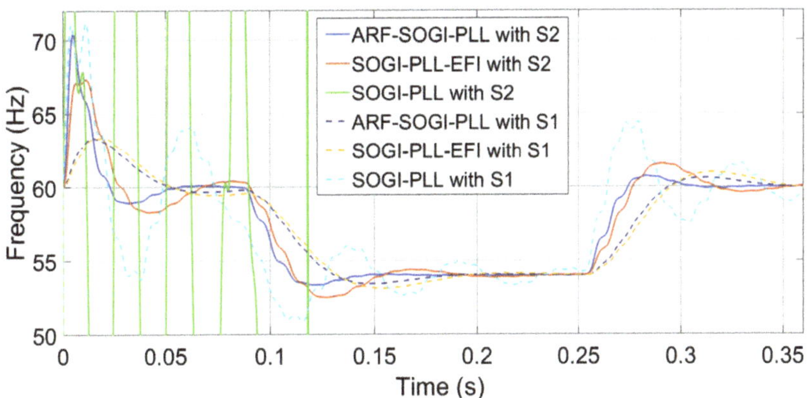

Figure 12. Frequency measured using a high bandwidth and a typical bandwidth in the PLLs in the face deviation of −6 Hz in the frequency of the line voltage.

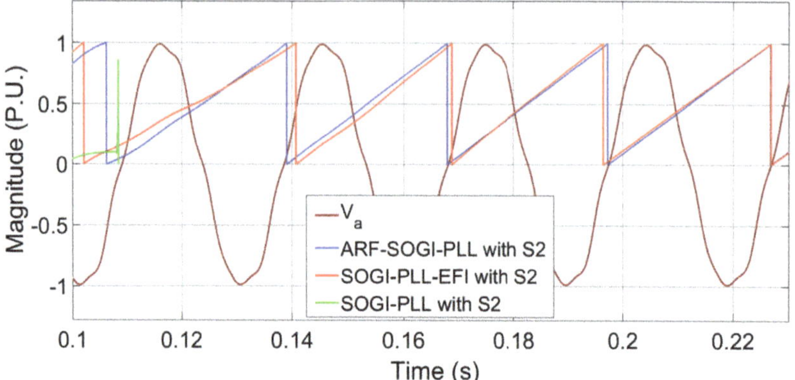

Figure 13. ωt measured during -6 Hz deviation to before 60 Hz recovery.

Another feature of the ARF-SOGI-PLL is that it is possible to almost match its overshoot level, for a large bandwidth loop filter setting, to that of a much smaller typical bandwidth reported by another PLL, such as that of Simulink [26] or any of the versions of SOGI-PLL considered here. In other words, faster PLL response times can be achieved without incurring a high overshoot.

With the tuning set S1, the SOGI-PLL still performs measurement and synchronization, but with high overshoots, and it very slowly reaches the steady state. The SOGI-PLL with the tuning set S2 is no longer able to perform frequency measurement and timing. On the other hand, Figure 13 shows that the ARF-SOGI-PLL synchronizes adequately before the SOGI-PLL. In addition, in the measured frequency, the SOGI-PLL can not synchronize and the signal is lost. This frequency deviation is within the allowed adjustment ranges given in [31,32]. The ARF-SOGI-PLL shows the best behavior, since it synchronizes three cycles before the SOGI-PLL. The ARF-SOGI-PLL is faster when reporting a frequency measurement, and it is more reliable, since its overshoot does not exceed ± 1 Hz. This gives greater assurance that the system will not trip due to an erroneous frequency measurement, compared to the other two versions of the SOGI-PLL considered here.

3.2. Response to Voltage Sag

Voltage sags are a tough test for synchronization algorithms [14,15]; they arise from disturbances in the electrical network, caused by, for example, the energization of transformers or large induction motors [12]. To compare the performance of the PLLs subjected to voltage sag, a test is applied where the grid voltage suddenly changes to 0.2 per unit. The gains of the ARF-SOGI-PLL and the other two PLLs are those of Tables 1 and 2 (tuning sets S1 and S2). The voltage sag causes a desynchronization of the PLLs; when the fault is cleared, the algorithms must resynchronize. As seen in Figure 14 for the case of a typical PLL bandwidth given by S1, during the resynchronization, the ARF-SOGI-PLL and the SOGI-PLL-EFI report a smooth transient frequency measurement with similar errors. It is also observed that the SOGI-PLL with the typical bandwidth (S1) incurs a large error in the transient frequency measurement and, therefore, the phase tracking error in the resynchronization process is greater than for the other PLLs. For the case of a large PLL bandwidth given by tuning S2, the frequency measurements by the ARF-SOGI-PLL and the SOGI-PLL-EFI are comparable, while the SOGI-PLL is no longer capable of resynchronizing and, therefore, cannot correctly measure the frequency either.

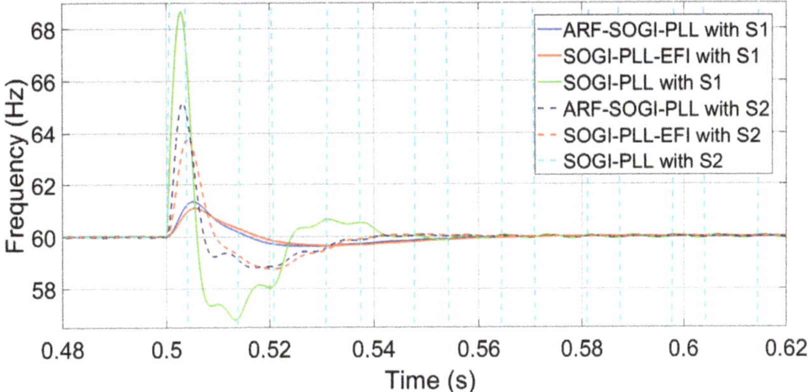

Figure 14. Frequency before, during and after a voltage sag of 80%.

3.3. Response to Ground Fault

Regarding ground faults, during the event, the synchronization is lost by all the algorithms; when the system recovers from the fault, all of the considered SOGI-PLLs re-synchronize after approximately four cycles, except for the standard SOGI-PLL with S2 tuning, as seen in Figure 15. The SOGI-PLL-EFI synchronizes on the third cycle after system recovery; however, it loses synchronization again, regaining it on the fifth cycle, while on the sixth cycle, its error in phase tracking increases. Figure 16 shows the responses for the SOGI-PLL-EFI and the ARF-SOGI-PLL, both using the S2 tuning, in greater detail. Figure 17 shows how, as the resynchronization evolves, the ARF-SOGI-PLL for both tunings (S1 and S2) reports the smallest phase tracking error, synchronizing correctly before the other PLLs. On the other hand, the SOGI-PLL with the S1 tuning can re-synchronize, but it does so slowly, compared to the other two, and reports the highest phase tracking error. For its part, the same PLL with the S2 tuning no longer achieves synchronization.

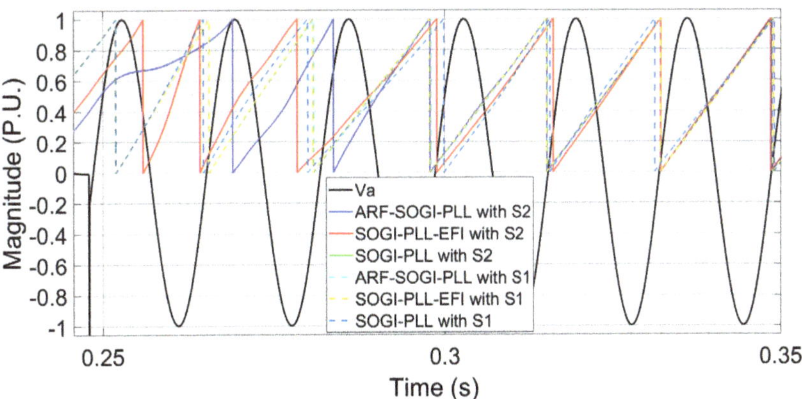

Figure 15. ωt during and after phase-to-ground short-circuit fault using a high and typical bandwidth in the control loop of the PLLs.

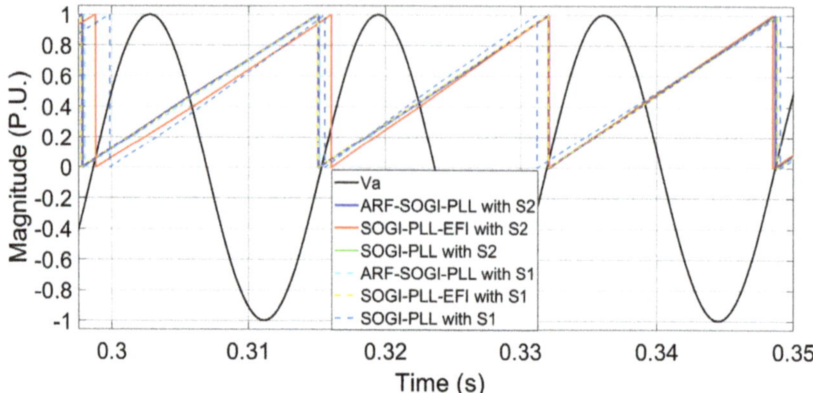

Figure 16. ωt during the synchronization process, from the third to the sixth cycle after system recovery following a phase-to-ground short-circuit.

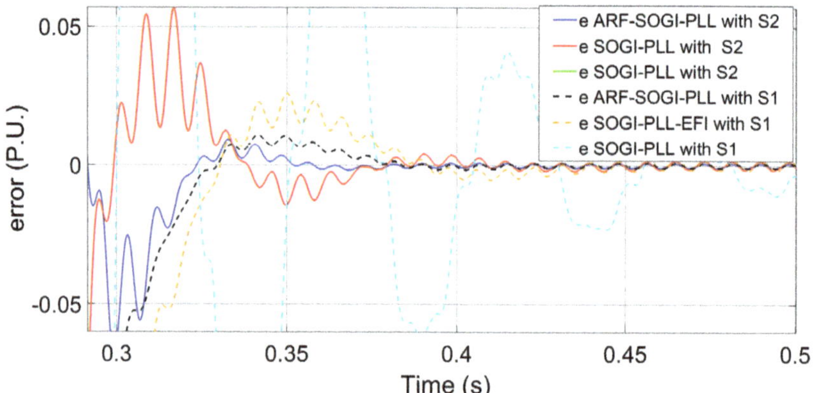

Figure 17. Phase tracking error from third to sixth cycle after system recovery from phase-to-ground short-circuit.

Moreover, in Figure 17, in the case of the S2 tuning, the ARF-SOGI-PLL reports the lowest error in the phase tracking. The SOGI-PLL-EFI presents an error of zero at $t = 0.29$ s (it is synchronized), but when it reaches the next cycle at approximately $t = 0.316$ s, it presents an error of 0.05 s (it loses the correct synchrony). Then, at $t = 0.332$ s, the error is reduced to 0.01268 pu; later, at $t = 0.3715$ s, it is synchronized with an error of only 0.000422 pu, and at $t = 0.3902$ s, the error again grows to 0.00424 pu.

Figure 18 shows that the ARF-SOGI-PLL is the first to report the correct value of the frequency for both tunings S1 and S2. It is also observed how the standard SOGI-PLL with the S1 tuning can measure the frequency of the system by recovering from the short-circuit; however, it is slower, and during the re-synchronization, its measurement has the largest error. The measurement of frequency by the standard SOGI-PLL with the S2 tuning is not seen because, in this case, the initial synchronization is not achieved.

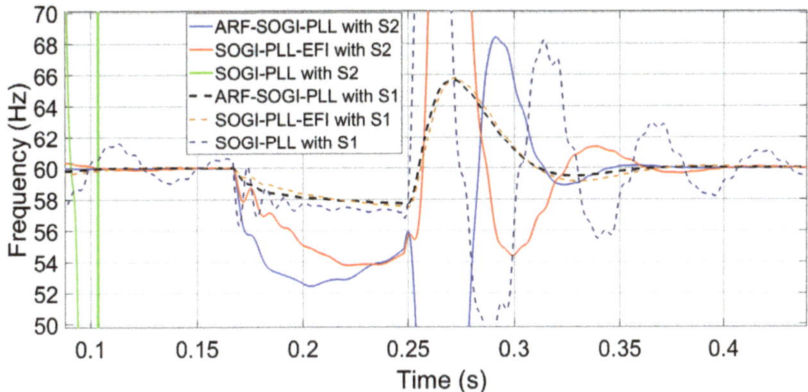

Figure 18. Frequency measured before, during, and after phase-to-ground short-circuit fault.

3.4. Harmonic Content Rejection Capability of Large Bandwidth SOGI Versions

The total rejection of the DC component in the orthogonal signal and the improved high-frequency harmonic filtering capability are key features of the SOGI-QSG [35]. For this reason, it is desirable to verify that the proposed ARF-SOGI-PLL does not lose these features. In this section, it is verified that, when using a small-bandwidth (Case A) or a typical-bandwidth (Case B) ARF-SOGI-PLL, the harmonic content rejection capacity is not lost. The loop filters of the PLLs have a common adjustment. The THD of the unit vectors is compared for an input THD of 4.99%.

Case A. The three PLLs considered are adjusted with the gains given in Table 1. Harmonic analysis reveals that the THD of the unit vectors of the PLLs meet the requirement of being less than 1% [36]. However, the THD for the ARF-SOGI-PLL is less than the corresponding value for the SOGI-PLL, and a is little greater than the value for the SOGI-PLL-EFI, as shown in Figure 19.

Case B. The values assigned to the gains of the PLLs are shown in Table 3. For the ARF-SOGI-PLL, THD_α=0.12% and THD_β=0.21%, which is lower than the corresponding values for the SOGI-PLL (THD_α=0.21%, THD_β=0.30%) and slightly higher than the values for SOGI-PLL-EFI (THD_α=0.10%, THD_β=0.17%). In Figures 19 and 20, the harmonic components and THD of the unit vectors α and β are compared.

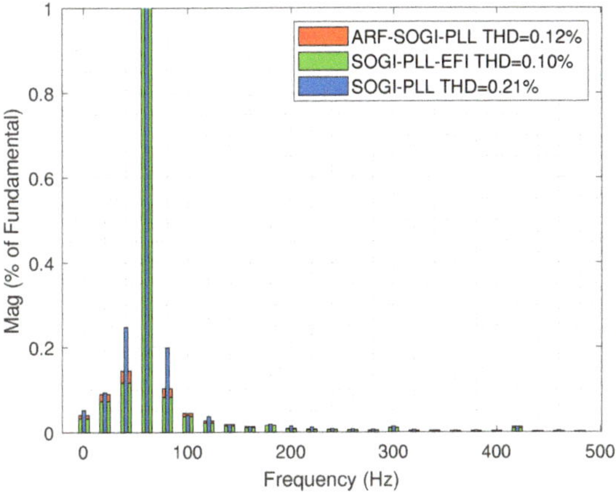

Figure 19. THD of the unit vectors $\alpha= \cos\left(\omega t'\right)$ for $k_{\alpha\beta} = 0.5$ and $k = 0.5$ and an input THD of 4.99%.

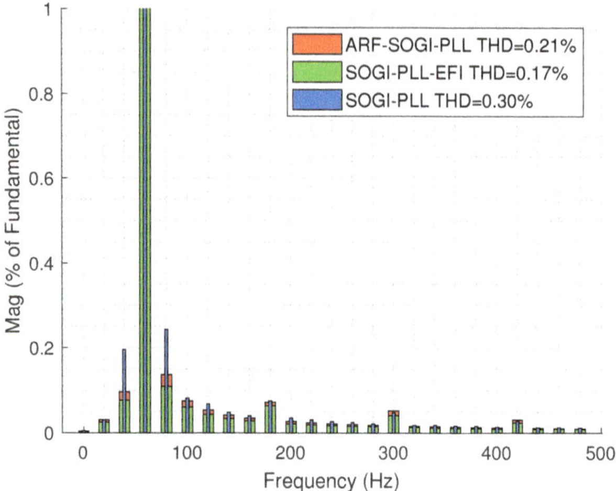

Figure 20. THD of the unit vectors $\beta = \sin\left(\omega t'\right)$ for $k_{\alpha\beta} = 0.5$ and $k = 0.5$ and an input THD of 4.99%.

3.5. Synchronization of a Single-Phase Inverter

In this section, the ARF-SOGI-PLL and SOGI-PLL-EFI are used to synchronize a single-phase full-bridge inverter connected to the grid. A comparative analysis of the influence of each PLL on the response of the synchronized inverter is made. Only these two PLLs are considered since they are the ones that have shown the best performance. Figure 21 shows the frequency measurement by the two algorithms using two different settings S1 and S2 during a deviation of -6 Hz. The gain values for settings S1 and S2 are listed in Tables 1 and 2. Figure 21 shows how the ARF-SOGI-PLL achieves a correct frequency before the SOGI-PLL-EFI. The figure also shows that the overshoots given by the ARF-SOGI-PLL at the beginning of the frequency deviations are smaller than those reported by the other PLL. In fact, the overshoot produced by the ARF-SOGI-PLL with a larger bandwidth in its PI is practically the same as that given by the SOGI-PLL-EFI with the smaller bandwidth. This indicates that it is possible to synchronize faster with the ARF-SOGI-PLL than with the SOGI-PLL-EFI.

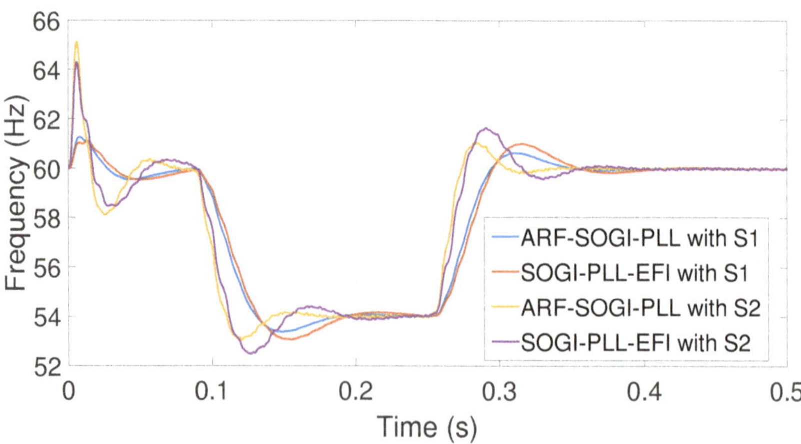

Figure 21. Frequency measured following -6 Hz deviation during 0.166 s, and its recovery, using a high bandwidth and a typical bandwidth in the PLLs control loop.

Figure 22 shows that the ARF-SOGI-PLL syncs 16.67 ms faster than the SOGI-PLL-EFI, and it can be seen that the error in the phase tracking is lower with the ARF-SOGI-PLL.

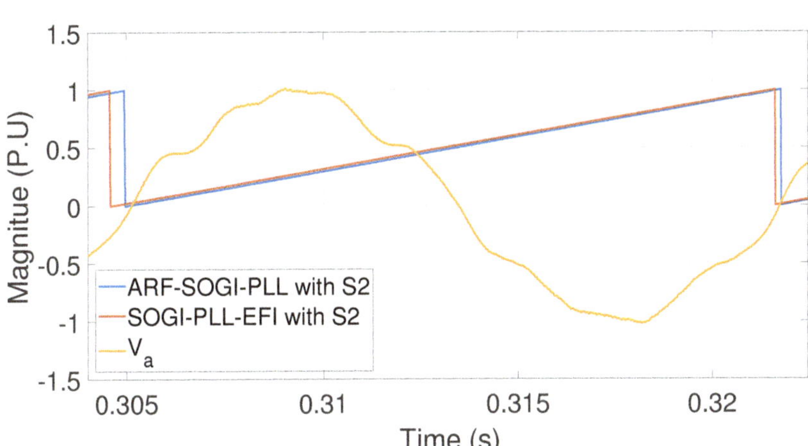

Figure 22. ωt in the final cycle in the synchronization of the ARF-SOGI-PLL.

3.6. Resources and Computing Times

The metrics shown in Table 4 are obtained using a processor-in-the-loop approach. The corresponding measurements are the time necessary for the processing of each of the algorithms in a cycle of the fundamental, and the memory demanded from the 262 KB available in the DSC F28335 processor. Further, in Table 4, it is observed that the flash memory needed by the SOGI-PLL and the SOGI-PLL-EFI is similar, and that it is slightly higher for the ARF-SOGI-PLL.

Table 4. Required memory and processing time of each algorithm in the DSC.

Algorithm	Flash Memory	Average Execution Time	Maximum Execution Time
SOGI-PLL	4030 bytes (1%)	12,861 ns	14,887 ns
SOGI-PLL-EFI	4013 bytes (1%)	15,136 ns	17,187 ns
ARF-SOGI-PLL in state space	4090 bytes (1%)	13,392 ns	15,307 ns
ARF-SOGI-PLL	4049 bytes (1%)	15,451 ns	17,787 ns

The memory required by the transfer fuction ARF-SOGI-PLL is greater than the memory required by the SOGI-PLL and SOGI-PLL-EFI versions, but less than the state space ARF-SOGI-PLL. This is seen in Table 4, and it is concluded that the increase in memory required by either of the two implementations of the ARF-SOGI-PLL, with respect to the existing versions of SOGI-PLL, is not a determining factor in a real implementation.

In Table 4, it is seen that the SOGI-PLL is processed faster than any of the others. After this comes the ARF-SOGI-PLL, and the slowest is the SOGI-PLL-EFI. During a run cycle, the ARF-SOGI-PLL resolves 1.744 µs faster than the SOGI-PLL-EFI, and 0.531 µs slower than the traditional SOGI-PLL. As can be seen in the same Table 4, the modifications made to the SOGI imply raising the processing time of the algorithm.

The ARF-SOGI-PLL implemented in the state space runs 2.059 µs faster per DSC processing cycle than the structure of Figure 3. This indicates that, although the memory required by the ARF-SOGI-PLL implemented in the state space is larger than that required by its equivalent in Figure 3, its processing time is lower.

4. Conclusions

The traditional SOGI-PLL is the base for the proposed ARF-SOGI-PLL. The SOGI-QSG transfer functions were brought into the state space representation in controllable canonical form with the intention of having direct access to the sections of the SOGI-QSG algorithm, where gain has an influence. The ARF-SOGI-QSG was proposed, which, unlike the standard SOGI-QSG, contains a second gain k_s that is located in the system matrix in the state space; moreover, a feedback of the output signal of the band-pass filter is added to the input of the modified QSG. These modifications allow for selecting the extra attenuation of the quadrature signals, while moving the QSG poles further to the left, thereby achieving greater stability in the ARF-SOGI-QSG compared to the standard SOGI-QSG.

The block diagram of the ARF-SOGI-QSG and its transfer functions were reported. A linear analysis of the ARF-SOGI-QSG was performed and compared with that of the standard SOGI-QSG, and it was shown that the ARF-SOGI-QSG is a quadrature signal generator with greater stability and a faster response time than that given by the SOGI-QSG.

The greater stability reported by the ARF-SOGI-QSG permits the increase of the bandwidth of the control loop filter, and with that, the response speed of the ARF-SOGI-PLL is increased without compromising the rejection of the harmonic content and without reaching the instability or impossibility of processing compared to the SOGI-PLL.

The complete scheme of a PLL based on the ARF-SOGI-QSG was reported; this proposal was called ARF-SOGI-PLL. To verify the performance of the proposed PLL, this was subjected to different disturbances, such as frequency deviations, phase jumps, voltage sags, phase-to-ground short-circuits, and harmonic contents. The response was compared with the response of two other versions of SOGI-PLL under two different tunings in the loop filter. Finally, a single-phase inverter was synchronized with the voltage of an electrical network. From the simulations, it is confirmed that a PLL based on the new ARF-SOGI-QSG is more stable and faster in its transient response without losing the ability to reject harmonic content.

The responses of the ARF-SOGI-PLL and the SOGI-PLL-EFI are quite similar; however, the ARF-SOGI-PLL can be adjusted in such a way that the overshoot in the frequency measurement is less than the overshoot reported by the SOGI-PLL-EFI. Even when the ARF-SOGI-PLL has a high bandwidth in its loop filter, the overshoot in its response is very similar in magnitude to the overshoot given by the SOGI-PLL-EFI with a lower bandwidth. This contributes to lowering the error in the transient frequency measurements, achieving a correct frequency measurement faster.

The ARF-SOGI-QSG implemented in the state space demonstrated a more agile computational processing. The rejection of the harmonic content in the ARF-SOGI-PLL is better than the rejection given by the traditional SOGI-PLL and it is slightly lower than that presented by the SOGI-PLL-EFI. Although the ARF-SOGI-PLL consumes 77 bytes more than the SOGI-PLL-EFI, the DSC processing of the ARF-SOGI-PLL is 1744 µs faster. The ARF-SOGI-PLL requires 60 bytes more than the traditional SOGI-PLL and requires only 0.531 µs more in processing time than the traditional SOGI-PLL.

Author Contributions: Conceptualization, G.A.H.-P. and E.M.-V.; methodology, G.A.H.-P. and J.d.J.C.; software, G.A.H.-P.; validation, J.d.J.C.; formal analysis, G.A.H.-P. and J.d.J.C.; investigation, G.A.H.-P.; resources, E.M.-V.; data curation, G.A.H.-P.; writing—original draft preparation, G.A.H.-P. and E.M.-V.; writing—review and editing, G.A.H.-P., E.M.-V. and J.d.J.C.; visualization, G.A.H.-P.; supervision, E.M.-V.; project administration, E.M.-V.; funding acquisition, E.M.-V. All authors have read and agreed to the published version of the manuscript.

Funding: This research was funded by Tecnológico Nacional de México, grant number 13392.21-P, and by Consejo Nacional de Ciencia y Tecnología, México (CONACyT).

Data Availability Statement: Not applicable.

Conflicts of Interest: The authors declare no conflict of interest.

Abbreviations

The following abbreviations are used in this manuscript:

ARF	Adjustable re-filtering
DG	Distributed generation
EPS	Electrical power system
PLL	Phase-locked loop
QSG	Quadrature signal generator
SOGI	Second-order generalized integrator
THD	Total harmonic distortion

References

1. Best, R.E. *Phase-Locked Loops: Design, Simulation, and Applications*, 5th ed.; McGraw-Hill: New York, NY, USA, 2003.
2. Timbus, A.; Liserre, M.; Teodorescu, R.; Blaabjerg, F. Synchronization methods for three phase distributed power generation systems—An overview and evaluation. In Proceedings of the 2005 IEEE 36th Power Electronics Specialists Conference, Recife, Brazil, 16 June 2005; pp. 2474–2481.
3. Best, R.E. *Phase-Locked Loops: Design, Simulation, and Applications*, 6th ed.; McGraw-Hill Education: New York, NY, USA, 2007.
4. Yazdani, A.; Iravani, R. *Voltage-Sourced Converters in Power Systems: Modeling, Control, and Applications*; John Wiley & Sons: Hoboken, NJ, USA, 2010.
5. Golestan, S.; Guerrero, J.M.; Vasquez, J.C. Three-Phase PLLs: A Review of Recent Advances. *IEEE Trans. Power Electron.* **2017**, *32*, 1894–1907. [CrossRef]
6. Rodríguez, P.; Teodorescu, R.; Candela, I.; Timbus, A.V.; Liserre, M.; Blaabjerg, F. New positive-sequence voltage detector for grid synchronization of power converters under faulty grid conditions. In Proceedings of the 2006 37th IEEE Power Electronics Specialists Conference, Jeju, Korea, 18–22 June 2006; pp. 1–7. [CrossRef]
7. Chung, S.K. Phase-locked loop for grid-connected three-phase power conversion systems. *IEE Proc.-Electr. Power Appl.* **2000**, *147*, 213–219. [CrossRef]
8. Gao, S.; Barnes, M. Phase-locked loop for AC systems: Analyses and comparisons. In Proceedings of the 6th IET International Conference on Power Electronics, Machines and Drives (PEMD 2012), Bristol, UK, 27–29 March 2012; pp. 1–6. [CrossRef]
9. Mahdian, H.; Hashemi, M.; Ghadimi, A.A. Improvement in the synchronization process of the voltage-sourced converters connected to the Grid by PLL in order to Detect and Block the Double Frequency Disturbance Term. *Indian J. Sci. Technol.* **2013**, *6*, 4940–4952. [CrossRef]
10. Jauch, C.; Matevosyan, J.; Ackermann, T.; Bolik, S. International comparison of requirements for connection of wind turbines to power systems. *Wind Energy* **2005**, *8*, 295–306. [CrossRef]
11. Rodriguez, P.; Luna, A.; Ciobotaru, M.; Teodorescu, R.; Blaabjerg, F. Advanced Grid Synchronization System for Power Converters under Unbalanced and Distorted Operating Conditions. In Proceedings of the IECON 2006—32nd Annual Conference on IEEE Industrial Electronics, Paris, France, 7–10 November 2006; pp. 5173–5178. [CrossRef]
12. Teodorescu, R.; Liserre, M.; Rodriguez, P. *Grid Converters for Photovoltaic and Wind Power Systems*; John Wiley & Sons: New Delhi, India, 2011.
13. Karimi Ghartemani, M.; Khajehoddin, S.A.; Jain, P.K.; Bakhshai, A. Problems of Startup and Phase Jumps in PLL Systems. *IEEE Trans. Power Electron.* **2012**, *27*, 1830–1838. [CrossRef]
14. Abdali Nejad, S.; Matas, J.; Martín, H.; de la Hoz, J.; Al-Turki, Y.A. New SOGI-FLL Grid Frequency Monitoring with a Finite State Machine Approach for Better Response in the Face of Voltage Sag and Swell Faults. *Electronics* **2020**, *9*, 612. [CrossRef]
15. Abdali Nejad, S.; Matas, J.; Elmariachet, J.; Martín, H.; de la Hoz, J. SOGI-FLL Grid Frequency Monitoring with an Error-Based Algorithm for a Better Response in Face of Voltage Sag and Swell Faults. *Electronics* **2021**, *10*, 1414. [CrossRef]
16. Bollen, M.H. Understanding power quality problems. In *Voltage Sags and Interruptions*; IEEE Press: Piscataway, NJ, USA, 2000.
17. Rodríguez, P.; Luna, A.; Muñoz-Aguilar, R.S.; Etxeberria-Otadui, I.; Teodorescu, R.; Blaabjerg, F. A Stationary Reference Frame Grid Synchronization System for Three-Phase Grid-Connected Power Converters Under Adverse Grid Conditions. *IEEE Trans. Power Electron.* **2012**, *27*, 99–112. [CrossRef]
18. Velasco, J.A.M. *Power System Transients: Parameter Determination*; CRC Press: London, UK, 2010.
19. Green, T.; Marks, J. Control techniques for active power filters. *IEE Proc.-Electr. Power Appl.* **2005**, *152*, 369–381. [CrossRef]
20. Subramanian, C.; Kanagaraj, R. Rapid Tracking of Grid Variables Using Prefiltered Synchronous Reference Frame PLL. *IEEE Trans. Instrum. Meas.* **2015**, *64*, 1826–1836. [CrossRef]
21. Chung, S.K. A phase tracking system for three phase utility interface inverters. *IEEE Trans. Power Electron.* **2000**, *15*, 431–438. [CrossRef]
22. Arricibita, D.; Marroyo, L.; Barrios, E.L. Simple and robust PLL algorithm for accurate phase tracking under grid disturbances. In Proceedings of the 2017 IEEE 18th Workshop on Control and Modeling for Power Electronics (COMPEL), Stanford, CA, USA, 9–12 July 2017; pp. 1–6. [CrossRef]

23. Nos, O.V.; Abramushkina, E.E.; Kharitonov, S.A. Control Design of Fast Response PLL for FACTS Applications. In Proceedings of the 2019 International Ural Conference on Electrical Power Engineering (UralCon), Chelyabinsk, Russia, 1–3 October 2019; pp. 301–305. [CrossRef]
24. Luhtala, R.; Alenius, H.; Roinila, T. Practical Implementation of Adaptive SRF-PLL for Three-Phase Inverters Based on Sensitivity Function and Real-Time Grid-Impedance Measurements. *Energies* **2020**, *13*, 1173. [CrossRef]
25. Golestan, S.; Ramezani, M.; Guerrero, J.M. An Analysis of the PLLs With Secondary Control Path. *IEEE Trans. Ind. Electron.* **2014**, *61*, 4824–4828. [CrossRef]
26. Cao, Y.; Yu, J.; Xu, Y.; Li, Y.; Yu, J. An Efficient Phase-Locked Loop for Distorted Three-Phase Systems. *Energies* **2017**, *10*, 280. [CrossRef]
27. Aldbaiat, B.; Nour, M.; Radwan, E.; Awada, E. Grid-Connected PV System with Reactive Power Management and an Optimized SRF-PLL Using Genetic Algorithm. *Energies* **2022**, *15*, 2177. [CrossRef]
28. Ciobotaru, M.; Teodorescu, R.; Blaabjerg, F. A new single-phase PLL structure based on second order generalized integrator. In Proceedings of the 2006 37th IEEE Power Electronics Specialists Conference, Jeju, Korea, 18–22 June 2006; pp. 1–6. [CrossRef]
29. Ciobotaru, M.; Agelidis, V.G.; Teodorescu, R.; Blaabjerg, F. Accurate and Less-Disturbing Active Antiislanding Method Based on PLL for Grid-Connected Converters. *IEEE Trans. Power Electron.* **2010**, *25*, 1576–1584. [CrossRef]
30. Guerrero-Rodríguez, N.; Rey-Boué, A.B.; Bueno, E.; Ortiz, O.; Reyes-Archundia, E. Synchronization algorithms for grid-connected renewable systems: Overview, tests and comparative analysis. *Renew. Sustain. Energy Rev.* **2017**, *75*, 629–643. [CrossRef]
31. *IEEE Std 1547a-2020 (Amendment to IEEE Std 1547-2018)*; IEEE Standard for Interconnection and Interoperability of Distributed Energy Resources with Associated Electric Power Systems Interfaces—Amendment 1: To Provide More Flexibility for Adoption of Abnormal Operating Performance Category III. IEEE: Piscataway, NJ, USA, 2020; pp. 1–16. [CrossRef]
32. *IEC61727*; Photovoltaic (PV) Systems-Characteristics of the Utility Interface. International Electrotechnical Commission: Geneva, Switzerland, 2004 .
33. Motahhir, S.; Ghzizal, A.E.; Sebti, S.; Derouich, A. MIL and SIL and PIL tests for MPPT algorithm. *Cogent Eng.* **2017**, *4*, 1378475. [CrossRef]
34. Krishna Srinivasan, M.; Daya John Lionel, F.; Subramaniam, U.; Blaabjerg, F.; Madurai Elavarasan, R.; Shafiullah, G.M.; Khan, I.; Padmanaban, S. Real-Time Processor-in-Loop Investigation of a Modified Non-Linear State Observer Using Sliding Modes for Speed Sensorless Induction Motor Drive in Electric Vehicles. *Energies* **2020**, *13*, 4212. [CrossRef]
35. Golestan, S.; Guerrero, J.M.; Vasquez, J.C. Single-Phase PLLs: A Review of Recent Advances. *IEEE Trans. Power Electron.* **2017**, *32*, 9013–9030. [CrossRef]
36. Kulkarni, A.; John, V. Analysis of Bandwidth–Unit-Vector-Distortion Tradeoff in PLL During Abnormal Grid Conditions. *IEEE Trans. Ind. Electron.* **2013**, *60*, 5820–5829. [CrossRef]

Article

Profit Maximization with Imbalance Cost Improvement by Solar PV-Battery Hybrid System in Deregulated Power Market

Ganesh Sampatrao Patil [1], Anwar Mulla [2], Subhojit Dawn [3,*] and Taha Selim Ustun [4,*]

[1] Department of Technology, Shivaji University, Kolhapur 416004, India; patilganesh84@gmail.com
[2] Dr. Daulatrao Aher College of Engineering, Shivaji University, Kolhapur 416004, India; ammaitp@rediffmail.com
[3] Department of Electrical & Electronics Engineering, Velagapudi Ramakrishna Siddhartha Engineering College, Vijayawada 520007, India
[4] Fukushima Renewable Energy Institute, AIST (FREA), Koriyama 963-0298, Japan
* Correspondence: subhojit.dawn@gmail.com (S.D.); selim.ustun@aist.go.jp (T.S.U.)

Abstract: The changeable nature of renewable sources creates difficulties in system security and stability. Therefore, it is necessary to study system risk in several power system scenarios. In a wind-integrated deregulated power network, the wind farm needs to submit the bid for its power-generating quantities a minimum of one day ahead of the operation. The wind farm submits the data based on the expected wind speed (EWS). If any mismatch occurs between real wind speed (RWS) and expected wind speed, ISO enforces the penalty/rewards to the wind farm. In a single word, this is called the power market imbalance cost, which directly distresses the system profit. Here, solar PV and battery energy storage systems are used along by the wind farm to exploit system profit by grasping the negative outcome of imbalance cost. Along with system profit, the focus has also been on system risk. The system risk has been calculated using the risk assessment factors, i.e., Value-at-Risk (VaR) and Cumulative Value-at-risk (CVaR). The work is performed on a modified IEEE 14 and modified IEEE 30 bus test system. The solar PV-battery storage system can supply the demand locally first, and then the remaining power is given to the electrical grid. By using this concept, the system risk can be minimized by the incorporation of solar PV and battery storage systems, which have been studied in this work. A comparative study has been performed using three dissimilar optimization methods, i.e., Artificial Gorilla Troops Optimizer Algorithm (AGTO), Artificial Bee Colony Algorithm (ABC), and Sequential Quadratic Programming (SQP) to examine the consequence of the presented technique. The AGTO has been used for the first time in the risk assessment and alleviation problem, which is the distinctiveness of this work.

Keywords: deregulated market; nodal pricing; system risk; imbalance cost; AGTO; ABC

Citation: Patil, G.S.; Mulla, A.; Dawn, S.; Ustun, T.S. Profit Maximization with Imbalance Cost Improvement by Solar PV-Battery Hybrid System in Deregulated Power Market. *Energies* **2022**, *15*, 5290. https://doi.org/10.3390/en15145290

Academic Editor: Juri Belikov

Received: 30 May 2022
Accepted: 19 July 2022
Published: 21 July 2022

Publisher's Note: MDPI stays neutral with regard to jurisdictional claims in published maps and institutional affiliations.

1. Introduction

According to the norm, electricity was a monopoly owned by regional powers that had both production and distribution [1]. Countries allow the status quo to exist in the equivalent exchange for being provided a cut in the cost of service. This system was adopted even though it had a huge flaw, which was the potential to influence state policymakers. To break down this monopoly environment, deregulation was introduced over a centralized action taken over many years. Deregulation refers to the breakdown of monopolies at the state level, and these monopolies are sold or transferred to third parties [2]. The regulation resulted in the monopoly of the production and distribution of electricity by electric utility companies, which, in extension, led to a monopoly over the wholesale market. Additionally, by introducing deregulation, the monopoly was reduced.

The introduction of deregulation was an unintended move on the part of the government. It started in the 1970s in the form of an unintentional act called the Public Utility Regulatory Policy Act (PURPA). PURPA started as an act to encourage alternative sources of energy [3]. After these initiatives, most countries are moving toward the implementation of deregulation in their electrical networks to provide more economical facilities to their citizens [4,5].

With the continuous deprivation in the quantity of coal and fossil fuels throughout the world, power-generating stations are thinking about unconventional sources [6]. The uncertain nature of renewable energy creates issues such as energy management [7] and protection [8] in renewable combined power systems. The renewable addition of the present thermal power plant in the day-ahead market is very complicated, but the consumer experiences the benefit [9]. Some work has been accomplished by researchers in this field in recent years.

Paper [10] portrays the significance of renewable additions in the electricity market by the lower structure disruptions. Xing et al. [11] demonstrated that variable renewable electricity (VRE) performs a vital role in global decarbonization. The incorporation expenses of solar and wind on the demand and source sides by the economic dispatch model are discussed by the author. Life cycle assessment based on the performance degradation of solar panels has been discussed in [12]. The authors discussed how installed capacity decreases after deployment in the field and how this affects overall finances. Shujin et al. [13] discussed the decrement of renewable resources due to the overconsumption of renewable electricity in day-to-day life. In [14], the author states that the available transfer capability (ATC) performs a significant part in the deregulated market, and knowing it in advance can help to use the transmission network more efficiently. The authors of [15] displayed an arrangement of conventional and non-conventional systems with energy storage equipment to learn the impression of renewable uncertainties. The work in [16] showed how virtual power plants can be utilized to collectively manage renewable energy-based resources for efficient use. Reddy et al. [17] presented a methodology of renewable combined systems to exploit system safety and economic profit. In [18,19], the advantages of CAES in the system economy have been presented for the electricity market. In [20], the work aimed to decouple the focused solar energy output by CAESs and to model the MCP with the proposed offering strategies.

Chang et al. [21] discussed the influence of wind turbine generators (WTG) on system operation using Evolutionary Particle Swarm optimization (EPSO). A risk-mitigation bidding plan considering CVaR has been elucidated in [22]. Matevosyan et al. [23] projected a bidding strategy to lessen the imbalance of pricing in the wind-integrated short-term deregulated power market. In [24], a technique is proposed by the author for evaluating the effect of the unpredictable nature of wind flow in a wind combined competitive electrical system.

The authors of [25] introduced a novel optimization system to regulate the optimum generator schedule and involve load response to reduce the risk of transmission overloading burden in the forecasting power market. Rubin et al. [26] illustrated an equilibrium modeling technique to examine the impression of integrating wind power in a deregulated market. Das et al. [27] proposed risk mitigation methods in a wind-incorporated competitive power system using flexible AC Transmission Systems (FACTS) devices. The authors of [28] showed the importance of wind power incorporation in the system economy in deregulated markets. Khamees et al. [29] presented a key method of optimum power flow in a wind-incorporated electrical system to optimize the system fuel cost. Paper [30] depicts a scheduling technique for the best capacity sharing of a solar PV, wind farm, and pumped hydro storage system. The works in [31,32] presented a method for risk curtailment using FACTS devices and pumped hydro storage plants simultaneously in a wind-incorporated system.

From the detailed studies, it can be seen that several risks and financial mitigation work have been completed earlier, but there are still some scopes that have been done in this paper.

In the electricity market, wind farms need to submit the power generation scenario for the next day to ISO, a minimum day ahead of operation. Based on the acquiesced bid for the wind power plant, ISO arranged the power generation scheduling for all present generating stations. Due to the vagueness of the wind, there is a chance of not satisfying the arrangement of power from the wind plant. The ISO imposes an imbalance cost on the wind farm when a violation occurs in the electricity market. When real wind power (RWP) is more than expected wind power (EWP), then ISO grants rewards to the wind plant for the extra power sourced to the grid. ISO enforces a penalty on the wind plant if the EWP is more than the RWP. The negative imbalance cost (i.e., penalty) minimizes the system profit. Therefore, it is necessary to reduce the damaging effect of imbalance costs by reducing the mismatching amount between real and expected power generation from the wind farm to maintain system profit. Solar PV and battery energy storage can play an important role in this situation by providing additional power. The key highlights of this work are:

- Twenty scenarios with different system abnormalities (i.e., bus failure, transmission line failure, generator failure, sudden load increment, etc.) have been created to verify the success of the presented work. The VaR and CVaR have been calculated for all scenarios based on two system parameters: nodal pricing (NP) and transmission line flow (LF).
- The wind farm placement has been performed to reduce the system risk and exploit the system economics.
- Solar PV and battery storage are used to maximize the system profit while minimizing the harmful consequences of imbalance costs in the system. A comparative study has been performed using AGTO, ABC, and SQP to check the success of renewable integration in the electricity market in terms of operating cost and system risk.
- The AGTO has been used for the first time in the risk assessment and alleviation problem, which is the distinctiveness of this work.

2. Mathematical Formulations

This unit contains detailed studies on the mathematical formulation of wind power and risk assessment tools.

2.1. Wind Power Quantity and Investment Cost

The wind flows are very uncertain. This is changing every moment. The quality of wind power generation depends on the air density (ρ), efficiency of wind turbine (η), swept area of the wind turbine (A), and wind speed (WS). The generated wind power (GWP) is formulated as follows [33]:

$$\text{GWP} = \frac{1}{2}\rho \cdot A \cdot \eta \cdot (\text{WS})^3 \tag{1}$$

All the parameters are fixed for a particular place; only wind speed varies at every moment. In this case, the considered wind farm parameters are as follows: $\rho = 1.225 \text{ kg/m}^3$, $\eta = 0.49$, wind turbine rotor radius (r) = 40 m. Real-time wind speed data are not obtainable at the height of the wind turbine. In India, real-time wind speed data are available at a height of 10 m from the ground level. Under maximum conditions, the wind turbine height is 120 m. Therefore, the calculation is required to determine the wind speed at the desired height [33]:

$$\frac{\text{WV}_h}{\text{WV}_{10}} = \left(\frac{h}{10}\right)^N \tag{2}$$

Here, WV_h and WV_{10} are the wind speed at heights 'h' and 10 m. N is the Hellman co-efficient (1/7).

2.2. Risk Assessment Parameters (VaR and CVaR)

VaR and CVaR have been chosen as the risk assessment parameters in this work. Other risk assessment tools can also be used, but based on the efficiencies and feasibilities in the field of power systems, these tools have been considered here. CVaR has greater numerical properties than other risk valuation tools. CVaR is also known as a coherent risk measurement tool. The CVaR of a set is a continuous function, whereas the other tools may be discontinuous. The CVaR deviation is a robust contestant to the standard deviation. Under maximum conditions, the standard deviation can be substituted by a CVaR deviation to obtain better results. In risk management, CVaR functions can be performed more efficiently than the other risk assessment tools. CVaR can be optimized with linear programming methods, whereas VaR and other risk-assessing tools are relatively difficult to enhance. CVaR delivers a suitable picture of risks replicated in extreme tails. This is a very significant property if extreme tail losses are properly projected. For these reasons, CVaR is chosen as the risk assessment tool in this work.

Both the considered assessment tools work based on probabilistic studies and the confidence level of assurance (ω). The confidence level of occurrence is 98% for measuring the values of VaR and CVaR. VaR represents the smallest loss with loss amount of $(1 - \omega)$ percentile but CVaR illustrates the average loss mechanisms. $m(x,y)$ is the loss mechanism related to the decision vector P, which is taken from a definite subset x of $\dot{Q}$ and the random vector y in $\dot{Q}$. The probability of loss components $m(x,y)$ is indicated by $n(y)$, which must be ranged with a threshold limit (ζ) [34]:

$$\beta(x,\zeta) = \int_{m(x,y)\leq\zeta} n(y)d\zeta \tag{3}$$

The assurance level-based VaR and CVaR are as follows [34]:

$$\zeta_p(x) = min\left\{\zeta\epsilon\dot{Q} : \beta(x,\zeta)\right\} \tag{4}$$

$$\theta_p(x) = \frac{1}{1-\omega}\left[\left(\sum_{c=1}^{C_a}(n_c-\omega)p_{c_a} + \sum_{c=c_a}^{T} n_c p_c\right] \tag{5}$$

Here, T is the number of trials composed under numerous conditions.

Figure 1 shows a graphical illustration of the risk assessment factors. The maximum negative values of the parameters portray the maximum system risk. Therefore, it is necessary to transfer toward the right-hand side to lessen system loss and system risk.

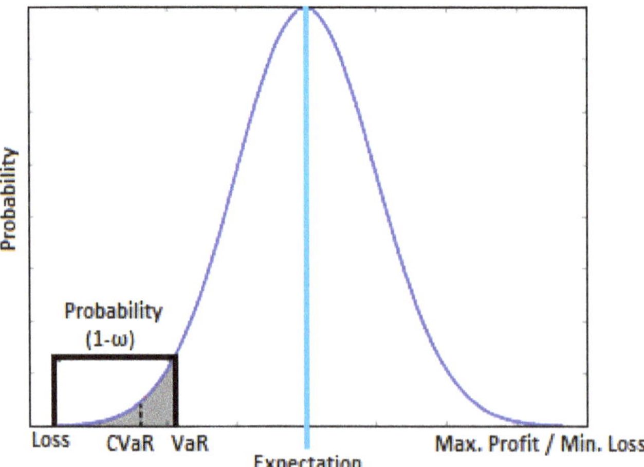

Figure 1. VaR and CVaR Representation.

3. Optimization Techniques

Some nature-inspired optimization techniques are popular in today's world due to their efficacy in solving stability issues for renewable incorporated systems [35]. These include frequency stability [36,37], cost minimization [38], and energy and storage otmimization [39–41]. This unit shows the particulars of the considered optimization techniques, i.e., AGTO, ABC, and SQP. Here, SQP is the linear optimization technique, whereas ABC and AGTO are advanced optimization tools. Other optimization techniques can also be used, but these algorithms are chosen randomly for comparative studies. The AGTO algorithm has been proposed in 2021. Therefore, its application has been displayed here to check the efficiencies of the proposed approach along with the relatively old optimization technique ABC and the linear optimization technique SQP.

The following features of metaheuristic algorithms have created interest among the researchers over the analytic methods: (i) the accuracy is advanced than that of the analytical approaches, (ii) the iteration number is less for metaheuristic methods, (iii) the processes can be simply adapted for the two diode models of solar PV systems with metaheuristic algorithms, and (iv) the recital of the parameters abstraction can be enhanced using meta heuristic algorithms.

The concept of AGTO has been taken from [42], whereas [43] provides a detailed concept of ABC optimization algorithms. Here, MATPOWER software has used to solve the OPF problem using SQP. The SQP is also the same as a simplification of the Newton Raphson Technique, in which non-linear controlled optimization difficulties are answered in steps wise to determine the OPF.

4. Problem Formulation

The aim of this work is to exploit the system profit and diminish the system risk by best location of wind farms, solar PV, and battery storage. Solar PV and batteries are considered backup power sources that are used to lessen the negative influence of imbalance costs in the electricity market environment. The second objective is to impact the valuation of imbalance costs on system profit in a wind-incorporated deregulated power network and to exploit the system profit using the best action of a solar PV-battery storage system.

Objective Function 1:

$$\text{Maximize, } P(t) = R(t) - GC(t) \tag{6}$$

Here, $P(t)$, $R(t)$ and $GC(t)$ are system profit, revenue earned and generation cost at time 't' 1 in \$/h.

$$R(t) = \sum_{m=1}^{NG} PG_r(m,t) \cdot \lambda_{loc}(m,t) \tag{7}$$

$$GC(t) = GC_{Th}(t) + GC_W(t) \tag{8}$$

$$GC_{Th}(t) = \sum_{m=1}^{N_g} \left(a_m + b_m \cdot PG_r(m,t) + c_m \cdot PG_r^2(m,t) \right) \tag{9}$$

Here, '$PGr(m,t)$' is the power generated capacity with RWS at time 't' for bus-m, '$\lambda_{loc}(m,t)$' is the retailing price of generator-m. The system generation cost has two parts: thermal generation cost ($GC_{Th}(t)$) and wind generation cost ($GC_W(t)$). NG is the generator number that is linked to the system. 'a_m', 'b_m' and 'c_m' are the cost co-efficient of generation units.

The scientific expression of the risk-associated objective function is as follows [34]:

$$\text{Max. } \xi_p(x) = min\left\{ \xi \epsilon \dot{Q} : \beta(x,\xi) \right\} \tag{10}$$

$$\text{Max. } \theta_p(x) = \frac{1}{1-\omega} \left[\left(\sum_{c=1}^{C_a} (n_c - \omega) p_{c_a} + \sum_{c=c_a}^{T} n_c p_c \right] \tag{11}$$

Here, the VaR and CVaR have not been included in the optimization techniques directly. After applying the optimization techniques for profit maximization, the system data has been collected. Then, VaR and CVaR are calculated using that data. From Figure 1, it is observed that the system risk will be diminished when VaR and CVaR rise. Therefore, this objective function is taken as a maximization problem.

***Objective Function 2*:**

In this objective function, the concept of imbalance has been introduced along with the first objective function. Now, the scientific expression of this objective is as:

$$\text{Maximize}, P(t) = R(t) + IC(t) - GC(t) \tag{12}$$

Here, $IC(t)$ is the system imbalance cost at time 't'. This is generated due to a mismatch in bidding quantities and actual generated quantities of wind power. ISO imposes a penalty on wind farms for their deficit power supply conditions. Furthermore, ISO provides a reward if surplus power has been supplied to the grid by the wind farm. Imposing a penalty on the wind farm creates '$-ve$' imbalance cost, and rewards provided to the wind firm create '$+ve$' imbalance cost. The solar PV and battery energy storing systems can play a dynamic role in this situation. By providing extra power to the wind farm at the required time, a solar PV-battery hybrid system can alleviate the negative effect of imbalance costs and can exploit the system profit. The mathematical expression of the system imbalance cost is as follows:

$$IC(t) = \sum_{m=1}^{NG}\left(R_S(t) + R_D(t)\cdot\left(\frac{PG_e(m,t)}{PG_r(m,t)}\right)^2\right)\cdot(PG_r(m,t) - PG_e(m,t)) \tag{13}$$

$$R_D(t) = (1+\beta)\cdot\lambda_{loc}(m,t), R_S(t) = 0 \; if \, PG_e(m,t) > PG_r(m,t) \tag{14}$$

$$R_S(t) = (1-\beta)\cdot\lambda_{loc}(m,t), R_D(t) = 0 \; if \, PG_e(m,t) < PG_r(m,t) \tag{15}$$

$$R_S(t) = R_D(t) = 0 \; otherwise \tag{16}$$

Here, '$R_S(t)$' and '$R_D(t)$' are surplus charge rate and deficit charge rate, '$PG_e(m,t)$', '$PG_r(m,t)$' are generated power quantities with expected and real wind speed respectively. 'β' is the system imbalance cost co-efficient. In this work, 'β' is presumed to be 0.8, as this value varies from 0 to 1 [33].

- ***Constraints:***

Equality and inequality constraints have been taken to solve the OPF problem.

$$\sum_{m=1}^{NG} PG_r + GWP - P_{loss} - P_L = 0 \tag{17}$$

$$P_{loss} = \sum_{n=1}^{NTL} G_n\left[|V_p|^2 + |V_q|^2 - 2|V_p||q|cos(\delta_p - \delta_q)\right] \tag{18}$$

$$P_m - \sum_{k=1}^{NB}|V_m V_k Y_{mk}|cos(\theta_{mk} - \delta_m + \delta_k) = 0 \tag{19}$$

$$Q_m + \sum_{k=1}^{NB}|V_m V_k Y_{mk}|sin(\theta_{mk} - \delta_m + \delta_k) = 0 \tag{20}$$

$$V_m^{min} \le V_m \le V_m^{max} \; m = 1,2,3\ldots NB \tag{21}$$

$$\varnothing_m^{min} \le \varnothing_m \le \varnothing_m^{max} \; m = 1,2,3\ldots NB \tag{22}$$

$$TL_l \le TL_l^{max} \; l = 1,2,3\ldots N_{TL} \tag{23}$$

$$P_{Gm}^{min} \le P_{Gm} \le P_{Gm}^{max} \; m = 1,2,3\ldots NB \tag{24}$$

$$Q_{Gm}^{min} \le Q_{Gm} \le Q_{Gm}^{max} \; m = 1,2,3\ldots NB \tag{25}$$

'P_{loss}' and 'PL' are transmission line loss and system loads. N_{TL} is the number of transmission lines. Y_{mk} and θ_{mk} are the magnitude and angle of the m x n-th element of bus admittance matrix. The voltage magnitude are $|V_p|$, $|V_q|$, and V_k for bus p, q, and k. P_m

and Q_m are active and reactive power at bus-m. Gn is transmission line conductance. The voltage angles are δm and δk for bus 'm' and 'k,' respectively. $\phi_m{}^{min}$ and $\phi_m{}^{max}$ are the least and most extreme angle limits at bus 'm'. $V_m{}^{min}$ and $V_m{}^{max}$ are lesser and greater voltage bounds. TL_l and $TL_l{}^{max}$ are actual and extreme line flows. $PG_m{}^{min}$, $PG_m{}^{max}$, $QG_m{}^{min}$, and $QG_m{}^{max}$ are lesser and higher real and reactive power limits. NB is the bus number.

The step-by-step process of the presented work is as follows:

Step 1: Read all system information for the considered test system.

Step 2: Generate 20 different scenarios for creating system congestion by bus outage, line outage, generator outage, and load increment.

Step 3: Calculate the system generation cost, revenue, profit, and system risk (based on NP and LF) without wind placement in the system.

Step 4: Choose the 2 most severe scenarios with the base case from the 20 scenarios based on the values of risk assessment tools.

Step 5: Collect hourly real and expected wind speed data from Kolhapur and Mumbai.

Step 6: Calculate wind power generation and wind power costs.

Step 7: Calculate the system generation cost, revenue, profit, and system risk (based on NP and LF) with wind placement in the system.

Step 8: Calculate the imbalance cost considering wind speed data and compare the system profit with and without the imbalance cost.

Step 9: Place solar PV and battery energy storage systems and check system risk and system profit.

Step 10: Compare system risk and system economy with different optimization techniques.

5. Implementation of the Proposed Approach

A modified IEEE 14-bus and modified IEEE 30-bus system is deliberated to explore the effect in this work. The base MVA is 100 for the system, and bus no. 1 is the reference bus for the IEEE 14-bus system [28,34]. SQP, AGTO, and ABC algorithms have been used to solve the optimal power flow problem. Different scenarios have been taken to examine system performance.

Case 1: Scenario Generation and Finding the Worst Case Based on System Risk (Modified IEEE 14-bus System)

Twenty different scenarios have been generated considering the different types of system instabilities, such as bus failure, generator failure, transmission line failure, and sudden increment of system load. System risk and system generation costs have been calculated on behalf of every chosen case using SQP. Table 1 shows the system economic parameters and system risk for the considered scenarios, along with the base case. The risk assessment tools (i.e., VaR and CVaR) are operated based on system nodal prices (NP) and power flow in the transmission lines (LF).

Table 1 shows that scenarios 9 and 10 are the most severe risky scenario due to their negative highest values of VaR and CVaR, which indicate the minimum profit and maximum losses of the system. The system generation cost depends on several system parameters, including transmission line congestion. When the system is riskier, then the generation cost is also high due to the high congestion cost.

Figure 2 depicts the relation between system generation cost with risk assessment tools for all considered cases. For further study of this work, the most 2 risky scenarios along with the base conditions have been considered. If solar PV and battery storage provide better results for the worst cases, then this method will also provide better results for other cases. Therefore, only three cases have been considered for further studies.

Table 1. Economic parameters and system risk of the system.

Scenario No.	Details	System Generation Cost ($/h)	Revenue ($/h)	Profit ($/h)	VaR on NP	CVaR on NP	VaR on LF	CVaR on LF
1	Base Case	899.09	1226.19648	327.1065	−0.3897	−0.5996	−0.7529	−0.7926
2	Line outage (2–3)	952.57	1368.904	416.334	−0.4719	−0.726	−0.9598	−1.0664
3	Line outage (4–5)	967.36	1405.72038	438.3604	−0.5295	−0.8146	−0.752	−0.8356
4	Generator outage (2)	1094.92	1784.70386	689.7839	−0.604	−0.829	−0.8012	−0.8433
5	Bus_4 (15%)	935.66	1286.4382	350.7782	−0.4009	−0.6168	−0.7536	−0.7932
6	Bus_6 (15%)	906.36	1235.63664	329.2766	−0.3896	−0.5993	−0.7527	−0.7923
7	Line outage (13–14)	1024.48	1583.71197	559.232	−0.8787	−1.3519	−0.9069	−1.0076
8	Bus_11 (15%)	1094.94	1783.79553	688.8555	−0.9822	−1.5111	−0.7702	−0.8108
9	Bus_14 (15%)	1115.49	1821.69013	706.2001	−10.978	−16.889	−0.9718	−1.0229
10	Bus_10 (15%)	1113.42	1817.68268	704.2627	−10.153	−15.623	−0.9901	−1.0422
11	Bus_9 (11%)	1156.23	1919.31698	763.087	−6.4408	−9.9089	−0.8375	−0.8816
12	Bus_2 (11%)	909.85	1242.168	332.318	−0.3817	−0.5872	−0.7532	−0.7929
13	Line outage (1–5)	918.75	1266.423	347.673	−0.4597	−0.7073	−0.7533	−0.837
14	Line outage (10–11)	986.68	1486.184	499.504	−0.8845	−1.3608	−0.9218	−1.0243
15	Line outage (6–12)	902.7	1237.829	335.129	−0.3932	−0.6049	−0.7476	−0.8307
16	Bus_12 (14%)	900.32	1227.485	327.165	−0.3917	−0.6026	−0.7837	−0.8249
17	Bus_3 (14%)	995.92	1405.286	409.366	−0.5162	−0.7941	−0.7536	−0.7933
18	Line outage (12–13)	897.04	1224.889	327.849	−0.3923	−0.6036	−0.7469	−0.8299
19	Line outage (9–14)	895.84	1218.687	322.847	−0.3919	−0.6029	−0.7595	−0.8439
20	Line outage (12–13)	936.58	1332.214	395.634	−0.5027	−0.7734	−0.9511	−1.0567

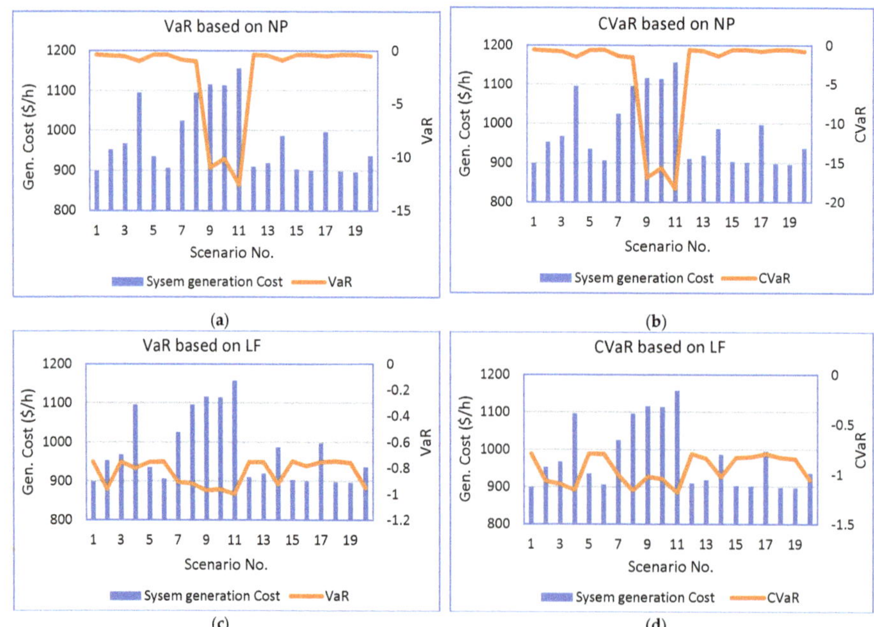

Figure 2. Relation between generation cost and risk assessment tools. VaR based on NP (**a**), CvaR basd on NP (**b**), VaR based on LF (**c**), CvaR based on LF (**d**).

Step 2: System Economy (without imbalance cost) and Risk with Wind Farm Integration (Modified IEEE 14-bus system)

Wind speed varies in every period throughout the world. Considering the flexible nature of wind flow, two places have been considered in India (i.e., Kolhapur and Mumbai) for real-time problem solving. The wind speed data has been taken for 4 different times (6 AM, 12 noon, 6 PM and 12 midnight) for both considered places. Both real and expected wind speed data have been considered for all cases.

Table 2 depicts the real-time wind speed data taken for both considered places with four different periods. The running cost or generation cost of wind power is zero; only the investment cost is present for the wind farm. The average lifetime of the wind farm is 20 years. The approximate wind power investment cost is 3.75 $/MWh, which was taken from [33]. Table 3 depicts the generated wind power quantity and the wind power cost for the considered wind speeds. In this work, it is assumed that the height of the wind turbine is 120 m. Therefore, at first, the wind speed at the considered height is calculated using Equation (2). Then, the generated wind power quantity is measured by Equation (1). Here, 50 wind turbines have been chosen for their series-connected operations.

Table 2. Real-time Wind Speed Data [44].

Sl. No.	Details	Kolhapur		Mumbai	
		RWS (km/h)	EWS (km/h)	RWS (km/h)	EWS (km/h)
1	Base Case_(00.00 h)	8	9	7	8
2	Base Case_(06.00 h)	9	11	8	9
3	Base Case_(12.00 h)	13	13	11	13
4	Base Case_(18.00 h)	8	7	9	9
5	Bus_14 (15%)_(00.00 h)	8	9	7	7
6	Bus_14 (15%)_(06.00 h)	9	11	8	9
7	Bus_14 (15%)_(12.00 h)	13	7	11	8
8	Bus_14 (15%)_(18.00 h)	8	8	9	13
9	Bus_10 (15%)_(00.00 h)	8	9	7	11
10	Bus_10 (15%)_(06.00 h)	9	9	8	7
11	Bus_10 (15%)_(12.00 h)	13	11	11	13
12	Bus_10 (15%)_(18.00 h)	8	11	9	9

Table 3. Wind Power Quantity and Wind Power Generation Cost.

WS at 10 m Height (km/h)	WS at 10 m Height (m/s)	WS at 120 m Height (m/s)	GWP for 1 Turbine (MW)	GWP for 50 Turbines (MW)	Wind Power Cost for 50 Turbines ($/h)
7	1.939	2.7661774	0.031914783	1.595739129	5.984021733
8	2.216	3.1613456	0.047639559	2.381977942	8.932417281
9	2.493	3.5565138	0.067830544	3.391527186	12.71822695
10	2.77	3.951682	0.093046013	4.652300667	17.4461275
11	3.047	4.3468502	0.123844244	6.192212188	23.2207957
12	3.324	4.7420184	0.160783511	8.039175553	30.14690832
13	3.601	5.1371866	0.204422091	10.22110457	38.32914212

Wind power has also been considered in this work as the secondary source of generation beside the thermal power, which is working as the primary energy source. Table 4 displays the system risk, along with the risk assessment parameters, after the placement of the wind farm at bus no. 4 for Kolhapur. Figure 3 shows risk assessment parameter values, which have been calculated based on system NP and transmission line power flow (LF) for base cases of Kolhapur. The system generation costs with wind farms for different considered cases in Kolhapur is shown in Figure 4. From these results, it can be concluded that the highest values of wind farm placement reduce system risk and system generation costs in higher quantities. This happens due to the additional power supply to the grid by the wind farm.

Similar to the previous case (i.e., Kolhapur), in this case (i.e., Mumbai), the impact of wind placement on the system risk has been obtained. The optimal location of a wind farm delivers extra protection to the electrical system by providing the additional power generated. The negative maximum values of VaR and CVaR deliver the minimum profit and maximum risk for the power system shown in Figure 1. It is necessary to shift the values of VaR and CVaR toward the right-hand side to deliver maximum profit. Figures 5–7 show the VaR and CVaR values after placement of the wind farm in modified IEEE 14-bus systems for Mumbai. Four different amounts of wind power are merged into the system to show the variable nature of wind power. From the results, it is understood that after the placement of maximum quantities of wind farms in the system, the system risk is minimized. The same scenario is also observed for system generation costs. The maximum quantities of wind power provide a minimum generation cost-based system. These results directly support the incorporation of wind farms with high capacity in a deregulated power system to mitigate system risks and maximize system profit.

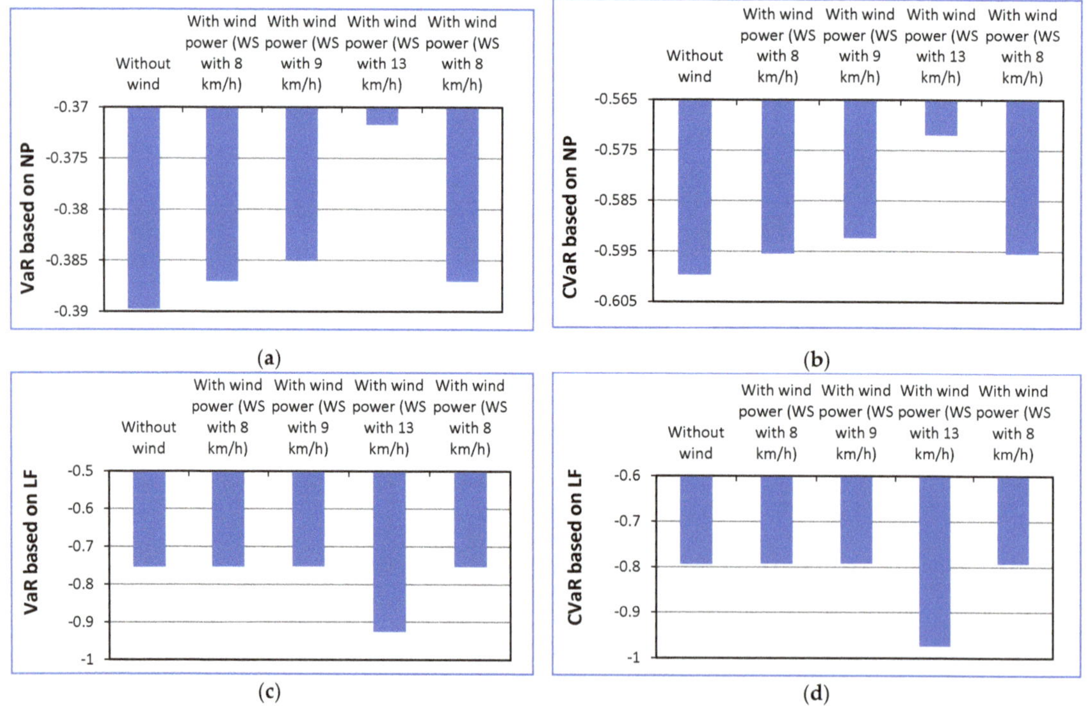

Figure 3. System Risk with Wind Power Integration (Vijayawada @Base Case). VaR based on NP (**a**), CvaR basd on NP (**b**), VaR based on LF (**c**), CvaR based on LF (**d**).

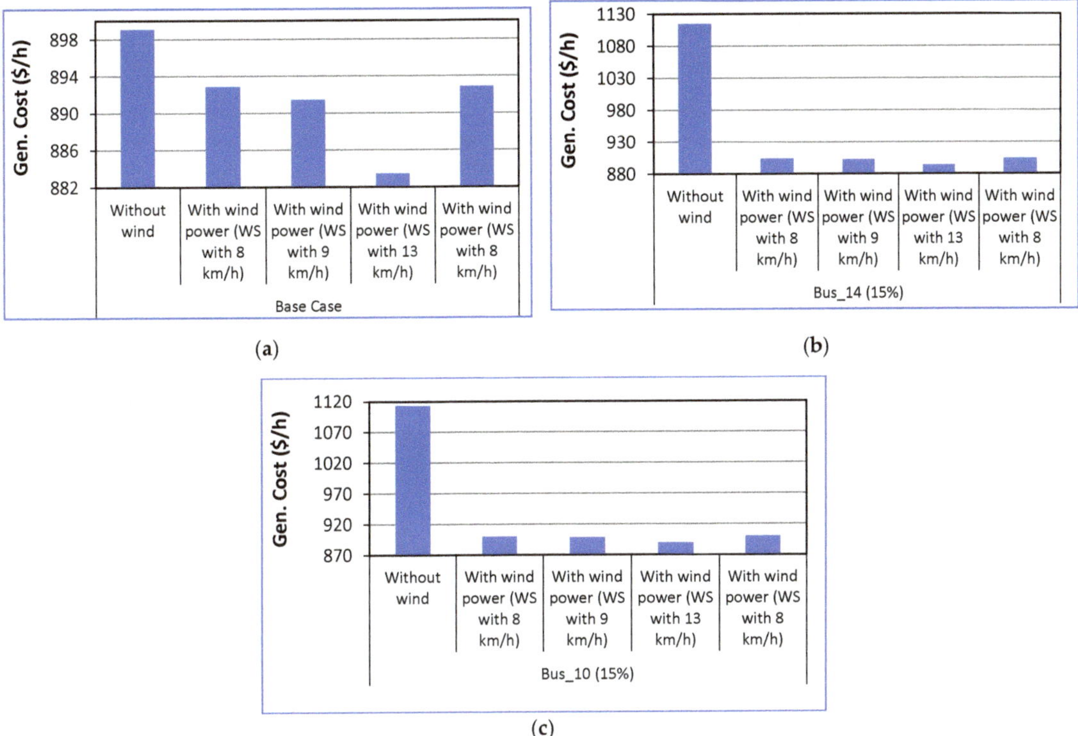

Figure 4. System Generation Costs ($/h) with Wind Power Integration (Kolhapur). Base case (**a**), Bus-14 15% (**b**), Bus_10 15% (**c**).

Table 4. System Risk and Profit with Wind Power Integration (Kolhapur).

Sl. No.	Details	Wind Power (km/h)	Revenue ($/h)	Generation Cost ($/h)	Profit ($/h)	NP		LF	
						VaR	CVaR	VaR	CVaR
1	Base Case_(0.0 h)	8	1212.317	892.902	319.415	−0.387	−0.5954	−0.752	−0.7915
2	Base Case_(6.0 h)	9	1208.188	891.468	316.72	−0.385	−0.5923	−0.7515	−0.7911
3	Base Case_(12.0 h)	13	1179.431	883.47	295.961	−0.3717	−0.5719	−0.9246	−0.9732
4	Base Case_(18.0 h)	8	1212.317	892.902	319.415	−0.387	−0.5954	−0.752	−0.7915
5	Bus_14 (15%)_(0.0 h)	8	1230.963	904.232	326.731	−0.3898	−0.5996	−0.7553	−0.7951
6	Bus_14 (15%)_(6.0 h)	9	1226.665	902.71	323.955	−0.3878	−0.5966	−0.755	−0.7947
7	Bus_14 (15%)_(12.0 h)	13	1196.225	894.14	302.085	−0.3746	−0.5763	−0.8833	−0.9298
8	Bus_14 (15%)_(18.0 h)	8	1230.963	904.232	326.731	−0.3898	−0.5996	−0.7553	−0.7951
9	Bus_10 (15%)_(0.0 h)	8	1226.94	900.482	326.458	−0.3903	−0.6005	−0.7521	−0.7917
10	Bus_10 (15%)_(6.0 h)	9	1222.536	898.931	323.605	−0.3883	−0.5974	−0.7517	−0.7913
11	Bus_10 (15%)_(12.0 h)	13	1191.6	890.13	301.47	−0.375	−0.577	−0.8823	−0.9287
12	Bus_10 (15%)_(18.0 h)	8	1226.94	900.482	326.458	−0.3903	−0.6005	−0.7521	−0.7917

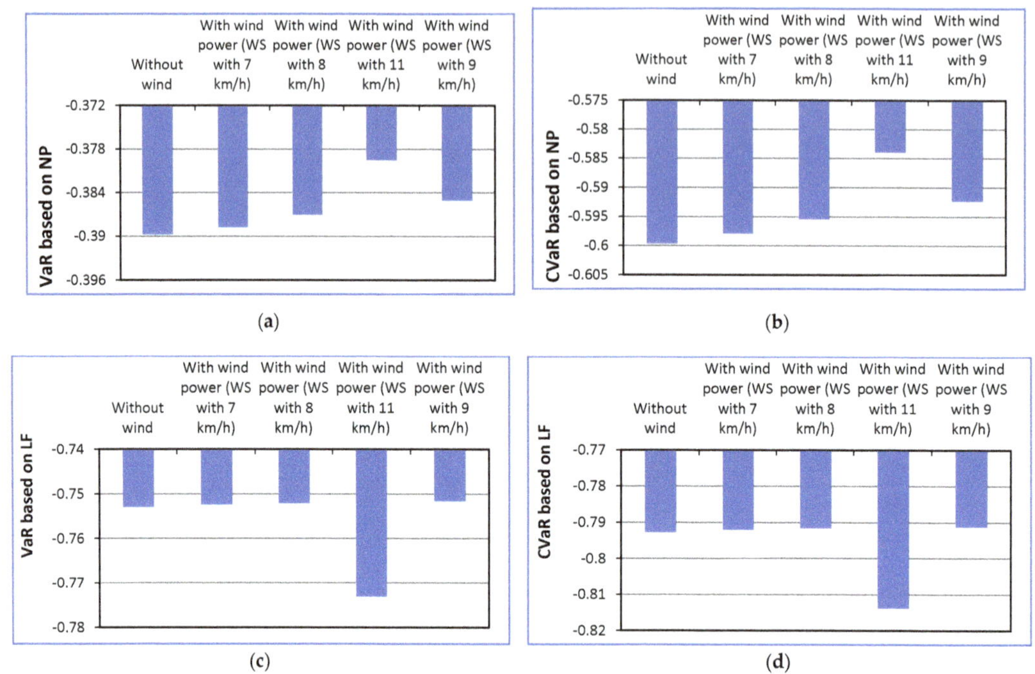

Figure 5. System Risk with Wind Power Integration (Mumbai @Base Case). VaR based on NP (**a**), CvaR basd on NP (**b**), VaR based on LF (**c**), CvaR based on LF (**d**).

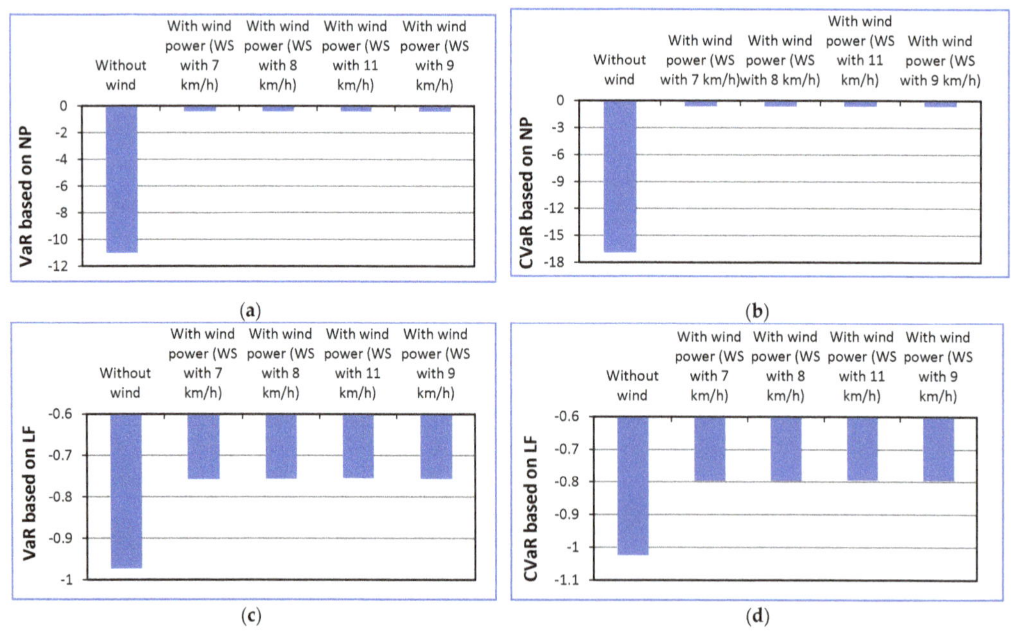

Figure 6. System Risk with Wind Power Integration (Mumbai @ Bus_14 (15%)). VaR based on NP (**a**), CvaR basd on NP (**b**), VaR based on LF (**c**), CvaR based on LF (**d**).

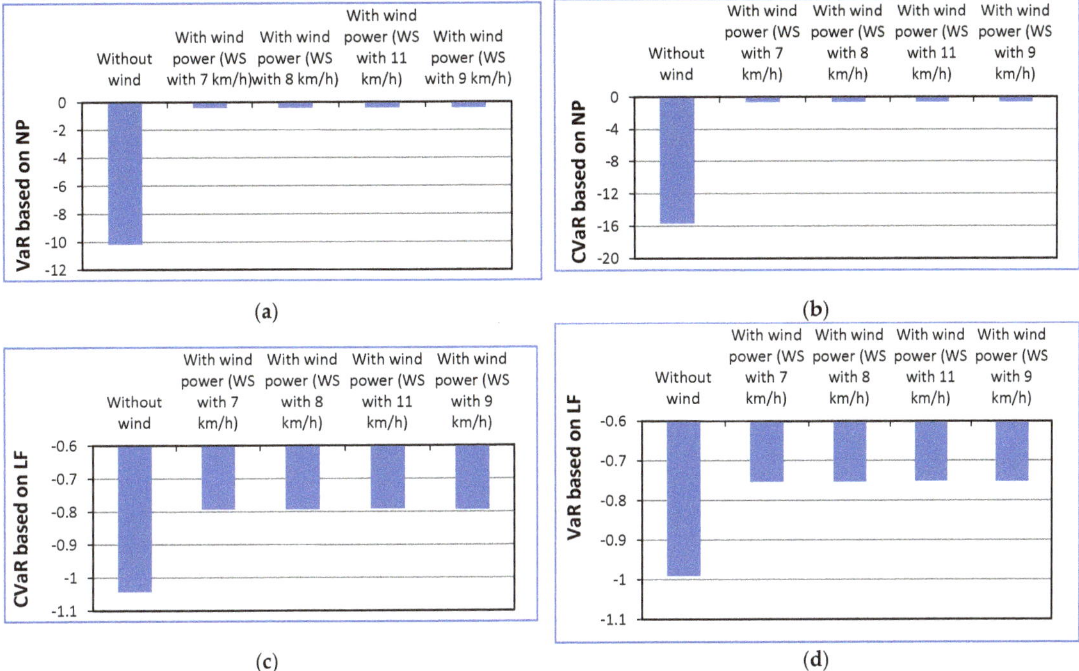

Figure 7. System Risk with Wind Power Integration (Mumbai @ Bus_10 (15%)). VaR based on NP (**a**), CvaR basd on NP (**b**), VaR based on LF (**c**), CvaR based on LF (**d**).

The comparative studies of system generation cost with and without placement of wind farms for the IEEE 14-bus system considering the cases of Mumbai are shown in Figure 8. It can be seen that the generation cost is reduced by a huge amount after the placement of the highest values of wind power in the system. Therefore, it can be concluded that the placement of WF provides risk minimization and generation cost minimization for any power system.

Case 3: With Wind Placement and Considering Imbalance Cost (Modified IEEE 14-bus System)

In the deregulated system, wind farms need to submit the future power generation scenario to the ISO before the date of operation. Based on their submitted data of power, ISO scheduled power generation from different generating stations. In reality, the wind farm cannot generate the scheduled power in maximum cases due to the uncertain nature of the wind flow. The violation of market contracts can impose an economic burden (i.e., imbalance cost) on the generating companies. The imbalance cost directly affects the system economy. When RWP is more than the EWP, then ISO gives rewards to the wind farm for their surplus power supply; however, ISO imposes a penalty if EWP is more than RWP. Thus, the adverse effect of imbalance costs directly disturbs the economic advancement of the market players. Here, the imbalance cost is calculated for every considered variation in expected and real wind speeds. The imbalance cost of the system reflects the mismatch between the predicted and real wind speed data. The imbalance cost is maximum when the difference between expected and real wind speed is maximum. When the expected wind speed is large compared to the real wind speed, the deficit charge rate arises, and when the real wind speed is larger than the expected wind speed, the surplus charge rate occurs. The deficit and surplus charge rates are zero for that case when the expected and real wind speeds are the same. Using the deficit and surplus charge rates, the total imbalance cost of the electrical system can be calculated. The imbalance cost is '−ve' when

ISO imposes the penalty on the generating station for their deficit supply of power from renewable sources. However, the imbalance cost is '+ve' when ISO provides the reward to the generating station for their surplus supply of power from renewable energy sources. Here, the imbalance cost is calculated for every variation in the expected and real wind speeds using the formula stated in Equations (13)–(16). Both expected and real wind speed data have been taken for Kolhapur and Mumbai, Maharashtra to check the effectiveness of the proposed method. Tables 5 and 6 depict the profit comparison considering the imbalance costs for Kolhapur and Mumbai, respectively.

The impact of imbalance cost on system profit has shown in Figures 9 and 10 for Kolhapur and Mumbai respectively. In the last considered case, the real wind speed is 8 km/h whereas the expected wind speed was 11 km/h. This is the maximum amount of mismatch in wind speed. Therefore, at this hour, the impact of imbalance cost is also high.

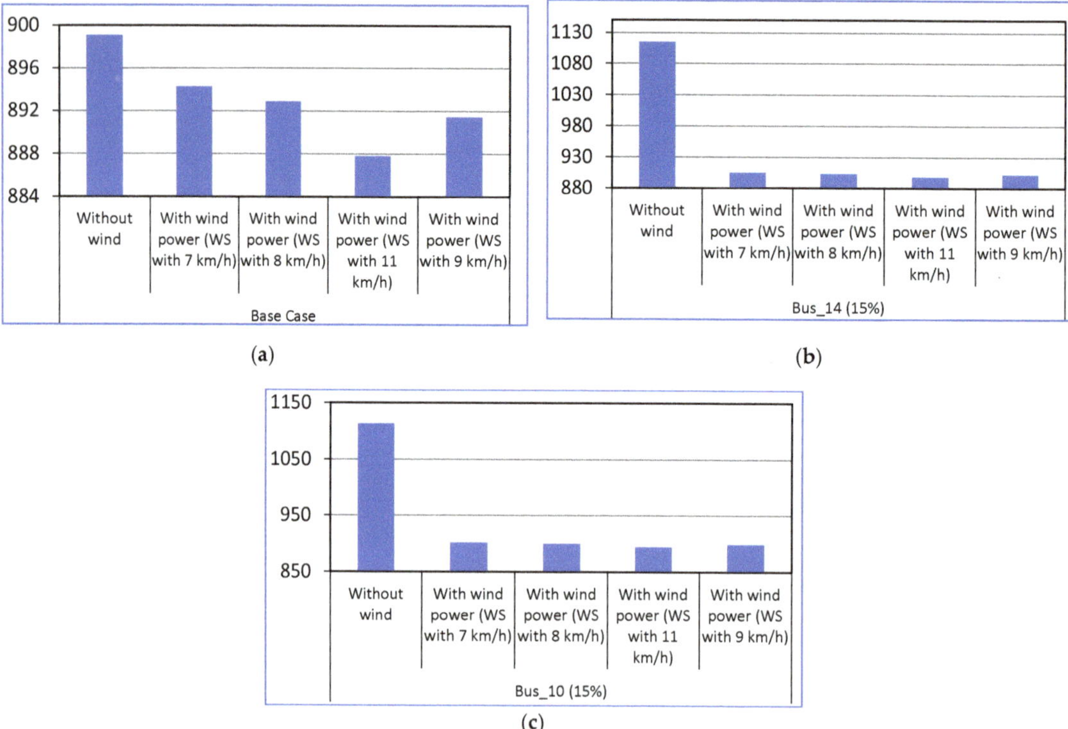

Figure 8. System Generation Cost ($/h) with Wind Power Integration (Mumbai). Base case (**a**), Bus-14 15% (**b**), Bus_10 15% (**c**).

Table 5. Profit Comparison with Imbalance Cost (Kolhapur).

Sl. No.	Details	Generation Cost ($/h)	Revenue ($/h)	Profit without Imbalance Cost ($/h)	Real Wind Power (km/h)	Expected Wind Power (km/h)	Imbalance Cost ($/h)	Profit with Imbalance Cost ($/h)
1	Base Case_(00.00 h)	892.902	1212.317	319.415	8	9	−9.601	309.814
2	Base Case_(06.00 h)	891.468	1208.188	316.72	9	11	−27.07	289.65
3	Base Case_(12.00 h)	883.47	1179.431	295.961	13	13	0	295.961
4	Base Case_(18.00 h)	892.902	1212.317	319.415	8	7	1.602	321.017
5	Bus_14(15%)_(00.00 h)	904.232	1230.963	326.731	8	9	−9.6017	317.1293
6	Bus_14(15%)_(06.00 h)	902.71	1226.665	323.955	9	11	−27.0724	296.8826
7	Bus_14(15%)_(12.00 h)	894.14	1196.225	302.085	13	7	3.0779	305.1629
8	Bus_14(15%)_(18.00 h)	904.232	1230.963	326.731	8	8	0	326.731
9	Bus_10(15%)_(00.00 h)	900.482	1226.94	326.458	8	9	−16.3831	310.0749
10	Bus_10(15%)_(06.00 h)	898.931	1222.536	323.605	9	9	0	323.605
11	Bus_10(15%)_(12.00 h)	890.13	1191.6	301.47	13	11	1.8957	303.3657
12	Bus_10(15%)_(18.00 h)	900.482	1226.94	326.458	8	11	−65.58	260.878

Table 6. Profit Comparison with Considering Imbalance Cost (Mumbai).

Sl. No.	Details	Generation Cost ($/h)	Revenue ($/h)	Profit without Imbalance Cost ($/h)	Real Wind Power (km/h)	Expected Wind Power (km/h)	Imbalance Cost ($/h)	Profit with Imbalance Cost ($/h)
1	Base Case_(00.00 h)	894.28	1216.005	314.295	7	8	−10.43	311.295
2	Base Case_(06.00 h)	892.902	1212.317	319.415	8	9	−12.89	306.525
3	Base Case_(12.00 h)	887.8	1196.124	308.324	11	13	−51.03	257.294
4	Base Case_(18.00 h)	891.468	1208.188	316.72	9	9	0	316.72
5	Bus_14(15%)_(00.00 h)	905.42	1234.303	328.883	7	7	0	328.883
6	Bus_14(15%)_(06.00 h)	904.232	1230.963	326.731	8	9	−13.33	313.401
7	Bus_14(15%)_(12.00 h)	898.83	1215.162	316.332	11	8	9.0246	325.3566
8	Bus_14(15%)_(18.00 h)	902.718	1226.655	323.937	9	13	−94.0949	229.8421
9	Bus_10(15%)_(00.00 h)	901.71	1230.382	328.672	7	11	−61.5737	267.0983
10	Bus_10(15%)_(06.00 h)	900.48	1226.94	326.46	8	7	2.6938	329.1538
11	Bus_10(15%)_(12.00 h)	894.98	1210.795	315.815	11	13	−57.936	257.879
12	Bus_10(15%)_(18.00 h)	898.93	1222.536	323.606	9	9	0	323.606

Case 4: Solar PV-Battery Operation with SQP, ABC, and AGTO (Modified IEEE 14-bus System)

After a detailed study of the first 3 cases, it is found that the system imbalance cost is very dangerous for the economic operation of generating stations. If the imbalance cost is positive, the profit of the generation unit increases, but the system profit is lower for the negative imbalance cost. In this scenario, solar PV and battery hybrid systems play a vital role in mitigating the mismatch between real and expected wind power and minimizing dependency on the thermal power plant. Environmental benefits can also be obtained by using a solar PV-battery system. To check the effectiveness of the presented method, three different optimization techniques have been used.

The solar PV-battery storage system has been placed on bus no. 9 with a supply capacity of 2 MW. Here, the modeling of solar PV-battery storage systems has not been considered. Only a fixed value of generated power from a solar PV-battery storage system has been chosen. The placement bus has been considered at 9 due to the large load connected to that particular bus. Tables 7 and 8 show the comparative studies of system profit and system risk with different optimization techniques. For risk assessment studies, some selected cases have been considered for Kolhapur. From both tables, it is observed that AGTO techniques provide the best results among all considered cases by providing accurate optimal settings to minimize the system risk and maximize the system profit. Therefore, it can be concluded that the solar PV-battery storage system can reduce the negative impact of imbalance costs in the system economy and minimize system risk.

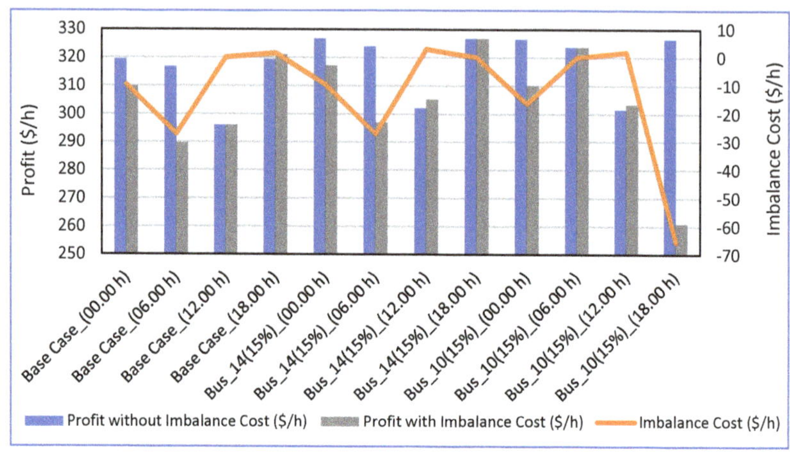

Figure 9. Profit Comparison with Imbalance Cost (Kolhapur).

Case 5: Solar PV-Battery Operation with SQP, ABC, and AGTO for Kolhapur (Modified IEEE 30-bus system)

Similar to the modified IEEE 14-bus system, the impact assessment of solar PV and battery storage systems has also been investigated for the modified IEEE 30-bus system. The system data has been taken from [33]. Only the base case of Kolhapur with four time intervals (i.e., 0.0 h, 6.0 h, 12.0 h, and 18.0 h) has been studied for the modified IEEE 30-bus system. The same wind speeds have been considered here as the 14-bus system. A combined amount of 2 MW of power from a solar PV-battery storage system has been chosen.

Tables 9 and 10 show comparative studies of system profit and system risk with different optimization techniques. For risk assessment studies, base cases have been considered for Kolhapur. From both tables, it is observed that AGTO techniques offer the finest results among all considered cases by providing accurate optimal settings. Therefore, it can be concluded that the solar PV-battery storage system can also reduce the negative impact of imbalance costs in the modified IEEE 30-bus system in terms of profit maximization. The system risk has also been minimized after the placement of solar PV and battery storage systems.

Table 7. System profit with different optimization techniques.

Sl. No.	Details	Kolhapur				Mumbai			
		Profit with Imbalance Cost ($/h) Using SQP without Solar PV-Battery	Profit with Imbalance Cost ($/h) Using SQP with Solar PV-Battery	Profit with Imbalance Cost ($/h) Using ABC with Solar PV-Battery	Profit with Imbalance Cost ($/h) Using AGTO with Solar PV-Battery	Profit with Imbalance Cost ($/h) Using SQP without Solar PV-Battery	Profit with Imbalance Cost ($/h) Using SQP with Solar PV-Battery	Profit with Imbalance Cost ($/h) Using ABC with Solar PV-Battery	Profit with Imbalance Cost ($/h) Using AGTO with Solar PV-Battery
1	Base Case_(00.00 h)	309.814	313.265	318.256	319.165	311.295	315.354	320.651	321.425
2	Base Case_(06.00 h)	289.65	293.698	299.032	300.168	306.525	310.265	315.321	316.521
3	Base Case_(12.00 h)	295.961	299.021	304.658	305.785	257.294	261.954	266.357	267.462
4	Base Case_(18.00 h)	321.017	325.964	330.254	331.457	316.72	320.547	325.835	327.125
5	Bus_14(15%)_(00.00 h)	317.1293	321.238	326.265	327.158	328.883	332.154	337.254	338.652
6	Bus_14(15%)_(06.00 h)	296.8826	300.982	305.954	307.054	313.401	317.561	322.647	323.851
7	Bus_14(15%)_(12.00 h)	305.1629	309.347	314.325	315.647	325.3566	329.617	334.247	335.324
8	Bus_14(15%)_(18.00 h)	326.731	330.657	335.931	337.265	229.8421	233.991	238.487	239.623
9	Bus_10(15%)_(00.00 h)	310.0749	314.0535	320.654	321.781	267.0983	271.364	276.954	278.126
10	Bus_10(15%)_(06.00 h)	323.605	327.835	333.254	334.438	329.1538	333.614	338.725	339.957
11	Bus_10(15%)_(12.00 h)	303.3657	307.215	312.657	313.981	257.879	261.983	266.587	267.754
12	Bus_10(15%)_(18.00 h)	260.878	264.325	270.258	271.435	323.606	327.751	332.652	333.723

Table 8. System risk with different optimization techniques (for Kolhapur).

Sl. No.	Details	NP							
		VaR				CVaR			
		With Wind Farm Using SQP	With Wind Farm-Solar PV-Battery Storage Using SQP	With Wind Farm-Solar PV-Battery Storage Using ABC	With Wind Farm-Solar PV-Battery Storage Using AGTO	With Wind Farm Using SQP	With Wind Farm-Solar PV-Battery Storage Using SQP	With Wind Farm-Solar PV-Battery Storage Using ABC	With Wind Farm-Solar PV-Battery Storage Using AGTO
1	Base Case_(12.0 h)	−0.3717	−0.3615	−0.3525	−0.3419	−0.5719	−0.5526	−0.5416	−0.5316
2	Bus_14 (15%)_(12.0 h)	−0.3746	−0.3634	−0.3548	−0.3439	−0.5763	−0.5586	−0.5474	−0.5357
3	Bus_10 (15%)_(12.0 h)	−0.375	−0.3647	−0.3559	−0.3442	−0.577	−0.5591	−0.5465	−0.5368

Table 9. System profit with different optimization techniques (for Kolhapur).

Sl. No.	Details	Kolhapur			
		Profit with Imbalance Cost ($/h) Using SQP without Solar PV-Battery	Profit with Imbalance Cost ($/h) Using SQP with Solar PV-Battery	Profit with Imbalance Cost ($/h) Using ABC with Solar PV-Battery	Profit with Imbalance Cost ($/h) Using AGTO with Solar PV-Battery
1	Base Case_(00.00 h)	324.268	329.658	335.627	336.751
2	Base Case_(06.00 h)	297.685	302.685	308.776	310.021
3	Base Case_(12.00 h)	307.247	312.654	318.168	319.685
4	Base Case_(18.00 h)	341.038	346.658	352.951	354.237

Table 10. System risk with different optimization techniques (for Kolhapur).

Sl. No.	Details	NP							
		VaR				CVaR			
		With Wind Farm Using SQP	With Wind Farm-Solar PV-Battery Storage Using SQP	With Wind Farm-Solar PV-Battery Storage Using ABC	With Wind Farm-Solar PV-Battery Storage Using AGTO	With Wind Farm Using SQP	With Wind Farm-Solar PV-Battery Storage Using SQP	With Wind Farm-Solar PV-Battery Storage Using ABC	With Wind Farm-Solar PV-Battery Storage Using AGTO
1	Base Case_(12.0 h)	−0.3735	−0.3667	−0.3521	−0.3434	−0.5753	−0.5567	−0.5423	−0.5334
2	Bus_14 (15%)_(12.0 h)	−0.3748	−0.3675	−0.3536	−0.3445	−0.5771	−0.5574	−0.5446	−0.5348
3	Bus_10 (15%)_(12.0 h)	−0.3759	−0.3686	−0.3571	−0.3462	−0.5792	−0.5585	−0.5475	−0.5359

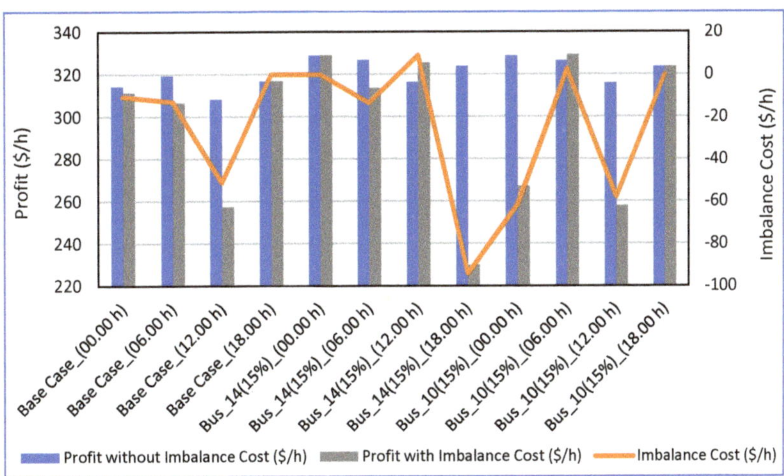

Figure 10. Profit Comparison with Imbalance Cost (Mumbai).

6. Conclusions

The hybrid effect of solar PV, wind farms, and battery storage systems on power system economics has been considered in this paper. To authenticate the task, a deregulated power market was considered. To evaluate system risk, Value-at-risk (VaR) and Cumulative Value-at-risk (CVaR) have been used here. The stability and safety of an electrical system can be improved by minimizing system risk. The placement of solar PV and battery storage systems minimizes the negative impact of system imbalance costs, which are developed due to the disparity between the bidding and running wind power quantities. The hybrid system minimizes system risk, which can further reduce the instability conditions of the system. To examine the effect of system risk considering wind farms with solar PV-battery storage systems under deregulated power systems, comparative studies are conducted using different optimization techniques, such as the Artificial Gorilla Troops Optimizer Algorithm (AGTO), Artificial Bee Colony Algorithms (ABC), and Sequential Quadratic Programming (SQP). The modified IEEE 14-bus and modified IEEE 30-bus systems have been used here to analyze the efficiency and robustness of the presented work. As evident from the results, the presence of a solar PV-battery storage system with a wind farm improves the economic parameters of the system by reducing the system risk. The Artificial Gorilla Troops Optimizer Algorithm (AGTO) has been used for the first time in this kind of risk mitigation problem, which is the uniqueness of this paper. This work can be performed with different renewable energy sources and energy storage devices in the near future for any small or large electrical system.

Author Contributions: Conceptualization, G.S.P., S.D. and A.M.; methodology, G.S.P., S.D., A.M. and T.S.U.; software, G.S.P.; validation, G.S.P., A.M. and T.S.U.; formal analysis, T.S.U.; investigation, G.S.P. and S.D.; resources, G.S.P., A.M. and T.S.U.; data curation, T.S.U.; writing—original draft preparation, G.S.P. and S.D.; writing—review and editing, T.S.U.; visualization, G.S.P., A.M. and T.S.U.; supervision, G.S.P.; project administration, G.S.P., A.M. and T.S.U.; funding acquisition, T.S.U. All authors have read and agreed to the published version of the manuscript.

Funding: This research received no external funding.

Conflicts of Interest: The authors declare no conflict of interest.

References

1. Ustun, T.S.; Aoto, Y. Analysis of Smart Inverter's Impact on the Distribution Network Operation. *IEEE Access* **2019**, *7*, 9790–9804.
2. Farooq, Z.; Rahman, A.; Hussain, S.S.; Ustun, T.S. Power Generation Control of Renewable Energy Based Hybrid Deregulated Power System. *Energies* **2022**, *15*, 517.
3. Public Utility Regulatory Policies Act of 1978 (PURPA), Office of Electricity, U.S. Department of Energy. Available online: https://www.energy.gov/oe/services/electricity-policy-coordination-and-implementation/other-regulatory-efforts/public (accessed on 1 March 2022).
4. Latif, A.; Paul, M.; Das, D.C.; Hussain, S.S.; Ustun, T.S. Price Based Demand Response for Optimal Frequency Stabilization in ORC Solar Thermal Based Isolated Hybrid Microgrid under Salp Swarm Technique. *Electronics* **2020**, *9*, 2209.
5. Ustun, T.S.; Aoto, Y.; Hashimoto, J.; Otani, K. Optimal PV-INV Capacity Ratio for Residential Smart Inverters Operating Under Different Control Modes. *IEEE Access* **2020**, *8*, 116078–116089.
6. Hussain, S.S.; Nadeem, F.; Aftab, M.A.; Ali, I.; Ustun, T.S. The Emerging Energy Internet: Architecture, Benefits, Challenges, and Future Prospects. *Electronics* **2019**, *8*, 1037.
7. GM Abdolrasol, M.; Hannan, M.A.; Hussain, S.S.; Ustun, T.S.; Sarker, M.R.; Ker, P.J. Energy Management Scheduling for Microgrids in the Virtual Power Plant System Using Artificial Neural Networks. *Energies* **2021**, *14*, 6507.
8. Ustun, T.S.; Ozansoy, C.; Zayegh, A. Extending IEC 61850-7-420 for distributed generators with fault current limiters. In Proceedings of the 2011 IEEE PES Innovative Smart Grid Technologies, Perth, Australia, 13–16 November 2011; pp. 1–8.
9. Kumar, K.K.P.; Soren, N.; Latif, A.; Das, D.C.; Hussain, S.S.; Al-Durra, A.; Ustun, T.S. Day-Ahead DSM-Integrated Hybrid-Power-Management-Incorporated CEED of Solar Thermal/Wind/Wave/BESS System Using HFPSO. *Sustainability* **2022**, *14*, 1169.
10. Arango-Aramburo, S.; Bernal-García, S.; Larsen, E.R. Renewable energy sources and the cycles in deregulated electricity markets. *Energy* **2021**, *223*, 120058.
11. Yao, X.; Yi, B.; Yu, Y.; Fan, Y.; Zhu, L. Economic analysis of grid integration of variable solar and wind power with conventional power system. *Appl. Energy* **2020**, *264*, 114706.
12. Ustun, T.S.; Nakamura, Y.; Hashimoto, J.; Otani, K. Performance analysis of PV panels based on different technologies after two years of outdoor exposure in Fukushima, Japan. *Renew. Energy* **2019**, *136*, 159–178.
13. Hou, S.; Yi, B.W.; Zhu, X. Potential economic value of integrating concentrating solar power into power grids. *Comput. Ind. Eng.* **2021**, *160*, 107554.
14. Kumar, A.; Bag, B. Incorporation of probabilistic solar irradiance and normally distributed load for the assessment of ATC. In Proceedings of the 2017 8th International Conference on Computing, Communication and Networking Technologies (ICCCNT) IEEE Transactions on Power Systems, Delhi, India, 3–5 July 2017; Volume 13, pp. 1–6.
15. Reddy, S.S. Optimal scheduling of thermal-wind-solar power system with storage. *Renew. Energy* **2017**, *101*, 1357–1368.
16. Nadeem, F.; Aftab, M.A.; Hussain, S.S.; Ali, I.; Tiwari, P.K.; Goswami, A.K.; Ustun, T.S. Virtual Power Plant Management in Smart Grids with XMPP Based IEC 61850 Communication. *Energies* **2019**, *12*, 2398.
17. Reddy, S.S.; Bijwe, P.R. Day-Ahead and Real Time Optimal Power Flow considering Renewable Energy Resources. *Electr. Power Energy Syst.* **2016**, *82*, 400–408.
18. Gope, S.; Goswami, A.K.; Tiwari, P.K. Transmission congestion management using a wind integrated compressed air energy storage system. *Eng. Technol. Appl. Sci. Res.* **2017**, *7*, 1746–1752.
19. Xu, X.; Ye, Z.; Qian, Q. Economic, exergoeconomic analyses of a novel compressed air energy storage-based cogeneration. *J. Energy Storage* **2022**, *51*, 104333. [CrossRef]
20. Sun, S.; Kazemi-Razi, S.M.; Kaigutha, L.G.; Marzband, M.; Nafisi, H.; Al-Sumaiti, A.S. Day-ahead offering strategy in the market for concentrating solar power considering thermoelectric decoupling by a compressed air energy storage. *Appl. Energy* **2022**, *305*, 117804. [CrossRef]
21. Chang, Y.C.; Lee, T.Y.; Chen, C.L.; Jan, R.M. Optimal power flow of a wind-thermal generation system. *Electr. Power Energy Syst.* **2014**, *55*, 312–320.
22. Xu, Z.; Hu, Z.; Song, Y.; Wang, J. Risk-Averse Optimal Bidding Strategy for Demand-Side Resource Aggregators in Day-Ahead Electricity Markets Under Uncertainty. *IEEE Trans. Smart Grid* **2017**, *8*, 96–105.
23. Matevosyan, J.; Soder, L. Minimization of Imbalance Cost Trading Wind Power on the Short-Term Power Market. *IEEE Trans. Power Syst.* **2016**, *21*, 1396–1404.
24. Dawn, S.; Tiwari, P.K.; Goswami, A.K. An approach for efficient assessment of the performance of double auction competitive power market under variable imbalance cost due to high uncertain wind penetration. *Renew. Energy* **2017**, *108*, 230–248.
25. Wu, J.; Zhang, B.; Jiang, Y.; Bie, P.; Li, H. Chance-constrained stochastic congestion management of power systems considering uncertainty of wind power and demand side response. *Electr. Power Energy Syst.* **2019**, *107*, 703–714.
26. Rubin, O.D.; Babcock, B.A. The impact of expansion of wind power capacity and pricing methods on the efficiency of deregulated electricity markets. *Energy* **2013**, *59*, 676–688.
27. Das, A.; Dawn, S.; Gope, S.; Ustun, T.S. A Strategy for System Risk Mitigation Using FACTS Devices in a Wind Incorporated Competitive Power System. *Sustainability* **2022**, *14*, 8069. [CrossRef]
28. Patil, G.S.; Mulla, A.; Ustun, T.S. Impact of Wind Farm Integration on LMP in Deregulated Energy Markets. *Sustainability* **2022**, *14*, 4354. [CrossRef]

29. Khamees, A.K.; Abdelaziz, A.Y.; Eskaros, M.R.; El-Shahat, A.; Attia, M.A. Optimal Power Flow Solution of Wind-Integrated Power System Using Novel Metaheuristic Method. *Energies* **2021**, *14*, 6117. [CrossRef]
30. Xu, Y.; Lang, Y.; Wen, B.; Yang, X. An Innovative Planning Method for the Optimal Capacity Allocation of a HybridWind–PV–Pumped Storage Power System. *Energies* **2019**, *12*, 2809. [CrossRef]
31. Dawn, S.; Tiwari, P.K.; Goswami, A.K.; Panda, R. An Approach for System Risk Assessment and Mitigation by Optimal Operation of Wind Farm and FACTS Devices in a Centralized Competitive Power Market. *IEEE Trans. Sustain. Energy* **2018**, *10*, 1054–1065. [CrossRef]
32. Singh, N.K.; Koley, C.; Gope, S.; Dawn, S.; Ustun, T.S. An Economic Risk Analysis in Wind and Pumped Hydro Energy Storage Integrated Power System Using Meta-Heuristic Algorithm. *Sustainability* **2013**, *13*, 13542. [CrossRef]
33. Dawn, S.; Tiwari, P.K.; Goswami, A.K. An approach for long term economic operations of competitive power market by optimal combined scheduling of wind turbines and FACTS controllers. *Energy* **2019**, *181*, 709–723.
34. Das, A.; Dawn, S.; Gope, S.; Ustun, T.S. A Risk Curtailment Strategy for Solar PV-Battery Integrated Competitive Power System. *Electronics* **2022**, *11*, 1251. [CrossRef]
35. Abdolrasol, M.G.M.; Hussain, S.M.S.; Ustun, T.S.; Sarker, M.R.; Hannan, M.A.; Mohamed, R.; Ali, J.A.; Mekhilef, S.; Milad, A. Artificial Neural Networks Based Optimization Techniques: A Review. *Electronics* **2021**, *10*, 2689.
36. Latif, A.; Hussain, S.S.; Das, D.C.; Ustun, T.S. Double stage controller optimization for load frequency stabilization in hybrid wind-ocean wave energy based maritime microgrid system. *Appl. Energy* **2021**, *282*, 116171.
37. Latif, A.; Hussain, S.S.; Das, D.C.; Ustun, T.S. Optimum Synthesis of a BOA Optimized Novel Dual-Stage PI − (1 + ID) Controller for Frequency Response of a Microgrid. *Energies* **2020**, *13*, 3446.
38. Singh, S.; Chauhan, P.; Aftab, M.A.; Ali, I.; Hussain, S.S.; Ustun, T.S. Cost Optimization of a Stand-Alone Hybrid Energy System with Fuel Cell and PV. *Energies* **2020**, *13*, 1295.
39. Dey, P.P.; Das, D.C.; Latif, A.; Hussain, S.S.; Ustun, T.S. Active Power Management of Virtual Power Plant under Penetration of Central Receiver Solar Thermal-Wind Using Butterfly Optimization Technique. *Sustainability* **2020**, *12*, 6979.
40. Chauhan, A.; Upadhyay, S.; Khan, M.T.; Hussain, S.S.; Ustun, T.S. Performance Investigation of a Solar Photovoltaic/Diesel Generator Based Hybrid System with Cycle Charging Strategy Using BBO Algorithm. *Sustainabiliy* **2021**, *13*, 8048.
41. Hussain, I.; Das, D.C.; Sinha, N.; Latif, A.; Hussain, S.S.; Ustun, T.S. Performance Assessment of an Islanded Hybrid Power System with Different Storage Combinations Using an FPA-Tuned Two-Degree-of-Freedom (2DOF) Controller. *Energies* **2020**, *13*, 5610.
42. Abdollahzadeh, B.; Gharehchopogh, F.S.; Mirjalili, S. Artificial gorilla troops optimizer: A new Nature-inspired Metaheuristic Algorithm for Global Optimization Problems. *Int. J. Intell. Syst.* **2021**, *36*, 5887–5958.
43. Karaboga, D.; Basturk, B. Artificial Bee Colony (ABC) Optimization Algorithm for Solving Constrained Optimization Problems. In Proceedings of the 12th International Fuzzy Systems Association World Congress on Foundations of Fuzzy Logic and Soft Computing, Cancun, Mexico, 18–21 June 2007; Springer: Berlin/Heidelberg, Germany, 2007; Volume 4529, pp. 789–798.
44. Database. World Temperatures & Weather around the World. Available online: www.timeanddate.com/weather/ (accessed on 29 May 2022).

energies

MDPI

Article

Arc Furnace Power-Susceptibility Coefficients

Zbigniew Olczykowski

Faculty of Transport, Electrical Engineering and Computer Science, Kazimierz Pulaski University of Technology and Humanities, Malczewskiego 29, 26-600 Radom, Poland; z.olczykowski@uthrad.pl

Abstract: The article presents the susceptibility coefficients active power k_p and reactive power k_q, as proposed by the author. These coefficients reflect the reaction of arc furnaces (change of the furnace operating point) to supply voltage fluctuations. The considerations were based on the model of the arc device in which the electric arc was replaced with a voltage source with an amplitude dependent on the length of the arc. In the case of voltage fluctuations, such a model gives an assessment of the arc device's behavior closer to reality than the model used, based on replacing the arc with resistance. An example of the application of the k_p and k_q coefficients in a practical solution is presented.

Keywords: arc furnaces; power–voltage characteristics; voltage fluctuations; power quality

Citation: Olczykowski, Z. Arc Furnace Power-Susceptibility Coefficients. *Energies* **2022**, *15*, 5508. https://doi.org/10.3390/en15155508

Academic Editors: Frede Blaabjerg and Ahmed Abu-Siada

Received: 13 June 2022
Accepted: 25 July 2022
Published: 29 July 2022

Publisher's Note: MDPI stays neutral with regard to jurisdictional claims in published maps and institutional affiliations.

1. Introduction

The article proposes power susceptibility factors: active power k_p and reactive power k_q, determined on the basis of quasi-static power–voltage characteristics. These coefficients were determined by replacing the arc of an electric arc furnace with a voltage with a value that depends on the length of the electric arc. In analytical research, many models of arc devices can be distinguished. These include classic models based on nonlinear differential equations using the Mayra and Cassie equations [1,2], models using a voltage source varying in time defined as a dependent nonlinear function on the length of the electric arc [3–5], models based on a series-connected resistor and inductance [6,7], and models using current–voltage characteristics [8–10].

In the work on arc devices, simulation studies are also using computer programs based, among others, on neural networks [11–20]. However, this requires the use of complex electric arc models and advanced simulation programs. This greatly limits the practical application of these algorithms. The article proposed a certain compromise resulting from the advancement of the model, the accuracy of the obtained results, and its practical application.

In the publication [21] to date, the assessment of the interaction of arc loads was based on the adoption of generally used static power–voltage characteristics marked in the article with the superscript (*):

$$Q^* = f(U^*) \text{ and } P^* = f(U^*) \tag{1}$$

For small changes in the supply voltage, the characteristics are, in the vicinity of the rated voltage U_n, approximately linear dependencies, determined by the slope coefficients of the static characteristics. These factors are also referred to as the receiver power susceptibility factors for active power k_p^* and reactive power k_q^* [21]:

$$k_p = \frac{dP^*}{dU^*} = \frac{dP}{dU} \cdot \frac{U_n}{P_n} \quad k_q = \frac{dQ^*}{dU^*} = \frac{dQ}{dU} \frac{U_n}{Q_n} \tag{2}$$

where P_n and Q_n mean active and reactive power consumed by the load under rated conditions (for an arc device, it means operation at rated voltage and current).

For practical considerations, the use of the active power susceptibility coefficients—k_p and reactive power—k_q allows for a direct assessment of the relative change in power consumed by the receiver with a relative voltage change. The exact values of the coefficients are obtained experimentally by determining the power–voltage static characteristics of active and reactive power. Carrying out the appropriate measurements requires a lot of work and is not always possible for technical reasons.

In the case of an arc device, for the most commonly used electrode control system, based on keeping the arc resistance constant (more precisely on maintaining a constant arc voltage to arc current ratio [22]), the entire circuit of the arc device has a constant impedance character. As a result, for slight voltage changes, one can take:

$$k_p^* = 2 \ \text{ and } \ k_q^* = 2 \tag{3}$$

The difficulties in determining the k_p and k_q coefficients are evidenced by the data presented in the publication [21], where, based on Canadian research, the values $k_p = 2.3$ and $k_q = 4.6$ were given for the electric arc furnace. Especially the last, unexpectedly high value raises reservations as to the correctness of assuming the constant arc resistance. The arc model adopted in the article makes it possible to explain this discrepancy.

2. Assumptions Adopted for the Calculation of Power–Voltage Characteristics

The electric arc supply system is presented by means of a simplified single-phase diagram of a substitute arc device together with the supply network. The circuit diagram is shown in Figure 1.

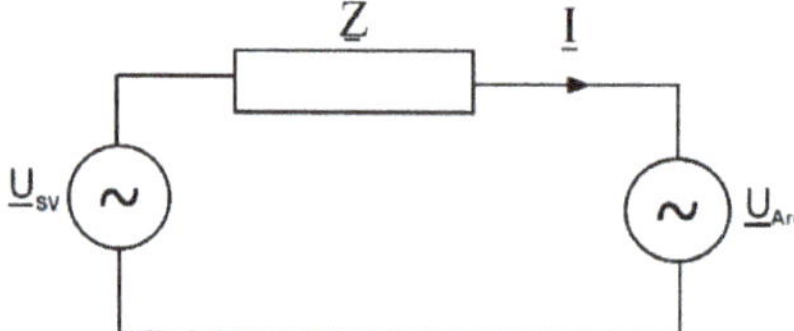

Figure 1. Simplified single-phase equivalent diagram of the arc device.

The phase voltage of the supply network (equivalent voltage supplying the arc) was taken as a reference value and marked as: $U'_{SV} = 100\%$ (it corresponds to the no-load voltage—with an interrupted arc). The index (') at the top of the symbols means operation with a supply voltage equal to 100%, (similarly marked—I', P'_{Arc}, Q'). The index (") used in the following text means operation at a voltage different from the rated $U''_{SV} \neq 100\%$ (marked as—I'', P''_{Arc}, Q''). The equivalent impedance of the arc supply network includes the parameters (resistances and reactances) of such elements as: a steelworks supply network with a power transformer, furnace transformer (with a choke), and high-current circuit consisting of a flexible part, bus bars, and electrodes: $Z = R + jX$ (it corresponds to the impedance of the arc supply circuit when the electrodes are short-circuited with the charge). The reactance value was assumed for the calculations in the amount of $X = 50\%$, in the amount given, e.g., in [21] for furnaces with a capacity of 50–200 Mg with transformers 21–80 MVA. The values of the assumed reactance and supply voltage correspond to the theoretical value of the operational short-circuit current (with the resistance determined from the omitted formula):

$$I'_{SC} = \frac{U'_S}{X} = \frac{100}{50} \cdot 100\% = 200\% \tag{4}$$

The electric arc was mapped using an ideal source of sinusoidal voltage with the amplitude value depending on the arc length. It is the fundamental harmonic of the square wave of the arc voltage. The proposed model derives from the most frequently proposed one in the literature, a nonlinear arc model in which the arc is represented by a voltage source with the value U_{Arc}, depending on the arc length l_{Arc}:

$$U_{Arc} = a + bl_{Arc} \tag{5}$$

where

a, b—denote constants,
l_{Arc}—arc length,
and with polarity according to the polarity of the arc current:

$$U_{Arc} = U_{Arc}sign(i_{Arc}) \tag{6}$$

As a result, a rectangular arc voltage waveform is obtained in each half of the period. The adopted arc model refers to the multi-voltage model presented by the author in the publication [23], consisting of mapping the arc with a system of series-connected voltage sources of higher harmonics:

$$u_{Arc} = u_{\sin}(t) + u_{rec}(t) = U_{\sin}\sin(\omega t) + U_{rec}sign(i) = bU_{Arc1}\sin(\omega t) + \frac{(1-b)\pi}{4}U_{Arc1}sign(i) \tag{7}$$

which, in relation to the amplitude of the fundamental harmonic of the arc voltage, and for $b = 1$, we obtain a sinusoidal waveform, adopted in the model proposed in the article:

$$u_{Arc} = bU_{Arc}\sin(\omega t) \tag{8}$$

For the assumed constant arc length (which may change, for example, as a result of electrode movement related to the operation of the electrode position regulator, sliding of the melted scrap, movement of the arc along the electrode and charge surface, etc.), the effective value of the arc voltage is determined depending on: voltage power supply—U_{SV}, resistance—R, and reactance—X of the arc supply circuit, and arc current I from the formula:

$$U'_{Arc} = \sqrt{U'^2_S - I^2 X^2} - IR \tag{9}$$

The relationship $U_{Arc} = f(I)$ is graphically illustrated in Figure 2, presenting the waveforms determined for $X = 50\%$ and two values of the resistance of the arc-supplying circuit in the amount of $R = 5\%$ and $R = 10\%$, as a function of the current consumed by the arc device expressed as a percentage of the rated current.

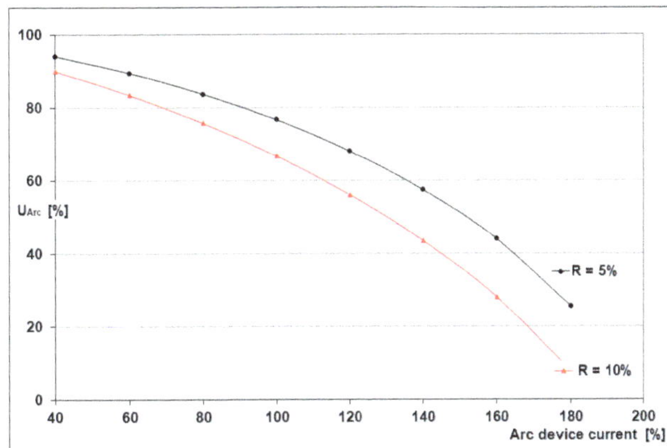

Figure 2. Changes in the arc voltage as a function of the arc device current for two different resistance values of the arc-supplying circuit.

3. Power–Voltage Characteristics of the Arc Device Taking into Account the Resistance of the Circuit Supplying the Arc

Figure 3 shows a diagram of the arc supply system, taking into account the resistance in the supply circuit.

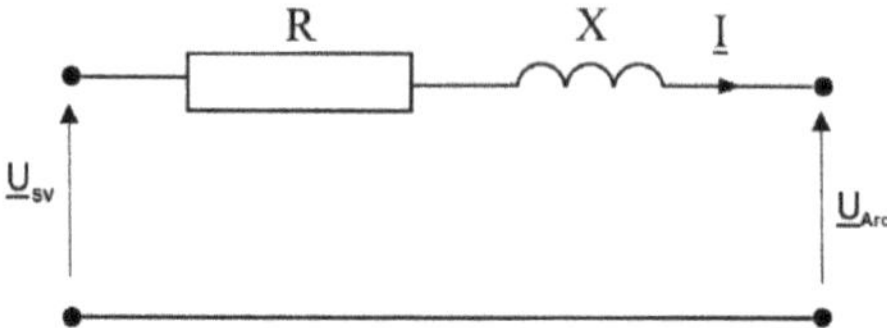

Figure 3. Single-phase equivalent diagram of the arc furnace taking into account the resistance of the arc-supplying circuit.

The resistance value was adopted as 0.1 and 0.2 of the supply reactance value, i.e., $R = 5\%$ and $R = 10\%$. For the resistances and reactances specified in this way, $X = 50\%$ (with $U'_{SV} = 100\%$), and the service short-circuit current I_{SC} is:

$$I_{SC5} = \frac{U'_{SV}}{\sqrt{(R^2 + X^2)}} = 199.00\% \text{ for } R = 5\% \tag{10}$$

$$I_{SC10} = \frac{U'_{SV}}{\sqrt{(R^2 + X^2)}} = 196.11\% \text{ for } R = 10\% \tag{11}$$

The arc voltage at the supply voltage $U'_{SV} = 100\%$, and current I (depending on the operating point of the arc device) is determined from the Formula (9).

For the rated current $U'_{SV} = 100\%$ (at rated voltage), it was determined:

- arc voltage U'_{Arc}:

$$U'_{Arc} = \sqrt{U'^2_S - I^2 X^2} - IR = 76.6026\% \tag{12}$$

- arc resistance r'_{Arc}:

$$r'_{Arcn} = \sqrt{\frac{U'^2_S}{I'^2_n} - X^2} - R = 76.60\% \tag{13}$$

- arc power P'_{Arcn}:

$$P'_{Arcn} = I'^2_n \cdot r'_{Arcn} = 76.6025\% \tag{14}$$

- active power drawn from the source P'_n:

$$P'_n = I'^2_n \cdot (r'_{Arcn} + R) = 86.6025\% \tag{15}$$

- reactive power Q'_n:

$$Q'_n = I'^2_n \cdot X = 50\% \tag{16}$$

- apparent power S'_n:

$$S'_n = I'_n \cdot U'_n = \sqrt{P'^2_n + Q'^2} = 100\% \tag{17}$$

The determined values will be helpful in determining the slope coefficients of the power–voltage characteristics of the active power and reactive power. Assuming that the value of the arc voltage depends only on its length, when changing the supply voltage from $U'_{SV} = 100\%$ to $U''_{SV} \neq 100\%$, the value of the current consumed by the arc device will change from I' to I'': Formula (18):

$$I'' = \frac{-R \cdot U'_{Arc} + \sqrt{(R^2 \cdot U'^2_{Arc} + (R^2 + X^2) \cdot (U''^2_{Arc} - U'^2_{Arc}))}}{(R^2 + X^2)} \tag{18}$$

was determined from the formula:

$$(U'_{Arc} + I'' \cdot R)^2 + (I'' X)^2 = U''^2_f \tag{19}$$

$$U'^2_{Arc} + 2U_L \cdot I'' \cdot R + (R^2 + X^2) I''^2 = U''^2_{SV} \tag{20}$$

Similarly, for the voltage $U''_{SV} \neq 100\%$, the following were determined:

- arc power P''_{Arc}:

$$P''_{Arc} = I'' \cdot U'_{Arc} \tag{21}$$

- active power losses $\Delta P''$:

$$\Delta P'' = I''^2 R \tag{22}$$

- active power supplied to the electric arc-supplying circuit P'':

$$P'' = P''_{Arc} + \Delta P'' \tag{23}$$

- reactive power supplied to the electric arc-supplying circuit Q''

$$Q'' = I''^2 \cdot X \tag{24}$$

As a result of changes in the current of the arc furnace, the supply voltage of the furnace changes. Figure 4 shows the dependence of changes in the supply voltage of the steel mill as a function of changes in the current of the arc furnace. Two exemplary levels of voltage changes between Point (H) and Point (L) are marked. The presented data was recorded in the power supply network of the steel plant. In the further part of the article, changes in the k_p and k_q coefficients resulting from changes in the supply voltage $\Delta U = U_{SVH} - U_{SVL}$ are considered. In Figure 4, blue is the color of the measured values of the supply voltage for a given value of the arc furnace current. The red color represents the trend line.

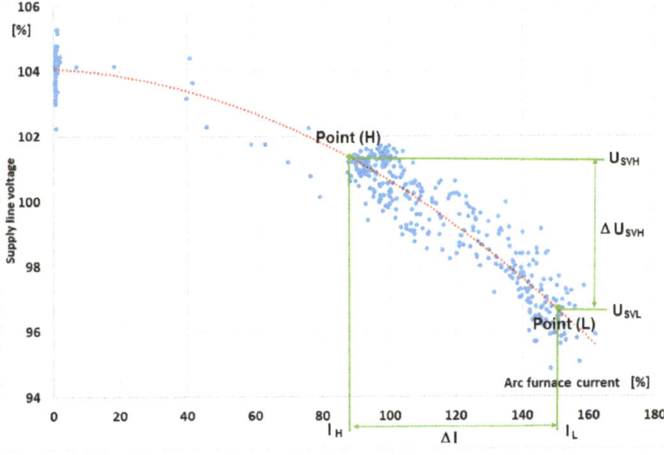

Figure 4. Voltage changes as a function of furnace current changes.

With changes in the supply voltage U_{SV} resulting from voltage fluctuations caused by arc devices—from:

$$U''_{SVB} = U'_{SV} + \Delta U/2 \tag{25}$$

to

$$U''_{SVA} = U'_{SV} - \Delta U/2 \tag{26}$$

the inclination coefficients of the power–voltage characteristics are as follows:

$$k'_p = \frac{\dfrac{\Delta P''}{P'}}{\Delta U} \tag{27}$$

and

$$k'_q = \frac{\dfrac{\Delta Q''}{Q'}}{\Delta U} \tag{28}$$

where $\Delta P'' = P_L - P_H$—active power difference, $\Delta Q'' = Q_L - Q_H$—reactive power difference.

The following part of the article presents an example of determining the k_p and k_q coefficients for the furnace current $I_M = 120\%$ (which corresponds to the charge-melting current $I_M = 1.2\,I_n$), reactance $X = 50\%$. and main resistance $R = 10\%$.

For the rated voltage $U'_{SV} = 100\%$, the following will be determined:

- arc voltage U'_{Arc}:

$$U'_{Arc} = \sqrt{U'^2_{SV} - I^2 X^2} - IR = 68\% \tag{29}$$

- electric arc power P'_{Arc}:

$$P'_{Arc} = I^2 \cdot r'_{Arc} = I^2\left(\sqrt{\frac{U'^2_{SV}}{I^2} - X^2} - R\right) = 81.6\% \tag{30}$$

- active power P':

$$P' = I^2 \cdot (r'_{Arc} + R) = 96\% \tag{31}$$

- reactive power Q':

$$Q' = I^2 \cdot X = 72\% \tag{32}$$

Assuming a constant value of the arc voltage with changes in the supply voltage (as a result of its fluctuations caused, for example, by a restless working receiver), Formulas (24) and (25), the current consumed by the arc device will change. For the assumed voltage fluctuations in the amount of: $\Delta U = 1\%$, $U''_{SVH} = 100 + 0.5\%$, and $\Delta U = 10\%$ $U''_{SVL} = 100 + 5\%$ (two ranges of voltage changes were introduced to analyze the influence of the fluctuations on the value of the k_p and k_q coefficients), the current will change, for $\Delta U = 1\%$—from:

$$I^{-0.5} = \frac{-R \cdot U'_{Arc} + \sqrt{(R^2 \cdot U'^2_{Arc} + (R^2 + X^2) \cdot (U''^2_{SV} - U'^2_{Arc}))}}{(R^2 + X^2)} = 118.681\% \tag{33}$$

to

$$I^{+0.5} = \frac{-R \cdot U'_{Arc} + \sqrt{(R^2 \cdot U'^2_{Arc} + (R^2 + X^2) \cdot (U''^2_{SV} - U'^2_{Arc}))}}{(R^2 + X^2)} = 121.313\% \tag{34}$$

The arc force values at these currents are:

$$P^{-0.5}_{Arc} = U_{Arc} \cdot I^{-0.5} = 80.704\% \text{ and } P^{+0.5}_{Arc} = U_{Arc} \cdot I^{+0.5} = 82.493\% \tag{35}$$

In addition, the changes in active power are:

$$P^{-0.5} = U_{Arc}I^{-0.5} + \left(I^{-0.5}\right)^2 R = 94.788\% \text{ and } P^{+0.5} = U_{Arc}I^{+0.5} + \left(I^{+0.5}\right)^2 R = 97.209\% \tag{36}$$

hence:

$$\Delta P^{\pm 0.5} = P^{+0.5} - P^{-0.5} = 2.421\% \tag{37}$$

Moreover, the changes in reactive power are:

$$Q^{-0.5} = \left(I^{-0.5}\right)^2 X = 70.427\% \quad Q^{+0.5} = \left(I^{+0.5}\right)^2 X = 73.584\% \tag{38}$$

hence:

$$\Delta Q^{\pm 0.5} = Q^{+0.5} - Q^{-0.5} = 3.157\% \tag{39}$$

The k_p and k_q coefficients corresponding to the above changes amount to:

$$k_p^{\pm 0.5} = \frac{\dfrac{\Delta P^{\pm 0.5}}{P'}}{\Delta U\%} 100\% = \frac{\dfrac{2.421}{96}}{1\%} 100\% = 2.522 \tag{40}$$

$$k_q^{\pm 0.5} = \frac{\dfrac{\Delta Q^{\pm 0.5}}{Q^o}}{\Delta U\%} 100\% = \frac{\dfrac{3.158}{72}}{1\%} 100\% = 4.386 \tag{41}$$

Correspondingly, for voltage fluctuations, $\Delta U_2 = 10\%$ of the current value will be: $I^{-5} = 106.552\%$ and $I^{+5} = 132.916\%$; arc power: $P^{-5Arc} = 72.455\%$ and $P^{+5Arc} = 90.383\%$; while changes in active power are $\Delta P^{+5} = P^{+5} - P^{-5} = 24.241\%$; and for reactive power are $\Delta Q^{+5} = Q^{+5} - Q^{-5} = 31.566\%$, while the resulting slope coefficients of the power–voltage characteristics are $k_p^{+5} = 2.525$ and $k_q^{+5} = 4.384$. The obtained values of the k_p and k_q coefficients are similar to the different ranges of voltage changes ($\Delta U_1 = 1\%$ and $\Delta U_2 = 10\%$) for a given average current (e.g., $I_M = 120\%In$) consumed by the arc device—Table 1.

Table 1. Summary of the inclination coefficients of the power–voltage characteristics for two different ranges of voltage fluctuations.

I	ΔU—1%U		ΔU—10%U	
[%]	k_p	k_q	k_p	k_q
40	13.15	25.24	13.79	23.93
60	7.22	13.58	7.33	13.43
80	4.66	8.57	4.69	8.54
100	3.31	5.94	3.32	5.94
120	2.52	4.39	2.53	4.38
140	2.03	3.39	2.03	3.39
160	1.72	2.72	1.72	2.72
180	1.59	2.25	1.59	2.25

Figure 5 summarizes the obtained values of the k_p and k_q coefficients as a function of the current consumed by the arc device.

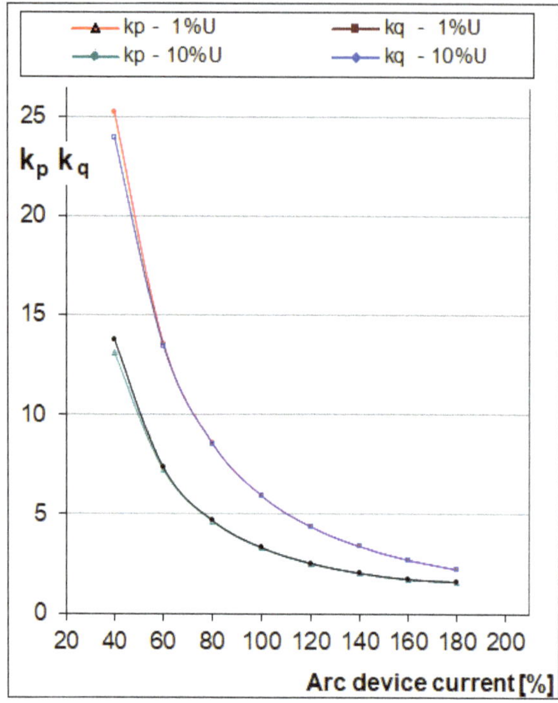

Figure 5. Changes in the coefficients of power–voltage characteristics as a function of the current consumed by the arc device.

In the manner described above, the k_p and k_q coefficients were calculated for different resistances of the arc-supplying circuit. Table 2 and Figure 6 show the obtained results for the resistance $R = 10\%$, which corresponds to the ratio $R/X = 0.2$ (at $X = 50\%$) and, additionally, $R = 20\%$, which corresponds to the ratio $R/X = 0.4$ with 1% changes in the supply voltage $\Delta U_1 = 1\%$. Table 2 and Figures 6 and 7 show the changes of k_p and k_q as a function of the current consumed by the arc furnace for $R = 0$ (resistance omitted—red).

Table 2. Summary of the inclination coefficients of the power–voltage characteristics for various resistances of the circuit supplying the arc.

I [%]	k_p			k_q		
	$R = 0$	$R = 10\%X$	$R = 20\%X$	$R = 0$	$R = 10\%X$	$R = 20\%X$
40	26.2650	13.1483	8.7790	50.0000	25.2444	16.8917
60	11.8736	7.2200	5.1802	22.2222	13.5822	9.7808
80	6.8505	4.6603	3.5230	12.5000	8.5716	6.5222
100	4.5229	3.3140	2.6075	8.0000	5.9417	4.7258
120	3.2682	2.5220	2.0479	5.5556	4.3859	3.6232
140	2.5385	2.0273	1.6869	4.0816	3.3899	2.8987
160	2.1307	1.7210	1.4550	3.1250	2.7174	2.4038
180	2.1030	1.5903	1.3365	2.4691	2.2511	2.0684

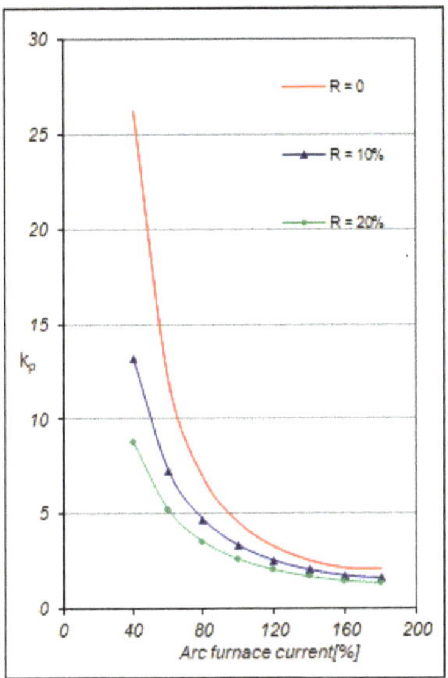

Figure 6. Changes in the kp factor of the power–voltage characteristics for various arc supply resistances, as a function of the current consumed by the arc device.

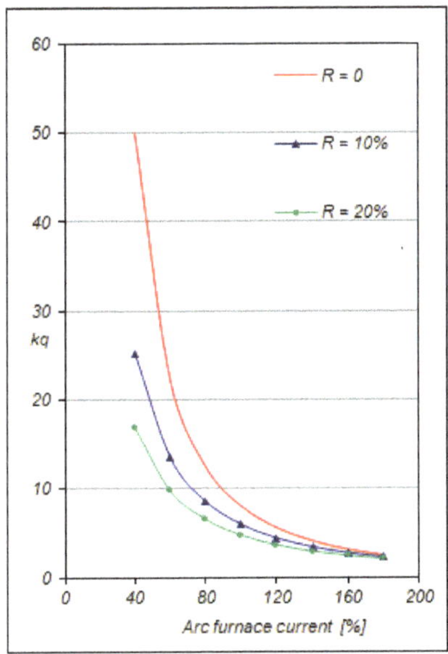

Figure 7. Changes in the k_q factor of the power–voltage characteristics for different arc supply resistances, as a function of the current consumed by the arc device.

Figures 6 and 7 confirm the obvious fact that voltage fluctuations are more suppressed in the LV and MV networks, where the resistance to reactance ratio is greater than in the case of the HV and LV networks.

4. Power–Voltage Characteristics of the Arc Device, Omitting the Resistance of the Circuit Supplying the Electric Arc of the Arc Furnace

In the following part, considerations are presented without the electric arc supply-circuit resistance. These assumptions correspond to the single-phase equivalent diagram of the arc device (furnace including mains) shown in Figure 8.

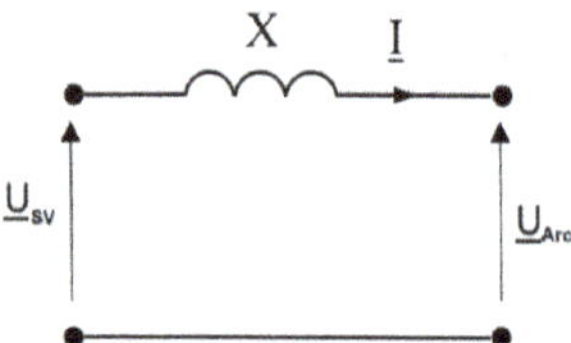

Figure 8. Simplified single-phase equivalent diagram of the arc device, omitting the resistance in the electric arc supply line.

The value of the current in the circuit during operation—at a certain value of UArc, is determined from the formula:

$$\underline{I} = \frac{U_{SV} - U_{ARC}}{jX} \text{ or } I = \frac{\sqrt{U_{SV}^2 - U_{Arc}^2}}{X} \tag{42}$$

the theoretical operational short-circuit current (electrode short-circuit with the charge) is determined by the relationship:

$$\underline{I}_{SC} = \frac{U_{SV}}{jX} \text{ or } I_{SC} = \frac{U_{SV}}{X} \tag{43}$$

Active power, disregarding the losses in the supply circuit ($R = 0$), will be the power released in the arc.

$$P \approx P_{Arc} = U_{Arc}I \tag{44}$$

and reactive power:

$$Q = I^2 X = \frac{U_{SV}^2 - U_{Arc}^2}{X} \tag{45}$$

The value of the current at the operational short circuit of the electrodes with the charge is:

$$I_{SC}' = \frac{U_{SV}'}{X} = 200\% \tag{46}$$

The rated current of the furnace transformer $I_n = 100\%$ and the melting current $I_M = 120\%I_n$ were assumed as the values of currents significant for determining the conditions of the rational operation of the arc furnace. Additionally, the value of the maximum efficiency current I_M, ensuring the maximization of the power released in the arcs and the maximum reduction in the time required for melting the scrap—even at the cost of overloading the arc device. W, this value, disregarding losses in the supply circuit $R = 0$, is determined by the relationship (for calculations, $I_M = 140\%I_n$ was assumed)—Table 3:

$$I_M' = \frac{I_{SC}'}{\sqrt{2}} = 141.42\% \tag{47}$$

The assumptions are made for the value of the arc voltage when working with rated current is:

$$U'_{Arcn} = \sqrt{U'^2_{SV} - I_n^2 X^2} = 86.603\% \tag{48}$$

power released in the arc:

$$P'_{Arcn} = I_n U'_{Arcn} = 86.60\% \tag{49}$$

and the reactive power of the supply circuit:

$$Q'_n = I_n^2 X = 50.0\% \tag{50}$$

Table 3. Changes in the value of the arc voltage—U_{Arc}, reactive power consumed by the arc-supplying circuit—Q, arc power—P_{Arc}, for different values of the current consumed by the arc device.

I	U_{Arc}	Q	P_{Arc}
[%]	[%]	[%]	[%]
0	100	0	0
20	99.5	0.2	1.99
40	97.98	0.8	3.92
60	95.39	1.8	5.72
80	91.65	3.2	7.33
100	86.6	5	8.66
120	80	7.2	9.6
140	71.41	9.8	10
160	60	12.8	9.6
180	43.59	16.2	7.85
200	0	20	0

Figure 9 shows the operating characteristics of the arc voltage as well as active and reactive power (for the supply voltage $U'_{SV} = 100\%$).

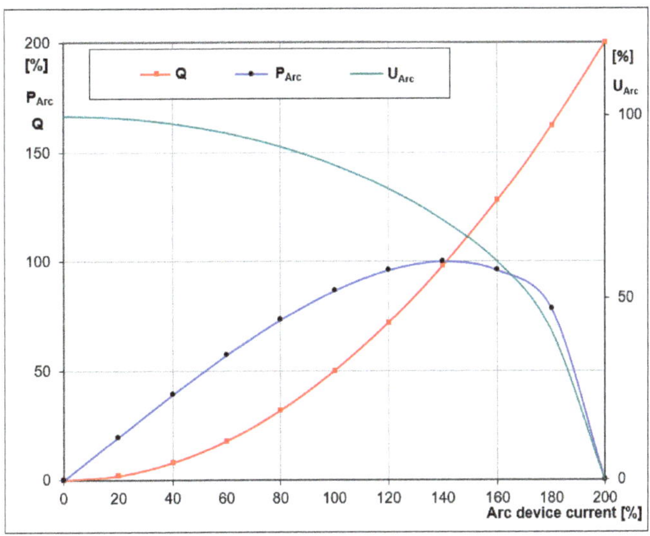

Figure 9. Operating characteristics of an ideal arc device: ($U'_{SV} = 100\%$, $X = 50\%$, $R = 0$).

In the event of changes in the supply voltage (from the value of U'_{SV} to the value of U''_{SV}), but while maintaining constant arc voltage, the arc current as well as the active and reactive power change.

The arc current is determined by the relationship:

$$I = \frac{\sqrt{U_{SV}^2 - U_{Arc}^2}}{X} \tag{51}$$

which at a constant value of the arc voltage U_{arc} = const, and a constant value of the circuit reactance X = const leads to the formulas determining the values of the currents: I' at rated voltage U'_{SV} and I'' at a voltage different from the rated U''_{SV}:

$$I' = \frac{\sqrt{U'_{SV} - U_{Arc}^2}}{X} \text{ and } I'' = \frac{\sqrt{U''^2_{SV} - U_{Arc}^2}}{X} \tag{52}$$

hence:

$$\frac{I''}{I'} = \frac{\sqrt{U''^2_{SV} - U_{Arc}^2}}{\sqrt{U'^2_{SV} - U_{Arc}^2}} \tag{53}$$

we have:

$$P'_{Arc} = U_{Arc} I' \text{ and } P''_{Arc} = U_{Arc} I'' \tag{54}$$

we obtain:

$$P''_{Arc} = U_{Arc} I'' = U_{Arc} I' \sqrt{\frac{U''^2_{SV} - U_{Arc}^2}{U'^2_{Arc} - U_{Arc}^2}} = P'_{Arc} \sqrt{\frac{U''^2_{SV} - U_L^2}{U'^2_{SV} - U_{Arc}^2}} \tag{55}$$

For reactive powers we have:

$$Q'' = I''^2 X = I'^2 X \left(\sqrt{\frac{U''^2_{SV} - U_{Arc}^2}{U'^2_{SV} - U_{Arc}^2}} \right)^2 = Q' \frac{U''^2_{SV} - U_{Arc}^2}{U'^2_{SV} - U_{Arc}^2} \tag{56}$$

The changed values are:

$$I'' = I' \sqrt{\frac{U''^2_{SV} - U_{Arc}^2}{U'^2_{SVf} - U_{Arc}^2}} \tag{57}$$

$$P''_{Arc} = P'_{Arc} \sqrt{\frac{U''^2_{SV} - U_{Arc}^2}{U'^2_{SV} - U_{Arc}^2}} \tag{58}$$

$$Q'' = Q' \frac{U''^2_{SV} - U_{Arc}^2}{U'^2_{SV} - U_{Arc}^2} \tag{59}$$

Figure 10 shows the power changes P'' and Q'' as a function of the AC supply voltage within $U''_{SV} = 80\% \dots 110\%$, for selected operating points defined by the arc voltages: $U_{Arcn} = 86.603\%$, $U_{ArcM} = 80\%$, $U_{ArcH} = 70.71\%$, and $U_{ArcSC} = 0$ (electrodes short circuit with the charge).

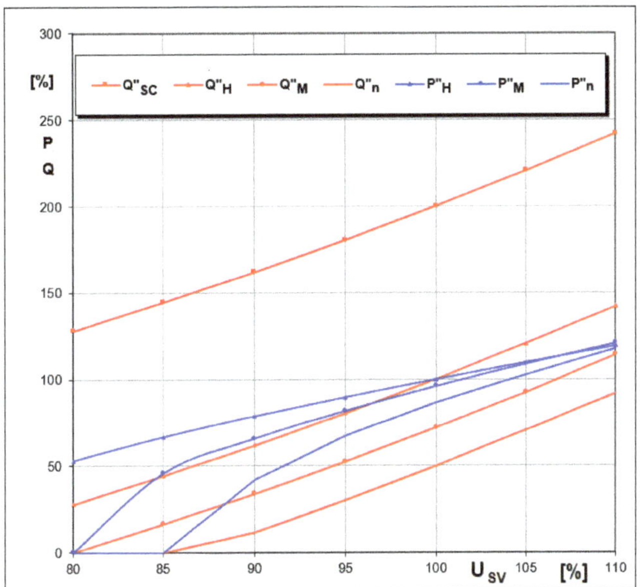

Figure 10. Changes of reactive and active power as a function of supply voltage—operation at a constant arc voltage value (U_{Arc} = const).

Table 4 shows the values (expressed as a percentage) of the power corresponding to the characteristics in Figure 11.

Table 4. Changes of reactive and active power of the arc device for different values of the supply voltage.

$U_{SV}^{''}$	$Q_{SC}^{''}$	$Q_{H}^{''}$	$Q_{M}^{''}$	$Q_{n}^{''}$	$P_{H}^{''}$	$P_{M}^{''}$	$P_{n}^{''}$
[%]	[%]	[%]	[%]	[%]	[%]	[%]	[%]
80	128	28	0	0	52.9	0	0
85	144.5	44.5	16.5	0	66.7	45.9	0
90	162	62	34	12	78.7	65.9	42.4
95	180.5	80.5	52.5	30.5	89.7	81.9	67.6
100	200	100	72	50	100	96	86.6
105	220.5	120.5	92.5	70.5	109.8	108.8	102.8
110	242	142	114	92	119.2	120.7	117.4

Figure 11 shows changes in k_p and k_q as a function of current. Attention should be paid to large values of these coefficients at lower currents, decreasing gradually with increasing current.

For the assessment of the interaction between the furnaces, it is important to estimate the influence of voltage changes on the reactive power consumption. Therefore, it is important to analyze the variability of the reactive power susceptibility coefficient k_q, which is determined for the voltage changing around the voltage $U_{SVH}^{''}$ from the value $U_{SVL}^{''}$ to the value $U_{SVH}^{''}$, with dependence:

$$k_q^{''} = \frac{\dfrac{(Q_H^{''} - Q_L^{''})}{Q''}}{\dfrac{(U_{SVH}^{''} - U_{SVL}^{''})}{U_{SV}^{''}}}\tag{60}$$

which allows to determine the changes of the reactive power $\Delta Q''$, depending on the changes of the supply voltage $U_{SV}^{''}$. Figure 12 shows the pie chart of the electric arc furnace.

In Figure 12, the blue color shows the measured values of active power at a given value of the reactive power of the electric arc furnace. Black represents the trend line. Changes in the ΔU supply voltage cause changes in the ΔP active power and changes in the reactive power ΔQ. The effect of voltage changes on power changes is defined by the coefficients k_p and k_q.

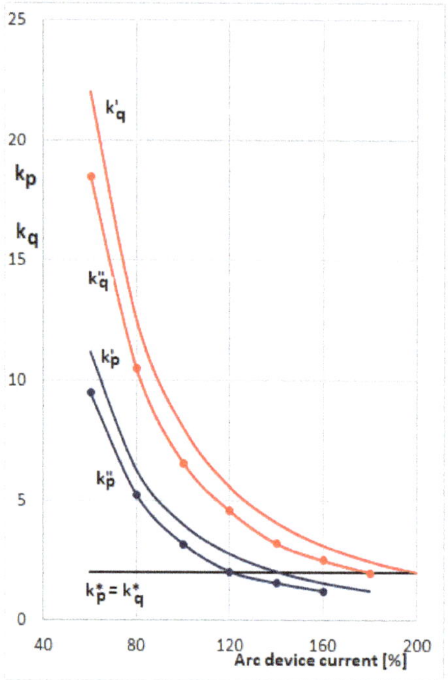

Figure 11. Changes of the k_p and k_q coefficients as a function of the arc device current (for the supply voltage $U'_{SV} = 100\%$ and $U''_{SV} \neq 100\%$).

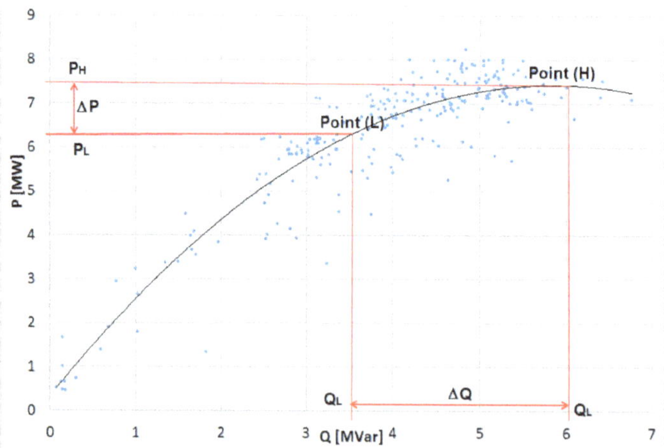

Figure 12. Dependence of changes in reactive power and active power of the arc furnace.

Formula (61) results from the definition of the k_q coefficient:

$$k_q'' = \frac{\dfrac{\Delta Q''}{Q''}}{\dfrac{\Delta U''}{U''}} = \frac{\dfrac{Q_H'' - Q_L''}{Q''}}{\dfrac{U_{SVH}'' - U_{SVL}''}{U_{SV}''}} \tag{61}$$

$$k_q'' \frac{\Delta U''}{U''} = \frac{\Delta Q''}{Q''} \tag{62}$$

$$\Delta Q'' = k_q'' Q'' \frac{\Delta U''}{U''} \tag{63}$$

For the given range of current variation, adequately for the melting process carried out, when supplied with a voltage of a constant value (U_{SV} = const), the following dependence was found:

$$k_q'' Q'' = const \tag{64}$$

For the supply voltage U_{SV}' = 100%, circuit reactance X = 50%, and currents I = 100–120–140–200%, we obtain:

- arc voltage: U_{Arc} = 86.6–80–71.41–0%

$$U_{Arc} = \sqrt{U_{SV}'^2 - I^2 X^2} \tag{65}$$

- reactive power value: Q' = 50.0–72.0–98.0–200.0%

$$Q' = (U_{SV}'^2 - U_{Arc}^2)/X \tag{66}$$

When the supply voltage changes from the upper level U_{SVH}'' = 105% to the lower U_{SVL}'' = 95%, i.e., when voltage fluctuations ΔU = 10%, they correspond to the reactive power values: Q_H = 70.5–92.5–118.5–220.5%

$$Q_H = (U_H^2 - U_{Arc}^2)/X \tag{67}$$

and: Q_L = 30.5–52.5–78.5–180.5%

$$Q_L = (U_L^2 - U_{Arc}^2)/X \tag{68}$$

As a result, the reactive power changes are: ΔQ = 40–40–40–40%,

$$\Delta Q = Q_H - Q_L \tag{69}$$

For a constant value of voltage changes ΔU = 10%, it gives the value of the coefficient: k_q'' = 8–5.55–4.08–2

$$k_q' = \Delta Q_\% / \Delta U_\% \tag{70}$$

and: ΔQ = 40–40–40–40%

$$\Delta Q = k_q' Q'' \Delta U_\% / 100 \tag{71}$$

So, it was obtained for different values of the current, different values of the coefficient of the slope of the reactive power characteristics k_q' (the arc mapped by a sinusoidal voltage with a constant RMS value depending on the arc length), and different values of the average reactive power Q' (for the supply voltage, U_{SV} = 100%), while the ratio (71) remains constant.

$$k_q' Q_{sc} \frac{\Delta U}{U'} \tag{72}$$

For the furnace operation at the service short-circuit current k_q^* = 2, it allows to determine the relationship, which is also important for other operating currents (k_q^* = 2—applies to an arc device with an arc represented by a resistance with a constant value, depending on the arc length):

$$k'_q Q'_{sc} \frac{\Delta U}{U_o} = k^*_q Q_{sc} \frac{\Delta U}{U'} \tag{73}$$

The above property of the arc (mapped with the voltage U_{Arc}) also results from the relationship:

$$Q_H - Q_L = \frac{U_H^2 - U_{Arc}^2}{X} - \frac{U_L^2 - U_{Arc}^2}{X} = \frac{U_H^2 - U_L^2}{X} = (U_H - U_L)\frac{U_H + U_L}{X} =$$
$$\Delta U \frac{2U'}{X} = 2\frac{\Delta U}{U'}\frac{U'^2}{X} = 2\frac{\Delta U}{U'}Q' \tag{74}$$

Similar dependencies can be given for the supply voltage $U'' \neq U'$ (then, an additional factor should be entered: $(U''/U')^2$).

$$Q''_{SC} = \frac{U''^2}{X} \text{ or } Q'_{SC} = \frac{U'^2}{X}\left(\frac{U''}{U'}\right)^2 \tag{75}$$

Formula (72) for the operational short-circuit state ($U_{Arc} = 0$) changes into the relationship

$$k''_q Q'' = k^*_q Q''_{SC} = k^*_q Q'_{SC}\left(\frac{U''}{U'}\right)^2 \tag{76}$$

Formula (76) is of great importance for solving the problem of interactions between furnaces, because it enables the determination of the fluctuations in reactive power consumption, resulting from voltage fluctuations on the bus bars of the steel mill, using the coefficient $k^*_q = 2$ and the short-circuit power of the furnace. This is a great simplification compared to the need to determine the coefficient k''_q and power Q'', depending on the operating point (this conclusion is correct, when the condition $R/X = 0$ is met).

5. Discussion

The proposed power susceptibility factors of active power—k_p and reactive power—k_q are applicable in practice. They were used to determine the increase in voltage fluctuations (and flicker of light) arising during the operation of several arc furnaces in relation to the fluctuations arising during the operation of a single furnace. The increase in fluctuations is determined by the K_N coefficient. The method of determining the increase in light flicker proposed by UIE takes into account only the melting phases in arc furnaces.

For the assessment of superposition of voltage fluctuations, the substitute parameter P_{st} obtained with the use of the light flicker meter is assumed, which is determined from the relationship [24].

$$P_{st} = \sqrt[m]{P_{st1}^m + P_{st2}^m + \ldots + P_{stn}^m} \tag{77}$$

where for arc furnaces, the following values of the m coefficient are adopted:

$m = 4$—used only for the summation of voltage changes, due to arc furnaces specifically run to avoid coincident melts;

$m = 2$—used where coincident stochastic noise is likely to occur, e.g., coincident melts on arc furnaces.

Analyzing the waveforms of fast-changing voltage fluctuations recorded in the networks supplying arc devices, with a different number of arc furnaces operating in parallel, it was found that the mutual influence of the furnaces should be taken into account. Switching on successive arc furnaces causes an increase in the amplitude of voltage fluctuations, as well as a decrease in the average value of the supply voltage to the furnaces—Figure 13. The developed factors k_p and k_q allow for this phenomenon to be taken into account.

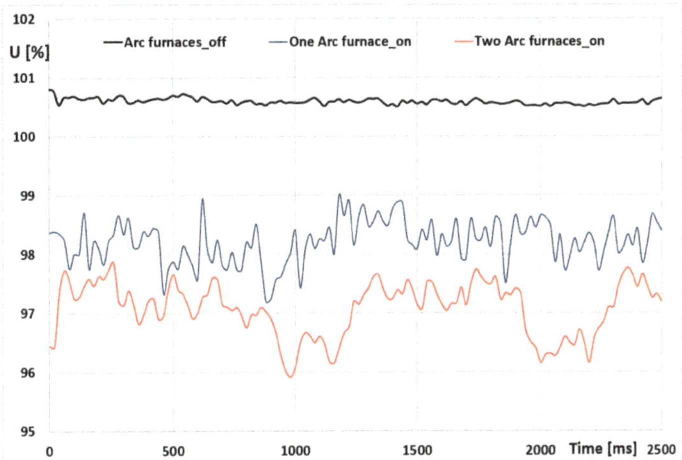

Figure 13. RMS value changes recorded in the steel network supplying the steelworks with different numbers of arc furnaces.

Taking into account the decrease in the average voltage value resulting from the connected successive furnaces, the short-circuit power of the supply network, the number of arc furnaces operating in parallel, and the power of transformers of arc devices, the formulas allow for determining the K_N coefficient

For the same arc devices in the scrap-melting phase, K_N is determined from the formula:

$$K_N = \frac{\sqrt{N}}{\sqrt{1 + \frac{(N-1)k_q'' \overline{Q_1}}{S_{SC}} \left(\frac{U_{SN}}{U_{S1}}\right)^2}} \tag{78}$$

and for arc devices with different powers of their furnace transformers, from the dependence:

$$K_N = \frac{\sqrt{\sum\limits_{i=1}^{N} \left(\frac{S_{ni}}{S_{n1}}\right)^2}}{\sqrt{1 + \sum\limits_{j=2}^{N} \left(\frac{k_q'' \overline{Q_j}}{S_{sc}}\right) \left(\frac{U_{SN}}{U_{S1}}\right)^2}} \tag{79}$$

where:
k_q''—slope coefficients of the power–voltage characteristic calculated at a constant arc voltage;
$\overline{Q_j}$—the mean reactive power drawn by j-th furnace;
S_{cc}—the short-circuit power on the bus-bars of the steelwork (in PCC furnaces);
S_{ni}—the nominal power drawn by i-th furnace.
U_{SN}; U_{S1}—voltage on the rails of the steel plant in the operation of N furnaces and in the operation of a single furnace (furnace with the highest power is the reference furnace).

The issues of voltage fluctuations and flickering caused by arc furnaces in detail have been discussed, among others, in publications [25,26].

In a steady state, both arc models lead to the same solution. The differences in the response of both models to voltage fluctuations are presented below. If the arc device works in a steady state at the rated current (I_n = 100%) and constant voltage (U_{SV} = 100%), when we only take into account the reactance of the supply network (X = 50%) for individual arcs models, we obtain:

R_{Arc} (electric arc replaced by resistance), U_{Arc} (electric arc replaced by a voltage source):

$$R_{Arcn} = \sqrt{\left(\frac{U_{SV}}{I_n}\right)^2 - X^2} = 86.602\% \quad U_{Arcn} = \sqrt{U_{SV}^2 - I^2 X^2} = 86.602\% \tag{80}$$

$$P_n = P_{Arcn} = I_n^2 R_{Arcn} I_n = 86.602\% \quad Q_n = I_n^2 X = 50\% \tag{81}$$

which proves that the models are equivalent to each other.

In case of voltage fluctuations: $U_{SV} = 100 \pm 5\%$ for, e.g., two-state voltage changes, will change between the low state L: $U_{SVL} = 95\%$ and the high state H: $U_{SVH} = 105\%$. Figure 14 shows a vector diagram showing the changes in voltage fluctuation between U_{SVH} and U_{SVH}.

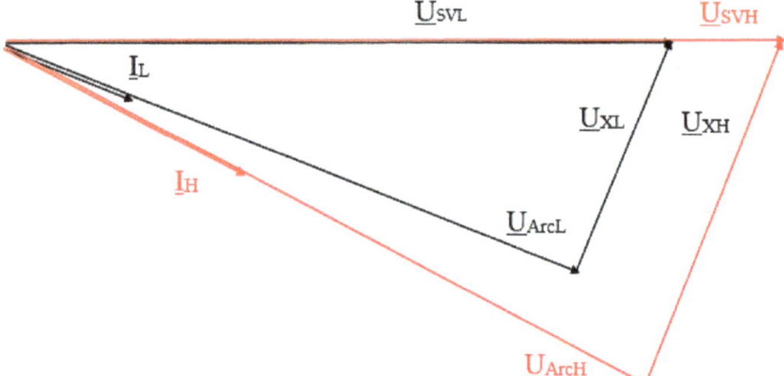

Figure 14. Vector diagram of the currents and voltages of the arc furnace.

For high level (H) $U_{SVH} = 105\%$, it can be written:
R_{Arc} (electric arc replaced by resistance), U_{Arc} (electric arc replaced by a voltage source):

$$I_H = \frac{U_{SVH}}{\sqrt{R_{Arc}^2 + X^2}} = 105\% \quad I_H = \frac{\sqrt{U_{SVH}^2 + U_{Arc}^2}}{X} = 118.74\% \tag{82}$$

$$U_{ArcH} = I_H R_{Arc} = 90.93\% \quad R_{ArcH} = \frac{U_{ArcH}}{I_H} = 70.93\% \tag{83}$$

$$P_{ArcH} = I_H^2 R_{Arc} = 95.48\% \quad P_{ArcH} = I_H^2 R_{Arc} = 102.83\% \tag{84}$$

$$Q_H = I_H^2 X = 55.12\% \quad Q_H = I_H^2 X = 70.5\% \tag{85}$$

and low level (L): $U_{SVL} = 95\%$

$$I_H = \frac{U_{SVH}}{\sqrt{R_{Arc}^2 + X^2}} = 95\% \quad I_H = \frac{\sqrt{U_{SVH}^2 + U_{Arc}^2}}{X} = 78.10\% \tag{86}$$

$$U_{ArcH} = I_H R_{Arc} = 82.24\% \quad R_{ArcH} = \frac{U_{ArcH}}{I_H} = 110.88\% \tag{87}$$

$$P_{ArcH} = I_H^2 R_{Arc} = 78.16\% \quad P_{ArcH} = I_H^2 R_{Arc} = 67.64\% \tag{88}$$

$$Q_H = I_H^2 X = 45.125\% \quad Q_H = I_H^2 X = 30.5\% \tag{89}$$

For the compared models of arches, the coefficients k_p and k_q are, respectively:

R_{Arc} (electric arc replaced by resistance), U_{Arc} (electric arc replaced by a voltage source):

$$k_p = \frac{\frac{\Delta P}{P_n}100}{\frac{\Delta U}{U}100} = \frac{20}{10} = 2 \quad k_p = \frac{\frac{\Delta P}{P_n}100}{\frac{\Delta U}{U}} = \frac{40.64}{10} = 4.06 \tag{90}$$

$$k_q = \frac{\frac{\Delta Q}{Q_n}100}{\frac{\Delta U}{U}100} = \frac{20}{10} = 2 \quad k_q = \frac{\frac{\Delta Q}{Q_n}100}{\frac{\Delta U}{U}100} = \frac{80}{10} = 8 \tag{91}$$

The presented calculations confirm that the reaction of the circuit to rapidly changing voltage fluctuations, using the constant resistance arc model, is different than the reaction to the same fluctuations when replacing the arc with a voltage source.

Using the k_p and k_q coefficients, it is possible to estimate the degree of suppression of disturbances generated by the arc devices. Figure 15 shows the changes in the k_p and k_q coefficients as a function of the ratio of the circuit resistance to the constant reactance ($X = 50\%$) for the rated current $I = 100\%$ and the U_{SV} supply voltage changes $= 1\%$. The damping effect of voltage fluctuations also depends on the power–voltage characteristics of low-voltage loads (e.g., devices for non-metallurgical processing of steels operating in a quiet manner—compared to scrap-metal arc furnaces or furnaces in the production phase).

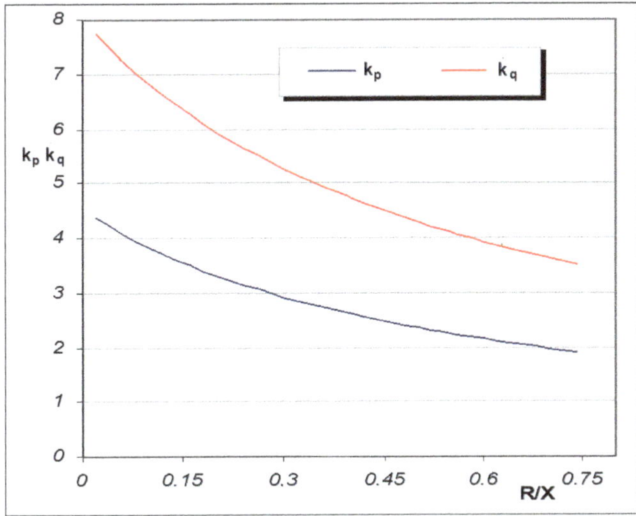

Figure 15. Changes in the inclination coefficients of the power–voltage characteristics as a function of the ratio of resistance to reactance supplying the arc device (operation at rated current).

Figure 15 confirms the known fact that voltage fluctuations are more absorbed in LV and MV grids, where the resistance to reactance ratio is greater than in the case of HV and VHV grids. The absorbing effect of voltage fluctuations also depends on the power–voltage characteristics of low-voltage loads (e.g., devices for non-metallurgical processing of steels operating in a quiet manner—compared to scrap-metal arc furnaces or furnaces in the production phase). In the event of difficulties in determining these dynamic characteristics, practical approximate calculations can use the knowledge of the static power–voltage characteristics.

6. Summary

In the case of slow changes in the voltage supplying arc devices, it is recommended to use static power–voltage characteristics. Then, the power susceptibility factors $k_p^* = 2$ and $k_q^* = 2$ should be used. In the case of fast-changing voltage fluctuations, one should use the power–voltage characteristics determined for the condition $U_{Arc} = $ const and assume the power susceptibility coefficients for active power k_p' or k_p'' and reactive power k_q' or k_q'' with a value depending on the operating point of the arc device. Then, the mutual influence of the arc devices on the supply conditions (changes in the supply voltage) is taken into account. For the characteristics using the developed k_p and k_q factors, the name of quasi-static power–voltage characteristics of arc devices is proposed.

Funding: This research received no external funding.

Institutional Review Board Statement: Not applicable.

Informed Consent Statement: Not applicable.

Conflicts of Interest: The author declares no conflict of interest.

Abbreviations

The following nomenclatures are used in this manuscript:

USV	supply voltage
U_{SV}'	rated supply voltage ($U_{SV}' = 100\%$)
U_{SV}''	supply voltage different from the rated voltage ($U_{SV}'' \neq 100\%$)
U_{SVL}''	the highest value of the supply voltage
U_{SVH}''	the smallest value of the supply voltage
U^*	arc voltage when the arc is replaced by resistance
U_{Arc}	electric arc voltage of the arc furnace
U_{Arc}'	arc voltage at rated supply voltage ($U_{SV}' = 100\%$)
U_{Arc}''	arc voltage at a supply voltage different from the rated voltage ($U_{SV}'' \neq 100\%$)
I	arc furnace current
I_M	arc furnace melting current
I_n	nominal arc furnace current
I_{SC}	arc furnace current when the electrodes are short-circuited with the charge
P_{Arc}	active power of the electric arc
P'	active power at rated supply voltage ($U_{SV}' = 100\%$)
Q'	reactive power at rated supply voltage ($U_{SV}' = 100\%$)
P''	active power at a voltage different from the rated supply voltage ($U_{SV}'' \neq 100\%$)
P^*	active power when the arc is replaced with resistance
Q''	reactive power at a voltage different from the rated supply voltage ($U_{SV}'' \neq 100\%$)
Q^*	reactive power when the arc is replaced with resistance
Q_{SC}	reactive power at the short-circuit of the electrodes with the charge
k_p^*, k_q^*	slope coefficients of the power–voltage characteristic calculated at a constant arc voltage and a constant arc resistance respectively
k_p', k_q'	slope coefficients of power–voltage characteristics determined at constant arc voltage at rated supply voltage ($U_{SV}' = 100\%$)
k_p'', k_q''	slope coefficients of power–voltage characteristics determined at constant arc voltage and supply voltage different from the rated one ($U_{SV}'' \neq 100\%$)
Z	impedance of the circuit supplying the electric arc
R	resistance of the circuit supplying the electric arc
X	reactance of the circuit supplying the electric arc
P_{st}	short-term flicker severity
K_N	coefficient determining the increase in flicker of light depending on the number of parallel operating arc furnaces in steel plant
P_{stN}	value of the short-term light flicker indicator recorded during the operation N of arc furnaces
P_{st1}	value of the short-term light flicker indicator recorded during the operation of a single arc furnace
N	number of parallel operating arc furnaces in steel plant
S_{scf}	short-circuit power capacity when shorting the electrodes with the scrap
S_{sc}	short-circuit power capacity of the network
Q_j	mean reactive power drawn by j-th furnace
U_{SN}	voltage on the bus-bars of the steelwork at the work of N furnaces
U_{SN1}	voltage on the bus-bars of the steelwork at the work of a one arc furnace

References

1. Tseng, K.J.; Wang, Y.; Vilathgamuwa, D.M. An experimentally verified hybrid Cassie–Mayr electric arc model for power electronics simulations. *IEEE Trans. Power Electron.* **1997**, *12*, 429–436. [CrossRef]
2. Samet, H.; Farjah, E.; Zahra Sharifi, Z. A dynamic, nonlinear and time-varying model for electric arc furnace. *Int. Trans. Electr. Energy Syst.* **2015**, *25*, 2165–2180. [CrossRef]
3. Pauna, H.; Willms, T.; Aula, M.; Echterhof, T.; Huttula, M.; Fabritius, T. Electric Arc Length-Voltage and Conductivity Characteristics in a Pilot-Scale AC Electric Arc Furnace. *Metall. Mater. Trans.* **2020**, *51*, 1646. [CrossRef]
4. Horton, R.; Haskew, T.A.; Burch, R.F. A time-domain ac electric arc furnace model for flicker planning studies. *IEEE Trans. Power Deliv.* **2009**, *24*, 1450–1457. [CrossRef]
5. Tang, L.; Kolluri, S.; McGranaghan, M.F. Voltage flicker prediction for two simultaneously operated AC arc furnaces. *IEEE Trans. Power Deliv.* **1997**, *12*, 985–992. [CrossRef]
6. Sadeghian, A.; Lavers, J.D. Dynamic reconstruction of nonlinear v–i characteristic in electric arc furnaces using adaptive neuro-fuzzy rule-based networks. *Appl. Soft Comput.* **2011**, *11*, 1448. [CrossRef]
7. Alonso, M.A.P.; Donsion, M.P. An improved time domain arc furnace model for harmonic analysis. *IEEE Trans. Power Deliv.* **2004**, *19*, 367–373. [CrossRef]
8. Cano Plata, E.A.; Tacca, H.E. Arc Furnace Modeling in ATP-EMTP. In Proceedings of the 6th International Conference on Power Systems Transients, Montreal, QC, Canada, 20–23 June 2005; pp. 19–23.
9. Odenthal, H.J.; Kemminger, A.; Krause, F.; Sankowski, L.; Uebber, N.; Vogl, N. Review on Modeling and Simulation of the Electric Arc Furnace (EAF). *Steel Res. Int.* **2018**, *89*, 1700098. [CrossRef]
10. Terzija, V.; Stanojevic, V. Power quality indicators estimation using robust Newton-typealgorithm. *IEE Proc. Gener. Transm. Distrib.* **2004**, *151*, 477–485. [CrossRef]
11. Acha, E.; Madrigal, M. *Power Systems Harmonics: Computer Modeling and Analysis*; John Wiley & Sons: Hoboken, NJ, USA, 2001.
12. Lee, C.; Kim, H.; Lee, E.J.; Baek, S.T.; Jae Woong Shim, J.W. Measurement-Based Electric Arc Furnace Model Using Ellipse Formula. *IEEE Access* **2021**, *9*, 155609–155621. [CrossRef]
13. Marulanda-Durango, J.; Escobar-Mejía, A.; Alzate-Gómez, A.; Álvarez-López, M. A Support Vector machine-Based method for parameter estimation of an electric arc furnace model. *Electr. Power Syst. Res.* **2021**, *196*, 107228. [CrossRef]
14. Lozynskyi, O.Y.; Paranchuk, Y.S.; Paranchuk, R.Y.; Matico, F.D. Development of methods and means of computer simulation for studying arc furnace electric modes. *Electr. Eng. Electromech.* **2018**, 28–36. [CrossRef]
15. Bhonsle, D.C.; Kelkar, R.B. Simulation of Electric Arc Furnace Characteristics for Voltage Flicker study using MATLAB, International Conference on Recent Advancements in Electrical. In Proceedings of the 2011 International Conference on Recent Advancements in Electrical, Electronics and Control Engineering, Sivakasi, India, 15–17 December 2011; pp. 174–181.
16. Ghiormez, L.; Prostean, O.; Panoiu, M.; Panoiu, C. GUI for studying the parameters influence of the electric arc model for a three-phase electric arc furnace. *IOP Conf. Ser. Mater. Sci. Eng.* **2017**, *163*, 012026. [CrossRef]
17. Mokhtari, H.; Hejri, M. A new three phase time-domain model for electric arc furnace using MATLAB. In Proceedings of the IEEE Asia-Pacific Transmission and Distribution Conference and Exhibition, Yokohama, Japan, 6–10 October 2002; Volume 3, pp. 2078–2083.
18. Gajic, D.; Savic-Gajic, I.; Savic, I.; Georgieva, O.; Di Gennaro, S. Modelling of electrical energy consumption in an electric arc furnace using artificial neural networks. *Energy* **2016**, *108*, 132–139. [CrossRef]
19. Lei, W.; Wang, Y.; Wang, L.; Cao, H. A Fundamental Wave Amplitude Prediction Algorithm Based on Fuzzy Neural Network for Harmonic Elimination of Electric Arc Furnace Current. *Math. Probl. Eng.* **2015**, *2015*, 268470. [CrossRef]
20. Kowalski, Z. *Unbalance in Electrical Power Systems*; PWN (Polish Scientific Publishers): Warsaw, Poland, 1987.
21. Kowalski, Z. *Voltage Fluctuations in Power Systems*; WNT: Warsaw, Poland, 1985. (In Polish)
22. Kurbiel, A. *Electrothermal Arc Devices*; WNT: Warszawa, Poland, 1988. (In Polish)
23. Olczykowski, Z. Arc Voltage Distortion as a Source of Higher Harmonics Generated by Electric Arc Furnaces. *Energies* **2022**, *15*, 3628. [CrossRef]
24. UIE. *Guide to Quality of Electical Supply for Industrial Installations. Part 5. Flicker and Voltage Fluctuations*; Power Quality, Working Group WG 2; UIE: North Andover, MA, USA, 1999.
25. Olczykowski, Z. Modeling of Voltage Fluctuations Generated by Arc Furnaces. *Appl. Sci.* **2021**, *11*, 3056. [CrossRef]
26. Olczykowski, Z.; Łukasik, Z. Evaluation of Flicker of Light Generated by Arc Furnaces. *Energies* **2021**, *14*, 3901. [CrossRef]

Article

The Influence of the Transformer Core Material on the Characteristics of a Full-Bridge DC-DC Converter

Krzysztof Górecki, Kalina Detka * and Krystian Kaczerski

Department of Marine Electronics, Faculty of Electrical Engineering, Gdynia Maritime University, Morska 83, 81-225 Gdynia, Poland
* Correspondence: k.detka@we.umg.edu.pl

Abstract: The paper analyzes the influence of the material from which the ferromagnetic core of a transformer is made on the characteristics of a full-bridge converter. Experimental investigations were carried out for three bridge converters containing transformers with ring cores made of various materials: iron powder, ferrite, and nanocrystalline material. The properties of the aforementioned converters were considered in a wide range of changes of input voltage, load resistance, as well as frequency and the duty cycle of the control signal. Based on the obtained measurements results of the relationship between the parameters of the used transformer core and the obtained values of the output voltage, the energy efficiency of the full bridge converter was discussed. The method of transformer modeling in the SPICE program for the analysis of the considered converter in this program was proposed. The correctness of this model was demonstrated for a converter containing a transformer with a powdered iron core.

Keywords: full-bridge converter; transformer; ferromagnetic core; measurements; computations; SPICE; modeling

Citation: Górecki, K.; Detka, K.; Kaczerski, K. The Influence of the Transformer Core Material on the Characteristics of a Full-Bridge DC-DC Converter. *Energies* **2022**, *15*, 6160. https://doi.org/10.3390/en15176160

Academic Editors: Juri Belikov, Eugenio Meloni, Alberto-Jesus Perea-Moreno, Iva Ridjan Skov, Giorgio Vilardi, Antonio Zuorro and José Carlos Magalhães Pires

Received: 21 July 2022
Accepted: 22 August 2022
Published: 25 August 2022

Publisher's Note: MDPI stays neutral with regard to jurisdictional claims in published maps and institutional affiliations.

1. Introduction

Contemporary electronic devices require power systems that supply electrical energy with given parameters [1]. Recently, the economic conversion of electrical energy and related technical, economic, and environmental issues have become of great importance [2].

Pulse DC-DC converters are commonly used to convert electrical energy [1–3]. In many cases, it is desirable to provide galvanic isolation between the input and output of such a converter. Then, isolated DC-DC converters are used, which include, e.g., bridge converters [4–6]. These converters are also used in the systems cooperating with wind farms or photovoltaic installations [7].

One of the tasks of designers of the considered class of DC-DC converters is the proper selection of a pulse transformer, in particular a ferromagnetic core. Manufacturers of these cores offer a wide range of products made of various ferromagnetic materials of various shapes and sizes. As shown in [8,9], the shape and size of the ferromagnetic core significantly affect thermal parameters of inductors with ferromagnetic cores. On the other hand, in [10,11] it was shown that the material used for the construction of the inductor core significantly influences the characteristics of buck and boost converters. As shown in the literature, the functional parameters of magnetic elements such as losses, operating temperature, size, weight, and material parameters have a significant impact on the properties of systems and converting devices [8,9,12].

For example, in [9], it was shown that the use of ferrite material for the construction of the inductor core operating in the boost converter resulted in a change of the operating mode of the mentioned system from CCM to DCM at load resistance above 1 kΩ, whereas the use of powder material for the construction of the inductor core resulted in a change in the operating mode of the converter from CCM to DCM at 100 Ω load resistance.

Meanwhile, the computer analyses and descriptions of properties of DC-DC converters typically ignore the imperfections of magnetic elements and models of the transformers contained in these converters that use linear coupled inductors [13,14]. This method of transformer modeling may lead to significant inaccuracies in calculations. Additionally, with such simplified analyses, it is not possible to take into account the properties of the considered elements. As is shown in [15–19], the magnetic materials used are characterized by different values of parameters, e.g., saturation of magnetic flux density, power losses, and the maximum value of the operating frequency. Of course, these materials are constantly being improved, but their development is not as dynamic as that of semiconductor devices.

Designers of electronic devices strive to reduce the size and weight of the designed devices and to improve their energy efficiency. However, the construction process must be preceded by the appropriate analyses that take into account the limitations of the used components [2,3]. The aforementioned strive to miniaturize converting devices that require, inter alia, increasing the switching frequency of switching elements and, consequently, searching for new technologies for the production of semiconductor devices and magnetic elements [4,20].

Before starting the construction of the aforementioned electronic systems, computer analyses are carried out to verify the properties of the designed systems. This allows eliminating some errors at the design stage. During the analysis of electronic circuits such as energy conversion systems, it is necessary to use the appropriate models and calculation methods [5,21,22]. In simulated systems, it is desirable to carry out a transient analysis, which is a very time-consuming process [15,23,24].

The paper [25] describes the modeling method and the phase-shifted bidirectional dual active bridge (PSBBAB) model. It was found out that the standard methodology for modeling such systems is inappropriate when the average value of the transformer current is zero. For the analysis of this type of systems, a discrete model with two time scales was proposed. This model performs a fast and slow dynamic analysis of the system, but the process is performed separately, which simplifies the calculations and reduces the simulation time. This model takes into account a number of factors, including: inductor core losses, dead-time of semiconductor devices, on-resistance, transformer winding losses, etc. Additionally, a phase-shifted bidirectional dual active bridge prototype was built. The conducted investigations show that the model proposed in [25] allows for obtaining similar characteristics to the characteristics of the real system when non-idealities of the transformer can be omitted.

As for now, little attention has been focused on the influence of the transformer core material on properties of DC-DC converters. A typical approach is to idealize properties of these elements [4,5,13,14]. Such transformers are modeled with the use of linearly coupled linear inductors. It can be expected that when taking into account non-idealities of the transformer, it is possible to more accurately model DC-DC converters containing such elements. Due to the differences in properties of different magnetic materials, some differences between characteristics of such converters can be also observed. The authors do not know any articles describing such a problem.

The aim of this paper is to analyze the influence of the selection of the ferromagnetic material used to build the transformer core on the characteristics of a full-bridge converter. The results of experimental investigations for the mentioned converters containing transformers with a core made of iron powder, ferrite, and nanocrystalline material are presented. Based on the obtained results of the measurements, the influence of the core material on the output voltage and energy efficiency of the tested converter operating at different values of the input voltage, load resistance, frequency, and the duty of the control signal was discussed. A transformer model for the SPICE program is proposed and its usefulness was demonstrated to determine the characteristics of the considered converter containing a transformer with a powdered iron core. The method of including the non-ideality of the transformer in the considered model is discussed. The converter operating conditions, in which selected cores ensure the best properties of the investigated converter, were indicated.

2. Investigated Circuit

The investigations were carried out for the full-bridge converter, whose topology is shown in Figure 1. According to the classic concept of operation of this system, presented among others in [4], transistors T_1 and T_4, as well as T_2 and T_3, are switched on in pairs, alternately allowing the current to flow through the primary winding of the transformer at the zero value of the average voltage on this winding. Two identical secondary windings alternately deliver energy to the load through diodes D_1 and D_2. The energy storage element is inductor L_4. Capacitor C_0 maintains the ripple of the output voltage V_{out} at an acceptable level.

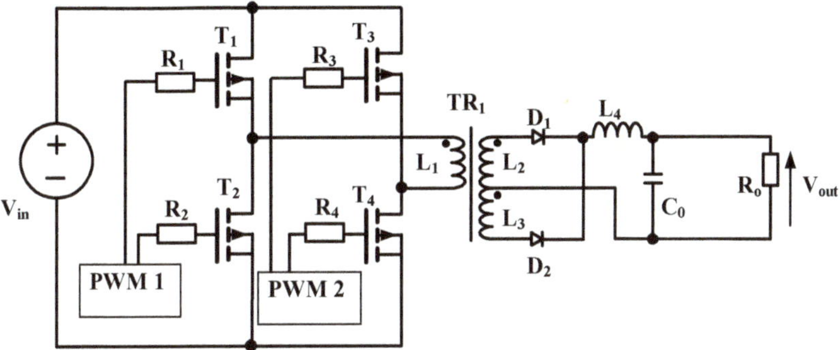

Figure 1. Diagram of the investigated full-bridge converter.

In the considered circuit, four MOSFET transistors of the IRF540 type [26], two Schottky diodes of the MBR10100 type [27] and passive elements with the following values were used: $R_1 = R_2 = R_3 = R_4 = 100\ \Omega$, $C_O = 47\ \mu F$, $L_1 = 80\ \mu H$. Resistor R_0 is the load of the converter, and the voltage source V_{in}—the power source. In the considered circuit, T_{R1} transformers with the cores described in Section 3 were used. The signal controlling the gates of the transistors is obtained from two drivers IR2111 [28] marked in the diagram as PWM1 and PWM2 controlled by the PWM square wave signal generator. This generator produces a signal with frequency f and the duty cycle d with a value not exceeding 0.5.

For the considered system, the measurements were performed for the dependence of the output voltage and energy efficiency with changes in the input voltage V_{in} in the range from 0 to 50 V, load resistance from 15 Ω to 1 kΩ, and the frequency of the control signal from 10 to 50 kHz. All the measured values were obtained at the steady state.

3. Investigated Transformers

The investigations were carried out for three transformers containing toroidal cores with similar external diameters of 27 mm, internal diameters of 15 mm, and heights of 10 mm. The investigated cores are made of various ferromagnetic materials:

- ferrite material, designated in this paper as SM-100 [29],
- nanocrystalline material, designated in this paper as RTN (material M-074) [30],
- powdered iron, designated in this paper as RTP (material -26) [31].

According to the information given by the producers [29] of the considered cores, the SM-100 material is dedicated to applications which need very high values of initial permeability and frequency up to 1 MHz. This material is characterized by a strong influence of temperature on saturation flux density. Its temperature coefficient is about 0.67%/K. In turn, the material -26 is dedicated to the cores of inductors and its permeability is nearly constant for frequency below 100 kHz [31]. It can operate at a very high values of flux density up to 1.2 T. Finally, the material M-074 is dedicated, e.g., for transformers in switch-mode, power supplies operate at frequencies up to 100 kHz [30].

Figure 2 shows the constructed transformers containing the mentioned cores. On each of them, 10 turns of enameled copper wire with a diameter of 0.9 mm were wound on the primary side. In turn, the split secondary winding contained 2×15 turns of enameled copper wire with a diameter of 0.8 mm. The most important material parameters of the cores used are listed in Table 1.

Figure 2. Toroidal transformers with the cores made of (**a**) SM-100, (**b**) RTN, (**c**) RTP.

Table 1. Selected material parameters of the used ferromagnetic cores.

Material	B_{sat} [T]	T_C [°C]	μ_i	P_V [kW/m³]	A_L [nH]
			Parameters		
SM-100	0.41	120	10,000	No data	10,200
RTN	1.2	600	80,000	37 @ B_m = 0.3 T, f = 25 kHz	30,000–102,000
RTP	1.38	250	75	6500 @ B_m = 0.3 T, f = 25 kHz	70

Table 1 shows that the RTP core has the highest value of the saturation of magnetic flux density, but the initial magnetic permeability μ_i of this core is up to 250 times lower than for the RTN core and 130 times lower than for the SM-100 core. In turn, the RTN core has the highest value of Curie temperature T_C. The parameter of losses P_V is not listed for the ferrite core, and for the RTN core, it is nominally 25% higher, even 175 time lower than for the RTP core. However, considering the influence of frequency on losses, it can be seen that under the same operating conditions, the losses in the RTN core are much lower than in the RTP core. It is worth noting that the parameter A_L, which is proportional to the inductance of the windings is the highest for the RTN core and the lowest (even over a thousand times) for the RTP core.

Such significant differences in the values of the cores parameters lead to the assumption that significant differences will also appear in the characteristics of the transformers containing these cores. Differences in the characteristics of the converters containing these transformers are also expected.

4. Results of Measurements

To assess the influence of the transformer core material on the characteristics of the full-bridge converter, a number of measurements of the characteristics of the mentioned converter were carried out. The figures in this section show only the measurement results obtained at the steady state. They illustrate the influence of the input voltage V_{in}, load resistance R_0, and the transformer core material on the measured values of the output voltage V_{out} and the energy efficiency η of the investigated converter.

Figure 3 shows the dependences V_{out} (V_{in}) and η (V_{in}) measured at frequency $f = 25$ kHz, the duty cycle $d = 0.48$, and load resistance $R_0 = 1$ kΩ.

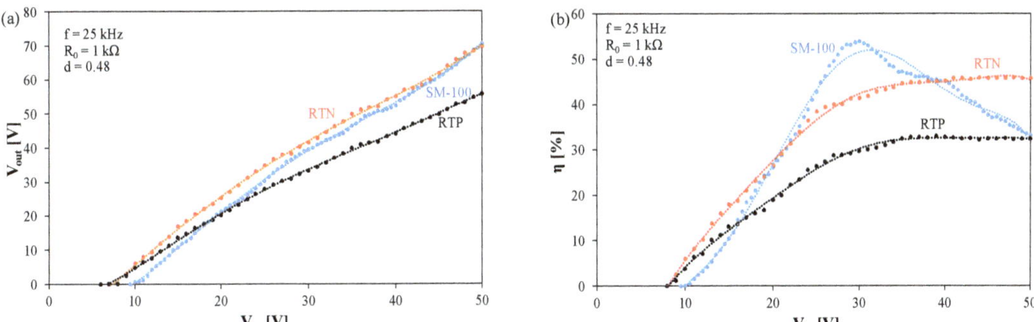

Figure 3. Measured dependences of the output voltage (**a**) and energy efficiency (**b**) of the tested DC-DC converter on its input voltage at $R_0 = 1$ kΩ.

As can be seen, for resistance $R_0 = 1$ kΩ, the character of the dependence V_{out} (V_{in}) is identical for all the considered transformers. This is approximately a linear relationship, but the slope of this characteristic changes as the transformer core changes. The highest values of the output voltage V_{out} were obtained for the converter with a transformer with the RTN core, and the lowest for the converter with a transformer with the RTP core. The differences between the obtained values of the output voltage V_{out} reach up to 20%. It is also worth noting that the output voltage only increases after the input voltage exceeds the value of a few volts.

In turn, in Figure 3b, it can be seen that the dependence η (V_{in}) is a monotonically increasing function for the converter with the a transformer with RTP and RTN cores, whereas for the SM-100 core, this dependence has a maximum at $V_{in} = 30$ V. It is worth noting that that the obtained energy efficiency values are low and reach a maximum of 32% for a transformer with the RTP core, 46%-with the RTN core, and 54%-with the SM-100 core. For higher values of the input voltage V_{in}, the highest energy efficiency is ensured by the use of a transformer with the RTN core, and in the range of the low input voltage values V_{in} a transformer with the SM-100 core.

Figure 4 shows the dependences V_{out} (V_{in}) and η (V_{in}) measured at frequency $f = 25$ kHz, the duty cycle $d = 0.48$, and load resistance $R_0 = 33$ Ω.

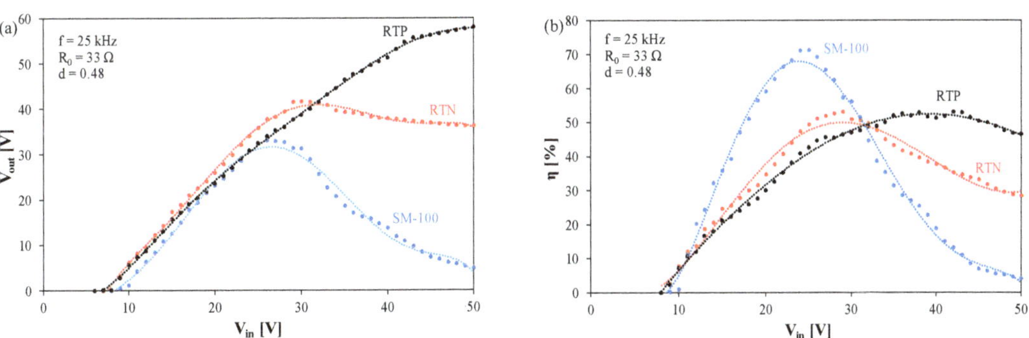

Figure 4. Measured dependences of the output voltage (**a**) and energy efficiency (**b**) of the tested DC-DC converter on its input voltage at $R_0 = 33$ Ω.

As can be seen, at $R_0 = 33\ \Omega$ only for the transformer with the RTP core, the monotonically increasing dependence of V_{out} (V_{in}) was obtained. In the other cases, the considered dependence has a maximum at voltage V_{in} equal to about 26 V. The observed character of the considered dependence results from the differences in the course of the magnetization curve for selected cores, which also cause the dependence of the winding inductance on the current of these windings.

In turn, Figure 4b shows that for all the considered transformers, the dependences η (V_{in}) have a maximum. The highest value of about 70% of the maximum is achieved for the transformer with the SM-100 core. For the remaining transformers, this maximum achieves 53%. The occurrence of this maximum is related to the limitation of the maximum value of the power that can be transferred by each of the considered transformers.

Figure 5 shows the dependences V_{out} (V_{in}) and η (V_{in}) measured at frequency $f = 25$ kHz, the duty cycle $d = 0.48$, and load resistance $R_0 = 15\ \Omega$.

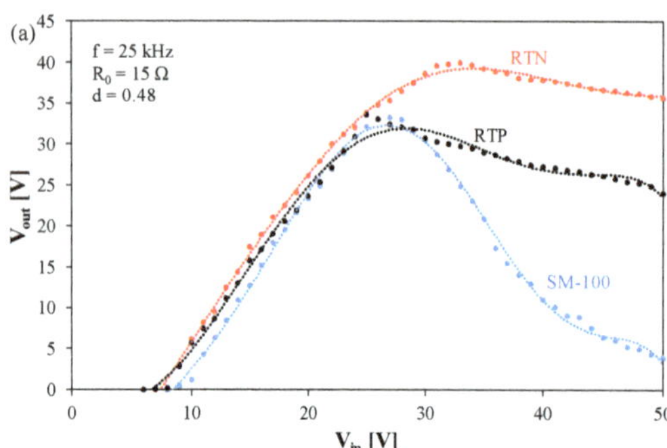
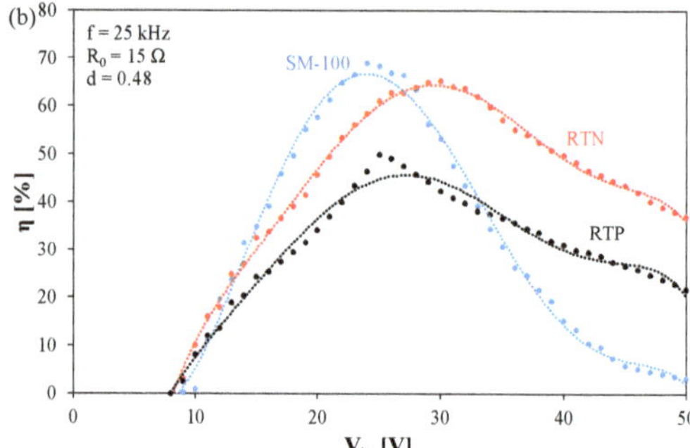

Figure 5. Measured dependences of the output voltage (**a**) and energy efficiency (**b**) of the tested DC-DC converter on its input voltage at $R_0 = 15\ \Omega$.

As can be seen in Figure 5a, at a low value of resistance R_0, V_{out} (V_{in}) characteristics obtained for each of the considered transformers have a maximum, but it is most visible for the transformer with the SM-100 core. For the input voltage $V_{in} < 25$ V, the influence of the core material on the output voltage V_{out} is small, and the obtained values of this voltage do not differ from each other by more than 3 V. In turn, for the highest of the considered input voltage values $V_{in} = 50$ V, the highest value of the output voltage V_{out} was obtained for the transformer with the RTN core. It is seven times higher than for the transformer with the SM-100 core.

Similarly, Figure 5b clearly shows the maxima of the dependences η (V_{in}) for each of the considered transformers. They occur at voltage V_{in} in the range from 25 to 30 V. The value of this maximum is the highest for the transformer with the SM-100 core and it exceeds 70%. For higher values of V_{in} voltage, the highest energy efficiency is obtained using the transformer with the RTN core, and the lowest with the SM-100 core.

Figure 6 shows the dependences V_{out} (V_{in}) and η (V_{in}) measured at frequency $f = 25$ kHz, the duty cycle $d = 0.48$, and load resistance $R_0 = 470\ \Omega$.

It is visible in Figure 6a that for resistance $R_0 = 470\ \Omega$, the dependence V_{out} (V_{in}) is approximately a linear relationship for all the considered cores. The highest values of the output voltage V_{out} were obtained for the converter with transformers containing the RTN and SM-100 cores, whereas the lowest were obtained for the converter with the transformer with the RTP core. The differences between the obtained values of the output voltage V_{out} reach up to 20%.

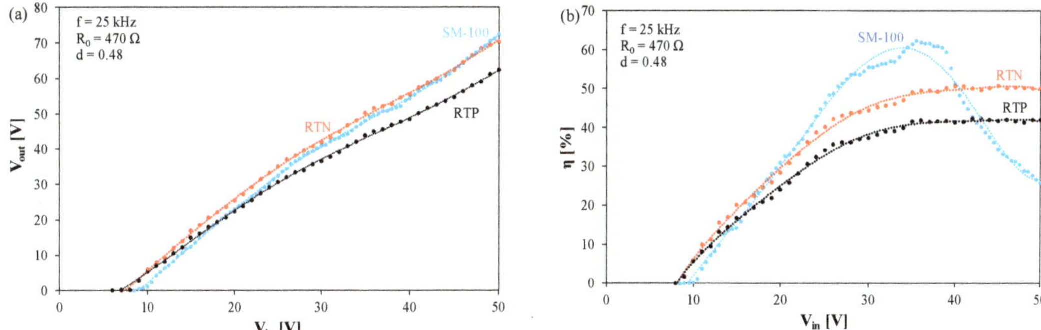

Figure 6. Measured dependences of the output voltage (**a**) and energy efficiency (**b**) of the tested DC-DC converter on its input voltage at $R_0 = 470\ \Omega$ and $d = 0.48$.

In Figure 6b, one can observe that the dependence $\eta\ (V_{in})$ is a monotonically increasing function for a converter with the transformer with RTP and RTN cores, whereas for the SM-100 core, this dependence has a maximum at $V_{in} = 35$ V. The obtained energy efficiency values do not exceed 42% for the transformer with the RTP core, 50% with the RTN core and 62% with the SM-100 core. For the input voltage V_{in} higher than 40 V, the highest energy efficiency is ensured by the use of the transformer with the RTN core, whereas for lower values of this voltage, it is the transformer with the SM-100 core.

Comparing the results presented in Figures 3–6, one can observe that both the ferromagnetic material used and load resistance visibly influence the characteristics V_{out} (V_{in}) and $\eta\ (V_{in})$ of the considered DC-DC converter. For higher values of load resistance, the dependences $V_{out}\ (V_{in})$ are nearly linear, but for lower values they are nonlinear and possess the maxima. The considered characteristics obtained for the RTP core are monotonically increasing functions in the widest range in R_0 values. It worth observing that for each considered value of R_0 the dependences, $V_{out}\ (V_{in})$ are linear for all the considered ferromagnetic materials and they lie very close to one another for low values of V_{in} voltage. This range of V_{in} voltage is smaller and smaller when the value of R_0 decreases. In turn, in the characteristics $\eta\ (V_{in})$, for all the ferromagnetic materials, a maximum can be observed which moves left when the value of R_0 decreases. In a wider range of change in V_{in} and R_0 values, a high value of energy efficiency can be obtained using the RTN core.

In addition, besides the load resistance, the characteristics of the considered converter also depend on the parameters of the control signal.

Figure 7 shows the dependences $V_{out}\ (V_{in})$ and $\eta\ (V_{in})$ measured at frequency $f = 25$ kHz, the value of the duty cycle $d = 0.25$, and load resistance $R_0 = 470\ \Omega$.

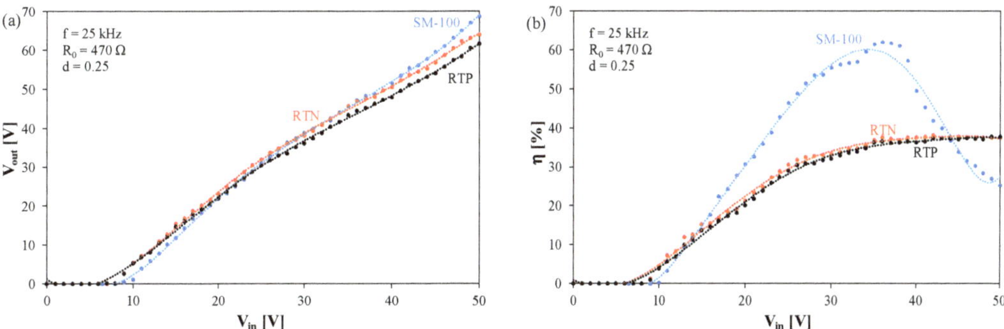

Figure 7. Measured dependences of the output voltage (**a**) and energy efficiency (**b**) of the tested DC-DC converter on its input voltage at $R_0 = 470\ \Omega$ and $d = 0.25$.

As can be seen in Figure 7a, for the considered operation and conditions of the full-bridge converter, the transformer core material does not significantly influence the course of V_{out} (V_{in}) characteristics. These dependences monotonically increase, and the maximum divergence between the output voltage values V_{out} obtained with the use of different transformer cores do not exceed 14%.

In turn, in Figure 7b, it can be seen that the dependences η (V_{in}) obtained for the transformers with the RTP and RTN cores are monotonically increasing functions, which do not practically differ from each other and reach a maximum value of less than 40%. For the SM-100 core, a maximum more than 60% can be seen at voltage V_{in} = 36 V.

Figure 8 shows the dependences V_{out} (V_{in}) and η (V_{in}) measured at frequency f = 50 kHz, the duty cycle d = 0.48, and load resistance R_0 = 470 Ω.

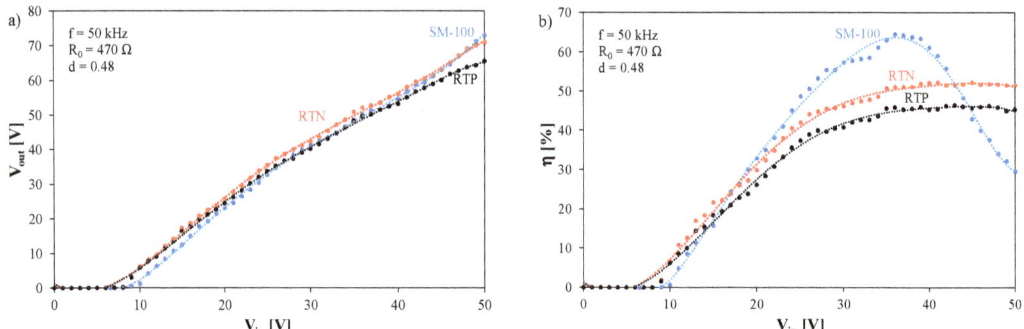

Figure 8. Measured dependences of the output voltage (**a**) and energy efficiency (**b**) of the tested DC-DC converter on its input voltage at f = 50 kHz and R_0 = 470 Ω.

As can be seen in Figure 8a, with an increased frequency of the control signal, the monotonically increasing characteristics V_{out} (V_{in}) are obtained. The highest values of voltage V_{out} were obtained for the transformer with the RTN core.

Figure 8b shows different values of the converter energy efficiency for the value obtained for the transformers with different cores. For the transformers with the RTN and RTP cores, the dependences η (V_{in}) are monotonically increasing functions, and for the transformer with the SM-100 core, the maximum occurs at V_{in} = 35 V. For the highest input voltage values V_{in}, the highest energy efficiency values were obtained for the transformer with the RTN core. The energy efficiency values obtained at f = 50 kHz are higher than the values obtained at f = 25 kHz.

Comparing the results presented in Figures 6–8, one can observe that the double change in switching frequency or the duty cycle very slightly influences characteristics V_{out} (V_{in}), but a change in the characteristics η (V_{in}) is visible. For the cores RTN and RTP, a decrease in the duty cycle causes a decrease in energy efficiency, whereas an increase in the frequency value causes a small increase in energy efficiency. For the core SM-100, the changes in the value of energy efficiency are very small.

The presented measurement results prove that the choice of the materials, from which the transformer core is made, significantly influences both the output voltage and the energy efficiency of the full-bridge converter. These differences are particularly noticeable at low values of load resistance R_0. Apart from the quantitative differences, one can also see qualitative differences in the shape of the obtained dependences V_{out} (V_{in}) and η (V_{in}).

5. Simulation Results

The measurement results shown in the previous section prove that the selection of the transformer core material can significantly affect the characteristics of the full-bridge converter. The strong influence of the input voltage and load resistance on the output voltage and energy efficiency of the considered converter is also visible. In order to investigate the influence of individual circuit parameters and the parameters of individual components on the properties of the tested DC-DC converter, it is convenient to use computer simulations [32].

This section presents the results of simulation tests of the considered converter made with the use of the SPICE program. In the computer simulations, the scheme of the tested circuit shown in Figure 1 and the models of the power MOS transistor and diodes built-in in the SPICE program were used [33]. The parameter values of these models were taken from the websites of the manufacturers of these components [34]. The passive elements were described with the use of linear models with the values given in the description of Figure 1. Each transistor is controlled from a voltage source generating a sequence of rectangular pulses with given values of frequency and the duty cycle.

The results presented in this section were determined for the converter with the transformer containing the powdered iron RTP core. First, calculations were performed for the classical transformer model in the form of linear coupled coils. The following values of the inductance of these coils were used in the calculations: $L_1 = 70$ µH, $L_2 = L_3 = 157.5$ µH and the coupling factor equal to 0.99. The calculated (lines) and measured (points) characteristics of V_{out} (V_{in}) and η (V_{in}) of the considered converter for the presented transformer model are shown in Figure 9.

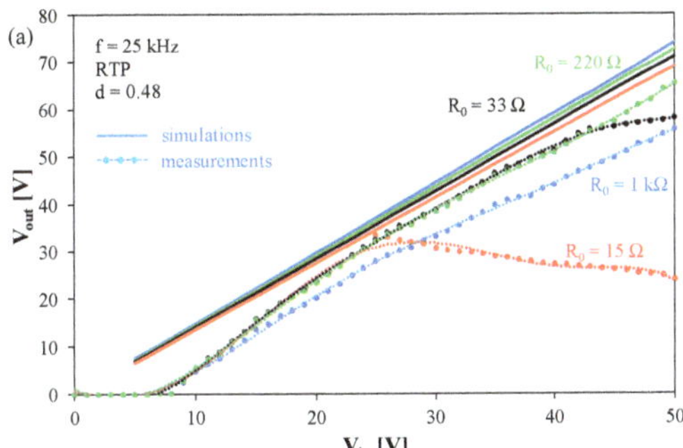

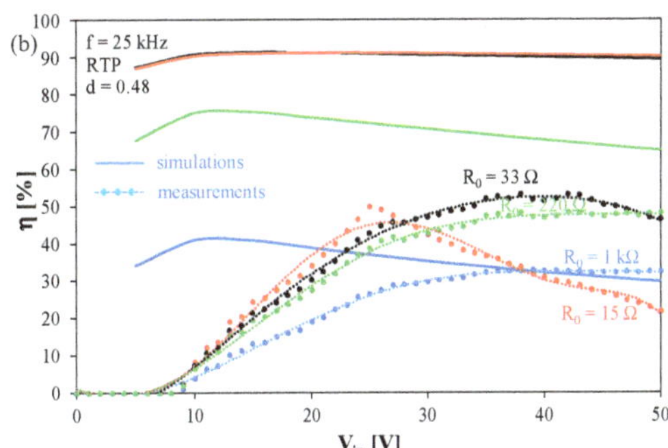

Figure 9. Measured and computed simulated with the model of an ideal transformer dependences of the output voltage (**a**) and energy efficiency (**b**) of the tested DC-DC converter on its input voltage at f = 25 kHz.

Figure 9a shows that when using the ideal transformer model, the calculation results significantly different from the measurement results are obtained. The calculated V_{out} (V_{in}) characteristics are linear functions, whose slope slightly decreases as the load resistance value decreases. Particularly big differences between the results of calculations and measurements occur at low values of voltage V_{in} and at a low value of resistance R_0.

In Figure 9b, it can be seen that the use of an ideal transformer model in the calculations causes significant errors in the calculations of the energy efficiency of the considered converter. In particular, for the lowest of the considered load resistance values, the calculated energy efficiency values are up to three times higher than the measurement results. Apart from the quantitative differences, there are also qualitative differences between the calculated and measured dependences η (V_{in}). In particular, it is worth noting that the calculations obtain the efficiency exceeding 90% over a wide range of V_{in} voltage variations, whereas the maximum measured value of this efficiency slightly exceeds 50%.

The results of calculations and measurements presented in Figure 9 clearly show that the application of the ideal transformer model in computer analyses of the full-bridge converter does not allow for obtaining reliable results. Therefore, the authors proposed a modification of the transformer model by taking into account undesirable phenomena, which are the source of power losses in this element [15,17]. The equivalent diagram of the modified transformer model is shown in Figure 10.

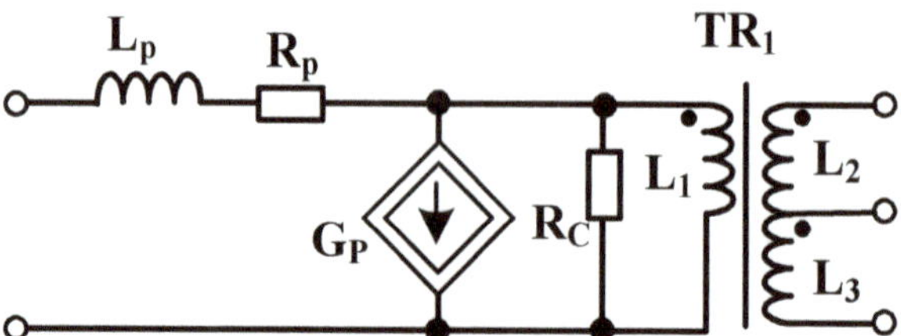

Figure 10. Network representation of the modified model of the transformer.

This model takes into account the dependence of the inductance of individual transformer windings (L_1, L_2 and L_3) on the average value of the current of these windings. These inductances are an increasing function of the load resistance of the converter. The values of these inductances at the load resistance tending to infinity are calculated on the basis of the value of parameter A_L of the used core and the number of turns in each winding. Furthermore, the value of the coupling coefficient between each pair of the windings was made depending on the input voltage and the load resistance of the DC-DC converter. R_C resistor represents the core loss. The value of this resistance depends on the load resistance of the tested converter. The value of this resistor strongly decreases when the load resistance of the converter decreases and it decreases in the range from 1 kΩ to about 10 Ω. If the simulations are performed at variable load resistances, this resistor should be replaced by the controlled current source of the output current depending on the converter output current. Resistor R_p represents the resistance of the primary winding and inductor L_P represents the dissipated inductance of this winding. The controlled current source G_P represents the no-load current of the transformer. The source current is a decreasing function of the input voltage and the load current. Of course, the values of the parameters occurring in the described model are different for each ferromagnetic core.

Using the presented transformer model, the dependences of the output voltage of the tested DC-DC converter and its energy efficiency were calculated on the input voltage at selected values of load resistance. The obtained calculation results (solid lines) were compared with the measurement results (points) in Figure 11.

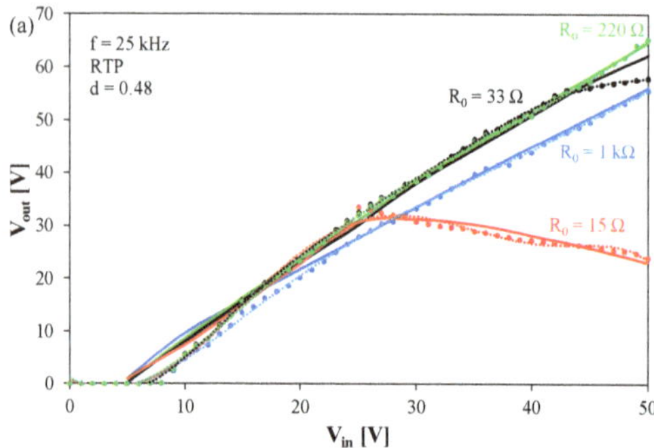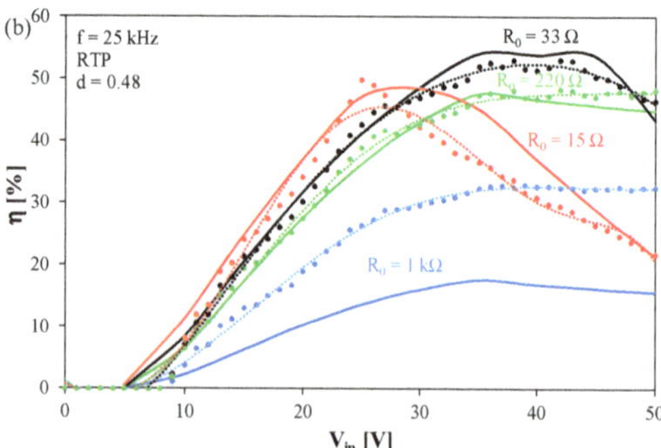

Figure 11. Measured and computed simulated with the proposed model of the transformer dependences of the output voltage (**a**) and energy efficiency (**b**) of the tested DC-DC converter on its input voltage at f = 25 kHz.

As can be seen, the use of the modified transformer model made it possible to significantly improve the accuracy of the calculations compared with the results obtained using the ideal transformer model shown in Figure 9.

In Figure 11a, it can be seen that depending on the value of resistance R_0, different shapes of the characteristic V_{out} (V_{in}) are obtained. In the range of low voltage V_{in}, the no-load current has a significant impact on the considered characteristics. In turn, in the range of high voltage V_{in} and low load resistances, a decrease in the value of the output voltage occurs due to the limitation of the power value that can be transmitted by the transformer. Modeling this phenomenon is possible using the dependence of the coupling coefficient between the windings on the input voltage and resistance R_0. Taking into account the non-ideality of the transformer also allowed for the correct description of the shape of the $\eta\,(V_{in})$ characteristics. It was shown that the highest value of the maximum efficiency was obtained with load resistance $R_0 = 33\ \Omega$.

Figure 12 illustrates the measured and simulated characteristics of the considered DC-DC converters operating at f = 25 kHz, d = 0.48, $R_0 = 220\ \Omega$, and different cores of the transformer.

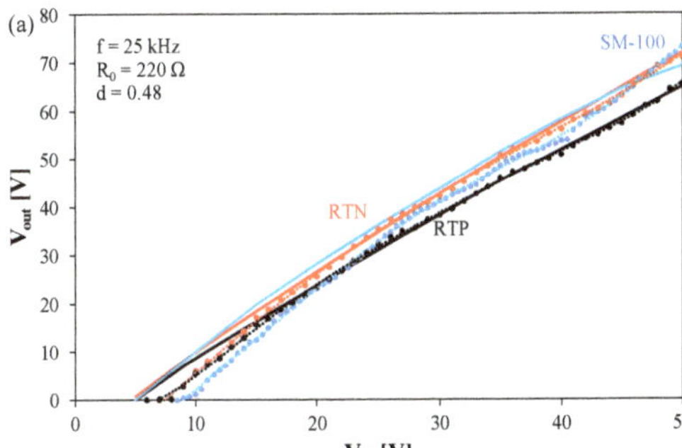

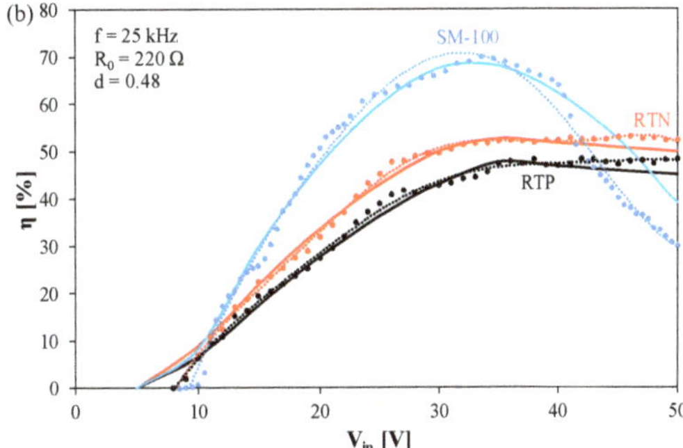

Figure 12. Measured and simulated with the proposed model of the transformer dependences of the output voltage (**a**) and energy efficiency (**b**) of the tested DC-DC converter on its input voltage at f = 25 kHz and $R_0 = 220\ \Omega$.

As is visible, a good agreement between the results of measurements and simulations is obtained for all cores. In the considered, operating conditions and the characteristics $V_{out}\,(V_{in})$ are monotonically increasing functions. Due to the biggest power losses in the RTP core, the values of the output voltage obtained for the converter with the transformer with this core are the lowest. In turn, the dependences $\eta\,(V_{in})$ have the maxima. The largest of them is obtained for the SM-100 core. At the selected value of load resistance, the high value of the output voltage and a high value of energy efficiency in a wide range of the input voltage can be obtained for the transformer with the RTN core.

6. Conclusions

The paper considers the problem of the influence of the ferromagnetic material used in the transformer contained in the full-bridge DC-DC converter on the characteristics of this converter. The measurements illustrating the influence of the input voltage and load resistance of the considered converter on its output voltage and energy efficiency were performed. The investigations were carried out for three different transformers with the same winding structure and different cores made of powdered iron, ferrite, and nanocrystalline material. It was proven that non-idealities of the transformer core can strongly influence the characteristics of full-bridge DC-DC converters. Additionally, in contrast to the classical models, the model of a transformer described in this article makes it possible to properly calculate the characteristics of the considered DC-DC converter.

The performed measurements showed that the material used for the transformer core significantly influences the obtained values of the output voltage and the energy efficiency of the bridge converter. There are also visible differences in the shape of the $V_{out}\,(V_{in})$ and $\eta\,(V_{in})$ characteristics. In a wide range of changes of load resistance and the output voltage, the most advantageous properties were shown by the converter containing the transformer with the nanocrystalline core. On the other hand, in the narrow range of the input voltage changes, the highest efficiency was achieved for the transformer with the ferrite core. Particularly visible differences between the output voltage values obtained for the converters containing different transformer cores occur at the low load resistance values. These differences are even sevenfold. Similarly, there are clear differences between the energy efficiency values of the considered converter, which can be even five times higher.

The computer simulations carried out with the use of the ideal transformer model and models embedded in the SPICE program of the other components showed that such simplified, but often used, calculations give significantly different results than the measurement results. The observed differences are not only quantitative, but also qualitative. The modified transformer model proposed by the authors made it possible to obtain a good agreement between the calculated and measured V_{out} (V_{in}) and η (V_{in}) characteristics. To achieve this compliance, it was necessary to take into account such non-idealities of the transformer as the dependence of the winding coupling coefficient on the input voltage and load resistance, taking into account at the same time the no-load current and resistances modeling the losses in the core and the primary winding.

The results of the investigations presented in this paper may be useful for designers of switch-mode power converter systems. In further work, the authors will develop a universal transformer model dedicated to the analysis of the considered class of converters which will take into account the properties of the magnetic materials used.

Author Contributions: Conceptualization, K.G. and K.D.; methodology, K.G. and K.D.; measurements, K.K.; computations, K.G. and K.D.; resources, K.G., K.D. and K.K.; writing—original draft preparation, K.G. and K.D.; writing—review and editing, K.G. and K.D.; visualization, K.G. and K.D.; supervision, K.G. All authors have read and agreed to the published version of the manuscript.

Funding: Project financed in the framework of the program by Ministry of Science and Higher Education called "Regionalna Inicjatywa Doskonałości" in the years 2019–2022, project number 006/RID/2018/19, the sum of financing 11,870,000 PLN.

Institutional Review Board Statement: Not applicable.

Informed Consent Statement: Not applicable.

Data Availability Statement: Data available for request.

Conflicts of Interest: The authors declare no conflict of interest.

References

1. Rashid, M.H. *Power Electronic Handbook*; Academic Press: Cambridge, MA, USA, 2007.
2. Janke, W. *Impulsowe Przetwornice Napięcia Stałego*; Wydawnictwo uczelniane Politechniki Koszalińskiej: Koszalin, Poland, 2014.
3. Billings, K.; Morey, T. *Switch-Mode Power Supply Handbook*; McGraw-Hill: New York, NY, USA, 2011.
4. Ericson, R.; Maksimovic, D. *Fundamentals of Power Electronics*; Kluwer Academic Publisher: Norwell, MA, USA, 2001.
5. Basso, C. *Switch-Mode Power Supply SPICE Cookbook*; McGraw-Hill: New York, NY, USA, 2001.
6. Mohan, N.; Undeland, T.M.; Robbins, W.P. *Power Electronics: Converters, Applications, and Design*, 3rd ed.; Wiley: Hoboken, NJ, USA, 2003.
7. Ademulegun, O.O.; Moreno Jaramillo, A.F. Power conversion in a grid-connected residential PV system with energy storage using fuzzy-logic controls. *Int. Trans. Electr. Energy Syst.* **2020**, *30*, e12659. [CrossRef]
8. Detka, K.; Górecki, K. Influence of the Size and Shape of Magnetic Core on Thermal Parameters of the Inductor. *Energies* **2020**, *13*, 3842. [CrossRef]
9. Detka, K.; Górecki, K. Influence of the size and the material of the magnetic core on thermal properties of the inductor. *Microelectron. Reliab.* **2022**, *129*, 114458. [CrossRef]
10. Detka, K.; Górecki, K. Wpływ samonagrzewania w dławiku na charakterystyki przetwornicy typu boost. *Przegląd Elektrotechniczny* **2014**, *90*, 19–21.
11. Górecki, K.; Detka, K.; Zarębski, J.A. Wierszyło: Wpływ rdzenia dławika na charakterystyki przetwornicy Buck. *Przegląd Elektrotechniczny* **2016**, *92*, 137–139.
12. Tumański, S. *Handbook of Magnetic Measurements*; Taylor and Francis Group: Boca Raton, FL, USA, 2011.
13. Ghadimi, H.; Rastegar, A. Keyhani: Development of Average Model for Control of a Full Bridge PWM DC-DC Converter. *Eng. J. Iran. Assoc. Electr. Electron. Engineers* **2007**, *4*, 52–59.
14. Ekici, A.; Najafi, F. Dirisaglik: 600W DC-DC Converter Design Using Flyback, Half Bridge, Full Bridge LLC Topologies and Comparison of Simulation Results. *Eurasia Proc. Sci. Technol. Eng. Math.* **2021**, *16*, 219–224. [CrossRef]

15. Górecki, K.; Detka, K.; Górski, K. Compact thermal model of pulse transformer taking into account nonlinearity of heat transfer. *Energies* **2020**, *13*, 2766. [CrossRef]
16. Van den Bossche, A.; Valchev, V. *Inductor and Transformers for Power Electronic*; CRC Press: Boca Raton, FL, USA, 2005.
17. Górecki, K.; Godlewska, M. Modelling characteristics of the impulse transformer in a wide frequency range. *Int. J. Circuit Theory Appl.* **2020**, *48*, 750–761. [CrossRef]
18. Wilson, P.R.; Ross, J.N.; Brown, A.D. Simulation of magnetics components models in electrics circuits including dynamic thermal effects. *IEEE Trans. Power Electr.* **2002**, *17*, 55–65. [CrossRef]
19. Lullo, G.; Sciere, D.; Vitale, G. Non-linear inductor modelling for a DC/DC Buck converter. In Proceedings of the International Conference on Renewable Energies and Power Quality (ICREPQ'17), Malaga, Spain, 21–25 March 2017. [CrossRef]
20. Kazimierczuk, M. *High-Frequency Magnetic Components*; Wiley: Hoboken, NJ, USA, 2014.
21. Maksimovic, D. Automatem steady-state analysis of switching power converters using a general-purpose simulation tool. In Proceedings of the IEEE Power Electronics Specialists Conference PESC, St. Louis, MO, USA, 27 June 1997; Volume 2, pp. 1352–1358.
22. Górecki, K.; Zarębski, J. Electrothermal analysis of the self-excited push-pull dc-dc converter. *Microelectron. Reliab.* **2009**, *49*, 424–430. [CrossRef]
23. Huang, B.; Hu, M.; Chen, L.; Jin, G.; Liao, S.; Fu, C.; Wang, D.; Cao, K. A Novel Electro-Thermal Model of Lithium-Ion Batteries Using Power as the Input. *Electronics* **2021**, *10*, 2753. [CrossRef]
24. Górecki, K.; Detka, K. Electrothermal model of choking-coils for the analysis of dc-dc converters. *Mater. Sci. Eng. B* **2012**, *177*, 1248–1253. [CrossRef]
25. Iqbal, M.T.; Maswood, A.I.; Tariq, M.; Iqbal, A.; Verma, V.; Urooj, S. A Detailed Full-Order Discrete-Time Modeling and Stability Prediction of the Single-Phase Dual Active Bridge DC-DC Converter. *IEEE Access* **2022**, *10*, 31868–31884. [CrossRef]
26. IRF540 Datasheet. Available online: https://pdf1.alldatasheet.com/datasheet-pdf/view/17799/PHILIPS/IRF540.html (accessed on 5 July 2022).
27. MBR10100 Datasheet. Available online: https://www.alldatasheet.com/datasheet-pdf/pdf/3128/MOTOROLA/MBR10100.html (accessed on 5 July 2022).
28. IR2111 Datasheet. Available online: https://www.alldatasheet.com/datasheet-pdf/pdf/68060/IRF/IR2111.html (accessed on 5 July 2022).
29. SM-100 Material Parameters. Available online: https://en.aet.com.pl/Shop/Type/ProductDetails/ProductID/R-040-0003/ProductName/Toroidal-ferrite-core-R40-24-16-SM-100-OR40X16-24HC-coated-RoHS (accessed on 5 July 2022).
30. RTN M-074 Material Parameters. Available online: https://feryster.pl/rdzenie-rtn (accessed on 5 July 2022).
31. RTP Material Parameters. Available online: https://feryster.pl/rdzenie-proszkowe-rtp (accessed on 5 July 2022).
32. Górecki, P.; Górecki, K. Methods of Fast Analysis of DC–DC Converters—A Review. *Electronics* **2021**, *10*, 2920. [CrossRef]
33. Wilamowski, B.; Jager, R.C. *Computerized Circuit Analysis Using SPICE Programs*; McGraw-Hill: New York, NY, USA, 1997.
34. SPICE Models, Mouser Electronics. Available online: https://eu.mouser.com/ProductDetail/onsemi/MBR10100?qs=3JMERSakeboFOtFAhNHcug%3D%3D (accessed on 18 July 2022).

Article

Alternative Simplified Analytical Models for the Electric Field, in Shoreline Pond Electrode Preliminary Design, in the Case of HVDC Transmission Systems

George J. Tsekouras [1,2,*], Vassiliki T. Kontargyri [1,2], John M. Prousalidis [3], Fotios D. Kanellos [4], Constantinos D. Tsirekis [1,5], Konstantinos Leontaritis [5], John C. Alexandris [5], Panagiota M. Deligianni [1,5], Panagiotis A. Kontaxis [1,2] and Antonios X. Moronis [1]

[1] Department of Electrical and Electronics Engineering, University of West Attica, 250 Thivon Str., Egaleo, 12241 Athens, Greece
[2] School of Electrical and Computer Engineering, National Technical University of Athens, Iroon Polytechniou 9, Zografou, 15780 Athens, Greece
[3] School of Naval Architecture and Marine Engineering, National Technical University of Athens, Iroon Polytechniou 9, Zografou, 15780 Athens, Greece
[4] School of Electrical and Computer Engineering, Technical University of Crete, University Campus, Akrotiri, 73100 Chania, Greece
[5] Hellenic Independent Power Transmission Operator, Dyrrachiou 89 & Kifissou, Sepolia, 10443 Athens, Greece
* Correspondence: gtsekouras@uniwa.gr; Tel.: +30-21-0538-1750

Citation: Tsekouras, G.J.; Kontargyri, V.T.; Prousalidis, J.M.; Kanellos, F.D.; Tsirekis, C.D.; Leontaritis, K.; Alexandris, J.C.; Deligianni, P.M.; Kontaxis, P.A.; Moronis, A.X. Alternative Simplified Analytical Models for the Electric Field, in Shoreline Pond Electrode Preliminary Design, in the Case of HVDC Transmission Systems. *Energies* **2022**, *15*, 6493. https://doi.org/10.3390/en15176493

Academic Editor: Juri Belikov

Received: 15 July 2022
Accepted: 30 August 2022
Published: 5 September 2022

Publisher's Note: MDPI stays neutral with regard to jurisdictional claims in published maps and institutional affiliations.

Abstract: In Greece, a new bi-polar high voltage direct current (HVDC) transmission system with a ground return was designed with nominal characteristics of ±500 kV, 1 GW, between Attica in the continental country and the island of Crete, which is an autonomous power system based on thermal diesel units. The interconnection line has a total length of about 380 km. The undersea section is 330 km long. In this paper, the use of the Aegean Sea as an active part of the ground return, based on shoreline pond electrodes, was proposed to avoid EUR 200 M of expenses. According to the general guidelines for HVDC electrode design by the International Council on Large Electric Systems (CIGRE) working group B4.61/2017, the electric field and ground potential rise of shoreline electrodes should be studied to analyze safety, electrical interference and corrosion impacts related to the operation of the electrodes. Two kinds of studies are available; one is a simplified approach based on a spherical/pointed electrode centered at the edge of the seashore and seabed, assuming it to be sloping to the horizontal, and the other is a detailed simulated model using a suitable electric field software package. The first approach usually gives more unfavorable results than the second one, especially in the near electric field, while it can not take into account obstacles, i.e., dams, near to electrode position. The second approach demands a detailed description of the wider installation area, which cannot be available during the preliminary study, significant computational time and considerable financial resources for the purchase of a reliable specialized software package. In this research, a two-step modification of the CIGRE simplified model was proposed. The first modification deals with the obstacles in the near electric field, and the second modification deals with the use of a linear current source (instead of a point one), which can give more accurate results. Additionally, the electric field for complex electrode formation is calculated by applying the superposition method, which can be easily achieved using a common software package, i.e., MATLAB. The proposed simplified approaches were applied on shoreline pond electrode locations for the Attica–Crete HVDC interconnection line (between Stachtoroi island in Attica and Korakia beach in Crete), allowing the preliminary study to be conducted swiftly, giving satisfactory results about electric field gradient, ground potential rise and resistance to remote earth of electrodes stations for the near and far electric field.

Keywords: analytical models; electric field; HVDC transmission system; shoreline pond electrode

---

1. Introduction

The history of high voltage direct current interconnections (HVDC) begins in New York with Th. Edison at about 1880 [1]. This was followed by individual HVDC interconnections [2–4]. Currently, their use is widespread in offshore wind parks [5–9]. Many technical guidelines have been written about HVDC interconnections, mainly by the International Council on Large Electric Systems (CIGRE) [10–22]; the Institute of Electrical and Electronics Engineers (IEEE) [23–30]; the International Electrotechnical Commission (IEC) [30–42]; Electrical Power Research Institute (EPRI) [43–46]; Energy Department—Oak Ridge National Laboratory, USA [47,48]; Det Norske Veritas (DNV) [49]; and the European Commission [50]. These technical guidelines solve issues such as the configuration of networks and their general characteristics [10,19,25,34,38,43]; the design of individual components, such as transformers [11,12,15,23,24,30,39–42], electrodes [17,34,45–47], switches [18], insulators [26], cables [27,37] and reactors [29]; the environmental/acoustic/electromagnetic effects [13,31,32,36,44,48]; the feasibility of the relevant interconnection projects [14]; the configuration of special purpose networks, such as wind farms [16,49,50]; control and protection techniques of the entire HVDC network [20,22,28]; and testing [29,33,37].

The Hellenic Independent Power Transmission Operator (IPTO) studied, in 2018–2019, the bi-directional interconnection between the island of Crete and mainland Greece in the region of Attica, where there is a 400 kV AC high voltage network, in order to reduce the operation of petroleum thermal power plants and increase the penetration of renewable energy sources in Crete. The required power of the interconnection is 1 GW with a length of at least 380 km, of which the largest part (330 km) is underwater, as shown in Figure 1 [51]. Therefore, the interconnection took place with HVDC in order to reduce the required reactive power due to the existence of cable capacities (where it is only required by the inverters) and to achieve the stability of the electric power system. An HVDC bipolar heteropolar configuration with a nominal power of 2 × 500 MW, at a nominal voltage of ±500 kV DC and voltage source converters, was selected. The return is made via land, as EUR 200 M is saved. Considering that seawater is a much better conductor than the land (at least 100 times lower resistivity), the ground return beyond the sea was proposed by placing two electrode stations in the sea. Near Attica, the island of Stachtoroi was chosen, where the nearest residential area is located at a distance of about 8 km, so that there is no nuisance to the inhabitants (especially from electrochemical erosions). The distance from the 400 kV High Voltage Substation where the inverters are installed is less than 20 km (Figure 2). Near the island of Crete, the electrode station is constructed on the deserted Korakia beach, which is accompanied by the respective converters in the area of Damasta (Figure 3). In addition, the two specific sites meet various criteria, such as geophysical, geological, hydrological, seismological, volcanic, exclusion zones, licenses, the possibility of construction, etc. In addition, electrochemical corrosion should be eliminated in nearby installations. Step and touch voltages should also be eliminated according to IEC 60749-1:2007 [17] (pp. 59–70). From the six usual types of electrodes (land–shallow horizontal, land vertical, land–deep well, sea, shore–beach, shore–pond) according to [17] (p. 19), the shoreline pond electrode was selected for which the determination of the electric field and the potential rise was calculated according to the guidelines of CIGRE B4.61 675:2017 [17]. Based on [17] (pp. 109–120), two methods can be applied during its preliminary design:

- *Simplified analytical method*: Electric current is injected at points, and it is considered that space is divided into a soil hemisphere and the area of air [17] (pp. 118–119), thus solving the problem with a simple application of electric field and potential equations;
- *Computational method*: Numerical methods are applied for solving electric field problems in order to calculate ground potential rise, electrode resistance, etc. [17] (pp. 119–120). The input data are the configuration, the electrical resistivity of the conductors in the area, especially the resistivity of the ground, which is determined by geophysical methods, such as electrical resistivity tomography and magnetotelluric tomography [52,53], which are extremely time-consuming and costly.

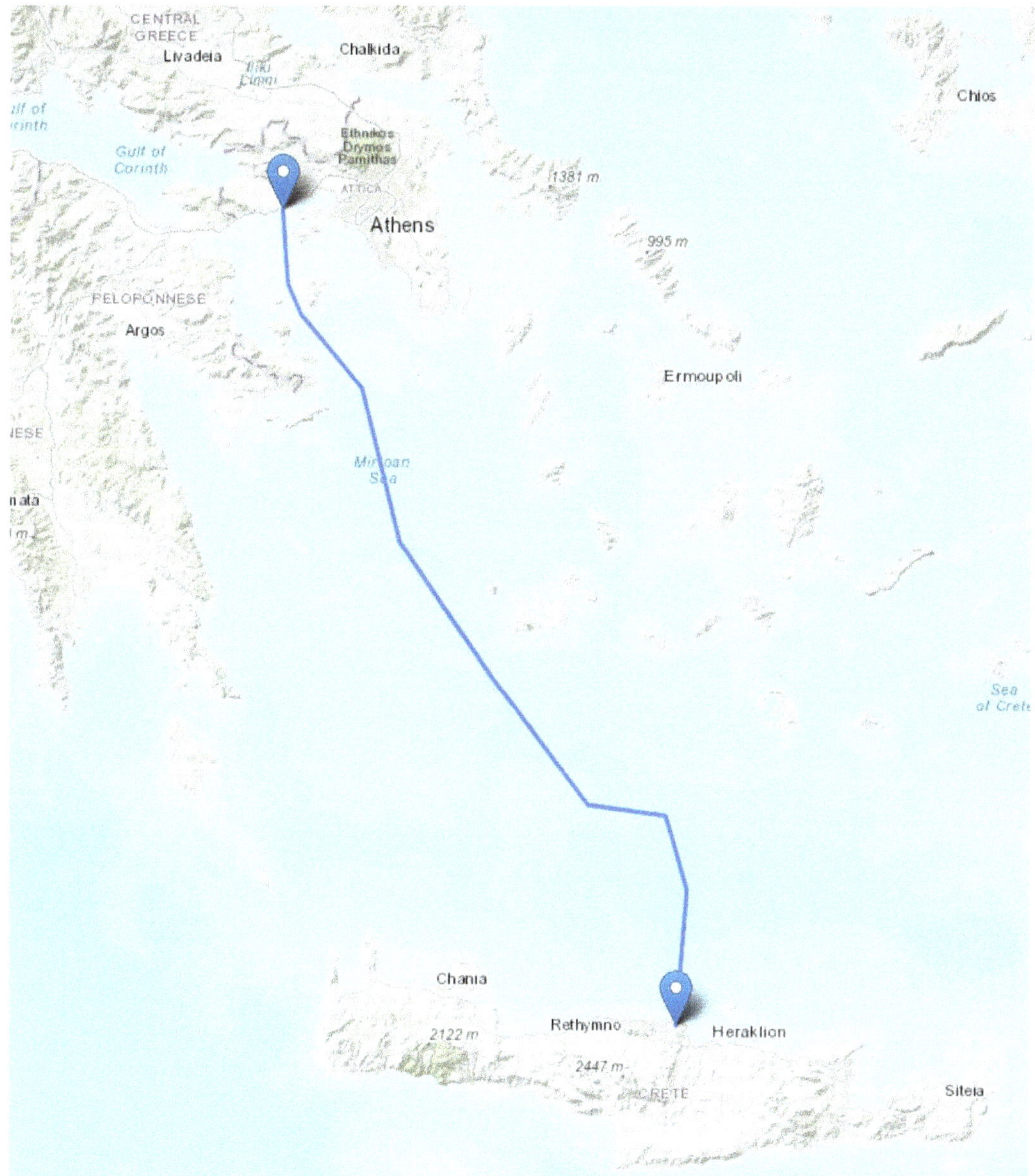

Figure 1. Geographical map of the interconnection between Crete (electrode station in Korakia beach, converter station in Damasta) and continental Greece power systems in Attica (electrode station in Stachtoroi island, converter station in Koumoundourou) [51].

An analytical method similar to CIGRE B4.61 675:2017 [17] is described in IEC TS 62344:2013 [34] (pp. 30–32), with the difference that the air occupies a hemisphere, the water forms a "wedge" of a specific angle, and the rest is homogeneous soil. It is based on Rusck methodology [54], which determined the electrode resistance, electric field strength and current density in the inland ground and seawater, which was repeated in [55] (pp. 465–476), [56–58]. In addition, there is the Uhlmann method [57,59] (pp. 267–272), which discusses the issue of sea electrodes at a distance from the shore, but considering the ground and the seabed completely insulated. The detailed solution of the field of a shoreline pond anode electrode is mentioned in technical reports about Italy–Greece Interconnection, such as [60].

Figure 2. Location of the electrode station for an HVDC transmission system in the region of Attica–Stachtoroi (with italics are the nearby residential areas under study).

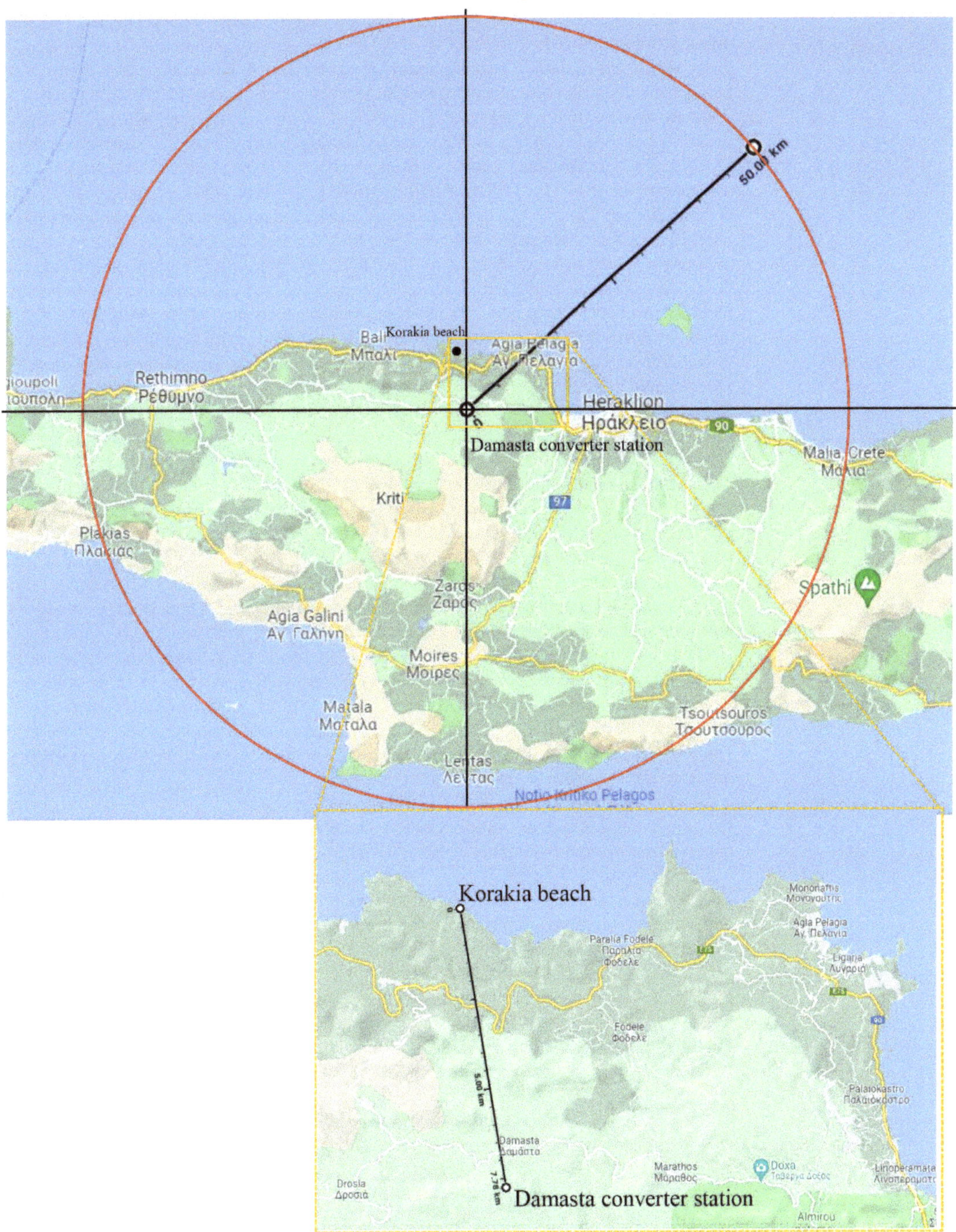

Figure 3. Location of the electrode station for an HVDC transmission system in the region of Crete.

EPRI proposes the use of numerical algorithms using resistivity data from geophysical methods [45], which are more widely used in grounding electrodes [61] but also in sea electrodes [58,62,63]. In [62], both near and far electric fields were analyzed, using the numerical methods of hemispheroidal, finite volume element and inclined layer, and finally proposes the use of the first two methods [64,65] that have finite limits on the volume of water for the calculation of the scalar potential and ground potential. In [63,66], the use of the finite element method for near-field analysis was suggested. Alternatively, semi-analytical methods can be used, such as the complex images method with a "generalized pencil of functions" interpolator, achieving numerical convergence and solution stability with satisfactory results in the HVDC electrode study [67], which was confirmed with the help of finite elements in grounding systems [68]. Numerical methods can also be used to draw useful conclusions about the problems that stray currents of marine electrodes can cause in tubes [69] and railways [70] using the CDEGS (current distribution, electromagnetic fields, grounding and solid structure) software package. A comparative cost study between metallic and earth return on HVDC transmission line was carried out using the COMSOL (cross-platform finite element analysis, solver and multiphysics simulation software) programming package [71]. In the most recent technical reports, simulations were proposed. The far-field behavior of the HVDC interconnection electrodes of the Lower Churchill Project in three areas (Gull Island, Soldiers Pond, New Brunwick) using three-dimensional finite elements over an area of 1,000 km × 1,200 km and at a depth of 5 km, taking into investigation appropriate geographical/geophysical data, is determined [72]. By using computational tools of mathematical analysis and having implemented geophysical studies, the possible locations in the respective areas of Gull Island and Soldiers Pond were examined in order to determine the GPR in areas of interest and the effects in the respective areas [73], while the near field is studied in [74]. In [75], the field behavior of the electrodes in the Fagelsundet–Forsmark area for the Fenno-Scan HVDC link was studied, taking the resistivity of the area up to a depth of 41 km as input data, which were obtained by methods such as geophysical–petrophysical–electric measurements in drillings, transient electromagnetic soundings, electric soundings, electric potential measurements in the sea. These data were utilized by DCIPF3D software from UBC-GIF based on finite differences. However, the comparison of the experimental data shows a significant deviation near the Forsmark power plant and near the HVAC substation at a distance of fewer than 25 km. Therefore, the estimated field behavior, either with analytical models or with computational models, deviates from the experimental measurements [60,75].

The guidelines from the Oak Ridge National Laboratory [47] (pp. 39–40) are descriptive, while in other regulations, they are practically non-existent [49,50]. In many scientific papers, there are general instructions for the selection of electric field solution methods [76].

In this paper, the preliminary design of the shoreline pond electrode station locations for the Attica–Crete HVDC interconnection line (between Stachtoroi island in Attica and Korakia beach in Crete) is studied using analytical models. The study of the electric field through a suitable electric field software package was not possible because it is time-consuming and costly. The proposed method was based on the general guidelines for HVDC electrode design analyzed by the CIGRE working group B4.61/2017 [17] using a simplified approach based on a spherical/pointed electrode centered at the edge of the seashore and seabed, assuming it to be sloping to the horizontal so that the electric field and ground potential rise of shoreline electrodes can be identified in order to analyze safety issues, electrical interference and corrosion impacts related to the operation of the electrode. The theoretical background of the electric field strength distribution was proven, generalizing the mathematical solutions proposed by both CIGRE B4.61 675: 2017 [17] (pp. 118–119) and by the IEC TS 62344:2013 [34] (pp. 30–32). Two modifications were proposed in this paper. The first modification deals with the obstacles in the near electric field, such as the existence of a dam. The second modification is relevant to the use not of a point current source but of a linear one, which can give more valuable results, especially in areas where the sea is shallow and has a relatively constant depth. In addition, when an electrode system is formed, the electric field strength can be calculated by applying the superposition method using a common software package, i.e., MATLAB. The proposed methods were applied to shoreline pond electrode locations for the Attica–Crete HVDC interconnection line allowing the preliminary study to be conducted swiftly, giving satisfactory results about electric field gradient, ground potential rise and resistance to remote earth of electrode stations for the near and far electric field.

2. Theoretical Background

2.1. Method "A"—Combination of Electric Field Distribution Methods by CIGRE B4.61 675:2017 and IEC TS 62344:2013

According to CIGRE B4.61 675:2017 [17] (pp. 118–119 and Figure 5.35), an electrode on the shore (or an electrode near the shore) is considered, which is placed in the center of the shore (or at the bottom of the sea in the shallows), while the seabed is assumed to be inclined to the horizon by an angle a. A sphere with radius r is considered around the electrode. The ground forms a hemisphere, the seawater is part of a sphere with the angle of a, and the rest part of the sphere is air (Figure 4a). According to IEC TS 62344:2013 [34] (pp. 30–32 and Figure 5), the assumption is the same, the only difference being that the soil and the seawater form a hemisphere (Figure 4b). In summary, the area around the electrode forms a sphere, which is divided into three parts: the homogeneous ground of electrical resistivity ρ_s with angle θ_s, the seawater of electrical resistivity ρ_w with angle θ_w, and the air (Figure 4c).

In spherical coordinates, the areas of the corresponding parts of the sphere of radius r for the ground of azimuth angle θ_s and for the sea of azimuth angle θ_w are, respectively:

$$S_S = \int_{-\pi/2}^{\pi/2} \left(r \times cos\varphi \times d\varphi \times \int_0^{\theta_s} r \times d\theta \right) = 2 \times \theta_s \times r^2 \tag{1}$$

$$S_w = \int_{-\pi/2}^{\pi/2} \left(r \times cos\varphi \times d\varphi \times \int_0^{\theta_w} r \times d\theta \right) = 2 \times \theta_w \times r^2. \tag{2}$$

The total current intensity I_{tot} passes through the ground and water, assuming that the air is an insulator of very high electrical resistivity. Due to uniform resistivity, the total current intensity passes through the ground and seawater sections radially and symmetrically. Considering the current intensity of the ground I_s and the current density of the ground J_s, the current intensity of the seawater I_w and the current density of the seawater J_w, the total current is equal to:

$$I_{tot} = I_s + I_w = \int_{S_s} \vec{J}_s \times \vec{n} \times dS + \int_{S_w} \vec{J}_w \times \vec{n} \times dS = J_s \times S_s + J_w \times S_w \tag{3}$$

Because the radial component of the electric field strength on the dividing surface is continuous, the radial electric field strength of the sea E_{rw} and the radial electric field strength of the ground E_{rs} are equal. Due to symmetry, there are no azimuth and polar components. In combination with Ohm's law, the electric field strength is given by:

$$E_{rs} = E_{rw} = \rho_s \times J_s = \rho_w \times J_w \Rightarrow J_s = \frac{\rho_w}{\rho_s} \times J_w \tag{4}$$

Combining Equations (1)–(4), the current densities and electric field intensities are determined as follows:

$$I_{tot} = \frac{\rho_w}{\rho_s} \times J_w \times 2 \times \theta_s \times r^2 + J_w \times 2 \times \theta_w \times r^2 = J_w \times 2 \times r^2 \times \left(\theta_w + \frac{\rho_w}{\rho_s} \times \theta_s \right) \Rightarrow J_w = \frac{I_{tot}}{2 \times r^2 \times \left(\theta_w + \frac{\rho_w}{\rho_s} \times \theta_s \right)} \tag{5}$$

$$J_s = \frac{\rho_w}{\rho_s} \times \frac{I_{tot}}{2 \times r^2 \times \left(\theta_w + \frac{\rho_w}{\rho_s} \times \theta_s \right)} = \frac{I_{tot}}{2 \times r^2 \times \left(\frac{\rho_s}{\rho_w} \times \theta_w + \theta_s \right)} \tag{6}$$

$$E_{rs} = E_{rw} = \rho_w \times \frac{I_{tot}}{2 \times r^2 \times \left(\theta_w + \frac{\rho_w}{\rho_s} \times \theta_s \right)} = \frac{I_{tot}}{2 \times r^2 \times \left(\frac{\theta_w}{\rho_w} + \frac{\theta_s}{\rho_s} \right)} \tag{7}$$

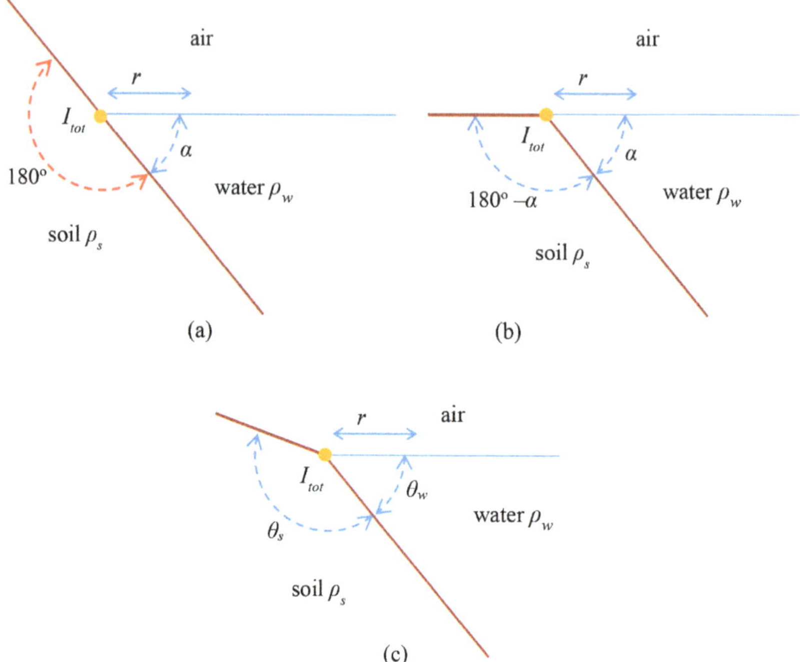

Figure 4. Simplified model for electrode placement on shore or in near-shore sea based on (**a**) Figure 5.35 by CIGRE B4.61 675:2017 [17], (**b**) Figure 5 by IEC TS 62344:2013 [34], (**c**) the proposed generalization.

Considering that the potential is zero at infinite distance, the absolute potential is determined as follows:

$$V(r) = \int_r^{\infty} \vec{E} \times \vec{d\ell} = \int_r^{\infty} E \times d\ell = \int_r^{\infty} \frac{I_{tot}}{2 \times \ell^2 \times \left(\frac{\theta_w}{\rho_w} + \frac{\theta_s}{\rho_s}\right)} \times d\ell \Rightarrow V(r) = \frac{I_{tot}}{2 \times \left(\frac{\theta_w}{\rho_w} + \frac{\theta_s}{\rho_s}\right)} \times \left(-\frac{1}{\ell}\right)\Big|_r^{\infty} = \frac{I_{tot}}{2 \times r \times \left(\frac{\theta_w}{\rho_w} + \frac{\theta_s}{\rho_s}\right)} \tag{8}$$

Dividing the potential difference between the surface of the electrode of radius r_{el} and remote earth (infinite distance) for a per unit current intensity, the resistance of remote earth is given by:

$$R_{el} = \frac{V(r_{el})}{I_{tot}} = \frac{1}{2 \times r_{el} \times \left(\frac{\theta_w}{\rho_w} + \frac{\theta_s}{\rho_s}\right)} \tag{9}$$

The above approach is quite simplified and includes the following assumptions as mentioned in ([17], p. 119):

- *Infinite shore level*: Electrodes are usually placed in protected areas, such as a cave or a shore, and full exposure to the beach is not available. Moreover, the straight part of the coast is limited;
- *Inclination*: The actual inclination differs radially and axially;
- *Uniform electrical resistivities of seawater and soil*: Due to the variation in the electrical resistivities, the isodynamic surfaces are not circular;
- *"Wedge" shape of the ground and "wedge" shape of the water*: The soil does not take the form of a wedge. Moreover, water does not form a uniform wedge, and its shape differs in different directions. However, it is a better approach than those of CIGRE B4.61 675:2017 and IEC TS 62344:2013.

Additionally, the above theoretical analysis shows the following:

- *Confirmation of CIGRE and mathematical errata*: Equations (5.5–8) through (5.5–10), (5.5–12), (5.5–13) in [17], on current densities, voltage and remote ground resistance present typographical errors and, in some cases, inconsistencies regarding measurement units (e.g., the voltage in V/m and resistance in Ω/m). From Equations (5)–(9), the correct quantities result, setting π rad (where θ_s) and α rad (where θ_w);
- *Confirmation of IEC*: Equation (13) in [34] results from Equation (8), in the present paper, by setting π-α rad (where θ_s) and α rad (where θ_w), whereby all other equations in [34] are directly confirmed;
- *Eliminating the hemisphere of soil*: The soil does not form a hemisphere or a wedge with the horizontal plane. The ground angle is no longer π rad according to CIGRE B4.61 675:2017 or (π-α) rad according to IEC TS 62344:2013. In addition, as suggested in CIGRE B4.61 675:2017 [17] (pp. 118–119), the shape of the water wedge along the coast is flat. Only a part (less than 180°) of the hemispherical part forms the wedge and, in some cases, is limited to a few degrees if the electrode is placed on a narrow beach. The analysis can be improved by taking different inclinations of the seabed and multiplying by a correction factor if the exposed side of the sea is limited to an angle φ (rad), so the multiplier by π/φ should be applied to the calculated distance of the remote earth;
- *Results in favor of safety*: By modifying the assumptions or always considering the worst-case scenario, the corresponding assessment can be made on the safe side, e.g., considering the average inclination as the distance of interest and not the initial inclination from the shore, which is usually relatively large or setting the ground resistivity infinite, as was performed with other analytical models [57,59], (pp. 267–272). In the last case, Equations (5)–(9) are simplified as follows:

$$J_{w,\rho_s=\infty} = \frac{I_{tot}}{2 \times r^2 \times \theta_w} \tag{10}$$

$$J_{s,\rho_s=\infty} = 0 \tag{11}$$

$$E_{rs,\rho_s=\infty} = E_{rw,\rho_s=\infty} = \frac{I_{tot} \times \rho_w}{2 \times r^2 \times \theta_w} \tag{12}$$

$$V(r)_{\rho_s=\infty} = \frac{I_{tot} \times \rho_w}{2 \times r \times \theta_w} \tag{13}$$

$$R_{el,\rho_s=\infty} = \frac{\rho_w}{2 \times r_{el} \times \theta_w} \tag{14}$$

2.2. Method "B"—Combination of Electric Field Distribution Methods by CIGRE B4.61 675:2017 and IEC TS 62344:2013 with the Addition of a Dam

The simplified methodology of Section 2.1 is extended by considering a dam of uniform electrical resistivity ρ_d constructed from stones or artificial blocks. Instead of the typical dam, the simplifying structure of Figure 5 was considered, where an electrode on the shore (or at the bottom of the sea in the shallows) is placed at the center of the respective coast, while the seabed is assumed to be inclined to the horizon by an angle θ_w. The distance of the electrode from the dam is r_1, and the thickness of the dam is d, so the outer surface of the dam has a radius of $r_2 = r_1 + d$. Seawater has the same electrical resistivity ρ_w on both sides of the dam. The homogeneous ground of electrical resistivity ρ_s has an angle θ_s, and the rest is air (Figure 5). The initial assumptions of Section 2.1 are applied, i.e., about infinite shore level, the uniform inclination of the seabed, etc. The electromagnetic field theory requires:

- Continuity of the vertical current density on the dividing surface:

$$J_{n1} = J_{n2} \tag{15}$$

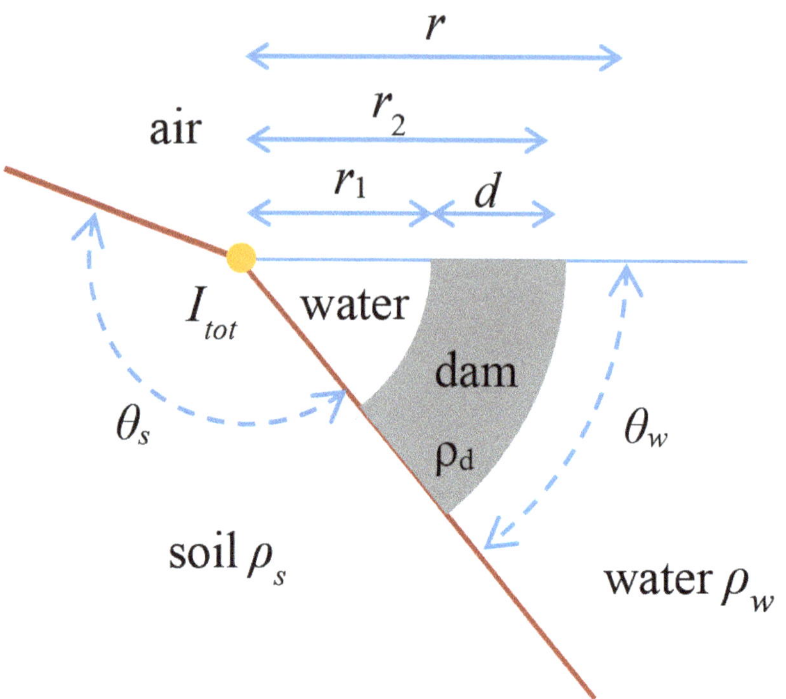

Figure 5. Simplified generalized model for placement of an electrode on the shore or in near-shore sea considering a point electrode, in spherical coordinates and using a dam–soil surface.

- Continuity of the tangential electric field strength on the dividing surface:

$$E_{t1} = E_{t2} \Leftrightarrow \rho_1 \times J_{t1} = \rho_2 \times J_{t2} \tag{16}$$

Because the resistivity of dam material ($\rho_d = \rho_2 = 100\ \Omega{\cdot}m$) is larger than the resistivity of seawater ($\rho_w = \rho_1 = 0.25\ \Omega{\cdot}m$), the ratio of the tangential current densities in a dam against water is limited significantly (1:400). Considering that the direction of the current density vector is primarily radial in spherical coordinates, the tangential/polar component is ignored in water–dam interfaces. Therefore, the areas of the corresponding parts of the sphere of radius r for the ground S_s with azimuth angle θ_w and for the sea S_w (or for the dam S_d) with azimuth angle θ_s (in rad) are given by the Equations (1) and (2), respectively.

Assuming that the air is an insulator of a high resistivity value, the total current intensity I_{tot} passes through the ground and water (or the dam). Due to uniform resistivity, the total current intensity passes through the ground and seawater mainly radially and symmetrically. Considering the electric current intensity of the soil I_s, the current density of the soil J_s, the current intensity of the seawater I_w and the current density of the seawater J_w, the current intensity of the dam I_d and the current density of the dam J_d, Equation (3) is applied to the seawater–ground complex, while the total current intensity for the seawater–dam is given by:

$$I_{tot} = I_s + I_d = \int_{S_s} \vec{J}_s \times \vec{n} \times dS + \int_{S_d} \vec{J}_d \times \vec{n} \times dS = J_s \times S_s + J_d \times S_d \tag{17}$$

Equation (15) becomes:

$$I_d = I_w = J_d \times S_d = J_w \times S_w \Rightarrow J_d = J_w\ : r = r_1\ and\ r = r_2 \tag{18}$$

Considering the continuity of the radial electric field strength on the water–soil interface, the radial electric field strength of the sea E_{rw} and the radial electric field strength of the soil E_{rs} are equal and practically the only components. It follows that:

$$E_{rs} = E_{rw} = \rho_s \times J_s = \rho_w \times J_w \Rightarrow J_s = \frac{\rho_w}{\rho_s} \times J_w : r < r_1 \; or \; r > r_2 \tag{19}$$

Considering the continuity of the radial electric field strength on the dam–soil interface, the radial electric field strength of the dam E_{rd} and the radial electric field strength of the soil E_{rs} are equal and practically the only components. It follows that:

$$E_{rs} = E_{rd} = \rho_s \times J_s = \rho_d \times J_d \Rightarrow J_s = \frac{\rho_d}{\rho_s} \times J_d : r_1 < r < r_2 \tag{20}$$

By combining Equations (1)–(3), (17), (19) and (20), the current densities and electric field strengths are determined as follows:

$$J_w = \frac{I_{tot}}{2 \times r^2 \times \left(\theta_w + \frac{\rho_w}{\rho_s} \times \theta_s \right)} : r < r_1 \; or \; r > r_2 \tag{21}$$

$$J_s = \frac{I_{tot}}{2 \times r^2 \times \left(\frac{\rho_s}{\rho_w} \times \theta_w + \theta_s \right)} : r < r_1 \; or \; r > r_2 \tag{22}$$

$$E_{rs} = E_{rw} = \frac{I_{tot}}{2 \times r^2 \times \left(\frac{\theta_w}{\rho_w} + \frac{\theta_s}{\rho_s} \right)} : r < r_1 \; or \; r > r_2 \tag{23}$$

$$J_d = \frac{I_{tot}}{2 \times r^2 \times \left(\theta_w + \frac{\rho_d}{\rho_s} \times \theta_s \right)} : r_1 < r < r_2 \tag{24}$$

$$J_s = \frac{I_{tot}}{2 \times r^2 \times \left(\frac{\rho_s}{\rho_d} \times \theta_w + \theta_s \right)} : r_1 < r < r_2 \tag{25}$$

$$E_{rs} = E_{rd} = \frac{I_{tot}}{2 \times r^2 \times \left(\frac{\theta_w}{\rho_d} + \frac{\theta_s}{\rho_s} \right)} : r_1 < r < r_2 \tag{26}$$

It was noted that Equation (18) does not apply because, in this case, the results of Equations (21) and (24) should be equal on the boundary surfaces $r = r_1$ and $r = r_2$, which, however, is not the case due to the fact that non-radial currents on the respective surfaces were ignored. Equations (21)–(23) are strictly valid for $r < r_1$. Further on, another approach is attempted, as was also the case with the initial assumptions in CIGRE B4.61 675:2017 [17] (pp. 118–119), with regards to the sea–soil interface, which (being vertical to the plane of Figure 4c) is not depicted.

The absolute value of potential considering zero potential at infinite distance is determined as follows:

$$V(r) = \int_r^\infty \vec{E} \times \vec{d\ell} = \int_r^{r_1} E_w \times d\ell + \int_{r_1}^{r_2} E_d \times d\ell + \int_{r_2}^\infty E_w \times d\ell$$

$$V(r) = \begin{cases} \frac{I_{tot}}{2 \times \left(\frac{\theta_w}{\rho_d} + \frac{\theta_s}{\rho_s} \right)} \times \left(\frac{1}{r_1} - \frac{1}{r_2} \right) + \frac{I_{tot}}{2 \times \left(\frac{\theta_w}{\rho_w} + \frac{\theta_s}{\rho_s} \right)} \times \left(\frac{1}{r_2} - \frac{1}{r_1} + \frac{1}{r} \right) : r < r_1 \\[4mm] \frac{I_{tot}}{2 \times \left(\frac{\theta_w}{\rho_d} + \frac{\theta_s}{\rho_s} \right)} \times \left(\frac{1}{r} - \frac{1}{r_2} \right) + \frac{I_{tot}}{2 \times \left(\frac{\theta_w}{\rho_w} + \frac{\theta_s}{\rho_s} \right)} \times \frac{1}{r_2} : r_1 < r < r_2 \\[4mm] \frac{I_{tot}}{2 \times \left(\frac{\theta_w}{\rho_w} + \frac{\theta_s}{\rho_s} \right)} \times \frac{1}{r} : r > r_2 \end{cases} \tag{27}$$

The resistance of remote earth is calculated through Equation (27) for $r = r_{el} < r_1$:

$$R_{el} = \frac{V(r_{el})}{I_{tot}} = \frac{1}{2 \times \left(\frac{\theta_w}{\rho_d} + \frac{\theta_s}{\rho_s}\right)} \times \left(\frac{1}{r_1} - \frac{1}{r_2}\right) + \frac{1}{2 \times \left(\frac{\theta_w}{\rho_w} + \frac{\theta_s}{\rho_s}\right)} \times \left(\frac{1}{r_2} - \frac{1}{r_1} + \frac{1}{r_{el}}\right) \quad (28)$$

2.3. Method "C"—Near Electric Field Distribution Method with a Linear Current Source

The two previous methods, which are based on CIGRE B4.61 675:2017 and IEC TS 62344:2013, with or without a dam, leading to very high electric field strengths near the electrode because a point current source is considered. One way to overcome this problem is to replace the point source with a linear current source. Particularly, instead of the typical dam, the simplifying structure of Figure 6 is considered only for the effective height L in cylindrical coordinates. On the "right" side of the section of Figure 6 the distance between the electrode and the dam is r_1, and the thickness of the dam is d, so the radius of the outer surface of the dam is $r_2 = r_1 + d$. On the "left" side of the section of Figure 6 the distance between the electrode and the soil is r_3, providing that the corresponding depth is ensured. The angle on the ground plan is θ. As mentioned, the electromagnetic field theory requires:

- Continuity of the vertical current density on the dividing surface according to Equation (15);
- Continuity of the tangential electric field strength on the dividing surface according to Equation (16).

Due to the multiple values of electrical resistivity of dam material ($\rho_d = \rho_2 = 100 \ \Omega\cdot m$) in relation to the resistivity of seawater ($\rho_w = \rho_1 = 0.25 \ \Omega\cdot m$), the ratio of the current densities is significantly reduced (1:400). Considering that the direction of the electric current density vector is mainly radial in the cylindrical coordinates, the tangential and vertical components in the water–dam, water–soil interfaces at a constant radius are ignored. Therefore, the areas of the respective parts of the cylinder of radius r for the horizontal soil S_s and the water reservoir S_{w_l}, with corresponding plan angle θ (in rad), and for the sea S_{w_r} and the dam S_d for plan angle ($2 \times \pi$-θ) are, respectively:

$$S_S = \theta \times r \times L : r \geq r_3 \quad (29)$$

$$S_{w\_l} = \theta \times r \times L : r < r_3 \, for \, left \, hand \, area \, (water - soil) \quad (30)$$

$$S_{w\_r} = (2 \times \pi - \theta) \times r \times L : r < r_1 \, or \, r > r_2 \ \, for \, right \, hand \, area \, (water - dam) \quad (31)$$

$$S_d = (2 \times \pi - \theta) \times r \times L : r_1 \leq r \leq r_2 \quad (32)$$

The total current I_{tot} passes through the soil and the seawater (or dam), assuming that air is an insulator of very high resistivity.

Due to the uniform electrical resistivity, the total current passes through the soil and seawater sections radially and symmetrically. If the current intensity and the soil current density are I_s and J_s, respectively, of the seawater in the right section; I_{w_r} and J_{w_r}, respectively, of the seawater in the left section; I_{w_l} and J_{w_l}, respectively, of the seawater in the area (for $r < min\{r_1, r_3\}$ or $r_2 < r < r_3$); I_w and J_w, respectively, of the dam; and I_d and J_d, respectively, based on the current densities, Kirchhoff's law and the existence of practically only one radial component, the total current is given by:

$$I_{tot} = I_s + I_{w\_r} = \int_{S_s} \vec{J}_s \times \vec{n} \times dS + \int_{S_{w\_r}} \vec{J}_{w\_r} \times \vec{n} \times dS = J_s \times S_s + J_{w\_r} \times S_{w\_r} : r > max\{r_2, r_3\} \ or \ r_3 < r < r_1 \quad (33)$$

$$\begin{aligned} I_{tot} = I_{w\_r} + I_{w\_l} &= \int_{S_{w\_r}} \vec{J}_{w\_r} \times \vec{n} \times dS + \int_{S_{w\_l}} \vec{J}_{w\_l} \times \vec{n} \times dS = J_{w\_r} \times S_{w\_r} + J_{w\_l} \times S_{w\_l} \\ &= J_w \times 2 \times \pi \times r \times L : r < min\{r_1, r_3\} \ or \ r_2 < r < r_3 \end{aligned} \quad (34)$$

$$I_{tot} = I_d + I_{w\_l} = \int_{S_d} \vec{J}_d \times \vec{n} \times dS + \int_{S_{w\_l}} \vec{J}_{w\_l} \times \vec{n} \times dS = J_d \times S_d + J_{w\_l} \times S_{w\_l} : r_1 < r < min\{r_2, r_3\} \quad (35)$$

$$I_{tot} = I_d + I_s = \int_{S_d} \vec{J}_d \times \vec{n} \times dS + \int_{S_s} \vec{J}_s \times \vec{n} \times dS = J_d \times S_d + J_s \times S_s : max\{r_1, r_3\} < r < r_2 \tag{36}$$

$$I_d = I_{w\_r} \Rightarrow J_d = J_{w\_r} : r = r_1 \; or \; r = r_2 \tag{37}$$

$$I_s = I_{w\_l} \Rightarrow J_s = J_{w\_l} : r = r_3 \tag{38}$$

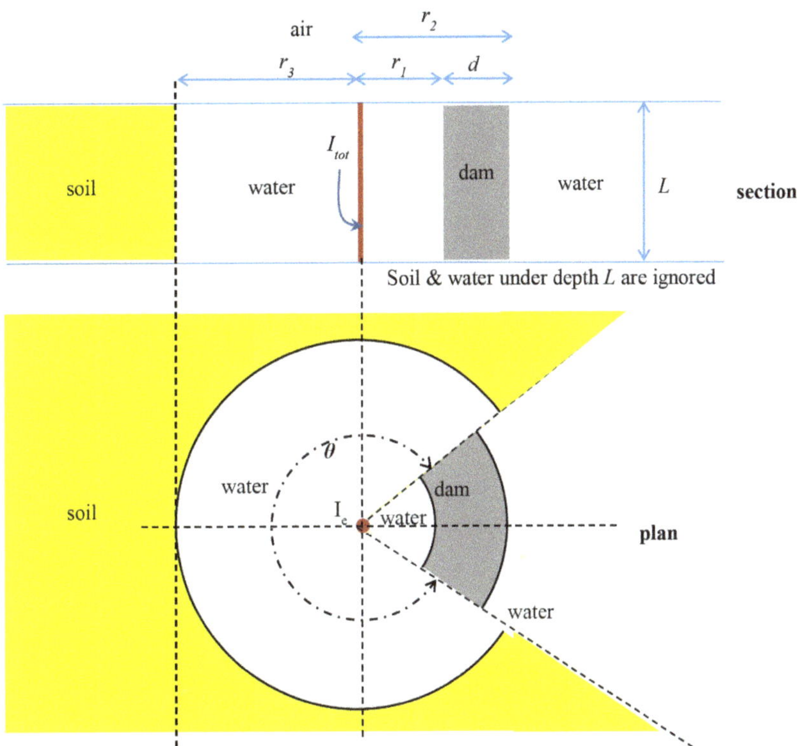

Figure 6. Simplified model for placement of an electrode on the shore or in near-shore sea considering a linear electrode, in cylindrical coordinates and using a dam–soil surface.

Due to the continuity of the radial electric field strength on the water–soil interface, the radial electric field strength of the right part of the sea E_{rw_r} and the radial electric field strength of the soil E_{rs} are equal and practically the only components. Combined with Ohm's law, it follows that:

$$E_{rs} = E_{rw\_r} \Leftrightarrow \rho_s \times J_s = \rho_w \times J_{w\_r} \Rightarrow J_s = \frac{\rho_w}{\rho_s} \times J_{w\_r} : r > max\{r_2, r_3\} \; or \; r_3 < r < r_1 \tag{39}$$

Due to the continuity of the radial electric field strength on the water–dam interface, the radial electric field strength of the left part of the sea E_{rw_l} and the radial electric field strength of the dam E_{rd} are equal and practically the only components. Combined with Ohm's law, it follows that:

$$E_{rd} = E_{rw\_l} \Leftrightarrow \rho_d \times J_d = \rho_w \times J_{w\_l} \Rightarrow J_d = \frac{\rho_w}{\rho_d} \times J_{w\_d} : r_1 < r < min\{r_2, r_3\} \tag{40}$$

Due to the continuity of the radial electric field strength on the dam–soil interface (Figure 6), the radial electric field strength of the dam E_{rd} and the radial electric field strength of the soil E_{rs} are equal and practically the only components. Combined with Ohm's law, it follows that:

$$E_{rd} = E_{rs} \Leftrightarrow \rho_d \times J_d = \rho_s \times J_s \Rightarrow J_s = \frac{\rho_d}{\rho_s} \times J_d : max\{r_1, r_3\} < r < r_2 \tag{41}$$

By combining Equations (29)–(36) and (39)–(41), the current densities and electric field strengths are determined as follows:

$$I_{tot} = \frac{\rho_w}{\rho_s} \times J_{w\_r} \times \theta \times r \times L + J_{w\_r} \times (2 \times \pi - \theta) \times r \times L \Rightarrow$$

$$J_{w\_r} = \frac{I_{tot}}{r \times L \times \left((2 \times \pi - \theta) + \frac{\rho_w}{\rho_s} \times \theta\right)} : r > max\{r_2, r_3\} \; or \; r_3 < r < r_1 \tag{42}$$

$$\Rightarrow J_s = \frac{I_{tot}}{r \times L \times \left(\frac{\rho_s}{\rho_w} \times (2 \times \pi - \theta) + \theta\right)} : r > max\{r_2, r_3\} \; or \; r_3 < r < r_1 \tag{43}$$

$$\Rightarrow E_{rs} = E_{rw\_r} = \frac{I_{tot}}{r \times L \times \left(\frac{2 \times \pi - \theta}{\rho_w} + \frac{\theta}{\rho_s}\right)} : r > max\{r_2, r_3\} \; or \; r_3 < r < r_1 \tag{44}$$

$$I_{tot} = J_w \times 2 \times \pi \times r \times L \Rightarrow J_w = \frac{I_{tot}}{2 \times \pi \times r \times L} : r < min\{r_1, r_3\} \; or \; r_2 < r < r_3 \tag{45}$$

$$\Rightarrow E_{rw} = \frac{\rho_w \times I_{tot}}{2 \times \pi \times r \times L} : r < min\{r_1, r_3\} \; or \; r_2 < r < r_3 \tag{46}$$

$$I_{tot} = J_{w\_l} \times \theta \times r \times L + \frac{\rho_w}{\rho_d} \times J_{w\_l} \times (2 \times \pi - \theta) \times r \times L \Rightarrow$$

$$J_{w\_l} = \frac{I_{tot}}{r \times L \times \left(\frac{\rho_w}{\rho_d} \times (2 \times \pi - \theta) + \theta\right)} : r_1 < r < min\{r_2, r_3\} \tag{47}$$

$$\Rightarrow J_d = \frac{I_{tot}}{r \times L \times \left((2 \times \pi - \theta) + \frac{\rho_d}{\rho_w} \times \theta\right)} : r_1 < r < min\{r_2, r_3\} \tag{48}$$

$$\Rightarrow E_{rd} = E_{rw\_l} = \frac{I_{tot}}{r \times L \times \left(\frac{2 \times \pi - \theta}{\rho_d} + \frac{\theta}{\rho_w}\right)} : r_1 < r < min\{r_2, r_3\} \tag{49}$$

$$I_{tot} = \frac{\rho_d}{\rho_s} \times J_d \times \theta \times r \times L + J_d \times (2 \times \pi - \theta) \times r \times L \Rightarrow$$

$$J_d = \frac{I_{tot}}{r \times L \times \left((2 \times \pi - \theta) + \frac{\rho_d}{\rho_s} \times \theta\right)} : max\{r_1, r_3\} < r < r_2 \tag{50}$$

$$\Rightarrow J_s = \frac{I_{tot}}{r \times L \times \left(\frac{\rho_s}{\rho_d} \times (2 \times \pi - \theta) + \theta\right)} : max\{r_1, r_3\} < r < r_2 \tag{51}$$

$$\Rightarrow E_{rd} = E_{rs} = \frac{I_{tot}}{r \times L \times \left(\frac{2 \times \pi - \theta}{\rho_d} + \frac{\theta}{\rho_s}\right)} : max\{r_1, r_3\} < r < r_2 \tag{52}$$

Be it noted that Equations (37) and (38) do not apply because if that were the case, the results of Equations (44), (46), (49) and (52) should be identical on the boundary surfaces $r = r_1$, $r = r_2$ and $r = r_3$, which is not observed, due to the fact that the non-radial currents on the respective surfaces were ignored. Equations (45) and (46) apply for $r < min\{r_1, r_3\}$. Additional approximations were made, as was also the case in the original assumption of CIGRE B4.61 675:2017 [17] (pp. 118–119):

- *An infinite layer of seawater–dam–soil of active thickness L:* The modeling layer practically grows significantly, e.g., here, a thickness of the order of meters is assumed, while depths at long distances reach tens of meters at Stachtoroi and hundreds of meters at Korakia. This is a safe assumption to make, as it ignores a large part of the conductive material making the model unsuitable for the far field unless one is referring to a water surface of constant depth, e.g., an artificial lake;
- *Uniform seawater and soil resistivities:* The resistivities vary; therefore, the equipotential surfaces are not circular. However, by using the most unfavorable values, the worst-case scenario for this equivalent linear electrode can be estimated;
- *Soil, dam and seawater cylindrical segment:* The respective materials do not form correspondingly uniform surfaces; their shape differs in different directions (especially that of the dam). However, the angle θ is the best approach, despite the fact that more interfaces are formed in radial directions, as shown in Figure 6.

For the calculation of the absolute potential, a zero potential at an infinite distance cannot be considered because, in the case of infinite distance, the application of Equation (44) to the outer side of the dam leads to a non-zero value. Therefore, in this study, the radius of the infinity r_∞ for the calculation of the absolute potential is taken as equal to half of the distance between the two electrode stations.

In the case of Figure 6 with $r_1 < r_2 < r_3$, Equation (52) does not apply, and the absolute potential is calculated as follows:

$$V(r) = \int_r^{r_\infty} \vec{E} \times \vec{d\ell} = \int_r^{r_\infty} E \times d\ell = \underbrace{\int_r^{r_1} E \times d\ell}_{\text{Equation (46)}} + \underbrace{\int_{r_1}^{r_2} E \times d\ell}_{\text{Equation (49)}} + \underbrace{\int_{r_2}^{r_3} E \times d\ell}_{\text{Equation (46)}} + \underbrace{\int_{r_3}^{r_\infty} E \times d\ell}_{\text{Equation (44)}} \Rightarrow$$

$$V(r) = \begin{cases} \left\{ \begin{aligned} & \frac{I_{tot}}{L \times \left(\frac{2 \times \pi - \theta}{\rho_d} + \frac{\theta}{\rho_w} \right)} \times ln\left(\frac{r_2}{r_1}\right) + \\ & \frac{\rho_w \times I_{tot}}{2 \times \pi \times L} \times ln\left(\frac{r_3}{r_2} \times \frac{r_1}{r}\right) + \frac{I_{tot}}{L \times \left(\frac{2 \times \pi - \theta}{\rho_w} + \frac{\theta}{\rho_s} \right)} \times ln\left(\frac{r_\infty}{r_3}\right) \end{aligned} \right\} & : r < r_1 \\[6pt] \left\{ \begin{aligned} & \frac{I_{tot}}{L \times \left(\frac{2 \times \pi - \theta}{\rho_d} + \frac{\theta}{\rho_w} \right)} \times ln\left(\frac{r_2}{r}\right) + \\ & \frac{\rho_w \times I_{tot}}{2 \times \pi \times L} \times ln\left(\frac{r_3}{r_2}\right) + \frac{I_{tot}}{L \times \left(\frac{2 \times \pi - \theta}{\rho_w} + \frac{\theta}{\rho_s} \right)} \times ln\left(\frac{r_\infty}{r_3}\right) \end{aligned} \right\} & : r_1 < r < r_2 \\[6pt] \frac{\rho_w \times I_{tot}}{2 \times \pi \times L} \times ln\left(\frac{r_3}{r}\right) + \frac{I_{tot}}{L \times \left(\frac{2 \times \pi - \theta}{\rho_w} + \frac{\theta}{\rho_s} \right)} \times ln\left(\frac{r_\infty}{r_3}\right) & : r_2 < r < r_3 \\[6pt] \frac{I_{tot}}{L \times \left(\frac{2 \times \pi - \theta}{\rho_w} + \frac{\theta}{\rho_s} \right)} \times ln\left(\frac{r_\infty}{r}\right) & : r > r_3 \end{cases} \tag{53}$$

Similarly, the resistance of the remote earth electrode is obtained by dividing the potential difference between the surface of the electrode of radius r_{el} and remote earth r_∞ for unary current intensity. It is calculated from Equation (53) for $r = r_{el} < r_1$ as follows:

$$R_{el} = \frac{V(r_{el})}{I_{tot}} = \frac{1}{L} \times \left(\frac{ln\left(\frac{r_2}{r_1}\right)}{\left(\frac{2 \times \pi - \theta}{\rho_d} + \frac{\theta}{\rho_w} \right)} + \frac{ln\left(\frac{r_3}{r_2} \times \frac{r_1}{r_{el}}\right)}{\frac{2 \times \pi}{\rho_w}} + \frac{ln\left(\frac{r_\infty}{r_3}\right)}{\left(\frac{2 \times \pi - \theta}{\rho_w} + \frac{\theta}{\rho_s} \right)} \right) \tag{54}$$

In the case of $r_1 < r_3 < r_2$, the absolute potential is calculated as follows:

$$V(r) = \int_r^{r_\infty} \vec{E} \times \vec{d\ell} = \int_r^{r_\infty} E \times d\ell = \underbrace{\int_r^{r_1} E \times d\ell}_{\text{Equation (46)}} + \underbrace{\int_{r_1}^{r_3} E \times d\ell}_{\text{Equation (49)}} + \underbrace{\int_{r_3}^{r_2} E \times d\ell}_{\text{Equation (52)}} + \underbrace{\int_{r_2}^{r_\infty} E \times d\ell}_{\text{Equation (44)}} \Rightarrow$$

$$
V(r) = \begin{cases}
\left\{ \begin{aligned}
&\frac{\rho_w \times I_{tot}}{2 \times \pi \times L} \times ln\!\left(\frac{r_1}{r}\right) + \frac{I_{tot}}{L \times \left(\frac{2 \times \pi - \theta}{\rho_d} + \frac{\theta}{\rho_w}\right)} \times ln\!\left(\frac{r_3}{r_1}\right) + \\
&\frac{I_{tot}}{L \times \left(\frac{2 \times \pi - \theta}{\rho_d} + \frac{\theta}{\rho_s}\right)} \times ln\!\left(\frac{r_2}{r_3}\right) + \frac{I_{tot}}{L \times \left(\frac{2 \times \pi - \theta}{\rho_w} + \frac{\theta}{\rho_s}\right)} \times ln\!\left(\frac{r_\infty}{r_2}\right)
\end{aligned} \right\} & : r < r_1 \\[2em]
\left\{ \begin{aligned}
&\frac{I_{tot}}{L \times \left(\frac{2 \times \pi - \theta}{\rho_d} + \frac{\theta}{\rho_w}\right)} \times ln\!\left(\frac{r_3}{r}\right) + \\
&\frac{I_{tot}}{L \times \left(\frac{2 \times \pi - \theta}{\rho_d} + \frac{\theta}{\rho_s}\right)} \times ln\!\left(\frac{r_2}{r_3}\right) + \frac{I_{tot}}{L \times \left(\frac{2 \times \pi - \theta}{\rho_w} + \frac{\theta}{\rho_s}\right)} \times ln\!\left(\frac{r_\infty}{r_2}\right)
\end{aligned} \right\} & : r_1 < r < r_3 \\[2em]
\frac{I_{tot}}{L \times \left(\frac{2 \times \pi - \theta}{\rho_d} + \frac{\theta}{\rho_s}\right)} \times ln\!\left(\frac{r_2}{r}\right) + \frac{I_{tot}}{L \times \left(\frac{2 \times \pi - \theta}{\rho_w} + \frac{\theta}{\rho_s}\right)} \times ln\!\left(\frac{r_\infty}{r_2}\right) & : r_3 < r < r_2 \\[1.5em]
\frac{I_{tot}}{L \times \left(\frac{2 \times \pi - \theta}{\rho_w} + \frac{\theta}{\rho_s}\right)} \times ln\!\left(\frac{r_\infty}{r}\right) & : r > r_2
\end{cases}
\tag{55}
$$

Similarly, the resistance of the remote earth electrode is obtained from Equation (55) for $r = r_{el} < r_1$ as follows:

$$
R_{el} = \frac{V(r_{el})}{I_{tot}} = \frac{1}{L} \times \left(\frac{ln\!\left(\frac{r_1}{r_{el}}\right)}{\frac{2 \times \pi}{\rho_w}} + \frac{ln\!\left(\frac{r_3}{r_1}\right)}{\left(\frac{2 \times \pi - \theta}{\rho_d} + \frac{\theta}{\rho_w}\right)} + \frac{ln\!\left(\frac{r_2}{r_3}\right)}{\left(\frac{2 \times \pi - \theta}{\rho_d} + \frac{\theta}{\rho_s}\right)} + \frac{ln\!\left(\frac{r_\infty}{r_2}\right)}{\left(\frac{2 \times \pi - \theta}{\rho_w} + \frac{\theta}{\rho_s}\right)} \right)
\tag{56}
$$

In the case of $r_3 < r_1 < r_2$, the absolute potential is calculated as follows:

$$
V(r) = \int_r^{r_\infty} \overrightarrow{E} \times \overrightarrow{d\ell} = \int_r^{r_\infty} E \times d\ell = \underbrace{\int_r^{r_3} E \times d\ell}_{\text{Equation (46)}} + \underbrace{\int_{r_3}^{r_1} E \times d\ell}_{\text{Equation (49)}} + \underbrace{\int_{r_1}^{r_2} E \times d\ell}_{\text{Equation (52)}} + \underbrace{\int_{r_2}^{r_\infty} E \times d\ell}_{\text{Equation (44)}} \Rightarrow
$$

$$
V(r) = \begin{cases}
\left\{ \begin{aligned}
&\frac{\rho_w \times I_{tot}}{2 \times \pi \times L} \times ln\!\left(\frac{r_3}{r}\right) + \frac{I_{tot}}{L \times \left(\frac{2 \times \pi - \theta}{\rho_d} + \frac{\theta}{\rho_w}\right)} \times ln\!\left(\frac{r_1}{r_3}\right) + \\
&\frac{I_{tot}}{L \times \left(\frac{2 \times \pi - \theta}{\rho_d} + \frac{\theta}{\rho_s}\right)} \times ln\!\left(\frac{r_2}{r_1}\right) + \frac{I_{tot}}{L \times \left(\frac{2 \times \pi - \theta}{\rho_w} + \frac{\theta}{\rho_s}\right)} \times ln\!\left(\frac{r_\infty}{r_2}\right)
\end{aligned} \right\} & : r < r_3 \\[2em]
\left\{ \begin{aligned}
&\frac{I_{tot}}{L \times \left(\frac{2 \times \pi - \theta}{\rho_d} + \frac{\theta}{\rho_w}\right)} \times ln\!\left(\frac{r_1}{r}\right) + \\
&\frac{I_{tot}}{L \times \left(\frac{2 \times \pi - \theta}{\rho_d} + \frac{\theta}{\rho_s}\right)} \times ln\!\left(\frac{r_2}{r_1}\right) + \frac{I_{tot}}{L \times \left(\frac{2 \times \pi - \theta}{\rho_w} + \frac{\theta}{\rho_s}\right)} \times ln\!\left(\frac{r_\infty}{r_2}\right)
\end{aligned} \right\} & : r_3 < r < r_1 \\[2em]
\frac{I_{tot}}{L \times \left(\frac{2 \times \pi - \theta}{\rho_d} + \frac{\theta}{\rho_s}\right)} \times ln\!\left(\frac{r_2}{r}\right) + \frac{I_{tot}}{L \times \left(\frac{2 \times \pi - \theta}{\rho_w} + \frac{\theta}{\rho_s}\right)} \times ln\!\left(\frac{r_\infty}{r_2}\right) & : r_1 < r < r_2 \\[1.5em]
\frac{I_{tot}}{L \times \left(\frac{2 \times \pi - \theta}{\rho_w} + \frac{\theta}{\rho_s}\right)} \times ln\!\left(\frac{r_\infty}{r}\right) & : r > r_2
\end{cases}
\tag{57}
$$

Similarly, the resistance of the remote earth electrode is obtained from Equation (57) for $r = r_{el} < r_3$ as follows:

$$
R_{el} = \frac{V(r_{el})}{I_{tot}} = \frac{1}{L} \times \left(\frac{ln\!\left(\frac{r_3}{r_{el}}\right)}{\frac{2 \times \pi}{\rho_w}} + \frac{ln\!\left(\frac{r_1}{r_3}\right)}{\left(\frac{2 \times \pi - \theta}{\rho_d} + \frac{\theta}{\rho_w}\right)} + \frac{ln\!\left(\frac{r_2}{r_1}\right)}{\left(\frac{2 \times \pi - \theta}{\rho_d} + \frac{\theta}{\rho_s}\right)} + \frac{ln\!\left(\frac{r_\infty}{r_2}\right)}{\left(\frac{2 \times \pi - \theta}{\rho_w} + \frac{\theta}{\rho_s}\right)} \right)
\tag{58}
$$

It is noted that the seabed slope bears no effect, as it is limited to a narrow zone equal to the active length of the electrode.

2.4. Extending the Application of Electric Field Distribution Methods When Using More Electrodes—Superposition Application

When studying the distribution of the electric field near the electrode station, a more accurate representation of the electrode from a point source is necessary. Therefore, in order to limit the electric current density of each electrode, as well as for reliability reasons, more than one electrode is used. According to CIGRE B4.61 675:2017 [17] (p. 69), the electric current density is proposed to be between 6 and 10 A/m^2 for beach and sea electrodes to reduce chlorine selectivity for elements in contact with water. However, higher values of current densities can be used in pond electrodes when inaccessible to animals and people. Of course, lower electric current density leads to lower chlorine selectivity and reduced electrode consumption. Accordingly, IEC TS 62344:2013 [34] (p. 32) suggests that the electric current density should again be between 6 and 10 A/m^2 for sea electrodes and pond electrodes alike so that the electric field strength near it is less than 1.25 to 2 V/m. If the electrodes work in free water, which is not accessible to people and marine fauna, the average electric current density can reach up to 100 A/m^2. Therefore, the manufacturer's instructions must be taken into account (e.g., ANOTEC suggests 10 A/m^2 for continuous operation [77]), as well as the mode of operation (e.g., IPTO suggests 20 A/m^2 for temporary use in case one pole is out of operation (because of failure or maintenance) for a few hours during the lifespan of the HVDC bipolar heteropolar link between Attica and Crete [78]).

However, from the moment that the steady state electric current density limit J_{steady} is selected and given the area of the peripheral surface of the electrode S_{p_el}, the number of necessary electrodes N_{min_el} is determined, based on the maximum nominal current I_{tot_steady} flowing through the electrode station:

$$I_{tot\_steady} = \int_{N_{min\_el} \times S_{p\_el}} \vec{J}_{steady} \times \vec{n} \times dS \Rightarrow N_{min\_el} = \frac{I_{tot\_steady}}{J_{steady} \times S_{p\_el}} \tag{59}$$

If the necessary number of electrodes is divided into v_{frame} frames with an equal number of electrodes, then the number of electrodes per frame N_{el_frame} is equal to:

$$N_{el\_frame} = round\left(\frac{N_{min\_el}}{v_{frame}}\right) \tag{60}$$

However, the electrode station consists of (v_{frame} + 1) frames, based on the relevant reliability criterion (n + 1), so the total number of electrodes in operation, under full load conditions N_{full_load} is equal to (v_{frame} + 1) $\times N_{el\_frame}$, while under conditions of periodic maintenance of a frame, $N_{maintenance}$ is equal to $v_{frame} \times N_{el\_frame}$, through which the respective electric current density (with which each individual electrode is electrified) is determined. Furthermore, due to a possible non-uniformity in the electric current distribution in the relevant Hatch report [73,74], IPTO has required, during the pre-study phase, to increase the electric current density by a factor β (in the present case of the order of 6%). Hence, the final values of current densities under full load conditions $J_{full\_load\_steady}$ and under periodic maintenance conditions $J_{maintenance\_steady}$ are, respectively, equal to:

$$J_{full\_load\_steady} = \frac{(1+\beta) \times I_{tot-steady}}{\left(v_{frame}+1\right) \times N_{el\_frame} \times S_{p\_el}} \tag{61}$$

$$J_{maintenance\_steady} = \frac{(1+\beta) \times I_{tot-steady}}{v_{frame} \times N_{el\_frame} \times S_{p\_el}} \tag{62}$$

For a transient state, the final values of electric current densities, under full load conditions $J_{full\_load\_transient}$ and under periodic maintenance conditions $J_{maintenance\_transient}$ are, respectively, calculated through Equations (61) and (62) for the respective current $I_{tot-transient}$ (instead of $I_{tot-steady}$).

Based on the positioning of the electrodes in space (e.g., linear arrangement of electrodes at fixed distances D_{el} in each frame of length D_{frame} of Figure 7 or linear arrangement of frames per distance d_{frames} of Figure 8), as well as on the typical load of each electrode, the total electric field strength is calculated as follows:

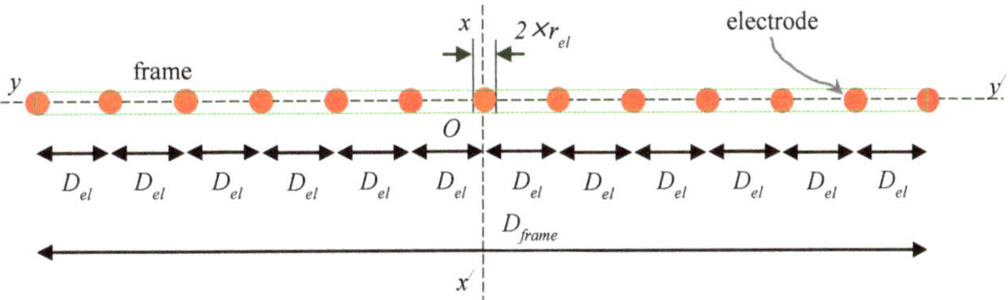

Figure 7. Typical frame view with 13 electrodes placed in series, vertically positioned, parallel to the yOy' axis.

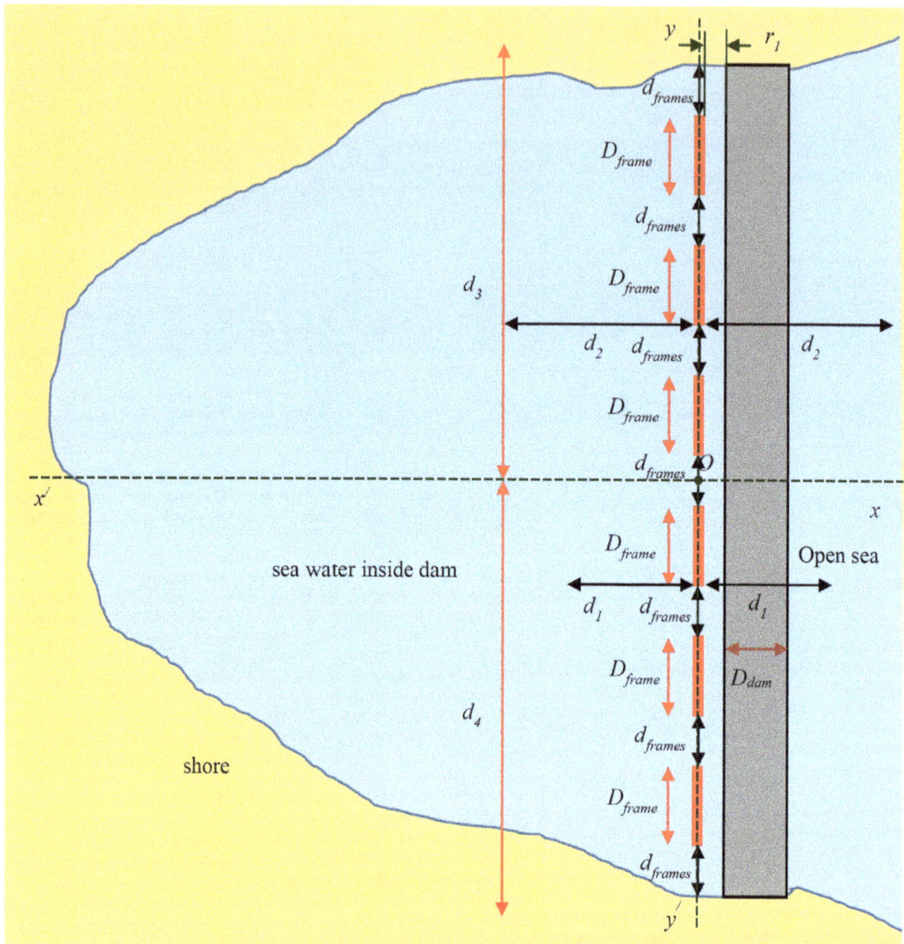

Figure 8. Indicative floor plan of six frames in a row, each frame parallel to the axis of the protective dam.

- The area of interest concerns the surface of the water and is configured in an appropriate system of two-dimensional Cartesian coordinates, where the respective electrode is placed in a specific position. A canvas of points, with a step of d_{step_x} and d_{step_y} is formed, at which the electric field strength is "measured". The respective step is not constant but variable when studying the far field (closer to electrodes, the canvas is denser, farther away, sparser and larger step);
- The radial electric field strength E_r is calculated based on the method applied, with the respective electrode as the reference point and with an electric current equal to the product of the electric current density $J_{full\_load\_transient}$ or $J_{maintenance\_transient}$ with the corresponding area of the peripheral surface of the electrode S_{p_el}. The electric field strength caused by the ℓ-th electrode was analyzed into the corresponding components $E_{x\text{-}\ell}$ and $E_{y\text{-}\ell}$ along the axes xOx' and yOy', as shown in Figure 9, with the help of the coordinates (x_ℓ, y_ℓ) of the electrode, (x,y) of the point of interest (canvas points) and the respective distance r_ℓ:

$$r_\ell = \sqrt{(x_\ell - x)^2 + (y_\ell - y)^2} \tag{63}$$

$$E_{x\text{-}\ell} = E(r_\ell) \times \frac{(x - x_\ell)}{r_\ell} \tag{64}$$

$$E_{y-\ell} = E(r_\ell) \times \frac{(y - y_\ell)}{r_\ell} \tag{65}$$

- By using the superposition, the individual electric field strengths of all the electrodes in the respective xOx′ and yOy′ directions are added:

$$E_x = \sum_{\ell=1}^{(v_{frame}+1) \times N_{el\_frame}} E_{x-\ell} \tag{66}$$

$$E_y = \sum_{\ell=1}^{(v_{frame}+1) \times N_{el\_frame}} E_{y-\ell} \tag{67}$$

$$E = \sqrt{E_x^2 + E_y^2} \tag{68}$$

- The calculation of the absolute electric potential was performed numerically along the main directions xOx′ and yOy′, at various points of the canvas, with respect to either "infinity":

$$V(x) = \int_x^\infty E_x \times dx \tag{69}$$

$$V(y) = \int_y^\infty E_y \times dy \tag{70}$$

- Similarly, from the respective potential difference for specific lengths, the respective average electric field strength values are calculated:

$$E_{mean-x}(x) = \frac{V(x) - V(x + 1 \times sign(x))}{1\,\text{m}} \tag{71}$$

$$E_{mean-y}(y) = \frac{V(y) - V(y + 1 \times sign(y))}{1\,\text{m}} \tag{72}$$

$$E_{mean}(x,y) = \sqrt{E_{mean-x}^2(x) + E_{mean-y}^2(y)} \tag{73}$$

In this way, all the necessary variables for the preliminary study are numerically determined.

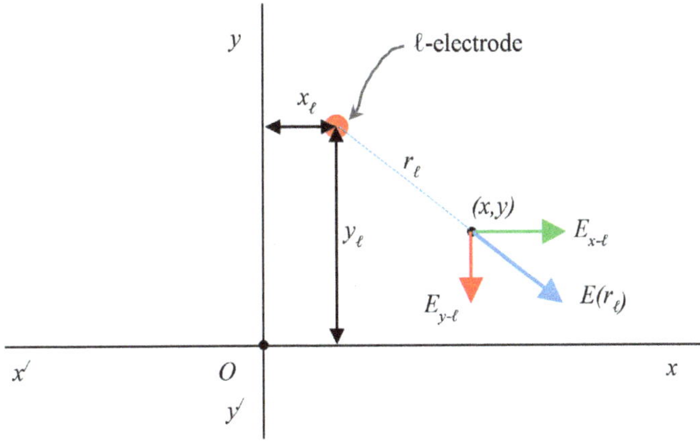

Figure 9. Basic principle of electric field strength analysis of an electrode in xOx′ and yOy′ axes.

2.5. Electric Field Strength and Voltage Limits According to CIGRE B4.61 675:2017 and IEC TS 62344:2013

The safety requirements for an electrode station can be summarized in this single objective: "The operation of the electrodes must not lead to unsafe conditions for people or animals in areas accessible either by the public or by authorized maintenance/operation personnel". In order to meet this, a complete list of possible operating states of the electrode needs to be studied, and possible conditions and procedures in the wider area of the electrode, which could be affected by its operation, need to be considered. The operating conditions are basically divided into two categories:

- Conditions lasting 10 s or more when, for safety reasons, the operation is considered to be continuous;
- Conditions lasting less than 10 s are considered to be transient operations (e.g., faults and short-time overloads). In the case of a transient fault on the HVDC transmission line, the transient current overload can be considered as short as 50 ms (this is the typical time required for the current protection of the line and its respective control device to cut off the fault current).

The criteria for these two different time frames lead to different requirements since tolerance to electric current by humans is time-dependent, according to IEC TS60479-1:2007, as shown in Table 4.1 and Figure 5.2 of CIGRE 675-B4. 61:2017 [17] (p. 52 and p.60 respectively).

According to the general guidelines of CIGRE 675-B4.61:2017 [17] (p. 65), the limit of the potential gradient for the protection of divers in marine electrodes amounts to 2.5 V/m for the continuous operating conditions (steady state), and 15 V/m for the transient operating conditions. Especially in IEC TS 62344:2013 [34] (p. 32), the value 1.25 V/m is taken for continuous operating conditions in order to ensure no effect on marine mammals, and in [17] (p. 93, section 5.4.3, 2nd paragraph), the same value is reported as the generally accepted value of potential difference, per unit length, from the electrode surface, directly accessible to humans and marine fauna. In addition, based on [79], the value of 4 V is reported as the effect value on buried metal tubes (pipe to soil potential difference), as it is considered to be the maximum potential value that can be applied by a cathodic protection device. Therefore, in the present case, the following conditions were applied:

- *Potential gradient/electric filed strength for continuous operating conditions in water:* E_{limit_S} = 1.25 V/m;
- *Potential gradient/electric filed strength for transient operating conditions in water:* E_{limit_T} = 15 V/m;
- *Absolute potential with respect to remote earth for continuous operating conditions:* V_{limit_S} = 4 V;
- *Step voltage, touch voltage and metal-to-metal touch voltage for continuous operating conditions:* E_{soil_S} = 5 V or greater, according to the equations of Figure 5.3 in [17] (p. 63);
- *Step voltage, touch voltage and metal-to-metal touch voltage for transient operating conditions:* E_{soil_T} = 30 V or greater, according to the equations of Figure 5.3 in [17] (p. 63).

In IEC TS 62344:2013 [34] (p. 17), no distinction was made in terms of different states regarding step voltage and touch voltage. Nonetheless, it must not exceed 70 V regardless, thus proving the IEC TS 62344:2013 requirements as evidently stricter [34] (p. 16).

In any case, the first three criteria are critical for the installation of the electrode station in each area due to the choice of a submerged electrode near the coast since all other features lie within an IPTO-controlled area.

Therefore, by considering the unified electric field distribution method, according to CIGRE B4.61 675:2017 and IEC TS 62344:2013, it follows that the distance from the electrode (at which the voltage has indicatively dropped below the limit V_{limit_S} for electric current intensity $I_{tot\text{-}steady}$), according to Equation (8), is equal to:

$$r_{limit1} = \frac{I_{tot-steady}}{2 \times \left(\frac{\theta_w}{\rho_w} + \frac{\theta_s}{\rho_s}\right) \times V_{limit\_S}} : r_{limit1} > r_2 \tag{74}$$

Accordingly, the distance from the electrode, at which the potential gradient in the water has fallen, on average, below the limit E_{limit_S} for continuous operating conditions, is calculated as follows:

$$E_{limit\_S} = \frac{V(r_{limit2}) - V(r_{limit2} + 1\text{ m})}{1\text{ m}} \tag{75}$$

$$\Rightarrow r_{limit2} = 0.5 \times \left(-1 + \sqrt{1 + 4 \times \frac{I_{tot-steady}}{2 \times \left(\frac{\theta_w}{\rho_w} + \frac{\theta_s}{\rho_s}\right) \times E_{limit\_S}}}\right) : r_{limit2} > r_2 \tag{76}$$

For transient operating conditions, the corresponding limit r_{limit3} is calculated from Equation (76) if the potential gradient amounts to E_{limit_T} (instead of E_{limit_S}) and with the respective electric current intensity being $I_{tot\text{-}transient}$ (instead of $I_{tot\text{-}steady}$).

Accordingly, the distance from the electrode, at which the potential gradient (electric field strength) in the water has spot-dropped E_{limit_S} for continuous operating conditions, according to Equation (7), is equal to:

$$r_{limit4} = \sqrt{\frac{I_{tot-steady}}{2 \times \left(\frac{\theta_w}{\rho_w} + \frac{\theta_s}{\rho_s}\right) \times E_{limit\_S}}} : r_{limit4} > r_2 \tag{77}$$

For transient operating conditions, the corresponding limit r_{limit5} is calculated from Equation (77) if the potential gradient amounts to E_{limit_T} (instead of E_{limit_S}) and with the respective electric current intensity being $I_{tot-transient}$ (instead of $I_{tot-steady}$).

The aforementioned equations are also valid when considering the use of a dam, provided that the corresponding distances extend beyond the dam ($r > r_2$). On the dam itself, it makes no sense to study the potential gradient for the protection of the diver and sea creatures, whilst from the point of view of absolute potential on the dam and according to Equation (27), it is true that:

$$r_{limit1} = \left[\frac{1}{r_2} - \frac{2 \times \left(\frac{\theta_w}{\rho_w} + \frac{\theta_s}{\rho_d}\right) \times V_{limits}}{I_{tot-steady}} - \frac{\frac{\theta_w}{\rho_w} + \frac{\theta_s}{\rho_d}}{\frac{\theta_w}{\rho_w} + \frac{\theta_s}{\rho_s}} \times \frac{1}{r_2}\right]^{-1} : r_1 < r_{limit1} < r_2 \tag{78}$$

Accordingly, if the electric field distribution method with a linear current source is used instead of a point source (as presented in Section 2.3), then the distance from the electrode, at which the voltage has indicatively fallen below the limit V_{limit_S}, for electric current intensity $I_{tot-steady}$ (on condition that it does not lie within the dam; i.e., $r > r_1$), is given by Equations (53), (55) or (57) and is equal to:

$$r_{limit1} = \begin{cases} r_\infty \times exp\left[-\frac{L \times V_{limit\_S}}{I_{tot}} \times \left(\frac{2\times\pi-\theta}{\rho_w} + \frac{\theta}{\rho_s}\right)\right] : r_{limit1} > max\{r_2, r_3\} \\[2ex] r_3 \times exp\left[-\frac{2\times\pi\times L}{\rho_w\times I_{tot}} \times \left(V_{limits} - \frac{I_{tot}\times ln\left(\frac{r_\infty}{r_2}\right)}{L\times\left(\frac{2\times\pi-\theta}{\rho_w}+\frac{\theta}{\rho_s}\right)}\right)\right] : r_2 < r_{limit1} < r_3 \\[2ex] \left\{r_2 \times exp\left[-\frac{\left(\frac{2\times\pi-\theta}{\rho_d}+\frac{\theta}{\rho_s}\right)\times L}{I_{tot}} \times \left(V_{limits} - \frac{I_{tot}\times ln\left(\frac{r_\infty}{r_2}\right)}{L\times\left(\frac{2\times\pi-\theta}{\rho_w}+\frac{\theta}{\rho_s}\right)}\right)\right]\right\} \\ \qquad\qquad : max\{r_1, r_3\} < r_{limit1} < r_2 \\[2ex] r_3 \times exp\left[\left(V_{limits} - \frac{\left(\frac{2\times\pi-\theta}{\rho_d}+\frac{\theta}{\rho_w}\right)\times L}{I_{tot}} \times \right.\right. \\ \qquad\left.\left. - \frac{I_{tot}\times ln\left(\frac{r_2}{r_3}\right)}{L\times\left(\frac{2\times\pi-\theta}{\rho_d}+\frac{\theta}{\rho_s}\right)} - \frac{I_{tot}\times ln\left(\frac{r_\infty}{r_2}\right)}{L\times\left(\frac{2\times\pi-\theta}{\rho_w}+\frac{\theta}{\rho_s}\right)}\right)\right] : r_1 < r_{limit1} < r_3 \end{cases} \tag{79}$$

Accordingly, the distance from the electrode, at which the potential gradient in the water has fallen, on average, below the limit E_{limit_S} for continuous operating conditions, since it lies outside the dam zone, results from Equation (75) and is equal to:

$$r_{limit2} = \begin{cases} \left[exp\left(\frac{L\times E_{limit\_S}}{I_{tot}} \times \left(\frac{2\times\pi-\theta}{\rho_w} + \frac{\theta}{\rho_s}\right)\right) - 1\right]^{-1} : r_{limit2} > max\{r_2, r_3\} \\[2ex] \left[exp\left(\frac{2\times\pi\times L\times E_{limit\_S}}{\rho_w\times I_{tot}}\right) - 1\right]^{-1} : r_2 < r_{limit2} < (r_3 - 1\,m) \\[2ex] \frac{(r_{limit2}+1\,m)^{\left(1/\left(\frac{2\times\pi-\theta}{\rho_w}+\frac{\theta}{\rho_s}\right)\right)}}{r_{limit2}^{\left(\frac{\rho_w}{2\times\pi}\right)}} \overset{solving}{=} \\ exp\left(\frac{L\times E_{limit\_S}}{I_{tot}} - \frac{\frac{\rho_w}{2\times\pi}\times\left(\frac{\theta}{\rho_s}-\frac{\theta}{\rho_w}\right)}{\left(\frac{2\times\pi-\theta}{\rho_w}+\frac{\theta}{\rho_s}\right)} \times ln(r_3)\right) \end{cases} : \left\{\begin{array}{l}(r_2 < r_{limit2})\,\wedge\\(r_3 < (r_{limit2}+1\,m))\end{array}\right\} \tag{80}$$

Within (or on) the dam, the potential combinations are many more but of no practical interest. Accordingly, the distance from the electrode, at which the potential gradient in the water has spot-dropped below the limit E_{limit_S}, for the continuous operating conditions, provided it lies outside the dam, is given by Equations (44) and (46) and is equal to:

$$r_{limit4} = \begin{cases} \frac{I_{tot}}{L\times E_{limit\_S}\times\left(\frac{2\times\pi-\theta}{\rho_w}+\frac{\theta}{\rho_s}\right)} : r_{limit4} > max\{r_2, r_3\} \\[2ex] \frac{\rho_w\times I_{tot}}{2\times\pi\times L\times E_{limit\_S}} : r_2 < r_{limit4} < r_3 \end{cases} \tag{81}$$

Likewise, the respective limits for the transient operating conditions can be obtained if the potential gradient amounts to E_{limit_T} (instead of E_{limit_S}) and the corresponding electric current intensity $I_{tot-transient}$ (instead of $I_{tot-steady}$).

By taking all electrodes, the electric field strength was calculated analytically for each one and numerically for the whole array through superposition, whilst the remaining quantities through numerical analysis as described in Section 2.4. The corresponding distance limits are calculated from the numerical results and not through analytical equations, such as (74)–(81).

Moreover, during the preliminary design, it can be considered a worst-case scenario that soil resistivity is infinite in relation to that of seawater.

3. Basic Preliminary Study Admissions for the Configuration of a Shoreline Pond Electrode Station of HVDC Link, between Attica and Crete, Greece

3.1. Breakwater Basic Structure

Based on [78], the relevant literature [1] and the Lower Churchill project [72–74] (which in terms of interconnection structure is quite similar to the structure of Crete—Attica), the configuration of coastal electrode stations in Stachtoroi and Korakia was initially proposed, with the formation of a pond and either a rubble mound breakwater (see Figure 10a), a concrete-block one (see Figure 10b) or with concrete caissons (see Figure 10c). The rubble mound breakwater, due to the openings between the rubble, acts as a filter and allows the renewal of water within the pond, while for the other two cases, this is achieved through appropriate openings.

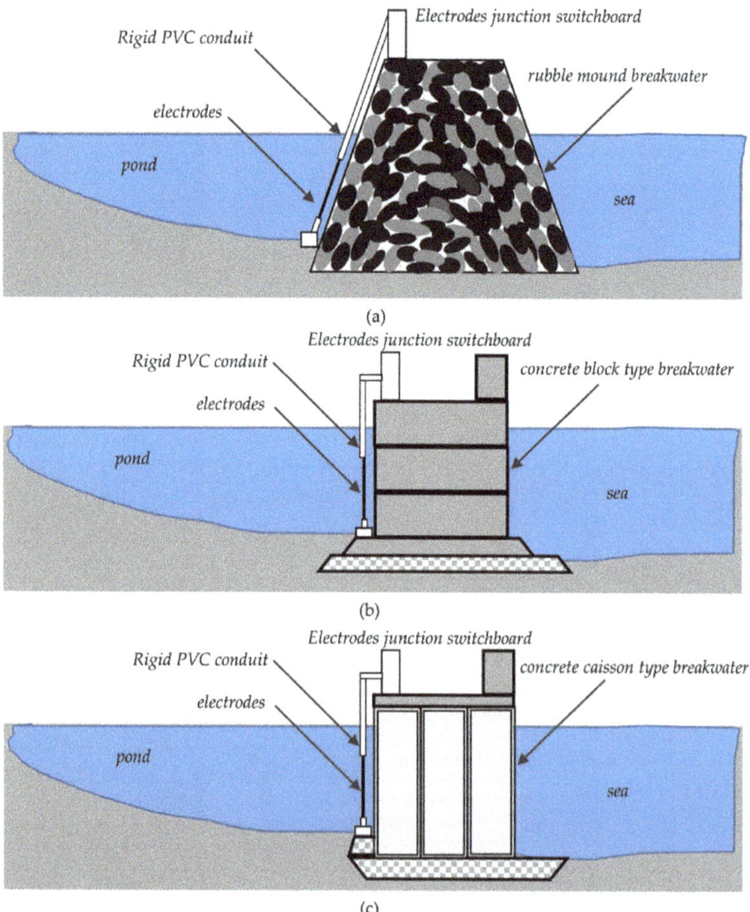

Figure 10. Shoreline electrode station with (**a**) rubble mound breakwater, (**b**) concrete block breakwater, (**c**) concrete caisson breakwater.

3.2. Basic Breakwater Layout at Stachtoroi, Attica

The corresponding layout of the electrode station for the area of Stachtoroi islet in the Argosaronic gulf, Attica, Greece, is shown in the plan view of Figure 11.

Figure 11. Shoreline electrode station on the islet of Stachtoroi, in the Argosaronic gulf, in the region of Attica [78].

3.3. Basic Breakwater Layout at Korakia, Crete

The corresponding layout of the electrode station for the area of Korakia beach in Crete is shown in the plan view of Figure 12.

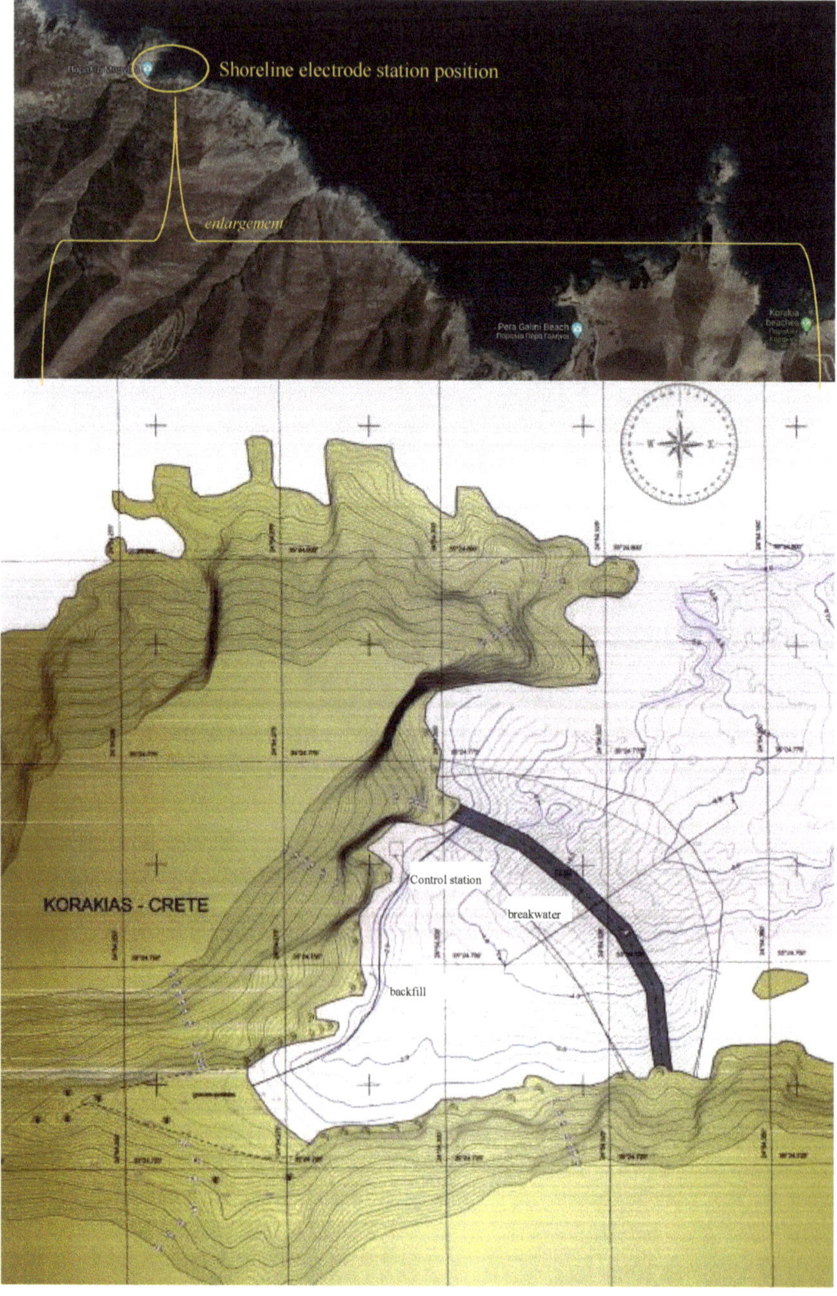

Figure 12. Shoreline electrode station at Korakia beach, Crete [78].

3.4. Technical Features of the Electrode Installation

The technical features of the electrode installation are set by IPTO as follows:

- *Operation mode*: Bipolar operation, where the currents between the two poles are theoretically equal and opposite. In emergency situations, the unipolar operation occurs due to a failure of the main pole, where the shoreline pond electrodes are used as a return conductor in cooperation with the corresponding DC medium-voltage protection conductor. In case of failure/maintenance of the converter of either pole, the corresponding pole conductor can be used without utilizing the shoreline pond electrodes;
- *Nominal current intensity at monopolar operation*: 1,000 A, as each pole has a nominal power of 500 MW under a nominal voltage of 500 kV;
- *Imbalance current intensity at bipolar operation*: 11–25 A, since it is not possible to achieve absolute synchronization between the AC/DC converters located at each conversion station, and there is a very small (asymmetrical) current, which flows through the electrodes in their normal operation and does not exceed 1–2% of the nominal current intensity of the converters;
- *Maximum short-time current intensity, under overload conditions*: 1,100 A, (i.e., +10% of nominal). Sizing of the electrodes for continuous operating conditions, as well as respective operation effects, were performed with this value;
- *Maximum transient fault current intensity*: peak value of 12,800 A at a maximum duration of 0.5 s, as required to clear the fault, due to the use of a voltage source converter;
- *Lifetime of the technical project*: 50 years;
- *Economic lifetime*: 25 years for electromechanical projects and 50 for civil engineering projects;
- *Load factor at monopolar operation*: No data are given, while in itself, it is a very complex problem because the load demand from the Attica–Crete interconnection depends, to a significant extent, not only on the load demand estimates but also on the penetration of R.E.S. in Crete, from the moment this interconnection is made. In the present case, the worst possible factor is assumed (i.e., 1);
- *Transmission line reliability and availability–forced pole outage rate*: According to CIGRE guideline 379 [80] (Table 11, p. 11 and Table 30, p. 66), the expected number of failures is 22 in 50 years or 0.433 failures per year;
- *Time interval of forced pole outages and time interval of scheduled pole outages for maintenance reasons*: In [78], an aggregate estimation is given that the duration of restoration and maintenance of one of the two poles amounts to 3 months every 5 years, without any additional data;
- *Annual electrode operational duty in Ah*: Assuming that the electrode station operates 3 months per 5 years at a maximum short-time current intensity of 1,100 A at monopolar operation and the rest time period at an imbalance current intensity of 25 A at bipolar operation, the average annual electrode operational duty is equal to 646,050 Ah;
- *Electrode station operation, during installation and commission acceptance*: During these phases, it is not expected to operate under the IPTO guidelines;
- *Reliability*: The configuration of the electrode station is performed in such a way that the necessary electrodes are divided into ν sections, with a reserve of $\nu + 1$. In the present case, the IPTO requirement is $\nu = 5$ so as to achieve a reserve of 20%;
- *Polarity*: The polarity of each earth electrode is fully reversible due to the possibility of power flow from Attica to Crete and vice versa, as well as due to the structure of the high voltage DC transmission system, where the flow of asymmetry current and monopolar operation current can reverse the operating polarity of the electrodes;
- *Electrode Materials*: IPTO recommends the use of high-silicon iron electrodes of the tubular form (indicatively by ANOTEC, Centertec Z series), conforming to ASTM A518 G3. Silicon content 14.20% ÷ 14.75%, chromium 3.25 ÷ 5.00%, carbon 0.70÷1.10%, manganese up to 1.50%, copper up to 0.50%, molybdenum up to 0.20% and the rest iron. The electrode is cast in a cooled die, with zinc connection in the center of the tube, with mechanical stress resistance equivalent to 1,000 kg, connection resistance of 1 mΩ, type 4884 SZ, weight 143 kg, diameter 122 mm ($=2 \cdot r_{el}$) and length 2130 mm ($=L_{el}$) [77]. In order to achieve reversible operation of the electrode, according to the manufacturer, the electric current density must be limited to 20 A/mm^2.

In addition, to determine the interactions and environmental impacts in each area, the following information is required:

- *Seawater electrical resistivity*: It generally depends on several factors, such as its salt content–salinity, the depth of the sea, the season, climatic conditions (e.g., prevailing temperatures) and so forth. From measurements carried out by the Hellenic Marine Research Centre, the following data were obtained regarding the seawater at Stachtoroi (Attica) and Korakia (Crete) areas:

- ➢ *Salinity*: The water salinity value varies:
  - ■ For the area of Stachtoroi, 38 ÷ 39 psu (practical salinity units or ‰ content), in water temperatures of 24 ÷ 29 °C, to a depth of approximately 90 m;
  - ■ For the area of Korakia, 38.9 ÷ 39.6 psu (practical salinity units or ‰ content), in water temperatures of 24 ÷ 26 °C, to a depth of approximately 90 m.
- ➢ *Electrical resistivity*: It takes the value of 0.167 ÷ 0.212 Ω·m, which means seawater is a medium of very good conductivity.

Therefore, in the present study, the seawater's electrical resistivity took a value of 0.25 Ω·m. In the case of breaking water (water foaming on the shore, jetties, etc., due to the air contained within and at a completely local level of a few meters from the shore or jetty), the respective value can be taken as equal to 2.0 Ω·m:

- *Soil electrical resistivity*: It is roughly determined through the types of soil, since at Korakia, there is slate, with an electrical resistivity of 20 and 1,000 Ω·m, and at Stachtoroi limestone, with a value of 1,000 to 10,000 Ω·m, as long as it does not have sediments. The seabed consists, in the best-case scenario, of sandy saline materials with an electrical resistivity that is estimated to be two orders of magnitude greater than that of seawater, i.e., about 10 Ω·m [81], and in the worst-case scenario, of rocks, with an electrical resistivity that is estimated at least at three to four orders of magnitude greater than that of seawater, i.e., >1,000 Ω·m [81]. In other words, the seabed is a path with a much higher electrical resistivity than seawater, so for the sake of saving time and financial resources, it is suggested, at the level of a preliminary study, not to measure electrical resistivity at minor and great depths;
- *Soil thermal characteristics of soil*: No measurements are needed concerning these characteristics since the electrodes are immersed in the sea;
- *Marine life around the electrodes*: There is no special type of marine fauna and flora in the electrode area other than protected birds at Stachtoroi and Posidonia Oceanica meadows (marine plants) near Korakia;
- *Salinity reduction due to freshwater inflow to seawater*: In the area of Stachtoroi, due to the small area of the islet, there is no question of changing the seawater salinity from the overflow of rainwater on the islet after rainfall. In the area of Korakia, two small dry rivers end in the small bay (as shown in Figure 12), which are not expected to cause a substantial change in seawater salinity from the runoff of rainwater as they are not rivers or underground sources of constant or significant flow. Moreover, the "rounding-up" of the seawater electrical resistivity to 0.25 instead of the maximum 0.212 Ω·m leaves a significant reserve margin, in case of changes in the electrical resistivity.

4. Application of Electric Field Distribution Methods Using an Equivalent Electrode

4.1. General Remarks

Any method based on an equivalent electrode through which all electric current flows is suitable for the determination of the electric field in the case of the far field. The method based on a linear current source is suitable for the calculation of the near electric field, but it is not suitable for the far field. This method is presented only for comparison purposes. In order to study the far-field distribution at each electrode station, nearby areas with houses or industries were studied. In the case of Stachtoroi, four characteristic areas were studied: Aegina, whose nearest coast with houses is at 7.8 km, as shown in Figure 2; Salamina at 9.5 km; Pachi-Megara at 17.5 km; and the natural gas unit at Revythousa at 16.4 km. The corresponding maximum depths and angles are given in Table 1. The exposed side to the sea near the coast is limited to 150°, while if one moves away from the island, it widens well beyond 180°. The corresponding correction factor for the electric field distribution method according to CIGRE B4.61 675:2017 and IEC TS62344:2013, as well as for the modified one with the addition of a dam, is 180°/150° = 1.20, very close to shore and less than 1 if one goes beyond 50 m from an electrode placement point. Therefore, in this case, no correction factor was necessary. In the case of Korakia, the nearest opposite coast is over 110 km on the island of Santorini. However, on the island of Crete, at a distance of about 1.5 km from the coast, there are seaside buildings, while at 0.8 km on land, there are farms. The exposed side towards the sea near the shore is limited to 112°, while, further away, it widens over 180°, according to Figure 3. Very close to shore, the corresponding correction factor is 180°/112° = 1.60, while 100 m away from an electrode, it is about 1. In this study, a correction factor of 1.30 was set. Table 1 lists the respective depth and angle values θ_w at near distances (50 m) and long distances (20 km). In addition, in both areas under study, it is considered that the ground has a small elevation in relation to the variations in the sea depth. Consequently, it can be considered as horizontal, i.e., $\theta_s = \pi\text{-}\theta_w$, approximating better the IEC standard TS62344:2013.

Table 1. Geometric characteristics (distances, depths and angles of seabed) of electrode station.

Electrode Station at Stachtoroi, Attica				Electrode Station at Korakia, Crete			
Area	Distance [m]	Depth [m]	θ_w [rad]	Area	Distance [m]	Depth [m]	θ_w [rad]
Aegina	7,800	37	0.004743554	Far	20,000	800	0.039978687
Salamina	9,500	80	0.008420854	Near	50	3	0.059928155
Revithousa	16,400	90	0.005487750				
Megara Pachi	17,500	100	0.005714224				

4.2. Application of Method "A"—Combination of Electric Field Distribution Methods by CIGRE B4.61 675:2017 and IEC TS 62344:2013

Applying the combination method of electric field distribution according to CIGRE B4.61 675:2017 and IEC TS 62344:2013 in the area of Stachtoroi, the results for the worst-case scenario, i.e., with infinite soil resistivity, and for the scenarios with the smallest and the highest possible electrical resistivities ($\rho_s = 1,000~\Omega{\cdot}m$ and $\rho_s = 10,000~\Omega{\cdot}m$) are presented in Table 2. It is found that the worst-case scenario and the scenario with the highest possible electrical resistivity give very similar results. In addition, for the worst-case scenario, the voltage is smaller than 4 V before reaching any residential shore, comparing the limit r_{limit1} to the corresponding distances in Table 1. Moreover, the electric field strength is limited to 1.25 V/m in a radius of 153 m from the position of the point electrode. In the case of the transient phenomenon, the distances are slightly shorter.

Similarly, by applying the corresponding method in the area of Korakia, the results for the worst-case scenario, i.e., with infinite soil resistivity, and for the scenarios with the smallest and the highest possible electrical resistivities ($\rho_s = 100~\Omega{\cdot}m$ and $\rho_s = 1,000~\Omega{\cdot}m$) are presented in Table 3. The initial values of Equations (8) and (9), (74), (76) and (77) were recorded, as well as after the implementation of the correction factor 1.30, where the respective values are increased. It was found that the scenario with the highest possible electrical resistivity gives slightly improved results compared to the worst-case scenario due to the fact that the resistivity is set at $1,000~\Omega{\cdot}m$. However, even for the worst-case scenario, the voltage is smaller than 4 V at a distance of 1,200 m (shorter distance than the distance to existing structures, roads, etc.). Only in the south–southeast is there an arable area without buildings at a distance of 850–1,200 m. In this area, no effects are expected because the voltage is smaller than 1.18 V at a distance of 350 m. The electric field strength is limited to 1.25 V/m for a radius of 70 m from the position of the point electrode (due to the effect of a near field, the exact position of the electrode must be taken into account). In the case of the transient phenomenon, the distances are slightly shorter.

The indicative value of the resistance of the electrode station in relation to remote earth for the worst-case scenario is extremely high in both cases (432 Ω for Stachtoroi and 67 Ω for Korakia, i.e., a total of 542 Ω). This value does not represent reality, as it was assumed that the whole electric current passes through an electrode with a radius of 0.061 m and with extremely high potential (475 kV for Stachtoroi and 73.3 kV for Korakia). The use of non-infinite values of resistivity improves the corresponding quantities by up to 21%, which, however, remain at high levels. The above data show that the method proposed by CIGRE B4.61 675:2017 and IEC TS 62344:2013 gives results of quite higher values, rendering the design more difficult.

Table 2. Minimum values of distances for absolute potential in relation to remote earth, for potential gradient/electric field strength in steady and transient state, absolute potential at electrode surface and equivalent electrode resistance in the area of Stachtoroi (Attica) in relation to neighboring coasts based on method "A" according to Equations (8) and (9), (74), (76) and (77).

Area	Worst-Case: $\rho_s = \infty$ $\Omega \cdot m$							Lower Expected: $\rho_s = 1{,}000$ $\Omega \cdot m$							Higher Expected: $\rho_s = 10{,}000$ $\Omega \cdot m$						
	r_{limit1} [m]	r_{limit2} [m]	r_{limit3} [m]	r_{limit4} [m]	r_{limit5} [m]	$V(r_{el})$ [kV]	R_{el} [Ω]	r_{limit1} [m]	r_{limit2} [m]	r_{limit3} [m]	r_{limit4} [m]	r_{limit5} [m]	$V(r_{el})$ [kV]	R_{el} [Ω]	r_{limit1} [m]	r_{limit2} [m]	r_{limit3} [m]	r_{limit4} [m]	r_{limit5} [m]	$V(r_{el})$ [kV]	R_{el} [Ω]
Aegina	7,246.7	151.78	149.46	152.28	149.96	475.2	431.99	6,218.6	140.57	138.41	141.07	138.91	407.8	370.71	7,128.8	150.54	148.23	151.04	148.73	467.5	424.97
Salamina	4,082.1	113.79	112.05	114.29	112.55	267.7	243.35	3,734.7	108.82	107.15	109.32	107.65	244.9	222.64	4,044.5	113.27	111.53	113.76	112.03	265.2	241.10
Revithousa	6,264.0	141.08	138.92	141.58	139.42	410.8	373.41	5,480.9	131.94	129.91	132.43	130.41	359.4	326.73	6,175.7	140.08	137.93	140.58	138.43	405.0	368.15
Megara Pachi	6,015.7	138.25	136.13	138.75	136.63	394.5	358.61	5,289.9	129.61	127.62	130.11	128.12	346.9	315.35	5,934.3	137.30	135.20	137.80	135.70	389.1	353.76

Note: Numbers in italics are the worst results.

Table 3. Minimum values of distances for absolute potential in relation to remote earth, for potential gradient/electric field strength in steady and transient state, absolute potential at electrode surface and equivalent electrode resistance in the area of Korakia (Crete) in relation to neighboring coasts based on method "A" according to Equations (8) and (9), (74), (76) and (77).

Area	Worst-Case: $\rho_s = \infty$ $\Omega \cdot m$							Lower Expected: $\rho_s = 100$ $\Omega \cdot m$							Higher Expected: $\rho_s = 1{,}000$ $\Omega \cdot m$						
	r_{limit1} [m]	r_{limit2} [m]	r_{limit3} [m]	r_{limit4} [m]	r_{limit5} [m]	$V(r_{el})$ [kV]	R_{el} [Ω]	r_{limit1} [m]	r_{limit2} [m]	r_{limit3} [m]	r_{limit4} [m]	r_{limit5} [m]	$V(r_{el})$ [kV]	R_{el} [Ω]	r_{limit1} [m]	r_{limit2} [m]	r_{limit3} [m]	r_{limit4} [m]	r_{limit5} [m]	$V(r_{el})$ [kV]	R_{el} [Ω]
Far	859.8	51.96	51.16	52.45	51.65	56.38	51.26	720.2	47.51	46.77	48.01	47.27	47.22	42.93	843.5	51.46	50.66	51.95	51.16	55.31	50.28
Near	573.6	42.35	41.69	42.84	42.19	37.61	34.19	508.3	39.83	39.22	40.33	39.71	33.33	30.30	566.3	42.07	41.42	42.57	41.92	37.14	33.76
Far *	1,117.8	67.54	66.50	68.19	67.15	73.30	66.63	936.2	61.76	60.81	62.41	61.45	61.39	55.81	1,096.5	66.89	65.86	67.54	66.51	71.90	65.37
Near *	745.7	55.05	54.20	55.70	54.85	48.90	44.45	660.7	51.78	50.98	52.43	51.63	43.33	39.39	736.2	54.70	53.85	55.34	54.50	48.28	43.89

Note: Numbers in italics are the worst results. (*)—After the implementation of the correction factor (=1.30).

4.3. Application of Method "B"—Combination of Electric Field Distribution Methods by CIGRE B4.61 675:2017 and IEC TS 62344:2013 (Modification Taking the Dam into Consideration)

By applying the modified method of electric field distribution according to CIGRE B4.61 675:2017 and IEC TS 62344:2013 with the addition of a dam in the area of Stachtoroi, it was assumed that the thickness of the dam is 16.0 m at a distance of 1.0 m from the electrode ($r_1 = 1.0$ m, $r_2 = 17.0$ m). Six scenarios were considered involving all possible combinations using infinite soil resistivity, the smallest and highest possible resistivities ($\rho_s = 1,000$ $\Omega \cdot$m and $\rho_s = 10,000$ $\Omega \cdot$m) and using the smallest and highest possible electrical resistance of a rubble mound or concrete dam with gaps filled with seawater ($\rho_d = 100$ $\Omega \cdot$m and $\rho_d = 120$ $\Omega \cdot$m).

The safety distances r_{limit1} for voltage V_{limit_S}, r_{limit2} for a steady-state mean value of electric field strength E_{limit_S}, r_{limit3} for a transient mean value of electric field strength E_{limit_T}, r_{limit4} for the steady-state point value of electric field strength E_{limit_S}, and r_{limit5} for the transient point value of electric field strength E_{limit_T} present no variations compared to the values in Table 2, because the corresponding distances are located outside the dam, and are therefore not affected by the dam resistivity. On the contrary, the values of the electrode station resistance and the corresponding high absolute potential are significantly increased due to the addition of the dam, as shown in Table 4. When the effect of the soil is ignored, the values are 23.9 times higher than the corresponding values of method "A" for dam resistivity $\rho_d = 100$ $\Omega \cdot$m, while, if the soil is taken into account, the variations are between 1.35 and 5.85 times higher achieving the smallest differences for the smallest possible soil resistivity and the widest angles θ_w. If the dam resistivity is increased by 20% ($\rho_d = 120$ $\Omega \cdot$m), then in the worst-case scenario (ignoring the soil effect), there is an increment of 19.2%, while when the soil effect is taken into account, then it is limited from 0.1% to 3% for resistivities ρ_s from 1,000 to 10,000 $\Omega \cdot$m.

Table 4. Absolute potential at electrode surface $V(r_{el})$ and equivalent electrode resistance R_{el} in the area of Stachtoroi, Attica, in relation to the neighboring coasts based on method "B" according to Equations (27) and (28) and variation in corresponding values in relation to method "A".

	Worst-Case: $\rho_s = \infty$ $\Omega \cdot$m			Lower Expected: $\rho_s = 1,000$ $\Omega \cdot$m			Higher Expected: $\rho_s = 10,000$ $\Omega \cdot$m		
$\rho_d = 100$ $\Omega \cdot$m /Area	$V(r_{el})$ [kV]	R_{el} [Ω]	Variation [-]	$V(r_{el})$ [kV]	R_{el} [Ω]	Variation [-]	$V(r_{el})$ [kV]	R_{el} [Ω]	Variation [-]
Aegina	*11,361*	*10,327.8*	*23.91*	546.9	497.21	1.341	1,874	1,703.7	4.009
Salamina	6,400	5,817.7	23.91	391.7	356.12	1.600	1,552	1,411.1	5.853
Revithousa	9,820	8,927.2	23.91	501.0	455.45	1.394	1,787	1,624.1	4.411
Megara Pachi	9,431	8,573.4	23.91	489.1	444.62	1.410	1,763	1,602.8	4.531
$\rho_d = 120$ $\Omega \cdot$m /Area	$V(r_{el})$ [kV]	R_{el} [Ω]	Variation [-]	$V(r_{el})$ [kV]	R_{el} [Ω]	Variation [-]	$V(r_{el})$ [kV]	R_{el} [Ω]	Variation [-]
Aegina	*13,543*	*12,311.9*	*28.50*	547.3	497.58	1.342	1,906	1,732.9	4.078
Salamina	7,629	6,935.4	28.50	392.4	356.76	1.602	1,600	1,454.4	6.032
Revithousa	11,706	10,642.3	28.50	501.5	455.87	1.395	1,822	1,656.6	4.500
Megara Pachi	11,243	10,220.5	28.50	489.6	445.06	1.411	1,800	1,636.3	4.625

Note: Numbers in italics are the worst results.

With the addition of a rubble mound dam in the area of Korakia, the assumption that the thickness of the dam is 18.0 m at a distance of 1.0 m from the electrode, i.e., $r_1 = 1.0$ m, $r_2 = 19.0$ m, is made. Six scenarios were considered, including all possible combinations using infinite soil resistivity, the lowest and highest possible soil resistivity ($\rho_s = 100$ $\Omega \cdot$m and $\rho_s = 1,000$ $\Omega \cdot$m) and using the lowest and highest possible resistivity of a rubble mound dam with seawater-filled gaps ($\rho_d = 100$ $\Omega \cdot$m and $\rho_d = 120$ $\Omega \cdot$m). The values of the respective safety distances r_{limit1}, r_{limit2}, r_{limit3}, r_{limit4} and r_{limit5} are the same compared to the values in Table 3. On the contrary, the values of the electrode station resistance with respect to remote earth and the corresponding high absolute potential increase significantly due to the addition of the dam, as shown in Table 5. Ignoring the soil effect, the values are 24.1 times higher than the corresponding values of method "A" for the case of dam resistivity $\rho_d = 100$ $\Omega \cdot$m, while, if the soil is taken into account, the variation is limited between 1.29 and 4.75 times higher achieving the smallest differences for the smallest possible soil resistivity. If the dam resistivity is increased by 20% ($\rho_d = 120$ $\Omega \cdot$m), then in the worst-case scenario, by ignoring the soil effect, there is an increment of 19.2%, while when the soil effect is taken into account, then it is limited from 0.1% to 2.2% for the electrical resistivities ρ_s from 100 to 1,000 $\Omega \cdot$m.

Table 5. Absolute potential at electrode surface $V(r_{el})$ and equivalent electrode resistance R_{el} in the area of Korakia (Crete), in relation to the neighboring coasts based on method "B" according to Equations (27) and (28) and variation in corresponding values in relation to method "A".

$\rho_d = 100\ \Omega\cdot m$	Worst-Case: $\rho_s = \infty\ \Omega\cdot m$			Lower Expected: $\rho_s = 100\ \Omega\cdot m$			Higher Expected: $\rho_s = 1{,}000\ \Omega\cdot m$		
/Area	$V(r_{el})$ [kV]	R_{el} [Ω]	Variation [-]	$V(r_{el})$ [kV]	R_{el} [Ω]	Variation [-]	$V(r_{el})$ [kV]	R_{el} [Ω]	Variation [-]
Far	1,357	1,233.1	24.06	61.08	55.527	1.293	200.9	182.66	3.633
Near	904.9	822.64	24.06	47.99	43.626	1.440	176.5	160.49	4.754
Far *	1,763	1,603.0	24.06	79.40	72.185	1.293	261.2	237.46	3.633
Near *	1,176	1,069.4	24.06	62.39	56.713	1.440	229.5	208.64	4.754
$\rho_d = 120\ \Omega\cdot m$ /Area	$V(r_{el})$ [kV]	R_{el} [Ω]	Variation [-]	$V(r_{el})$ [kV]	R_{el} [Ω]	Variation [-]	$V(r_{el})$ [kV]	R_{el} [Ω]	Variation [-]
Far	1,617	1,470.1	28.68	61.12	55.559	1.294	203.8	185.28	3.685
Near	1,079	980.72	28.68	48.04	43.674	1.441	180.5	164.08	4.860
Far *	2102	1,911.1	28.68	79.45	72.227	1.294	265.0	240.87	3.685
Near *	1,402	1,274.9	28.68	62.45	56.776	1.441	234.6	213.31	4.860

Note: Numbers in italics are the worst results. (*)—After the implementation of the correction factor (=1.30).

Of course, in both cases examined (Stachtoroi and Korakia), the values of the resistance of the electrode station with respect to remote earth for the worst-case scenario are extremely high, much higher than those of method "A". This is not the actual case, as it was assumed that all the current intensity passes through an electrode of a radius of 0.061 m and that this electrode is also surrounded by a high resistivity dam of thickness d (16 m for Stachtoroi and 18 m for Korakia) and by the soil of infinite resistivity. The use of non-infinite resistivity values regarding the soil improves the respective values by up to 95%, which, however, remain at high levels. The modified point source method by CIGRE B4.61 675:2017 and IEC TS 62344:2013 with the addition of a dam eventually gives quite unfavorable results, especially regarding the resistance of the electrode station with respect to remote earth and the corresponding developed absolute potential.

4.4. Application of Method "C"—Near Electric Field Distribution Method with Linear Current Source

By applying the near electric field distribution method, with a linear current source and with the addition of a rubble mound breakwater, the inclination of the seabed and of ground is not used, but the water zone of height/active electrode length L. In the case of a rubble mound breakwater with an inclination λ (base:height), as in Figure 10a, the water zone L is determined through the geometric length of the electrode L_{el} equal to:

$$L = L_{el} \times cos(atan(\lambda)) \tag{82}$$

If the electrode is placed vertically, as in Figure 10b,c, then $L = L_{el}$.

In the case of Stachtoroi, based on Figure 11, the angle θ of the top view of Figure 6 is 210°, and the thickness of the dam at the point where the electrode is placed is 16.0 m, so the radius r_2 is equal to 17.0 m, while the radius r_3 (for which it is ensured the same depth as the lower end of the electrode, along the entire length of the dam) is equal to 10.0 m. The sea distance between the two electrode stations is equally divided, thus setting r_∞ equal to 150 km. The rubble mound breakwater inclination λ amounts to 3:2, so the respective electrode inclination is the same; thus, the active length L of the ANOTEC electrode (see Section 3.4) is equal to 1.1815 m. In the case of a special composition concrete breakwater, the electrode is placed vertically with a respective active length of 2.13 m. Similar to the application of method "B", the six scenarios are examined, which include all possible combinations, using three soil electrical resistivity and two breakwater electrical resistivity values for the two different active electrode lengths (3:2 inclination and vertical). Table 6 lists the respective results. In the case of the 3:2 electrode inclination, it is found that the voltage drops below 4 V, with respect to "infinity", at 143.5 km, which fully displays the inability to describe the far field since it studies only a narrow water zone, equal to the active length L (=1.1815 m). However, the electric field strength below 1.25 V/m is limited to a radius of 71 m from the position of this linear electrode. In the case of the transient phenomenon, slightly shorter distances result. Effect of the electrical resistivity of the breakwater material does not exist at the distance limits because they lie beyond the breakwater area and is limited to large soil electrical resistivities, increasing the maximum electrode resistance to remote earth from 16.0 Ω to 21.5 Ω, and the absolute potential on the electrode from 17.6 kV to 23.7 kV. Here, the effect of the linear electrode on the calculation of the electrode resistance to remote earth and on the voltage on the electrode is additionally observed, as there is a significant reduction compared to the results of Tables 2 and 4. Therefore it seems that beyond the distance of the far-field effects (where the model's failure was expected from the start), in the remaining quantities, it gives values that represent the near field much better. Additionally, if the electrode is placed vertically and not with a 3:2 inclination, with a suitable suspension device, then a significant reduction in the electric field strength is observed, below 1.25 V/m, as it is limited to a radius of 40 m from the position of the linear electrode, for an active length of 2.13 m, which basically states—in relation to the 16 m breakwater thickness—that at an external distance from the breakwater, of the order of 23 m, there will be no problem for swimmers and sea creatures/mammals, dropping at 26% of the value of methods "A" or "B". Moreover, significant reductions occur regarding the maximum electrode resistance to the remote earth (dropping to 12.0 Ω) and absolute potential across on electrode (to 13.1 kV), even for infinite soil electrical resistivity.

Table 6. Minimum values of distances for absolute potential in relation to remote earth, for potential gradient/electric field strength in steady and transient state, absolute potential at electrode surface and equivalent electrode resistance in the area of Stachtoroi (Attica) in relation to neighboring coasts based on the method "C" according to Equations (55) and (56), (79)–(81).

L [m]	ρ_d [$\Omega \cdot$m]	ρ_s [$\Omega \cdot$m]	r_{limit1} [m]	r_{limit2} [m]	r_{limit3} [m]	r_{limit4} [m]	r_{limit5} [m]	$V(r_{el})$ [kV]	R_{el} [Ω]
		∞	*143,401*	70.625	68.470	*71.124*	68.969	19.927	18.116
	100	1,000	143,399	70.600	68.446	71.099	68.945	17.610	16.009
1.1815		10,000	143,401	70.623	68.468	71.122	68.966	19.667	17.879
(slope 3:2)		∞	*143,401*	70.625	68.470	*71.124*	68.969	*23.702*	*21.547*
	120	1,000	143,399	70.600	68.446	71.099	68.945	20.444	18.586
		10,000	143,401	70.623	68.468	71.122	68.966	23.327	21.207
		∞	*138,314*	38.955	37.759	*39.452*	38.257	11.054	10.049
	100	1,000	138,310	38.941	37.746	39.439	38.244	9.768	8.880
2.1300		10,000	138,313	38.953	37.758	39.451	38.256	10.909	9.917
(vertical)		∞	*138,314*	38.955	37.759	*39.452*	38.257	*13.147*	*11.952*
	120	1,000	138,310	38.941	37.746	39.439	38.244	11.340	10.309
		10,000	138,313	38.953	37.758	39.451	38.256	12.940	11.763

Note: Numbers in italics are the worst results.

Respectively, in the case of Korakia, based on Figure 12, the angle θ of the plan view of Figure 6 is 245°, and the thickness of the breakwater at the electrode position is 18.0 m, so the radius r_2 is equal to 19.0 m, while the radius r_3 is equal with 25.0 m. The rest of the geometric features are the same as those at Stachtoroi. Similar to the application of method "B", the six scenarios are examined, which include all possible combinations, using three soil electrical resistivity and two breakwater electrical resistivity values for the two different active electrode lengths (3:2 inclination and vertical). Table 7 lists the respective results. In the case of the 3:2 electrode inclination, it was found that the voltage drops below 4 V, with respect "infinity", at 145 km, which fully displays the inability to describe the far-field since it studies only a narrow water zone, equal to the active length L (=1.1815 m). However, the electric field strength, below 1.25 V/m, is limited to a radius of 93 m from the position of this linear electrode. In the case of the transient phenomenon, slightly shorter distances result. Effect of the electrical resistivity of the breakwater material does not exist at the distance limits because they lie beyond the breakwater area and is limited to large soil electrical resistivities, increasing the maximum electrode resistance to remote earth from 1.161 Ω to 1.166 Ω, and the absolute potential on the electrode from 1.277 kV to 1.283 kV. Here, the effect of the linear electrode on the calculation of the electrode resistance to the remote earth and on the voltage on the electrode is additionally observed, as there is a significant reduction compared to the results in Tables 3 and 5. Therefore it seems that beyond the distance the far field affects, in the remaining quantities, it gives values that represent the near field much better. Additionally, if the electrode is placed vertically with a suitable suspension device, then a significant reduction in the electric field strength is observed, below 1.25 V/m, as it is limited to a radius of 51.5 m from the position of the linear electrode for an active length of 2.13 m, which basically states—in relation to the 18 m breakwater thickness—that at an external distance from the breakwater, of the order of 32 m, there will be no problem for swimmers and sea creatures/mammals, dropping at 76% of the value of methods "A" or "B". Respectively, significant reductions occur regarding the maximum electrode resistance to remote earth (dropping to 0.65 Ω) and absolute potential across on electrode (to 0.71 kV), even for infinite soil electrical resistivity.

Table 7. Minimum values of distances for absolute potential in relation to remote earth, for potential gradient/electric field strength in steady and transient state, absolute potential at electrode surface and equivalent electrode resistance in the area of Korakia (Crete) in relation to neighboring coasts based on the method "A" according to Equations (53) and (54), (79)–(81).

L [m]	ρ_d [$\Omega\cdot$m]	ρ_s [$\Omega\cdot$m]	r_{limit1} [m]	r_{limit2} [m]	r_{limit3} [m]	r_{limit4} [m]	r_{limit5} [m]	$V(r_{el})$ [kV]	R_{el} [Ω]
1.1815 (slope 3:2)	100	∞	144,914	92.271	89.460	92.770	89.959	1.2827	1.1661
		100	144,888	91.780	88.984	92.279	89.483	1.2773	1.1612
		1,000	144,911	92.222	89.412	92.721	89.911	1.2821	1.1656
	120	*∞*	*144,914*	*92.271*	*89.460*	*92.770*	*89.959*	*1.2827*	*1.1661*
		100	144,888	91.780	88.984	92.279	89.483	1.2774	1.1612
		1,000	144,911	92.222	89.412	92.721	89.911	1.2822	1.1656
2.1300 (vertical)	100	∞	140,956	50.961	49.402	51.460	49.900	0.7115	0.6468
		100	140,910	50.689	49.138	51.187	49.636	0.7085	0.6441
		1,000	140,952	50.934	49.375	51.432	49.874	0.7112	0.6465
	120	*∞*	*140,956*	*50.961*	*49.402*	*51.460*	*49.900*	*0.7115*	*0.6468*
		100	140,910	50.689	49.138	51.187	49.636	0.7086	0.6441
		1,000	140,952	50.934	49.375	51.432	49.874	0.7112	0.6466

Note: Numbers in italics are the worst results.

From the study of the two cases (Stachtoroi and Korakia), in terms of the effect of the values of the electrical resistivities of soil and breakwater, there are no large differences despite their range of values. This is due to the low seawater electrical resistivity and the water zones formed. On the contrary, the active length of the electrode has a significant effect, leading to significantly improved results.

The data above demonstrate that the proposed model is suitable for the near field since the electrode is considered a linear rather than a point source. On the contrary, due to the assumption of a constant conduction zone of constant depth, much smaller than that in which the electric current is diffused over long distances, it is unsuitable for the far field.

4.5. Comparison of Methods and Combined Utilization

In order to compare the proposed methods, the corresponding results of the electric field strength, with respect to the distance from the one concentrated electrode, for the case of Stachtoroi, Attica, are presented in detail. The geometric dimensions of the layout of Figures 4 and 6 are the following: $\theta_w = 0.2718° = 0.004743554$ rad (the worst-case scenario concerning Aegina in Table 1), $\theta = 210°$, $r_1 = 1.0$ m, $r_2 = 17.0$ m, $r_3 = 10.0$ m, $r_\infty = 150$ km. The electrical resistivity of the soil amounts to $\rho_s = 1,000$ $\Omega\cdot$m and of the rubble-mound or concrete with seawater-filled gaps breakwater to $\rho_d = 100$ $\Omega\cdot$m. Figure 13 shows the change in the electric field strength on a semi-logarithmic scale in relation to the distance of up to 150 km since, on a linear scale, the respective changes would not be easy to read. It can be seen that, for long distances, the "C" method of the linear current source gives much higher electric field strength values, as it disregards the seabed slope and the increasing depth of the seawater, which reaches up to 37 m, remaining at a layer of water of constant thickness of 1.1815 m. Additionally, the results of methods "A" and "B" are identical.

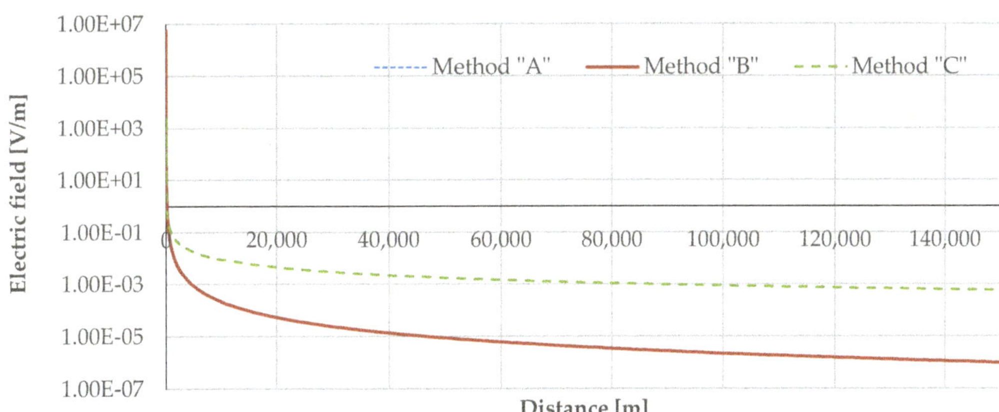

Figure 13. Semi-logarithmic diagram of electric field strength, with respect to distance (for up to 150 km) with three methods, "A", "B" and "C", for the case of Stachtoroi ($\theta_w = 0.2718°$, $\theta = 210°$, $r_1 = 1.0$ m, $r_2 = 17.0$ m, $r_3 = 10.0$ m, $r_\infty = 150$ km, $\rho_s = 1{,}000$ $\Omega \cdot$m, $\rho_d = 100$ $\Omega \cdot$m).

Figure 14 shows the respective distribution of the electric field strength within the area of the breakwater (i.e., for distances between r_{el} and r_1), resulting in the values of methods "A" and "B", by using a point current source, being identical and giving much higher values than method "C" and much less approximate the near field behavior. This is reinforced by the results of Figure 15 for the electric field strength in the breakwater area, where the values are extremely high due to the breakwater electrical resistivity. This demonstrates the weakness of the original "A" method, of electric field distribution, according to CIGRE B4.61 675:2017 and IEC TS 62344:2013, which does not take the breakwater into account. However, method "B", also leads to very high values, at close distances, due to the use of a point source. On the contrary, it is at this point that method "C" of electric field strength distribution is advantageous by using a linear current source, which better approximates reality, giving smoother changes in electric field strength. In particular, the area between the distances $r_3 = 10.0$ m and $r_2 = 17.0$ m gives high values due to the existence of only soil and breakwater. From Figure 16, which concerns a water area outside the breakwater, it is clearly seen that method "C" of the linear current source gives smaller electric field strength values at close distances, utilizing, in essence, the larger water zone, with respect to methods "A" and "B", whose results are identical, but use a small seawater wedge of angle θ_w. This is attenuated at longer distances since Figure 17 demonstrates that at about 240 m and beyond, the results are identical. Of course, for very long distances, the values of methods "A" and "B" provide more favorable results (as was seen in Figure 13), as the seawater wedge, of angle θ_w, grows larger than the horizontal layer of water, defined by method "C" of the linear power source.

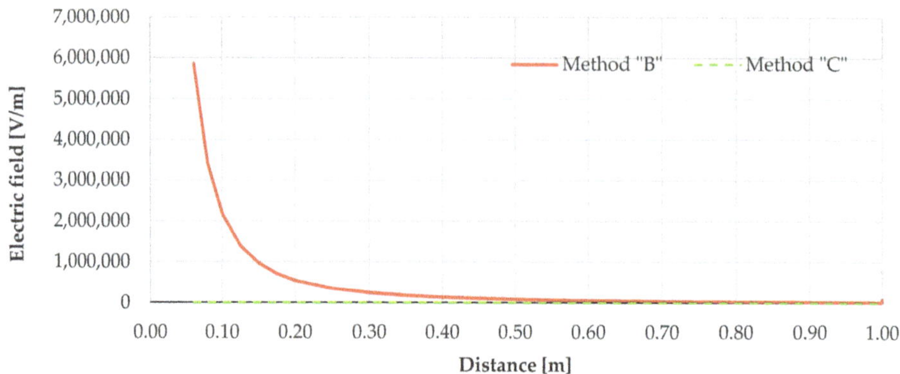

Figure 14. Electric field strength diagram, with respect to distance within the breakwater ($r < 1.0$ m), with three methods, "A", "B" and "C", for the case of Stachtoroi ($\theta_w = 0.2718°$, $\theta = 210°$, $r_1 = 1.0$ m, $r_2 = 17.0$ m, $r_3 = 10.0$ m, $r_\infty = 150$ km, $\rho_s = 1,000$ $\Omega{\cdot}$m, $\rho_d = 100$ $\Omega{\cdot}$m).

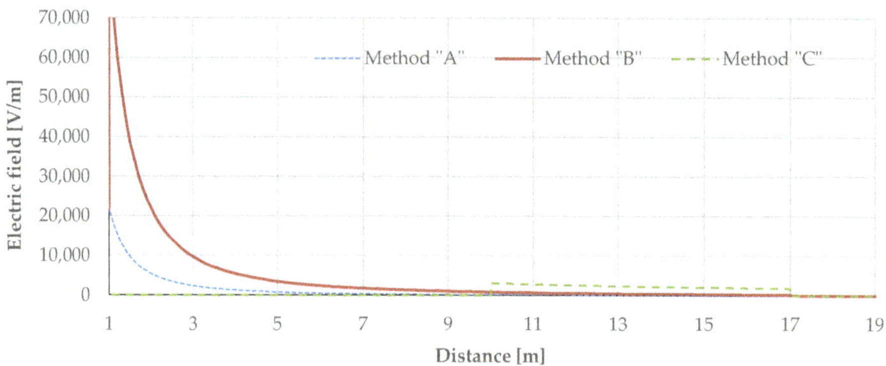

Figure 15. Electric field strength diagram, with respect to distance in the breakwater area ($r_1 = 1$ m $< r < r_2 = 17$ m), with three methods, "A", "B"and "C", for the case of Stachtoroi ($\theta_w = 0.2718°$, $\theta = 210°$, $r_1 = 1.0$ m, $r_2 = 17.0$ m, $r_3 = 10.0$ m, $r_\infty = 150$ km, $\rho_s = 1,000$ $\Omega{\cdot}$m, $\rho_d = 100$ $\Omega{\cdot}$m).

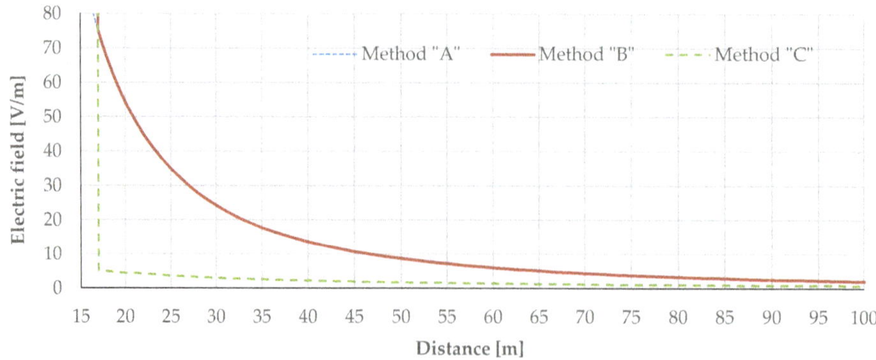

Figure 16. Electric field strength diagram, with respect to distance in the area very closely to the exterior of the breakwater ($r_2 = 17$ m $< r < 100$ m), with three methods, "A", "B"and "C", for the case of Stachtoroi ($\theta_w = 0.2718°$, $\theta = 210°$, $r_1 = 1.0$ m, $r_2 = 17.0$ m, $r_3 = 10.0$ m, $r_\infty = 150$ km, $\rho_s = 1,000$ $\Omega{\cdot}$m, $\rho_d = 100$ $\Omega{\cdot}$m).

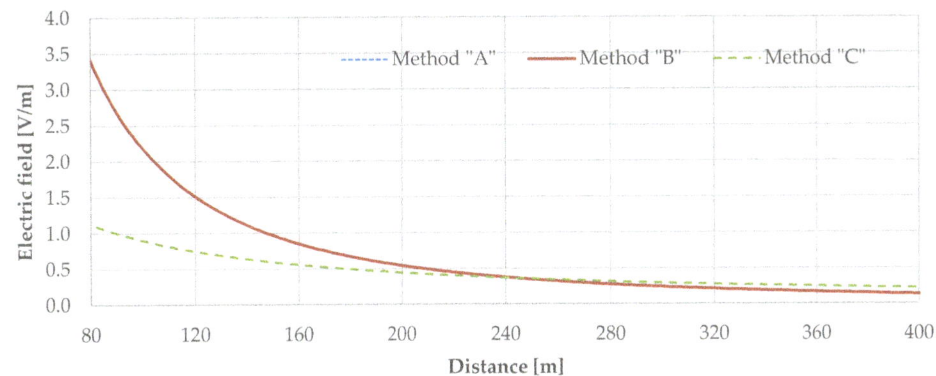

Figure 17. Electric field strength diagram, with respect to distance in the near outer area of the breakwater (80 m < r < 400 m), with three methods, "A", "B" and "C", for the case of Stachtoroi ($\theta_w = 0.2718°$, $\theta = 210°$, $r_1 = 1.0$ m, $r_2 = 17.0$ m, $r_3 = 10.0$ m, $r_\infty = 150$ km, $\rho_s = 1{,}000\ \Omega\cdot$m, $\rho_d = 100\ \Omega\cdot$m).

Therefore, method "C" of the linear current source is recommended within the breakwater area, inside the breakwater, and on the outer area, for up to the distance $r_{C\to A}$, beyond which method "A" (or "B", since in that region they are identical) of the point current source, gives smaller values of electric field strength, as the wedge of seawater of angle θ_w better approximates the seawater mass than does the limited horizontal water layer, defined by the method "C" (of the linear current source) at long distances. The smooth transition between the two methods takes place at a distance where the respective electric field strengths of methods "A" and "C" (for the external area of the breakwater) are equal. Therefore, by equalizing Equations (7) or (23) and (44), the distance $r_{C\to A}$ results as equal to:

$$r_{C\to A} = \frac{L}{2} \times \frac{\left(\frac{2\times\pi-\theta}{\rho_w} + \frac{\theta}{\rho_s}\right)}{\left(\frac{\theta_w}{\rho_w} + \frac{\theta_s}{\rho_s}\right)} \times c_f \tag{83}$$

where c_f is the correction factor of the electric field strength, due to limited exposure of a point electrode to the sea, at an angle φ, smaller than 180° by methods "A" and "B", equal to π/φ, where φ expressed in rad. In the case of Stachtoroi, the correction factor is 1.0, while in the case of Korakia, it is 1.3. In both cases, the angle of the soil layer θ_s is equal to $2 \times \pi\text{-}\theta_w$ in Figure 4 or Figure 5.

Therefore, to calculate the safety distances r_{limit2}, for an average value of electric field strength E_{limit_S} (continuous operating conditions); r_{limit3}, for an average value of electric field strength E_{limit_T} (transient operating conditions); r_{limit4}, for a point value of electric field strength E_{limit_S} (continuous operating conditions); and r_{limit5}, for a point value of electric field strength E_{limit_T} (transient operating conditions), it is suggested to use the method "C", of the linear current source, due to its suitability for short distances (of the order of tens of meters). On the contrary, for the calculation of the safety distance r_{limit1} for voltage V_{limit_S}, with respect to infinity, it is recommended to use method "A" (or "B", since in that area they are identical) due to its suitability for long distances (of the order of a few km).

For the calculation of electric field strength, the absolute potential on the surface of an electrode, with respect to infinity (remote earth), and the equivalent electrode station resistance, the combined application of the methods was proposed. More specifically, method "C" of electric field strength distribution, through a linear current source and a water zone of constant thickness, was applied from the surface of the electrode up to the distance where the electric field strengths of methods "A" and "C" are equalized (i.e., $r_{el} < r < r_{C\to A}$), whereas the unified method "A", of electric field strength distribution through a point current source, according to CIGRE B4.61 675:2017 and IEC TS 62344:2013, was applied from the distance where the electric field strengths of methods "A" and "C" are equalized, towards infinity (i.e., $r_{C\to A} < r < \infty$). From the calculation of the electric field strength, the absolute potential, with respect to infinity (remote earth), can be calculated and, subsequently, the respective equivalent electrode resistance. Hence, for the case of Stachtoroi, Attica (since $r_1 < r_3 < r_2 < r_{C\to A}$), according to Equation (83), the absolute potential is calculated as follows:

$$V(r) = \int_r^{r_\infty} \vec{E} \times \vec{d\ell} = \underbrace{\int_r^{r_1} E \times d\ell}_{\text{Equation (46)}} + \underbrace{\int_{r_1}^{r_3} E \times d\ell}_{\text{Equation (49)}} + \underbrace{\int_{r_3}^{r_2} E \times d\ell}_{\text{Equation (52)}} + \underbrace{\int_{r_2}^{r_{C \to A}} E \times d\ell}_{\text{Equation (44)}} + \underbrace{\int_{r_{C \to A}}^{\infty} E \times d\ell}_{\text{Equation (8)}} \Rightarrow$$

$$V(r) = \begin{cases} \begin{cases} \frac{\rho_w \times I_{tot}}{2 \times \pi \times L} \times ln\left(\frac{r_1}{r}\right) + \frac{I_{tot}}{L \times \left(\frac{2 \times \pi - \theta}{\rho_d} + \frac{\theta}{\rho_w}\right)} \times ln\left(\frac{r_3}{r_1}\right) + \\ \frac{I_{tot}}{L \times \left(\frac{2 \times \pi - \theta}{\rho_d} + \frac{\theta}{\rho_s}\right)} \times ln\left(\frac{r_2}{r_3}\right) + \\ \frac{I_{tot}}{L \times \left(\frac{2 \times \pi - \theta}{\rho_w} + \frac{\theta}{\rho_s}\right)} \times ln\left(\frac{r_{C \to A}}{r_2}\right) + \frac{I_{tot} \times c_f}{2 \times r_{C \to A} \times \left(\frac{\theta_w}{\rho_w} + \frac{\theta_s}{\rho_s}\right)} \end{cases} & : r < r_1 \\[6pt]
\begin{cases} \frac{I_{tot}}{L \times \left(\frac{2 \times \pi - \theta}{\rho_d} + \frac{\theta}{\rho_w}\right)} \times ln\left(\frac{r_3}{r}\right) + \frac{I_{tot}}{L \times \left(\frac{2 \times \pi - \theta}{\rho_d} + \frac{\theta}{\rho_s}\right)} \times ln\left(\frac{r_2}{r_3}\right) + \\ \frac{I_{tot}}{L \times \left(\frac{2 \times \pi - \theta}{\rho_w} + \frac{\theta}{\rho_s}\right)} \times ln\left(\frac{r_{C \to A}}{r_2}\right) + \frac{I_{tot} \times c_f}{2 \times r_{C \to A} \times \left(\frac{\theta_w}{\rho_w} + \frac{\theta_s}{\rho_s}\right)} \end{cases} & : r_1 < r < r_3 \\[6pt]
\begin{cases} \frac{I_{tot}}{L \times \left(\frac{2 \times \pi - \theta}{\rho_d} + \frac{\theta}{\rho_s}\right)} \times ln\left(\frac{r_2}{r}\right) + \\ \frac{I_{tot}}{L \times \left(\frac{2 \times \pi - \theta}{\rho_w} + \frac{\theta}{\rho_s}\right)} \times ln\left(\frac{r_{C \to A}}{r_2}\right) + \frac{I_{tot} \times c_f}{2 \times r_{C \to A} \times \left(\frac{\theta_w}{\rho_w} + \frac{\theta_s}{\rho_s}\right)} \end{cases} & : r_3 < r < r_2 \\[6pt]
\frac{I_{tot}}{L \times \left(\frac{2 \times \pi - \theta}{\rho_w} + \frac{\theta}{\rho_s}\right)} \times ln\left(\frac{r_{C \to A}}{r}\right) + \frac{I_{tot} \times c_f}{2 \times r_{C \to A} \times \left(\frac{\theta_w}{\rho_w} + \frac{\theta_s}{\rho_s}\right)} & : r_2 < r < r_{C \to A} \\[6pt]
\frac{I_{tot} \times c_f}{2 \times r \times \left(\frac{\theta_w}{\rho_w} + \frac{\theta_s}{\rho_s}\right)} & : r > r_{C \to A} \end{cases} \tag{84}$$

Accordingly, the electrode station resistance of remote earth results from Equation (85) (for $r = r_{el} < r_1$) as follows:

$$R_{el} = \frac{V(r_{el})}{I_{tot}} = \frac{1}{L} \times \left(\frac{ln\left(\frac{r_1}{r_{el}}\right)}{\frac{2 \times \pi}{\rho_w}} + \frac{ln\left(\frac{r_3}{r_1}\right)}{\left(\frac{2 \times \pi - \theta}{\rho_d} + \frac{\theta}{\rho_w}\right)} + \frac{ln\left(\frac{r_2}{r_3}\right)}{\left(\frac{2 \times \pi - \theta}{\rho_d} + \frac{\theta}{\rho_s}\right)} + \frac{ln\left(\frac{r_{C \to A}}{r_2}\right)}{\left(\frac{2 \times \pi - \theta}{\rho_w} + \frac{\theta}{\rho_s}\right)} + \frac{L \times c_f}{2 \times r_{C \to A} \times \left(\frac{\theta_w}{\rho_w} + \frac{\theta_s}{\rho_s}\right)} \right) \tag{85}$$

If the inequality $r_1 < r_3 < r_2 < r_{C \to A}$ does not apply, then it is examined whether the inequality $r_1 < r_3 < r_{C \to A} < r_2$ is valid, determining the equalization distance of the electric field strengths of methods "B" and "C", through Equations (26) and (52), hence it follows that:

$$r_{C \to A} = \frac{L}{2} \times \frac{\left(\frac{2 \times \pi - \theta}{\rho_d} + \frac{\theta}{\rho_s}\right)}{\left(\frac{\theta_w}{\rho_d} + \frac{\theta_s}{\rho_s}\right)} \times c_f \tag{86}$$

Here, method "B" is necessarily used instead of method "A", since the respective distance lies on the breakwater. Consequently, the absolute potential is calculated, similarly to Equation (85), and from this, the electrode station resistance of remote earth results as equal to:

$$R_{el} = \frac{1}{L} \times \left(\frac{ln\left(\frac{r_1}{r_{el}}\right)}{\frac{2 \times \pi}{\rho_w}} + \frac{ln\left(\frac{r_3}{r_1}\right)}{\left(\frac{2 \times \pi - \theta}{\rho_d} + \frac{\theta}{\rho_w}\right)} + \frac{ln\left(\frac{r_{C \to A}}{r_3}\right)}{\left(\frac{2 \times \pi - \theta}{\rho_d} + \frac{\theta}{\rho_s}\right)} + \frac{L \times c_f}{2 \times \left(\frac{\theta_w}{\rho_w} + \frac{\theta_s}{\rho_s}\right)} \times \left(\frac{1}{r_{C \to A}} - \frac{1}{r_2}\right) + \frac{L \times c_f}{2 \times r_2 \times \left(\frac{\theta_w}{\rho_w} + \frac{\theta_s}{\rho_s}\right)} \right) \tag{87}$$

It is noted that the equalization distance of the electric field strengths of methods "A" and "C" by applying Equation (83) is greater than that resulting from applying Equation (86) under the practical condition that the electrical resistivity of the soil ρ_s is much greater than the respective seawater ρ_w. In particular, the corresponding condition for $\theta_s = 2 \times \pi \text{-} \theta_w$ is $(2 \times \pi \text{-} \theta)/\theta < \rho_w/\rho_s$, which is practically the case.

Likewise, for the case of Korakia, Crete, if it is true that $r_1 < r_2 < r_3 < r_{C \to A}$, where the equalization distance of the electric field strengths of methods "A" and "C" (for the external area of the breakwater) results from Equation (83), then the absolute potential is calculated in a similar way and from this electrode station resistance of remote earth results as equal to:

$$R_{el} = \frac{1}{L} \times \left(\frac{ln\left(\frac{r_2}{r_1}\right)}{\left(\frac{2 \times \pi - \theta}{\rho_d} + \frac{\theta}{\rho_w}\right)} + \frac{ln\left(\frac{r_3}{r_2} \times \frac{r_1}{r_{el}}\right)}{\frac{2 \times \pi}{\rho_w}} + \frac{ln\left(\frac{r_{C \to A}}{r_3}\right)}{\left(\frac{2 \times \pi - \theta}{\rho_w} + \frac{\theta}{\rho_s}\right)} + \frac{L \times c_f}{2 \times r_{C \to A} \times \left(\frac{\theta_w}{\rho_w} + \frac{\theta_s}{\rho_s}\right)} \right) \tag{88}$$

If the inequality $r_1 < r_2 < r_3 < r_{C \to A}$ does not apply, then it is examined whether the inequality $r_1 < r_2 < r_{C \to A} < r_3$ is valid, determining the equalization distance of the electric field strengths of methods "A" or "B" and "C", through Equations (7) or (23) and (46), hence it follows that:

$$r_{C \to A} = L \times \frac{\frac{\pi}{\rho_w}}{\left(\frac{\theta_w}{\rho_w} + \frac{\theta_s}{\rho_s}\right)} \times c_f \tag{89}$$

Consequently, the absolute potential is calculated, similarly to Equation (85), and from this, the electrode station resistance of remote earth results as equal to:

$$R_{el} = \frac{1}{L} \times \left(\frac{ln\left(\frac{r_1}{r_{el}} \times \frac{r_{C \to A}}{r_2}\right)}{\frac{2 \times \pi}{\rho_w}} + \frac{ln\left(\frac{r_2}{r_1}\right)}{\left(\frac{2 \times \pi - \theta}{\rho_d} + \frac{\theta}{\rho_w}\right)} + \frac{L \times c_f}{2 \times r_{C \to A} \times \left(\frac{\theta_w}{\rho_w} + \frac{\theta_s}{\rho_s}\right)} \right) \tag{90}$$

It is noted that, if during the calculation of the equalization distance of the electric field strengths of methods "A" and "C", by applying Equation (83), the condition $r_1 < r_2 < r_{C \to A} < r_3$ is met, instead of $r_1 < r_2 < r_3 < r_{C \to A}$, and, respectively, by applying Equation (89), the condition $r_1 < r_2 < r_3 < r_{C \to A}$ is met, instead of $r_1 < r_2 < r_{C \to A} < r_3$, then distance r_3 is considered the corresponding equalization distance of the electric field strengths of methods "A" or "B" and "C", at which the strength of the electric field presents a discontinuity.

Each case, different from those of Stachtoroi and Korakia, should be examined, particularly in terms of finding analytical relations because due to the numerical values (regarding distances r_1, r_2, r_3 and $r_{C \to A}$), different equations could result in calculating the absolute potential (with respect to infinity) and the electrode station resistance (with respect to remote earth), but corresponding to those resulting from Equations (84) and (85).

In the present case, from the relevant calculations, for the most unfavorable conditions of Stachtoroi and Korakia, the respective results of Table 8 were obtained. From the comparison of the results with the corresponding ones in Tables 6 and 7, a small improvement between 1.7% and 2.7% was observed for the area of Stachtoroi, and a great one between 60% and 68% for the area of Korakia. This occurs because the equalization distance of the electric field strengths between methods "A" or "B" and "C", in the case of Korakia, is much shorter compared to that of Stachtoroi, which is mainly attributable to the much greater seabed slope used in methods "A" and "B". However, the total equivalent electrode station resistance, with respect to remote earth, has been significantly reduced compared to the values determined either by method "A" (smaller by 20 to 100 times) or by method "B" (smaller by 1,000 to 2,000 times). Even under the worst-case scenario, for a 3:2 inclination of the electrode, the total resistance of the two electrodes through remote earth amounts to 21.5 Ω and the potential difference between them to 23.69 kV, while for a vertical electrode, it is further reduced to 12.0 Ω and at 13.21 kV, respectively, demonstrating the superiority of the combined use of methods "A" or "B" and "C" to determine an upper limit in terms of the absolute potential on an electrode surface, as well as in terms of the equivalent electrode station resistance, with a relatively easy analytical mathematical procedure of direct calculation.

Table 8. Equalization distance of the electric field strengths between methods "A" or "B" and "C"; absolute potential on the electrode surface and equivalent electrode station resistance, in the areas of Stachtoroi, Attica and Korakia, Crete, with combined utilization of the methods under the worst-case scenario, in terms of geometric dimensions.

		Stachtoroi (Aegina: $\theta_w = 0.00474355$ rad)				Korakia (Far *: $\theta_w = 0.03997869$ rad)			
L [m]	ρ_d [Ω·m]	ρ_s [Ω·m]	$r_{C \to A}$ [m]	$V_{(rel)}$ [kV]	R_{el} [Ω]	ρ_s [Ω·m]	$r_{C \to A}$ [m]	$V_{(rel)}$ [kV]	R_{el} [Ω]
1.1815 (slope 3:2)	100	∞	326.0	19.471	17.701	∞	38.56	0.4401	0.4001
		1,000	245.1	17.128	15.571	100	27.88	0.4018	0.3652
		10,000	315.6	19.208	17.462	1,000	37.13	0.4356	0.3960
	120	∞	326.0	23.245	21.132	∞	38.56	0.4401	0.4001
		1,000	245.1	19.963	18.148	100	27.88	0.4018	0.3653
		10,000	315.6	22.868	20.789	1,000	37.13	0.4356	0.3960
2.1300 (vertical)	100	∞	587.8	10.830	9.845	∞	69.51	0.2820	0.2564
		1,000	441.8	9.530	8.664	100	50.26	0.2606	0.2369
		10,000	569.0	10.684	9.712	1,000	66.93	0.2795	0.2541
	120	∞	587.8	12.923	11.748	∞	69.51	0.2820	0.2564
		1,000	441.8	11.102	10.093	100	50.26	0.2606	0.2369
		10,000	569.0	12.714	11.558	1,000	66.93	0.2795	0.2541

Note: (*)—After the implementation of the correction factor (=1.30).

Regarding the safety distance from the breakwater, the use of method "C" was proposed, with the relevant results as analyzed in Section 4.4, where the most unfavorable results in the present cases result from the activation of the limits of the electric field strength at steady state (see Tables 6 and 7). On the contrary, the minimum distance, regarding absolute potential with respect to remote earth at a steady state, was achieved through methods "A" or "B", where for Stachtoroi, it was located outside all inhabited areas, whilst for Korakia, it was limited to a distance of 1.12 km from the electrode station, even for the most unfavorable combination of geometric dimensions and electrical resistivities.

5. Application of Electric Field Distribution Methods Using Superposition for Near Field Analysis

5.1. General Remarks

As already mentioned in Section 2.4, in this study, an ANOTEC electrode [77] was chosen, and it was suggested by IPTO that the electric current density limit value J_{st} be equal to 20 A/m^2, while for reliability reasons, the number of linear frames v_{frame} be 5 with an additional reserve of 1. Thus, for a total electric current intensity I_{tot_steady}, at a steady state (under overload conditions) equal to 1,100 A from the application of Equation (59), it follows that the number of necessary electrodes N_{min_el} is equal to 67. Therefore, by applying Equation (60), it follows that in each frame, the number of necessary electrodes N_{el_frame} is 13, while in total (including the reserve frame), 78 electrodes were used.

By taking into account the increment factor β equal to 6.1%, the final values of current densities, under full load conditions $J_{full\_load\_steady}$ and under periodic maintenance conditions $J_{maintenance\_steady}$ were calculated as equal to 18.33 A/m^2 and 22.00 A/m^2, through Equations (61) and (62), respectively. Similarly, for the transient state, for a respective current intensity $I_{tot\text{-}transient}$ equal to 12.8 kA, the final values of current densities, under full load conditions $J_{full\_load\_transient}$ and under periodic maintenance conditions $J_{maintenance\_transient}$ are calculated as equal to 213.28 A/m^2 and 255.93 A/m^2, respectively.

Based on Sections 4.4 and 4.5, for the study of the electric field distribution at short distances (near the breakwater), the most suitable method is "C" using the linear current source. For the sake of simplifying the calculation process, the effect of the soil (considering its electrical resistivity to be infinite), the water zone of a respective arc of angle θ on the plan view of Figure 6, as well as of the breakwater, were all ignored, therefore from Equations (44), (46), (49) and (52), the electric field strength results as follows:

$$E_{rw\_r} = \frac{\rho_w \times I_{tot}}{(2 \times \pi - \theta) \times r \times L} \tag{91}$$

The first two simplifying admissions led to much more unfavorable results, whilst not considering the breakwater only has an effect on the calculation of the electric field strength when one is on the breakwater, as well as on the calculation of the absolute potential for radii within or up to the breakwater (i.e., for $r < r_2$). In the present case, of interest is the electric field strength mainly beyond the breakwater, while regarding the electric field strength on it E_{rd_r} by simplifying, its value can be approximated by multiplying the respective seawater electric field strength E_{rw_r} by the ratio of the electrical resistivities ρ_d/ρ_w, that is:

$$E_{rd\_r} = \frac{\rho_d}{\rho_w} \times E_{rw\_r} \tag{92}$$

Equation (91) was practically applied; for each electrode located at position (x_ℓ, y_ℓ) of Figure 9 and through Equations (63) to (68), the superposition theorem was applied to determine the electric field strength resultant, in the case of the frame of Figure 7 and in the case of the array of linear frames placed parallel to the axis of the protective breakwater of Figure 8. Consequently, the relative analysis was made in a step-by-step process, emphasizing the area of Korakia, Crete, considering the arc of the "right" water–breakwater zone ($2 \times \pi$-θ) of Figure 6 as equal to 112°, which is smaller than the respective arc of 150° at Stachtoroi, in seawater of electrical resistivity of 0.25 Ω·m. In addition, emphasis is placed on the study of current intensity densities under conditions of periodic maintenance at a steady state because they generally give the most unfavorable results, in terms of safety distance, with respect to the point electric field strength criterion of 1.25 V/m.

5.2. Case of an Electrode at Maximum Electric Current Density, under Periodic Maintenance Conditions

Initially, the study was made of an electrode at a steady state, electrified with a current density of 22 A/m^2 (under conditions of periodic maintenance), on a canvas of dimensions 5 m (Ox semi-axis) by 10 m (yOy′ axis), with a step of 0.05 m, in the area of Korakia (with a computational mesh of $101 \times 201 = 20{,}301$ points). Due to the 3:2 inclination of the electrode, the effective length L was taken equal to 1.1835 m. Through this simulation, the electric field strength on the Oxy plane (Figure 18), as well as the area of the electric field strength, with values greater than 1.25 V/m (Figure 19), were obtained.

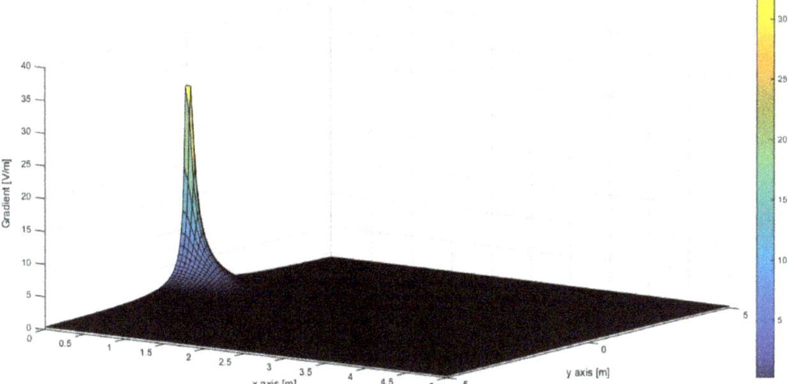

Figure 18. Electric field strength for method "C", of linear current source, for the area of Korakia ($\rho_S = \infty$, without dam, $L = 1.1815$ m, $\rho_w = 0.25$ Ω·m, $\theta = 248°$, $J_{maintenance\_steady} = 22$ A/m^2).

The maximum electric field strength was calculated as equal to 31.87 V/m, which is fully consistent with the respective electric field strength of Equation (91), while the distance where it stays below the limit of 1.25 V/m was determined at 1.555 m. These quantities are significantly improved in relation to those that would be obtained by method "A" (15,091.5 V/m and 6.703 m, respectively). Essentially, this means that there is no electric field strength greater than 1.25 V/m beyond the dam, as the latter has a crest width of at least 4.0 m. Additionally, at a distance of 1.0 m from the electrode, the dam begins, so in this case, the corresponding electric field strength in seawater is equal to 1.9441 V/m, which is equivalent to $\rho_d = 100 \div 120$ Ω·m, according to Equation (92), with an electric field strength E_{rd_r}, on the dam, ranging between 777.64 and 933.168 V/m. Therefore, protection measures need to be taken for step voltage, etc. If the same process is repeated with the electrode vertically suspended, with an effective length L equal to 2.13 m, then the maximum electric field strength is calculated to equal to 17.69 V/m, and the distance where it stays below the limit of 1.25 V/m equal to 0.863 m (i.e., within the area of the dam), while the electric field strength E_{rd_r} on the dam (1.0 m from the electrode) is between 107.84 and 113.232 V/m (need for protection measures for step voltage, etc.).

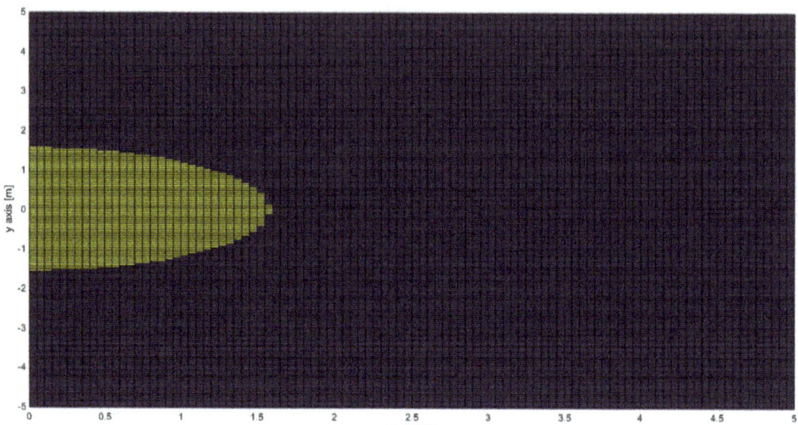

Figure 19. Area of electric field strength, with values higher (yellow) and lower (blue) than the limit of 1.25 V/m for method "C", of linear current source, for the area of Korakia ($\rho_S = \infty$, without dam, $L = 1.1815$ m, $\rho_w = 0.25$ Ω·m, $\theta = 248°$, $J_{maintenance\_steady} = 22$ A/m^2).

5.3. Case of a Frame of 13 Electrodes, at Maximum Current Density, under Periodic Maintenance Conditions

Initially, a straight frame was formed consisting of 13 electrodes, which are at a steady state, electrified with a current density (under periodic maintenance conditions) of 22 A/m^2, at a distance D_{el} of 1.0 m from each other. This frame is placed on a canvas of 30 m (Ox semi-axis) by 60 m (yOy′ axis—parallel to the axis of the frame), with a step of 0.05 m, with the 7th electrode placed at point O in the area of Korakia (with a computational mesh of 601 × 1,201 = 721,801 points). Due to the 3:2 inclination of the electrode, the effective length L is taken equal to 1.1835 m. From this simulation, the electric field strength on the Oxy plane (Figure 20), and the area of electric field strength, with values greater than 1.25 V/m (Figure 21), were obtained. The maximum electric field strength E_{max} was calculated as equal to 37.76 (against the 31.87 V/m that was the case with one electrode) due to the superposition of the electric fields of the 13 electrodes. The distance d_1, where it stays below the limit of 1.25 V/m along the Ox axis, was determined at 19.53 m, and the distance $d_1{}'$ along the Oy axis at 20.91 m, with respect to the beginning of the axes. Given that the frame is placed along the perceived axis yOy′ (i.e., parallel to the dam) with its center at point O (from −6.00 m to 6.00 m), it follows that from the end of the frame, the respective required distance d_{frames} is 20.91 − 6.00 = 14.91 m along the Oy axis, so the required frame length ℓ_k amounts to 12.0 + 2 × 14.91 = 41.82 m. These sizes are significantly improved compared to the values of the corresponding sizes that would be obtained with method "A" ($d_{frames-A}$ = 50.15 m and E_{max-A} = 15,152 V/m, respectively). It was also found that, compared to the single electrode (where the required distance would be 1.555 m), here it increases significantly due to the superposition effect of the electric field strengths of the neighboring fields, by approximately ten times, while, as expected, the symmetry around the center of the frame is no longer circular but elliptical.

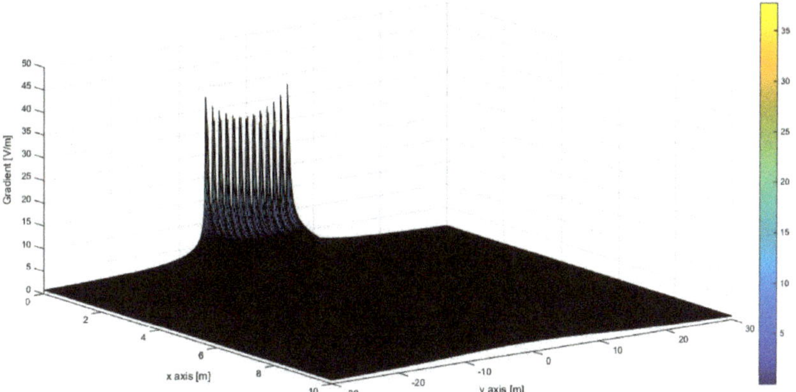

Figure 20. Electric field strength for method "C", of linear current source, for a linear frame of 13 electrodes, with D_{el} = 1.0m for the area of Korakia (ρ_S = ∞, without dam, L = 1.1815 m, ρ_w = 0.25 Ω·m, θ = 248°, $J_{maintenance\_steady}$ = 22 A/m²).

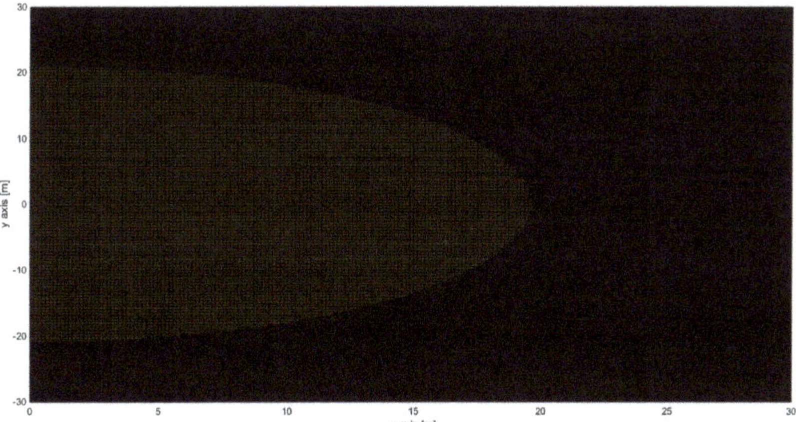

Figure 21. Area of electric field strength with values higher (yellow) and lower (blue) than the limit of 1.25 V/m for method "C" of linear current source for a linear frame of 13 electrodes with D_{el} = 1.0 m for the area of Korakia (ρ_S = ∞, without dam, L = 1.1815 m, ρ_w = 0.25 Ω·m, θ = 248°, $J_{maintenance\_steady}$ = 22 A/m²).

Then the respective distance between the electrodes and rods D_{el} changes, from 0.20 m to 1.50 m, and the critical electric field strength areas, with values greater than 1.25 V/m, were calculated (in addition to the total frame length D_{frame} and the maximum electric field strength of the array E_{max}), via the following:

- The width of the critical frame zone d_1 perpendicular to the dam consisted of 13 electrodes on the semi-axis Ox;
- The total length of the critical frame zone ℓ_k on the dam consisted of 13 electrodes (on the yOy′ axis);
- The area of rectangular zone arrangement that ensures the critical frame zone S_k (taking the width d_1 for both sides);
- The estimated distance between successive frames on the dam that ensures the critical zone for diver dropping for repairs d_{frames} (on yOy′);
- The estimated total length includes the critical zone of 6 frames ℓ_t on the dam (on yOy′);
- The area of the rectangular zone arrangement ensures the critical zone of 6 frames S_t (taking the width d_1 for both sides).

Table 9 lists the results for an effective length of 1.1815 m, with an inclination according to that of the dam, while Table 10 lists the results for an effective length of 2.13 m, with vertical suspension.

Table 9. Simulation results with method "C", of linear current source for a linear frame of 13 electrodes for the area of Korakia ($\rho_S = \infty$, without dam, $L = 1.1815$ m, $\rho_w = 0.25$ $\Omega \cdot$m, $\theta = 248°$, $J_{maintenance\_steady} = 22$ A/m², $E_{limit\_S} = 1.25$ V/m).

D_{el} [m]	D_{frame} [m]	E_{max} [V/m]	d_1 [m]	ℓ_k [m]	S_k [m²]	d_{frames} [m]	ℓ_t [m]	S_t [m²]
	point	414.32	20.219	40.439	1635.29	20.219	141.535	5,723.51
0.2	2.4	58.83	20.190	40.500	1635.40	19.050	147.748	5,966.17
0.3	3.6	50.48	20.156	40.561	1635.14	18.481	150.965	6,085.82
0.4	4.8	46.08	20.108	40.658	1635.11	17.929	154.303	6,205.48
0.5	6.0	43.37	20.045	40.784	1635.04	17.392	157.744	6,323.99
0.6	7.2	41.53	19.968	40.935	1634.83	16.868	161.274	6,440.75
0.7	8.4	40.19	19.878	41.113	1634.51	16.356	164.895	6,555.68
0.8	9.6	39.18	19.773	41.319	1634.04	15.860	168.618	6,668.27
0.9	10.8	38.39	19.655	41.553	1633.45	15.377	172.436	6,778.46
1.0	12.0	37.76	19.527	41.813	1632.92	14.906	176.344	6,886.81
1.1	13.2	37.23	19.373	42.097	1631.12	14.448	180.339	6,987.55
1.2	14.4	36.80	19.210	42.409	1629.40	14.005	184.433	7,086.03
1.3	15.6	36.42	19.034	42.748	1627.39	13.574	188.620	7,180.56
1.4	16.8	36.10	18.841	43.112	1624.58	13.156	192.892	7,268.72
1.5	18.0	35.83	18.634	43.503	1621.31	12.752	197.261	7,351.70

Table 10. Simulation results with method "C", of linear current source for a linear frame of 13 electrodes for the area of Korakia ($\rho_S = \infty$, without dam, $L = 2.13$ m, $\rho_w = 0.25$ $\Omega \cdot$m, $\theta = 248°$, $J_{maintenance\_steady} = 22$ A/m², $E_{limit\_S} = 1.25$ V/m).

D_{el} [m]	D_{frame} [m]	E_{max} [V/m]	d_1 [m]	ℓ_k [m]	S_k [m²]	d_{frames} [m]	ℓ_t [m]	S_t [m²]
	point	229.82	11.215	22.430	503.12	11.215	78.506	1,760.92
0.2	2.4	32.63	11.165	22.530	503.12	10.065	84.856	1,894.92
0.3	3.6	28.00	11.103	22.655	503.07	9.528	88.294	1,960.60
0.4	4.8	25.56	11.015	22.829	502.90	9.014	91.900	2,024.52
0.5	6.0	24.06	10.901	23.051	502.55	8.525	95.678	2,085.96
0.6	7.2	23.04	10.762	23.322	501.99	8.061	99.628	2,144.41
0.7	8.4	22.30	10.596	23.634	500.87	7.617	103.720	2,198.08
0.8	9.6	21.74	10.404	24.005	499.49	7.203	108.018	2,247.59
0.9	10.8	21.30	10.184	24.415	497.30	6.807	112.452	2,290.49
1.0	12.0	20.94	9.936	24.869	494.20	6.435	117.043	2,325.87
1.1	13.2	20.65	9.660	25.367	490.10	6.084	121.785	2,352.94
1.2	14.4	20.41	9.353	25.906	484.61	5.753	126.672	2,369.57
1.3	15.6	20.20	9.016	26.487	477.63	5.443	131.704	2,374.96
1.4	16.8	20.03	8.648	27.106	468.81	5.153	136.871	2,367.25
1.5	18.0	19.87	8.245	27.763	457.84	4.882	142.172	2,344.54

From the respective study of the results, the following conclusions emerge:

- As the electrode spacing increases, so do the necessary length on the dam along the yOy′ axis, ℓ_k, and the area of the rectangular zone arrangement, which ensures the critical zone of six frames, S_t, (but not monotonously, since for $L = 2.13$ m and $D_{el} > 1.3$ m the area S_t begins to decrease slowly). On the other hand, the maximum electric field strength E_{max} decreases slightly, as well as the width of the critical zone (in front of the location of the dam d_1) and the estimated distance between successive frames d_{frames}, as can be seen from the results of Tables 9 and 10;
- By taking into account that, in order to be able to repair each electrode, a distance of 0.50 m between them is practically required, then the required length of the dam is 96 m for $L = 2.13$ m, which is greater than the available (about 70 m) according to Figure 12. Of course, if from the beginning the array is allowed to move marginally closer to the coast, with the appropriate deepening and by reducing the distance between the frames to 6.5 m, then the required length

reaches 68.5 (=6 × 6 + 5 × 6.5), which is feasible. Additionally, the critical zone of the dam marginally extends outside by 5 m (with a crest width of 5 m and a suspension distance of at least 1 m). With the appropriate vertical suspension arrangement, at 6 m from the inner side of the dam, the electric field strength can be less than 1.25 V/m on the outside;

- In relation to the results of Table 3, it was found that the equivalent point source of method "A" would require a circular zone of about 52.5 m without the correction factor and 68.2 m with the correction factor, whilst, according to Table 7, the equivalent linear source requires 51.5 m, in contrast to the present case of the frame which requires a zone of 10.9 m along the Ox axis and 23.1 m along the Oy axis (for an electrode spacing of 0.50 m and L = 2.13 m). Respectively, the reserved area for the point source amounts to 8,659 m^2 (without a correction factor) and 14,612 m^2 (with a correction factor), while for the linear source amounts to 8,832 m^2, against 1,895 to 2,370 m^2 of the linear frames on the dam.;
- Based on this consideration, the vertical suspension of the electrodes is more suitable; however, due to the phenomenon of the diffusion of the electric current in the seawater, at a short distance from the frame, the behavior of the inclined electrodes tends to approximate that of the vertical ones.

For the sake of completeness, the same procedure was carried out for the area of Stachtoroi, with θ = 210°, L = 2.13 m and ρ_w = 0.25 Ω·m. From the respective change in the electrodes—rod spacing D_{el}, from 0.20 m up to 1.50 m—the values of the respective geometric and field parameters were calculated (which determine the critical areas of electric field strength higher than 1.25 V/m) and listed in Table 11.

Table 11. Simulation results with method "C", of linear current source, for a linear frame of 13 electrodes, for the area of Stachtoroi, Attica (ρ_S = ∞, without dam, L = 2.13 m, ρ_w = 0.25 Ω·m, θ = 210°, $J_{maintenance\_steady}$ = 22 A/m^2, E_{limit_S} = 1.25 V/m).

D_{el} [m]	D_{frame} [m]	E_{max} [V/m]	d_1 [m]	ℓ_k [m]	S_k [m^2]	d_{frames} [m]	ℓ_t [m]	S_t [m^2]
	point	171.60	8.374	16.748	280.50	8.374	58.618	981.73
0.2	2.4	24.37	8.307	16.882	280.49	7.241	65.086	1,081.39
0.3	3.6	20.91	8.223	17.049	280.39	6.724	68.670	1,129.39
0.4	4.8	19.09	8.155	17.279	281.82	6.240	72.478	1,182.10
0.5	6.0	17.96	7.951	17.576	279.50	5.788	76.515	1,216.80
0.6	7.2	17.20	7.763	17.934	278.44	5.367	80.768	1,254.01
0.7	8.4	16.65	7.538	18.353	276.67	4.977	85.236	1,284.93
0.8	9.6	16.23	7.274	18.831	273.95	4.615	89.908	1,307.96
0.9	10.8	15.90	6.971	19.367	270.01	4.283	94.783	1,321.49
1.0	12.0	15.64	6.627	19.956	264.49	3.978	99.844	1,323.35
1.1	13.2	15.42	6.240	20.596	257.02	3.698	105.085	1,311.39
1.2	14.4	15.24	5.807	21.286	247.22	3.443	110.500	1,283.38
1.3	15.6	15.09	5.325	22.021	234.54	3.210	116.072	1,236.28
1.4	16.8	14.95	4.793	22.799	218.55	3.000	121.797	1,167.51
1.5	18.0	14.84	4.204	23.617	198.58	2.808	127.659	1,073.43

From the respective study of the results, the following conclusions emerge:

- As before, as the electrodes spacing increases, so do the necessary length on the dam, along the yOy' axis ℓ_k (monotonously) and the area of the rectangular zone arrangement that ensures the critical zone of six frames S_t (but not monotonously, as for D_{el} > 1.1 m it begins to decrease slowly). Accordingly, the maximum electric field strength E_{max} is slightly reduced, along with the width of the critical zone, in front of the location of the dam d_1 and the estimated distance between successive frames d_{frames}, as can also be seen from the results of Table 11;
- The required length of the dam (at least 76.5 m) is greater than the available (about 55 m), according to Figure 12. Of course, if from the beginning the array is allowed to move marginally closer to the coast, with the appropriate deepening and by reducing the distance between the frames to 4.5 m, then the required length reaches 58.5 (=6 × 6 + 5 × 4.5), which is feasible. Additionally, the critical zone of the dam marginally extends outside by 2 m (with a crest width of 5 m and a suspension distance of at least 1 m). With the appropriate vertical suspension arrangement, at 3 m from the inner side of the dam, the electric field strength can be less than 1.25 V/m on the outside;

- In relation to the results of Table 2, it was found that the equivalent point source requires a zone of around 152.3 m without the correction factor; according to Table 6, the equivalent linear source would require 39.5 m, in contrast to the present case of the frame, which requires a zone of 8.0 m, on the Ox axis and 17.6 m, on the Oy axis, for an electrode spacing of 0.50 m and $L = 2.13$ m. Respectively, the reserved area for the point source amounts to 72,870 m^2, and for the linear source, it amounts to 4,901 m^2, against 1,070 to 1,325 m^2 of the linear frames on the dam.

5.4. Case of an Arrangement of 6 Linear Frames of 13 Electrodes Placed in a Row, Parallel to the Protective Dam, at Maximum Current Density under Normal Operation or Periodic Maintenance Conditions

Initially, an arrangement of six linear frames is formed, each consisting of 13 electrode rods with a distance between rods D_{el} equal to 0.50 m and a total length D_{frame} equal to 6.00 m. The distance between frames d_{frames} is 6.5 m (against 8.52 m in Table 10), which suggests that the total estimated required length of the protective dam, behind which the array of frames is placed, is 81.5 m (=6 × 6.00 + 7 × 6.5), of which 6.5 m, on either outer side, can be considered onshore to be covered by the plan view of the preliminary study dam of Figure 12. The case of loading the electrode station with the maximum nominal load $I_{total\text{-}steady}$ equal to 1,100 A is considered, for six-frame or five-frame operation, with respective current densities under full load conditions $J_{full\_load\_steady}$ and under periodic maintenance conditions $J_{maintenance\_steady}$, respectively. The entire arrangement is placed on a canvas of 60 m (Ox semi-axis) by 160 m (yOy' axis—parallel to the axis of the frame) and with a simulation step of 0.10 m by 0.10 m (with a computational mesh of 601 × 1,601 = 962,201 points), as seen in Figure 8. From the corresponding simulation, the result of the data of Table 12, with uniform loading of the six-frame or five-frame electrode station, with the sixth, fifth or fourth frame off, as well as the electric field strength graphs, on the Oxy plane in Figures 22–25 and the area of electric field strength, with values greater than $E_{limit\_S} = 1.25$ V/m in Figures 26–29, respectively. The results include the following, depending on the mode of operation:

- The electric current density J_{steady} with respect to the peripheral surface;
- The width d_2 of the electrode station critical zone, perpendicular to the dam (on the xOx' axis);
- The length of the electrode station critical zone d_3 on the dam above the center of the dam (on the yOy' axis);
- The length of the electrode station critical zone d_4 on the dam below the center of the dam (on the yOy' axis);
- The distance of the lowermost rod of the electrode station $\ell_{b\text{-}c}$, on the dam from the center of the dam that is connected to the power supply (on yOy');
- The distance of the uppermost rod of the electrode station $\ell_{u\text{-}c}$, on the dam from the center of the dam, connected to the power supply (by yOy');
- The distance between the lowermost rod of the electrode station $s_{b\text{-}c}$ on the dam that is connected to the power supply and the nearest point of protection (dam end, electrode not connected to power supply for maintenance purposes);
- The distance between the uppermost rod of the electrode station $s_{u\text{-}c}$ on the dam, which is connected to the power supply, and the nearest point of protection (dam end, electrode not connected to power supply for maintenance purposes);
- The distance y_p between the nearest point of protection (dam edge, electrode not connected to power supply for maintenance purposes) from the center of the dam that is connected to the power supply (on yOy');
- The safety margin Dy_p in relation to an initial preliminary study of a frame under the same conditions (where negative values indicate a requirement for a greater safety distance);
- The maximum electric field strength of the arrangement E_{max}.

Table 12. Simulation results with method "C", of a linear current source, for the case of an arrangement of 6 linear frames, each consisting of 13 electrodes, for the area of Korakia, Crete ($\rho_S = \infty$, without dam, L = 2.13 m, D_{el} = 0.50 m, d_{frames} = 6.50 m, ρ_w = 0.25 $\Omega \cdot$m, θ = 248°, E_{limit_S} = 1.25 V/m).

Supply Method	J_{steady} [A/m²]	d_2 [m]	d_3 [m]	d_4 [m]	ℓ_{b-c} [m]	ℓ_{u-c} [m]	s_{b-c} [m]	s_{u-c} [m]	y_p [m]	Δy_p [m]	E_{max} [V/m]
Operation of 6 frames	18.33	47.67	63.97	−63.97	−34.25	34.25	29.72	29.72	40.75	−23.22	21.91
Operation of 5 frames (except no. 6)	22.00	50.29	55.31	−67.81	−34.25	21.75	33.56	33.56	28.25	−27.06 *1	26.09
Operation of 5 frames (except no. 5)	22.00	48.01	18.62	−66.98	−34.25	9.25	32.73	9.37	15.75 −40.75	−2.87 *2 −26.23	26.04
			61.59	25.98	28.25	34.25	2.27	27.34	21.75 40.75	4.23 −20.84	
Operation of 5 frames (except no. 4)	22.00	45.66	2.97	−65.86	−34.25	−3.25	31.61	6.22	3.25 −40.75	0.28 −25.11	25.95
			64.24	11.60	15.75	34.25	4.15	29.99	9.25 40.75	2.35 −23.49	

Note: (*1)—Maximum electric field strength at the area of a nonoperating frame 3.13 V/m; (*2)—Maximum electric field strength at the area of a nonoperating frame 1.98 V/m.

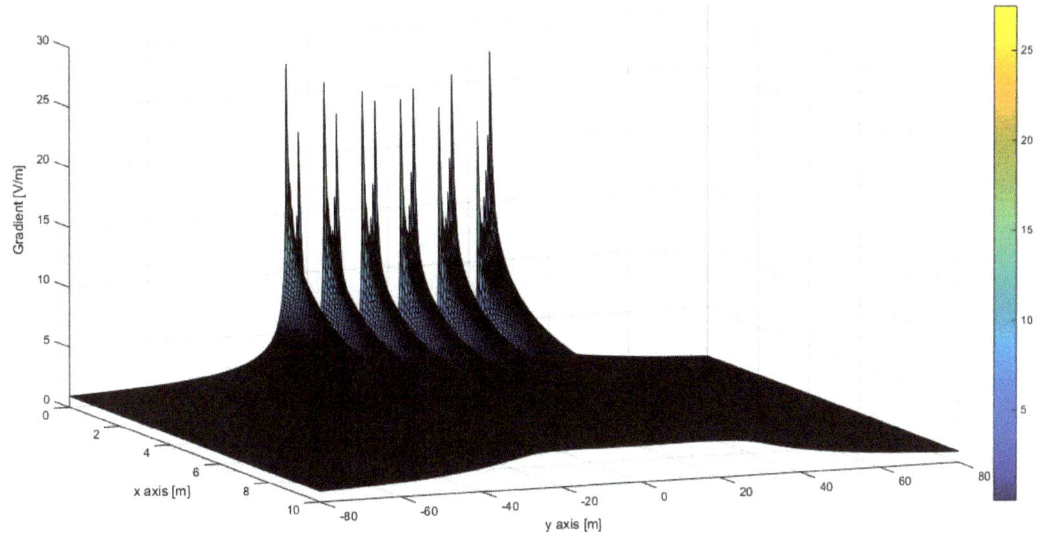

Figure 22. Electric field strength for method "C", of linear current source, for an electrode station of 6 linear frames in a row with d_{frames} = 6.50 m; each frame consists of 13 electrodes of active length L = 2.13 m, D_{el} = 0.50 m, ρ_w = 0.25 $\Omega \cdot$m, θ = 248° (area of Korakia), $J_{full\_load\_steady}$ = 18.33 A/m² (steady state and operation of all 6 frames).

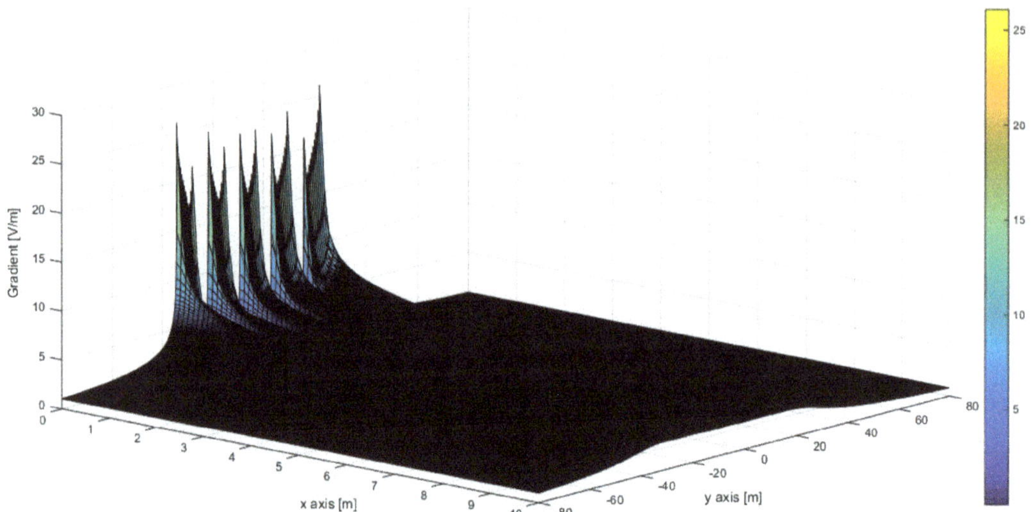

Figure 23. Electric field strength for method "C", of linear current source, for an electrode station of 6 linear frames in a row with d_{frames} = 6.50 m; each frame consists of 13 electrodes of active length L = 2.13 m, D_{el} = 0.50 m, ρ_w = 0.25 Ω·m, θ = 248° (area of Korakia), $J_{full\_load\_steady}$ = 22 A/m^2 (steady state and operation of 5 frames, except no. 6).

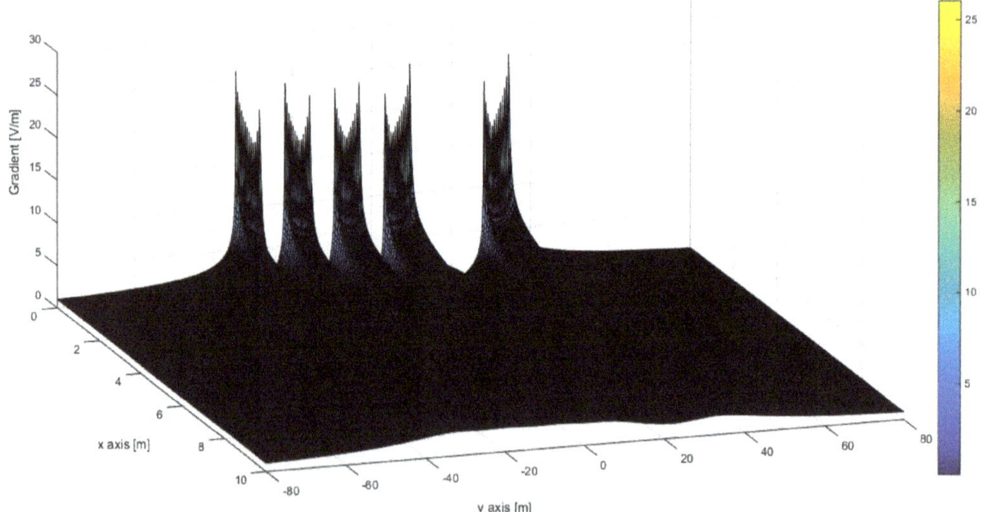

Figure 24. Electric field strength for method "C", of linear current source, for an electrode station of 6 linear frames in a row with d_{frames} = 6.50 m; each frame consists of 13 electrodes of active length L = 2.13 m, D_{el} = 0.50 m, ρ_w = 0.25 Ω·m, θ = 248° (area of Korakia), $J_{full\_load\_steady}$ = 22 A/m^2 (steady state and operation of 5 frames, except no. 5).

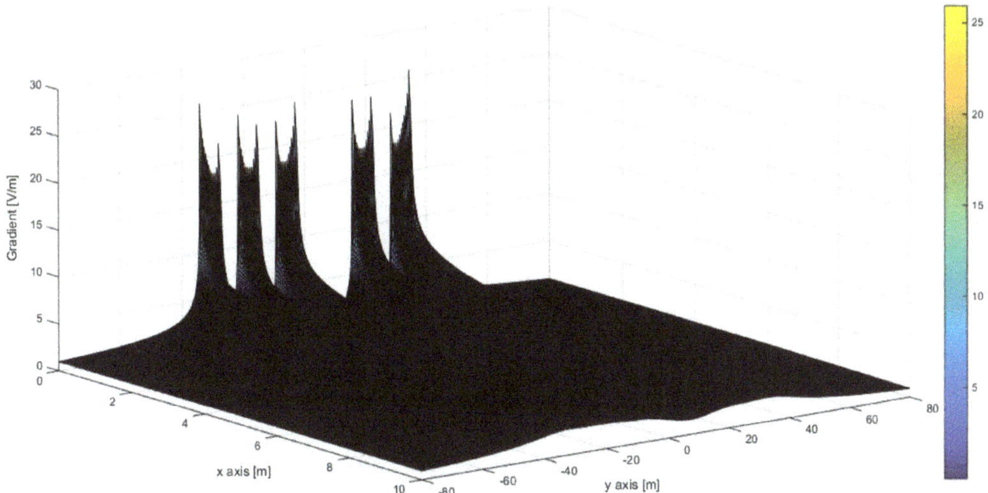

Figure 25. Electric field strength for method "C", of linear current source, for an electrode station of 6 linear frames in a row with d_{frames} = 6.50 m; each frame consists of 13 electrodes of active length L = 2.13 m, D_{el} = 0.50 m, ρ_w = 0.25 Ω·m, θ = 248° (area of Korakia), $J_{full\_load\_steady}$ = 22 A/m^2 (steady state and operation of 5 frames, except no. 4).

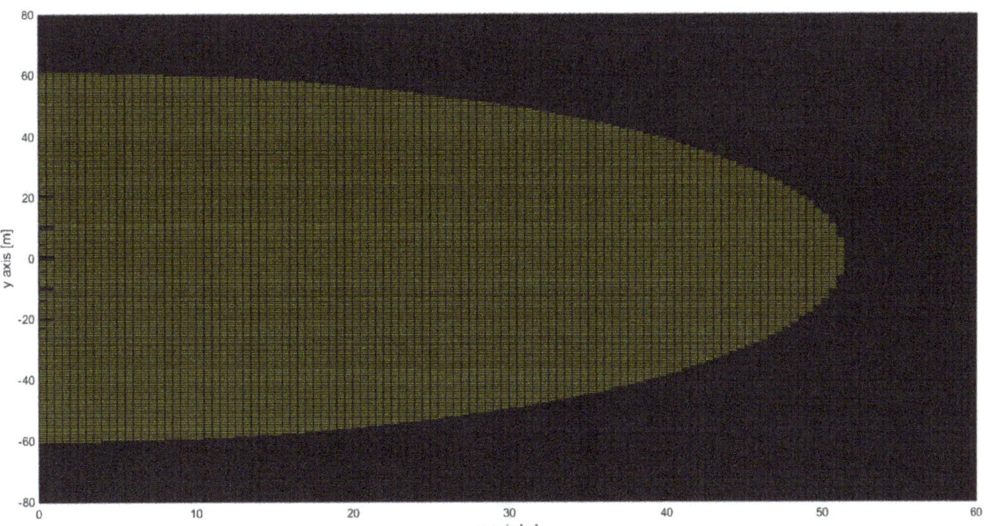

Figure 26. Area of electric field strength, with values higher (yellow) and lower (blue), than the limit of 1.25 V/m, for method "C", of linear current source, for an electrode station of 6 linear frames in a row with d_{frames} = 6.50 m; each frame consists of 13 electrodes of active length L = 2.13 m, D_{el} = 0.50 m, ρ_w = 0.25 Ω·m, θ = 248° (area of Korakia), $J_{full\_load\_steady}$ = 18.33 A/m^2 (steady state and operation of all 6 frames).

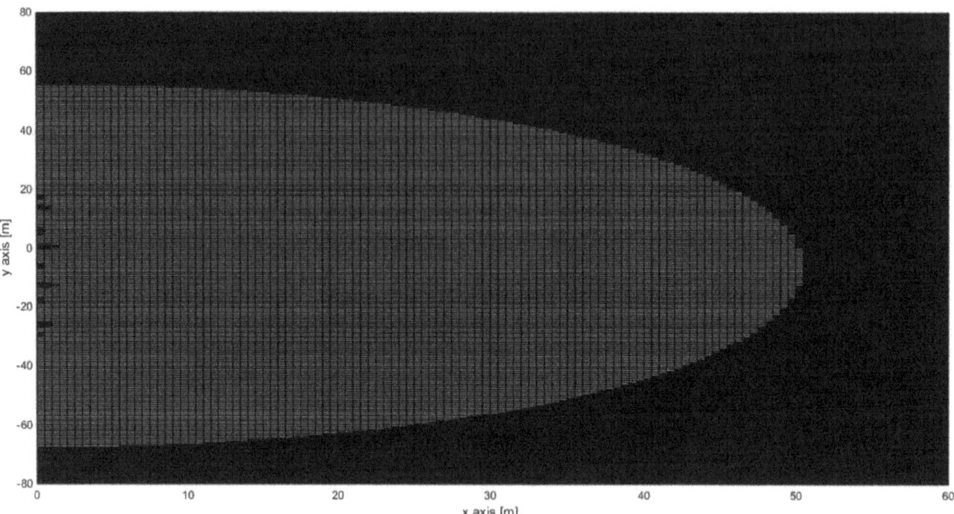

Figure 27. Area of electric field strength, with values higher (yellow) and lower (blue) than the limit of 1.25 V/m for method "C" of linear current source, for an electrode station of 6 linear frames in a row with d_{frames} = 6.50 m; each frame consists of 13 electrodes of active length L = 2.13 m, D_{el} = 0.50 m, ρ_w = 0.25 Ω·m, θ = 248° (area of Korakia), $J_{full\_load\_steady}$ = 22 A/m^2 (steady state and operation of 5 frames, except no. 6).

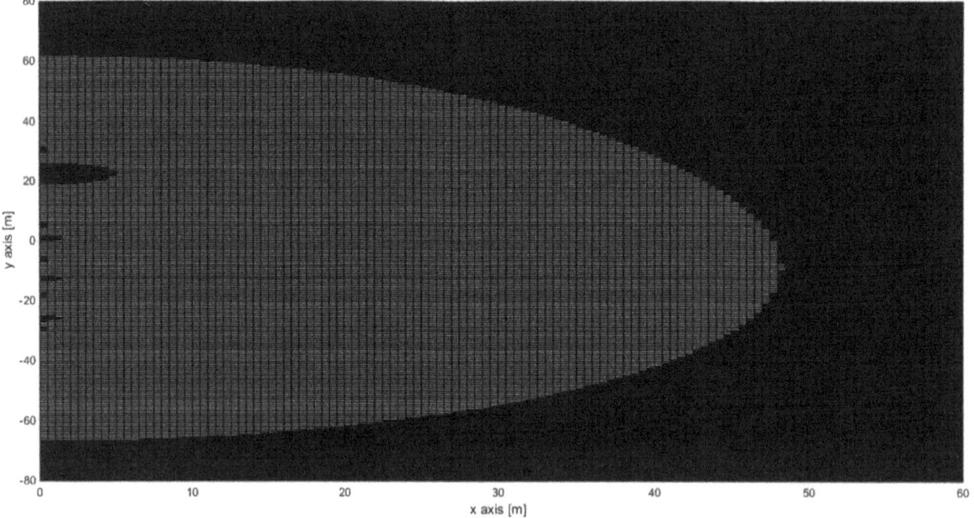

Figure 28. Area of electric field strength, with values higher (yellow) and lower (blue) than the limit of 1.25 V/m for method "C" of linear current source, for an electrode station of 6 linear frames in a row with d_{frames} = 6.50 m; each frame consists of 13 electrodes of active length L = 2.13 m, D_{el} = 0.50 m, ρ_w = 0.25 Ω·m, θ = 248° (area of Korakia), $J_{full\_load\_steady}$ = 22 A/m^2 (steady state and operation of 5 frames, except no. 5).

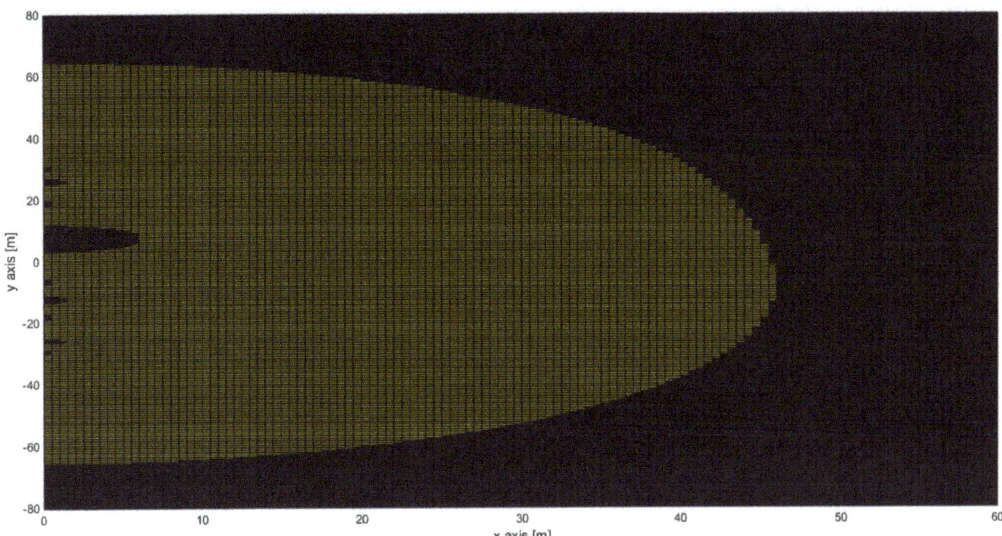

Figure 29. Area of electric field strength, with values higher (yellow) and lower (blue), than the limit of 1.25 V/m, for method "C", of linear current source, for an electrode station of 6 linear frames in a row, with d_{frames} = 6.50 m, each frame consisting of 13 electrodes of active length L = 2.13 m, D_{el} = 0.50 m, ρ_w = 0.25 Ω·m, θ = 248° (area of Korakia), $J_{full\_load\_steady}$ = 22 A/m^2 (steady state and operation of 5 frames, except no. 4).

From the study of the relevant results in Table 12, the following conclusions emerge:

- The deviation at the ends of the arrangement reaches up to 27.0 m depending on the electrifying method (especially when electrifying five consecutive panels). However, in the area of the frame that does not operate, the respective electric field strength is marginally above 2.5 V/m, so there is no safety issue during maintenance as long as the diver takes the appropriate measures;
- The deviation of the maximum developing electric field strength of the electrode station against that of a single frame (26.09 V/m against 24.06 V/m) is of the order of 8.4%, which is quite large, but expected, given the fact that, instead of 8.5 m, the distance between of frames decreased to 6.5 m;
- The critical zone of the dam extends outside the dam by 44 m (with a crest width of 5 m and a suspension distance of at least 1 m) on the xOx' axis (vertical to the dam), where the initial estimation of Table 10 has failed;
- The critical zone of the dam extends beyond the dam by 66 m (33 m on either side) on the yOy' axis (parallel to the axis of the dam) in the shore area. The reason is that the distance between the frames was significantly reduced.

If the previous procedure is repeated, setting the intermediate distance between the frames at 8.5 m, according to Table 10, the total estimated required length of the protective dam behind which the array is placed must be 95.5 m (=6 × 6.00 + 7 × 8.5), of which 8.5 m on either outer side can be considered on the coast. From the corresponding simulation, the results of Table 13 are obtained, from the study of which the following conclusions are drawn:

- The deviation at the ends of the electrode station reaches up to 23.0 m depending on the electrifying method (especially when electrifying five consecutive panels). However, in the area of the frame that does not operate, the respective electric field strength is marginally above 2.5 V/m, so there is no safety issue during maintenance;
- The deviation of the maximum developing electric field strength of the electrode station against that of a single frame (25.84 V/m against 24.06 V/m) is of the order of 7.4%, which is quite large, despite the fact that the distance between frames is 8.5 m, as set from the beginning. This happens because, in method "C", the constant effective length causes the field effect to decrease more slowly;

- The critical zone of the dam extends outside the dam by 42 m (with a crest width of 5 m and a suspension distance of at least 1 m) on the xOx' axis (vertical to the dam), where the initial assessment of Table 10 has failed. However, it is limited to 83% of the most favorable value, resulting from concentrated source methods;
- The critical zone of the dam extends beyond the estimated dam (according to Section 5.3) by 46 m (23 m on either side) on the yOy' axis (parallel to the dam axis) on the shore area. Due to the large size, the respective area (13,622 m^2) approaches the respective area of method "A", using a correction factor;
- From the comparison of Tables 12 and 13, it emerged that there is no substantial benefit, in the present case, from increasing the distance between the frames from 6.5 to 8.5 m, so the spacing of 6.5 m can be applied. It is estimated that the electric field strength drops below the value of 2.5 V/m (with respect to 3.1 V/m) because, in the present case, an active zone as long as the height of the electrode was assumed by simplification, and another 2.0 m of seawater depth in the nearby area was ignored, as well as the mass of water "behind" the dam in Figure 6.

Table 13. Simulation results with method "C", of a linear current source, for the case of an arrangement of 6 linear frames, in a row, each consisting of 13 electrodes, for the area of Korakia, Crete ($\rho_S = \infty$, without dam, $L = 2.13$ m, $D_{el} = 0.50$ m, $d_{frames} = 8.50$ m, $\rho_w = 0.25$ $\Omega\cdot$m, $\theta = 248°$, $E_{limit\_S} = 1.25$ V/m).

Supply Method	J_{steady} [A/m^2]	d_2 [m]	d_3 [m]	d_4 [m]	ℓ_{b-c} [m]	ℓ_{u-c} [m]	s_{b-c} [m]	s_{u-c} [m]	y_p [m]	Δy_p [m]	E_{max} [V/m]
Operation of 6 frames	18.33	44.50	66.64	−66.64	−39.25	39.25	27.39	27.39	47.75	−18.89	21.68
Operation of 5 frames (except no. 6)	22.00	48.23	56.11	−70.61	−39.25	24.75	31.36	31.36	33.25	−22.86 *3	25.84
Operation of 5 frames (except no. 5)	22.00	45.46	20.58	−69.80	−39.25	10.25	30.55	10.33	18.75 −47.75	−1.83 *4 −22.05	25.79
			64.18	30.42	33.25	39.25	2.83	24.93	24.75 47.75	5.67 −16.43	
Operation of 5 frames (except no. 4)	22.00	42.14	2.65	−68.68	−39.25	−4.25	29.43	6.90	4.25 −47.75	1.60 −20.93	25.72
			67.03	14.07	18.75	39.25	4.68	27.78	10.25 47.75	3.82 −19.28	

Note: (*3)—Maximum electric field strength at the area of a nonoperating frame 2.61 V/m, (*4)—Maximum electric field strength at the area of a nonoperating frame 1.60 V/m.

For the sake of completeness, by applying the respective procedure for the case of Stachtoroi, an arrangement of six linear frames was formed, each consisting of 13 electrodes with a distance between bars D_{el} equal to 0.50 m and a total length D_{frame} equal to 6.00 m. Considering that the arc of the "right" water–dam zone of Figure 6 is equal to 150° and that the distance between the frames d_{frames} is equal to 4.5 m (with respect to 5.78 m of Table 11), consequently, the total required length of protective dam (behind which the array is placed) is estimated to be 67.5 m (=6 × 6.0 + 7 × 4.5), of which 9.0 m, on either outer side, can be considered on the shore, so as to be covered by the plan view of the preliminary study dam of Figure 11. Same as before, the case of loading the electrode station with the maximum nominal load $I_{total\text{-}steady}$, equal to 1,100 A, was examined, for six-frame or five-frame operation, with the respective electric current densities under full load conditions $J_{full\_load\_steady}$ and under periodic maintenance conditions $J_{maintenance\_steady}$, respectively. The arrangement was placed on a canvas of 50 m (Ox semi-axis) by 120 m (yOy' axis—parallel to the axis of the frame) and with a simulation step of 0.10 m by 0.10 m (with a computational mesh of 501 × 1,201 = 601,701 points), as seen in Figure 8. From the respective simulation, the results of Table 14 are obtained, from the study of which the following emerge:

- The deviation at the ends of the electrode station reaches up to 18.6 m depending on the electrifying method (especially when electrifying five consecutive panels). However, in the area of the frame that does not operate, the respective electric field strength is marginally above 2.5 V/m, so there is no safety issue during maintenance;

- The deviation of the maximum developing electric field strength of the electrode station against that of a single frame (19.73 V/m against 17.96 V/m) is of the order of 10%, which is quite large but expected because the distance between frames dropped to 4.5 m, instead of 5.8 m;
- The critical zone of the dam extends outside the dam by 30 m (with a crest width of 5 m and a suspension distance of at least 1 m) on the xOx' axis (vertical to the dam), where the initial assessment of Table 11 failed. However, it is limited to 92% of the most favorable value, resulting from concentrated source methods;
- The critical zone of the dam extends beyond the estimated dam (according to Section 5.3) by 37.1 m (18.6 m on either side) on the yOy' axis (parallel to the dam axis) on the shore area, attributable to the significant reduction in the distances between frames. Due to the large size, the respective area (7,602 m^2) is larger by 55% than that of method "C", with a concentrated current source, but larger only by 10% compared to that of method "A".

Table 14. Simulation results with method "C", of a linear current source, for the case of an arrangement of 6 linear frames, in a row, consisting of 13 electrodes, for the area of Stachtoroi, Attica ($\rho_S = \infty$, without dam, $L = 2.13$ m, $D_{el} = 0.50$ m, $d_{frames} = 4.50$ m, $\rho_w = 0.25$ $\Omega{\cdot}$m, $\theta = 210°$, $E_{limit\_S} = 1.25$ V/m).

Supply Method	J_{steady} [A/m^2]	d_2 [m]	d_3 [m]	d_4 [m]	$\ell_{b\text{-}c}$ [m]	$\ell_{u\text{-}c}$ [m]	$s_{b\text{-}c}$ [m]	$s_{u\text{-}c}$ [m]	y_p [m]	Δy_p [m]	E_{max} [V/m]
Operation of 6 frames	18.33	33.74	49.29	−49.29	−29.25	29.25	20.04	20.04	33.75	−15.54	16.59
Operation of 5 frames (except no. 6)	22.00	36.34	41.80	−52.30	−29.25	18.75	23.05	23.05	23.25	−18.55 *5	19.73
Operation of 5 frames (except no. 5)	22.00	34.35	15.10	−51.69	−29.25	8.25	22.44	6.85	12.75 −33.75	−2.35 *6 −17.94	19.68
			47.33	21.61	23.25	29.25	1.64	18.28	18.75 33.75	2.86 −13.78	
Operation of 5 frames (except no. 4)	22.00	32.02	2.17	−50.85	−29.25	−2.25	21.60	4.42	2.25 −33.75	0.08 −17.10	19.61
			49.62	9.88	12.75	29.25	2.87	20.37	8.25 33.75	1.63 −15.87	

Note: (*5)—Maximum electric field strength at the area of a nonoperating frame 2.95 V/m, (*6)—Maximum electric field strength at the area of a nonoperating frame 1.95 V/m.

If the respective process is repeated with the respective transient behavior currents and the respective electric field limits at transient behavior, the resulting requirements would be slightly smaller, so they were not recorded further.

5.5. Estimation of Maximum Absolute Electric Potential and Equivalent Remote Earth Resistance for an Electrode Station of 6 Linear Frames, in a Row, Each Consisting of 13 Electrodes, Parallel to the Protective Dam, at Maximum Current Density, under Conditions of Normal Operation or Periodic Maintenance

Based on the conclusions of Section 4.5, for the determination of the absolute electric potential and the equivalent resistance of each electrode station, both method "A" and method "C" must be used. The former is that of a point current source (based on the guidelines of CIGRE B4.61 675:2017 and IEC TS 62344:2013), which is suitable for calculating the electric field strength in the far field (considering a wedge-shaped sea zone, using the slope of the bottom), and the latter is of the linear current source, which is suitable for the near field near the electrodes (because it treats the electrodes as linear rather than point current sources), but at the cost of taking into account a small water zone of constant depth. The limit of switch between methods is the equalization distance of the electric field strengths provided by methods "A" and "C" in the area beyond the dam with the admissions made in Section 5.1 (ignoring the effect of the soil, the water zone in the respective arc of angle θ on the plan view of Figure 6, of the dam itself), using Equations (12) and (91), respectively, to calculate the strength of each electrode, at the points of the canvas under study in Figure 8 and then by superpositioning the total strength results through Equations (63)–(68), by determining the smaller of the two methods. The strength values of the dam area are calculated through method "C" (a nearby area) and are corrected by multiplying by the ratio of dam resistivity to seawater resistivity (ρ_d/ρ_w) (i.e., through Equation (92)). Subsequently, from the appropriate integration of the electric field strength, with respect to infinity (i.e., at 150 km, where the absolute potential is considered null), the total absolute potential was determined. In the present case, the integration was performed perpendicular to the dam axis (yOy' axis), of the row of frames on the semi-axis Ox, according to Equation (69).

In the area of Korakia, the arrangement of six linear frames was formed with a distance between them d_{frames} = 6.5 m. Each frame consists of 13 electrodes with a distance between them, D_{el} = 0.50 m and a total length D_{frame} = 6.00 m, vertically placed with an effective length L = 2.13 m. The geometrical features are θ = 248° in relation to Figure 6, θ_w = 2.29° = 0.039978687 rad (most unfavourable seabed inclination from Table 1), with respect to Figure 5. Similar to before, the case of loading the electrode station with the maximum nominal load $I_{total\text{-}steady}$ of 1,100 A was considered, for a six-frame or five-frame operation (sixth, fifth or fourth frame out of operation), with respective current densities under full load conditions, $J_{full\_load\_steady}$, and under periodic maintenance conditions, $J_{maintenance\_steady}$, respectively. The array of frames was placed on a canvas of 150 km (Ox axis) by 160 m (yOy' axis—parallel to the frame axis) and with a simulation step of 0.10 m up to 100 m, 1.0 m from 100 to 200 m, 5.0 m from 200 to 1,000 m, 10 m from 1,000 to 10,000 m, 100 m from 10 km to 150 km on the Ox axis and by 0.10 m, on the yOy' (with a computational mesh of 3,561 × 1,601 = 5,701,161 points), as seen in Figure 8. From the respective simulation, the data of Table 15 were obtained, additionally listing the current density J_{steady} in terms of the peripheral surface, the maximum value of the absolute potential V_{rel_max}, and the resistance between the electrode station and remote earth R_{el}. The graphs of the absolute potential $V_{max}(y)$ (on the perceived axis of the electrode station (x = 0)), of the electric field strength (perpendicular to the axis yOy', for that y at which the maximum value of all $V_{max}(y)$ occurs), and the respective value of absolute potential on this axis, are indicatively shown in Figures 30–32, respectively, for the case of non-operation of frame no. 6, which is the most unfavorable of all.

Table 15. Results of determination of absolute electric potential and resistance between electrode station and remote earth by applying method "A" for the far field ($\rho_S = \infty$) and method "C" for the near field, with simplifying admissions ($\rho_S = \infty$) and compensation for the presence of a dam; for an electrode station of 6 linear frames in a row, d_{frames} = 6.50 m (area of Korakia) and 4.50 m (area of Stachtoroi), each frame consisted of 13 electrodes of active length L = 2.13 m, D_{el} = 0.50 m, ρ_d = 100 Ω·m, ρ_w = 0.25 Ω·m, θ_w = 2.29°, θ = 248° (area of Korakia) and θ_w = 0.272°, θ = 210° (area of Stachtoroi).

		Korakia		Stachtoroi	
Supply Method	J_{steady} [A/m²]	V_{rel_max} [V]	R_{el} [Ω]	V_{rel_max} [V]	R_{el} [Ω]
Operation of 6 frames	18.33	19,527	16.713	14,899	12.766
Operation of 5 frames (except no. 6)	22.00	22,772	19.512	17,300	14.824
Operation of 5 frames (except no. 5)	22.00	22,124	18.957	16,772	14.371
Operation of 5 frames (except no. 4)	22.00	21,199	18.164	16,015	13.722

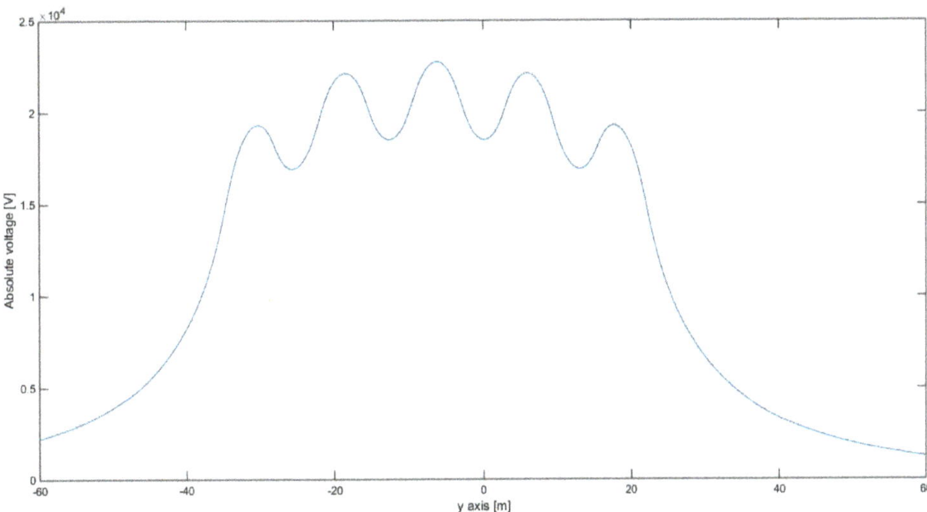

Figure 30. Absolute electric potential on the perceived axis of the electrode station (for $x = 0$ on the yOy' axis, applying method "A" for the far field ($\rho_S = \infty$) and method "C" for the near field, with simplifying admissions ($\rho_S = \infty$), with compensation for the presence of a dam), for an electrode station of 6 linear frames in a row, with $d_{frames} = 6.50$ m, each frame consisted of 13 electrodes of active length $L = 2.13$ m, $D_{el} = 0.50$ m, $\rho_d = 100$ $\Omega{\cdot}$m, $\rho_w = 0.25$ $\Omega{\cdot}$m, $\theta_w = 2.29°$, $\theta = 248°$ (area of Korakia), $J_{maitenance\_steady} = 22$ A/m^2 (steady state and operation of 5 panels, except no. 6).

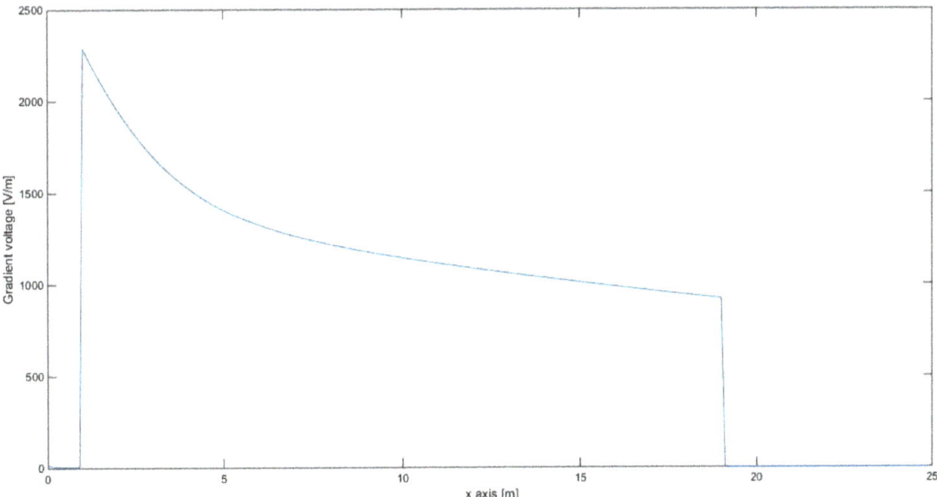

Figure 31. Electric field strength on the Ox semi-axis, perpendicular to the perceived axis of the electrode station, at the point of the maximum value of all absolute potentials, applying method "A" for the far field ($\rho_S = \infty$) and method "C" for the near field, with simplifying admissions ($\rho_S = \infty$), with compensation for the presence of a dam; for an electrode station of 6 linear frames in a row, with $d_{frames} = 6.50$ m, each frame consisted of 13 electrodes of active length $L = 2.13$ m, $D_{el} = 0.50$ m, $\rho_d = 100$ $\Omega{\cdot}$m, $\rho_w = 0.25$ $\Omega{\cdot}$m, $\theta_w = 2.29°$, $\theta = 248°$ (area of Korakia), $J_{maitenance\_steady} = 22$ A/m^2 (steady state and operation of 5 panels, except no. 6).

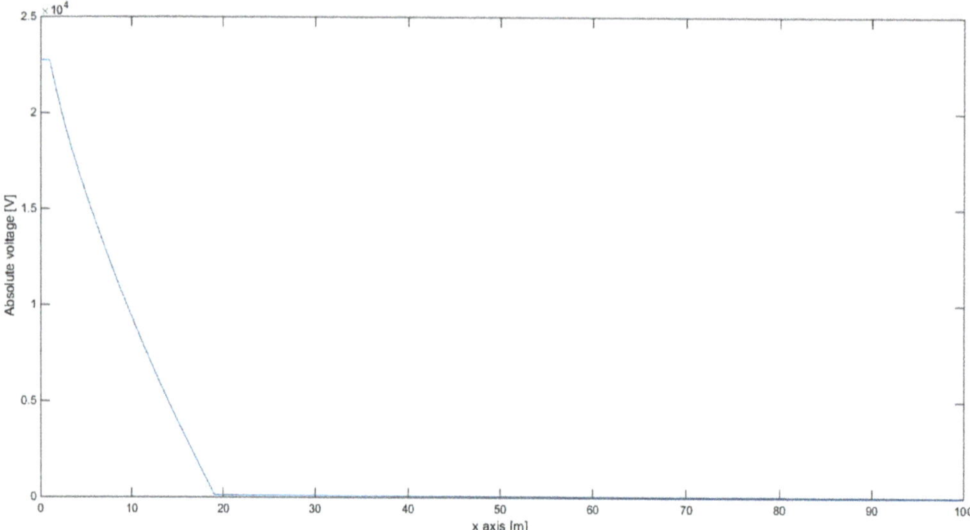

Figure 32. Absolute electric potential on the Ox semi-axis, perpendicular to the perceived axis of the electrode station, at the point of the maximum value of all absolute potentials, applying method "A" for the far field ($\rho_S = \infty$) and method "C" for the near field, with simplifying admissions ($\rho_S = \infty$), with compensation for the presence of a dam; for an electrode station of 6 linear frames in a row, with d_{frames} = 6.50 m, each frame consisted of 13 electrodes of active length L = 2.13 m, D_{el} = 0.50 m, ρ_d = 100 Ω·m, ρ_w = 0.25 Ω·m, θ_w = 2.29°, θ = 248° (area of Korakia), $J_{maitenance\_steady}$ = 22 A/m^2 (steady state and operation of 5 panels, except no. 6).

The same procedure is repeated for the area of Stachtoroi, with the difference that the six linear frames are arranged in a row at a distance between them d_{frames} = 4.5 m, while the geometrical features are θ = 210° with respect to Figure 6, θ_w = 0.272°= 0.004743554 rad (less favorable seabed inclination from Table 1), with respect to Figure 5.

The following conclusions emerge from the respective study:

- *For the case of Korakia*: The developed absolute potentials and the respective values of the resistance of the electrode station, with respect to remote earth (22.77 kV, 19.512 Ω) are much lower compared to those of method "A" (73.3 kV, 66.63 Ω, according to Table 3) and of method "B" (1,763 kV, 1,603 Ω, according to Table 5, considering the same dam material), and bigger compared to those of method "C" (0.712 kV, 0.6468 Ω, according to Table 7, considering the same dam material), under the conditions infinite soil resistivity. The latter is due to the fact that during the application of method "C" of Table 7, the effect of the water of the formed pond was also taken into account, whilst here, its part from the electrode station to the shore was practically ignored, giving much more unfavorable results. In any case, much smaller values than those in Table 15 are expected, while the effect of the dam on the development of the absolute potential is extremely important (as can be seen in Figure 31) due to the significant increase in the electric field strength, according to Figure 32. If the effect of the dam was ignored, an absolute potential of the order of 200 V (instead of 22 kV) would have resulted;
- *For the case of Stachtoroi*: The developed absolute potentials and the respective values of the resistance of the electrode station, with respect to remote earth (17.30 kV, 14.824 Ω) are much lower compared to those of method "A" (475.2 kV, 431.99Ω, according to Table 2), and of method "B" (11,361 kV, 10,328 Ω, according to Table 4, considering the same dam material) and bigger compared to those of method "C" (11.054 kV, 10.049 Ω, according to Table 6, considering the same dam material), under the conditions of infinite soil resistivity by disregarding the effect of the water of the pond formed;
- *General remarks*: The presence of the dam, the thickness of the dam, and resistivity all play an important role in the final value of the developed absolute potential. Taking the average thickness of the dam at the average immersion height of the electrodes and ignoring the upper

and lower water zones from the effective length of the electrode in method "C" leads to quite unfavorable results and in favor of safety. Analytical simulations with 3D field models would lead to significantly lower values of electric field strength, maximum absolute potential and electrode station resistance with respect to remote earth. Finally, it was clarified that the respective values are calculated at the average height of the electrodes; thus, towards the surface, reduced values are obtained due to non-ideal dam materials and water in terms of electrical conductivity.

6. Conclusions

The purpose of this paper was to study the distribution of the electric field strength at a shoreline pond electrode station with the aid of analytical methods. In particular, based on the analytical methods of CIGRE B4.61 675:2007 [17] (pp. 118–119) and IEC TS 6234:2003 [34] (pp. 30–32) standards, the following were developed:

- *Method "A"*: It was based on an equivalent point current source, with the formation of a sphere, where the homogeneous soil of electrical resistivity ρ_s occupies an angle θ_s, the water of electrical resistivity ρ_w occupies an angle θ_w and the rest of the space is occupied by non-conductive air, according to Figure 4c, thus unifying the two pre-existing analytical methods of the aforementioned standards [17,34];
- *Method "B"*: It was an extension of method "A", as a dam of thickness d and of electrical resistivity ρ_d (which extends from radius r_1 to radius $r_2 = r_1 + d$, occupying an angle θ_w, such as the seawater), is added inside the water, according to Figure 5;
- *Method "C"*: It was based on an equivalent linear current source, which approximates the structure of a rod-shaped electrode much better, corresponding to a water zone of thickness/effective length L (in the vertical sense), extending around the electrode in the form of a cylinder, according to Figure 6. The soil of electrical resistivity ρ_s and thickness L extends from a radius r_3 to infinity, occupying an arc of angle θ, and the dam of resistivity ρ_d, of the same thickness L, extends from radius r_1 to r_2, occupying an arc of angle $2 \times \pi\text{-}\theta$, whilst the remaining space of the same thickness L contains water of resistivity ρ_w. Above and below this zone of thickness L lies electrically non-conductive material.

For all methods, the necessary mathematical background was developed, and the theoretical assumptions and weaknesses of each method were commented on. The final purpose was to ensure that no high potential differences between two points develop (which can lead to electrochemical corrosion of metal structures, etc.), as well as no hazardous electric field strengths (near the electrode station), with regard to humans and other living beings, at steady and transient states. The above are expressed as safety distances r_{limit1} against voltage V_{limit_S} with respect to infinity, r_{limit2} against average steady-state electric field strength E_{limit_S}, r_{limit3} against average transient-state electric field strength E_{limit_T}, r_{limit4} against steady-state point value of electric field strength E_{limit_S}, and r_{limit5} against transient-state point value of electric field strength E_{limit_T}. Limits V_{limit_S}, E_{limit_S} and E_{limit_T} are 4 V, according to [79] (although considering specific points and not infinity); 1.25 V/m to 2 V/m, according to IEC TS 6234:2003, [34] (p. 32) or 2.5 V/m, according to CIGRE B4.61 675:2007 [17]; and 15 V/m according to CIGRE B.4.61 675:2007 [17], respectively, applying, in the present case, the most unfavorable values. In addition, the determination of the absolute electric potential, with respect to infinity (remote earth) and the equivalent ohmic resistance of the electrode station, with respect to remote earth, is of interest because they constitute basic criteria for dimensioning the insulation material of the switching devices and the return conductor of the HVDC interconnection. From the relative development of the mathematical background and also from the application of the above methods for the electrode stations at Stachtoroi, Attica and Korakia, Crete, for the new ±500 kV, 1 GW bi-polar HVDC transmission system with ground return between Attica and Crete, resulted in the following main conclusions regarding said methods:

- Regarding the safety distances, against average and point electric field strengths at steady and transient states, method "C" is more suitable, as these distances are located in the near field; the model of this method, with a water zone of constant thickness L, better approximates the real conditions near the dam. In addition, from the existing numerical simulations of the two regions, the most critical distance is that of the point electric field strength at a steady state, with an allowable limit value of 1.25 V/m (r_{limit4});
- Concerning the safety distance, with regards to potential difference, with respect to infinity, methods "A" or "B" are more suitable, as these methods are characterized by a more realistic representation of space in the far field by forming a water wedge of angle θ; that is, a larger

space with respect to the cylindrical water zone of constant thickness L for long distances from the electrode station. Moreover, at distances of some km, the electrode station of a size of some tens of m would appear as a "point". Furthermore, because this safety distance lies outside the dam, the results of methods "A" and "B" are identical;

- Regarding the absolute electric potential and the equivalent ohmic resistance of the electrode station, with respect to remote earth, the combined use of methods "A" and "C" is recommended. In particular, initially, the equalization distance of the electric field strengths of methods "A" and "C" for the external area of the dam is determined. Then (through the appropriate integration of the electric field strength), the corresponding values of the "C" method are used, from the surface of the electrode to the distance equalizing the strengths of the two methods (near field), and the values of the "A" method, from a distance, equalizing the strengths of the two methods towards infinity (far field). In this way, the advantages of these two methods are utilized, and the disadvantages curtailed;

- The calculation of the corresponding quantities (safety distances, absolute electric potential and equivalent ohmic resistance of the electrode station with respect to remote earth) is performed through analytical relations directly, even with a scientific calculator, providing the respective limits. Furthermore, should some parameters be unknown (such as soil electrical resistivity), they can be omitted through appropriate admissions (e.g., assumed infinite), leading to more unfavorable results, providing, nonetheless, an upper limit on the sizes, which is extremely critical for the designer of the electrode station at the preliminary study, in a cost-effective and swift way.

However, in order to better study the electric field distribution in the near field, so as to limit the respective safety distances, this paper also proposed the use of superpositioning to simulate the individual electrodes that constitute the electrode station instead of a concentrated one, as initially described by methods "A", "B" and "C". In particular, the total intensity of the electric current was distributed to the individual electrodes, including a corrective incrementation factor, due to the uneven distribution of the electric current among the electrodes. A dense, two-dimensional, orthogonal canvas was formed, where the electrodes of the station are appropriately placed, and the electric field strength is calculated separately for each electrode with the appropriate method at the respective points of the canvas and analyzed in the two components of the axes xOx' and yOy'. Consequently, the respective components of the two axes were then added/superimposed separately for all the electrodes, and then the total integrated electric field strength was formed, at each point of the canvas, allowing for the calculation of the safety distances concerning the limits of the point electric field strength. Through numerical integration, with respect to the xOx' and yOy' directions, of the individual electric field strength components, the absolute electric potential, with respect to remote earth, was approximated, and subsequently, both the ohmic resistance of the electrode station and the average electric field strength. For the calculation of the safety distances, in terms of electric field strengths, the application of method "C" was proposed, where, in the present study, for reasons of simplification and easy numerical simulation, the effects of the ground, the water zone (in the arc of angle θ, of Figure 6) and the dam were ignored. The first two admissions lead to more unfavorable results, while the third does not affect them since the safety distances are outside the area defined by the dam. The values on the dam are approximated by the initial values of the electric field strength, multiplied by the ratio of the dam resistivity over the water resistivity. Regarding the absolute electric potential and the equivalent ohmic resistance of the electrode station, with respect to remote earth, the combined application of methods "A" and "C" was proposed through the appropriate equalization of the electric field strengths and following numerical integration of the electric field strength (in the present case on the Ox semi-axis) through the method "C", from the perceived axis where the electrodes are placed (in the present case yOy' axis) to the point of equalization of the strengths and method "A" from the point of equalization of the strengths to infinity (in the present case 150 km).

The following main conclusions emerged, from the development of the relevant software in the MATLAB programming environment and from its application for the electrode stations at Stachtoroi, Attica and Korakia, Crete:

- Regarding the safety distances, with respect to average and point electric field strengths, at steady and transient states, the respective values are reduced by at least 10% compared to the respective values of the methods of concentrated sources despite the unfavorable admissions. However, the corresponding area occupied is comparable to or even greater than the one of concentrated sources since now the electrode station occupies a significant area instead of a single point on the plane;

- Regarding the safety distance, with respect to potential differences to infinity, no examination is conducted due to the long distances and the suitability of the methods of concentrated current sources;
- Regarding the absolute electric potential and the equivalent ohmic resistance of the electrode station with respect to remote earth, the respective values are much smaller compared to the respective methods "A" and "B" and larger compared to method "C". The latter is due to the fact that, during the application of method "C", the effect of the water of the formed pond was also taken into account (which, in this analysis, its part from the electrode station to the coast was ignored);
- The results are in favor of safety, as the upper and lower water zones and the rest of the lower ground/seaned are ignored, with respect to the active length of the electrode, in method "C";
- The calculation of the corresponding quantities (safety distances, absolute electric potential and equivalent ohmic resistance of the electrode station with respect to remote earth) was performed through simple software that can be developed in any computer programming platform (such as MATLAB, etc.), through a few dozen lines of code. Thus, it can be relatively easily implemented without the requirement of purchasing specialized software packages and training in them at the cost of numerical accuracy. In addition, this computational method does not require detailed data of the area under study through expensive and time-consuming geophysical methods. Therefore, at the preliminary study stage, it is considered suitable for implementation;
- In the case of analytical simulation with three-dimensional field models, significantly smaller values are expected in the quantities of the electric field strength, the maximum absolute potential and the resistance of the electrode station with respect to remote earth, given that the total mass of water, the soil, the seabed and the dam (with its possible openings) was included with greater precision.

Author Contributions: Conceptualization, G.J.T., J.M.P., C.D.T., K.L. and J.C.A.; methodology, G.J.T., V.T.K., F.D.K., J.M.P. and A.X.M.; software, G.J.T., V.T.K., F.D.K. and P.M.D.; validation, G.J.T., V.T.K., F.D.K. and P.M.D.; formal analysis, G.J.T., V.T.K., F.D.K. and P.A.K.; investigation, G.J.T. and V.T.K.; resources, G.J.T., V.T.K., K.L. and J.C.A.; data curation, C.D.T., K.L. and J.C.A.; writing—original draft preparation, G.J.T., V.T.K., F.D.K. and P.A.K.; writing—review and editing, G.J.T., V.T.K., F.D.K. and P.A.K.; visualization, G.J.T., V.T.K. and P.M.D.; supervision, G.J.T.; project administration, G.J.T.; funding acquisition, C.D.T., K.L. and J.C.A. All authors have read and agreed to the published version of the manuscript.

Funding: This research was funded by IPTO, grant number 191102.

Acknowledgments: The authors would like to thank the IPTO for the data availability of the initial shoreline electrode station design. The authors are grateful to Spyridon Gialampidis for his work on the original text version.

Conflicts of Interest: The authors declare no conflict of interest. The funders (IPTO) had no role in the design of the study; in the collection, analyses, or interpretation of data; in the writing of the manuscript, or in the decision to publish the results.

References

1. Sutton, S.J.; Lewin, P.L.; Swingler, S.G. Review of global HVDC subsea cable projects and the application of sea electrodes. *Electr. Power Energy Syst.* **2017**, *87*, 121–135. [CrossRef]
2. Longatt, F.G. High Voltage Direct Current (HVDC). Seminar in Norway, 25 April 2019. Available online: https://www.researchgate.net/publication/332642386_Serminar_High_Voltage_Direct_Current_HVDC (accessed on 19 December 2019). [CrossRef]
3. Sutton, S.J.; Swingler, S.J.; Lewin, P.L. *HVDC Subsea Cable Electrical Return Path Schemes: Use of Sea Electrodes and Analysis of Environmental Impact*, 1.1st ed.; HubNet: Manchester, UK, 2016; pp. 1–51.
4. Grid Systems HVDC. The Early HVDC Development, the Key Challenge in the HVDC Technique. Available online: www.abb.com/hvdc (accessed on 19 December 2019).
5. Rahman, S.; Khan, I.; Alkhammash, H.I.; Nadeem, M.F. A comparison review on transmission mode for onshore integration of offshore wind farms: HVDC or HVAC. *Electronics* **2021**, *10*, 1489. [CrossRef]
6. Holtsmark, N.; Bahirat, H.J.; Molinas, M.; Mork, B.A.; Hoidalen, H.K. An all-DC offshore wind farm with series-connected turbines: An alternative to the classical parallel AC model? *IEEE Trans. Ind. Electron.* **2013**, *60*, 2420–2428. [CrossRef]
7. Liljestrand, L.; Sannino, A.; Breder, H.; Thorburn, S. Transients in collection grids of large offshore wind parks. *Wind Energy* **2008**, *11*, 45–61. [CrossRef]

8. Xu, L.; Andersen, B.R. Grid connection of large offshore wind farms using HVDC. *Wind Energy* **2006**, *9*, 371–382. [CrossRef]
9. Akhmatov, V.; Callavik, M.; Franck, C.M.; Rye, S.E.; Ahndorf, T.; Bucher, M.K.; Müller, H.; Schettler, F.; Wiget, R. Technical guidelines and prestandardization work for first HVDC grids. *IEEE Trans. Power Deliv.* **2014**, *29*, 327–335. [CrossRef]
10. CIGRE Working Group B4.33. *HVDC and FACTS for Distribution Systems*, 1st ed.; CIGRE: Paris, France, 2005; Volume 280, pp. 1–63.
11. CIGRE Working Group A2/B4.28. *HVDC Converter Transformers—Design Review, test procedures, ageing evaluation and reliability in service*, 1st ed.; CIGRE: Paris, France, 2010; Volume 406, pp. 1–38.
12. CIGRE Working Group A2/B4.28. *HVDC Converter Transformers—Guidelines for Conducting Design Reviews for HVDC Converter Transformers*, 1st ed.; CIGRE: Paris, France, 2010; Volume 407, pp. 1–22.
13. CIGRE Working Group B4.44. *HVDC Environmental Planning Guidelines*, 1st ed.; CIGRE: Paris, France, 2012; Volume 508, pp. 1–59.
14. CIGRE Working Group B4.52. *HVDC Grid Feasibility Study*, 1st ed.; CIGRE: Paris, France, 2013; Volume 533, pp. 1–189.
15. CIGRE Working Group B4.04. *HVDC LCC Converter Transformers—Converter Transformer Failure Survey from 2003 to 2012*, 1st ed.; CIGRE: Paris, France, 2015; Volume 617, pp. 1–54.
16. CIGRE Working Group B4.55. *HVDC Connection of Offshore Wind Power Plants*, 1st ed.; CIGRE: Paris, France, 2021; Volume 619, pp. 1–100.
17. CIGRE Working Group B4.61. *General Guidelines for HVDC Electrode Design*, 1st ed.; CIGRE: Paris, France, 2017; Volume 675, pp. 1–150.
18. CIGRE Working Group A3/B4.34. *Technical Requirements and Specifications of State-of-the-Art HVDC Switching Equipment*, 1st ed.; CIGRE: Paris, France, 2017; Volume 683, pp. 1–240.
19. CIGRE Working Group B4/C1.65. *Recommended Voltages for HVDC Grids*, 1st ed.; CIGRE: Paris, France, 2017; Volume 684, pp. 1–67.
20. CIGRE Working Group B4.58. *Control Methodologies for Direct Voltage and Power Flow in a Meshed HVDC Grid*, 1st ed.; CIGRE: Paris, France, 2017; Volume 699, pp. 1–61.
21. CIGRE Working Group B4.60. *Designing HVDC Grids for Optimal Reliability and Availability Performance*, 1st ed.; CIGRE: Paris, France, 2017; Volume 713, pp. 1–128.
22. CIGRE Working Group B4.59. *Protection and Local Control of HVDC-Grids*, 1st ed.; CIGRE: Paris, France, 2018; Volume 39, pp. 1–96.
23. IEEE Working Group WGI10. *Guide for Commissioning High-Voltage Direct-Current (HVDC) Converter Stations and Associated Transmission Systems*, 2nd ed.; IEEE: New York, NY, USA, 2008; Volume 1378–1997, pp. 1–31.
24. IEEE Working Group WGI10. *IEEE Guide for the Evaluation of the Reliability of HVDC Converter Stations*, 2nd ed.; IEEE: New York, NY, USA, 2012; Volume 1240–2000, pp. 1–67.
25. IEEE Working Group WGI10. *IEEE Guide for Analysis and Definition of DC Side Harmonic Performance of HVDC Transmision Systems*, 2nd ed.; IEEE: New York, NY, USA, 2010; Volume 1240–2003, pp. 1–104.
26. IEEE Working Group WG-HVDC-CPI—Working Group for Establishing Standard Specifications of HVDC Composite Post Insulators. *IEEE Standard for High-Voltage Direct-Current (HVDC) Composite Post Insulators*, 1st ed.; IEEE: New York, NY, USA, 2017; Volume 1898–2016, pp. 1–35.
27. IEEE Working Group HVDC WG—Working Group for HVDC Cable Systems (Cables, Joints and Teminations) (DEI/SC/HVDC Cable Systems). *IEEE Recommended Practice for Space Charge Measurements on High-Voltage Direct-Current Extruded Cables for Rated Voltages up to 550 kV*, 1st ed.; IEEE: New York, NY, USA, 2017; Volume 1732–2017, pp. 1–36.
28. IEEE Working Group WG-UHVDC-TCP—Working Group for Establishing Basic Requirements for Ultra High-Voltage Direct-Current (UHVDC) Transmission Control and Protection. *IEEE Guide for Establishing Basic Requirements for High-Voltage Direct-Current Transmission Protection and Control Equipment*, 1st ed.; IEEE: New York, NY, USA, 2017; Volume 1899–2017, pp. 1–47.
29. IEEE Working Group HVConv-WG1277—HV Converter TR & Reactors—Req. & Test Code for HVDC Smoothing Reactors Working Group. *IEEE Standard General Requirements and Test Code for Dry-Type and Oil-Immersed Smoothing Reactors and for Dry-Type Converter Reactors for DC Power Transmission*, 1st ed.; IEEE: New York, NY, USA, 2020; Volume 1277–2020, pp. 1–90.
30. IEEE/IEC. *IEC/IEEE International Standard—Power transformers—Part 57–129: Transformers for HVDC Applications*, 1st ed.; IEEE: New York, NY, USA, 2017; Volume 60076-57-129-2017, pp. 1–58.
31. IEC Technical Commitee TC115—High Voltage Direct Current (HVDC) Transmission for DC Voltages above 100 kV. *High Voltage Direct Current (HVDC) Substation Audible Noise*, 1st ed.; IEC: Geneva, Switzerland, 2012; Volume IEC TR 61973:2012, pp. 1–82.
32. IEC Technical Commitee TC115—High Voltage Direct Current (HVDC) Transmission for DC Voltages above 100 kV. *Amendment 1: High Voltage Direct Current (HVDC) Substation Audible Noise*, 1st ed.; IEC: Geneva, Switzerland, 2019; Volume IEC TR 61973:2012/AMD1:2019, pp. 1–5.
33. IEC Technical Commitee SC22F—Power Electronics for Electrical Transmission and Distribution Systems. *High-Voltage Direct Current (HVDC) Installations—System Tests*, 1st ed.; IEC: Geneva, Switzerland, 2010; Volume IEC 61975:2010, pp. 1–165.
34. IEC Technical Commitee TC115—High Voltage Direct Current (HVDC) Transmission for DC Voltages above 100 kV. *Design of Earth Electrode Stations for High-Voltage Direct Current (HVDC) Links—General Guidelines*, 1st ed.; IEC: Geneva, Switzerland, 2013; Volume IEC 62344:2013, pp. 1–89.
35. IEC Technical Commitee TC115—High Voltage Direct Current (HVDC) Transmission for DC Voltages above 100 kV. *Reliability and Availability Evaluation of HVDC Systems*, 1st ed.; IEC: Geneva, Switzerland, 2018; Volume IEC TR 62672:2018, pp. 1–45.

36. IEC Technical Commitee TC115—High Voltage Direct Current (HVDC) Transmission for DC Voltages above 100 kV. *Electromagnetic Performance of High Voltage Direct Current (HVDC) Overhead Transmission Lines*, 1st ed.; IEC: Geneva, Switzerland, 2014; Volume IEC TR 62681:2014, pp. 1–92.
37. IEC Technical Commitee TC 20—Electric Cables. *High Voltage Direct Current (HVDC) Power Transmission—Cables with Extruded Insulation and Their Accessories for Rated Voltages up to 320 kV for Land Applications—Test Methods and Requirements*, 1st ed.; IEC: Geneva, Switzerland, 2017; Volume IEC TR 62895:2017, pp. 1–136.
38. IEC Technical Commitee TC115—High Voltage Direct Current (HVDC) Transmission for DC Voltages above 100 kV. *HVDC installations—Guidelines on asset management*, 1st ed.; IEC: Geneva, Switzerland, 2017; Volume IEC TR 62978-1:2017, pp. 1–60.
39. IEC Technical Commitee TC115—High Voltage Direct Current (HVDC) Transmission for DC Voltages above 100 kV. *High Voltage Direct Current (HVDC) Power Transmission—System Requirements for DC-Side Equipment—Part 1: Using Line-Commutated Converters*, 1st ed.; IEC: Geneva, Switzerland, 2018; Volume IEC TR 63014-1:2018, pp. 1–87.
40. IEC Technical Commitee TC115—High Voltage Direct Current (HVDC) Transmission for DC Voltages above 100 kV. *Guidelines for Operation and Maintenance of Line Commutated Converter (LCC) HVDC Converter Station*, 1st ed.; IEC: Geneva, Switzerland, 2017; Volume IEC TR 63065-1:2017, pp. 1–50.
41. IEC Technical Commitee TC115—High Voltage Direct Current (HVDC) Transmission for DC Voltages above 100 kV. *Guideline for the System Design of HVDC Converter Stations with Line-Commutated Converters*, 1st ed.; IEC: Geneva, Switzerland, 2020; Volume IEC TR 63127-1:2020, pp. 1–64.
42. IEC Technical Commitee TC115—High Voltage Direct Current (HVDC) transmission for DC voltages above 100 kV. *Guideline for Planning of HVDC Systems—Part 1: HVDC Systems with Line-Commutated Converters*, 1.0st ed.; IEC: Geneva, Switzerland, 2020; Volume IEC TR 63179-1:2020, pp. 1–28.
43. EPRI. *Life Extension Guidelines for HVDC Systems*, 1st ed.; EPRI: California, CA, USA, 2006; Volume ID 1012516, p. 26.
44. EPRI. *Electrical Effects of HVDC Transmission Lines*, 1st ed.; EPRI: California, CA, USA, 2010; Volume ID 1020118, p. 322.
45. EPRI. *HVDC Ground Electrode Overview*, 1st ed.; EPRI: California, CA, USA, 2010; Volume ID 1020116, p. 66.
46. EPRI. *HVDC Ground Electrode Design*, 1st ed.; EPRI: California, CA, USA, 1981; Volume EL-2020, Research Project 1467-1.
47. Halt, R.J.; Debkowski, J.; Hauth, R.L. *HVDC Power Transmission Electrode Siting and Design*, 1.0st ed.; Oak Ridge National Laboratory: Oak Ridge, TN, USA, 1997; Volume ORL/Sub/95-SR893/3, pp. 1–138.
48. Bailey, W.H.; Weil, D.E.; Stewart, J.R. *HVDC Power Transmission Environmental Issues Review*, 1st ed.; Oak Ridge National Laboratory: Oak Ridge, TN, USA, 1997; Volume ORL/Sub/95-SR893/2, pp. 1–128.
49. DVN-GL. *Recommended Practice: Qualification Procedure for Offshore High-Voltage Direct Current (HVDC) Technologies*, 1st ed.; EPRI: Oslo, Norway, 2014; Volume DNVGL-RP-0046:2014-08, p. 47.
50. Antoine, O.; Papangelis, L.; Michels Alfaro, S.; Guittonneau, A.; Bertinato, A. Technical requirements for connection to offshore HVDC grids in the North Sea. *Eur. Comm. Dir. Gen. Energy Intern. Energy Mark.* **2020**, 1–89.
51. Available online: https://www.admie.gr/erga/erga-diasyndeseis/diasyndesi-tis-kritis-me-tin-attiki (accessed on 10 December 2018).
52. Manglik, A.; Verma, S.K.; Muralidharan, D.; Sasmal, R.P. Electrical and electromagnetic investigations for HVDC ground electrode site in India. *Phys. Chem. Earth* **2011**, *36*, 1405–1411. [CrossRef]
53. Da Fonseca Freire, P.E.; Pereira, S.Y.; Padilha, A.L. Adjustment of the geoelectric model for a ground electrode design—The case of the Rio Madeira HVDC transmission system, Brazil. *Sci. Eng.* **2021**, *20*, 137–151.
54. Rusck, S. HVDC power transmission: Problems relating to earth return. *Direct Curr.* **1962**, 290–300.
55. Kimbark, E.W. *Direct Current Transmission*, 1st ed.; Wiley Interscience: New York, NY, USA, 1971; pp. 1–496.
56. Kovarsky, D.; Pinto, L.J.; Caroli, C.E.; Santos, N. Soil surface potentials induced by Itaipu HVDC ground return current. I. Theoretical evaluation. *IEEE Trans. Power Deliv.* **1988**, *3*, 1204–1210. [CrossRef]
57. Girdinio, P.; Molfino, P.; Nervi, M.; Rossi, M.; Bertani, A.; Malgarotti, S. Technical and compatibility issues in the design of HVDC sea electrodes. In Proceedings of the International Symposium on Electromagnetic Compatibility—EMC EUROPE, Rome, Italy, 17–21 September 2012; IEEE Press: New York, NY, USA; pp. 1–5.
58. Marzinotto, M.; Mazzanti, G.; Nervi, M. Ground/sea return with electrode systems for HVDC transmission. *Int. J. Electr. Power Energy Syst.* **2018**, *100*, 222–230. [CrossRef]
59. Uhlmann, E. *Power Transmission by Direct Current*, 1st ed.; Springer-Verlag: New York, NY, USA, 1975; pp. 1–389.
60. Pirelli Cavi & Sistemi SpA. *Italy-Greece Interconnection Anode—Evaluation of the Corrosion Effects*, 1st ed.; Pirelli Cavi & Sistemi SpA: Zona Asi, Italy, 1999; p. 6.
61. Hao, J.; Teng, W.; Zhang, Y.; Liu, W. Research on distribution characteristics of DC potential near the UHVDC grounding electrode. *IEEE Access* **2020**, *8*, 122360–122365. [CrossRef]
62. Hajiaboli, A.; Fortin, S.; Dawalibi, F.P. Numerical techniques for the analysis of HVDC sea electrodes. *IEEE Trans. Ind. Appl.* **2015**, *51*, 5175–5181. [CrossRef]
63. Molfino, P.; Nervi, M.; Rossi, M.; Malgarotti, S.; Odasso, A. Concept design and development of a module for the construction of reversible HVDC submarine deep-water sea electrodes. *IEEE Trans. Power Deliv.* **2017**, *32*, 1682–1687. [CrossRef]
64. Ma, J.; Dawalibi, F.P. Analysis of grounding systems in soils with finite volumes of different resistivities. *IEEE Trans. Power Deliv.* **2002**, *17*, 596–602. [CrossRef]

65. Hajiaboli, A.; Fortin, S.; Dawalibi, F.P.; Zhao, P.; Ngoly, A. Analysis of grounding systems in the vicinity of hemi-spheroidal heterogeneities. *IEEE Trans. Ind. Appl.* **2015**, *51*, 5070–5077. [CrossRef]
66. Pompili, M.; Cauzillo, B.A.; Calcara, L.; Codino, A.; Sangiovanni, S. Steel reinforced concrete electrodes for HVDC submarine cables. *Electr. Power Syst. Res.* **2018**, *163*, 524–531. [CrossRef]
67. Brignone, M.; Karimi Qombovani, A.; Molfino, P.; Nervi, M. An algorithm for the semianalytical computation of fields emitted in layered ground by HVDC electrodes. In Proceedings of the 19th Edition of the Power Systems Computation Conference-PSCC 2016, Genova, Italy, 20–24 June 2016; IEEE Press: New York, NY, USA; pp. 1–5.
68. Freschi, F.; Mitolo, M.; Tartaglia, M. An effective semianalytical method for simulating grounding grids. *IEEE Trans. Ind. Appl.* **2013**, *49*, 256–263. [CrossRef]
69. Charalambous, C.A. Interference activity on pipeline systems from VSC-based HVDC cable networks with earth/sea return: An Insightful review. *IEEE Trans. Power Deliv.* **2021**, *36*, 1531–1541. [CrossRef]
70. Charalambous, C.A.; Dimitriou, A.; Gonos, I.F.; Papadopoulos, T.A. Modeling and assessment of short-term electromagnetic interference on a railway system from pole-to-ground faults on VSC-HVDC cable networks with sea electrodes. *IEEE Trans. Ind. Appl.* **2021**, *57*, 121–129. [CrossRef]
71. Bouzid, M.A.; Flazi, S.; Stambouli, A.B. A cost comparison of metallic and earth return path for HVDC transmission system case study: Connection Algeria-Europe. *Electr. Power Syst. Res.* **2019**, *171*, 15–25. [CrossRef]
72. Hatch-Statnett. *Newfoundland and Labrador Hydro—Lower Churchill Project—DC1110 Electrode Review—Gull Island & Soldiers Pond*, 1.1st ed.; Muskrat Falls Project—CE-09 Rev.1; Hatch: Mississauga, NL, Canada, 2008; p. 73.
73. Hatch. *Nalcor Energy—Lower Churchill Project—DC1250 Electrode Review Types and Locations*, 1st ed.; Muskrat Falls Project—CE-11; Hatch: Mississauga, NL, Canada, 2010; p. 277.
74. Hatch. *Nalcor Energy—Lower Churchill Project—DC1500 Electrode Review Confirmation of Types and Site Locations*, 1.1st ed.; Muskrat Falls Project—CE-12 Rev.1; Hatch: Mississauga, NL, Canada, 2010; p. 319.
75. Thunehed, H.; GeoVista, A.B. *Compilation and Evaluation Earth current measurements in the Forsmark area*, 1.1st ed.; AB, R-14–34; Svensk Karnbranslehantering: Stockholm, Sweden, 2017; p. 46.
76. Molfino, P.; Nervi, M.; Malgarotti, S. On the choice of the right HVDC Electrode type. In Proceedings of the 2019 AEIT HVDC International Conference, Florence, Italy, 9–10 May 2019; IEEE Press: New York, NY, USA; pp. 1–6.
77. Datasheet ANOTEC. *High Silicon Iron Tubular Anodes Centertec Z-Series*; bulletin 04-14/06.02.28; ANOTEC: Langley, BC, Canada, 2018.
78. Independent Power Transmission Operator, S.A. *Technical Description for Shoreline Electrodes for HVDC Link Attica-Creta*, 3rd ed.; Independent Power Transmission Operator S.A.: Athens, Greece, 2018; pp. 1–5.
79. Freire, P.E.; Fihlo, J.N.; Nicola, G.L.; Borin, P.O.; Perfeito, M.D.; Bartelotti, M.; Estrella, M.; Pereira, S.Y. Electrical interference of the Bipole I Ground Electrode from Rio Madeira HVDC Transmission System on the Bolivia-Brazil gas pipeline—Preliminary calculations and field measurements. In Proceedings of the 19th Edition of the Power Systems Computation Conference—PSCC 2016, Rio de Janeiro, Brazil, 22–24 September 2015; pp. 1–8.
80. CIGRE Working Group B1.10. *Update of Service Experience of HV Underground and Submarine Cable Systems*, 1st ed.; CIGRE: Paris, France, 2009; Volume 379, pp. 1–86.
81. Dorf, C. *The Electrical Engineering Handbook*, 2nd ed.; CRC Press LLC: Boca Raton, FL, USA, 2000; pp. 1–2976.

Article

A Novel Dynamic Event-Triggered Mechanism for Distributed Secondary Control in Islanded AC Microgrids

Boyang Huang [1], Yong Xiao [1], Xin Jin [1], Junhao Feng [1], Xin Li [2] and Li Ding [2,*]

[1] Electric Power Research Institute of China Southern Power Grid Co., Ltd., Guangzhou 510700, China
[2] Department of Artificial Intelligence and Automation, School of Electrical Engineering and Automation, Wuhan University, Wuhan 430072, China
* Correspondence: liding@whu.edu.cn

Abstract: In this paper, the frequency/voltage restoration and active power sharing problems of islanded AC microgrids are studied. A novel distributed dynamic event-triggered secondary control scheme is proposed to reduce the communication burden. The continuous monitoring of event-triggered conditions and Zeno behavior can be fundamentally avoided by periodically evaluating event-triggered conditions. In addition, by introducing an adaptive coefficient related to the system deviations, the control performance can be improved. Sufficient conditions to ensure the stability of the system are provided through a Lyapunov function. Lastly, the effectiveness of our proposed secondary control scheme is verified in a MATLAB/SimPowerSystems environment.

Keywords: distributed secondary control; dynamic event-triggered; islanded AC microgrid

1. Introduction

Microgrids (MGs) are small-scale power systems consisting of loads, energy storage systems, distributed generations (DGs), and controllers that help in fully utilizing renewable energy [1]. Renewable energy generators represented by photovoltaic (PV) generators have great development potential, and MGs based on PV generators will play an important role [2,3]. Since the output of PV generators is the DC voltage, three phase inverters are needed to connect PV generators to MGs. Therefore, the control of AC MGs based on voltage source inverters is particularly important.

A hierarchical control framework [4] including primary, secondary, and tertiary controls is widely adopted in the control of MGs. In islanded mode, the primary control ensures power sharing through decentralized droop control. The secondary control is introduced to compensate for frequency and voltage deviations caused by the primary control. Distributed secondary control is regarded to be a superior alternative to the traditional centralized secondary control strategy because it does not require a central controller, and only neighboring information is needed to compute the control inputs [5]. With the technique of feedback linearization, the secondary control problem can be transformed into a one- or two-order consensus problem [6,7]. In this distributed control framework, important issues such as communication delay [8] in MG control are studied. In addition, the hierarchical distributed control strategy considering optimal economic dispatch problem has been a research hotspot in recent years [9–11].

With the increased number of DGs, such distributed control schemes face the limitation of communication bandwidth. Recently, dynamic event-triggered (ET) control has been proposed to alleviate the communication burden [12,13]. Compared with conventional ET control [14], an internal dynamic variable is added in the ET conditions (ETCs) in dynamic ET control, which increases the difficulty of meeting ETCs and thus further reduces ET times. However, most current papers applying dynamic ET to MG control design ET function in a continuous form [15–17], which is energy-consuming in practical implementation. Similar to periodic ET (PET) control [18], continuous monitoring of ETCs and Zeno behavior can

Citation: Huang, B.; Xiao, Y.; Jin, X.; Feng, J.; Li, X.; Ding, L. A Novel Dynamic Event-Triggered Mechanism for Distributed Secondary Control in Islanded AC Microgrids. *Energies* **2022**, *15*, 6883. https://doi.org/10.3390/en15196883

Academic Editor: Juri Belikov

Received: 29 August 2022
Accepted: 16 September 2022
Published: 20 September 2022

Publisher's Note: MDPI stays neutral with regard to jurisdictional claims in published maps and institutional affiliations.

be fundamentally avoided through periodically checking ETCs. That is, dynamic periodic ET (DPET) control [19]. There are few works applying the DPET technique in the field of MGs. The authors in [20] proposed a DPET strategy to address the output consensus problem of DC MGs. The time-varying communication delay problem was solved with a DPET strategy proposed in [21].

On the other hand, the design of ETCs usually only ensures the stability of the system. For practical MG systems, when system frequency or voltage deviates from an acceptable range, the first consideration is the control performance, not the communication burden. The reduction in communication burden may sacrifice control performance. On the basis of the above considerations, in this paper, an adaptive coefficient relating to system deviations is introduced to balance these two aspects. To the best of our knowledge, this idea has not been studied in the secondary control of AC MGs. The major contributions of our work are listed below:

- The secondary control problem is addressed under the DPET control structure, which further reduces the communication burden, and fundamentally avoids Zeno behavior and the continuous monitoring of ETCs.
- By introducing adaptive coefficients related to system deviation, the proposed control scheme can take into account both the communication burden and the control performance.

The rest of this paper is organized as follows. Section 2 provides preliminary knowledge and the model of MG primary droop control. Section 3 elaborates the proposed novel distributed DPET secondary control scheme. Section 4 provides the simulation results and a discussion on related works. Lastly, this work is concluded in Section 5.

2. Preliminaries and Problem Formulation

2.1. Graph Theory

The communication topology of a networked MG system with N DGs can be represented by a graph $\mathcal{G} = (\mathcal{V}, \mathcal{E})$, where $\mathcal{V} = \{v_1, \ldots, v_N\}$ is the set of vertexes and $\mathcal{E} \subseteq \mathcal{V} \times \mathcal{V}$ is the set of edges. Adjacency matrix $\mathcal{A} = (a_{ij}) \in \mathbb{R}^{N \times N}$ represents the connection relationship between vertexes. If $(v_i, v_j) \in \mathcal{E}$, $a_{ij} = 1$. Otherwise, $a_{ij} = 0$. The set of neighbors of vertex i is denoted by $N_i = \{v_j | (v_i, v_j) \in \mathcal{E}, i \neq j\}$. The degree matrix is defined by $\mathcal{D} = diag\{d_i\} \in \mathbb{R}^{N \times N}$, where $d_i = \sum_{j=1}^{j=N} a_{ij}$. The Laplacian matrix of $\mathcal{G}$ is defined as $\mathcal{L} = \mathcal{D} - \mathcal{A}$. The maximal eigenvalue of matrix P is denoted by $\lambda_n(P)$.

2.2. AC MG System and Primary Droop Control

In general, the power controller, nested voltage and current controllers, and the pulse width modulator comprise the primary control layer. More details about the primary control structure can be found in [7]. The decentralized droop control is widely applied in the primary control of MGs. The droop characteristic of the i-th DG is given as follows.

$$\omega_i = \omega_{ni} - m_{pi}P_i, \tag{1}$$

$$v_{odi} = V_{ni} - n_{qi}Q_i, v_{oqi} = 0, \tag{2}$$

where ω_{ni} and V_{ni} are the primary frequency and voltage reference signals, respectively; m_{pi} and n_{qi} are the droop coefficients associated with DG i's capacity; and P_i and Q_i are the measured output active and reactive power, respectively.

2.3. Problem Formulation

The secondary control adjusts primary control reference signals to compensate for frequency or voltage deviations. Differentiating (1) and (2) yields

$$\dot{\omega}_i = \dot{\omega}_{ni} - m_{pi}\dot{P}_i, \tag{3}$$

$$\dot{v}_{odi} = \dot{V}_{ni} - n_{qi}\dot{Q}_i. \tag{4}$$

Define $\dot{\omega}_i = u_i^\omega$, $m_{pi}\dot{P}_i = u_i^P$, $\dot{v}_{odi} = u_i^V$ and $n_{qi}\dot{Q}_i = u_i^Q$. The reference signals sent to the primary control can be given by $\omega_{ni} = \int (u_i^\omega + u_i^P)dt$ and $V_{ni} = \int (u_i^V + u_i^Q)dt$, where the four controllers u_i^ω, u_i^P, u_i^V, and u_i^Q need to be designed in the secondary control scheme.

3. Distributed DPET Secondary Control Design

In this part, a novel distributed DPET secondary control scheme is proposed, and the stability of the control scheme is analyzed.

3.1. Distributed DPET Frequency and Active Power Controllers

The operating frequency of each DG should be restored to the nominal value ω_{ref}. Therefore, the frequency control objective for the i-th DG can be expressed as

$$\lim_{t \to \infty} \omega_i(t) - \omega_{ref} = 0, \; \forall i \in \mathcal{V}. \tag{5}$$

The distributed frequency controller for the i-th DG was designed as follows.

$$u_i^\omega = k_\omega c_{\omega i}(t), k_\omega > 0, \tag{6}$$

$$c_{\omega i}(t) = \sum_{j \in N_i} \left(\tilde{\omega}_j(t) - \tilde{\omega}_i(t) \right) + b_i \left(\omega_{ref} - \tilde{\omega}_i(t) \right), \tag{7}$$

where k_ω denotes the control gain, $c_{\omega i}(t)$ denotes the ET coordination error, b_i represents whether DG i can receive ω_{ref}, and $\tilde{\omega}_i$ denotes the frequency state at the ET instant. That is, $\tilde{\omega}_i(t) = \omega_i(t_{k_i}^\omega), t \in [t_{k_i}^\omega, t_{k_i+1}^\omega)$.

The ET instant $t_{k_i+1}^\omega$ is determined by the frequency ETC, whose evaluating period is defined as h_w. Thus, the ETC evaluating moment is denoted by $t = lh_w, l \in \mathbb{N}$. Obviously, $t_{k_i+1}^\omega$ is an integer multiple of h_ω determined by

$$t_{k_i+1}^\omega = \inf\{t = lh_w | t > t_{k_i}^\omega, e_{\omega i}^2(t) - \frac{\alpha_i}{4\sigma^2}c_{\omega i}^2(t) - \beta_{\omega i}(\omega)\chi_i^\omega(t) \geq 0\}, \tag{8}$$

where $\alpha_i \in (0,1)$, $\sigma = \max\{d_i + b_i/2\}$, and $e_{\omega i}(t)$ is the ET measurement error defined as

$$e_{\omega i}(t) = \tilde{\omega}_i(t) - \omega_i(t). \tag{9}$$

Adaptive coefficient $\beta_{\omega i}(\omega)$ need to be designed later, and the internal dynamic variable $\chi_i^\omega(t)$ is calculated with

$$\dot{\chi}_i^\omega(t) = -\gamma_i\chi_i^\omega(t), \gamma_i > 0, \chi_i^\omega(0) > 0. \tag{10}$$

Theorem 1. *Assume that the communication topology $\mathcal{G}$ of the MG system is connected, and at least one DG could receive leader information ω_{ref}. If evaluating period h_w satisfies*

$$h_w < \frac{1 - \bar{\alpha}}{2k_\omega \lambda_n(\mathcal{L} + \mathcal{B})}, \; \bar{\alpha} = \max\{\alpha_i\}, \mathcal{B} = diag\{b_i\}, \; \forall i \in \mathcal{V}, \tag{11}$$

and $\beta_{\omega i}(\omega)$ and γ_i satisfy

$$2\sigma^2 k_\omega \beta_{\omega i}(\omega) - \gamma_i \leq 0, \tag{12}$$

frequency control objective (5) can be achieved under Control Law (6)–(7) and the ETCs (8)–(12).

Proof. The frequency tracking error is denoted by

$$\zeta_{\omega i} = \omega_i - \omega_{ref}. \tag{13}$$

The compact form of (7), (9), and (13) can be derived as follows.

$$c_\omega(t) = -\mathcal{H}(\tilde{\omega}(t) - \mathbf{1}_N \omega_{ref}), \tag{14}$$

$$e_\omega(t) = \tilde{\omega}(t) - \omega(t), \tag{15}$$

$$\zeta_\omega(t) = \tilde{\omega}(t) - \mathbf{1}_N \omega_{ref}, \tag{16}$$

where $c_\omega = [c_{\omega 1}, \ldots, c_{\omega N}]^T \in \mathbb{R}^{N \times 1}$, $\omega = [\omega_1, \ldots, \omega_N]^T \in \mathbb{R}^{N \times 1}$, $\tilde{\omega} = [\tilde{\omega}_1, \ldots, \tilde{\omega}_N]^T \in \mathbb{R}^{N \times 1}$, $e_\omega = [e_{\omega 1}, \ldots, e_{\omega N}]^T \in \mathbb{R}^{N \times 1}$, $\mathbf{1}_N = [1, .., 1]^T \in \mathbb{R}^{N \times 1}$ and $\mathcal{H} = \mathcal{L} + \mathcal{B}$.

Consider a Lyapunov candidate function:

$$V = V_1 + V_2 = \frac{1}{2}\zeta_\omega^T(t)\mathcal{H}\zeta_\omega(t) + \sum_{i=1}^{n} \chi_i^\omega(t). \tag{17}$$

Using (14)–(16), the time derivative of V_1 can be derived as follows.

$$\dot{V}_1 = k_\omega c_\omega^T(t)c_\omega(t) - k_\omega c_\omega^T(t)\mathcal{H}e_\omega(t). \tag{18}$$

For any $t \in [lh_\omega, (l+1)h_\omega) \subseteq [t_{k_i}^\omega, t_{k_i+1}^\omega)$, we have $\tilde{\omega}_i(t) = \tilde{\omega}_i(lh_\omega)$. Thus, $\dot{V}_1$ can be derived as

$$\dot{V}_1 = k_\omega c_\omega^T(lh_\omega)c_\omega(lh_\omega) - k_\omega c_\omega^T(lh_\omega)\mathcal{H}e_\omega(t). \tag{19}$$

According to (15), we have $e_\omega(t) = e_\omega(lh_\omega) - (t - lh_\omega)\dot{\omega}(t)$. Applying inequality $t - lh_\omega \le h_\omega$, (19) can be bounded as

$$\dot{V}_1 \le -k_\omega c_\omega^T(lh_\omega)c_\omega(lh_\omega) - k_\omega c_\omega^T(lh_\omega)\mathcal{H}e_\omega(lh_\omega) + k_\omega^2 h_\omega c_\omega^T(lh)\mathcal{H}c_\omega(lh_\omega). \tag{20}$$

For simplicity, $c_\omega(lh_\omega)$ and $e_\omega(lh_\omega)$ are abbreviated to c_ω and e_ω in the following steps. Since $\mathcal{H}$ is symmetric and positive definite [22], (20) can be bounded as

$$\dot{V}_1 = -k_\omega c_\omega^T c_\omega - k_\omega c_\omega^T \mathcal{H}e_\omega + k_\omega^2 h_\omega c_\omega^T \mathcal{H}c_\omega$$
$$\le -k_\omega \sum_{i=1}^{n} c_{\omega i}^2 - k_\omega \sum_{i=1}^{n}(d_i + g_i)c_{\omega i}e_{\omega i} + k_\omega \sum_{i=1}^{n}\sum_{j \in N_i} c_{\omega i}e_{\omega j} + k_\omega^2 h_\omega \lambda_n(\mathcal{H})c_\omega^T c_\omega. \tag{21}$$

According to Young's Inequality, $x^2/(2a) + (ay^2)/2 \ge xy, \forall x, y \ge 0, a > 0$, it can be obtained that

$$-k_\omega \sum_{i=1}^{n}(d_i + g_i)c_{\omega i}e_{\omega i} \le \frac{1}{4\sigma}k_\omega \sum_{i=1}^{n}(d_i + g_i)c_{\omega i}^2 + \sigma k_\omega \sum_{i=1}^{n}(d_i + g_i)e_{\omega i}^2 \,,$$

$$k_\omega \sum_{i=1}^{n}\sum_{j \in N_i} c_{\omega i}e_{\omega j} \le \frac{1}{4\sigma}k_\omega \sum_{i=1}^{n} d_i c_{\omega i}^2 + \sigma k_\omega \sum_{i=1}^{n}\sum_{j \in N_i} e_{\omega j}^2 \,, \tag{22}$$

where $\sigma = \max\{d_i + g_i/2\}$. Since $\mathcal{H}$ is symmetric, we have

$$\sigma k_\omega \sum_{i=1}^{n}\sum_{j \in N_i} e_{\omega j}^2 = \sigma k_\omega \sum_{i=1}^{n}\sum_{j \in N_i} e_{\omega i}^2 = \sigma k_\omega \sum_{i=1}^{n} d_i e_{\omega i}^2. \tag{23}$$

Combining (11), (21)–(23), $\dot{V}_1$ can be bounded as

$$\dot{V}_1 = -k_\omega \sum_{i=1}^{n} c_{\omega i}^2 + \frac{1}{2\sigma}k_\omega \sum_{i=1}^{n}(d_i + g_i/2)c_{\omega i}^2 + 2\sigma k_\omega \sum_{i=1}^{n}(d_i + g_i/2)e_{\omega i}^2 + k_\omega^2 h_\omega \lambda_n(H)c_\omega^T c_\omega$$

$$\le -k_\omega \sum_{i=1}^{n} c_{\omega i}^2 + \frac{1}{2}k_\omega \sum_{i=1}^{n} c_{\omega i}^2 + 2\sigma^2 k_\omega \sum_{i=1}^{n} e_{\omega i}^2 + k_\omega \frac{1 - \bar{\alpha}}{2} \sum_{i=1}^{n} c_{\omega i}^2 \tag{24}$$

$$= -k_\omega \frac{\bar{\alpha}}{2} \sum_{i=1}^{n} c_{\omega i}^2 + 2\sigma^2 k_\omega \sum_{i=1}^{n} e_{\omega i}^2.$$

Using the ET inequality $e_{\omega i}^2(lh_\omega) \leq \alpha_i c_{\omega i}^2(lh_\omega)/4\sigma^2 + \beta_{\omega i}(\omega)\chi_i^\omega(t)$, and combining (10), (12) and (24), $\dot{V}$ can be bounded as

$$
\begin{aligned}
\dot{V} &= -k_\omega \frac{\bar{\alpha}}{2} \sum_{i=1}^n c_{\omega i}^2 + 2\sigma^2 k_\omega \sum_{i=1}^n e_{\omega i}^2 - \sum_{i=1}^n \gamma_i \chi_i^\omega(t) \\
&\leq -k_\omega \frac{\bar{\alpha}}{2} \sum_{i=1}^n c_{\omega i}^2 + 2\sigma^2 k_\omega \sum_{i=1}^n \left(\frac{\alpha_i}{4\sigma^2} c_{\omega i}^2 + \beta_{\omega i}(\omega)\chi_i^\omega(t) \right) - \sum_{i=1}^n \gamma_i \chi_i^\omega(t) \\
&= k_\omega \frac{\alpha_i - \bar{\alpha}}{2} \sum_{i=1}^n c_{\omega i}^2 + \sum_{i=1}^n \left(2\sigma^2 k_\omega \beta_{\omega i}(\omega) - \gamma_i \right) \chi_i^\omega(t) \\
&\leq 0.
\end{aligned}
\tag{25}
$$

Therefore, the system frequency can reach consensus and restore to ω_{ref} under Control Law (6) and (7) and ETCs (8)–(12). The proof is completed. $\square$

Next, the value of the adaptive coefficient $\beta_{\omega i}(\omega)$ is discussed. Usually, coefficient $\beta_{\omega i}$ is fixed in ETCs. In practical implementation of the secondary control scheme, when inverter operating frequency deviates from an acceptable range, the controller should restore system frequency to the acceptable range as soon as possible instead of first prioritizing the reduction in communication burden. Obviously, an adaptive coefficient related to frequency deviations is more effective in balancing control performance and communication burden. Thus, adaptive coefficient $\beta_{\omega i}(\omega)$ is determined by

$$
\beta_{\omega i}(\omega) = \begin{cases} \frac{1}{R_\omega |\omega_i - \omega_{ref}|}, & \text{if } |\omega_i - \omega_{ref}| > \Delta_\omega \text{ and } g_i = 1, \\ \beta_{\omega 0}, & \text{otherwise}, \end{cases}
$$

where Δ_ω is an acceptable threshold, and $\beta_{\omega 0}$ and R_ω are constants to guarantee that $\beta_{\omega i}(\omega)$ always satisfies Condition (12). A larger frequency deviation leads to a smaller value of $\beta_{\omega i}(\omega)$ when the operating frequency deviates from the safety range, which renders ETCs easier to be met, and data exchange and control input updates are then more frequent. When the frequency deviation is within the acceptable range, $\beta_{\omega i}(\omega)$ is fixed to further reduce the communication burden.

The control objective of active power can be expressed as follows.

$$
\lim_{t \to \infty} (m_{pi}P_i(t) - m_{pj}P_j(t))0, \ \forall i \neq j, i, j \in \mathcal{V}.
\tag{26}
$$

The active power sharing control law of the i-th DG was designed as follows.

$$
u_i^P = k_P c_{Pi}(t), k_P > 0,
\tag{27}
$$
$$
c_{Pi}(t) = -\sum_{j \in N_i} (\tilde{P}_{m_i}(t) - \tilde{P}_{m_j}(t)),
\tag{28}
$$

where $P_{m_i}(t) \overset{\Delta}{=} m_{pi}P_i(t)$, k_P is the control gain, and $c_{Pi}(t)$ represents the ET coordination error of active power ratios. $\tilde{P}_{m_i}(t)$ is updated by $\tilde{P}_{m_i}(t) = P_{m_i}(t_{k_i}^P), t \in [t_{k_i}^P, t_{k_i+1}^P)$. $t_{k_i+1}^P$ represents the ET instant of the active power controller, which is determined by

$$
t_{k_i+1}^P = \inf\{t = lh_p | t > t_{k_i}^P, e_{Pi}^2(t) - \frac{\alpha_i}{4\sigma^2} c_{Pi}^2(t) - \beta_{Pi}(P)\chi_i^P(t) \geq 0\},
\tag{29}
$$

where h_p denotes the active power evaluating period, $\alpha_i \in (0,1)$, and $\sigma = \max\{d_i\}$. The ET measurement error is calculated with $e_{Pi}(t) = \tilde{P}_{m_i}(t) - P_{m_i}(t)$, and internal dynamic variable $\chi_i^P(t)$ is calculated with $\dot{\chi}_i^P(t) = -\gamma_i \chi_i^P(t), \gamma_i > 0, \chi_i^P(0) > 0$.

Theorem 2. *Assume that communication topology $\mathcal{G}$ is connected. Control Law (27)–(28) and ETC (29) can achieve control objective (26) if h_p satisfies*

$$h_p < \frac{1 - \bar{\alpha}}{2k_P \lambda_n(\mathcal{L})}, \quad \bar{\alpha} = \max\{\alpha_i\}, \ \forall i \in \mathcal{V},$$

and $\beta_{Pi}(P)$ and γ_i satisfy $2\sigma^2 k_P \beta_{Pi}(P) - \gamma_i \leq 0$.

Proof. The detailed proof is similar to that of Theorem 1 and is omitted here. $\square$

Considering that the frequency and active power controller jointly generate primary frequency signal ω_{ni}, the value of $\beta_{Pi}(P)$ was set to

$$\beta_{Pi}(P) = \begin{cases} \beta_{\omega i}(\omega), & \text{if } g_i = 1, \\ \beta_{P0}, & \text{otherwise.} \end{cases}$$

The constraints of active power can also be designed according to actual requirements.

3.2. Distributed DPET Voltage and Reactive Power Controllers

Since accurate voltage restoration and reactive power sharing are two contradictory control objectives, this paper mainly considers the voltage restoration objective, that is, $\lim_{t \to \infty} v_{odi}(t) - v_{ref} = 0$, $\forall i \in \mathcal{V}$. Reactive power controller u_i^Q was adopted from low-pass filter [7] and is omitted here.

Similar to the secondary frequency controller, the voltage control law of the i-th DG was designed as follows,

$$u_i^V = k_v c_{vi}(t), k_v > 0, \tag{30}$$
$$c_{vi}(t) = -\sum_{j \in N_i} (\tilde{v}_{odi}(t) - \tilde{v}_{odj}(t)) + b_i(v_{ref} - \tilde{v}_{odi}(t)), \tag{31}$$

where k_v is the control gain, b_i represents whether DG i could receive v_{ref}, $c_{vi}(t)$ represents the ET coordination error of the voltage, and $\tilde{v}_{odi}(t)$ is updated by $\tilde{v}_{odi}(t) = v_{odi}(t_{k_i}^v)$, $t \in [t_{k_i}^v, t_{k_i+1}^v)$. $t_{k_i+1}^v$ represents the ET instant of the voltage controller, which is generated by dynamic ETCs, similar to the frequency controller. For simplicity, the design details are not provided here, but the simulation results of voltage restoration are provided in Section 4.

In addition, the output voltage of inverters should fall in an acceptable range as much as possible. To balance control performance and communication burden, adaptive coefficient $\beta_{Vi}(v)$ is defined as follows.

$$\beta_{Vi}(v) = \begin{cases} \frac{1}{R_v |v_i - v_{ref}|}, & \text{if } |v_i - v_{ref}| > \Delta_v \text{ and } g_i = 1, \\ \beta_{V0}, & \text{otherwise,} \end{cases}$$

where Δ_v is threshold, and β_{V0} and R_v are constants to guarantee that $\beta_{Vi}(v)$ satisfies the above constraints all the time.

If DGs receiving reference signals crash, frequency and voltage controllers (6)–(7) and (30)–(31) may encounter failures, since the reference signals are not available for the system. Some leaderless control structures were proposed to cope with this problem [23–26] where the reference signals are treated as global quantity. Thus, these control structure can overcome leader outage or failure. There are many works addressing communication failures or node hijacking issues under the leader–follower structure, and this topic deserves further study.

4. Simulation Results

The effectiveness of the proposed distributed DPET secondary control scheme was verified in MATLAB/SimPowerSystems. A 60 Hz/380 V 4-DG test MG system was built. The diagram of communication and electrical topology of the MG is shown in Figure 1. The electric and control parameters are provided in Table 1.

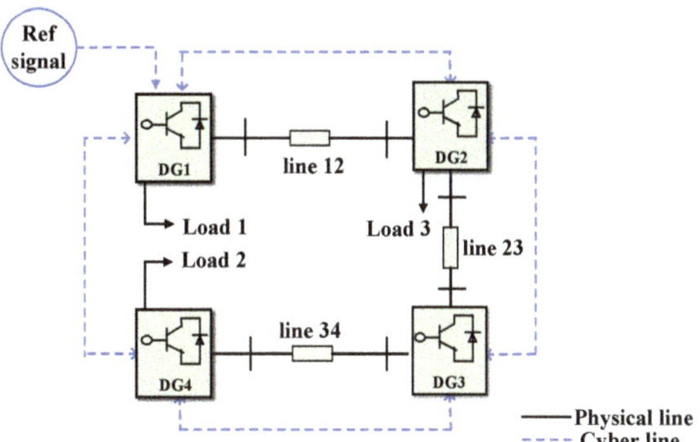

Figure 1. Diagram of the communication and electrical topology of the MG.

Table 1. Electric and control parameters of the MG system.

DGs	Mp	Np	Rf (Ω)	Lf (mH)	Cf (μF)
DG1 and 2	9.4×10^{-5}	1.3×10^{-3}	0.1	1.35	50
DG3 and 4	12.5×10^{-5}	1.5×10^{-3}	0.1	1.35	50
Lines	**Lines 12 and 34**	**line 23**	**Loads**	**Loads 1 and 3**	**Load 2**
Rt (Ω)	0.23	0.35	P (kW)	45.9	36
Lt (mH)	0.318	1.847	Q (kVar)	22.8	36
Controller	k	h	β_0	Δ/R_0	γ
Frequency	5	0.02	0.016	0.05/1250	1
Voltage	5	0.02	0.016	5/12.5	1
Active power	1	0.06	0.125	-	1

4.1. Case Studies

The MG system operates in islanded mode at $t = 0$ s. Initially, only the primary control is activated. The timeline of the study cases is described below.

1. Case 1: At $t = 1$ s, the secondary control is activated.
2. Case 2: At $t = 10$ s, load 3 is connected to the MG.
3. Case 3: At $t = 20$ s, load 3 is disconnected. The communication link between DG 1 and DG 4 fails at $t = 21$ s and restores at $t = 34$ s.
4. Case 4: At $t = 35$ s, DG 3 is disconnected from the MG..
5. Case 5: At $t = 40$ s, DG 3 is connected to the MG.

The simulation results of frequency, active power ratio, and voltage under the proposed distributed DPET secondary control scheme are shown in Figure 2. During $0 < t < 1$ s, in the absence of secondary control, the system frequency and voltage deviated from ω_{ref} and v_{ref}. When the proposed secondary control activated at $t = 1$ s, the system frequency and voltage gradually recovered to 60 Hz and 380 V, and the active power ratios reached consensus. Moreover, in Cases 2–5, when load changes, communication-link failures, and plug-and-play events occurred, the system frequency and voltage could still recover to their nominal values while achieving the active power sharing objective. The simulation results of power mismatch between the total supplied and the rated active/reactive power are shown in Figure 3. The power mismatch was under an acceptable range. The controller ET interval of DG 1 is shown in Figure 4, which is usually larger than the evaluating period. Overall, the DPET secondary control scheme proposed in this paper could achieve the secondary control objectives while reducing the communication burden, and the design of dynamic ETCs was reasonable.

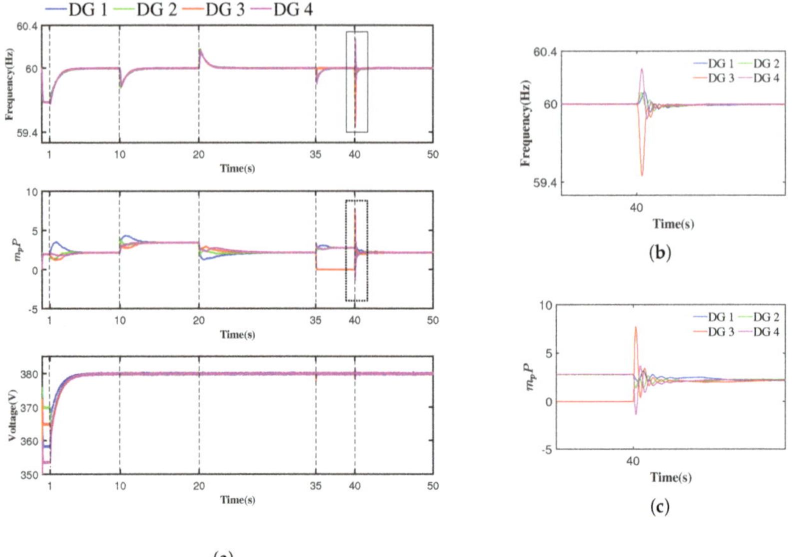

Figure 2. Simulation results of distributed DPET secondary control scheme (the time period of different cases is separated with dashed lines). (**a**) Frequency, active power ratio, and voltage response curves. (**b**) Enlarged view of the solid rectangular area on the left. (**c**) Enlarged view of the dotted rectangular area on the left.

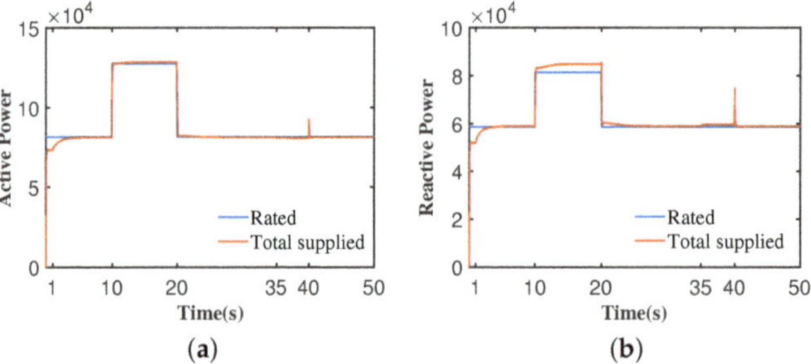

Figure 3. Power mismatch of active/reactive power. (**a**) Active power. (**b**) Reactive power.

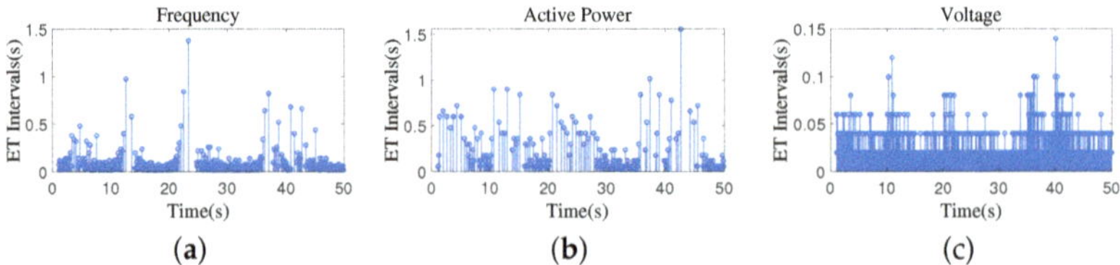

Figure 4. ET interval of the frequency, active power, and voltage controllers of DG 1. (**a**) Frequency. (**b**) Active power. (**c**) Voltage.

4.2. Comparisons

The proposed distributed DPET secondary control scheme is compared with a traditional PET secondary control scheme from the perspective of communication burden. The controller ET times of the above two schemes are shown in Figure 5. Our proposed scheme had fewer ET times than the PET scheme did. The average ET times of frequency, voltage, and active power controllers of the DPET scheme were 768, 1905 and 204, respectively, which were smaller than the 1372, 2158 and 413 of the PET scheme, respectively.

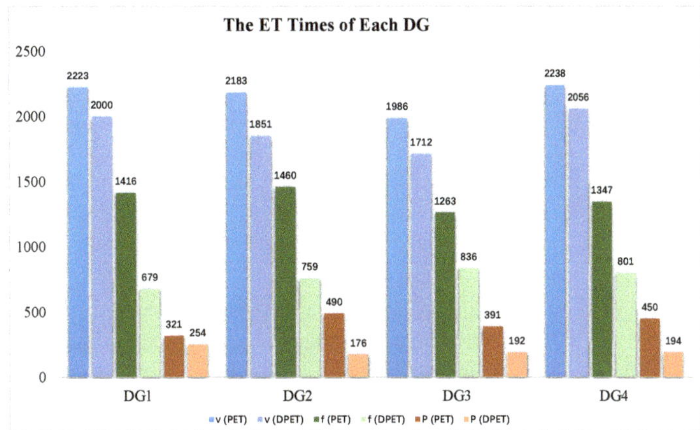

Figure 5. ET times of frequency, voltage, and active power controllers of each DG under the PET and DPET schemes (for each DG, the blue, green, and orange columns represent the ET times of voltage, frequency, and active power controllers under PET and DPET schemes, respectively).

The advantages of adaptive coefficient β_i are discussed. When a DG is plugged in the system, the system state faces large oscillations that could even lead to system instability. Thus, the plug-in event is considered to compare the control performance of the adaptive β_i and fixed β_i strategies. The frequency controller ET instants and the frequency response curves of the above two methods are shown in Figures 6 and 7, respectively.

The total ET times of frequency controllers were reduced from 52 for the fixed $\beta_{\omega 0}$ to 40 for the adaptive $\beta_{\omega i}$. As shown in Figure 7, under a fixed $\beta_{\omega 0}$, the system frequency oscillated from 58.61 to 60.69 Hz, which was larger than 58.66 to 60.65 Hz under the adaptive $\beta_{\omega i}$. Since the oscillation of the system itself was relatively small, and only DG 1 could adaptively change its $\beta_{\omega i}$, the advantage of adaptive $\beta_{\omega i}$ in narrowing the oscillation range is not very obvious, but it did narrow the oscillation amplitude of the system to some extent. As shown in Figure 6, the triggering events of the adaptive $\beta_{\omega i}$ were more frequent than those of the fixed $\beta_{\omega 0}$, since the adaptive $\beta_{\omega i}$ was smaller than $\beta_{\omega 0}$ at the beginning. Due to the rapid improvement of the system oscillations, the subsequent ET times were reduced, resulting in a decrease in the total ET times and an improvement in control performance.

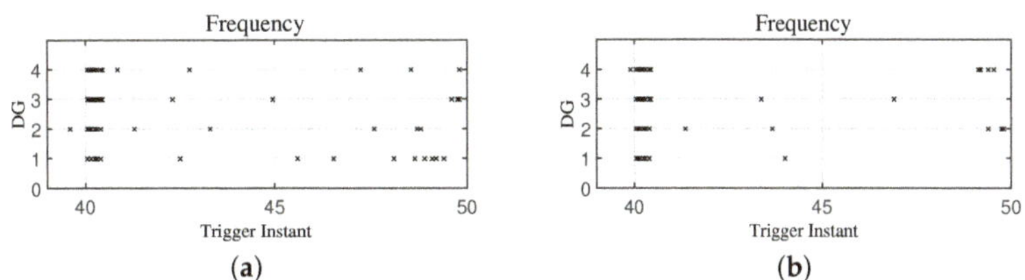

Figure 6. Frequency controller ET instants of each DG. ('x' represents the ET instant). (**a**) Fixed $\beta_{\omega 0}$. (**b**) Dynamic $\beta_{\omega i}$.

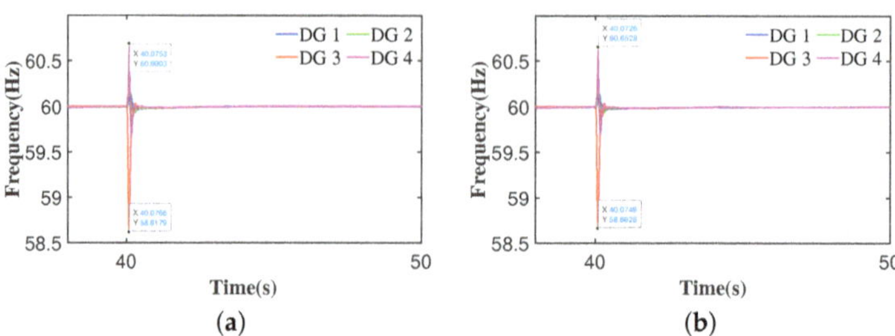

Figure 7. Frequency response curves under plug-and-play event. (**a**) Fixed $\beta_{\omega 0}$. (**b**) Dynamic $\beta_{\omega i}$ (comparison experiments were performed with $k_\omega = k_v = 2$, $k_p = 1$, $h_\omega = h_v = 0.05$, $h_p = 0.06$, $\gamma = 0.5$, $\beta_{\omega 0} = \beta_{v0} = 0.02$, $\beta_{P0} = 0.06$, $R_{\omega 0} = 1000$ and $R_{v0} = 10$).

4.3. Performance under Communication Delay

In this section, the effects of communication delay on our proposed frequency, voltage, and active power controllers are discussed. The communication delay is defined as τ. Take case 1 in Section 4.1 as an example. Simulation conditions were $\tau = h_\omega/10 = h_v/10 = 1$ ms, $\tau = h_\omega = h_v = 20$ ms and $\tau = 2h_\omega = 2h_v = 40$ ms. As shown in Figure 8, with the increase in τ, the response curves of frequency, voltage, and active power ratios had more oscillations. When τ was much larger than the evaluating period, for example, $\tau = 40$ ms, the system failed to achieve consensus and became unstable.

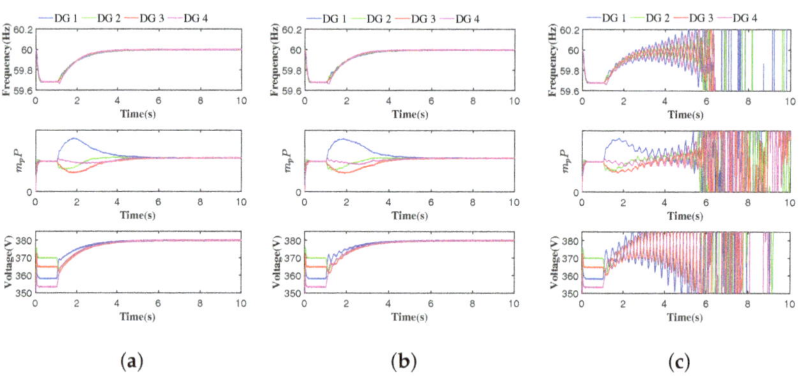

Figure 8. The effect of communication delay on frequency, active power, and voltage controllers. (**a**) $\tau = 1$ ms. (**b**) $\tau = 20$ ms. (**c**) $\tau = 40$ ms.

5. Conclusions

In this paper, a novel distributed DPET secondary control scheme for islanded AC MGs was proposed to achieve the control objectives of frequency or voltage restoration and active power sharing. In our case studies, the ET times of frequency, active power, and voltage controllers in the proposed DPET scheme were reduced by 604, 253 and 209 times, respectively, compared to the traditional PET scheme. Compared with the commonly used fixed coefficient, frequency controllers with an adaptive coefficient reduced the ET times by 12 times. The frequency oscillations in the plug-and-play events were also reduced by about 0.05 Hz. In general, the proposed DPET secondary control scheme had advantages in reducing the communication burden, and the introduction of adaptive coefficients helps in further reducing the communication burden and improving the control performance. The effectiveness of the proposed DPET secondary control scheme was verified with the simulation results of the MG system under the events of island operation, load changes, link failures, and plug and play.

Author Contributions: B.H.: methodology, writing—original draft, funding acquisition. Y.X.: conceptualization, writing—original draft, methodology. X.J.: validation, supervision. J.F.: validation, supervision. X.L.: investigation. L.D.: investigation. All authors have read and agreed to the published version of the manuscript.

Funding: this research was funded by the National Natural Science Foundation of China under grant 62173256.

Institutional Review Board Statement: Not applicable.

Informed Consent Statement: Not applicable.

Data Availability Statement: Not applicable.

Conflicts of Interest: the authors declare no conflict of interest.

References

1. Hatziargyriou, N.; Asano, H.; Iravani, R.; Marnay, C. Microgrids. *IEEE Power Energy Mag.* **2007**, *5*, 78–94. [CrossRef]
2. Razmi, D.; Lu, T. A Literature Review of the Control Challenges of Distributed Energy Resources Based on Microgrids (MGs): Past, Present and Future. *Energies* **2022**, *15*, 4676. [CrossRef]
3. Kim, B.Y.; Oh, K.K.; Moore, K.L.; Ahn, H.S. Distributed coordination and control of multiple photovoltaic generators for power distribution in a microgrid. *Automatica* **2016**, *73*, 193–199. [CrossRef]
4. Bidram, A.; Davoudi, A. Hierarchical Structure of Microgrids Control System. *IEEE Trans. Smart Grid* **2012**, *3*, 1963–1976. [CrossRef]
5. Olfati-Saber, R.; Fax, J.A.; Murray, R.M. Consensus and Cooperation in Networked Multi-Agent Systems. *Proc. IEEE* **2007**, *95*, 215–233. [CrossRef]
6. Bidram, A.; Davoudi, A.; Lewis, F.L.; Guerrero, J.M. Distributed Cooperative Secondary Control of Microgrids Using Feedback Linearization. *IEEE Trans. Power Syst.* **2013**, *28*, 3462–3470. [CrossRef]
7. Bidram, A.; Davoudi, A.; Lewis, F.L.; Qu, Z. Secondary control of microgrids based on distributed cooperative control of multi-agent systems. *IET Gener. Transm. Distrib.* **2013**, *7*, 822–831. [CrossRef]
8. Ullah, S.; Khan, L.; Sami, I.; Ullah, N. Consensus-Based Delay-Tolerant Distributed Secondary Control Strategy for Droop Controlled AC Microgrids. *IEEE Access* **2021**, *9*, 6033–6049. [CrossRef]
9. Lu, X.; Xia, S.; Sun, G.; Hu, J.; Zou, W.; Zhou, Q.; Shahidehpour, M.; Chan, K.W. Hierarchical distributed control approach for multiple on-site DERs coordinated operation in microgrid. *Int. J. Electr. Power Energy Syst.* **2021**, *129*, 106864. [CrossRef]
10. Pullaguram, D.; Rana, R.; Mishra, S.; Senroy, N. Fully distributed hierarchical control strategy for multi-inverter-based AC microgrids. *IET Renew. Power Gener.* **2020**, *14*, 2468–2476. [CrossRef]
11. Nguyen, T.L.; Wang, Y.; Tran, Q.T.; Caire, R.; Xu, Y.; Gavriluta, C. A Distributed Hierarchical Control Framework in Islanded Microgrids and Its Agent-Based Design for Cyber–Physical Implementations. *IEEE Trans. Ind. Electron.* **2021**, *68*, 9685–9695. [CrossRef]
12. Girard, A. Dynamic Triggering Mechanisms for Event-Triggered Control. *IEEE Trans. Autom. Control* **2015**, *60*, 1992–1997. [CrossRef]
13. Ge, X.; Han, Q.L.; Ding, L.; Wang, Y.L.; Zhang, X.M. Dynamic Event-Triggered Distributed Coordination Control and its Applications: A Survey of Trends and Techniques. *IEEE Trans. Syst. Man Cybern. Syst.* **2020**, *50*, 3112–3125. [CrossRef]
14. Dimarogonas, D.V.; Frazzoli, E.; Johansson, K.H. Distributed Event-Triggered Control for Multi-Agent Systems. *IEEE Trans. Autom. Control* **2012**, *57*, 1291–1297. [CrossRef]
15. Yang, C.; Yao, W.; Fang, J.; Ai, X.; Chen, Z.; Wen, J.; He, H. Dynamic event-triggered robust secondary frequency control for islanded AC microgrid. *Appl. Energy* **2019**, *242*, 821–836. [CrossRef]
16. Qian, Y.Y.; Premakumar, A.V.P.; Wan, Y.; Lin, Z.; Shamash, Y.A.; Davoudi, A. Dynamic Event-Triggered Distributed Secondary Control of DC Microgrids. *IEEE Trans. Power Electron.* **2022**, *37*, 10226–10238. [CrossRef]
17. Han, F.; Lao, X.; Li, J.; Wang, M.; Dong, H. Dynamic event-triggered protocol-based distributed secondary control for islanded microgrids. *Int. J. Electr. Power Energy Syst.* **2022**, *137*, 107723. [CrossRef]
18. Heemels, W.P.M.H.; Donkers, M.C.F.; Teel, A.R. Periodic Event-Triggered Control for Linear Systems. *IEEE Trans. Autom. Control* **2013**, *58*, 847–861. [CrossRef]
19. Deng, C.; Che, W.W.; Wu, Z.G. A Dynamic Periodic Event-Triggered Approach to Consensus of Heterogeneous Linear Multiagent Systems with Time-Varying Communication Delays. *IEEE Trans. Cybern.* **2021**, *51*, 1812–1821. [CrossRef]
20. Geng, Y.; Ji, J.; Hu, B. The Output Consensus Problem of DC Microgrids With Dynamic Event-Triggered Control Scheme. *Front. Energy Res.* **2021**, *9*, 446. [CrossRef]
21. Lian, Z.; Deng, C.; Wen, C.; Guo, F.; Lin, P.; Jiang, W. Distributed Event-Triggered Control for Frequency Restoration and Active Power Allocation in Microgrids With Varying Communication Time Delays. *IEEE Trans. Ind. Electron.* **2021**, *68*, 8367–8378. [CrossRef]

22. Barooah, P.; Hespanha, J.P. Graph Effective Resistance and Distributed Control: Spectral Properties and Applications. In Proceedings of the Proceedings of the 45th IEEE Conference on Decision and Control, San Diego, CA, USA, 13–15 December 2006; pp. 3479–3485. [CrossRef]
23. Li, Z.; Cheng, Z.; Si, J.; Li, S. Distributed Event-Triggered Hierarchical Control to Improve Economic Operation of Hybrid AC/DC Microgrids. *IEEE Trans. Power Syst.* **2022**, *37*, 3653–3668. [CrossRef]
24. Ullah, S.; Khan, L.; Sami, I.; Ro, J.S. Voltage/Frequency Regulation With Optimal Load Dispatch in Microgrids Using SMC Based Distributed Cooperative Control. *IEEE Access* **2022**, *10*, 64873–64889. [CrossRef]
25. Li, Z.; Cheng, Z.; Liang, J.; Si, J.; Dong, L.; Li, S. Distributed Event-Triggered Secondary Control for Economic Dispatch and Frequency Restoration Control of Droop-Controlled AC Microgrids. *IEEE Trans. Sustain. Energy* **2020**, *11*, 1938–1950. [CrossRef]
26. Ullah, S.; Khan, L.; Sami, I.; Hafeez, G.; Albogamy, F.R. A Distributed Hierarchical Control Framework for Economic Dispatch and Frequency Regulation of Autonomous AC Microgrids. *Energies* **2021**, *14*, 8408. [CrossRef]

Article

Impact of Spotted Hyena Optimized Cascade Controller in Load Frequency Control of Wave-Solar-Double Compensated Capacitive Energy Storage Based Interconnected Power System

Arindita Saha [1], Puja Dash [2], Naladi Ram Babu [3], Tirumalasetty Chiranjeevi [4,*], Bathina Venkateswararao [5] and Łukasz Knypiński [6,*]

[1] Department of Electrical Engineering, Regent Education & Research Foundation Group of Institutions, Kolkata 700121, West Bengal, India
[2] Department of Electrical and Electronics Engineering, Gayatri Vidya Parishad College of Engineering (Autonomous), Visakhapatnam 530048, Andhra Pradesh, India
[3] Department of Electrical & Electronics Engineering, Aditya Engineering College, Surampalem 533437, Andhra Pradesh, India
[4] Department of Electrical Engineering, Rajkiya Engineering College Sonbhadra, Sonbhadra 231206, Uttar Pradesh, India
[5] Department of EEE, V R Siddhartha Engineering College, Vijayawada 520007, Andhra Pradesh, India
[6] Institute of Electrical Engineering and Electronics, Faculty of Automatic Control, Robotic and Electrical Engineering, Poznan University of Technology, 60-965 Poznan, Poland
* Correspondence: chirupci479@gmail.com (T.C.); lukasz.knypinski@put.poznan.pl (Ł.K.)

Citation: Saha, A.; Dash, P.; Babu, N.R.; Chiranjeevi, T.; Venkateswararao, B.; Knypiński, Ł. Impact of Spotted Hyena Optimized Cascade Controller in Load Frequency Control of Wave-Solar -Double Compensated Capacitive Energy Storage Based Interconnected Power System. *Energies* **2022**, *15*, 6959. https://doi.org/10.3390/en15196959

Academic Editor: Juri Belikov

Received: 22 August 2022
Accepted: 15 September 2022
Published: 22 September 2022

Publisher's Note: MDPI stays neutral with regard to jurisdictional claims in published maps and institutional affiliations.

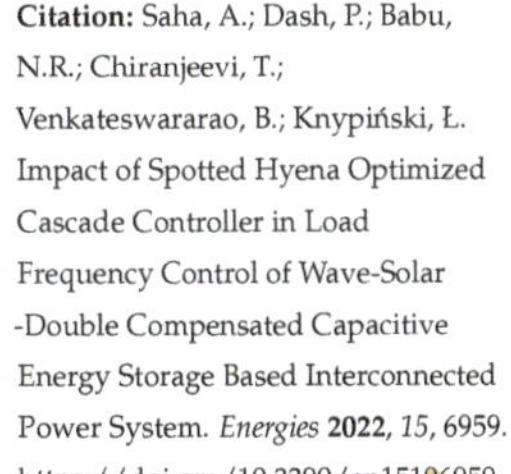

Abstract: The concept of automatic generation control has an immense role in providing quality power in an interconnected system. To obtain quality power by controlling the oscillations of frequency and tie-line power, a proper controller design is necessary. So, an innovative endeavor has been undertaken to enforce a two-stage controller with the amalgamation of a proportional-derivative with filter (PDN) (integer-order) and a fractional order integral-derivative (FOID), i.e., PDN(FOID). In an effort to acquire the controller's gains and parameters, a bio-inspired meta-heuristic spotted hyena optimizer is applied. Various examinations manifest the excellence of PDN(FOID) over other controllers such as integral, proportional–integral, proportional–integral-derivative filter, and fractional order PID from perspectives concerning the diminished amount of peak anomaly oscillations, and the instant of settling for a three-area system. The system includes thermal–bio-diesel in area-1; a thermal–geothermal power plant in area-2; and a thermal–split-shaft gas turbine in area-3. It is also observed that the presence of renewable sources such as wave power plants and photovoltaics makes the system significantly better compared to the base system, when assessed individually or both together. Action in a combination of capacitive energy storage with duple compensation is also examined using the PDN(FOID) controller, which provides a noteworthy outcome in dynamic performance. Moreover, PDN(FOID) parameter values at a nominal condition are appropriate for the random patterns of disturbance needed for optimization.

Keywords: Archimedes wave energy conversion; automatic generation control; bio-diesel plant; capacitive energy storage; geothermal power plant; PDN(FOID) controller; PV; spotted hyena optimizer; wave power plant

1. Introduction

The principle of automatic generation control (AGC) is to maintain the balance between power generation and power demand along with losses [1–3]. If this equilibrium is not maintained then it will lead to excessive fluctuations from the nominal values of frequency and tie-line power connecting areas. Back in earlier days, most of the literature in AGC learning highlighted work in isolated systems [4–6]. Later works were reported on interconnected systems for two-area, and even five-area, thermal systems [7–12]. Nowadays, the literature reflects the usage of multiple sources as generating units such as hydro, gas, and diesel along with thermal as a base generating unit [13–16]. However, far fewer works have been reported on another form of gas turbine, which is the split-shaft gas turbine (Ss (GT)) [17]. So, many studies can be conducted on thermal-Ss (GT) systems.

The increasing use of the same conventional sources is extensively depleting them. Conventional sources also affect the environment with their many by-products, which calls for the association of renewable sources with conventional means. The most common forms of renewable sources that are readily available are solar and wind. Many works have been reported in the literature about the involvement of solar and wind in AGC learning in a single area as well as in interconnected systems. Arya [18] reported on the use of a photovoltaic (PV) system in a hydro-thermal system. In addition to these, geothermal and bio-diesel are also coming into the picture. Geothermal energy is a type of thermal energy that is stored by the earth itself. Thus, this type of energy can be extracted from the earth's crust. Tasnin et al. [19] reported on the application of geothermal in AGC learning. Bio-diesel plants utilize bio-diesel to drive generators. Bio-diesel is produced from oil that has been extracted from various plants such as sunflower, palm, or soybean. The most common form is the use of palm oil. Bio-diesel is a type of renewable fuel. Barik et al. [20] highlighted the use of bio-diesel in an isolated system. In addition, wave power plants (WavePPts) have found minimal consideration in AGC. WavePPts have Archimedes energy translation parts that convert wave energy into electrical energy. Hasanien et al. [21] united a WavePPt with AGC knowledge for a dual-arena thermal scheme. The amalgamation of a geothermal power plant (GPP) and a bio-diesel plant along with a WavePPt and photovoltaic (PV) in AGC learning has not yet been reflected in the literature. Thus, a thermal-Ss (GT) system incorporating GPP, bio-diesel, a WavePPt, and PV calls for further extensive assessments.

The perception of AGC leads to a great effort to decrease the anomaly of frequency along with tie-line power, interlinking diverse areas from their basic value. However, periodically, a state may ascend when oscillations grow to an excessive amount so that a scheme might bring uncertainty. In this circumstance, if the scheme is involved with an energy storage unit, such as capacitive energy storage (C^{ES}), then it can avoid such an alteration. As such, C^{ES} will draw an extra amount of power, which indicates less usage of kinetic energy to subdue small load needs. C^{ES} [22] has found its application in AGC learning. It can be used in the existence of duple compensation or in the absence. The influence of the contrast of C^{ES} with/without duple compensation on scheme dynamics is hitherto to be discovered.

In the scheme of AGC knowledge, there is a dual diverse sort of control similar to a primary control and subordinate control. A major consideration in AGC knowledge concerning control is the appropriate choice of secondary controllers. Numerous categories of subordinate controllers such as integer order (InO), fractional order (FrO), and cascade controllers are described in the literature associated with AGC. Diverse authors have described numerous InO controllers, such as integral (I) [23], proportional–integral (PI) [24], and proportional–integral-derivative with filter (PIDN) [25], in AGC. Dual [26] or trio [27] higher grade of freedom subordinate controllers have also been examined in this arena of learning, and correspond to the InO sort. The few FrO controllers, which initiate its application, are FOPI [28] and FOPIDN [29]. The AGC knowledge literature reflects the practice of InO order two-stage controllers PD-PID [30], and FrO two-stage controllers FOPI-FOPD [19], as well as a grouping of InO and FrO controllers (PIDN-FOPD) [31]. A dual-stage controller with the amalgamation of InO PDN with FrO FOID, termed PDN(FOID), is never hitherto specified in AGC works. Furthermore, the utilization of the PDN(FOID) subordinate controller in this trio-area thermal–bio-diesel–GPP-Ss (GT) scheme along with WavePPt, PV, and CES has not been stated beforehand, so it claims the necessity of examination.

The performance of each subservient controller is exceptionally solitary if the finest amount of gain and related constraints are appropriately favored. These could be executed with the assistance of typical or optimization measures. However, the usage of typical methods such as straight pursuit, arbitrary pursuit, incline pursuit, and numerous others is pretty arduous and delivers substandard consequences, as well as craving a great number of repetitions to deliver outcomes. The optimization procedures that have previously been found in AGC knowledge are whale optimization algorithm (WOA) [17], bacterial foraging optimization (BFO) [24], cuckoo search (CS) [26], differential evolution (DE) [32], particle swarm optimization (PSO) [33], firefly algorithm (FA) [34], grey wolf optimization (GWO) [35], imperialist competitive algorithm (ICA) [36], flower pollination algorithm (FPA) [37], honey badger algorithm [38], AdaBoost algorithm [39], and improved moth-flame algorithm [40]. A newly developed bio-inspired meta-heuristic algorithm titled spotted hyena optimizer (SHO) [41] is obtainable from the literature. SHO was established from the behavioral nature of spotted hyenas, which portrays the social bond between the spotted hyena and their collaborative deeds. To our great surprise, the implementation of SHO has not been identified in AGC learning for its ability to obtain the best values of controller gains and parameters, and this demands a complete investigation.

With reference to the above-mentioned discussions, the prime purpose of the present article is as follows:

(a) Formulation of a three-area scheme with a thermal–bio-diesel in area-1, thermal–GPP in area-2, and thermal–Ss (GT) in area-3;

(b) The gains of I/PI/PIDN/PDN(FOID) are simultaneously optimized individually using the SHO algorithm in order to obtain an excellent controller;

(c) The scheme stated in (a) is combined with WavePPt in area-1, and its impact on the system dynamics is assessed;

(d) The scheme stated in (a) is combined with PV in area-3, and its impact on the system dynamics is assessed;

(e) The scheme stated in (a) is combined with WavePPt in area-1 and PV in area-3 together, and their impact on the system dynamics is assessed;

(f) The scheme stated in (e) is combined with C^{ES} with/without duple compensation separately, and their impact on the system dynamics is studied on an individual basis;

(g) Sensitivity investigation is undertaken to examine the toughness of the superlative 'controller's gains when subjected to a random pattern of load disturbance.

For ease of understanding, the present article is schematically represented in Figure 1.

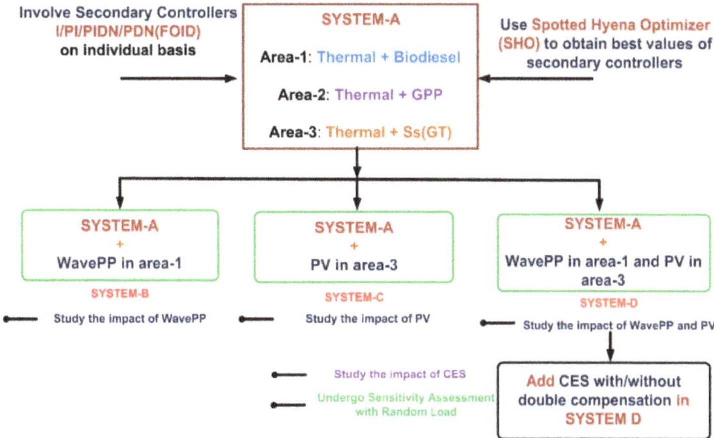

Figure 1. Schematic representation of the entire work.

2. Structure Portrayal

2.1. Overall Portrayal of Structure

A trio-area scheme of uneven sort is contemplated for scrutiny, confining the area size ratio in the arrangement of 2:3:4. The scheme encompasses a bio-diesel–thermal plant in area-1. In the same manner, thermal–geothermal power plants (GPP) in area-2, and thermal–split-shaft gas turbine (Ss (GT)) in area-3. The parameters values are provided in Appendix A. Typical diesel plants are currently being substituted by bio-diesel plants since they are nonpoisonous, as well being ecologically friendly, and pretty bulky with curtailed viscidity; additionally, they emit a relatively small amount of carbon monoxide. Thus, they could be employed as a reserve for power origination. The following comprises the valve controller and ignition engine. The first order transfer functions (*Tfn*) of the valve controller and ignition engine of a bio-diesel plant are detailed by (1) and (2) on an individual basis.

$$Tf_{valve\ regulator}^{Bio-diesel} = \frac{K^{VR}}{1 + sT^{VR}}, \tag{1}$$

K^{VR} and T^{VR} are the bio-diesel plant valve regulator's gain and time constants individually.

$$Tf_{Combustion\ engine}^{Bio-diesel} = \frac{K^{CE}}{1 + sT^{CE}} \tag{2}$$

K^{CE} and T^{CE} are the bio-diesel plant combustion engine's gain and time constants individually.

Geothermal energy is a potential renewable source (RWS) of energy where underground thermal energy is transformed into electricity. The *Tfn* modeling of *GPP* is similar to thermal plants, but it does not have a boiler for reheating steam [19]. The first order *Tfn* of the governor and turbine of *GPP* is given by (3) and (4), respectively.

$$Tf_n^{G_{GPP}} = \frac{1}{1 + sG^{GPP}}, \tag{3}$$

$$Tf_n^{T_{GPP}} = \frac{1}{1 + sT^{GPP}} \tag{4}$$

G^{GPP} and T^{GPP} are varied constants of *GPP*, independently. These values are obtained by the optimization technique SHO within the prescribed limits [19]. The participation factors (pf) of each generating unit of the respective areas are $pf_{11} = 0.7$, $pf_{12} = 0.3$ in area-1; $pf_{21} = 0.6$, $pf_{22} = 0.4$ in area-2; and $pf_{31} = 0.65$, $pf_{32} = 0.35$ in area-3. This is supposed to be scheme-1. Afterward, the structure is unified with a wave power plant (WavePPt) in area-1. This is supposed to be scheme-2. Afterward, structure-1 will be involved with the photovoltaic (PV) system in arena-3. This is scheme-3. Next, both the WavePPt and PV are integrated into scheme-1 with the WavePPt in area-1 and PV in area-3. This is scheme-4. When the WavePPt and PV are both present in the system, then the *pf*'s are: $pf_{11} = 0.7$, $pf_{12} = 0.3$ in arena-1; $pf_{21} = 0.6$, $pf_{22} = 0.4$ in arena-2; and $pf_{31} = 0.5$, $pf_{32} = 0.3$, and $pf_{33} = 0.2$ in arena-3. After that, the energy storing component, namely, capacitive energy storage (C^{ES}) is included in all areas. This is treated as scheme-5. Again, structure-5 is provided with C^{ES} having duple compensation in all areas. This is scheme-6. The representation and transfer function (*Tf_n*) model of the arrangements is replicated in Figure 2. The *Tfn* model of Ss (GT) is obtained from [17]. The elementary values of structure parameters are specified in the addendum. The best values of controller gains and correlated constraints are attained with the assistance of the spotted hyena optimizer algorithm by taking into account the integral squared error as a performance index (*Pi*_{ISE}) specified by (5)

$$Pi_{ISE} = \int_0^T \left\{ (\Delta f_1)^2 + (\Delta f_2)^2 + (\Delta f_3)^2 + \left(\Delta P_{tie_{1-2}}\right)^2 + \left(\Delta P_{tie_{2-3}}\right)^2 + \left(\Delta P_{tie_{1-3}}\right)^2 \right\} dt. \tag{5}$$

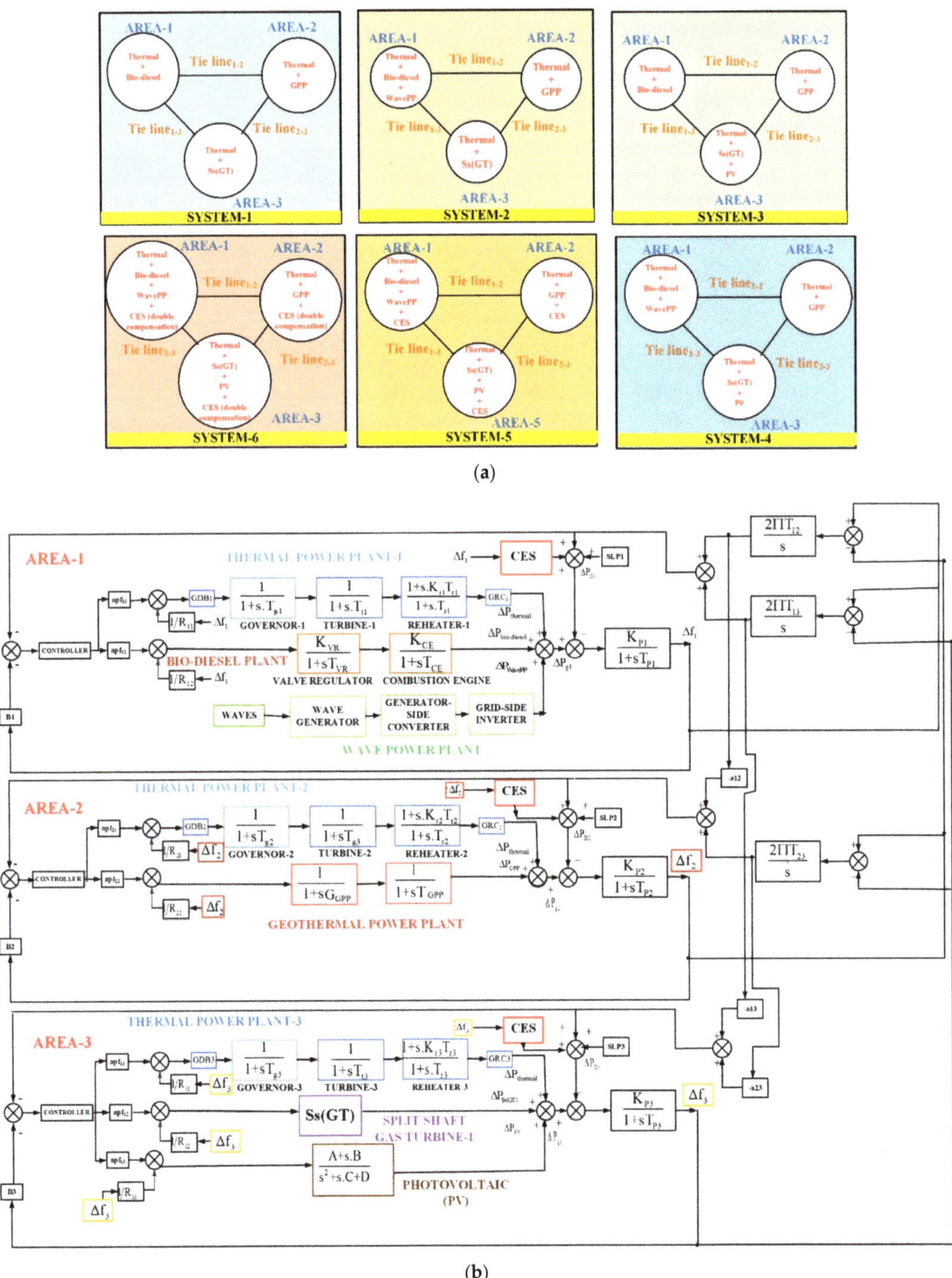

Figure 2. The representation and *Tfn* model of an unequal three-area system with thermal–biodiesel–WavePPt in area-1, thermal–GPP in area-2, and thermal–Ss (GT)–PV in area-3: (**a**) Schematic diagram of six different systems in a step-by-step method, (**b**) *Tfn* model of the ultimate system.

2.2. RWS—Wave Power Plant (WavePPt)

The power of the WavePPt is attained from sea surf. For accomplishing this rendition, Archimedes-wave swing (AdWS), which is a sort of translation segment pooled with a permanent magnet synchronous generator (PtMSg), converts sea surf mechanical to electrical energy. The Tf_n of AdWS attached to PtMSg is established. Here, the WavePPt is unified to area-1 through the assistance of a Converter I generator sideways/grid sideways inverter. All components of studied system were modelled MATLAB R2020a software. The converter present near a generator is employed in order to attain situations of extreme power point trailing. Mutually, the converter and inverter are intended for a gain value of 1 and a period constant of 0.01 s. The first order Tf_n prototype of AdWS of the WavePPt is specified by (6)

$$Tfn_{PPt_{AWS}}{}^{Wave} = \frac{K^{Wave}{}_{PPt_{AWS}}}{1 + sT^{Wave}{}_{PPt_{AWS}}}, \tag{6}$$

$K^{Wave}{}_{PPtAWS}$ and $T^{Wave}{}_{PPtAWS}$ are the gain and time constants of the AdWS of the *WavePPt*, respectively.

2.3. Energy Storage Device—Capacitive Energy Storage (C^{ES})

An energy storage device such as a capacitive energy storage device (C^{ES}) is equipment that usually employs a capacitor for storage along with a power adaptation segment, which is connected to the AC network with the assistance of a rectifier/inverter. The C^{ES} unit responds instantly to the system in the case of instant recurrent or current drift. Subsequently, any manner of unpredictability is moderated; C^{ES} yet again reestablishes the initial voltage amount in the plates of the capacitor by employing the additional energy obtainable in the scheme.

The Tfn of C^{ES} is specified by (7)

$$Tfn^{CES} = \frac{K^{CES}}{1 + sT^{CES}}, \tag{7}$$

K^{CES} is the gain and T^{CES} are the time constants of C^{ES}.

The C^{ES} plays the role of a frequency mediator in the case of a twofold compensation technique. The additive revision in power yield of C^{ES} with twofold compensation is detailed by (8)

$$\Delta P_{CES(duple\ compensation)} = \left[\frac{K^{CES(duple\ compensation)}}{1 + sT^{CES(duple\ compensation)}} \right] \left[\frac{1 + sT^1}{1 + sT^2} \right] \left[\frac{1 + sT^3}{1 + sT^4} \right] \Delta f_i(s), \tag{8}$$

$K^{CES(duple\ compensation)}$ and $T^{CES(duple\ compensation)}$ are the C^{ES} with duple compensation gain and time constants, individually. T^1, T^2, T^3, and T^4 are varied time factors of the recompensed segment of C^{ES}.

3. The Proposed Approach

3.1. Problem Declaration

The emphasis of the present learning is on the frequency excursion approach with the utilization of an innovative metaheuristic method to optimize the InO and FrO amalgamated controller in a renewable source integrated power system structure. The main aim is to obtain a zero error for aberration in frequency and tie-line power by interconnecting different areas using a suitable control input.

3.2. Commended Controller

The commended controller is an aggregate of integer order (InO) together through a fractional order (FrO) controller. The commended controller is an I/O proportional-derivative with filter (PDN) with FrO integral-derivative (FOID), hence, PDN(FOID). The arrangement of PDN(FOID) is substantiated in Figure 3. Segment-1 (B1) and Segment-2 (B2) are the layouts of PDN and FOID one-to-one. Rs_i (s) is the antecedent signal and Os_i (s) is the outcome signal for the PDN(FOID) controller. The *Trfn* of $B1_i$ (s) is manifested by (9)

$$B1_i(s) = \frac{K_{Pi}s + K_{Di}N_i}{(s + N_i)}.$$
(9)

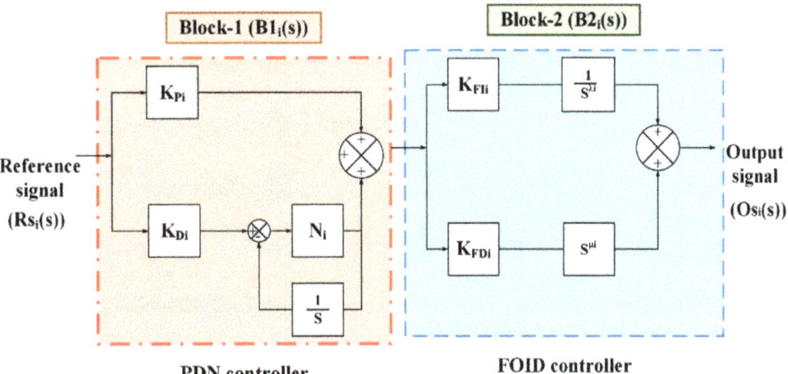

Figure 3. Layout of the proposed controller (PDN(FOID)).

The InO proportional gain is symbolized as K_{Pi} and the derivative gain as K_{Di} for the *i*-th suggested area, N_i is PDN controller's filter. Summarization of Riemann–Liouville for the FrO integrator and derivative are obtainable from (10) and (11) [42–44]

$$\alpha D_t^{-\alpha} f(t) = \frac{1}{\Gamma(n)} \int_\alpha^t (t-\tau)^{\alpha-1} f(\tau)d\tau, \ n-1 \le \alpha < n, \text{n is an integer}$$
(10)

$$\alpha D_t^\alpha f(t) = \frac{1}{\Gamma(n-\alpha)} \frac{d^n}{dt^n} \int_\alpha^t (t-\tau)^{n-\alpha-1} f(\tau)d\tau$$
(11)

αD_t^α is the fractional operator, and $\Gamma(.)$ is the Euler's gamma function. The alteration of the Fro integral and derivative in the Laplace domain is given by (12)

$$L\{\alpha D_t^\alpha f(t)\} = s^\alpha F(s) - \sum_{k=0}^{n-1} s^k \alpha D_t^{\alpha-k-1} f(t)|_{t=0}$$
(12)

The detriment of boundless computation of poles and zeros by virtue of absolute resemblance is manifested by Oustaloup, Mathieu, and Lanusse (1995) [45]. Here, a convenient Trfn is propounded that can approximate FrO derivatives together with integrators by dint of recursive distribution around poles and zeros substantiated by (13)

$$s^\alpha = K \prod_{n=1}^M \frac{1 + \left(\frac{s}{\omega_{Z,n}}\right)}{1 + \left(\frac{s}{\omega_{p,n}}\right)}$$
(13)

Suppose, attuned gain $K = 1$, gain = 0 dB through 1 rad/s frequency, M = Count of poles along with zeros (fixed beforehand), and frequencies choice for poles and zeros are manifested by (14)–(18).

$$\omega_{Z,l} = \omega_l \sqrt{n}$$
(14)

$$\omega_{p,n} = \omega_{Z,n}\varepsilon, \ n = 1, \ldots, M$$
(15)

$$\omega_{Z,n+1} = \omega_{p,n} \sqrt{\eta} \tag{16}$$

$$\varepsilon = \left(\frac{\omega_h}{\omega_l}\right)^{\frac{v}{M}} \tag{17}$$

$$\eta = \left(\frac{\omega_n}{\omega_l}\right)^{\frac{(1-v)}{M}} \tag{18}$$

The Trfn of $B2_i$ (s) is manifested by (19).

$$B2_i(s) = \frac{K_{FIi}}{s^{\lambda_i}} + s^{\mu_i} K_{FDi} \tag{19}$$

In the above expressions, λ is the FrO integrator's fragment and μ is the FrO derivative's fragment. K_{FIi} is the FrO's integral fragment gain, and K_{FDi} is the FrO's derivative fragment gain of the suggested area.

3.3. Objective Function

The main purpose of the controller design is the proper optimization task including the minimization of a particular cost function considering the constraints of controller gains and parameters with its confines. Here, in the present AGC learning, an integral squared error (ISE) is involved as the cost function. The mathematical expression of ISE as a performance index is provided by (20).

$$Pi_{ISE} = \int_0^T \left\{ (\Delta f_i)^2 + \left(\Delta P_{tie_{i-j}}\right)^2 \right\} dt. \tag{20}$$

Here, i and j are number of areas, where $i = 1, 2, 3$, and $i \neq j$.

In (20), the independent variables are K_P, K_D, N, K_{FI}, λ, K_{FD}, and μ, and the assumed constraints are provided in (21).

$$K_P^{min} \leq K_P \leq K_P^{max},\ K_D^{min} \leq K_D \leq K_D^{max},\ N^{min} \leq N \leq N^{max}, K_{FI}^{min} \leq K_{FI} \leq K_{FI}^{max}, \lambda^{min} \leq \lambda \leq \lambda^{max},\ K_{FD}^{min} \leq K_{FD} \leq K_{FD}^{max}, \mu^{min} \leq \mu \leq \mu^{max} \tag{21}$$

4. Spotted Hyena Optimizer

Spotted Hyenas Optimizers (SHOs) are known as competent chasers. They are the bulkiest of the hyena breed. The spotted hyenas are also renowned as laughing hyenas since they sound like humans laughing. They are highly complex, brainy, and hugely communal creatures. The SHs trail their victims by their vision, auditory, and odor features. This nature of SH inspired Dhiman et al. [41] to develop the meta-heuristic algorithm SHO. The authors have outlined the mathematical design of the communal acquaintance of SH and collegial agility to undergo optimization. The trio events allied with SHO are tracking down capture, encompassing, and conspicuous capture.

1. Encompassing capture: To develop the numerical prototype, it is assumed that the present finest contender is the destined capture, which is closest to the optimum given that the chase arena was not known previously. The remaining chase agents will seek to renew their spot with reference to the response of the finest contender about the finest location. The numerical prototype is manifested by (22) and (23)

$$\vec{D}_h = \left| \vec{B} \cdot \vec{P}_p(x) - P(x) \right| \tag{22}$$

$$\vec{P}(x+1) = \vec{P}_p(x) - \vec{E} \cdot \vec{D}_h \tag{23}$$

In the above expressions, $\vec{D}_h$ is the stretch between the Pr of the hunt, and $\vec{P}$ is the spot vector of SH. $\vec{B}$ and $\vec{E}$ are computed as in (24)–(26)

$$\vec{B} = 2\vec{r}\vec{d}_1 \tag{24}$$

$$\vec{B} = 2\vec{h}\vec{r}\vec{d}_2 - \vec{h} \tag{25}$$

$$\vec{h} = 5 - \left(iteration * \left(\frac{5}{\max_{\text{iteration}}} \right) \right), \ iteration = 1,\ 2,\ 3\ldots,\max_{\text{iteration}} \tag{26}$$

For appropriately corresponding the exploration and exploitation, $\vec{h}$ in straight line declined from 5 to 0 over the duration of the maximum iteration. Additionally, this execution indorses extra exploitation as the count value rises. However, $\vec{r}_{d1}$ and $\vec{r}_{d2}$ are arbitrary vectors within [0, 1].

2. Trapping: In order to characterize the conduct of SH numerically, it is assumed that the finest chase agent has information regarding the spot of the hunt. The remaining chase agents form groups toward the finest chase agent and save the finest results attained so far to restore their spots according to the following Equations (27)–(29)

$$\vec{D}_h = \left| \vec{B} \cdot \vec{P}_h(x) - P(x) \right| \tag{27}$$

$$\vec{P}_k = \vec{P}_h - \vec{E} \cdot \vec{D}_h \tag{28}$$

$$\vec{C}_h = \vec{P}_k + \vec{P}_{k+1} + \ldots + \vec{P}_{k+N} \tag{29}$$

$\vec{P}_h$ describes the spot of initial finest SH, and $\vec{P}_k$ describes the spot of further SH. At this time, N designates the figure of SH, which is figured as follows:

$$N = count_{nos}(\vec{P}_h,\ \vec{P}_{h+1},\ \vec{P}_{h+2}, \ldots,\ (\vec{P}_h + \vec{M})) \tag{30}$$

$\vec{M}$ is the arbitrary vector in [0.5, 1], the numbers outline the figure of the results and the totality of all the contender results, afterward adding $\vec{M}$, which is far from comparable to the finest ideal result in the specified hunt space, and $\vec{C}_h$, which is an assembly of N figure of ideal results.

3. Encroaching hunt (exploitation): To numerically model for invading the hunt, the $\vec{h}$ value is lessened. The disparity in $\vec{E}$ is also reduced from 5 to 0 in due course of the count. $|E| < 1$ forces the assembly of SH to attack on the way to the hunt. The numerical design for invading the hunt is

$$\vec{P}(x+1) = \frac{C_h}{N} \tag{31}$$

$\vec{P}(x+1)$ stores the finest result and revises the spot of further chase agents as per the spot of the finest chase agent.

4. Hunt for target (exploration): SH mostly chase the hunt, as per the spot of the assembly of the SH that exist in $\vec{C}_h$. They shift apart from one another to chase and to combat for the hunt. Then, they utilize $\vec{E}$ with arbitrary values >1 or <−1 to compel the chase agents to shift far away from the hunt. This mechanism permits the SHO algorithm to hunt in a wide-reaching manner. The SHO's flowchart is provided in Figure 4.

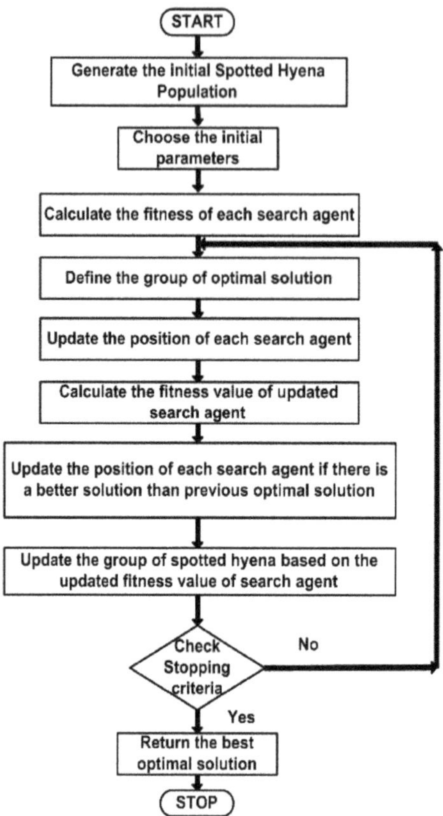

Figure 4. Flowchart of the SHO algorithm.

5. Methodology

A three-area system with a capacity ratio of 2:3:4 is considered for investigation. The investigated system comprises of thermal, bio-diesel, and wave power plant in area-1; thermal and geothermal power plant in area-2; and thermal, split-shaft gas turbine, and solar photovoltaic in area-3. System is also incorporated with an energy storage device. Investigations are carried out considering PDN(FOID) controller whose parameters are obtained by the SHO method with ISE as the performance index. The optimization technique is coded in MATLAB R2022a software, and the investigated system is modeled in Simulink with the FOMCON toolbox.

Studies are carried out for: (1) selection of best controller; (2) selection of appropriate performance index; (3) selection of best optimization method; (4) influence of wave power plant, PV, individually and both together; (5) influence of C^{ES} with or without duple compensation; (6) sensitivity assessment.

6. Outcomes and Valuation

6.1. Evaluation of Dynamic Outcome for the Choice of Superlative Controller

The scheme under evaluation embraces thermal including bio-diesel in area-1, GPP along with thermal in area-2, and Ss (GT) along with thermal in area-3 (Scheme-1). This scheme is familiarized with I/PI/PIDN/FOPID/PDN(FOID) controllers on a specific base. Evaluation is accomplished considering a 1% disturbance of the step content in area-1. The finest obtainable values of individual controller gains and related parameters are attained by employing SHO by means of Pi_{ISE}. The scheme is initially familiarized through a I controller to attain its gain values and also the parameters of GPP using SHO. The governor and turbine time constants obtained are 0.1 s, respectively. These values of G_{GPP} and T_{GPP} are kept the same for the remainder of the work. Subsequently, PI, PIDN, and FOPID in addition to PDN(FOID) controllers are cast off independently. The finest conceivable values are assembled in Table 1, and dynamic outcomes are associated, as revealed in Figure 5. Significant interpretation of each outcome articulates around the fineness of the PDN(FOID) overtop additional controllers concerning the diminished stage of peak_overshoot (Pk_O), extent-of-oscillations, and peak_undershoot (Pk_U), besides the duration of settling (S_Time). Thus, it is revealed that the SHO-optimized PDN(FOID) controller outperforms other controllers in terms of lessened Pk_O ($\Delta f_1 = 0.0007$ Hz, $\Delta f_2 = 0$ Hz, $\Delta Ptie_{1-2} = 0$ Hz, and $\Delta Ptie_{1-3} = 0$ Hz), Pk_U ($\Delta f_1 = 0.0112$ Hz, $\Delta f_2 = 0.0038$ Hz, $\Delta Ptie_{1-2} = 0.0039$ Hz, and $\Delta Ptie_{1-3} = 0.0039$ Hz), and S_Time ($\Delta f_1 = 25.81$ s, $\Delta f_2 = 35.31$ s, $\Delta Ptie_{1-2} = 21.71$ s, and $\Delta Ptie_{1-3} = 21.04$ s). In Table 2, the matching outcomes of Pk_O, Pk_U, and S_Time values are recorded, which imitates the improved conduct of the PDN(FOID) overtop remainder of the subordinate controllers.

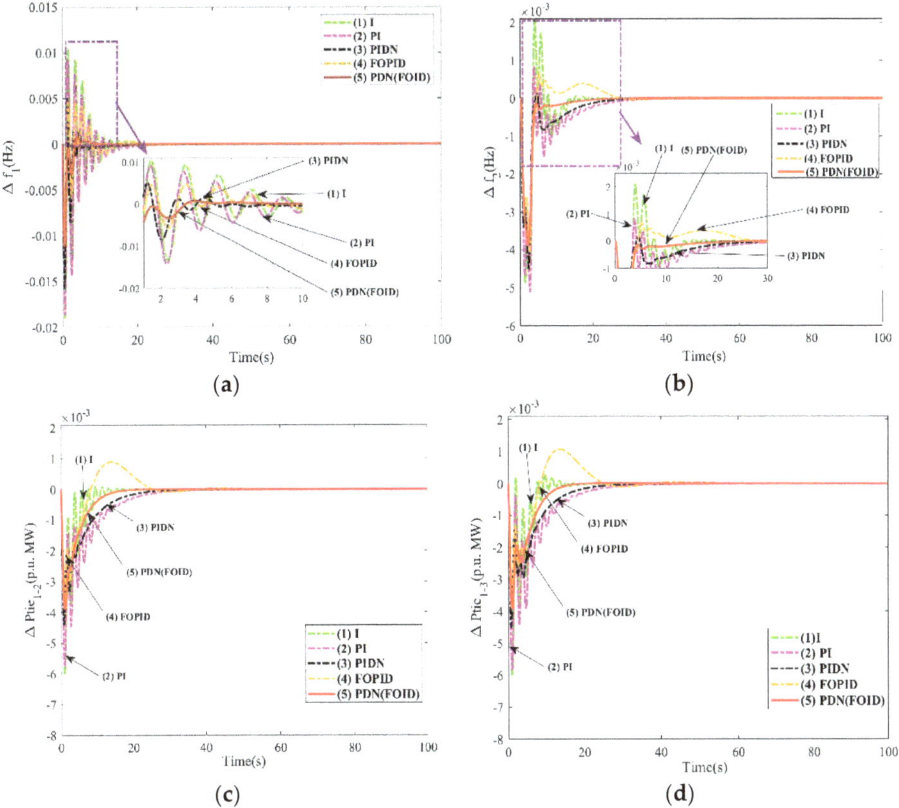

Figure 5. Evaluation of outcomes of subordinate controllers for scheme-1 for 1% step load disorder contrast time: (**a**) Anomaly of frequency of area-1, (**b**) Anomaly of frequency of area-2, (**c**) Anomaly of Tie-line power interlinking area-1 and area-2, (**d**) Anomaly of Tie-line power interlinking area-1 and area-3.

Table 1. Finest values of gains and related parameters of subordinate controllers for scheme-1.

Name of Controllers	Corresponding Gains and Correlated Parameters	Area-1	Area-2	Area-3
I	K_{Ii} *	0.9885	0.9897	0.9876
PI	K_{Pi} *	0.3565	0.5533	0.5538
	K_{Ii} *	0.4745	0.6497	0.7584
PIDN	K_{Pi} *	0.6975	0.9728	0.9739
	K_{Ii} *	0.5982	0.9836	0.9878
	K_{Di} *	0.6315	0.3140	0.3041
	N_i *	10.77	11.41	11.25
FOPID	K_{FPi} *	0.0094	0.0095	0.0096
	K_{FIi} *	0.3765	0.8354	0.9379
	λ_i	1.4352	1.1876	1.1587
	K_{FDi} *	0.4549	0.6357	0.5366
	μ_i *	1.0477	0.0768	0.1585
PDN(FOID)	K_{Pi} *	0.8686	0.5188	0.6875
	K_{Di} *	0.5796	0.7362	0.9421
	N_i *	55.58	68.83	79.77
	K_{FIi} *	0.9976	0.9041	0.7454
	λ_i *	1.0099	0.9454	0.9710
	K_{FDi} *	0.8508	0.8261	0.8554
	μ_i *	0.7853	0.2804	0.7216

* Signify the optimum values.

Table 2. Peak anomaly and duration of settling for outcomes in Figure 5 in the case of system-1 employing SHO-optimized I/PI/PIDN/FOPID/PDN(FOID) controllers.

Responses	Name of Controllers	Pk_O	Pk_U	S_Time (in Seconds)
Δf_1 (Figure 5a)	I	0.0103	0.0191	39.42
	PI	0.0092	0.0186	35.53
	PIDN	0.0051	0.0156	31.42
	FOPID	0.0045	0.0113	27.84
	PDN(FOID)	0.0007	0.0112	25.81
Δf_2 (Figure 5b)	I	0.0021	0.0048	43.52
	PI	0.0008	0.0051	42.23
	PIDN	0.0001	0.0045	39.63
	FOPID	0.0006	0.0041	39.97
	PDN(FOID)	0	0.0038	35.31
$\Delta Ptie_{1-2}$ (Figure 5c)	I	0.0001	0.0059	39.45
	PI	0	0.0057	39.41
	PIDN	0	0.0044	34.24
	FOPID	0.0008	0.0041	33.87
	PDN(FOID)	0	0.0039	21.71
$\Delta Ptie_{1-3}$ (Figure 5d)	I	0.0002	0.0062	42.23
	PI	0	0.0058	34.88
	PIDN	0	0.0044	34.24
	FOPID	0.0011	0.0041	37.02
	PDN(FOID)	0	0.0039	21.04

6.2. Nomination of Performance Index

The excellent outcome of the performance index (Pi) among the integral squared error (Pi_{ISE}), integral time squared error (Pi_{ITSE}), integral absolute error (Pi_{IAE}), and integral time absolute error (Pi_{ITAE}) are procured by assisting system-1 with each of the Pi on individual terms using the PDN(FOID) controller. The premium standards of PDN(FOID) controller gains and related parameters are attained using the SHO algorithm. The expressions for Pi_{ITSE}, Pi_{IAE}, and Pi_{ITAE} are given by (32)–(34), respectively, and for Pi_{ISE}, it is given by (5)

$$Pi_{ITSE} = \int_0^T \left\{ (\Delta f_1)^2 + (\Delta f_2)^2 + (\Delta f_3)^2 + \left(\Delta P_{tie_{1-2}}\right)^2 + \left(\Delta P_{tie_{2-3}}\right)^2 + \left(\Delta P_{tie_{1-3}}\right)^2 \right\} t\, dt \tag{32}$$

$$Pi_{IAE} = \int_0^T \left\{ |\Delta f_1| + |\Delta f_2| + |\Delta f_3| + |\Delta P_{tie_{1-2}}| + |\Delta P_{tie_{2-3}}| + |\Delta P_{tie_{1-3}}| \right\} dt \tag{33}$$

$$Pi_{ITAE} = \int_0^T \left\{ |\Delta f_1| + |\Delta f_2| + |\Delta f_3| + |\Delta P_{tie_{1-2}}| + |\Delta P_{tie_{2-3}}| + |\Delta P_{tie_{1-3}}| \right\} t\, dt \tag{34}$$

With the help of illustrious values accomplished for PDN(FOID) controller's gains and correlated parameters in each case, the dynamic responses are contrasted in Figure 6a–c, and the corresponding Pk_O, Pk_U, and S_Time values are marked down in Table 3. A critical view of the responses says that responses considering Pi_{ISE} as Pi have a better performance compared to others with respect to lessened Pk_O, Pk_U, S_Time, and oscillations. Thus, it is revealed that the SHO-optimized PDN(FOID) controller using ISE as a performance index outperforms other Pi in terms of lessened Pk_O (Δf_1 = 0.0007 Hz, Δf_2 = 0 Hz, and $\Delta Ptie_{1-2}$ = 0 Hz), Pk_U (Δf_1 = 0.0112 Hz, Δf_2 = 0.0038 Hz, and $\Delta Ptie_{1-2}$ = 0.0039 Hz), and S_Time (Δf_1 = 25.81 s, Δf_2 = 35.31 s, and $\Delta Ptie_{1-2}$ = 21.71 s). Further, the values of Pi are Pi_{ISE} = 0.00021, Pi_{ITSE} = 0.00084, Pi_{IAE} = 0.1053, and Pi_{ITAE} = 0.7233, which reveals the better performance of the system with Pi_{ISE}. The convergence characteristics for system-1 using different Pi's is reflected in Figure 6d. It is observed that convergence characteristics using ISE as Pi converge faster in fewer iterations than other Pi's.

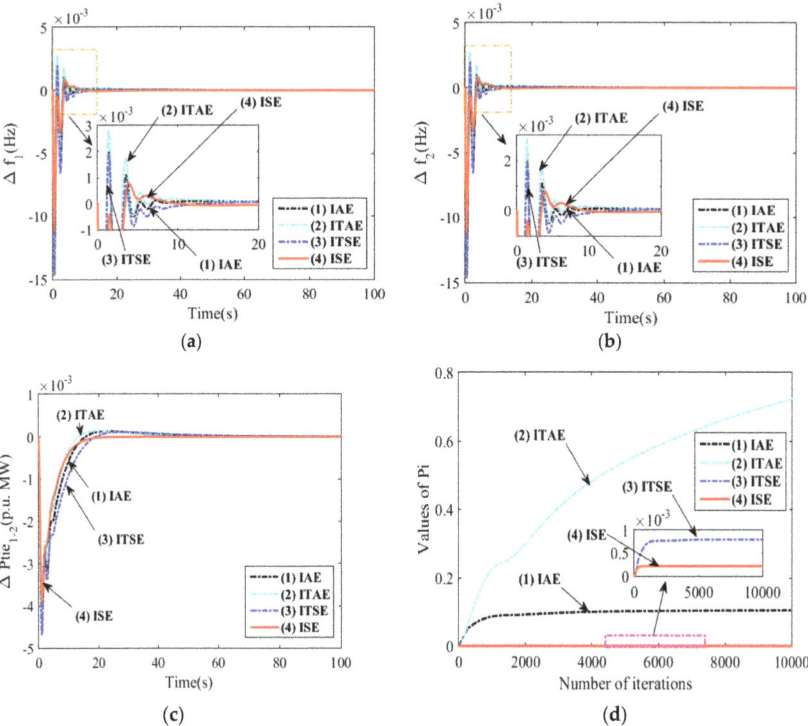

Figure 6. Evaluation for the recommendation of the Pi amongst IAE, ITAE, ITSE, and ISE for employing PDN(FOID) controller in the case of scheme-1 contrast time: (**a**) Anomaly of frequency of area-1, (**b**) Anomaly of frequency of area-2, (**c**) Anomaly of Tie-line power interlinking area-1 and area-2, (**d**) Convergence curve for the system with different Pi.

Table 3. Peak anomaly and duration of settling for outcomes in Figure 6a–c in the case of scheme-1 employing SHO-optimized PDN(FOID) controller for diverse performance indices.

Responses	Name of Performance Indices	Pk_O	Pk_U	S_Time (in Seconds)
Δf_1 (Figure 6a)	IAE	0.0021	0.0142	29.57
	ITAE	0.0027	0.0145	27.16
	ITSE	0.0019	0.0146	26.65
	ISE	0.0007	0.0112	25.81
Δf_2 (Figure 6b)	IAE	0.0019	0.0144	36.37
	ITAE	0.0028	0.0147	37.72
	ITSE	0.0020	0.0146	36.01
	ISE	0	0.0038	35.31
$\Delta Ptie_{1-2}$ (Figure 6c)	IAE	0.0013	0.0045	44.73
	ITAE	0.0013	0.0046	42.74
	ITSE	0.0001	0.0047	50.24
	ISE	0	0.0039	21.71

6.3. Nomination of Algorithm

For the nomination of the algorithm, system-1 is provided with different algorithms, separately, using PDN(FOID) controllers. The algorithms used here are the firefly algorithm (FA), cuckoo search algorithm (CS) [46], particle swarm optimization (PSO) [47], whale optimization algorithm (WOA), and SHO. The tuned values for FA are $\beta 0 = 0.3$, $\alpha = 0.5$, $\gamma = 0.4$, count of fireflies = 50, and maximum number of generations = 100. For CS, nests count = 50, rate of discovery = 0.5, exponent of levy = 1.5, maximum generation = 100, and count of dimensions = 10. For PSO, the tuned parameters values are w = 1, wdamp = 0.99, $c_1 = 1.4$, $c_2 = 1.98$, population size = 50, and maximum generation number = 100. For WOA, figure of hunt agents = 50, and number of repetitions = 100. For each of the algorithms, the best values for the PDN(FOID) controller are obtained. The values are not provided here. With these values, the responses of different algorithms are contrasted in Figure 7a–c. The corresponding responses of Pk_O, Pk_U, and S_Time values are given in Table 4. In Table 4, it can be seen that Pk_O, Pk_U, and S_Time values obtained by the SHO-optimized PDN(FOID) controller are much better compared to other algorithms. Thus, it is revealed that the SHO-optimized PDN(FOID) controller outperforms other algorithms in terms of lessened Pk_O ($\Delta f_1 = 0.0007$ Hz, $\Delta f_3 = 0$ Hz, and $\Delta Ptie_{1-3} = 0$ Hz), Pk_U ($\Delta f_1 = 0.0112$ Hz, $\Delta f_3 = 0.0052$ Hz, and $\Delta Ptie_{1-3} = 0.0039$ Hz), and S_Time ($\Delta f_1 = 25.81$ s, $\Delta f_3 = 23.53$ s, and $\Delta Ptie_{1-3} = 21.04$ s). Further, the supremacy is judged by the convergence curve provided in Figure 7d, where it is observed that the response with the SHO-optimized PDN(FOID) controller converges faster and has the least value of Pi_{ISE}. Therefore, further analyses are carried out using the SHO algorithm.

Table 4. Peak aberration and duration of settling for outcomes in Figure 7a–c, in the case of system-1 using FA/CS/PSO/WOA/SHO-optimized PDN(FOID) controller on an individual basis.

Responses	Name of Algorithms	Pk_O	Pk_U	S_Time (in Seconds)
Δf_1 (Figure 7a)	FA	0.0026	0.01201	32.95
	CS	0.0029	0.0113	33.79
	PSO	0.0018	0.0013	33.04
	WOA	0.0021	0.0118	29.19
	SHO	0.0007	0.0112	25.81
Δf_3 (Figure 7b)	FA	0.0011	0.0053	36.03
	CS	0.0013	0.0053	30.06
	PSO	0.0008	0.0053	38.04
	WOA	0.0009	0.0055	26.84
	SHO	0	0.0052	23.53

Table 4. *Cont.*

Responses	Name of Algorithms	Pk_O	Pk_U	S_Time (in Seconds)
$\Delta Ptie_{1-3}$ (Figure 7c)	FA	0.0012	0.0041	33.76
	CS	0.0010	0.0041	36.09
	PSO	0.0011	0.0040	37.15
	WOA	0.0006	0.0041	30.23
	SHO	0	0.0039	21.04

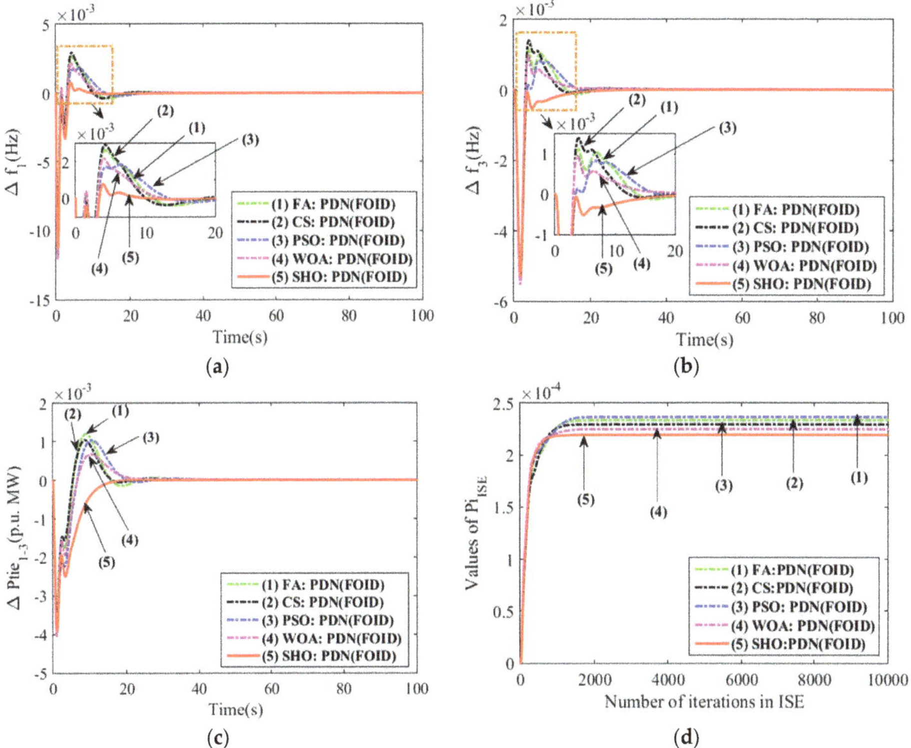

Figure 7. Evaluation for the recommendation of algorithm amid FA/CS/PSO/WOA/SHO techniques for obtaining finest values of PDN(FOID) controller for scheme-1 contrast time: (**a**) Anomaly of frequency in area-2, (**b**) Anomaly of Tie-line power interlinking area-1 and area-2, (**c**) Anomaly of Tie-line power interlinking area-1 and area-3, (**d**) Convergence curve.

6.4. Evaluation of Influence of the Wave Power Plant on Dynamics of System

In the preceding Section 6.1, it is perceived that the SHO-optimized PDN(FOID) is equipped with admirable pursuance for scheme-1. Now, scheme-1 is integrated with an additional RWS, namely, a wave power plant (WavePPt) in area-1, observed as scheme-2. Forthwith, scheme-2 is integrated with PDN(FOID) as the subservient controller, and the interrelated finest measure of gains in addition to related parameters is accomplished by employing the SHO. The premium values are furnished in Table 5. The assessment is performed for a step change in the WavePPt. With the finest values attained, the evaluation is accomplished for the impact of the WavePPt on the dynamic system by contrasting responses for the system with and without the WavePPt, as in Figure 8. In the outcomes in Figure 8, the vast decline in the values of Pk_U and S_Time is evidently noticeable.

Even Pk_O shows slightly lessened values. With the WavePPt, the values of Δf_2 have shown improvement in Pk_U and S_Time from 0.0038 to 0.00186 Hz (Pk_U) and 35.31 to 30.12 s (in comparison to the system without WavePPt), respectively. Similarly, for Δf_3, the improvement is 0.0052 to 0.00287 Hz (Pk_U) and 23.53 to 21.96 s; for $\Delta Ptie_{1-3}$, the improvement is 0.0039 to 0.00197 Hz and 21.04 to 20.22 s; and for $\Delta Ptie_{2-3}$, the improvement is 0.00077 to 0.000381 Hz (Pk_O), 0.00052 to 0.00027 Hz (Pk_U), and 24.65 to 20.09 s. Thus, the presence of the WavePPt improved the dynamics of the scheme.

Table 5. Finest values of gains and related parameters of PDN(FOID) controllers for scheme-2.

Name of Controller	Corresponding Gains and Correlated Parameters	Area-1	Area-2	Area-3
	K_{Pi} *	0.9898	0.9687	0.9603
	K_{Di} *	0.9785	0.4252	0.9232
	N_i *	72.24	56.07	78.85
PDN(FOID)	K_{Fli} *	0.5725	0.5575	0.8418
	λ_i *	1.0484	0.3997	1.0965
	K_{FDi} *	0.9632	0.3326	0.6901
	μ_i *	0.5911	0.7509	0.8949

* Signify optimum values.

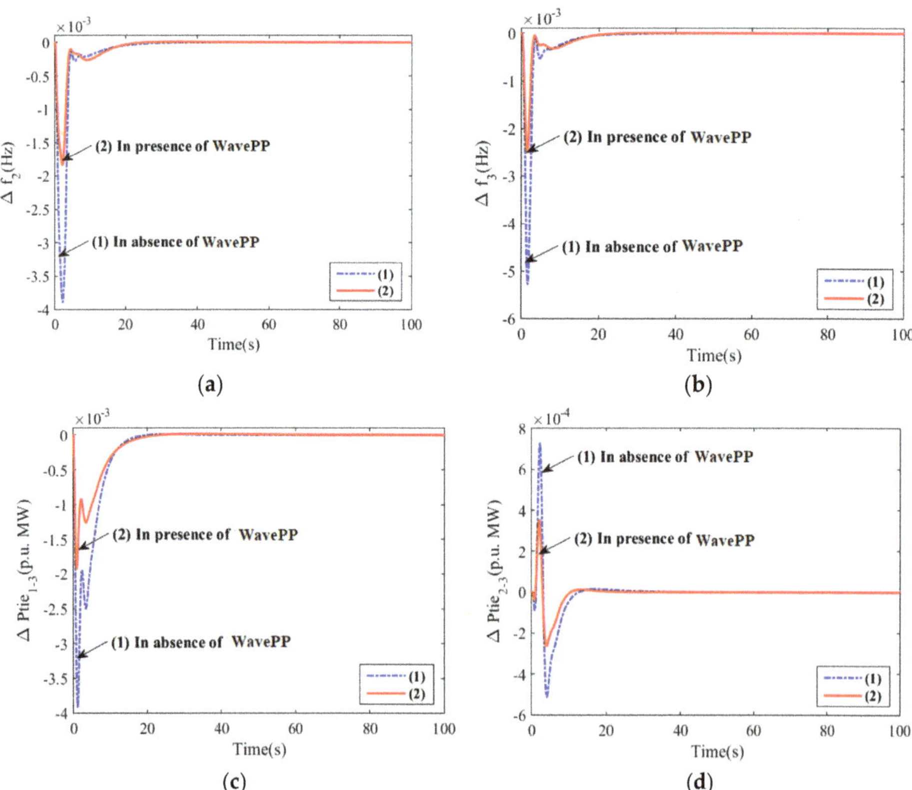

Figure 8. Evaluation of the influence of WavePPt on system changing aspects employing SHO-optimized PDN(FOID) controller in view of 1% step load disturbances for system-2 contrast time: (**a**) Anomaly of frequency in area-1, (**b**) Anomaly of frequency in area-3, (**c**) Anomaly of Tie-line power interlinking area-1 and area-3, (**d**) Anomaly of Tie-line power interlinking area-2 and area-3.

6.5. Evaluation of Influence of PV on Scheme Dynamics

From the previous analysis in Section 6.1, it can be concluded that the SHO-optimized PDN(FOID) provides with the best performance for system-1 over the I/PI/PIDN/FOPID controllers. Now, system-1 is involved with another RWS, namely, a photovoltaic (PV) unit in area-3, viewed as scheme-3. Then, scheme-3 is included with PDN(FOID) as the subordinate controller in all areas and its finest values of gains and the related parameters are accomplished via the SHO. The finest values are shown in Table 6. Employing the finest values, the evaluation is accomplished for the impact of the PV on a dynamic system by contrasting responses for the system with and without the PV unit, as in Figure 9. In the outcomes in Figure 9, the massive diminution in the values of Pk_O, Pk_U, and S_Time is undoubtedly noticeable. With PV, the values of Δf_1 have shown improvement in Pk_O, Pk_U, and S_Time from 0.0007 to 0 Hz (Pk_O), 0.0112 to 0.00109 Hz (Pk_U), and 25.81 to 20.11 s (in comparison to the system without PV). Similarly, for Δf_2, the improvement is 0.0038 to 0.00278 Hz (Pk_U) and 35.31 to 33.09 s; for Δf_3, the improvement is 0.0052 to 0.00298 Hz and 23.53 to 22.54 s; and for $\Delta Ptie_{2-3}$, the improvement is 0.00077 to 0.00011 Hz (Pk_O), 0.00052 to 0.000181 Hz (Pk_U), and 24.65 to 20.01 s. Thus, the presence of PV has improved the dynamics of the system.

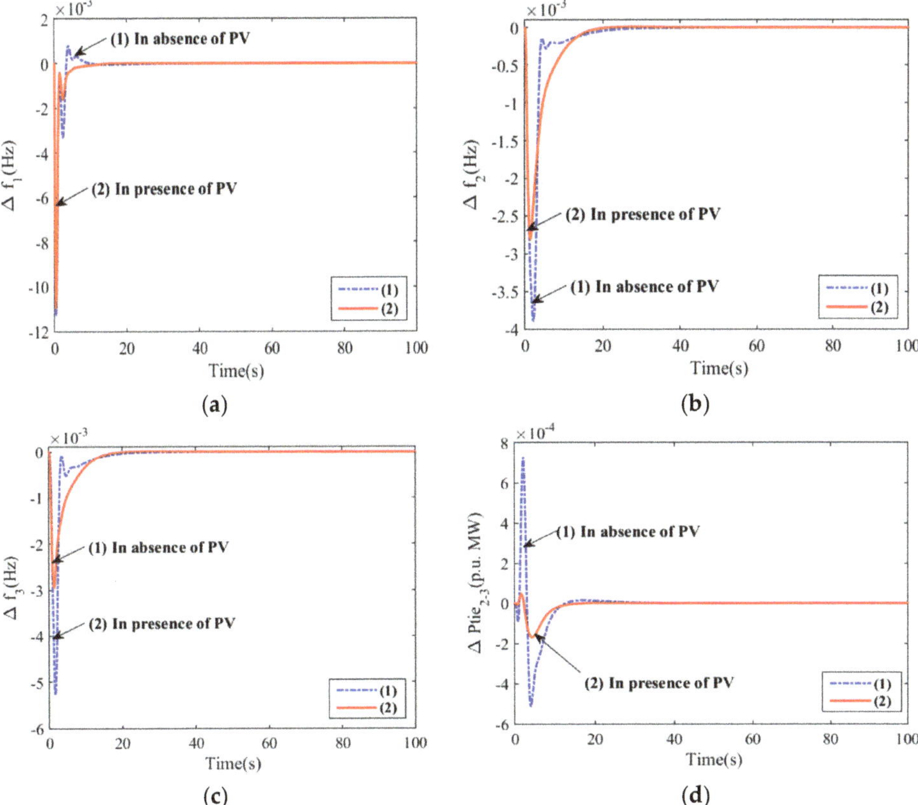

Figure 9. Valuation of influence of PV on system changing aspects employing SHO-optimized PDN(FOID) controller given 1% step load disturbances for system-3 contrast time: (**a**) Anomaly of frequency in area-1, (**b**) Anomaly of frequency in area-2, (**c**) Anomaly of frequency in area-3, (**d**) Anomaly of Tie-line power interlinking area-2 and area-3.

Table 6. Finest values of gains and related parameters of PDN(FOID) controllers for scheme-3.

Name of Controller	Corresponding Gains and Correlated Parameters	Area-1	Area-2	Area-3
PDN(FOID)	K_{Pi} *	0.9898	0.6324	0.7416
	K_{Di} *	0.7992	0.4900	0.4163
	N_i *	44.42	65.94	44.87
	K_{FIi} *	0.7752	0.4882	0.6698
	λ_i *	1.0305	0.6213	0.5595
	K_{FDi} *	0.8085	0.8341	0.7768
	μ_i *	0.6545	0.6640	0.7541

* Signify optimum values.

6.6. Assessment of Impact of Both WavePPt and PV on Scheme Dynamics

In the above Sections 6.4 and 6.5, it is observed that RWSs WavePPt and PV, respectively, have a noteworthy impact on the dynamics of the system when assessed individually. Now, system-1 is incorporated with both WavePPt and PV. WavePPt is associated with area-1 and PV in area-3. Thus, this system with both WavePP and PV is considered as system-4. In this case, the SHO-optimized PDN(FOID) controller is also used to analyze the impact of both WavePPt and PV together on the dynamics of the system. The premium standards of gains and related parameters of the PDN(FOID) controller are in Table 7. The dynamic responses for the system with and without both WavePPt and PV are contrasted in Figure 10. Critical assessment in Figure 10 reveals the better performance of the system with both WavePPt and PV with regard to the lessened values of Pk_O, Pk_U, and S_Time. With both WavePPt and PV, the values of Δf_1 showed improvement in Pk_O, Pk_U, and S_Time from 0.0007 to 0.00001 Hz, 0.0112 to 0.0063 Hz, and 25.81 to 22.65 s (in comparison to the system without WavePPt and PV). Similarly, for Δf_2, the improvement is 0.0038 to 0.00125 Hz (Pk_U) and 35.31 to 27.98 s; for $\Delta Ptie_{1-2}$, the improvement is 0.0039 to 0.00178 Hz and 21.71 to 19.81 s; and for $\Delta Ptie_{1-3}$, the improvement is 0.0039 to 0.00161 Hz and 21.04 to 20.01 s. Thus, the presence of both WavePPt and PV has a noteworthy impact on the system's performance, so further analysis is carried out for both WavePPt and PV.

Table 7. Best values of gains and related parameters of PDN(FOID) controllers for scheme-4 (thermal–bio-diesel–GPP-Ss (GT) system including both WavePPt, PV).

Name of Controller	Corresponding Gains and Correlated Parameters	Area-1	Area-2	Area-3
PDN(FOID)	K_{Pi} *	0.9899	0.8510	0.6775
	K_{Di} *	0.5957	0.4620	0.4404
	N_i *	39.08	50.13	42.96
	K_{FIi} *	0.4607	0.8522	0.5740
	λ_i *	1.0916	0.9588	0.7276
	K_{FDi} *	0.9379	0.8374	0.9869
	μ_i *	0.9947	0.7315	0.7899

* Signify optimum values.

6.7. Evaluation of the Influence of C^{ES} through/without Duple Compensation on Scheme Changing Aspects

In order to stabilize the system energy storing device, C^{ES} is incorporated into the system. Initially, C^{ES} without duple compensation is integrated into system-4 in all areas. Hence, this is system-5. The best-obtained controller PDN(FOID) from previous sections is used here to assess the impact of the energy-storing device. In this case, the best values of gains and parameters PDN(FOID) obtained using SHO are marked down in Table 8. Next, system-4 is integrated with C^{ES} with duple compensation in all areas. This is considered as system-6. The gains and related parameters of the PDN(FOID) controller for the scheme through C^{ES} involving duple compensation are provided in Table 9. With the values obtained in Tables 8 and 9, the dynamic responses are plotted and contrasted against the system response without having any forms of CES. This contrast is reflected in Figure 11. In Figure 11, it is observed that any form of C^{ES} has a better impact. However, if the responses for the system with/without C^{ES} with duple compensation are viewed, then they articulate the better performance of the system in existence with C^{ES} with duple compensation in terms of diminished Pk_U and S_Time. With C^{ES}, the values of Δf_2 showed improvement in Pk_U and S_Time from 0.00125 to 0.0008 Hz and 27.98 to 21.82 s (in comparison to the system without energy storage). Similarly, for Δf_3, the improvement is 0.00125 to 0.00078 Hz and 19.51 to 18.65 s; for $\Delta Ptie_{1-2}$, the improvement is 0.00178 to 0.00145 Hz and 19.81 to 17.55 s; for $\Delta Ptie_{1-3}$, the improvement is 0.00161 to 0.00143 Hz and 20.01 to 18.65 s. Next, with C^{ES} with duple compensation, the values of Δf_2 showed improvement in Pk_U and S_Time from 0.0008 to 0.0006 Hz and 21.82 to 13.65 s (in comparison to the system with C^{ES}). Similarly, for Δf_3, the improvement is 0.00078 to 0.00075 Hz and 18.65 to 13.75 s; for $\Delta Ptie_{1-2}$, the improvement is 0.00145 to 0.00123 Hz and 18.65 to 16.61 s; for $\Delta Ptie_{1-3}$, the improvement is 0.00143 to 0.00132 Hz and 18.65 to 16.58 s. So, C^{ES} with duple compensation will serve better in damping out oscillations and stabilizing the system when there are any fluctuations in demand.

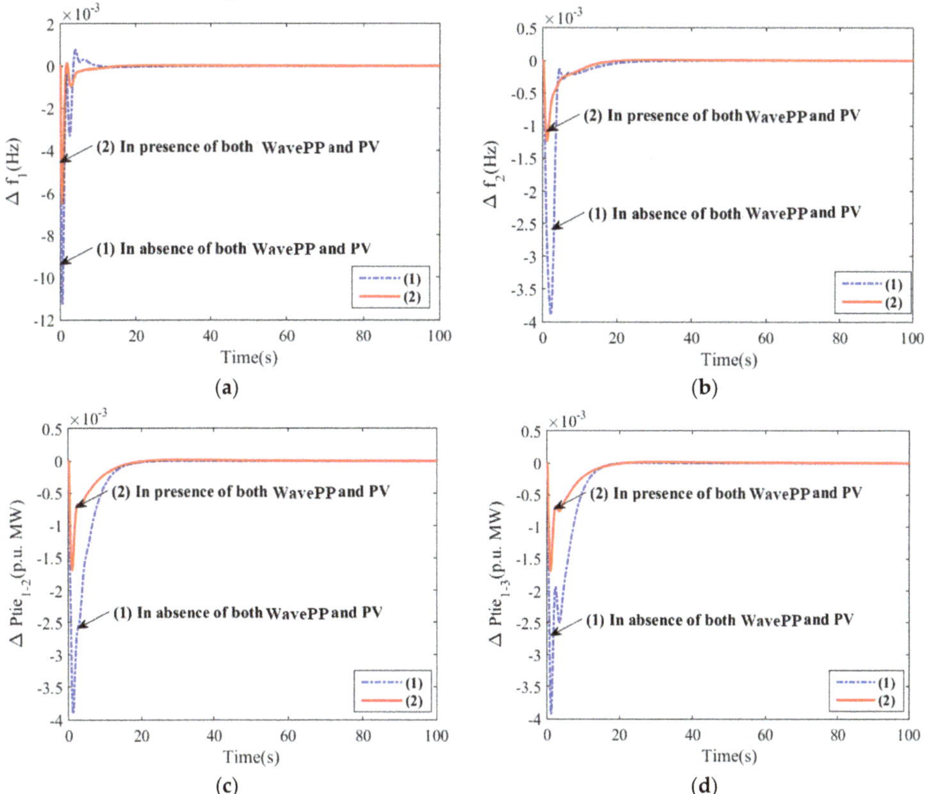

Figure 10. Assessment of the impact of both WavePPt for system-4 contrast time: (**a**) Anomaly of frequency in area-1, (**b**) Anomaly of frequency in area-2 vs. time, (**c**) Anomaly of Tie-line power interlinking area-1 and area-2, (**d**) Anomaly of Tie-line power interlinking area-1 and area-3.

Table 8. Finest values of gains and related parameters of PDN(FOID) controllers for scheme-5 (thermal–bio-diesel–GPP-Ss (GT) system including WavePPt, PV, and C^{ES} without duple compensation).

Name of Controller	Corresponding Gains and Correlated Parameters	Area-1	Area-2	Area-3
	K_{Pi} *	0.9897	0.9769	0.2093
	K_{Di} *	0.6681	0.5151	0.8276
	N_i *	74.90	63.37	93.11
PDN(FOID)	K_{FIi} *	0.5953	0.9368	0.5812
	λ_i *	1.0814	1.2086	0.9863
	K_{FDi} *	0.8434	0.9893	0.7160
	μ_i *	0.5678	0.0027	0.6173

* Signify optimum values.

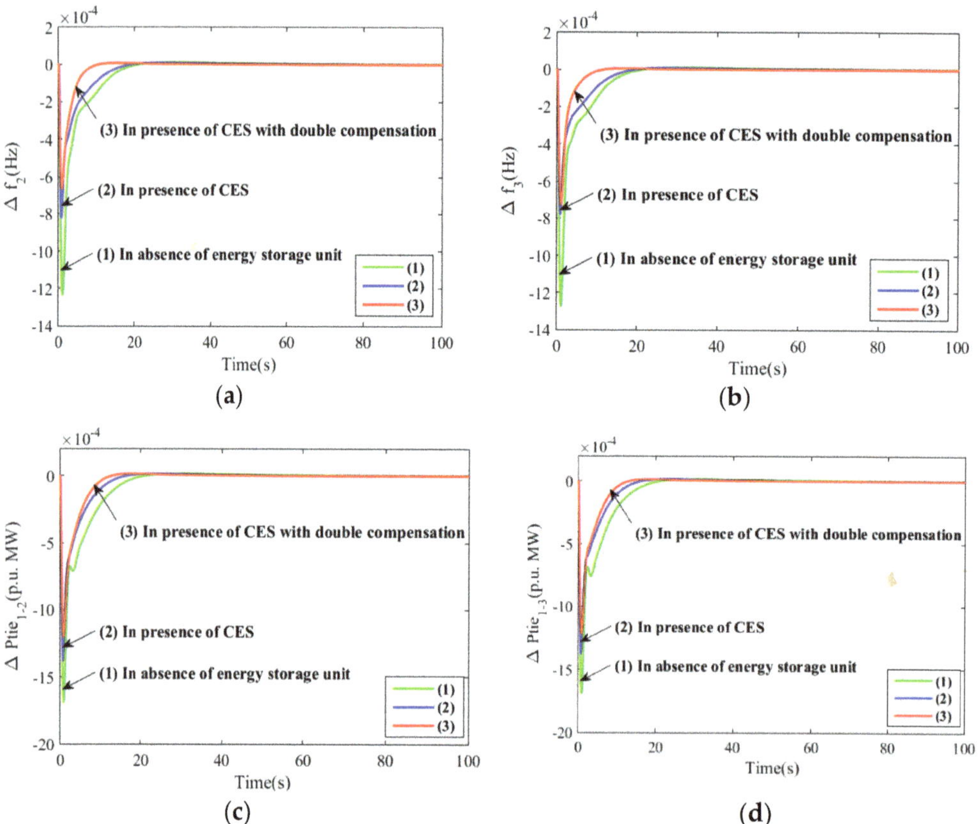

Figure 11. Valuation of influence of C^{ES} with/without duple compensation on scheme dynamics employing SHO-optimized PDN(FOID) controller taking into account 1% step load disturbances for system-5 and system-6 individually contrast time: (**a**) Anomaly of frequency in area-2, (**b**) Anomaly of frequency in area-3, (**c**) Anomaly of Tie-line power interlinking area-1 and area-2, (**d**) Anomaly of Tie-line power interlinking area-1 and area-3.

Table 9. Finest values of gains and related parameters of PDN(FOID) controllers for scheme-6 (thermal–bio-diesel–GPP-Ss (GT) system including WavePPt, PV, and C^{ES} with duple compensation) (* Signify optimum values).

Name of Controller	Corresponding Gains and Correlated Parameters	Area-1	Area-2	Area-3
PDN(FOID)	K_{Pi} *	0.9899	0.9288	0.1632
	K_{Di} *	0.9785	0.4533	0.4110
	N_i *	44.19	54.55	43.39
	K_{FIi} *	0.8900	0.9529	0.4464
	λ_i *	1.0914	0.0095	0.1938
	K_{FDi} *	0.9632	0.9630	0.7367
	μ_i *	0.7989	0.7919	0.8705

6.8. Sensitivity Determination When Subjected to Random Disturbance

The exploration of sensitivity is executed to perceive the heftiness of SHO-augmented PDN(FOID) controller gains traced at the basic event to comprehensive variance in the structure state of the thermal–bio-diesel–WavePPt in area-1, thermal–GPP in area-2, and thermal–Ss (GT)–PV in area-3 schemes along with C^{ES} with duple compensation. Here, the inspected scheme is stated with a random pattern of disturbance from a basic 1% step load disturbance. The augmented gains of the PDN(FOID) controller reflected in Table 10 are attained by retaining the SHO. The dynamic outcomes for the finest values analogous to the foundation and diverse outcomes are distinguished in Figure 12. In Figure 12a–d, it is indicated that all responses obtained with optimized values for the proposed controller (from Table 9) are stable for random load disturbance. Evaluation affirms that the outcomes are moderately identical, which claims that there are no circumstances for the added retuning of the finest values for modification.

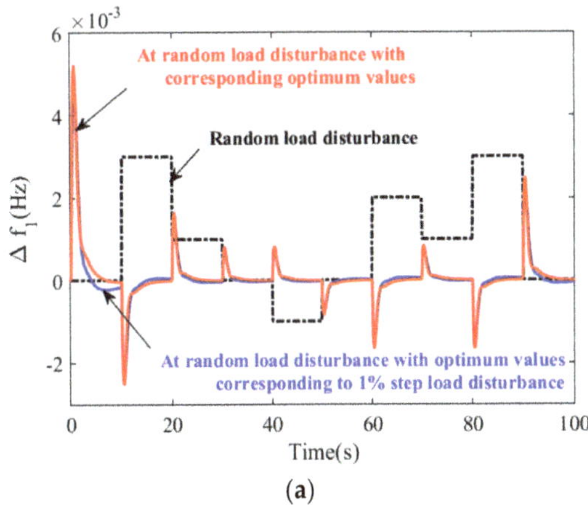

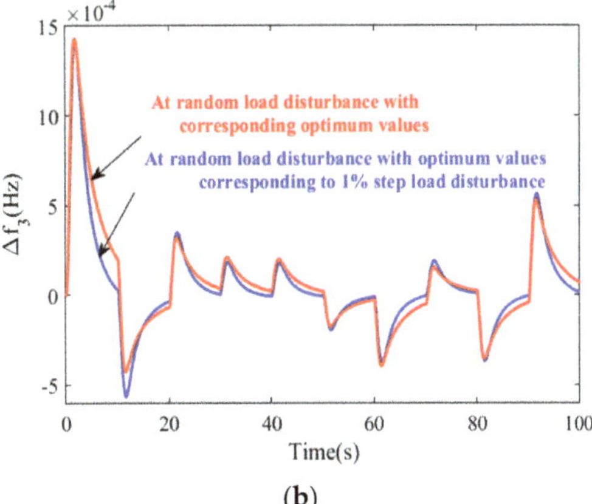

Figure 12. *Cont.*

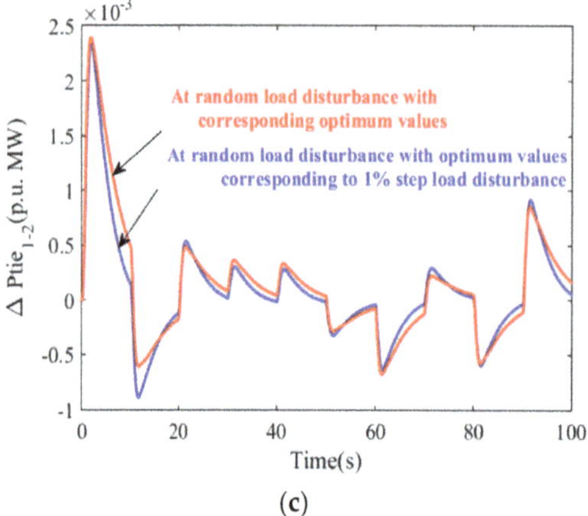

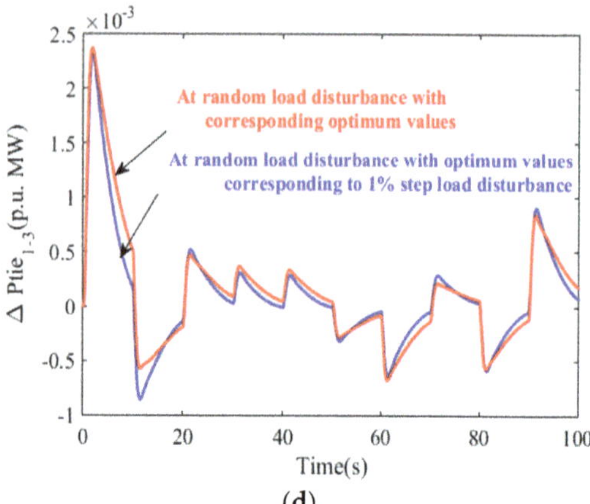

(c)   (d)

Figure 12. Evaluation of system-6 outcomes employing SHO-optimized PDN(FOID) controller after exposure to random load disturbance contrast time: (**a**) Anomaly of frequency in area-1, (**b**) Anomaly of frequency in area-3, (**c**) Anomaly of Tie-line power interlinking area-1 and area-2, (**d**) Anomaly of Tie-line power interlinking area-1 and area-3.

Table 10. Finest values of gains and correlated parameters of PDN(FOID) controllers for scheme-6 when conditional to a random pattern of disturbance.

Name of Controller	Corresponding Gains and Correlated Parameters	Area-1	Area-2	Area-3
PDN(FOID)	K_{Pi} *	0.8095	0.8965	0.2588
	K_{Di} *	0.8921	0.5145	0.5147
	N_i *	48.36	62.36	51.36
	K_{FIi} *	0.8478	0.9147	0.5147
	λ_i *	1.0549	0.0089	0.2147
	K_{FDi} *	0.8956	0.8899	0.8144
	μ_i *	0.6897	0.7514	0.8566

* Signify optimum values.

7. Conclusions

An innovative endeavor was placed to put into effect a two-stage controller through the amalgamation of a proportional-derivative with filter (PDN) (InO) and integral-derivative (FOID) (FrO) in AGC. A lately established biologically influenced meta-heuristic algorithm articulated as the spotted hyena optimizer (SHO) was proficiently employed for attaining gains and additional related parameters of diverse controllers. The superiority of PDN(FOID) was realized over additional classical controllers for the thermal–bio-diesel–geothermal power plant (GPP)–split-shaft gas turbine (Ss (GT)) system. The assessment shows that when the basic system was incorporated with either the wave power plant (WavePPt) or photovoltaic (PV) unit, the system responses improved in terms of lessened peak_overshoot, peak_undershoot, and settling_time using the PDN(FOID) controller. Much better responses were obtained when the system was involved with both the WavePPt and PV. Analysis also revealed the better performance of capacitive energy storage (C^{ES}) with duple compensation over the system with C^{ES} without duple compensation and systems without any form of C^{ES}. The robustness of the PDN(FOID) controller was undertaken by examining the system (having WavePPt, PV, and C^{ES} with duple compensation) when subjected to a random load of disturbance. It demonstrated that the finest values attained, considering PDN(FOID) gains and additional related constraints, were satisfactory and sufficient, and there were not, by any means, alterations under random load.

Author Contributions: Conceptualization, A.S. and P.D.; methodology, N.R.B.; software, T.C.; validation, A.S., P.D. and N.R.B.; formal analysis, N.R.B.; investigation, A.S.; resources, T.C. and B.V.; data curation, T.C.; writing—original draft preparation, A.S.; writing—review and editing, A.S., T.C., B.V. and Ł.K.; visualization, P.D.; supervision, N.R.B.; adaptation of optimization procedure, Ł.K. All authors have read and agreed to the published version of the manuscript.

Funding: Optimization procedure research was partially supported by the Poznan University of Technology, grant number (0212/SBAD/0568).

Institutional Review Board Statement: Not applicable.

Informed Consent Statement: Not applicable.

Data Availability Statement: Not applicable.

Conflicts of Interest: The authors declare no conflict of interest.

Nomenclature

f	Balance point valuation of frequency measured in Hertz (Hz).
*	Best aggregate suggested by exponent.
k	Count of areas suggested.
B_k	Portion of frequency bias of areas engaged.
T_{kj}	Portion of synchronization.
T	Total moment of simulation measured in seconds.
Δf_k	Modification of frequency of areas engaged.
ΔP_{Dk}	Degree of load modification of areas engaged.
H_k	Degree of inertia constant of areas engaged.
D_k	Degree of load modification of areas engaged (p.u. MW)/Modification of frequency of areas engaged (Hz).
R_k	Factor related to governor's speed regulation of area suggested.
β_k	Attributions of frequency outcome of area suggested.
K_{pk}	Gain constant of power system representation.
T_{pk}	Time constant of power system prototypes.
P_{rk}	Considerable rated power of area suggested.
a_{kj}	Considerable rated power of area-k/Considerable rated power of area-j
pf	Area contribution factor.
T_g, T_t, T_r	Time constants of thermal generating parts in seconds (s).
K_r	Reheater's gain.

Appendix A

1. Basic Power System: f = 60 Hz; Primary loading = 50%, K_{pm} = 120 Hz/p.u. MW, T_{pm} = 20 s, T_m = 0.086 s, H_m = 5 s; D_m = 0.00833 p.u. MW/Hz; β_m = 0.425 p.u. MW/Hz; R_m = 2.4 Hz/p.u MW
2. Thermal: T_{rm} = 10 s, K_{rm} = 5, T_{tm} = 0.3 s, T_{gm} = 0.08 s;
3. Bio-diesel: Kvr = 1, Tvr = 0.05 s, Kce = 1, Tce = 0.5 s;
4. Ss(GT): L_{max} = 1, T_3 = 3 s, $T_1 = T_2$ = 1.5 s, K_T = 1, FOVmax = 1, FOVmin = −0.02, Dtur = 0 p.u;
5. CES: KCES = 0.3, TCES = 0.0352 s;
6. CES with duple compensation: KCES (duple compensation) = 0.3, TCES (duple compensation) = 0.046 s, T_1 = 0.280, T_2 = 0.025, T_3 = 0.0411, T_4 = 0.39;
7. PV: A = 900, B = −18, C = 100, D = 50.

References

1. Elgerd, O.I. *Electric Energy Systems Theory: An Introduction*; McGraw-Hill: New Delhi, India, 2007.
2. Kundur, P. *Power System Stability and Control*; McGraw-Hill: New Delhi, India, 1993.
3. Ibraheem Kumar, P.; Kothari, D.P. Recent philosophies of automatic generation control strategies in power systems. *IEEE Trans. Power Syst.* **2005**, *20*, 346–357. [CrossRef]

4. Das, D.; Aditya, S.K.; Kothari, D.P. Dynamics of diesel and wind turbine generators on an isolated power system. *Electr. Power Energy Syst.* **1999**, *21*, 183–189. [CrossRef]

5. Das, D.C.; Roy, A.K.; Sinha, N. GA based frequency controller for solar thermal–diesel–wind hybrid energy generation/energy storage system. *Int. J. Electr. Power Energy Syst.* **2012**, *43*, 262–279. [CrossRef]

6. Ismayil, C.; Sreerama, K.R.; Sindhu, T.K. Automatic generation control of single area thermal power system with fractional order PID (PIλDμ) controllers. In Proceedings of the 3rd International Conference on Advances in Control and Optimization of Dynamical Systems, Kanpur, India, 13–15 March 2014; pp. 552–557.

7. Elgerd, O.I.; Fosha, C.E. Optimum Megawatt-Frequency Control of Multiarea Electric Energy Systems. *IEEE Trans. Power Appar. Syst.* **1970**, *89*, 556–563. [CrossRef]

8. Bhatt, P.; Roy, R.; Ghoshal, S.P. GA/particle swarm intelligence based optimization of two specific varieties of controller devices applied to two-area multi-units automatic generation control. *Int. J. Electr. Power Energy Syst.* **2010**, *32*, 299–310. [CrossRef]

9. Nanda, J.; Saikia, L.C. Comparison of performances of several types of classical controller in automatic generation control for an interconnected multi-area thermal system. In Proceedings of the Power Engineering Conference, 2008. AUPEC '08. Australasian Universities, Sydney, Australia, 14–17 December 2008.

10. Nanda, J.; Mishra, S.; Saikia, L.C. Maiden application of bacterial foraging based optimization technique in multi area automatic generation control. *IEEE Trans. Power Syst.* **2009**, *24*, 602–609. [CrossRef]

11. Golpira, H.; Bevrani, H.; Golpira, H. Application of GA optimization for automatic generation control design in an interconnected power system. *Energy Convers. Manag.* **2011**, *52*, 2247–2255. [CrossRef]

12. Dhamanda, A.; Dutt, A.; Bhardwaj, A.K. Automatic Generation Control in Four Area Interconnected Power System of Thermal Generating Unit through Evolutionary Technique. *Int. J. Electr. Eng. Inform.* **2015**, *7*, 569–583.

13. Saikia, L.C.; Nanda, J.; Mishra, S. Performance comparison of several classical controllers in AGC for multi-area interconnected thermal system. *Int. J. Electr. Power Energy Syst.* **2011**, *33*, 394–401. [CrossRef]

14. Nanda, J.; Mangla, A.; Suri, S. Some new findings on automatic generation control of an interconnected hydrothermal system with conventional controllers. *IEEE Trans. Energy Convers.* **2006**, *21*, 187–194. [CrossRef]

15. Saikia, L.C.; Chowdhury, A.; Shakya, N.; Shukla, S.; Soni, P.K. AGC of a multi area thermal system using firefly optimized IDF controller. In Proceedings of the 2013 Annual IEEE India Conference (INDICON), Mumbai, India, 13–15 December 2013.

16. Pradhan, P.C.; Sahu, R.K.; Panda, S. Firefly algorithm optimized fuzzy PID controller for AGC of multi-area multi-source power systems with UPFC and SMES. *Eng. Sci. Technol. Int. J.* **2016**, *19*, 338–354. [CrossRef]

17. Saha, A.; Saikia, L.C. Renewable energy source-based multiarea AGC system with integration of EV utilizing cascade controller considering time delay. *Int. Trans. Electr. Energy Syst.* **2019**, *29*, e2646. [CrossRef]

18. Arya, Y. AGC of PV-thermal and hydro-thermal energy systems using CES and a new multistage FPIDF-(1+PI) controller. *Renew. Energy* **2019**, *134*, 796–806. [CrossRef]

19. Tasnin, W.; Saikia, L.C. Performance comparison of several energy storage devices in deregulated AGC of a multi-area system incorporating geothermal power plant. *IET Renew. Power Gener.* **2018**, *12*, 761–772.

20. Barik, A.K.; Das, D.C. Expeditious frequency control of solar photobiotic/biogas/biodiesel generator based isolated renewable microgrid using Grasshopper Optimisation Algorithm. *IET Renew. Power Gener.* **2018**, *12*, 1659–1667.

21. Hasanien, M.H. Whale optimisation algorithm for automatic generation control of interconnected modern power systems including renewable energy sources. *IET Gener. Transm. Distrib.* **2018**, *12*, 607–614.

22. Dhundhara, S.; Verma, Y.P. Capacitive energy storage with optimized controller for frequency regulation in realistic multisource deregulated power system. *Energy* **2018**, *147*, 1108–1128.

23. Pathak, N.; Verma, A.; Bhatti, T.S.; Nasiruddin, I. Modeling of HVDC tie-links and their utilization in AGC/LFC operations of multi-area power systems. *IEEE Trans. Ind. Electron.* **2019**, *66*, 2185–2197. [CrossRef]

24. Kumari, N.; Malik, N.; Jha, A.N.; Mallesham, G. Design of PI Controller for Automatic Generation Control of Multi Area Interconnected Power System using Bacterial Foraging Optimization. *Int. J. Eng. Technol.* **2016**, *8*, 2779–2786. [CrossRef]

25. Jagatheesan, K.; Anand, B.; Samanta, S.; Dey, N.; Ashour, A.S.; Balas, V.E. Design of a proportional-integral-derivative controller for an automatic generation control of multi-area power thermal systems using firefly algorithm. *IEEE/CAA J. Autom. Sin.* **2019**, *6*, 503–515.

26. Dash, P.; Saikia, L.C.; Sinha, N. Comparison of performances of several FACTS devices using Cuckoo search algorithm optimized 2DOF controllers in multi-area AGC. *Electr. Power Energy Syst.* **2015**, *65*, 316–324. [CrossRef]

27. Rahman, A.; Saikia, L.C.; Sinha, N. AGC of dish-stirling solar thermal integrated thermal system with biogeography based optimised three degree of freedom PID controller. *IET Renew. Power Gener.* **2016**, *10*, 1161–1170. [CrossRef]

28. Reddy, P.J.; Kumar, T.A. AGC of three-area hydro-thermal system in deregulated environment using FOPI and IPFC. In Proceedings of the 2017 International Conference on Energy, Communication, Data Analytics and Soft Computing (ICECDS), Chennai, Indian, 1–2 August 2017; pp. 2815–2821. [CrossRef]

29. Pan, I.; Das, S. Fractional order AGC for distributed energy resources using robust optimization. *IEEE Trans. Smart Grid* **2016**, *7*, 2175–2186. [CrossRef]

30. Dash, P.; Saikia, L.C.; Sinha, N. Automatic generation control of multi area thermal system using Bat algorithm optimized PD–PID cascade controller. *Int. J. Electr. Power Energy Syst.* **2015**, *68*, 364–372. [CrossRef]

31. Saha, A.; Saikia, L.C. Utilisation of ultra-capacitor in load frequency control under restructured STPP-thermal power systems using WOA optimised PIDN-FOPD controller. *IET Gener. Transm. Distrib.* **2017**, *11*, 3318–3331. [CrossRef]
32. Sahu, R.K.; Panda, S.; Rout, U.K.; Sahoo, D.K. Teaching learning based optimization algorithm for automatic generation control of power system using 2-DOF PID controller. *Int. J. Elect. Power Energy Syst.* **2016**, *77*, 287–301. [CrossRef]
33. Haroun, A.H.G.; Li, Y.-Y. A novel optimized hybrid fuzzy logic intelligent PID controller for an interconnected multi-area power system with physical constraints and boiler dynamics. *ISA Trans.* **2017**, *71*, 364–379. [CrossRef]
34. Padhan, S.; Sahu, R.K.; Panda, S. Application of firefly algorithm for load frequency control of multi-area interconnected power system. *Electr. Power Compon. Syst.* **2014**, *42*, 1419–1430. [CrossRef]
35. Sharma, Y.; Saikia, L.C. Automatic generation control of a multi-area ST—thermal power system using Grey Wolf Optimizer algorithm based classical controllers. *Electr. Power Energy Syst.* **2015**, *73*, 853–862. [CrossRef]
36. Kumar, N.; Kumar, V.; Tyagi, B. Multi area AGC scheme using imperialist competition algorithm in restructured power system. *Appl. Soft Comput.* **2016**, *48*, 160–168. [CrossRef]
37. Dash, P.; Saikia, L.C.; Sinha, N. Flower pollination algorithm optimized PI-PD cascade controller in automatic generation control of a multi-area power system. *Int. J. Electr. Power Energy Syst.* **2016**, *82*, 19–28. [CrossRef]
38. Khan, M.H.; Ulasyar, A.; Khattak, A.; Zad, H.S.; Alsharef, M.; Alahmadi, A.A.; Ullah, N. Optimal Sizing and Allocation of Distributed Generation in the Radial Power Distribution System Using Honey Badger Algorithm. *Energies* **2022**, *15*, 5891. [CrossRef]
39. Sun, S.; Zhang, Q.; Sun, J.; Cai, W.; Zhou, Z.; Yang, Z.; Wang, Z. Lead–Acid Battery SOC Prediction Using Improved AdaBoost Algorithm. *Energies* **2022**, *15*, 5842. [CrossRef]
40. Zhu, L.; He, J.; He, L.; Huang, W.; Wang, Y.; Liu, Z. Optimal Operation Strategy of PV-Charging-Hydrogenation Composite Energy Station Considering Demand Response. *Energies* **2022**, *15*, 5915. [CrossRef]
41. Dhiman, G.; Kumar, V. Spotted hyena optimizer: A novel bio-inspired based metaheuristic technique for engineering applications. *Adv. Eng. Softw.* **2017**, *114*, 48–70. [CrossRef]
42. Chiranjeevi, T.; Biswas, R.K. Discrete-Time Fractional Optimal Control. *Mathematics* **2017**, *5*, 25. [CrossRef]
43. Chiranjeevi, T.; Biswas, R.K. Closed-Form Solution of Optimal Control Problem of a Fractional Order System. *J. King Saud Univ. –Sci.* **2019**, *31*, 1042–1047. [CrossRef]
44. Chiranjeevi, T.; Biswas, R.K.; Devarapalli, R.; Babu, N.R.; Marquez, P.G. On optimal control problem subject to fractional order discrete time singular systems. *Arch. Control. Sci.* **2021**, *31*, 849–863.
45. Oustaloup, A.; Mathieu, B.; Lanusse, P. The CRONE Control of Resonant Plants: Application to a Flexible Transmission. *Eur. J. Control.* **1995**, *1*, 113–121. [CrossRef]
46. Knypiński, Ł.; Kuroczycki, S.; Márquez, F.P.G. Minimization of Torque Ripple in the Brushless DC Motor Using Constrained Cuckoo Search Algorithm. *Electronics* **2021**, *10*, 2299.
47. Devarapalli, R.; Kumar Sinh, N.; Bathinavenkateswara, R.; Knypiński, Ł.; Jaya Naga, N.; Marquez, F.P.G. Allocation of real power generation based on computing over all generation cost: An approach of Salp Swarm Algorithm. *Arch. Electr. Eng.* **2021**, *70*, 337–349. [CrossRef]

Article

Evaluation of Technical Solutions to Improve Transient Stability in Power Systems with Wind Power Generation

Giuseppe Marco Tina *, Giovanni Maione and Sebastiano Licciardello

Department of Electrical, Electronic and Computer Engineering, University of Catania, 95125 Catania, Italy
* Correspondence: giuseppe.tina@unict.it

Abstract: Reliability and safety must be carefully considered in today's power systems, which are rapidly evolving toward ever higher penetration of renewable, inverter-based generation units. Power systems are constantly stressed by active power disturbances, which can be exacerbated by wind and solar systems that are subject to rapid fluctuations in primary energy. In this framework, a comparative technical analysis of solutions to improve transient stability, both rotor angle stability and frequency stability, is carried out. These solutions can be adopted by the transmission system operator (e.g., an additional parallel transmission line), by the generation companies (e.g., a fast excitation system), or by both, such as SVC (static VAR compensator) and STATCOM (static synchronous compensator). Sensitivity analyses were carried out to assess the impact of the location of the wind turbines in the buses of the grid on their rated power and level of production. On the basis of these analyses, the worst-case fault was considered, and the critical fault recovery time was determined as an engineering parameter to compare the different solutions. For the numerical analysis, a modified IEEE 9-bus system was considered, and the PowerWorld software tool was used. Rotor angle and frequency stability analyses were performed.

Keywords: power system simulation; power system stability; wind power generation; static VAr compensators

Citation: Tina, G.M.; Maione, G.; Licciardello, S. Evaluation of Technical Solutions to Improve Transient Stability in Power Systems with Wind Power Generation. *Energies* **2022**, *15*, 7055. https://doi.org/10.3390/en15197055

Academic Editor: Juri Belikov

Received: 31 August 2022
Accepted: 21 September 2022
Published: 26 September 2022

Publisher's Note: MDPI stays neutral with regard to jurisdictional claims in published maps and institutional affiliations.

1. Introduction

The energy model that has been introduced and developed in the last century to enable economic development is no longer sustainable under these new climate and environmental conditions. For instance, in the current Electrical Power System (EPS) model, thermoelectric generation is strongly predominant; this current source of energy today is the main cause of greenhouse gas emissions, including CO_2. The increase in these emissions has highlighted the need for evident and rapid decarbonization and efficiency of all energy sectors to limit their impacts on the environment and achieve a more sustainable system.

To this end, several national and international agreements have been reached, such as the PNIEC (Integrated National Plan for Energy and Climate 2030), the Clean Energy Package, and the European Green Deal, which set various strategic goals to counteract the increase in global average temperature. One of the most challenging goals is the strong increase in energy generation from renewable energy sources, i.e., wind and photovoltaic.

The generation of Renewable Energy Sources (RESs), has been identified as one of the main enablers of the transition. Thanks to their operation principles, these types of energy generation allow the production of energy without CO_2 emissions, and they help reduce energy prices, which (ultimately) benefits the consumers.

The power plants based on RES, specifically wind and solar generators, are characterized by nonprogrammable production profiles, which means that the electricity produced by such plants does not follow the dynamics of energy demand for consumption, but the dynamic characteristics of each energy source. Moreover, this type of technology is mainly based on inverter technology, i.e., systems connected to the grid through static components, which have a lower tendency to comply with the variation of power system the basic parameters for the safe operation of the grid than the traditional groups, that interface through the use of rotating machines [1,2].

These properties have the following implications for the management of the system [3,4]:

- Reduction in the number of generation resources capable of providing frequency regulation (active power regulation) and voltage regulation (reactive power regulation) services;
- Reduction in electromechanical inertia, which worsens the dynamic response after disturbances;
- Reduction in short-circuit power, which worsens the quality of service and the risk of propagation of voltage disturbances;
- Reduction in the margin to cover peak loads, which can be demonstrated in hours to a low production of RES;
- Increasing periods of overgeneration (production greater than demand) and line congestions, which without sufficient storage capacity or reserve can cause cuts in the energy produced;
- Increase in steepness of the evening ramp of the residual load, due to the sudden drop in the production of solar plants in the darkest hours, so that a fast response from flexible systems is essential;
- Increase in reserve requirements due to the increasing presence of nonprogrammable renewable energy sources (NP-RES)-based power plants and their randomness.

Looking at these impacts, the increase in the share of RES generation brings the electricity system to conditions of lower reliability, safety, and stability. In this context, stability, particularly transient stability, plays an important role as an index of robustness of the system, subject by its nature to events such as faults and disturbances [5].

Transient stability represents the ability of the system to withstand transient events caused by a disturbance, generally important, e.g., a short-circuit, and to reach the nominal operating conditions. For this reason, it has been the subject of many studies and is already described in the literature [6,7]. A part of this also applies to the impact on transient stability caused by wind turbine integration in the current traditional electric power systems [8,9].

This article first examines the impact of RES generation (in this case, wind energy) on the transient stability of the system when wind energy is added to existing traditional generation and when it replaces some of that generation. The main objective of this work is to analyze and compare some of the solutions to improve the transient stability, such as SVC (static VAR compensator), STATCOM (static synchronous compensator), fast excitation system, and the doubling of a transmission line, as performed in [6]. In this way, it is possible to evaluate and compare the effects of these solutions in a system with RES generation in two different scenarios and then compare them with those of a traditional system.

The PowerWorld Simulator is used for this paper. The IEEE 9-bus system used as a case study in [6] is modified by adding wind turbines. The decision to start with the IEEE 9-bus system is due to the widespread use of this grid, which allows for comparison, and its simplicity, which allows for various sensitivity analyses thanks to the small number of buses; this type of analysis would indeed be complicated and time-consuming for larger grids. As in [6], the system is modeled using component that are already tested and present in the IEEE or WECC lists and that are present in most power system simulator libraries. This is implemented to reach general and reproducible results. However, specific studies about new wind turbine controllers are present in the literature (e.g., [10,11]), which investigate new solutions to face adequately the stability issues in power system with low inertia.

The contributions of this paper are as follows:

- Sensitivity analysis to identify the generation bus in which to insert the wind generator, so as to define the worst operating condition in terms of transient stability following the insertion of the wind plant;

- Sensitivity analysis on the share of installed and generated wind power, to evaluate the impacts of the different share on the network following a failure;
- An evaluation of the impact on transient rotor angle and frequency stability, through the installation of SVC and STATCOM. These solutions provide TSOs (transmission system operators) and GENCOs (generation companies) with dynamic control during contingencies.

The paper is organized as follows: Section 2 presents the solutions to improve the transient stability. The studied system is described in Section 3. In Section 4, sensitivity to wind position, base power, and injected power is investigated. In Section 5, the improvement solutions are tested in two different scenarios. Conclusions are presented in Section 6, where the improvement solutions are summarized, and their main features are highlighted.

2. Description of Enhancement Solutions

According to [12], the solutions used to enhance the transient stability in power system can be classified as conventional and RES-based. Here, the focus is on the conventional techniques. The re-dispatching technique is not considered because no optimal power flow study is performed.

This study considers two types of solutions, as suggested in [12]:

- A preventive one, via the reduction in system reactance. To do this, the addition of a parallel transmission lines is considered;
- An emergency one, where the following solutions can be adopted:
  - The fast excitation system helps to reduce oscillations in the electrical system by controlling the generator voltage through the field-excitation circuit, under transient conditions;
  - SVC and STATCOM can regulate the voltage at a selected bus by varying their reactive power, and they can help improve transient stability by increasing the synchronization power flow between generators.

The two solutions mentioned above belong to the FACTS (flexible alternating current transmission systems) family. Their main features are as follows:

- The SVC (static VAR compensator) consists of passive elements, capacitors, and inductances, controlled by static devices, such as thyristors. Their performance depends on the voltage of the connection bus;
- The STATCOM (static synchronous compensator), unlike the SVC, does not need passive elements. It is basically a voltage source converter, whose performance is independent of the voltage at the connection bus.

3. IEEE 9-Bus System Description

The test system chosen for this study is the IEEE 9-bus system, which is part of a series of test systems often used for studies on the stability, steady state, and dynamics of power systems [8,13,14].

Here, a modified version of the IEEE 9-bus system is proposed. Starting from the modified version illustrated in [6], one more change is made, i.e., the addition of a type 3 wind turbine generator at bus 7. This is implemented to consider the effect of RES generation in power systems. The location of the wind turbine generator in this node is justified later.

The modified IEEE 9-bus system used in this article is shown in Figure 1, where the load flow results are shown for the case where the proposed solutions are not used or are switched off. Except for the wind generator, which is always connected, the proposed solutions are alternatively connected. In Figure 1, the corresponding connections are shown as empty squares. Their connections are also made to study their impact on the system in the different proposed simulations and scenarios.

As in [6], the transient stability analysis of the proposed system is performed by simulating a three-phase short-circuit. The frequency response of the load is not considered.

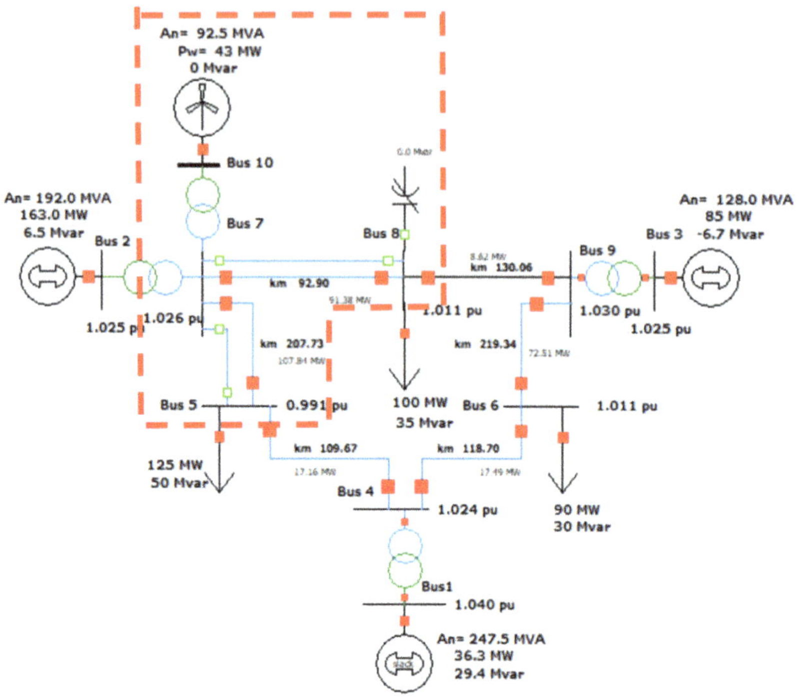

Figure 1. Modified IEEE 9-bus system with wind power generator and transient stability solutions (portion of the system identified by the dashed red line).

According to [15], the models used, in the PowerWorld Simulator, for implementing the type 3 wind turbine generator, i.e., DFIG (doubly-fed induction generator) wind generator, are shown in Table 1. Table 2 indicates the parameters set needed for the chosen control mode.

Table 1. Wind type 3 models.

Power Plant Parts	Type 3 Model
Plant controller	REPC_A
Generator/converter	REGC_A
Electrical control	REEC_A
Turbine	WTGT_A
Aerodynamics	WTGAR_A
Pitch controller	WTGPT_A
Torque controller	WTGTRQ_A

Table 2. Wind type 3 control mode selection.

Functionality	Models	PfFlag	Vflag	Qflag	RefFlag
Plant level V control + local coordinated V/Q control	REEC_A, REPC_A	0	1	1	1

Here, the analysis follows the steps shown below.

- Load-flow analysis is performed to initialize the system;
- A three-phase symmetrical short-circuit fault is applied to a transmission line at 1 s;
- The fault clearing time, FCT, is set to 0.300 s;

- The duration of the simulation is 10 s;
- The simulation time step is 5 ms.

4. Improved System Description

4.1. Sensitivity Analysis of Wind Generator Location

This analysis was performed to determine which wind generator installation bus had the greatest effect on the transient stability of the system. For this purpose, the wind turbine generator was installed in the three generator buses, i.e., bus 4, 7, and 9, and the maximum rotor angle values assumed by generators 2 and 3 (synchronous generators) were evaluated. The rated power of the wind turbine generator was 50 MVA, and its operative active power production was 25 MW. These two values were chosen considering the percentage of wind turbines installed in Italy (about 9.5%) and the energy produced by them (about 7.4%) for the year 2021, as indicated in [16].

The results of this analysis are presented in Table 3, where the maximum rotor angle is also given for the case where no wind generator is installed.

Table 3. Maximum rotor angle values for different wind bus positioning.

Gen.	No Wind	Wind@Bus 4	Wind@Bus 7	Wind@Bus 9
Gen 2	119.86°	126.49°	136.60°	136.33°
Gen 3	51.74°	55.71°	61.68°	61.38°

From Table 3 it can be seen that the maximum rotor angle for generators 2 and 3 was reached when the wind generator was installed at bus 7. Therefore, this system configuration was used going forward. In [6], the focus was on generator 2, which was also the most loaded generator in this case.

4.2. Sensitivity Analysis of Wind Generator Rated Power

Another sensitivity analysis was performed taking as parameter the rated power, An (in MVA), of the wind generator. The objective of this sensitivity analysis was to evaluate the impact of the MVA base power on the transient stability of the system. Five values were chosen, starting with 50 MVA and increasing by 25 MVA. The active power generated was kept constant at 25 MW.

To measure the effects on the transient stability, two parameters were considered: the rotor angle (at FCT = 0.300 s) and the critical clearing time (CCT). The variation of these parameters with the power base is shown in Table 4. Figure 2, on the other hand, shows a comparison of the rotor angle behavior for the five MVA basis values, including the case without wind generation.

Table 4. Rotor angle values for different rated power of the wind generator.

An (MVA)	50	75	100	125	150
Rotor angle	136.60°	134.33°	130.08°	123.58°	116.25°
CCT (s)	0.352	0.354	0.367	0.389	0.415

Table 4 and Figure 2 show that the rotor angle of generator 2 assumed decreasing maximum values as the installed wind power increased. The case characterized by an installed wind power of 150 MVA deserves special attention, resulting in a maximum rotor angle lower than that obtained in the case without wind power generation. The CCT took increasing values as the installed wind power increased.

If the two parameters are considered, it is possible to see that the stability of the system increased as the size of the installed wind generator increased.

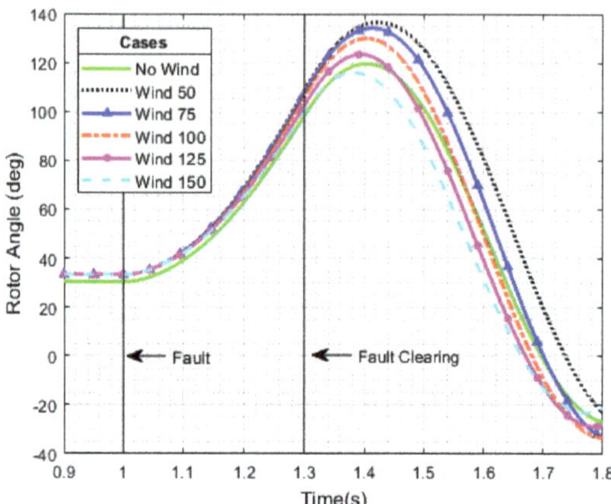

Figure 2. Comparison of rotor angle values for different rated power of wind generator.

4.3. Sensitivity Analysis of Wind Generator Produced Active Power

A sensitivity analysis was then performed for the power production of the wind generator. For this purpose, starting from the sensitivity study of the installed power, the sizes 50 MVA and 150 MVA (smallest and largest, respectively) were taken, as shown in Table 5.

Table 5. Rotor angle values for different wind power generation.

An (MVA)	50			150		
Pw (MW)	25	35	50	25	40	60
Rotor angle	136.60°	144.81°	159.78°	116.25°	136.17°	169.06°
CCT (s)	0.352	0.339	0.318	0.415	0.355	0.309

Once the basic values were established, the power values generated in the network were modified. Three values of generated power were evaluated (Figures 3 and 4), taking as the initial value the percentage of wind power in Italy in 2021 (7.4%) [16], while the final value was chosen to be very close to the system instability.

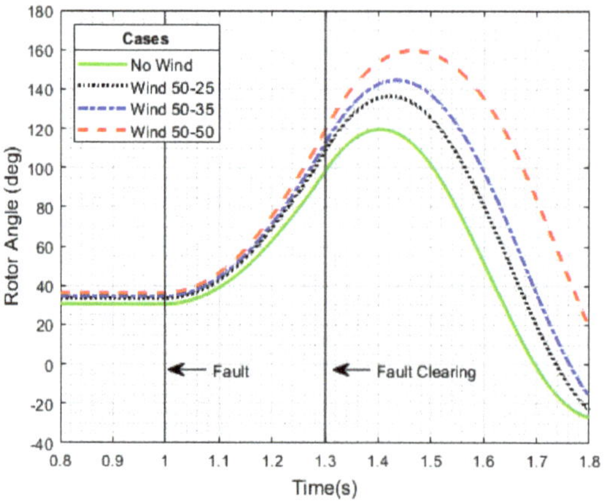

Figure 3. Comparison of rotor angle values for different wind power generation with 150 MVA wind power base.

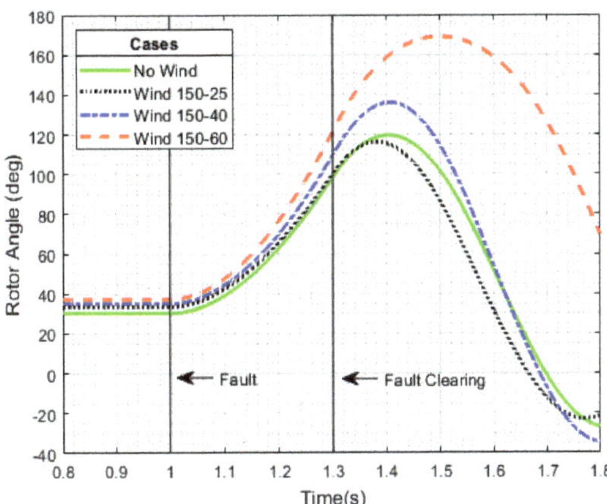

Figure 4. Comparison of rotor angle values for different wind power generation with 50 MVA wind power base.

Looking at Table 5, it is possible to note that, for both the installed power values and the increase in the power generated, there was an increase in the value of rotor angle and a decrease in the CCT. Specifically, it is possible to observe that the operating condition characterized by 150 MVA of installed wind power and 25 MW of generated power led to lower rotor angle values than those obtained in the scenario without wind generation.

5. Simulation on Scenarios

The aim of this study was to assess the impact of wind power generation in terms of stability and the effectiveness of the mentioned solutions to mitigate stability problems caused by the integration of the wind plant. For this purpose, an installed power value of 92.5 MVA was considered within the in-use test system. This value was chosen by applying to our test system the same percentage of wind generation expected from Terna to 2030 within the Italian electricity system [4].

With the same criterion, a power input value of 42.7 MW was chosen from this wind generator.

Once the generation and production quotas from wind have been defined, the impacts on transient stability related to the use of systems for their enhancement were studied, as seen in [6].

This assessment was carried out considering two different scenarios:

- Scenario A, which provides for the addition of the wind generator without the decrease in the share of conventional generation within the test system, so as not to decrease the inertia value present in the system;
- Scenario B, which involves the replacement of a share of conventional installed power equal to 92.5 MVA with wind power, so as to evaluate the impact of the system inertia reduction following the insertion of the wind generator (Figure 5).

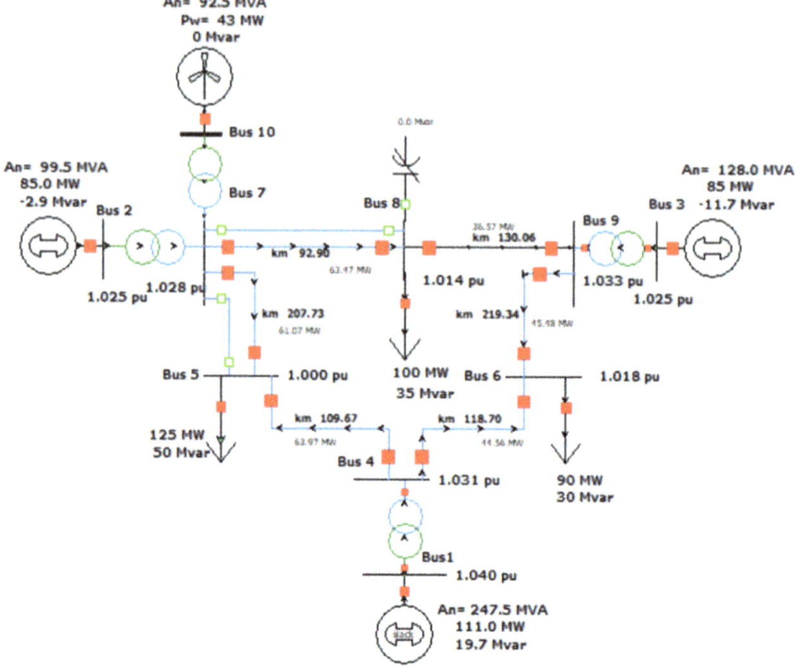

Figure 5. Modified IEEE 9-bus system for Scenario B.

5.1. Scenario A—Wind Added without Replacing Conventional Generation

In order to assess the device effectiveness in enhancing stability within the wind-added scenario, the system characterized by the presence of only the wind generator was taken as the base scenario.

Later comparisons were made in terms of rotor angle and CCT with data obtained from scenarios characterized by the presence of a double line, an SVC, a STATCOM, and a fast excitation system EXST4B, as carried out in [6] and shown in Tables 6 and 7, while the behavior of the rotor angle and the frequency is shown in Figures 6 and 7.

Table 6. Scenario A: maximum rotor angle values considering different stability improvement solutions.

	Base	**SVC**	**STATCOM**	**EXST4B**	**Line**
Rotor angle	150.98°	148.13°	148.07°	145.28°	126.76°

Table 7. Scenario A: CCT values considering different stability improvement solutions.

	Base	**SVC**	**STATCOM**	**EXST4B**	**Line**
CCT (s)	0.329	0.333	0.334	0.340	0.374

The results show that the addition of a new parallel line is the most effective solution to contain the rotor angle variations, as shown in Table 6 and Figure 6; moreover, the stability margin was increased by reducing the initial rotor angle, thanks to its effect on the system admittance matrix. SVC and STATCOM (see Figure 6) had a limited effect on the rotor angle stability, while the fast-acting exciter EXST4B anticipated the maximum rotor angle instant, while also reducing its value.

Figure 7 shows that the oscillations occurring in the base caused by the three-phase short-circuit case were mitigated using the solutions considered.

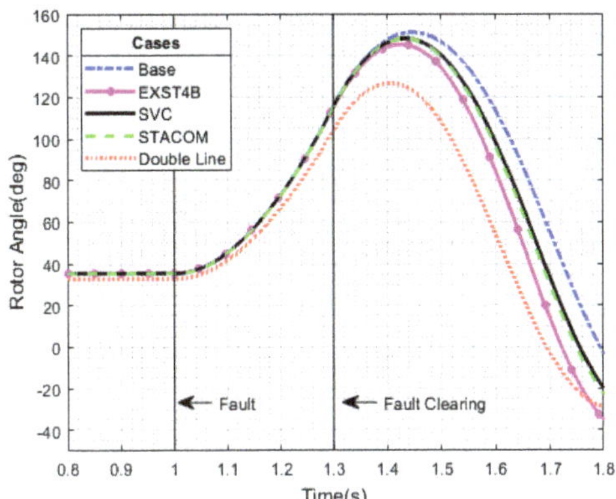

Figure 6. Scenario A: rotor angle curves considering different stability improvement solutions.

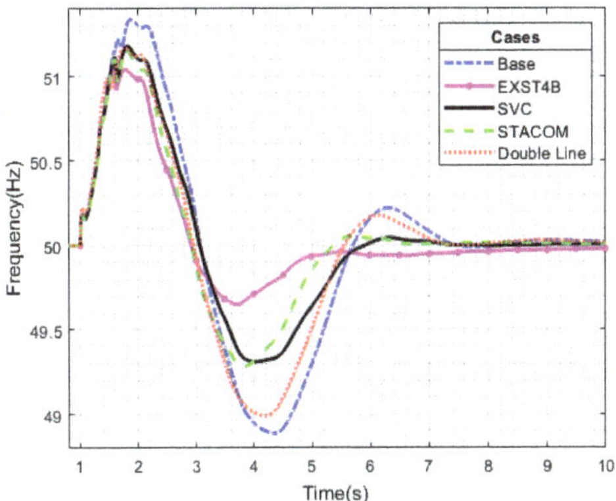

Figure 7. Scenario A: frequency curves considering different stability improvement solutions.

The addition of a parallel line represents a structural and preventive solution, with a small effect on the frequency stability. Emergency solutions such as SVC and STATCOM, as well as the fast-acting exciter, help the system frequency stability, especially during the underfrequency period. This is because these devices operate under dynamic conditions, which allows the control of electrical quantities such as voltage, which is a crucial quantity during a transient caused by a short-circuit. Comparing the results, the fast exciter had the best impact on frequency stability, resulting in the lowest oscillations, reaching the lowest point of 49.65 Hz.

5.2. Scenario B—Wind Added with Replacement of Conventional Generation

The same analyses described in the previous paragraph were evaluated under the scenario that involved replacing the traditional generation quota with wind generation, using the same percentages described. In this case, the effects of the tested devices were also evaluated as a function of the rotor angle and frequency values during the fault, as shown in Tables 8 and 9 and Figures 8 and 9.

Table 8. Scenario B: maximum rotor angle values considering different stability improvement solutions.

	Base	SVC	STATCOM	EXST4B	Line
Rotor angle	96.09°	94.43°	94.51°	95.43°	84.97°

Table 9. Scenario B: CCT values considering different stability enhancement solutions.

	Base	**SVC**	**STATCOM**	**EXST4B**	**Line**
CCT (s)	0.455	0.457	0.462	0.452	0.523

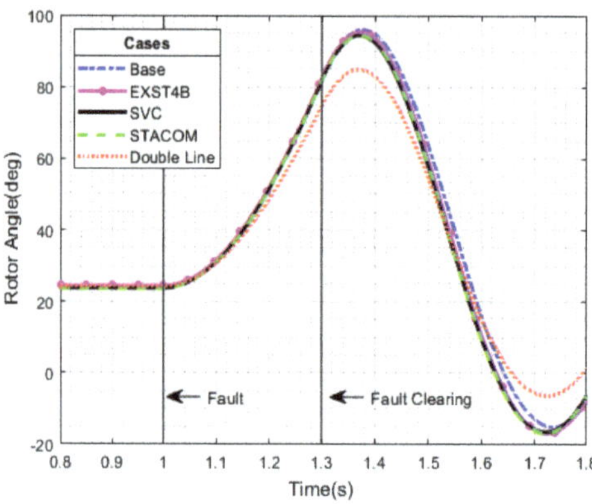

Figure 8. Scenario B: rotor angle curves with different stability improvement solutions.

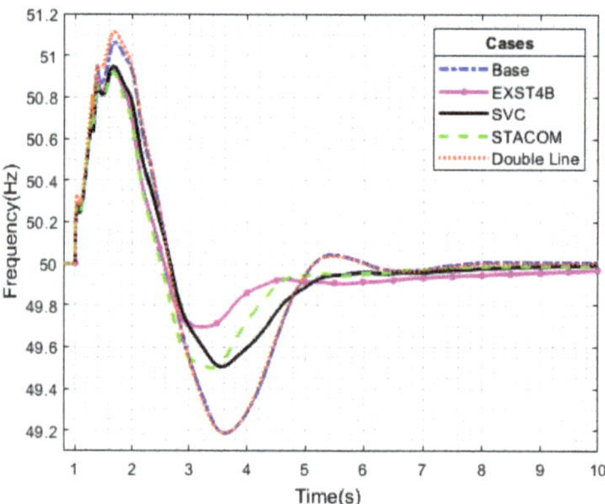

Figure 9. Scenario B: frequency curves with different stability improvement solutions.

A look at the rotor angle values obtained in Scenario B (see Table 8) shows that the maximum rotor angle for each solution was smaller than the rotor angle value obtained for the same solution in Scenario A. The same trend can be observed for CCT in Table 9. This is due to a different operating point of Generator 2, because the presence of the wind generator allows Generator 2 to feed a lower value of MW to Bus 2.

In the substitute scenario, the best solution for the stability of the rotor angle was the double line and not the other devices, which had only a small effect. The frequency oscillation in the replacement scenario behaved similarly to the addition scenario, confirming that EXST4B was the best solution in terms of frequency stability, reaching a minimum value of 49.7 Hz, as shown in Figure 9.

6. Conclusions

The analyses presented in this manuscript showed that the presence of a wind generator leads to lower system stability, due to inertia reduction, when limited to the grid under consideration. This is true until a very large amount of wind generation is installed (e.g., 150 MVA of wind capacity over 567.5 MVA of programmable generation capacity).

Injected power also affects stability, as an injected power close to the maximum power of the wind generator leads to lower stability margins in rotor angles.

In both the scenario with additional wind turbines and the scenario with replacement wind turbines, the addition of a parallel line (preventive control of transient stability) was the optimal solution to improve rotor angle stability, while the impact on frequency stability was irrelevant, due to the absence of dynamic contribution. In the case of emergency control of transient stability, the fast-acting exciter is the best solution, especially in terms of frequency stability, since it acts faster than any other solution. On the other hand, for rotor angle stability, the fast-acting exciter is useful to improve transient stability, but its contribution is limited with respect to the addition a new parallel line.

SVC and STATCOM showed limited capacity in helping the system to enhance its rotor angle stability, while they represent desirable solutions to improve frequency stability, allowing a better containment of frequency fluctuations, while avoiding the tripping of load shedding relays.

The results can be summarized as follows:

- SVC and STATCOM devices provide superior support for improving transient stability. These two solutions show similar behaviors; however, due to their unique features, the STATCOM solution was found to perform at higher standards than SVC—in the case of a weak grid or in a case characterized by a low percentage of synchronous generation (e.g., high penetration of renewable energy sources);
- From a monetary/financial point of view, SVC is a cheaper solution than STATCOM, due to its ease of allocation and maintenance, and the larger power range. Investment in these solutions can be economically incentivized for utilities in case of remuneration for reactive power compensation. Focusing on generator companies, an economic incentive, e.g., a remunerated voltage regulation service, could be useful to refurbish their plants and to increase performance through the use of fast excitation systems, as shown in the analyses.

Looking at the next generation of power system structures, specifically characterized by a high presence of RES generation [8,14,17] (which can strongly impact the system stability), it can be useful to reinforce the power grid structure by doubling key lines. This is surely useful to prevent critical congestion problems and significantly improve grid stability.

This paper provides a preliminary examination of the impact of new converter-based technologies on improving the security of modern power systems. Future developments will include the following topics: the impact of different control techniques for thermoelectric generators, concerning only generators with fast valving and brake resistors to mitigate the transient following a contingency. New market products are necessary to push not only utilities to provide services based on new technology, but also TSO to install new devices to manage the future power system safely. This can be achieved, for example, by introducing new ancillary services.

Author Contributions: Conceptualization, G.M.T.; Data curation, G.M. and S.L.; Investigation, G.M. and S.L.; Methodology, G.M.T.; Software, G.M. and S.L.; Supervision, G.M.T. All authors have read and agreed to the published version of the manuscript.

Funding: This research received no external funding.

Conflicts of Interest: The authors declare no conflict of interest.

References

1. Gautam, D.; Vittal, V.; Harbour, T. Impact of increased penetration of DFIG-based wind turbine generators on transient and small signal stability of power systems. *IEEE Trans. Power Syst.* **2009**, *24*, 1426–1434. [CrossRef]
2. Chamorro, H.R.; Ghandhari, M.; Eriksson, R. Wind Power Impact on Power System Frequency Response. In Proceedings of the 2013 North American Power Symposium (NAPS), Manhattan, KS, USA, 22–24 September 2013; pp. 1–6.
3. Terna. *Piano di Sviluppo 2020*; Terna S.p.A.: Italy, Rome, 2020.
4. Terna. *Piano di Sviluppo 2021*; Terna S.p.A.: Italy, Rome, 2021.

5.   Flynn, D.; Rather, Z.; Ardal, A.; D'Arco, S.; Hansen, A.D.; Cutululis, N.A.; Wang, Y. Technical impacts of high penetration levels of wind power on power system stability. *Wiley Interdiscip. Rev. Energy Environ.* **2017**, *6*, e216. [CrossRef]
6.   Tina, G.M.; Maione, G.; Licciardello, S.; Stefanelli, D. Comparative Technical-Economical Analysis of Transient Stability Improvements in a Power System. *Appl. Sci.* **2021**, *11*, 11359. [CrossRef]
7.   Kundur, P.; Paserba, J.; Ajjarapu, V.; Andersson, G.; Bose, A.; Canizares, C.; Vittal, V. Definition and classification of power system stability IEEE/CIGRE joint task force on stability terms and definitions. *IEEE Trans. Power Syst.* **2004**, *19*, 1387–1401.
8.   Kusumo, S.A.; Putranto, L.M. Transient Stability Study in Grid Integrated Wind Farm. In Proceedings of the 2018 5th International Conference on Information Technology, Computer, and Electrical Engineering (ICITACEE), Semarang, Indonesia, 27–28 September 2018; pp. 61–66.
9.   Salem, Q.; Altawil, I. Transient Stability Enhancement of Wind Farm Connected to Grid Supported with FACTS Devices. *Int. J. Electr. Energy* **2014**, *2*, 154–160. [CrossRef]
10.   Li, P.; Xiong, L.; Wu, F.; Ma, M.; Wang, J. Sliding mode controller based on feedback linearization for damping of sub-synchronous control interaction in DFIG-based wind power plants. *Int. J. Electr. Power Energy Syst.* **2019**, *107*, 239–250. [CrossRef]
11.   Li, P.; Xiong, L.; Ma, M.; Huang, S.; Zhu, Z.; Wang, Z. Energy-shaping L2-gain controller for PMSG wind turbine to mitigate subsynchronous interaction. *Int. J. Electr. Power Energy Syst.* **2022**, *135*, 107571. [CrossRef]
12.   Pertl, M.; Weckesser, T.; Rezkalla, M.; Marinelli, M. Transient stability improvement: A review and comparison of conventional and renewable-based techniques for preventive and emergency control. *Electr. Eng.* **2018**, *100*, 1701–1718. [CrossRef]
13.   Kaur, R.; Kumar, D. Transient stability improvement of IEEE 9 bus system using power world simulator. *MATEC Web Conf.* **2016**, *57*, 01026. [CrossRef]
14.   Katsivelakis, M.; Bargiotas, D.; Daskalopulu, A. Transient Stability Analysis in Power Systems Integrated with a Doubly-Fed Induction Generator Wind Farm. In Proceedings of the 2020 11th International Conference on Information, Intelligence, Systems and Applications (IISA), Piraeus, Greece, 15–17 July 2020; pp. 1–7.
15.   WECC Renewable Energy Modeling Task Force. *WECC Wind Plant Dynamic Modeling Guidelines*; Western Electricity Coordinating Council Modeling and Validation Work Group: Salt Lake City, UT, USA, 2014; p. 17.
16.   Terna. Transparency Report. Available online: https://www.terna.it/en/electric-system/transparency-report/actual-generation (accessed on 31 August 2022).
17.   Erlich, I.; Wilch, M.; Feltes, C. Reactive Power Generation by DFIG Based Wind Farms with AC Grid Connection. In Proceedings of the 2007 European Conference on Power Electronics and Applications, Aalborg, Denmark, 2–5 September 2007; pp. 1–10.

Review

Changes in Energy Sector Strategies: A Literature Review

Adam Sulich [1,2,*] and Letycja Sołoducho-Pelc [3,*]

[1] Schulich School of Business, York University, 4700 Keele Str., Toronto, ON M3J 1P3, Canada
[2] Department of Advanced Research in Management, Faculty of Business Management, Wroclaw University of Economics and Business, Komandorska Str. 118/120, 53-345 Wroclaw, Poland
[3] Strategic Management Department, Faculty of Business Management, Wroclaw University of Economics and Business, Komandorska Str. 118/120, 53-345 Wroclaw, Poland
* Correspondence: adam.sulich@ue.wroc.pl (A.S.); letycja.soloducho-pelc@ue.wroc.pl (L.S.-P.)

Abstract: Sustainable development (SD) can indicate the direction of the development of modern organizations' transition and transformation strategies in the energy sector. Currently, in most countries, the main challenge for the energy sector's strategies is to deal with energy security. The implementation of SD induces changes both in strategy and technology. The strategies are based on the technological transition toward renewable energy sources (RES). The aim of this paper is to explore business management literature dedicated to the transformation and transition strategies in the energy sector. The adopted methods are a systematic literature review (SLR) accompanied by a classical literature review (CLR) in Scopus database exploration. A literature review is developed in VOSviewer software and keyword co-occurrences analysis allowed to identify the main changes of direction in energy sector transformation strategies. The literature was explored by the 26 queries which resulted with 11 bibliometric maps. The analysis of the bibliometric maps was a challenge due to the cross-disciplinary strategic directions of development presented in indexed publications in the Scopus database. The identification of the changes in energy sector strategies is important because of its reliance on depleting resources and natural environment degradation. As a result of this paper, there is a visible shift of the trend in explored scientific publication from not only technological-based solutions but also towards managerial and organizational practices to achieve sustainability in the energy sector. This paper, besides the results, presents the theoretical contribution and managerial recommendations for business practices and addresses future research avenues. There are discussed implications of the presented analysis for further research.

Keywords: alternative energy; energy policy; energy transition; renewable energy; sustainable development; strategy classification

Citation: Sulich, A.; Sołoducho-Pelc, L. Changes in Energy Sector Strategies: A Literature Review. *Energies* **2022**, *15*, 7068. https://doi.org/10.3390/en15197068

Academic Editor: Juri Belikov

Received: 7 September 2022
Accepted: 22 September 2022
Published: 26 September 2022

Publisher's Note: MDPI stays neutral with regard to jurisdictional claims in published maps and institutional affiliations.

1. Introduction

Business has always depended on and had an impact on the natural environment [1,2]. However, through a proper strategy, it is possible to reverse or mitigate the negative influence of business [2,3]. The energy sector, together with other economic areas' processes, negatively influences the environment [4,5], despite this sector being a condition of the development of civilization [6,7]. Energy generation worldwide is based on fossil fuel combustion, which especially increases negative climate changes [8,9]. Therefore, energy sector producers have been implementing strategies to change their business models [10,11], technologies [12], and sources of energy generation towards renewable energy sources (RES) [13]. These transformation strategies are anchored in the sustainable development (SD) concept, but this idea is relatively new in business [14]. There have been two sides to the discussion about SD since the late 1980s [15]. The first is focused on the formulation of general laws and setting a formalized direction and strategic goals for business organizations [12]. The second side of this discussion addresses the question about the translation of all theoretical ideas related to sustainability into business practice, especially in the operations of manufacturing companies [16]. The SD paradigm is focused on quality of life and refers not only to the balance between social [17,18] and economic aspects but also the environmental dimension [19]. The number of household energy-saving appliances is constantly growing, though the intention of their producers is to reduce energy demand [20]. This contradiction is more visible in times when access to energy is a criterion of wealth, as it determines economic and social development [21]. On the other hand, ecological issues became an important determinant of the changes in the energy sector due to legal pressure [6,22]. For the energy sector, for a long time, the natural environment was the source of goods and waste reservoirs [23]. Despite public administrations and businesses putting pressure on the energy sector, economic development is still considered through the prism of natural resources exploitation [24]. The use of electricity still negatively influences natural resources and degrades the environment [25]. The lack of expected results from changes in the energy sector can be caused by a non-strategic approach to management [26]. Moreover, the contemporary business environment is characterized by high complexity and volatility, which affects the ways of conducting business in the energy sector [27]. Energy sector companies often struggle to achieve sustainability [9]. Still, the world is more concerned with the delivery of cheap and accessible electricity [28] than its impact on the natural environment [29]. For consumers, less important are the sources of energy compared to the energy prices.

This paper aims to explore the scientific literature indexed in the Scopus database and is dedicated to transformation and transition strategies in the energy sector. This research goal covers both theoretical and empirical research gaps revolving around the transition of the SD idea into practice. Understanding the process of business development in the energy sector and strategies transformation is notably not easy. However, exploration of the context of strategies transformation is the most important research agenda. The contributions of individual scientific papers can often appear minimal. Even when the study of the strategies transformation in the energy sector is viewed in terms of the collective endeavor, the various publications cannot easily be distilled into a consensus that would meet the standards of evidence routinely applied. Therefore, this scientific article employs two literature review methods.

There are two adopted literature review methods. The first is a systematic literature review (SLR) supported by the bibliometric maps generated in the VOSviewer on a query basis. The second method is a classical literature review (CLR), explaining the results obtained from the SLR method and providing an overview of the transformation strategies in the energy sector. This second method applies with special force in the identification of empirically salient transformation strategy determinants.

With this purpose of study, this scientific paper is structured as follows. In the first place, in Section 2, the paper develops a theoretical framework using the two literature review methods. There are limitations of this research caused by the adopted methods. In the third section, there are literature review results that cover two subjects: (1) changes in energy sector strategy directions among energy suppliers and (2) proposed typologies among energy sector strategies. The fourth chapter of this paper is a discussion of the presented results together with the limitations of the used methods. In the conclusions section, the scientific and practical contributions are presented together with the future promising research avenues.

2. Materials and Methods

The Scopus database's indexed scientific publications were the subject of the adopted methods in this research paper [30,31]. Scopus is a multidisciplinary repository that indexes different types of scientific publications [32]. The Scopus bibliographic database is an organized digital collection of references to published scientific literature, including journal articles, conference proceedings, patents, books, etc. [30]. The Scopus has been selected due to the popularity and recognizability of this database among researchers [33,34]. Scientific publications indexed in the Scopus database are considered prestigious. This database secures its scientific content due to rigorous conditions for the indexing of the title sources.

There are two literature review methods employed in this study, used as a collage of methods [35]. The first method is the SLR combined with the CLR, which is the second method used in this research. The purpose of a synthesis of these literature review methods is to identify, integrate, and evaluate research on a selected topic, based on clearly defined criteria [36]. The use of a combined literature review methodology is intended to provide three main benefits: (I) the literature review will include all research results on a given topic, (II) research results that for some reason do not correspond to the researcher's intentions or views will not be omitted, and (III) the possibility of verifying the relevance of the review will be created by enabling its replication. The literature review process [37,38] underpinning the operationalization and measurement of transformation strategies in the energy sector proceeded in three main stages [39]. Stage one was the selection of the literature base. This was followed by the selection of papers included in the analyzed set of results in stage two. Stage three consisted of critical content analysis [40]. Critical to the integrity of the literature review process is the thorough verification of the publications included within the Scopus literature base [41]. This requires reading each publication and preparing a concise analysis report [42]. Based on such work, it was possible to complement SLR with CLR prepared in this step's reports results [43].

The SLR method is used in this study as a tool for the identification of knowledge gaps, and its advanced applications allow the identification of the most common as well as the most desirable directions for further research, their specific assumptions, the operationalization of variables (items), the measurement of constructs, the research method adopted in seminal publications, and even the method of analyzing the raw data. The SLR method variation is supported by the research queries exploring the Scopus database [39]. Before embarking on the process of identifying academic publications in the field of transformation strategies in the energy sector [44], it was necessary to define automated search conditions for papers in the explored Scopus academic database [30,33]. The automation of queries in the Scopus scientific database raises the problem of overfitting [45,46]. In addition to analyzing the total results, an increasingly common approach is to use the most important journals in a given subject area. These are usually the most cited scientific publications indexed in the database being explored [30,47]. The query result sets created are often too large for a detailed analysis of the texts without software support [48,49].

Though linguistic differences are not the subject of this research, the choice of the proper form of a keyword or its synonym has an impact on the achieved results [21,30]. Due to this fact, the choice of keywords exploration for the queries was also limited. This is because if the query in the Scopus database is general, it yields too many results to indicate the proper direction of exploration [50,51]. Omitting the broader categories and successive limitations with the aim to restrict the occurrence of a keyword anywhere in the text is justified, because a keyword in the body of a research paper indexed in the Scopus database may appear accessory. This assumption focuses attention on articles in which the keyword reflects a research category that is relevant and not an accessory to the paper.

There was no initial assumption toward a literature exploration of energy sector transformation and transition strategies. There are some events which highly influence energy sector; however, they were not included into analyzed scientific publications. The factors coming from the geographical distribution of the resources or geopolitical situation were not a limitation of this study. In this study, there was no assumption about the sectoral or business strategy level.

There were multiple research assumptions decided during the Scopus database exploration with queries presented in Table 1. First, the area of the research was settled based on the initial query with the following syntax (TITLE-ABS-KEY ("power sector")) AND ("energy sector") with 1380 results representing all scientific publications indexed in the Scopus database. The syntax element AND is not only used by the Scopus search algorithm as conjunction "and" operator but also as an alternative "or" syntax element.

However, the two presented keywords "power sector" and "energy sector" were also researched independently. Finally, the "energy sector" was selected as most numerous with the highest number of publications indexed in Scopus, and it was the most popular keyword. The keyword "energy sector" initial query in TITLE-ABS-KEY fields in Scopus gave 15663 results compared to 6078 results for "power sector". Therefore, this sector is explored in query syntaxes presented in Table 1. There are two types of query elements in syntaxes, the specified and non-specified, respectively [40,45]. The specified part is marked by the citation marks, and non-specified searched keywords do not have them. In Table 1, there are presented queries used for the calibration and choice of more detailed queries. The queries are numbered continuatively in ascending order across Tables 1–3. Queries presented in the tables were used for studying the Scopus database on 20 August 2022, with different numerical results depending exact syntax of each formulated query.

Table 1. Syntaxes used in queries calibration for the Scopus scientific database exploration.

No.	Query Syntax	No. of Results (20 August 2022)
1	(TITLE-ABS-KEY ("energy sector")) AND (transition AND strategies)	2290
2	(TITLE-ABS-KEY ("energy sector")) AND (transformation AND strategies)	1400
3	(TITLE-ABS-KEY ("energy sector")) AND (transition AND transformation AND strategy)	830
4	(TITLE-ABS-KEY ("energy sector")) AND (transition AND strategy) AND (LIMIT-TO (SUBJAREA, "BUSI") OR LIMIT-TO (SUBJAREA, "DECI"))	340
5	(TITLE-ABS-KEY ("energy sector")) AND (transformation AND strategy) AND (LIMIT-TO (SUBJAREA, "BUSI") OR LIMIT-TO (SUBJAREA, "DECI"))	248
6	(TITLE-ABS-KEY ("energy sector")) AND (transformation AND transition AND strategy) AND (LIMIT-TO (SUBJAREA, "BUSI") OR LIMIT-TO (SUBJAREA, "DECI"))	134
7	(TITLE-ABS-KEY ("energy sector")) AND (transformation AND transition AND strategy)) AND (change) AND (LIMIT-TO (SUBJAREA, "BUSI") OR LIMIT-TO (SUBJAREA, "DECI"))	121
8	(TITLE-ABS-KEY ("energy sector")) AND ((transformation AND transition AND strategy)) AND (shift) AND (LIMIT-TO (SUBJAREA, "BUSI") OR LIMIT-TO (SUBJAREA, "DECI"))	26
9	(TITLE-ABS-KEY ("energy sector")) AND ("transformation strategy")	23
10	TITLE-ABS-KEY ("energy sector strategies")	15
11	((TITLE-ABS-KEY ("energy sector")) AND ("energy sector strategies")) AND (LIMIT-TO (SUBJAREA, "BUSI") OR LIMIT-TO (SUBJAREA, "DECI"))	1
12	((TITLE-ABS-KEY ("energy sector")) AND (transformation AND transition AND strategy)) AND (type AND typology AND classification AND category) AND (LIMIT-TO (SUBJAREA, "BUSI") OR LIMIT-TO (SUBJAREA, "DECI"))	0
13	((TITLE-ABS-KEY ("energy sector")) AND (transition AND transformation AND strategies)) AND ("startegy type") AND (LIMIT-TO (SUBJAREA, "BUSI") OR LIMIT-TO (SUBJAREA, "DECI"))	0
14	((TITLE-ABS-KEY ("energy sector")) AND (transition AND transformation AND strategies)) AND (strategy AND type) AND (LIMIT-TO (SUBJAREA, "BUSI") OR LIMIT-TO (SUBJAREA, "DECI"))	0

Source: authors' elaboration.

Table 2. Queries focused on typologies used in the Scopus scientific database exploration.

No.	Query Syntax	No. of Results (20 August 2022)
15	(TITLE-ABS-KEY ("energy sector")) AND ((typology)) AND (strategy)	281
16	(TITLE-ABS-KEY ("energy sector")) AND (transition AND transformation AND strategy) AND (type)	217
17	(TITLE-ABS-KEY ("energy sector")) AND (transition AND transformation AND strategy) AND (typology)	105
18	((TITLE-ABS-KEY ("energy sector")) AND (transformation AND transition AND strategy)) AND (type) AND (LIMIT-TO (SUBJAREA, "BUSI") OR LIMIT-TO (SUBJAREA, "DECI"))	28
19	((TITLE-ABS-KEY ("energy sector")) AND (transformation AND transition AND strategy)) AND (typology) AND (LIMIT-TO (SUBJAREA, "BUSI") OR LIMIT-TO (SUBJAREA, "DECI"))	21
20	((TITLE-ABS-KEY ("energy sector")) AND (transformation AND transition AND strategy)) AND (category) AND (LIMIT-TO (SUBJAREA, "BUSI") OR LIMIT-TO (SUBJAREA, "DECI"))	14
21	((TITLE-ABS-KEY ("energy sector")) AND (transformation AND transition AND strategy)) AND (classification) AND (LIMIT-TO (SUBJAREA, "BUSI") OR LIMIT-TO (SUBJAREA, "DECI"))	13

Source: authors' elaboration.

Table 3. Queries developed upon Query 2 from Table 1 used in the Scopus database exploration.

No.	Query Syntax	No. of Results (20 August 2022)
22	((TITLE-ABS-KEY ("energy sector")) AND (transformation AND strategies)) AND ("climate change")	724
23	((TITLE-ABS-KEY ("energy sector")) AND (transformation AND strategies)) AND (("climate change")) AND ("renewable energy sources")	229
24	((TITLE-ABS-KEY ("energy sector")) AND (transformation AND strategies)) AND ((("climate change")) AND ("renewable energy sources")) AND ("alternative energy")	56
25	((TITLE-ABS-KEY ("energy sector")) AND (transformation AND strategies)) AND (((("climate change")) AND ("renewable energy sources")) AND ("alternative energy")) AND ("energy policy")	53
26	((TITLE-ABS-KEY ("energy sector")) AND (transformation AND strategies)) AND ((((("climate change")) AND ("renewable energy sources")) AND ("alternative energy")) AND ("energy policy")) AND ("sustainable development")	40

Source: authors' elaboration.

There are explorative queries presented in Table 2 formulated to search title, abstract, and keywords indexed in the Scopus database of scientific publications related to the energy sector. Queries 15, 16, and 17 were used in unspecified science areas with higher numbers of the results than those achieved with Queries 18, 19, 20, and 21, respectively. The aim of this paper was the main reason for the construction of the queries presented in Table 2. Therefore, the queries used to explore Scopus were limited to the subject area of business management "BUSI" and decision sciences "DECI" due to their scope and specific interest in strategic management [52,53]. Then, initial queries were developed to explore the Scopus database in a selected field of science [54]. The full syntaxes of the used queries are presented in Table 1 and consist of syntax with an element indicating the choice of the science area, as follows: AND (LIMIT-TO (SUBJAREA, "BUSI") OR LIMIT-TO (SUBJAREA, "DECI"). Then, Query 6 presented in Table 1 was developed into five more specified Queries, 18, 19 20, and 21, in Table 2.

Queries 18, 19, 20, and 21 independently search type, typology, category, and classification of transformation or transition strategies in the energy sector. The exploration settled in the areas of business management and decision sciences. The results of Queries 18, 19, 20, and 21 were the subject of the VOSviewer software (version 1.6.18; Centre for Science and Technology Studies, Leiden University: Leiden, The Netherlands) for further analyses, and they are presented in the Results section. The choice of the query construction was also based on the other tested queries and logic choice of its construction. Authors checked also query syntax consisting of the following elements: (TITLE-ABS-KEY ("strategy type")) AND ("classification") AND (LIMIT-TO (SUBJAREA, "ENER")) to discover that there were 6 results, and two of them belong to the authors of this paper. Therefore, such queries' results were excluded from the further analysis, though they helped in further queries' development.

In Table 3, there are queries developed based on Query 2 in Table 1. To the syntax of Query 2, the indexed keywords recognized in the VOSviewer bibliometric map analysis of Query 2's results were added. The choice of keywords was based on the bibliometric map with the most dominant clusters (presented as Figure S1 in Supplementary Materials File S1). The keywords from these clusters were used to propose more specified Queries, 22, 23, 24, 25, and 26, in Table 3 to explore Scopus scientific database. Five queries differ from each other in the narrowing scope of added specified indexed keywords and the number of the obtained results. To achieve focused Scopus database exploration, there were more specific keywords added to the query syntaxes as presented in Table 3. These keywords were "climate change" (as a dominant keyword in the subnetwork dedicated to climate issues and their solutions, which consists of 96 items), "renewable energy sources" (indicated as the most dominant keyword in the second cluster with 87 items, dedicated mostly to technical aspects and business practice of energy transition), "alternative energy", "energy policy", and "sustainable development". The order of the additional keywords suggested by the VOSviewer software used in the extended Query 2 research presented in Table 1 was confirmed on the indicated keywords' significance by the bibliometric program used.

Obtained results from each query presented in Tables 1 and 2 were downloaded in .csv file format from the Scopus database as a separate set of data. However, during the export, all fields on the publication were marked. For the Scopus database, the following fields were selected for export: citation information, bibliographic information, abstract, keywords, funding details, and other information. Further analyses were carried out on the most recent data. The exported data were used for analyses in VOSviewer (version 1.6.18), and the results are shown in bibliometric maps. The choice of the number of keyword co-occurrences determines the result obtained in its graphical presentation and bibliometric map clarity. Additionally, during the VOSviewer bibliometric map preparation, some indexed keywords were excluded from the set proposed by the software. The excluded keywords were names or abbreviations for continents, international institutions, countries, and regions. The next groups besides their names were names of methods (regression analysis, climate model, or numerical model), publication types (article, review), chemical elements or compounds (carbon dioxide), and names of scientific disciplines (agriculture, chemistry, economics). Plural forms were excluded when the more numerous single form keyword was proposed by the VOSviewer software. Despite so many exclusions of keywords, the number of nodes in the most numerous queries was significant. Maps created, visualized, and explored using VOSviewer include items. Items are the objects of interest—the co-occurring keywords, which in the bibliometric map are nodes. Between nodes are edges that represent the relations between co-occurring keywords. In the bibliometric map, items are grouped into clusters. A cluster is a subnetwork, a set of items included in a bibliometric map. In the visualization of a map, items with higher importance are shown more prominently than items with lower significance.

The methods presented above for the purpose of the literature review were information analysis and synthesis, focusing on findings. The combined methods summarize the substance of the literature and conclude it. Traditional literature review adopts a critical approach, which assesses theories by critically examining the described sources, methods, and results with an emphasis on background and contextual material. On the other hand, the SLR is a review with a clearly stated purpose, questions, or queries and a defined search approach, stating inclusion and exclusion criteria and producing a qualitative appraisal for articles. The method combination can be supported by bibliometric analysis, a method that includes graphical and statistical analysis of published articles and citations therein to measure their impact. Bibliometric maps unveil pivotal articles and objectively illustrate linkages between and among articles focused on transformation strategies in the energy sector.

3. Results

Strategies in the energy sector have been subject to scientific interest since 1975; however, in the 19'80s, transition or transformation strategies emerged [55,56]. There are queries presented in Tables 1–3 along with the numerical results of the formulated queries.

Queries 1 and 2 were proposed to explore the "energy sector" for transition strategies and transformation strategies, respectively (Table 1). The main difference between Query 1 and Query 2 is in the number of results. Query 1 yielded 2290 results, while Query 2 gave 1400 results. This can be related to there being more technology related to transition. On the other hand, a transformation is related to business management and decision sciences areas. This observation is supported by the comparison of the 10 most cited publications in Tables S1 and S2 placed in Supplementary Materials File S1. Each query result in the Scopus database can be analyzed on the full-time horizon by the selection of the "Analyze search results" option. The results of Query 1 are presented as a chart in Figure 1. The number of Query 1 results yielded 2290 scientific publications. Figure 1 proves that transition strategies have been a subject of scientific interest since 1984. However, rapid growth in the number of those publications occurred in 2008 and surpassed 500 in 2021. The year 2022 was excluded from this analysis.

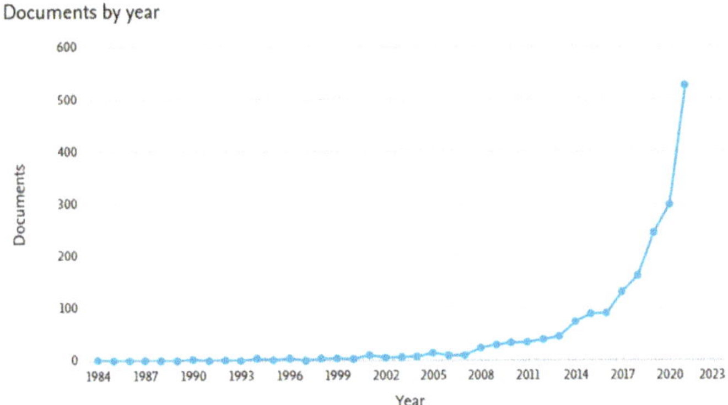

Figure 1. Scopus-indexed publications dedicated to transition strategies in the energy sector as Query 1's (Table 1) results. Source: Scopus database.

The graphical analysis of the number of other query results (Table 1) follows the same pattern of rapid growth since the year 2008 (Figure 2). However, the number of publications in the case of Query 2's results surpassed 300 publications in 2021. Another slight difference is that the first publication among the Query 2 results was published in 1983. The observation of the similarities in Figures 1 and 2 was the reason for the decision to combine them in a single query: Query 3 (Table 2).

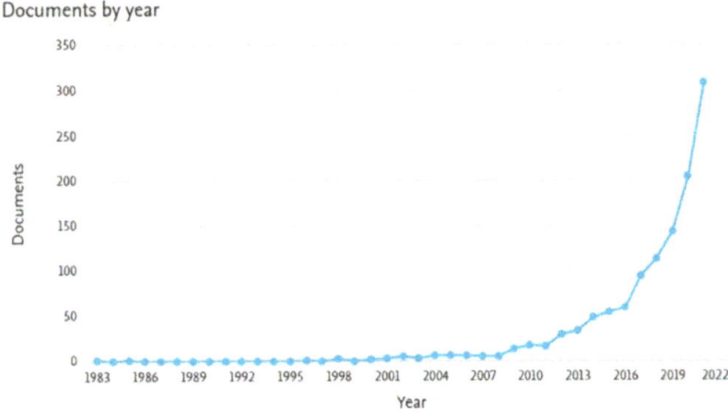

Figure 2. Scopus-indexed publications dedicated to transition strategies in the energy sector as Query 2's results (Table 1). Source: Scopus database.

Query 3, with syntax covering technological solutions ("transition") and organizational or management aspects of the changes towards SD (keyword "transformation"), was analyzed in the VOSviewer software. The chart illustrating the Query 3 results in Scopus also follows the pattern presented in Figures 1 and 2, indicating a lower number of publications when Queries 1 and 2 are combined. This observation stands with the descending number of results presented in Table 1. Query 3 was proposed to cover both indexed keywords of transition and transformation in a specified field of "energy sector". The combination of the two researched independently in Query 1 and 2's keywords yielded 830 results. These results only partially covered previously obtained similarities with the scientific publications resulting from Queries 1 and 2. The publications presented in

Table S3 in Supplementary Materials File S1 were not related to the energy sector despite the specified "energy sector" query syntax element.

Queries 4, 5, and 6 are developed versions of previous queries presented in Table 1. Queries 4 and 5's (Table 1) results in the analysis were the reason for their combination and the formulation of Query 6. There are 134 scientific publications in the Scopus database covering Query 6's results. The obtained results from Queries 4 and 5 were less numerous due to the introduced scientific areas limitation, and such exclusion has also influenced the quality of the most cited works researched from Query 6 that are presented in Table S4 in Supplementary Materials File S1.

Queries 7 and 8 were proposed as the development of Query 6. The scientific area remained the same as in Queries 4, 5, and 6. In Queries 7 and 8, additional keywords' impact on the obtained results is analyzed. Therefore, a broad approach to the topic of transformation strategy in the energy sector also required the consideration of alternative words and synonyms for the word "transition" and "transformation". The "change" or "shift" keyword can be seen as a synonym for transformation or transition words [57,58]. These keywords, however, initially gave very general results. Therefore, there was a careful choice of the synonyms to be explored in the further queries presented in Table 1. Then, the typing of search criteria was carried out starting from the identification of criteria related to the research model and the adopted cognitive context [41,57], i.e., the transformation strategies in the energy sector. In Table 1, there are also queries checking the difference between grammar forms and direct and unspecified queries. The difference between "strategy and strategies" was also tested, and the usage of single or plural forms does not affect the results number in Scopus. For example, query syntax (TITLE-ABS-KEY ("energy sector")) AND (changes) gave 7549 indexed publications as results.

Queries 9 and 10 in Table 1 are specified queries used to search scientific areas not defined in the Scopus database. The number of the results is significantly lower than the number obtained in Queries 1–6. The most cited 10 scientific publications from Queries 9 and 10 are presented in Table S5, and the most cited work results of Query 10 are presented in Tables S5 and S6 in Supplementary Materials File S1. Query 9 yielded 23 results due to its specified syntax: (TITLE-ABS-KEY ("energy sector")) AND ("transformation strategy").

Changes in several indexed publications covering the transformation strategies in the energy sector are presented in Figure 3. The first publication was indexed in the Scopus database in 1998, and the subject was again undertaken in the scientific literature in 2011 and 2015. There has been a growing interest in transformation strategies in the energy sector since 2017.

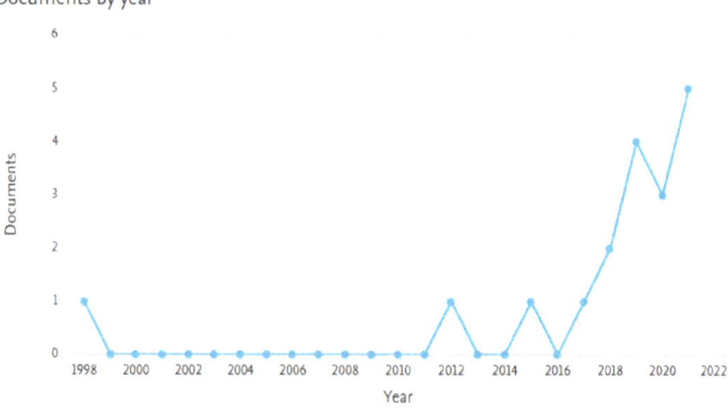

Figure 3. Scopus-indexed publications dedicated to transition strategies in the energy sector as Query 9's results (Table 1). Source: Scopus database.

Query 11 was very specified in its syntax and gave one result. This single scientific publication was titled "How to reform the power sector in Mexico? Insights from a simulation model" by R. Fuentes-Bracamontes [42], cited by seven other publications. R. Fuentes-Bracamontes in his paper addressed "the question of how a developing country, like Mexico, can reform its electricity industry at the same time as addressing climate change issues" [42].

Queries 12, 13, and 14 were too specific and detailed in the indicated area and gave no results to be analyzed. On the other hand, **Queries 15, 16, and 17**, although dedicated to the type or topology research, gave numerous numbers of results of publications (Table 2) that were developed in their syntaxes into **Queries 18, 19, and 20**. These queries, although less numerous, provided better insight into researched types, typologies, or categories of transformation strategies. In Table 2, there is also **Query 21**, the syntax of which consists of all combined searched elements from the previous three queries. Queries 18–21 results were then analyzed in the VOSviewer software (version 1.6.18), shown in detail below.

Query 15's results were explored in the VOSviewer software as a bibliometric map. The analysis of co-occurrences among indexed keywords with the full counting method was selected. The minimum number of co-occurrences of a keyword was 10, and then, among the total of 1573 keywords, 32 met the threshold. In the bibliometric map, three clusters with 32 nodes and 373 links were identified. The three clusters automatically identified were colored by the VOSviewer software and presented as the bibliometric map in Figure 4. This bibliometric map presents the co-occurrences of keywords related to the energy sector strategy typology. In the final dialogue box in the VOSviewer program, none of the keywords were deselected; however, the duplicates were removed.

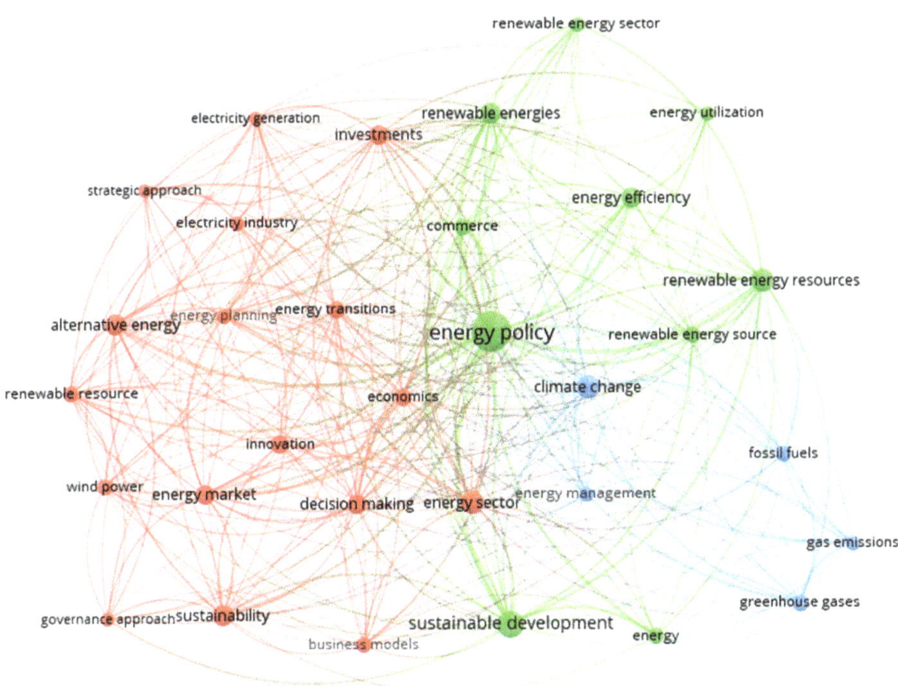

Figure 4. Bibliometric map of indexed keyword co-occurrence results from Scopus based on Query 15. Source: authors' elaboration performed in VOSviewer (version 1.6.18).

Based on the 281 scientific publications distinguished by Query 15, as presented in Table 2, the bibliometric map was proposed in the VOSviewer software. Although the keyword "typology" was included in the Query 15 syntax, there is no keyword related to this specific term. The keywords which were selected and used by the VOSviewer program are gathered in Table 4 and are separated by semicolons. Despite the original writing form, the keywords in Table 4 are written in small letters as in the VOSviewer software's unification procedure. In Table 4, there are clusters identified by colors, as in Figure 4, established by bibliometric software automatically. The order of clusters presented in Table 4 is caused by the number of keywords identified by the VOSviewer and represented as nodes in Figure 4.

Table 4. Clusters of index keyword co-occurrences in Figure 4 for Scopus Query 15.

Cluster	Color	Keywords
1	Red	alternative energy, business models, decision making, economics, electricity generation, electricity industry, energy market, energy planning, energy sector, energy transitions, governance approach, innovation, investments, renewable resources, strategic approach, sustainability, wind power
2	Green	commerce, energy, energy efficiency, energy policy, energy utilization, renewable energies, renewable energy resources, renewable energy sector, renewable energy sources, sustainable development
3	Blue	climate change, energy management, fossil fuels, gas emissions, greenhouse gases

Source: authors' elaboration.

The clusters presented graphically in Figure 4 and their nodes described in Table 4 as keywords indicate the three directions of the energy sector's strategies. The red cluster is the general overview of the energy sector's strategies. The green cluster is related to RES and revolves around sustainable future solutions. The last, blue-colored cluster is related to fossil fuels and energy management and is a subnetwork of the previous green cluster.

Query 18's results were used to propose Figure 5 as a bibliometric map. To draw this map, the minimum number of indexed keyword co-occurrences was selected as two, and then among 174 keywords, 14 met the threshold. The Query 18 syntax was constructed to explore transition and transformation strategies in the energy sector with the limitation to the science areas of business management and decision sciences. The obtained results are the scientific publications which are the edges between the nodes presented in figure's bibliometric map. The size of the node is related to the importance of the keyword. The keywords distinguished in Query 18 are gathered in Table 5.

In Figure 5, there are four clusters. The first is the red cluster with four keywords. This cluster is dedicated to management and economic sciences. The second cluster, also with four keywords, is related to the changes and innovation in the energy sector. There is also a blue cluster pointing at strategies toward renewable energies and energy conservation. There is also a yellow cluster visible in Figure 5, which represents the strategies and energy policies based on alternative energy (relation between two keywords). There is a shift among the presented keywords presented in Figure 5 and Figure S2 in Supplementary Materials File S1. This shift represents the change in scientific interests of publication authors dealing with transformation and transition strategy types in the energy sector (see Query 18). The bibliometric analysis of Query 18's results covered the years 2015–2021. The first keywords presented in Figure S2 were "energy policy", "alternative energy", and "sustainable development". The newest keywords from the analyzed scientific publications are "decision making", "commerce", and "economic and social effects".

The clusters presented in Table 5 are interrelated, and the picture presented in Figure 5 is complicated. However, there are three main directions of strategies presented in clusters two (green) and three (blue). There are strategies in the energy sector dedicated to innovation, transformation, and conservation. There is no connection (edge) between nodes representing innovation and "renewable energies" in Figure 5.

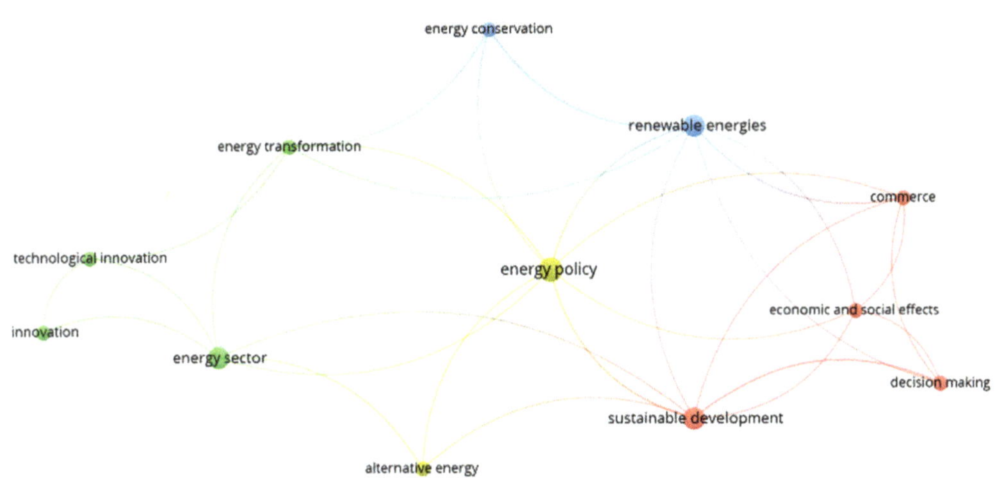

Figure 5. Bibliometric map of indexed keyword co-occurrence results from Scopus based on Query 18. Counting method: full counting. Minimum keyword co-occurrence is 10. Source: authors' elaboration performed in VOSviewer (version 1.6.18).

Table 5. Clusters of index keyword co-occurrences in Figure 5 for Scopus.

Cluster	Color	Keywords
1	Red	commerce, decision making, economic and social effects, sustainable development
2	Green	the energy sector, energy transformation, technological innovation, innovation
3	Blue	energy conservation, renewable energies
4	Yellow	alternative energy, energy policy

Source: authors' elaboration.

Query 18's results were analyzed in VOSviewer by all-keywords exploration. The results of this analysis are presented in Figure 6 and Table 6. The difference between Figures 5 and 6 is caused by the bigger number of all keywords. Therefore, there are five clusters identified by the VOSviewer software.

Table 6. Clusters of all keyword co-occurrences in Figure 6 for Scopus.

Cluster	Color	Keywords
1	Red	bioeconomy, biogas technologies, convergence, patent analysis
2	Green	energy conservation, energy policy, energy transformation, renewable energies,
3	Blue	energy sector, innovation, modeling, technological innovation
4	Yellow	alternative energy, commerce, decision making, sustainable development
5	Violet	coal industry, economic and social effects, renewable energy

Source: authors' elaboration.

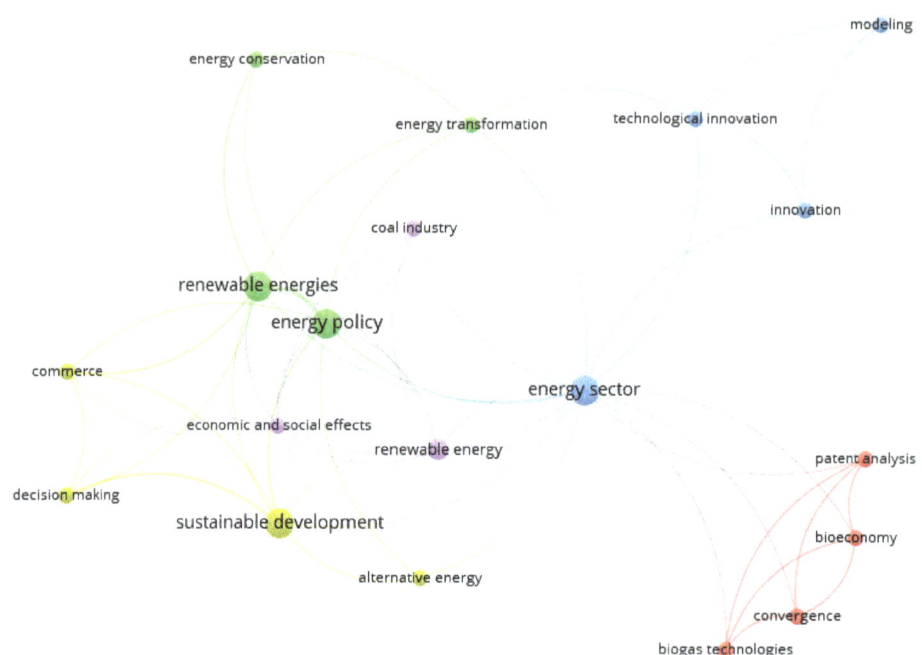

Figure 6. Bibliometric map of all keyword co-occurrence from Scopus based on Query 18's results. Source: authors' elaboration performed in VOSviewer (version 1.6.18).

In Figure 6, all keywords were used, and among 248 keywords, only 22 met the threshold of two minimum keyword co-occurrences. The method used in VOSviewer software was full counting. There were excluded keywords, specifically "india", "europe", and "carbon dioxide", despite their original writing form (the keywords in Table 2 are written in small letters as in the VOSviewer program). In Figure 6, there are clusters distinguished in color due to their common direction. The identified clusters are related to the different areas of the scientific interest of the analyzed scientific publications' authors. Figure 6 presents the graphical results of Query 18 used in the Scopus database exploration. There is a red cluster identified in Table 6. This cluster is related to the "bioeconomy", "biogas technologies", "convergence", and "patent analysis", which are common fields in business practice. The other clusters are more related to theory and are in opposition to the red cluster. The green cluster with four keywords revolves around the strategic direction of energy sector development. The third, the blue sector, is dedicated to technology and innovation. The yellow cluster describes the management sciences involved in alternative energy management in the energy sector. The last sector is violet and describes the choice between fossil fuels and renewable energy. This choice in the energy sector is intertwined with the economic and social effects of such a decision.

The scientific publications explored in Query 18 in both methods of analysis indicate the three main strategy directions; however, with the all-indexed keywords, the practical aspects are more visible.

Query 19's results were the basis for the analyses performed in VOSviewer and presented in Figure 7 and Table 7. These 21 results are scientific publications indexed in the Scopus database. To analyze them, the minimum number of indexed keyword co-occurrences was two, and among 174 keywords searched in the title, abstract, and keywords field, only 24 met the threshold when the full counting method was applied. In this research, no keyword was excluded. The graphical analysis result as the bibliometric map presented in Figure 7.

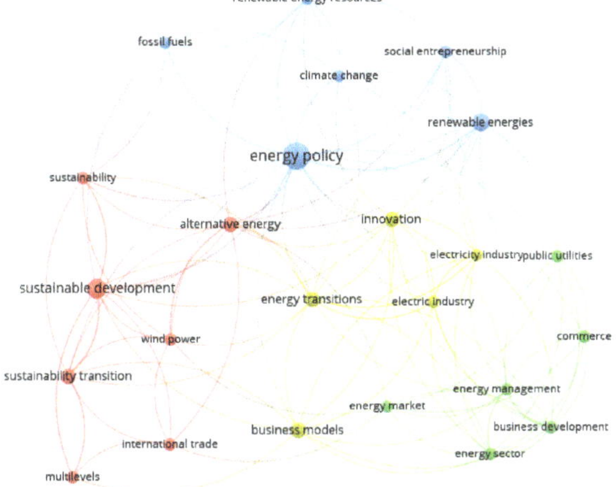

Figure 7. Bibliometric map of index keyword co-occurrence results from Scopus based on Query 19. Source: authors' elaboration performed in VOSviewer (version 1.6.18).

Table 7. Clusters of keyword co-occurrences in Figure 8 for Scopus Query 19.

Cluster	Color	Keywords
1	Red	alternative energy, international trade, multilevels, sustainability, sustainability transition, sustainable development, wind power
2	Green	business development, commerce, energy management, energy market, energy sector, public utilities
3	Blue	climate change, energy policy, fossil fuels, renewable energies, renewable energy resources, social entrepreneurship
4	Yellow	business models, electric industry, electricity industry, energy transitions, innovation

Source: authors' elaboration.

These four clusters are presented in Figure 7 as subnetworks of the presented bibliometric map. Then, there are four directions of scientific interest revealed in Query 19 (Table 2) dedicated to the typology of transformation and transition strategies in the energy sector limited to the subject areas "business, management, and accounting" and "decision sciences" in Scopus. The first and most numerous clusters are colored in red with seven keywords. This cluster's main node is oriented to SD-related topics represented by the keywords "sustainable transition" and "sustainability". The whole red cluster revolves around the development subject. The second subnetwork is the green-colored cluster dedicated to energy sector management. The blue cluster reflects the two main directions of development in the energy sector (pivotal keyword). These two directions are related to fossil fuels and RES. These research subjects are connected with climate change and social entrepreneurship. In Table 7, there is also a yellow cluster. This cluster is focused on management aspects of the electricity industry in the energy sector. Those aspects are "business models" and "innovation". Also visible in Figure 7 is a shift of the scientific interest among keywords in the result of Query 19 and analyzed in Figure S3 in Supplementary File S1. The newest keywords used in 2022 in scientific publications were those from the green and yellow clusters (Table 7): "energy sector", "energy transitions", "energy management", "energy market", "business development", "electric industry", and "electric industry". These keywords in Figure S3 in Supplementary File S1 are collected in the intertwined close yellow network.

The clusters presented in Table 7 are not randomly indexed keyword co-occurrences in the analyzed publications. The order of the clusters presented in Table 7 was caused by the number of keywords identified by VOSviewer and represented as nodes in Figure 7. The presented results were obtained from Query 19's syntax in Boolean style and were explored in two bibliometric maps presented in Figures 6 and 7. The initial number of indexed keywords grouped into clusters in Table 6 was lower than all the keywords analyzed in Table 7.

Query 20 presented in Table 2 the explored categories of transformation and transition strategies among scientific publications indexed in the Scopus database, limited to areas of "business, management and accounting" and "decision sciences". Although the "category" was the main subject explored by Query 20 in Scopus, this keyword is not present in Figure 8 and Table 8. The keywords which were selected and used by the VOSviewer program are gathered in Table 8. Keywords are separated by semicolons. In Table 8, there are clusters identified by colors, as in Figure 8. Despite the original writing form, the keywords in Table 8 are written in small letters as in the VOSviewer program. The order of clusters presented in Table 8 is caused by the number of keywords identified by VOSviewer. The identified clusters are related to the different areas of the scientific interest of the analyzed scientific publications' authors. Figure 8 presents the graphical results of Query 20 used for the Scopus exploration. In Figure 8, the minimum two co-occurrences of keywords are represented as a node in the bibliometric map. Among the 86 identified keywords in the VOSviewer software, only eight keywords met the threshold. In the bibliometric program, no keyword was excluded from the analysis. The edges of the network represent the explored co-occurrences between keywords in the data obtained from Scopus.

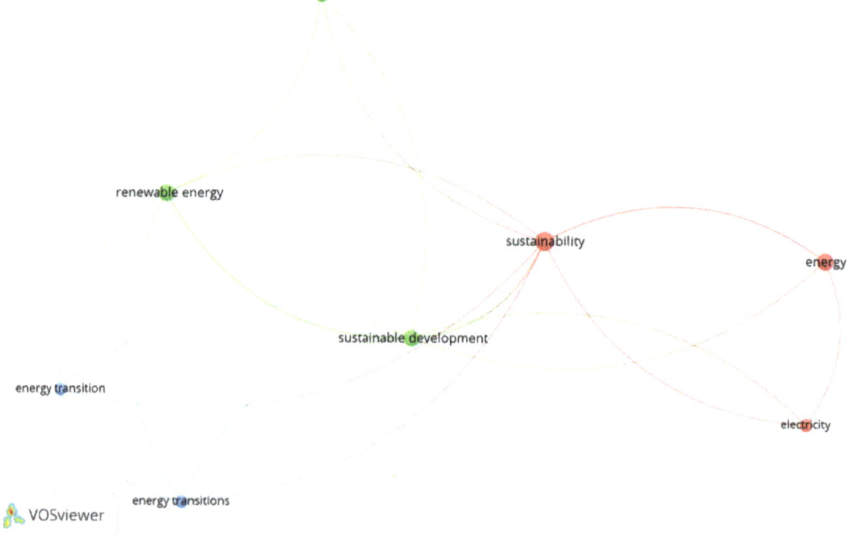

Figure 8. Bibliometric map of index keyword co-occurrences from Scopus based on Query 20's results. Source: authors' elaboration performed in VOSviewer (version 1.6.18).

Table 8. Clusters of keyword co-occurrences in Figure 8 for Scopus Query 20.

Cluster	Color	Keywords
1	Red	energy, electricity, sustainability
2	Green	business model, renewable energy, sustainable development
3	Blue	energy transition, energy transitions

Source: authors' elaboration.

In Figure 8, there are three clusters automatically identified by three colors as clusters in Table 8 and ordered by the VOSviewer program. In comparison to Queries 18 and 19, the number of identified keywords is significantly lower. The first cluster is marked in color red in Figure 8 and consists of co-occurring keywords related to general strategic orientations, such as "energy", "electricity", and "sustainability". The second cluster of keywords revolves around development visible in strategies in the energy sector and business models. Another group of scientific publications gathered in the third blue cluster created papers dedicated to the energy transition. Together with the performed Figure 8 analysis, there was an analysis of the keywords of interest in the VOSviewer software presented in Figure S4 in Supplementary Materials File S1. There was a shift towards keywords related to "sustainability", "electricity", and "energy" visible in the years 2018–2022.

Query 21's results were analyzed in the VOSviewer program in Figure 9 and Table 9. Analysis in bibliometric software was based on two minimum number of occurrences of a keyword among 49 indexed keywords. Only three indexed keywords met the threshold, and no keyword was excluded. These three keywords are included in the wider pool of all keywords, as proved in Query 18's analysis. Therefore, analysis of all keywords from Query 21's results was analyzed, and only nine of all keywords met the threshold of two minimum co-occurring keywords with the full counting method. In Figure 9, two clusters distinguished automatically by the VOSviewer program are visible. The results of Query 21 indicate both the importance and the low number of pieces of literature dealing with the classification of transformation and transition strategies in the energy sector.

Figure 9. Bibliometric map of all keyword co-occurrence results from Scopus based on Query 21. Source: authors' elaboration performed in VOSviewer (version 1.6.18).

Table 9. Clusters of keyword co-occurrences in Figure 9 for Scopus Query 21.

Cluster	Color	Keywords
1	Red	energy transitions, renewable energy, renewable energies, sustainability, sustainable development
2	Green	bioeconomy, biogas technologies, convergence, patent analysis

Source: authors' elaboration.

In Figure 9, there are two clusters visible as connected into a single network. The nodes are all the keywords co-occurring in the analyzed Query 21 results. The edges are the scientific publications that consist of these keywords. The sizes of the nodes represent the higher or lower number of co-occurrences. There are only two edges combining "sustainability" and "sustainable development" in the red cluster with "bioeconomy" in the green cluster. The first cluster is red, and it is the most numerous, indicating the direction of strategies oriented toward RES and SD. There is also a green cluster representing business practice visible in convergence and patent analysis. Among publications gathered in the green cluster, there are also those dedicated to biogas technologies and bioeconomy. The co-occurring keyword on the left side of the green cluster is more general than the keywords on the right side of this cluster. The centrally located keywords present in Figure 9 are also more general and theoretical than those placed in the bibliometric map's peripheries. There are the same keywords as presented in the red cluster in Figure 6 and Table 6.

Queries 22, 23, 24, and 25 were developed upon Query 2 from Table 1 and used in the Scopus database exploration. These queries consecutively limited their scope, reflected in the number of results presented in Table 3.

Query 26's results were analyzed in the VOSviewer program. The method adopted in the bibliometric program was full counting. The minimum number of indexed keyword co-occurrences was five, and among 378 keywords, only 16 met the threshold. There was no keyword excluded in the analysis procedure. There are three clusters presented in Figure 10 based on the 40 publications from Scopus. There is a bibliometric map of co-occurrences of keywords related to the transformation and transition strategies in the energy sector associated with "sustainable development", "energy policy", "alternative energy", "renewable energy sources", and "climate change". Despite such being specified in Query 26, only a few of them are present in Figure 10. The central node of the presented bibliometric map for Query 26 is "energy policy". This node is connected with all three clusters. First is the red cluster, organized around energy policy and decision making in the field of renewable energy sources and resources. The second, the green cluster, is revolving around sustainable development and management in the energy sector. There is also a third, blue cluster characterized by the environmental impact of technological solutions.

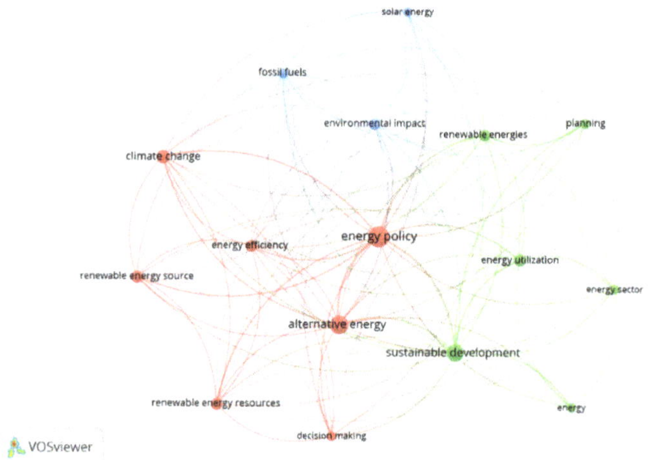

Figure 10. Bibliometric map of index keyword co-occurrence results from Scopus based on Query 26. The authors' elaboration was performed in VOSviewer (version 1.6.18).

The order of clusters presented in Table 10 is caused by the number of keywords identified by the VOSviewer. The identified clusters are related to the different areas of scientific interest of the analyzed scientific publications' authors.

Table 10. Clusters of keyword co-occurrences in Figure 10 for Scopus Query 26.

Cluster	Color	Keywords
1	Red	energy policy, alternative energy, climate change, decision making, energy efficiency, renewable energy sources, renewable energy resources
2	Green	energy, energy sector, energy utilization, planning, sustainable development
3	Blue	fossil fuels, environmental impact, solar energy

Source: authors' elaboration.

Figure 11 presents the division of the co-occurring keywords in the analyzed publication in the years 2019–2021. In this short period, there is a visible shift in scientific interest in publications reflected in the change of keywords in the bibliometric map. The results of Query 26 are publications with the oldest keyword, identified as "climate change" on the left side of Figure 11, and when going toward the right side there are the new and newest keywords in publications colored in yellow that have gained importance. When analyzing the general network of connections in Figure 11, it can be seen that the authors extended the areas of interest that were the focus three years ago. This means that these darker issues have not lost their importance, although the emphasis has slightly shifted toward issues related to planning and renewable energies in the energy sector. This further means that the lens is focused on the concept of energy policy as the equivalent of energy strategy, and this keyword has a central place in the presented bibliometric map.

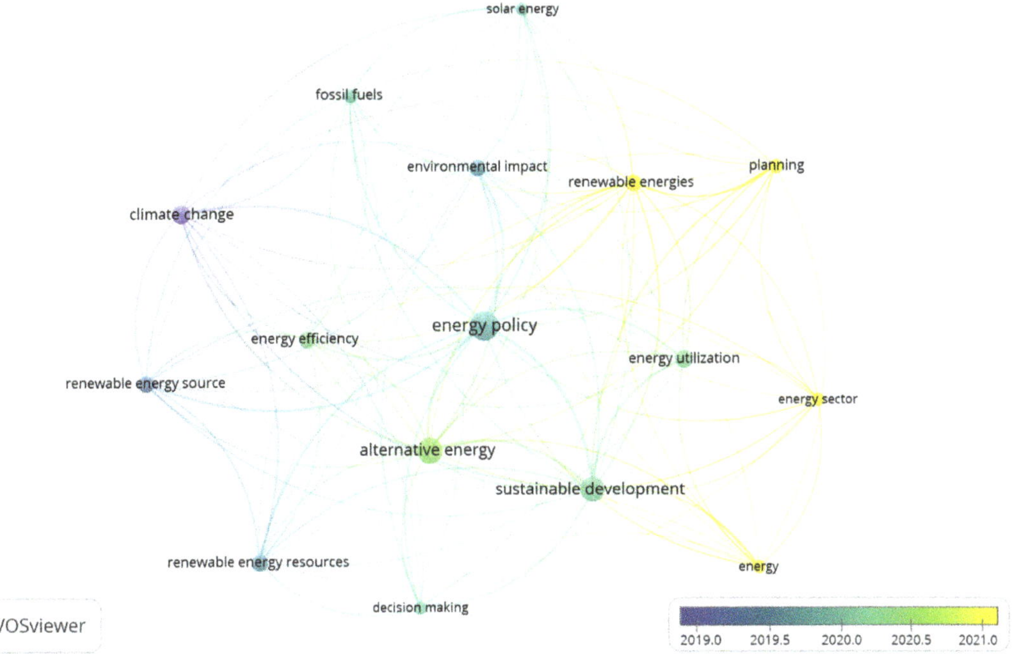

Figure 11. Bibliometric map of index keyword co-occurrence results from Scopus based on Query 26. Counting method: full counting. Minimum keyword co-occurrence is 10. Source: authors' elaboration performed in VOSviewer (version 1.6.18).

The exploration of the texts allowed for the identification of one more area of consideration. Based on Tables 4, 5 and 7–10, it can be noticed that there are more and less common keywords co-occurring in the publications indexed in Scopus. These tables were analyses of Queries 15, 18, 19, 20, 21, and 26, respectively. The maximum number of common keywords is six due to the number included in the analysis of six tables as sources. On the other hand, if the keyword is occurring only once among these tables, it can be stated as a difference in a strategic approach among analyzed strategies based on technology and those based on management and social change strategies in the energy sector. Additionally, the keyword "energy policy" is localized in all presented bibliometric maps in the central place.

In the analysis of the research results collected in Tables 4, 5 and 7–10, the keywords from the clusters were summarized. In this way, knowledge was gained about the keywords that are most often recurring and constitute a common element for most clusters of the resource. Also extracted were those keywords that are rarely realized research topics, presented in the last row of Table 11. Then, the gathered keywords in Table 11 represent the shift of the strategic direction and context in the third column, from the general perspective to the most specific solutions.

Table 11. Common and different keywords among analyzed Queries 15, 18, 19, 20, 21, and 26.

No of Occurrences	Keywords	Strategy Direction and Context
6	sustainable development	Strategic aim and priorities
5	energy transition *	Technological change
4	alternative energy; energy policy; energy sector; renewable energy *; sustainability	New energy sources
3	climate change; business model *; commerce; decision making; fossil fuels; innovation; renewable energy resources	Mitigation of negative changes in natural environment
2	electricity industry; energy; energy efficiency; energy management; energy market; energy utilization; renewable energy; renewable energy sources; wind power	Energy management and increase the energy efficiency
1	bioeconomy; biogas technologies; energy convergence; economic and social effects; economics; electric industry; electricity; electricity generation; energy conservation; energy planning; energy transformation; environmental impact; gas emissions; governance approach; greenhouse gases; international trade; investments; multilevels; patent analysis; planning; public utilities; renewable energy sector; renewable resources; social entrepreneurship; solar energy; strategic approach; sustainability transition; technological innovation	Future development and innovative directions

Source: authors' elaboration; * keywords occurring also in plural form.

The results of the keyword research are shown in Table 11. Taking the frequency of use in scientific articles as a criterion, the keywords were divided into six groups. The classification presents the keywords in descending order, beginning with the most frequently used. The author's elaboration is also presented in our strategic direction and context column.

The keyword that most often appears, analyzed in Table 11's queries, is "Sustainable Development". The context for the use of SD in energy sector research can be described as the main strategic aim and priorities. The second keyword that appears most often in the research is "energy transition", defining future development direction. The third group of keywords most frequently identified are "alternative energy", "energy policy", "energy sector", "renewable energy", and "sustainability". These keywords refer to research related to the future development direction of the energy sector in the context of technological change. Other keywords appeared three, two, and a single time in the surveyed texts of scientific publications, as indicated in the first column of Table 11. The first group of keywords consists of "climate change", "business model", "commerce", "decision making", "fossil fuels", "innovation", and "renewable energy resources", used in the context of the energy sector's future development direction. Twice the keywords appeared in the surveyed publications, "electricity industry", "energy", "energy efficiency", "energy management", "energy market", "energy utilization", "renewable energy", "renewable energy sources", and "wind power", pointing in the direction of research energy management. Keywords occurring a single time were listed in the last row of Table 11. To summarize the keywords in the scientific texts studied, they mainly refer to energy management, strategic concepts, and future development directions.

The directions of the changes in energy sector strategies can be generalized and became universal organization profiles implemented in the other economic sectors. In Table 12, there is a universal typology which can be assigned to energy sector companies illustrating their changes as progress between types I–IV. Those changes in strategies are confirmed by the bibliometric maps analysis, which illustrates shifts of scientific interest in explored publications.

Table 12. The energy sector strategy groups.

Source	Organizations' Profiles According to Transformation Strategy Types			
	I	II	III	IV
James, 1992 [59]	Non-compliance	Compliance	Compliance-plus	Excellence
Bostrum and Poysti, 1992 [60]	Resistant	Passive	Reactive	Innovative
Welford, 1994 [61]	Cowards	Laggards	Thinkers	Doers
Meima, 1994 [62]	Problem Fixer	Obstacle Jumper	Opportunist	Activist
Roome, 1994 [63]	Indifference	Offensive	Defensive	Innovative
Loknath and Abdul Azeem, 2017 [64]	Stable	Reactive	Anticipatory	Entrepreneurial
Sulich and Grudziński, 2019 [26]	Isolation	Redundancy	Adaptation	Cooperation
Zhou et al., 2019 [65]	Avoidant	Reverse	Pioneering	Aggressing

Source: author's own elaboration based on [64].

In the researched literature, there are also different divisions based on corporate responses to environmental pressures. The first group in Table 12 is the energy companies that have been forced to improve their environmental performance as a result of some well-known failures. The second group is the energy sector companies that have been able to exploit the opportunity created by green technology to gain a competitive advantage [66,67]. The third group includes energy companies that have moved beyond compliance and have incorporated their environmental strategy into their overall business strategy [64,66]. There is also the fourth group, which gathers the companies that exceed those in previous categories. However, among the presented groups, the authors focused on the different aspects of the organizations' profiles and their strategies proposing their classifications.

Although there are models of strategic options which include more than four strategy types, these four groups are the most popular and are used interchangeably by different authors (Table 12). These four categories are: indifferent, offensive, defensive, and innovative, proposed by Roome [63]. These classifications can be used also in other economy sectors. Indifferent companies are those that have low environmental risk and even less environmentally based opportunities for growth [63]. Offensive companies are those that have considerable potential for exploiting environmentally related market opportunities and include companies that manufacture pollution control equipment [63]. Those adopting a defensive strategy are companies such as chemical companies, which have high environmental risk and cannot afford to ignore environmental issues, or their very survival could be at stake [63]. The innovators are those that have high environmental risk and also a lot of environmentally based opportunities for growth.

Loknath and Abdul Azeem in their conference article presented a review of types of green management strategies, and among them they list four other element categorizations of strategy types, which were compared in Table 12. The first group is organizations that not only assume that concern for the environment is a passing phase and their impact on the environment is negligible but also assume that their competitors feel the same and hence carry out nothing to conserve the environment. The second group consists of organizations that are aware of the environmental challenges facing them but are unable to combat those challenges because of cost constraints, lack of trained manpower, and lack of knowledge. The third group is organizations that are aware of problems but are still waiting for others to show the way forward. Then, there are organizations that have proceeded to put their assumptions into action, and they make up the fourth group. In the science, there has been a shift in viewpoint from theorizing and successive divisions of strategy to more concrete solutions from management science in working out strategy. Then, strategies and types of strategies have ceased to be the center of academic interest.

The implementation of the transformation strategies by organizations operating in the energy sector began in the previous century, and it was a part of technological progress, which has provided a new eco-friendly solution. This technological change forced companies to leave fossil fuels consumption. In the new century, the energy sector should become an eco-centric one. Then, it must become focused on sustainability despite investments, risk, and time [68]. Therefore, an ever-greater number of organizations have begun to notice that the idea of sustainability is becoming a natural element of their action.

4. Discussion

In the presented bibliometric maps and tables, there is less interest among publications' authors in developing strategy theory (typology, type, or classification) in explored publications. Despite the formulated queries consisting of the part dedicated to strategy types, the results proved that there is focus on solving specific practical problems related to the energy sector [69]. This may be because of the strategic importance of the energy sector to economic and social development. On the other hand, the difficult situation in this sector is accompanied by its close synergy with the natural environment's depleting resources [70] and rapid change in the natural environment's deteriorating conditions [71]. In the results of the performed study, there was no specified sectoral strategy identified, neither in bibliometric maps in the SLR method nor in the CLR results. There were general change directions and types of energy sector strategies identified.

In the result tables associated with the bibliometric maps, there are results identified by the authors' strategic directions and contexts. There are then two categories of the results. The first group consists of the keywords identified by the VOSviewer program. The second group of results is formulated by the authors as the research directions. The identified clusters also reflect the shift from general perspective to the more specific solutions in the energy sector [72]. The identified clusters in bibliometric maps are explored as three main groups: strategies and management; technological innovation; and change in business practice. Among the identified clusters in the analyzed publications in the Scopus scientific database, the shift from protection and conservation toward environmental management is visible [73,74]. In addition, within the separate clusters there are subnetworks reflecting very specific subjects of the explored publications in Scopus. This is also a change of the paradigm which influences the strategies in the energy sector. Despite of the focus on the energy sector and the management and economics sciences' issues, there are also other subjects of rapidly growing attention to scientists: "social entrepreneurship" [75], "innovation" [76], "bioeconomy" [77], "biogas technologies" [78,79], "convergence" [80], and "patent analysis" [81,82]. These topics are related to the decision-making process in light of the shift from fossil fuels to RES.

The maps and results do not show clear divisions and ordering of strategy classifications, despite the fact that such an inquiry was formulated in the queries. The keywords lack direct mention of classification and strategy. It can be assumed that a typology of strategies can be found in the content of the scientific articles but is not explicitly called a classification. These maps show some strategy directions other than the typologies. The bibliometric research results indicate that companies operating in the energy sector attach great importance to strategy planning and business policies [83,84]. This observation is supported by the pivotal role of the "energy policy" keyword in presented bibliometric maps [85,86]. Therefore, the energy sector companies are looking for the most significant and influential activities [57,87].

This paper's goal was achieved by the exploration of the business management literature dedicated to the transformation and transition strategies in the energy sector. The adopted methods are a systematic literature review (SLR) accompanied by the classical literature review (CLR) in Scopus database exploration.

The limitation of this study's methods is an institutional access to the Scopus database. The researchers without the provided access to the Scopus, for example, cannot achieve reproducibility of this study. Another limitation of this study method could be related to the question of the research area limited by the subject area in Scopus or the presentation of the 10 most contributing scientific publications (presented in Tables S1–S6, Supplementary Materials File S1). Another limitation is also the fact that Scopus is an international database where English is a dominant language in science; therefore, keywords were only searched in this language. The restriction to the English language is an expression of the selection of only those works that have been internationally evaluated and circulated internationally [5]. On the other hand, there are no limits in terms of time span, publication accessibility, scope or type, geographical area, or author-related information in the presented results. Another limitation of this research is based on the aggregation of the plural forms from the keywords list. Another limitation is the authors' interpretation of the keywords collected in the clusters. However, there are two main macro streams in the explored publication in the Scopus database, the first related to SD and the second to strategies in the energy sector [65,88]. They reflect the clear division between more theoretical and empirical publications.

In the bibliometric study results, it is not surprising that SD is most often indicated as a keyword. The idea of SD is close to politicians, business leaders, entrepreneurs, and societies, but its implementation causes many problems [89]. SD should build a common future based on the energy sector.

Sustainable development builds on the concept of quality of life in an unlimited time span because its assumptions are timeless and universal [90,91]. Therefore, planning is the most crucial element of the strategies in the energy sector. This raises the question of how companies competing in the energy sector are to shape a high quality of life as part of their strategy. These assumptions must take into account the strategies of companies in the energy sector where diverse environmental, economic, and social problems are considered in a broad context. Thus, business sustainability is based on the ability to manage or mitigate pollution and depletion of non-renewable resources. This can be achieved by usage of technological innovations, recognized as the main groups of issues in the Scopus exploration. However, decision making and management have to initiate and coordinate those actions.

Publications analyzed in this research proved then that the lack of rapid changes and results in implemented strategies in energy sector companies constitute a premise for researching this area. Problems may be strategically conditioned and result from inadequate management and the lack of implementation of these strategies. The selection of strategic goals that positively impact the environment is essential when it is translated into the strategy's implementation by energy companies. The result of this bibliometric research is that current energy sector strategies are not studied from a theoretical point of view but as planning in the energy sector to achieve practical goals related primarily to SD.

The objectives of energy sector strategies are aimed at involving the improvement of the environment or minimizing anthropo-pressure. There is also complexity of the decision-making process in strategy field results from the growing number of elements (participants) of the business environment system and the ties between them, which are the result of unexpected events (discontinuities) that determine volatility, technological breakdowns, changes in consumer habits, economic transformations, and crises.

5. Conclusions

In this paper the authors assumed that scientists, who are authors of the explored publications indexed in Scopus, cooperate with business. Therefore, their works reflect the changes in the energy sector's strategies. There is close cooperation between academia and the business environment to achieve sustainability in the energy sector. To illustrate this close relationship and to perform a literature review, two methods of review were used. The combined methods of CLR and SLR were used to identify the state of the art and integrate and synthesize the existing body of literature but also explore the main directions for further research on transformation strategies in the energy sector.

The main conclusion of this paper is that there are the changes in the energy sector, but the pace those changes is too slow in relation to the real actions. This disparity is caused by the imbalance between struggles to improve the quality of life and stop climate change and especially the rise in levels. Energy sector transformation strategies aim to overcome the disadvantages caused by localization and gain energetic independence and diversification in energy generation. Therefore, there are multiple directions among those strategies. There are different driving forces behind the strategic changes in the energy sector; however, they have not appeared in the results of this bibliometric study despite the full-time analyzation period, and words such as "war", "armed conflict", or "pandemic" were not excluded.

The methodological contribution of this paper is based on the comparison of the two methods employed in this literature review. The SLR method has a clear advantage over the CLR method, as it avoids many of the typical inadequacies and at the same time opens the way for replication of bibliometric research, strengthening the development of management science. With the growing practice or dissemination of the standard of systematic literature review in the preparation of promotional works and empirical studies, problems specific to this method arise. The bibliometric study based on the SLR with queries can be an initial step in the exploration of the research problem. The results of this SLR method provided only initial recognition of the general directions of strategies and their contexts. The subject of the changes in energy sector strategies was deeply explored in the CLR method, which allowed the identification of missing typologies and strategy classifications. Better results are obtained only when these two methods are combined.

The scientific contribution of this study is based on the exploration of the research gaps. The first is the taxonomic research gap, which is satisfied in this study by the proposed order in bibliometric maps and their clusters. The second is the knowledge gap about the direction of the changes in energy sector strategies, which is answered in this study. In this study, the methodic research gap is also covered in terms of the combination of the two adopted methods. The last research gap addressed in this study is based on rare literature reviews and slender publications dedicated to the changes in the energy sector's strategies. However, these research gaps indicate promising research avenues which can be developed in future research.

The practical contributions of this paper can interest researchers, managers, and policymakers interested in energy sector strategies. The presentation of the directions and context of the strategies combined with their typologies allow one to assess the current situation of a company in the energy sector. The bibliometric maps along with their descriptions provide valuable information about promising scientific research which can combine subjects identified as outlier keywords. The direct implication for policymakers is the visible creation of the biogas and hydrogen power generation sectors of the bioeconomy. This specific direction of the development of strategies in the energy sector involves both managerial and technological solutions.

Among the publications, there are multiple shifts reflecting the authors' scientific interest changes. The dimensions of those shifts are, among others, change from a theoretical to empirical perspective, change from protection and conservation towards environmental management, and there is also growing interest in RES as technological innovation over fossil fuels. Among the results, there was no visible discussion about clear energy production or nuclear energy generation as alternatives and renewable energy sources. However, the RES advantages are listed in many publications as diversification solutions in strategies, providing energetic independence. Those advantages are presented more often than disadvantages in the explored research indexed in Scopus papers. Along with the positive image of the strategies in the energy sector, their influence on the labor market is present in scientific publication. The elements of the strategies employed by the energy sector's companies are also introduced in other sectors. The reason for an absence of classifications in the energy sector's strategies can be a lack of typologies and a lack of divisions in the bibliometric maps. Among the publications, there is ongoing generalization of the strategic-management-related topics and presentation of the new subjects than exploration of the well-explored and known strategy elements.

The novelty of this bibliometric research is based on the identification of the changes in energy sector strategies and is important because of its reliance on resources and natural environment degradation. Another novelty feature of this research is the presentation of the existing research gaps and identification of main directions of ongoing research.

New possibilities such as information and communication technologies, especially the easy access to scientific information, have led to the rapid development of systematic methods of knowledge management [92]. In recent years, there has been a particular development of systematic literature review methodology, which, by using a replicable and well-described research procedure, allows for the search, selection, and evaluation of all available scientific evidence in a given field of inquiry. The method allows for the synthesis of this evidence, through which it is provided in a user-friendly way, with reliable information from an up-to-date evidence base. Systematic literature reviews are used by researchers preparing new research projects. A review of the literature allows for an understanding of a given issue and for the identification of a body of knowledge that needs to be supplemented in subsequent research. As a prerequisite for increasing the effectiveness of public decision making is to base it on sound and reliable scientific evidence, it is worthwhile for politicians to use systematic reviews. Given the usefulness of systematic literature review methods, there is a need to disseminate knowledge about what the reviews are and how they can be used.

Supplementary Materials: The following supporting information can be downloaded at: https://www.mdpi.com/article/10.3390/en15197068/s1.

Author Contributions: Conceptualization, A.S. and L.S.-P.; methodology, A.S. and L.S.-P.; formal analysis, A.S. and L.S.-P.; investigation, A.S. and L.S.-P.; writing—original draft preparation, A.S. and L.S.-P.; writing—review and editing, A.S. and L.S.-P.; visualization, A.S. and L.S.-P.; supervision, A.S.; project administration, A.S. and L.S.-P.; funding acquisition, A.S. and L.S.-P. All authors have read and agreed to the published version of the manuscript.

Funding: (A.S.) The project was financed by the National Science Centre in Poland under the program "Business Ecosystem of the Environmental Goods and Services Sector in Poland", implemented in 2020–2022; project number 2019/33/N/HS4/02957; total funding amount PLN 120,900.00. (L.S.-P.) The project was financed by the Ministry of Education and Science in Poland under the program "Regional Initiative of Excellence" 2019–2023 project number 015/RID/2018/19 total funding amount 10 721 040,00 PLN.

Institutional Review Board Statement: Not applicable.

Informed Consent Statement: Not applicable.

Data Availability Statement: Not applicable.

Acknowledgments: The authors would like to thank Kinga Zmigrodzka-Ryszczyk and Agnieszka Dramińska (Wroclaw University of Economics and Business) and Alexandra Wang (York University Libraries) for the Scopus database exploration, data visualization and bibliometric workshops, and secondary data acquisition.

Conflicts of Interest: The authors declare no conflict of interest. The funders had no role in the design of the study; in the collection, analyses, or interpretation of data; in the writing of the manuscript or in the decision to publish the results.

References

1. Nagypál, N.C. Corporate social responsibility of hungarian SMEs with good environmental practices. *J. East Eur. Manag. Stud.* **2014**, *19*, 327–347. [CrossRef]
2. Franek, J.; Svoboda, O. Productivity and efficiency of automotive companies in the Czech Republic: A DEA approach. In Proceedings of the SMSIS 2019—13th International Conference on Strategic Management and Its Support by Information Systems, Ostrava, Czech Republic, 21–22 May 2019; pp. 87–98.
3. Aragón-Correa, J.A.; Rubio-López, E.A. Proactive Corporate Environmental Strategies: Myths and Misunderstandings. *Long Range Plann.* **2007**, *40*, 357–381. [CrossRef]
4. Sulich, A.; Sołoducho-Pelc, L. Renewable Energy Producers' Strategies in the Visegrád Group Countries. *Energies* **2021**, *14*, 3048. [CrossRef]
5. Gostkowski, M.; Rokicki, T.; Ochnio, L.; Koszela, G.; Wojtczuk, K.; Ratajczak, M.; Szczepaniuk, H.; Bórawski, P.; Bełdycka-Bórawska, A. Clustering Analysis of Energy Consumption in the Countries of the Visegrad Group. *Energies* **2021**, *14*, 5612. [CrossRef]

6.   Kozar, Ł.; Paduszyńska, M. Change Dynamics of Electricity Prices for Households in the European Union between 2011 and 2020. *Finans. Prawo Finans.* **2021**, *4*, 97–115. [CrossRef]

7.   Szaruga, E.; Załoga, E. Qualitative–Quantitative Warning Modeling of Energy Consumption Processes in Inland Waterway Freight Transport on River Sections for Environmental Management. *Energies* **2022**, *15*, 4660. [CrossRef]

8.   Fic, M.; Malinowski, M. Communication Infrastructure and Enterprise Effectiveness. In *The Essence and Measurement of Organizational Efficiency*; Dudycz, T., Osbert-Pociecha, G.B.B., Eds.; Springer Proceedings in Business and Economics; Springer Science and Business Media B.V.: Cham, Switzerland, 2016; pp. 39–63.

9.   Kozar, Ł. Energy Sector and the Challenges of Sustainable Development Analysis of Spatial Differentiation of the Situation in the EU based on selected indicators. *Probl. World Agric./Probl. Rol. Swiat.* **2018**, *18*, 173–186. [CrossRef]

10. Sullivan, K.; Thomas, S.; Rosano, M. Using industrial ecology and strategic management concepts to pursue the Sustainable Development Goals. *J. Clean. Prod.* **2018**, *174*, 237–246. [CrossRef]

11. Marinakis, V.; Koutsellis, T.; Nikas, A.; Doukas, H. Ai and data democratisation for intelligent energy management. *Energies* **2021**, *14*, 4341. [CrossRef]

12. Gallo, A.B.; Simões-Moreira, J.R.; Costa, H.K.M.; Santos, M.M.; Moutinho dos Santos, E. Energy storage in the energy transition context: A technology review. *Renew. Sustain. Energy Rev.* **2016**, *65*, 800–822. [CrossRef]

13. Connolly, D.; Lund, H.; Mathiesen, B.V.; Leahy, M. A review of computer tools for analysing the integration of renewable energy into various energy systems. *Appl. Energy* **2010**, *87*, 1059–1082. [CrossRef]

14. Csigéné Nagypál, N.; Görög, G.; Harazin, P.; Péterné Baranyi, R. "Future Generations" and Sustainable Consumption. *Econ. Sociol.* **2015**, *8*, 207–224. [CrossRef]

15. Gerasimova, K. *An Analysis of Brundtland Commission's Our Common Future*; Macat Library: London, UK, 2017; ISBN 9781912281220.

16. Olszak, C.M.; Mach-Król, M. A conceptual framework for assessing an organization's readiness to adopt big data. *Sustainability* **2018**, *10*, 3734. [CrossRef]

17. Chung, M.H. Comparison of Economic Feasibility for Efficient Peer-to-Peer Electricity Trading of PV-Equipped Residential House in Korea. *Energies* **2020**, *13*, 3568. [CrossRef]

18. Biard, G.; Nour, G.A. Industry 4.0 Contribution to Asset Management in the Electrical Industry. *Sustainability* **2021**, *13*, 10369. [CrossRef]

19. Sine, W.D.; Lee, B.H. Tilting at Windmills? The Environmental Movement and the Emergence of the U.S. Wind Energy Sector. *Adm. Sci. Q.* **2009**, *54*, 123–155. [CrossRef]

20. Lattanzio, R.K. *The World Bank Group Energy Sector Strategy*; Nova Science Publishers, Inc.: Hauppauge, NY, USA, 2012; ISBN 9781620818428.

21. Suriyankietkaew, S.; Petison, P. A Retrospective and Foresight: Bibliometric Review of International Research on Strategic Management for Sustainability, 1991–2019. *Sustainability* **2019**, *12*, 91. [CrossRef]

22. Kern, F.; Howlett, M. Implementing transition management as policy reforms: A case study of the Dutch energy sector. *Policy Sci.* **2009**, *42*, 391–408. [CrossRef]

23. Norton, T.A.; Ayoko, O.B.; Ashkanasy, N.M. A Socio-Technical Perspective on the Application of Green Ergonomics to Open-Plan Offices: A Review of the Literature and Recommendations for Future Research. *Sustainability* **2021**, *13*, 8236. [CrossRef]

24. Finkbeiner, M.; Schau, E.M.; Lehmann, A.; Traverso, M. Towards Life Cycle Sustainability Assessment. *Sustainability* **2010**, *2*, 3309–3322. [CrossRef]

25. Eurostat. *Sustainable Development in the European Union—Monitoring Report on Progress towards the SDGs in an EU Context*, 2018th ed.; Eurostat: Brussels, Belgium, 2018.

26. Sulich, A.; Grudziński, A. The analysis of strategy types of the renewable energy sector. In Proceedings of the Enterprise and Competitive Environment: 22nd Annual International Conference, Brno, Czech Republic, 21–22 March 2019; Kapounek, S., Vránová, H., Eds.; Mendel University in Brno: Brno, Czech Republic, 2019; p. 59.

27. Małecka, A.; Maciej, M.; Mróz-Gorgoń, B.; Pfajfar, G. Adoption of collaborative consumption as sustainable social innovation: Sociability and novelty seeking perspective. *J. Bus. Res.* **2022**, *144*, 163–179. [CrossRef]

28. González-Eguino, M. Energy poverty: An overview. *Renew. Sustain. Energy Rev.* **2015**, *47*, 377–385. [CrossRef]

29. Sawhney, A. Striving towards a circular economy: Climate policy and renewable energy in India. *Clean Technol. Environ. Policy* **2021**, *23*, 491–499. [CrossRef]

30. Zema, T.; Sulich, A. Models of Electricity Price Forecasting: Bibliometric Research. *Energies* **2022**, *15*, 5642. [CrossRef]

31. Visser, M.; van Eck, N.; Waltman, L. Large-scale comparison of bibliographic data sources: Web of Science, Scopus, Dimensions, and CrossRef. In Proceedings of the 17th International Conference on Scientometrics and Informetrics, ISSI 2019, Rome, Italy, 2–5 September 2019; Volume 2, pp. 2358–2369.

32. Fortunato, S.; Bergstrom, C.T.; Börner, K.; Evans, J.A.; Helbing, D.; Milojević, S.; Petersen, A.M.; Radicchi, F.; Sinatra, R.; Uzzi, B.; et al. Science of science. *Science* **2018**, *359*, eaao0185. [CrossRef] [PubMed]

33. Lyulyov, O.; Pimonenko, T.; Kwilinski, A.; Dzwigol, H.; Dzwigol-Barosz, M.; Pavlyk, V.; Barosz, P. The impact of the government policy on the energy efficient gap: The evidence from Ukraine. *Energies* **2021**, *14*, 373. [CrossRef]

34. Dobrowolski, Z.; Drozdowski, G.; Panait, M.; Babczuk, A. Can the Economic Value Added Be Used as the Universal Financial Metric? *Sustainability* **2022**, *14*, 2967. [CrossRef]

35. Gerstenblatt, P. Collage Portraits as a Method of Analysis in Qualitative Research. *Int. J. Qual. Methods* **2013**, *12*, 294–309. [CrossRef]
36. Denyer, D.; Tranfield, D.; van Aken, J.E. Developing Design Propositions through Research Synthesis. *Organ. Stud.* **2008**, *29*, 393–413. [CrossRef]
37. Denyer, D.; Tranfield, D. Using qualitative research synthesis to build an actionable knowledge base. *Manag. Decis.* **2006**, *44*, 213–227. [CrossRef]
38. Briner, R.B.; Denyer, D. Systematic Review and Evidence Synthesis as a Practice and Scholarship Tool. In *The Oxford Handbook of Evidence-Based Management*; Oxford University Press: Oxford, UK, 2012; pp. 112–129. ISBN 9780199968879; 0199763984; 9780199763986.
39. Waltman, L. A review of the literature on citation impact indicators. *J. Inform.* **2016**, *10*, 365–391. [CrossRef]
40. Philips, R.; Guttman, I. A new criterion for variable selection. *Stat. Probab. Lett.* **1998**, *38*, 11–19. [CrossRef]
41. Arslan, A.; Haapanen, L.; Hurmelinna-Laukkanen, P.; Tarba, S.Y.; Alon, I. Climate change, consumer lifestyles and legitimation strategies of sustainability-oriented firms. *Eur. Manag. J.* **2021**, *39*, 720–730. [CrossRef]
42. Fuentes-Bracamontes, R.; Fuentes-Bracamontes, R. How to reform the power sector in Mexico? Insights from a simulation model. *Int. J. Energy Sect. Manag.* **2012**, *6*, 438–464. [CrossRef]
43. Vogel, R.; Güttel, W.H. The Dynamic Capability View in Strategic Management: A Bibliometric Review. *Int. J. Manag. Rev.* **2012**, *15*, 426–446. [CrossRef]
44. Wróblewski, P.; Drożdż, W.; Lewicki, W.; Miązek, P. Methodology for Assessing the Impact of Aperiodic Phenomena on the Energy Balance of Propulsion Engines in Vehicle Electromobility Systems for Given Areas. *Energies* **2021**, *14*, 2314. [CrossRef]
45. Lamers, W.S.; Boyack, K.; Larivière, V.; Sugimoto, C.R.; van Eck, N.J.; Waltman, L.; Murray, D. Investigating disagreement in the scientific literature. *eLife* **2021**, *10*, e72737. [CrossRef]
46. Boyack, K.; Glänzel, W.; Gläser, J.; Havemann, F.; Scharnhorst, A.; Thijs, B.; van Eck, N.J.; Velden, T.; Waltmann, L. Topic identification challenge. *Scientometrics* **2017**, *111*, 1223–1224. [CrossRef]
47. van Eck, N.J.; Waltman, L. Software survey: VOSviewer, a computer program for bibliometric mapping. *Scientometrics* **2010**, *84*, 523–538. [CrossRef]
48. Vansovits, V.; Petlenkov, E.; Tepljakov, A.; Vassiljeva, K.; Belikov, J. Bridging the Gap in Technology Transfer for Advanced Process Control with Industrial Applications. *Sensors* **2022**, *22*, 4149. [CrossRef]
49. Kaczmarczyk, K.; Hernes, M. Financial decisions support using the supervised learning method based on random forests. *Procedia Comput. Sci.* **2020**, *176*, 2802–2811. [CrossRef]
50. Colavizza, G.; Boyack, K.W.; van Eck, N.J.; Waltman, L. Exploring the similarity of articles co-cited at different levels. In Proceedings of the ISSI 2017—16th International Conference on Scientometrics and Informetrics, Wuhan, China, 16–20 October 2017; pp. 549–560.
51. Waltman, L.; van Eck, N.J.; Wouters, P. Counting publications and citations: Is more always better? In Proceedings of the ISSI 2013—14th International Society of Scientometrics and Informetrics Conference, Vienna, Austria, 15–19 July 2013; Volume 1, pp. 455–467.
52. Organa, M.; Sus, A. Exploiting Opportunities in Strategic Business Nets—The Case Study. In *Innovation Management and Education Excellence through Vision 2020, Proceedings of the 31st International Business Information Management Association Conference (IBIMA), Milan, Italy, 25–26 April 2018*; Soliman, K.S., Ed.; International Business Information Management Association (IBIMA): Seville, Spain, 2018; pp. 5380–5390.
53. Niemczyk, J.; Sus, A.; Bielińska-Dusza, E.; Trzaska, R.; Organa, M. Strategies of European Energy Producers. Directions of Evolution. *Energies* **2022**, *15*, 609. [CrossRef]
54. van Eck, N.J.; Waltman, L.; Dekker, R.; van den Berg, J. A comparison of two techniques for bibliometric mapping: Multidimensional scaling and VOS. *J. Am. Soc. Inf. Sci. Technol.* **2010**, *61*, 2405–2416. [CrossRef]
55. Tcvetkov, P. Climate Policy Imbalance in the Energy Sector: Time to Focus on the Value of CO2 Utilization. *Energies* **2021**, *14*, 411. [CrossRef]
56. Janik, A.; Ryszko, A.; Szafraniec, M. Greenhouse Gases and Circular Economy Issues in Sustainability Reports from the Energy Sector in the European Union. *Energies* **2020**, *13*, 5993. [CrossRef]
57. Kochanek, E. The Energy Transition in the Visegrad Group Countries. *Energies* **2021**, *14*, 2212. [CrossRef]
58. Schaeffer, G.J. Energy sector in transformation, trends and prospects. *Procedia Comput. Sci.* **2015**, *52*, 866–875. [CrossRef]
59. James, P. The Corporate Response. In *Greener Marketing*; Charter, M., Ed.; Greenleaf Publishing: Sheffield, UK, 1992.
60. Bostrum, T.; Poysti, E. *Enviornmental Startegy in the Enterprise*; Helsinki School of Economics: Helsinki, Finland, 1992.
61. Welford, R. *Cases in Enviornmental Management and Business Startegy*; Rtman Publishing: London, UK, 1994.
62. Meima, R. Making Sense of Environmental Management Concepts and Practices: A critical exploration of Emerging Paradigms. In Proceedings of the 1994 Business Startegy and the Environment Conference, The University of, Nottingham, Nottingham, UK, 1994.
63. Roome, N. Business Strategy, R&D Management and Environmental Imperatives. *R&D Manag.* **1994**, *24*, 065–082. [CrossRef]
64. Loknath, Y.; Bepar, A.A. Green Management-Concept and Strategies. In Proceedings of the National Conference on Marketing and Sustainable Development, 13–14 October 2017; pp. 688–702.

65. Zhou, J.; He, P.; Qin, Y.; Ren, D. A selection model based on SWOT analysis for determining a suitable strategy of prefabrication implementation in rural areas. *Sustain. Cities Soc.* **2019**, *50*, 101715. [CrossRef]

66. Simpson, A. *The Greening of Global Investment—How the Environement, Ethics and Politics are Reshaping Strategies*; The Economicst Publications: London, UK, 1991.

67. Lyu, W.; Liu, J. Soft skills, hard skills: What matters most? Evidence from job postings. *Appl. Energy* **2021**, *300*, 117307. [CrossRef]

68. Martinot, E. Renewable energy investment by the World Bank. *Energy Policy* **2001**, *29*, 689–699. [CrossRef]

69. Balsalobre-Lorente, D.; Leitão, N.C.; Bekun, F.V. Fresh Validation of the Low Carbon Development Hypothesis under the EKC Scheme in Portugal, Italy, Greece and Spain. *Energies* **2021**, *14*, 250. [CrossRef]

70. Wróblewski, P.; Drożdż, W.; Lewicki, W.; Dowejko, J. Total Cost of Ownership and Its Potential Consequences for the Development of the Hydrogen Fuel Cell Powered Vehicle Market in Poland. *Energies* **2021**, *14*, 2131. [CrossRef]

71. Brodny, J.; Tutak, M.; Bindzár, P. Assessing the Level of Renewable Energy Development in the European Union Member States. A 10-Year Perspective. *Energies* **2021**, *14*, 3765. [CrossRef]

72. Tănasie, A.V.; Năstase, L.L.; Vochița, L.L.; Manda, A.M.; Boțoteanu, G.I.; Sitnikov, C.S. Green Economy—Green Jobs in the Context of Sustainable Development. *Sustainability* **2022**, *14*, 4796. [CrossRef]

73. Światowiec-Szczepańska, J.; Stępień, B. Drivers of Digitalization in the Energy Sector—The Managerial Perspective from the Catching Up Economy. *Energies* **2022**, *15*, 1437. [CrossRef]

74. Wróblewski, P.; Kupiec, J.; Drożdż, W.; Lewicki, W.; Jaworski, J. The Economic Aspect of Using Different Plug-In Hybrid Driving Techniques in Urban Conditions. *Energies* **2021**, *14*, 3543. [CrossRef]

75. Meek, W.R.; Pacheco, D.F.; York, J.G. The impact of social norms on entrepreneurial action: Evidence from the environmental entrepreneurship context. *J. Bus. Ventur.* **2010**, *25*, 493–509. [CrossRef]

76. Gabaldón-Estevan, D.; Peñalvo-López, E.; Alfonso Solar, D. The Spanish Turn against Renewable Energy Development. *Sustainability* **2018**, *10*, 1208. [CrossRef]

77. Colacicco, M.; Ciliberti, C.; Agrimi, G.; Biundo, A.; Pisano, I. Towards the Physiological Understanding of Yarrowia lipolytica Growth and Lipase Production Using Waste Cooking Oils. *Energies* **2022**, *15*, 5217. [CrossRef]

78. Senkus, P.; Glabiszewski, W.; Wysokińska-Senkus, A.; Cyfert, S.; Batko, R. The Potential of Ecological Distributed Energy Generation Systems, Situation, and Perspective for Poland. *Energies* **2021**, *14*, 7966. [CrossRef]

79. Paglini, R.; Gandiglio, M.; Lanzini, A. Technologies for Deep Biogas Purification and Use in Zero-Emission Fuel Cells Systems. *Energies* **2022**, *15*, 3551. [CrossRef]

80. Vavrek, R.; Chovancová, J. Energy Performance of the European Union Countries in Terms of Reaching the European Energy Union Objectives. *Energies* **2020**, *13*, 5317. [CrossRef]

81. Miśkiewicz, R.; Rzepka, A.; Borowiecki, R.; Olesiński, Z. Energy Efficiency in the Industry 4.0 Era: Attributes of Teal Organisations. *Energies* **2021**, *14*, 6776. [CrossRef]

82. Mitchell, D.; Blanche, J.; Harper, S.; Lim, T.; Gupta, R.; Zaki, O.; Tang, W.; Robu, V.; Watson, S.; Flynn, D. A review: Challenges and opportunities for artificial intelligence and robotics in the offshore wind sector. *Energy AI* **2022**, *8*, 100146. [CrossRef]

83. Karakosta, C. A Holistic Approach for Addressing the Issue of Effective Technology Transfer in the Frame of Climate Change. *Energies* **2016**, *9*, 503. [CrossRef]

84. Sarmiento, L.; Burandt, T.; Löffler, K.; Oei, P.-Y. Analyzing scenarios for the integration of renewable energy sources in the Mexican energy system—An application of the Global Energy System Model (GENeSys-MOD). *Energies* **2019**, *12*, 3270. [CrossRef]

85. Uğurlu, E. Impacts of Renewable Energy on CO 2 Emission: Evidence from the Visegrad Group Countries. *Polit. Cent. Eur.* **2022**, *18*, 295–315. [CrossRef]

86. Majeed, A.; Ye, C.; Chenyun, Y.; Wei, X. Muniba Roles of natural resources, globalization, and technological innovations in mitigation of environmental degradation in BRI economies. *PLoS ONE* **2022**, *17*, e0265755. [CrossRef]

87. Brodny, J.; Tutak, M. Analyzing Similarities between the European Union Countries in Terms of the Structure and Volume of Energy Production from Renewable Energy Sources. *Energies* **2020**, *13*, 913. [CrossRef]

88. Akinyele, D.; Belikov, J.; Levron, Y. Battery Storage Technologies for Electrical Applications: Impact in Stand-Alone Photovoltaic Systems. *Energies* **2017**, *10*, 1760. [CrossRef]

89. Sachs, J.D. From Millennium Development Goals to Sustainable Development Goals. *Lancet* **2012**, *379*, 2206–2211. [CrossRef]

90. Hwangbo, Y.; Shin, W.-J.; Kim, Y. Moderating Effects of Leadership and Innovation Activities on the Technological Innovation, Market Orientation and Corporate Performance Model. *Sustainability* **2022**, *14*, 6470. [CrossRef]

91. Medvedeva, Y.Y.; Luchaninov, R.S.; Poluyanova, N.V.; Semenova, S.V.; Alekseeva, E.A. The Stakeholders' Role in the Corporate Strategy Creation for the Sustainable Development of Russian Industrial Enterprises. *Economies* **2022**, *10*, 116. [CrossRef]

92. Czarnecka, M.; Kinelski, G.; Stefańska, M.; Grzesiak, M.; Budka, B. Social Media Engagement in Shaping Green Energy Business Models. *Energies* **2022**, *15*, 1727. [CrossRef]

Article

Integration of CSP and PV Power Plants: Investigations about Synergies by Close Coupling

Javier Iñigo-Labairu [1,*], Jürgen Dersch [1] and Luca Schomaker [2]

[1] German Aerospace Center (DLR), Institute of Solar Research, Linder Höhe, 51147 Köln, Germany
[2] Dornier Suntrace GmbH, Grosse Elbstrasse 145 c, 22767 Hamburg, Germany
* Correspondence: javier.inigolabairu@dlr.de

Abstract: Photovoltaic (PV) - concentrated solar power (CSP) hybrid power plants are an attractive option for supplying cheap and dispatchable solar electricity. Hybridization options for both technologies were investigated, combining their benefits by a deeper integration. Simulations of the different systems were performed for seven different sites by varying their design parameters to obtain the optimal configurations under certain boundary conditions. A techno-economic analysis was performed using the levelized cost of electricity (LCOE) and nighttime electricity fraction as variables for the representation. Hybrid power plants were compared to pure CSP plants, PV-battery plants, and PV plants with an electric resistance heater (ERH), thermal energy storage (TES), and power block (PB). Future cost projections were also considered.

Keywords: hybrid PV-CSP power plant; techno-economic analysis; thermal storage; battery energy storage system; electric heater; simulation tools

1. Introduction

The generation of electricity with photovoltaic (PV) systems has undergone a significant cost reduction in recent years. The global expansion of the technology has already resulted in PV electricity becoming the cheapest form of power generation in many markets. However, it is also known that a strong expansion of PV in the grid means that at times of very high irradiance, a considerable proportion of the possible production has to be curtailed because the generation would otherwise exceed the demand. The storage of larger amounts of PV electricity in batteries is not yet economically viable (2022). As a result, PV electricity utilization is not optimal when the installed production capacity is expanded, since the capacity of the grid is not sufficient at certain times, especially at midday. On the other hand, concentrated solar power (CSP) technology offers a proven and cost-effective storage option in the form of thermal storage, usually using molten salt stored in large tanks. So, it is possible to generate solar power even after sunset and at times of low irradiance. Thermal energy storage (TES) is economical and, due to its easy scale-up, has low marginal cost. The power block (PB) can work efficiently in a wide range of part-load conditions and is relatively flexible in terms of dispatchability (generation on-demand). At the same time, CSP solar fields can generate heat very efficiently. Compared to PV, CSP technology presents higher electricity generation costs, which can be attributed, among other things, to the fact that the installed capacity does not yet enable mass production. Nevertheless, there has been a significant cost reduction in CSP technology in recent years, which may lead to electricity production costs being below the level of conventional electricity generation if capacities are expanded further. Globally, by the end of 2021, approximately 6.4 GW of the CSP capacity was installed while there was about 843 GW of the PV installed capacity [1].

Due to the specific advantages of each technology, it makes sense to combine them. There are currently approaches to integrate both types of power plants and thus use the respective advantages. Some publications show the research carried out on hybrid PV-CSP plants in recent years. DLR participated in the THERMVOLT project [2], where the

Citation: Iñigo-Labairu, J.; Dersch, J.; Schomaker, L. Integration of CSP and PV Power Plants: Investigations about Synergies by Close Coupling. *Energies* **2022**, *15*, 7103. https://doi.org/10.3390/en15197103

Academic Editor: Juri Belikov

Received: 13 September 2022
Accepted: 23 September 2022
Published: 27 September 2022

combination of CSP, PV, TES, battery energy storage system (BESS), and backup combined cycle power plants was examined. The focus was on fulfilling the representative load curves while at the same time limiting the specific CO_2 emissions in kg/kWh of produced electricity. Grid coupling of the systems was also investigated but without as deep an integration as in the current paper. Starke et al. [3] conducted a techno-economic analysis to evaluate the performance of hybrid PV-CSP plants in northern Chile. The parametric analysis and optimization showed high potential in this region for the installation of hybrid PV-CSP plants due to the high levels of irradiation. Zhai et al. [4] proposed a thermal storage PV-CSP plant and calculated the annual hourly performance of the system. Two different dispatch strategies were studied: a conventional strategy in which the PV and CSP system operated independently and a constant-output strategy that integrated both systems, producing some synergies. Lower values of the levelized cost of electricity (LCOE) were presented by the constant-output strategy, which resulted in a more fluid and stable power output. Zurita et al. [5] performed a techno-economic evaluation of a hybrid PV-CSP plant integrated with thermal energy storage and large-scale battery energy storage for base load generation in northern Chile. LCOE and the capacity factor were used to evaluate the performance of the plant, identifying the configurations with a minimum LCOE. The results showed that the battery costs should be reduced by approximately 60–90% to attain competitive LCOEs in comparison to hybrid plants. Hamilton et al. [6] published a paper about dispatch optimization of hybrid PV-CSP-BESS systems using mixed-integer linear programming. They did not consider an electric resistance heater (ERH) and the focus was on the optimization algorithm. They examined two sites (northern Chile and Sierra Nevada) with different solar resources and two price curves for the electricity. The authors concluded that the hybrid systems outperform the CSP-only plants dramatically in terms of the capacity factor and economics (21–33% lower LCOE). Schöniger et al. [7] compared CSP with thermal storage, a photovoltaic system with a battery energy storage system (PV-BESS), and PV with thermal storage for a Spanish site with up to 24 h of storage capacity. They used simplified performance models for the subsystems and assumed a constant load of 100 MW. PV-BESS showed the lowest electricity costs for storage capacities lower than 4 h. Hassani et al. [8] performed a techno-economic analysis of hybrid PV-CSP plants in eastern Morocco and found that hybrid PV-CSP plants provide dispatching energy at a lower LCOE than standalone CSP plants and that the capacity factor could reach a value of more than 90%. Zurita et al. [9] investigated two sites in Chile with different solar resources (high and moderate). The TRNSYS and MATLAB models were used to create a surrogate model, which was then used for multi-objective optimization. Four load scenarios, including the baseload, were considered. The authors showed LCOE over the sufficiency factor (fulfilment of the load curve). They did not investigate the close integration with ERH. Only solar tower systems were simulated but no parabolic trough systems. Riffelmann et al. [10] studied different options to combine the benefits of PV and CSP technologies. A small PV plant that supplied its own consumption of the CSP plant, hybrid CSP-PV plants with an optional electric heater, and a PV system with an additional electric heater were simulated in different scenarios. For the Spanish location, the most economic configuration combined the CSP plant, which charged the thermal storage during daytime and produced electricity during nighttime, and a PV system, which delivered electricity during day hours, additionally charging the thermal storage with an electric heater. In contrast, the calculation of the best configuration in a high direct normal irradiance (DNI) site such as South Africa resulted in a system without an electric heater. For both locations, the electricity produced with the CSP plant and thermal storage was more economic than the one that originated from the PV system with an electric heater. Mata-Torres et al. [11] assessed the impact of solar irradiation and plant location for a hybrid PV-CSP plant integrated with a multi-effect distillation plant for simultaneous power generation and seawater desalination. A techno-economic analysis was performed at different locations to determine the most suitable sites for this kind of plant, showing that inland locations with a considerable increase in DNI with respect to the coast (over 300 kWh/m^2-year), distances from the sea of no more than 60 km, and altitudes up to 750–1000 m presented the most appropriate conditions. Richter et al. [12] defined a predictive storage strategy for the optimal design of hybrid PV-CSP plants with an immersion heater. It was proven that the strategy had a significant impact on the output of the plant and the subsystem sizing. Lui at al. [13] published a study about the optimization of hybrid PV-CSP-systems with ERH and BESS for a Chinese site. A 100 MW constant load was assumed for all systems. They showed LCOE over the loss of power supply probability (LPSP), which is the remaining fraction of the electricity demand that cannot be satisfied by the system. The lowest LCOE was found for the systems with TES capacities of about 16–18 h. The combination of PV, ERH, and CSP can reduce costs and lead to high fulfilment of the load curve.

Moreover, some hybrid plants are already under construction or operational. Phase 4 of MBR Solar Park in Dubai is under construction and will combine a 600 MW (three units of 200 MW each) parabolic trough, 100 MW molten salt solar power tower, and 250 MW photovoltaic plant [14]. Noor Midelt 1 in Morocco will have a total installed capacity of 800 MW and hybridize concentrated solar power and photovoltaic technologies with a 5 h thermal storage, using electrical heaters to increase the storage temperature and convert excess PV energy to heat [15]. Cerro Dominador was inaugurated in Chile in 2021 and has a 110 MW central tower and a 100 MW photovoltaic field [16]. Additionally, four projects that combine CSP and PV have been started in China. Each project will include 100 MW of CSP and 900 MW of PV [17].

The literature review shows that hybrid CSP-PV power plants can offer cost-effective solar power generation with high security of the supply and high solar coverage. A full integration of both technologies, which hybridizes them on the system level or, in other words, "inside the fence", was studied in this paper. These plants have only one common grid-access point and the plant operator is responsible for controlling the separate subsystem outputs, their interactions, and their contribution to the grid power (and ancillary services) supply. Electrical and thermal energy is enabled to flow between subsystems in order to provide charging power to energy storage systems or for unidirectional supply of auxiliary power. A fully integrated solar hybrid power plant benefits from multiple synergies but also comes with more challenges due to the higher number of degrees of freedom in the system design and operating strategy.

The combination of both systems would not lead to lower electricity generation costs compared to standalone PV plants. However, the generation of solar power at times of the day when the sun does not shine cannot be achieved by a PV plant without storage. In this respect, the storage of energy in the heat storage tank should be viewed in comparison to an alternative storage concept. Joint optimization of PV and CSP to a specific power generation profile will therefore lead to electricity generation costs that will lie between those of pure PV (without battery) and a pure CSP power plant. A shift in power generation from the CSP power plant to the evening hours can result, for example, in this CSP power plant having a smaller solar field than the pure CSP plant that also supplies electricity during the day. In this case, the storage size and the power plant unit remain basically unchanged. Consequently, savings are primarily made in the solar field, which may account for about 50% of the investment costs of the CSP power plant (this is given here only as an indicative value, since the cost fraction of the solar field depends on the storage size).

The impact of boundary conditions on the optimum configurations was another fundamental aspect of this investigation. Different locations with different radiation and latitude conditions were simulated. The most suitable plant types and configurations for the given boundary conditions were considered. The hybrid systems analyzed were the parabolic trough collectors in combination with a PV plant and the central receiver system also in combination with a PV plant. These systems were optimized and compared with other technologies such as pure CSP plants, PV with battery storage and photovoltaic system plus an electric resistance heater, thermal energy storage, and power block (PV+ERH+PB). This optimization was performed not only to achieve the lowest solar electricity costs for the systems but also to increase the proportion of electricity delivered to the grid during night hours.

The deeper integration of technologies to increase synergies, the representation of results considering the nighttime electricity fraction instead of the capacity factor, and the greater number of locations used for the calculation differentiate this work from previous publications.

A final report of the project [18] was also written and includes additional analyses such as the influence of the demand curve profile and systematic sensitivity analyses. More detailed information and explanations of the results are also included.

2. Technologies and Plant Concepts

Different plant concepts using CSP or PV technology were considered for this study.

- Pure CSP plants (Figure 1): Standalone parabolic trough and tower power plants (as "classical CSP" references):
    - There is no PV system.
    - Two tank molten-salt thermal energy storage (TES).
    - The CSP power block (steam turbine) runs during the day and night.
    - Cost-effective electricity during nighttime.

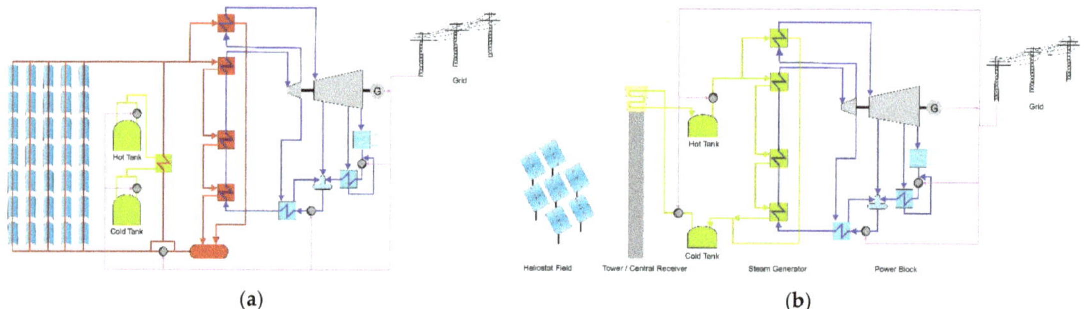

Figure 1. Pure concentrated solar power (CSP) plants: (**a**) Parabolic trough (PT); (**b**) Central receiver tower (CRT).

- PV-CSP hybrid power plants (Figure 2): Parabolic trough and central receiver tower:
    - PV injects electricity into the grid during the day.
    - The power block generates electricity at night using heat from the TES.
    - Parabolic trough: ERH is connected in series to the CSP field and works as a booster for the parabolic trough cycle, increasing the TES inlet temperature and thus improving the PB efficiency and increasing the TES energy capacity. PV power is firstly used to cover the auxiliary consumption of the solar field and, secondly, to power ERH to increase the molten salt temperature from the nominal outlet temperature of the solar field (385 °C) to the hot tank nominal temperature (565 °C), and, finally, excess electricity is delivered to the grid. The TES upper heat transfer fluid (HFT) temperature is equal to the central receiver plant temperature levels (565 °C). The steam generator is always operated with hot salt in this variant. The direct use of thermal oil to generate steam was not considered as both the steam cycle efficiency and TES capacity would be significantly reduced, with consequent impacts on the plant economics.
    - Tower: PV power is used in the first place to cover the auxiliary requirements of the solar field (pumps, heliostat tracking, PB auxiliaries) and then delivered to the grid. If there is excess electricity, it is used for heating up the molten salt from the cold tank temperature to the hot tank temperature. ERH is connected in parallel to the CSP field. The use of ERH as a booster is not considered for tower systems,

since they heat the molten salt directly and can easily bring it to its maximum temperature, which is determined by the physical limits of the molten salt.

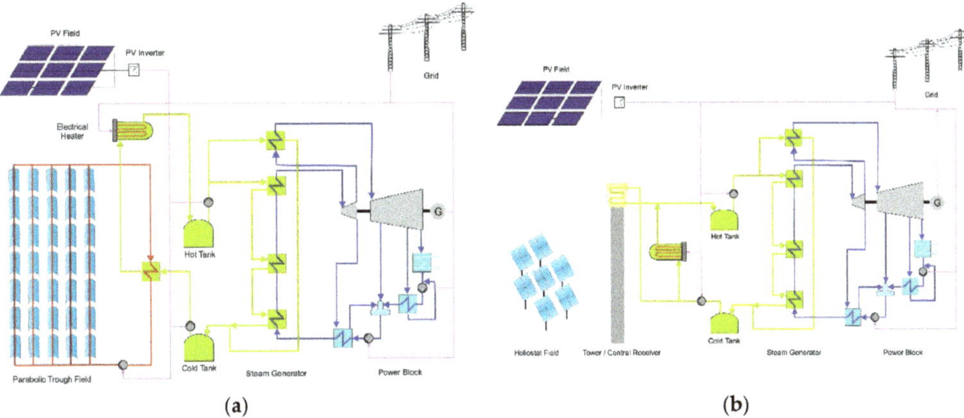

Figure 2. Hybrid photovoltaic (PV)-CSP power plants: (**a**) Hybrid parabolic trough; (**b**) Hybrid tower.

- PV systems (Figure 3): PV with battery and PV with an electric heater and power block:
    - There is no CSP field.
    - PV with battery: PV injects electricity into the grid during the day and the excess power is stored in the battery, which generates electricity during the night. Low-cost electricity can be used with flexibility and BESS provides highly efficient and flexible electricity storage.
    - PV with ERH, TES, and PB: Nighttime electricity is prioritized. PV power is used in the first place to cover the auxiliary demand of the PB and TES. Then, it is used by an ERH to charge the TES during the day. Excess PV power is injected into the grid during the day. PB generates electricity at night from the TES.
    - A standalone PV system was also calculated as a reference system.

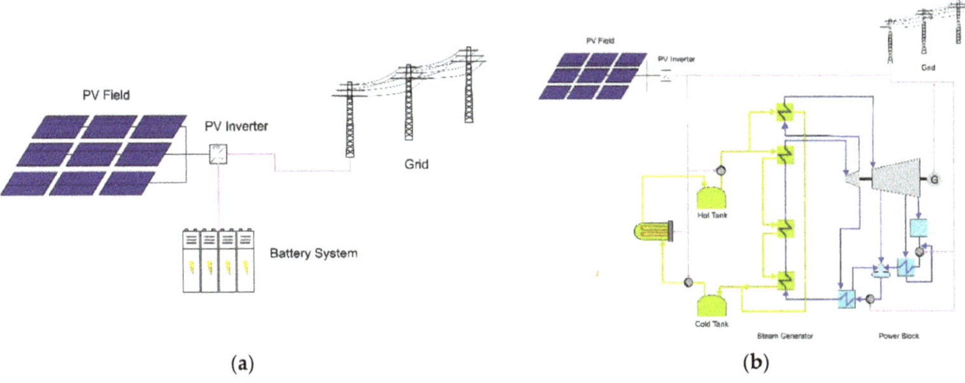

Figure 3. PV systems: (**a**) Photovoltaic system with a battery energy storage system (PV-BESS); (**b**) Photovoltaic system plus an electric resistance heater, thermal energy storage, and power block (PV+ERH+PB).

Some additional technical assumptions were made for the components and are shown in Table 1:

Table 1. Technical assumptions.

Technology	Technical Assumptions
Parabolic Trough	Heat transfer fluid (HTF): Therminol VP-1 Nominal field outlet temperature: 395 °C Nominal field inlet temperature: 310 °C Degradation: 0.4% per year
Central receiver tower	HTF: Solar Salt (60% $NaNO_3$, 40% KNO_3) Nominal tower outlet temperature: 565 °C Nominal tower inlet temperature: 300 °C Degradation: 0.4% per year
Thermal energy storage (TES)	Storage Medium: Solar Salt (60% $NaNO_3$, 40% KNO_3) Nominal hot temperature: 565 °C (380 °C for standalone PT) Nominal cold temperature: 300 °C (287 °C for standalone PT) Thermal loss: 1% per hour
PV	Bifacial-monocrystalline PV modules Single-Axis Tracking Systems DC/AC ratio: 1.3 PV panels nominal efficiency: 19% PV inverter nominal efficiency: 98.6% Degradation: 0.4% per year Assumption: Optimized standalone PV configuration will also lead to highest benefits in integrated hybrid plants. A representative system with one inverter was designed and in the hybrid plants, many of these systems were used to reach the required nominal power.
Power2 Heat technology: ERH	HTF: Solar Salt (60% $NaNO_3$, 40% KNO_3) Conversion efficiency: 99%
Battery energy storage system (BESS)	Technology: Lithium-Ion Nickel Manganese Cobalt Round-trip efficiency: 85% related to heating, ventilation and air conditioning, self-discharge, battery management system, power conversion system efficiency, etc. Lifetime (Warranty duration): 15 years (BESS has to be completely replaced after 15 years, operations and maintenance (O&M) cost assumption includes a share to build up an O&M budget for the replacement) Degradation: ≈2% of nominal capacity per year (lost BESS capacity has to be restored by adding additional batteries on a regular basis to keep up the BESS functionality, O& M cost includes a share to compensate degradation)
Power block (PB)	Gross efficiency: 46.5%

3. Materials and Methods

The software greenius was used for the simulations with an hourly time step resolution for typical years. DLR developed greenius several years ago as a software tool for the annual yield calculation and economic evaluation of renewable energy systems, with an emphasis on CSP plants. An official version is available for free from the DLR website [19] in addition to a manual with further details about the models used. For the evaluation of the results, the LCOE and the nighttime fraction of electricity were used. The LCOE is not sufficient for evaluating and comparing hybrid power plants. With LCOE as the only criteria, a standalone PV plant would be the least-cost solution, but it would just provide electricity during sunshine hours, providing no dispatchability. Hybrid solar power plants are instead capable of providing solar electricity even after sunset, often at elevated costs, and the aim of this project was to find cost-optimized combinations. Therefore, at least a second parameter had to be fixed to make the systems comparable on the same basis. This parameter was the nighttime electricity fraction.

Blended LCOE was used for the evaluation of the results. This is made of the tariff for direct feed-in and the tariff for electricity generated from the storage. Equation (1) shows how it was calculated. Equation (2) shows, on the other hand, the formula for the nighttime electricity:

$$\text{LCOE} = \frac{\text{Total Investment Costs} + \sum_{t=1}^{t_{ges}} \frac{\text{Annual Running Costs}_t}{(1+r)^t}}{\sum_{t=1}^{t_{ges}} \frac{\text{Annual Electricity Solar}_t \times (1-d)^{t-1}}{(1+r)^t}}, \tag{1}$$

where r is the interest rate, t is the year within the period of use (1, 2, … t_{ges}), t_{ges} is the period of use (system lifetime in years) [a], and d is the yearly degradation rate:

$$\text{Night time electricity fraction} = \frac{\text{annual night electricity production}}{\text{total annual electricity production}} \tag{2}$$

A parameter variation in greenius was used to identify the system configurations with the least LCOE. The steps and limits of this variation are shown in Table 2. The configurations with the lowest LCOE for each storage size were selected as the representation of the results.

Table 2. Steps and limits of the parameter variation.

Technology	PV Field Size [MW]	Storage Net Capacity [h]	CSP Field Nominal Power [MW]	ERH Nominal Power [MW]
Standalone tower	-	3–12 (3 h step)	400–1150 (50 MW step)	-
Standalone trough	-	3–12 (3 h step)	650–1850 (100 MW step)	-
Hybrid tower	150–300 (25 MW step)	3–12 (3 h step)	200–600 (50 MW step)	0–400 (variable step)
Hybrid trough	150–650 (50 MW step)	3–12 (3 h step)	20–196 (22 MW step)	45–405 (variable step)
PV with battery	150–525 (25 MW step)	3–12 (3 h step)	-	-
PV with ERH and PB	150–750 (50 MW step)	3–12 (3 h step)	-	5–600 (variable step)

The costs of each component used for the simulations were assumed by DLR and Dornier Suntrace GmbH, based on project experience, market analysis, and published information, and are shown in Table 3 for 2021 and 2030. The CSP costs were taken from [20] and BESS costs from [21] and [22]. According to an analysis conducted by DLR, the costs of the parabolic trough (PT) and central receiver tower (CRT) are expected to decrease by 12% between 2021 and 2030. The BESS cost is expected to decrease by only 20% for power and 23% for the capacity due to strong competition with the automotive sector, limited raw material availability, and high demand. The ERH cost values were assumed from DLR projects, with a 10 percent cost reduction expected from 2021 to 2030. The cost of the PV system was determined by Dornier Suntrace GmbH project experience and current market scenarios, and it is expected to fall by only 23% between 2021 and 2030. The storage costs for both TES and BESS refer to the net capacity. The costs for electricity supplied by the grid were adjusted for each site using the prices from [23]. The land costs were set to zero because of their negligible influence on the LCOE (assuming low-cost desert sites). Engineering, procurement, and construction (EPC) surcharges were assumed to be different for each subsystem because of the different maturity levels of the technologies and were related to the capital expenditure (CAPEX). Increased operation and maintenance costs for BESS were used to account for the replacement or addition of batteries in order to enable the lifetime of 25 years assumed for all systems. For all systems, a 5% debt interest rate was assumed.

Table 3. Cost assumptions.

Technology	Component	Value 2021	Value 2030	Unit
PT	Solar field	202	178	$/m^2
	Thermal storage	38	33	$/kWh_{th}$
	Power block	930	860	$/kW_{el}$
	EPC	0.2	0.2	of CAPEX
	O&M	0.015	0.015	of CAPEX
CRT	Heliostat field	127	112	$/m^2
	Tower	88000	77440	$/m
	Receiver	122	107	$/kW_{th}$
	Thermal storage	26	23	$/kWh_{th}$
	Power block	966	863	$/kW_{el}$
	EPC	0.2	0.2	of CAPEX
	O&M	0.015	0.015	of CAPEX
BESS	Cost per power	245	196	$/kW_{el}$
	Cost per energy capacity	246	189	$/kWh_{el}$
	EPC	0.235	0.235	of CAPEX
	O&M	0.045	0.045	of CAPEX
ERH	Cost per kW	100	90	$/kW
	EPC	0.2	0.2	of CAPEX
	O&M	0.01	0.01	of CAPEX
PV system	Inverter	53	41	$/kW_{ac}$
	PV field	454	350	$/kW_{dc}$
	EPC	0.1	0.1	of CAPEX
	O&M	0.005	0.005	of CAPEX

The grid injection limit was fixed at 150 MW$_{el}$ for all the systems (power block and battery net power) to allow the comparability of all concepts. In this way, it was also possible to avoid difficulties in the design of the PV plant. This limit might be caused by an upper limit of the grid connection of the solar power plant as well. It was necessary to set the nominal gross output of the CSP power block to 160 MW$_{el}$ because this plant should be able to deliver the required net power of 150 MW$_{el}$ to the grid and cover the auxiliary demand of the plant even under hot ambient conditions. A constant load curve for the whole day ("baseload") was assumed. This means that the required load at night was also 150 MW for each system. The power block and CSP solar field auxiliary requirements were modeled as being dependent on the ambient temperature. Stand-by auxiliaries (solar fields, PV, BESS) were considered with fixed values. Grid connection, substations, and transmission lines were not included in the project scope to allow the comparison of plants independently of their location.

The techno-economic evaluation was performed for different locations to consider different boundary conditions (e.g., DNI and global horizontal irradiance (GHI) values) and latitudes. The widest possible variety of locations was sought. Depending on the characteristic values of the sites, some locations are more suitable for CSP plants or for PV plants. Seven sites were used, for which high-quality meteorological datasets were available. Unfortunately, it was not possible to find sites with almost identical parameters but differing in only one parameter. Table 4 shows the main parameters of the selected sites.

The definition of the night hours was carried out for each month and location separately. An hour was considered to be nighttime if the PV output of a sunny day in the middle of the month (days with clear sky conditions and an almost ideal DNI curve) did not reach 25% of its nominal value for that hour. A time resolution of one hour was adopted for the meteorological datasets.

Table 4. Main parameters of the evaluated sites.

Site	Annual DNI (TMY) [kWh/m^2]	Annual GHI (TMY) [kWh/m^2]	Average Temperature [°C]	Latitude [°N]
Riyadh (Saudi Arabia)	2275	2236	26	24.9
De Aar (South Africa)	2712	2040	17.5	−30.7
PSA (Spain)	2207	1860	16.6	37.1
Diego de Almagro (Chile)	3477	2449	15.8	−26.3
Ouarzazate (Morocco)	2518	2123	18.8	30.9
Daggett (USA)	2723	2090	19.7	34.9
Dunhuang (China)	2158	1755	10.6	40.2

4. Results and Discussion

The results of the simulations were compared for each technology and site. This paper only shows the technology comparison diagram for Riyadh (Figure 4). Figures 5–9 compare the results of each site for every technology (standalone tower, hybrid tower, hybrid parabolic trough, PV with battery, and PV plus ERH, TES, and PB).

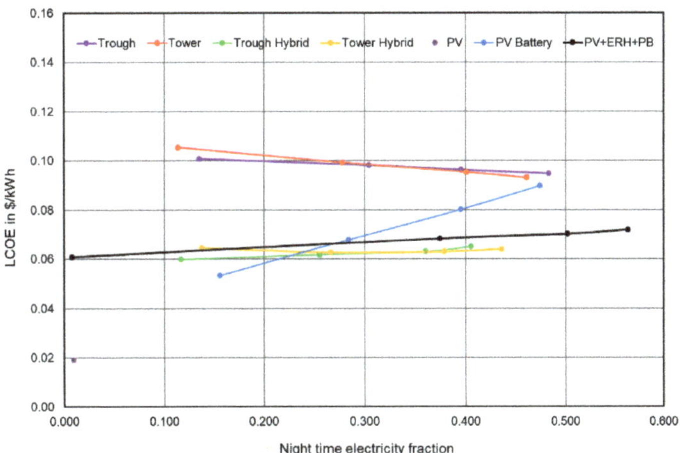

Figure 4. Lowest levelized cost of electricity (LCOE) configurations in Riyadh.

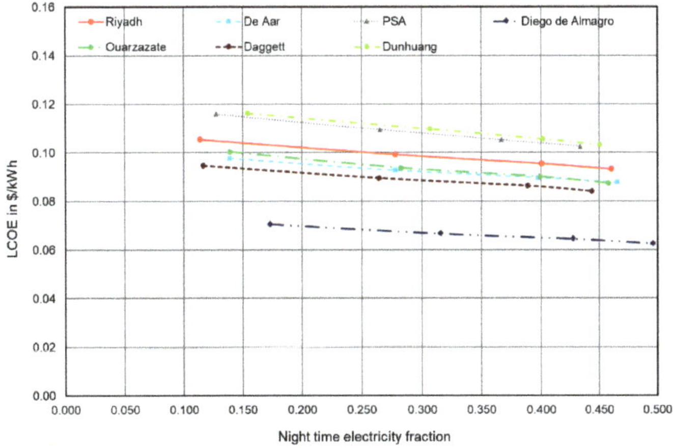

Figure 5. Comparison of locations: standalone tower.

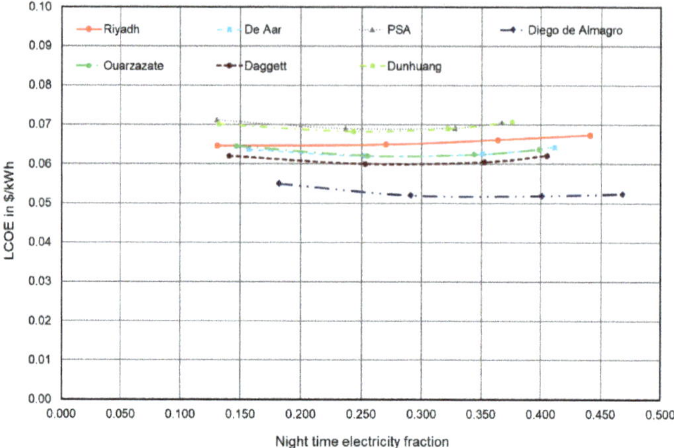

Figure 6. Comparison of locations: tower hybrid.

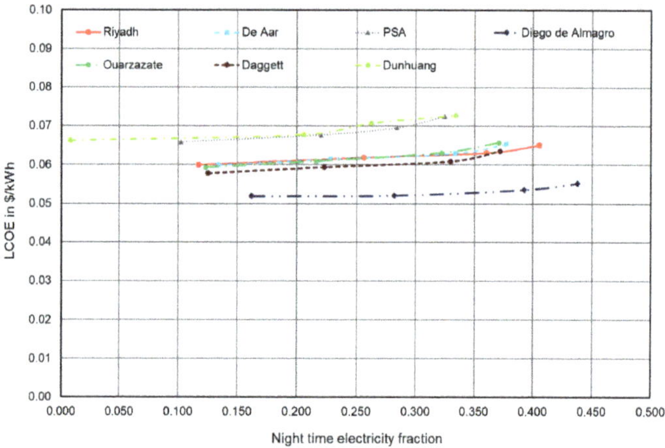

Figure 7. Comparison of locations: trough hybrid.

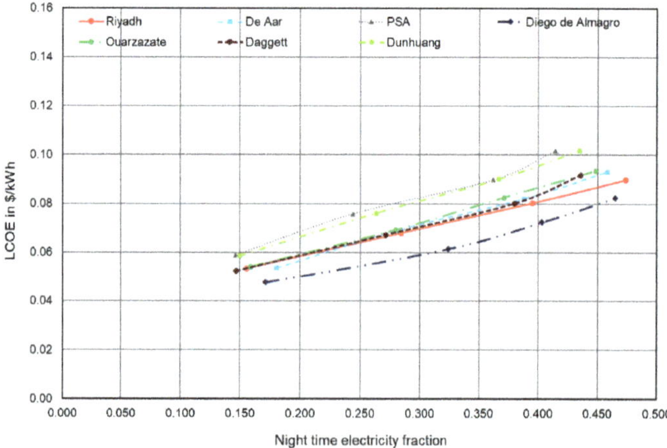

Figure 8. Comparison of locations: PV battery.

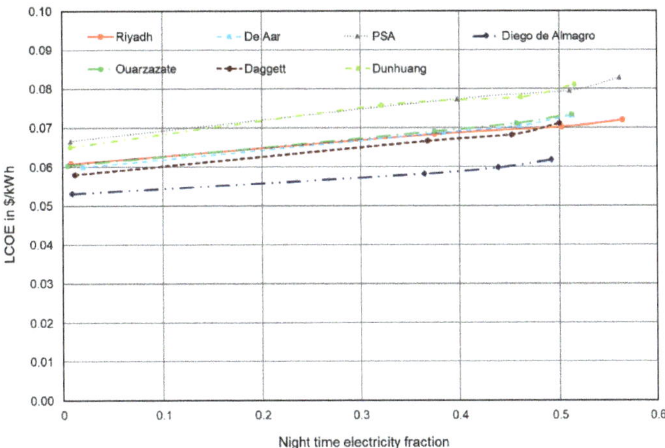

Figure 9. Comparison of locations: PV+ERH+PB.

Figure 4 shows the LCOE values versus the nighttime electricity fraction of the optimal configurations for different storage sizes of different topologies in Riyadh. Each point stands for an LCOE-optimized configuration for a certain storage capacity (3, 6, 9, and 12 h duration of discharge at the full power block output). Cases without storage are not displayed with one exception, which is the standalone PV system as a reference.

The top lines show the results for pure CSP systems. These plants have the highest LCOE since they do not benefit from the low electricity cost of PV. They are shown as a reference and their LCOE is higher than that of the hybrid systems. The LCOE decreases with increasing nighttime electricity fraction, which is caused by the cheap TES and better utilization of the PB (which is kept at a constant nominal power but reaches more operating hours with the increasing TES size).

The higher storage cost of BESS compared to TES results in a different dependency of LCOE with the increasing nighttime electricity fraction. PV with a battery is the technology that provides the lowest LCOE for nighttime fractions below 20–25% (approximately 4–5 h of storage capacity). However, from the 20–25% nighttime fraction upwards, the costs start to increase and this difference grows linearly as the storage size increases. For a storage size of 12 h, the LCOE is similar to standalone CSP plants. Hybrid PV-CSP systems, in contrast, show less dependency of LCOE on the nighttime electricity fraction and they are almost constant as shown in Figure 4.

Hybrid plants show a significant decrease in costs compared to standalone CSP plants. LCOE remains relatively constant regardless of the nighttime fraction (storage capacity). There is better utilization of the TES as it increases in size, but there is also an increase in the dumping and component costs. Both hybrid plants show almost the same LCOE, with a small advantage of the parabolic trough hybrid system for a low storage capacity and a small advantage of hybrid tower systems for a higher storage capacity.

Another interesting question is whether a system using PV, ERH, TES, and PB could be cheaper than a hybrid PV-CSP plant and how much nighttime electricity it can offer. Figure 4 shows that the LCOE values and night fractions of the PV+ERH+PB system are higher than those in hybrid plants. Note that the operating strategy of this kind of system prioritizes TES charging over direct grid injection. This operating strategy was used for direct comparison since, as in the case of hybrid systems, storage charging is prioritized. It can be seen that for a three-hour storage, the night share tends towards zero because the small storage capacity does not justify the cost of the ERH. The model considers fixed PB and TES costs even without any ERH. The optimized three-hour storage system would have storage and PB that would not be used at all. It would basically be a standalone PV plant but with a shared LCOE as a fixed PB and TES cost item is included. This changes for higher storage capacities, where the electric heater and storage tank would be used. Although the LCOE is higher than that for hybrid systems, the nighttime electricity fractions are much higher for large storage capacities. This is also of great importance since not only is reducing costs one of the objectives of the plant but also being able to provide a greater amount of electricity during the night.

Table 5 shows the main parameters of the optimized systems with 6 h of storage. The parabolic hybrid plant has the smallest CSP field. This is due to the greater flexibility in the design of this plant. The hybrid tower plant has a minimum CSP field design size of 200 MW, which represents the lower field size limit considered in the parametric study. Due to the significantly lower daytime electricity costs with PV, the slightly lower night costs with PV and ERH, and the relatively small storage size, a larger PV field is advantageous. For this reason, in the parabolic trough hybrid power plant, there is a significant difference in the size between the module area of the PV and the aperture area of the CSP fields. The PV field of the hybrid parabolic trough is also larger than the field of the PV battery system. The hybrid trough additionally uses PV electricity to charge the TES and due to the lower efficiency of the PB (compared to the battery system), a larger PV field is required. This optimized design is very sensitive to the cost assumptions. Slightly lower CSP costs would change this figure. The thermal capacity of the storage tank for the standalone parabolic trough powerplant is higher than that for the hybrid plants since the PB has a lower efficiency due to the lower temperature level and therefore requires more energy to deliver 150 MW$_{el}$ for 6 h.

Table 5. Parameters of the six-hour optimized configurations.

Technology	Trough	Tower	Hybrid Trough	Hybrid Tower	PV-BESS	PV+ERH+PB
Storage capacity (h)	6	6	6	6	6	6
TES capacity (MWh$_{th}$)	2494	2066	2066	2066	-	2066
BESS capacity (MWh$_{el}$)	-	-	-	-	900	-
CSP field aperture (km^2)	1.5	1.6	0.14	0.3	-	-
CSP field nominal output (MW$_{th}$)	978	850	86	200	-	-
PV field module area (km^2)	-	-	2.4	1.9	2.1	3.1
PV field nominal output (MW$_{ac}$)	-	-	350	275	300	450
ERH nominal power (MW$_{el}$)	-	-	160	100	-	200
PB nominal output (MW$_{el}$)	160	160	160	160	-	160
Total land area (km^2)	5.8	11.0	7.4	7.0	5.9	8.9
LCOE ($/kWh)	0.0982	0.0993	0.0619	0.0627	0.0680	0.0684
Nighttime fraction	0.305	0.278	0.256	0.267	0.284	0.374

Table 6 shows the parameters of the optimized systems with 12 h of storage. The sizes of the components are considerably increased to cover greater nighttime demand, allowing, at the same time, a higher load coverage during the daytime. The optimization of the hybrid tower prioritizes an increase in the PV field because of the lower heat production costs with PV and ERH. The ERH works in parallel, and does not require the CSP field to increase. In the case of parabolic trough hybrid plants, there is a greater increase in the CSP field aperture for higher storage capacities because the booster system requires an increase in both fields (PV and CSP) due to the fixed ratio of heat provided by each system. For the non-hybrid plants, the size of their main components is increased to make use of the additional storage capacity. The nominal power of the tower reaches 1150 MW, which is the upper limit available in greenius for the simulation.

Table 6. Parameters of the twelve-hour optimized configurations.

Technology	Trough	Tower	Hybrid Trough	Hybrid Tower	PV-BESS	PV+ERH+PB
Storage capacity (h)	12	12	12	12	12	12
TES capacity (MWh$_{th}$)	4989	4132	4132	4132	-	4132
BESS capacity (MWh$_{el}$)	-	-	-	-	1800	-
CSP field aperture (km^2)	2.1	2.2	0.24	0.3	-	-
CSP field nominal output (MW$_{th}$)	1304	1150	152	200	-	-
PV field module area (km^2)	-	-	3.1	3.3	3.1	4.8
PV field nominal output (MW$_{ac}$)	-	-	450	475	450	700
ERH nominal power (MW$_{el}$)	-	-	290	300	-	400
PB nominal output (MW$_{el}$)	160	160	160	160	-	160
Total land area (km^2)	7.7	19.1	9.7	10.9	8.9	13.8
LCOE ($/kWh)	0.0949	0.0932	0.0651	0.0640	0.0898	0.0720
Nighttime fraction	0.0482	0.460	0.405	0.436	0.473	0.563

There is a large difference in the land area of the standalone tower and the parabolic trough plant. Tower plants have a greater land use factor, and it increases with the increase in the system size. Heliostats far away from the central tower require a larger separation from their neighbors to limit shading of each other. This shows how the total land area of the tower plant increases significantly for plants with larger storage sizes. This considerable land use difference does not occur for hybrid plants because the CSP field of the hybrid tower is not as large as that of standalone tower plants. PV with ERH and PB also shows a significantly larger land use because of the greater PV capacity required. Land area costs were not included in the calculation.

Figure 5 shows LCOEs for the standalone solar tower plants at each site. The order in which the curves are placed corresponds to the magnitude of the annual DNI sums. The site with the lowest DNI is represented by the upper line and the site with the highest radiation by the lowest line. PSA has a higher LCOE than expected considering its DNI sum. This is due to the higher latitude of this site, which reduces the solar field performance in winter to a greater extent. Having a similar DNI to Riyadh, the worse distribution of radiation during the year for PSA implies a higher LCOE due to the lower efficiency of the CSP field. There is also a slightly higher than expected LCOE for De Aar for the CSP technologies. This is mainly due to the intercept heat, which must be dumped to protect the receiver. DNI exceeds 1000 W/m^2 in De Aar for several hours during the year, which causes greater dumping, increasing the LCOE. This explains why, despite having an annual DNI similar to Daggett, the LCOE is higher. The results at Diego de Almagro are outstanding, despite the dumping, for exceeding the maximum absorbable radiation for many hours. On the other hand, the LCOEs at Dunhuang and PSA exceed the LCOEs of the other sites by almost two cents. Most sites have values between 8 and 10 cents per kWh. There are no significant differences in the nighttime fraction, except for Diego de Almagro.

The positions in which the hybrid tower curves are arranged is shown in Figure 6 and can also be explained by the annual DNI and GHI sums, with the first still being very important even though the PV fields provide the majority of energy. The LCOE advantage for Diego de Almagro is not as big as for the standalone tower, but it is still significant. The hybridization causes a significant overall decrease in the costs compared to the standalone CSP systems.

The same dependence on solar resources can be observed for trough hybrid plants in Figure 7. There is a slight cost increase as the storage capacity increases because of the higher component costs and dumping. It should also be noted that the nighttime fraction is higher for Riyadh and Diego de Almagro due to the greater proximity to the equator.

Parabolic troughs suffer from low sun angles and sites close to the equator show smaller differences in the sun angles between summer and winter. Furthermore, the day length is also rather constant compared to sites with higher latitudes. This is an advantage as storage charging and discharging is more equally balanced over the year. More pronounced seasons with variable day lengths are not ideal, leading to storage and ERH over dimensioning for winter days. Moreover, there is an additional clear correlation of the nighttime electricity fraction and the solar resource. The plant with 12 h of thermal storage at Dunhuang only shows a nighttime fraction of 0.33 while the plant with the same storage size at Diego the Almagro shows a nighttime fraction of 0.43. For all sites, LCOE is below 8 cents per kWh. Due to the larger PV field, GHI increases its influence compared to the DNI. PSA clearly shows lower LCOE values than Dunhuang because of the higher GHI. Riyadh, the site with the second highest GHI, also displays a clear improvement, with this technology showing lower LCOE than Ouarzazate and De Aar for large storage capacities.

Figure 8 shows a similar trend for PV-BESS. This trend is dependent on the solar irradiation too. Similar values are observed for the De Aar, Riyadh, Ouarzazate, and Daggett sites.

The PV+ERH+PB systems (Figure 9) show some similar effects to the PV-BESS systems. The higher impact of GHI causes a proportionally lower LCOE and higher nighttime electricity fraction in Riyadh in relation to the other technologies because PV is less susceptible to low DNI as long as GHI stays high. Daggett shows lower LCOEs compared to Riyadh and Ouarzazate, although GHI is lower.

From the results, it can be concluded that Riyadh is particularly suited for PV technology and De Aar and Daggett are better suited for the introduction of CSP technologies. The design of the plant is thus highly dependent on the site, resulting in different optimums for each case. The results show that DNI and GHI are the determinant factors but not the only ones.

A financial sensitivity analysis of the electric heater cost was performed to reliably determine its impact on the LCOE. Figure 10 shows the results of varying the price of the heater from 278 (solid line) to 140 (dashed line) and 70 $/kW (dotted line). This cost reduction implies a decrease in LCOE in the order of 1 cent per kWh for the PV+EH+PB system. A similar order of magnitude is observed for the hybrid parabolic power plant. This cost reduction would also mean a significant shift to the left of the cut-off point with the battery PV system. All this proves the significant influence of this cost on the final techno-economic analysis and the final design of the hybrid power plants.

An interesting aspect to consider (though hard to determine) are cost projections for the future. There was active discussion about the cost values to be taken in the simulations for 2030. The costs of the components are continuously evolving depending on numerous factors and there is uncertainty in estimating their future values. The PV and BESS markets are currently (2022) in a crisis that has multiple causes, with a general raw material shortage, high demands for renewable energy technologies, and shortages in logistic capacities being only a few examples. The projection of costs for 2030 therefore has to be balanced between conservative cost assumptions reflecting the current crisis and allowing some optimism for market growth and potential cost reduction that can be expected as a result. The presented results are based on cost assumptions that are balanced between both extremes considering cost development forecasts from before the crisis and current medium-term price increases specifically for PV and BESS components. Table 3 shows the assumed costs. A general decrease in LCOE can be seen in Figure 11 for all technologies by 2030. It is interesting to note how the cut-off point of PV-BESS plants with hybrid plants shifts from the 4.5–5 h of storage expected in 2021 to approximately 6 h or more observed for 2030. For PV+ERH+PB systems, the distance to hybrid plants is becoming smaller due to the overall greater cost decrease expected for PV compared to the CSP components.

With regards to other publications, the results are overall comparable. As in [7], PV-BESS provides the lowest LCOE for storage capacities lower than 4 h and this break-even point will shift to larger storage capacities if battery costs decrease significantly. Hybrid plants' electricity costs are lower than those of standalone CSP plants as described in [6,8]. The outstanding performance of hybrid PV-CSP plants was calculated for high DNI sites in Chile in this paper, as reported in previous articles [3,6,9].

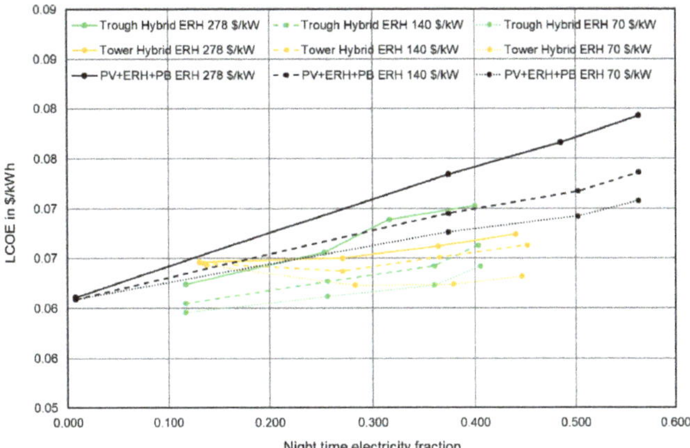

Figure 10. Variation in the ERH costs at Riyadh.

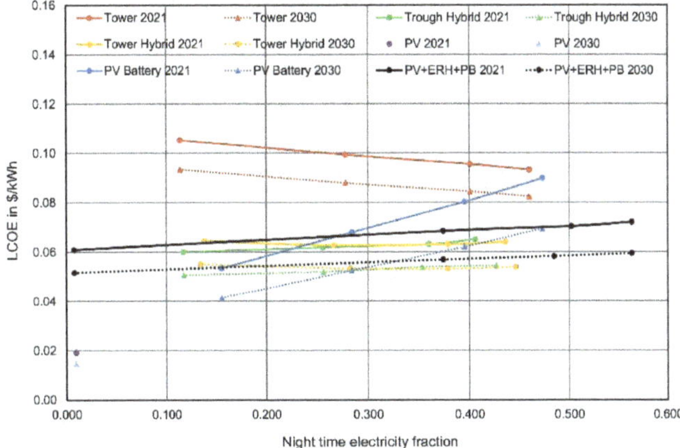

Figure 11. Projections for Riyadh 2030.

5. Conclusions

PV-CSP hybrid power plants were optimized by parameter variation and compared to standalone CSP plants, PV-battery plants, and PV plants with an electric heater, thermal storage, and power block. Hybrid plants show a significant cost decrease compared to pure CSP plants because of lower PV electricity costs. Hybrid PV-CSP plants show lower LCOE than PV-BESS for nighttime electricity fractions above 20–25% (corresponding to an approximately 4–5 h storage capacity) and different dependencies of LCOE for increasing nighttime electricity fraction (or storage capacity). Hybrid plants show a relatively constant LCOE for increasing storage capacities, but PV-BESS system costs grow significantly due to the higher storage costs. Heat production costs by CSP and by PV plus an electric heater are comparable and very sensitive to cost assumptions. The LCOE values for the PV system with an electric heater, thermal storage, and power block are higher than for hybrid systems but nighttime fractions are significantly higher as well, which is considered to be a really important factor in the design and optimization of the plant. Hybrid plants present the lowest LCOE values for nighttime electricity fractions above 20–25% and PV with a battery for lower fractions.

Close coupling and real integration offer additional benefits for hybrid plants. Parabolic trough systems using an ERH as a booster (in series to the solar field) show comparable LCOE and nighttime electricity fractions to molten salt solar tower systems using an ERH in parallel to the solar receiver.

To investigate the impact of latitude and irradiance, the systems were simulated and optimized for seven different sites. Considering nighttime fractions above 25%, sites with high direct normal irradiation are more suitable for higher CSP field shares in hybrid plants whereas sites with low direct normal irradiation are more suitable for higher PV field shares in hybrid plants. This trend was seen for all sites.

Costs projections for 2030 were made and the LCOE calculations were repeated for all technologies. A cost decrease in the next years may shift the break-even point of hybrid plants and PV-battery towards higher nighttime electricity fractions of around 25–30% (about 6 h storage capacity). This will depend on the actual decrease rates. PV with thermal storage and, particularly, with batteries may benefit to a greater extent than systems with CSP due to the higher cost reduction projected for their components.

Author Contributions: Conceptualization, J.D. and L.S.; methodology, J.I.-L., J.D., and L.S.; software, J.I.-L., J.D., and L.S.; validation, J.I.-L., J.D., and L.S.; formal analysis, J.I.-L., J.D., and L.S.; investigation, J.I.-L., J.D., and L.S.; resources, J.I.-L., J.D., and L.S.; data curation, J.I.-L., J.D., and L.S.; writing—original draft preparation, J.I.-L.; writing—review and editing, J.I.-L., J.D., and L.S.; visualization, J.I.-L., J.D., and L.S.; supervision, J.D. and L.S.; project administration, J.D. and L.S.; funding acquisition, J.D. and L.S. All authors have read and agreed to the published version of the manuscript.

Funding: This research was funded by the German Ministry of Economic Affairs and Climate Action, Project title: IntegSolar, Reference number 03 EE50011.

Data Availability Statement: Not applicable.

Acknowledgments: The authors gratefully acknowledge the funding of the work presented in this paper by the German Ministry of Economic Affairs and Climate Action. The responsibility for the content is solely by the authors.

Conflicts of Interest: The authors declare no conflict of interest. The funders had no role in the design of the study; in the collection, analyses, or interpretation of data; in the writing of the manuscript; or in the decision to publish the results.

Abbreviations

BESS	Battery Energy Storage System
CAPEX	Capital expenditure
CRT	Central Receiver Tower
CSP	Concentrated Solar Power
DNI	Direct Normal Irradiance
EPC	Engineering, Procurement and Construction
ERH	Electric Resistance Heater
GHI	Global Horizontal Irradiance
HTF	Heat Transfer Fluid
LCOE	Levelized Cost of Electricity
LPSP	Loss of Power Supply Probability
O&M	Operations and Maintenance
PB	Power Block
PT	Parabolic Trough
PV	Photovoltaic
PV-BESS	Photovoltaic System with Battery Energy Storage System
PV+ERH+PB	Photovoltaic System plus Electric Resistance Heater, Thermal Energy Storage and Power Block
TES	Thermal Energy Storage

References

1. IRENA. *Renewable Energy Statistics 2022*; The International Renewable Energy Agency: Abu Dhabi, United Arab Emirates, 2022.
2. Guiliano, S.; Puppe, M.; Hirsch, T.; Schenk, H.; Kern, J.; Moser, M.; Fichter, T. THERMVOLT Project: Systemvergleich von Solarthermischen und Photovoltaischen Kraftwerken Für die Versorgungssicherheit, Abschlussbericht (BMWI, 2014–2016). Available online: https://www.tib.eu/en/suchen/id/TIBKAT:100051305X/ (accessed on 1 March 2021).
3. Starke, A.; Cardemil, J.; Escobar, R.; Colle, S. Assessing the performance of hybrid CSP+PV plants in northern Chile. *Sol. Energy* **2016**, *138*, 88–97. [CrossRef]
4. Zhai, R.; Liu, H.; Chen, Y.; Wu, H.; Yang, Y. The daily and annual technical-economic analysis of the thermal storage PV-CSP system in two dispatch strategies. *Energy Convers. Manag.* **2017**, *154*, 56–67. [CrossRef]
5. Zurita, A.; Mata-Torres, C.; Valenzuela, C.; Felbol, C.; Cardemil, J.; Guzmán, A.; Escobar, R. Techno-economic evaluation of a hybrid CSP + PV plant integrated with thermal energy storage and a large-scale battery energy storage system for base generation. *Sol. Energy* **2018**, *173*, 1262–1277. [CrossRef]
6. Hamilton, W.; Husted, M.; Newman, A.; Braun, R.; Wagner, M. Dispatch optimization of concentrating solar power with utility-scale photovoltaics. *Optimization Eng.* **2020**, *21*, 335–369. [CrossRef]
7. Schöniger, F.; Thonig, R.; Resch, G.; Lilliestam, J. Making the sun shine at night: Comparing the cost of dispatchable concentrating solar power and photovoltaics with storage. *Energy Sources Part B Econ. Plan. Policy* **2021**, *16*, 55–74. [CrossRef]
8. Hassani, S.; Ait Lahoussine Oualia, H.; Moussaouia, M.; Mezrhab, A. Techno-Economic Analysis of a Hybrid CSP/PV Plants in the Eastern Region of Morocco. *Appl. Sol. Energy* **2021**, *57*, 297–309. [CrossRef]
9. Zurita, A.; Mata-Torres, C.; Cardemil, J.; Guédez, R.; Escobar, R. Multi-objective optimal design of solar power plants with storage systems according to dispatch strategy. *Energy* **2021**, *237*, 121627. [CrossRef]
10. Riffelmann, K.J.; Weinrebe, G.; Balz, M. Hybrid CSP-PV Plants with Integrated Thermal Storage. *AIP Conf. Proc.* **2022**, *2445*, 30020. [CrossRef]
11. Mata-Torres, C.; Palenzuela, P.; Alarcón-Padilla, D.; Zurita, A.; Cardemil, J.; Escobar, R. Multi-objective optimization of a Concentrating Solar Power + Photovoltaic + Multi-Effect Distillation plant: Understanding the impact of the solar irradiation and the plant location. *Energy Convers. Manag. X* **2021**, *11*, 100088. [CrossRef]
12. Richter, P.; Trimborn, T.; Aldenhoff, L. Predictive storage strategy for optimal design of hybrid CSP-PV plants with immersion heater. *Sol. Energy* **2021**, *218*, 237–250. [CrossRef]
13. Liu, T.; Yang, J.; Yang, Z.; Duan, Y. Techno-economic feasibility of solar power plants considering PV/CSP with electrical/thermal energy storage system. *Energy Convers. Manag.* **2022**, *255*, 115308. [CrossRef]
14. Fourth Phase of Mohammed Bin Rashid Al Maktoum Solar Park is the First CBI Certified Renewable Energy Project Financing in the GCC Region. Available online: https://www.acwapower.com/news/fourth-phase-of-mohammed-bin-rashid-al-maktoum-solar-park-is-the-first-cbi-certified-renewable-energy-project-financing-in-the-gcc-region/ (accessed on 28 July 2022).
15. Morocco Pioneers PV with Thermal Storage at 800 MW Midelt Concentrated Solar Power Project. Available online: https://helioscsp.com/morocco-pioneers-pv-with-thermal-storage-at-800-mw-midelt-cocentrated-solar-power-project/ (accessed on 25 August 2022).
16. Cerro Dominador CSP Plant in Chile Officially Opens. Available online: https://www.acciona.com/updates/news/cerro-dominador-csp-plant-chile-officially-opens/ (accessed on 23 August 2022).
17. 4 Concentrated Solar Power Projects Started & 2 CSP Projects Confirmed EPC Last Week in China. Available online: https://helioscsp.com/4-concentrated-solar-power-projects-started-2-csp-projects-confirmed-epc-last-week-in-china/ (accessed on 25 August 2022).
18. Dersch, J.; Schomaker, L.; Iñigo Labairu, J.; Mhaske, S.; Pfundmaier, L.; Schlecht, M.; Zimmermann, F. *Integration of CSP and PV Power Plants*; Final Report of the Research Project "IntegSolar" No. 03EE5011; Cologne, Germany, 2022; Available online: https://elib.dlr.de/188367/ (accessed on 12 September 2022).
19. DLR. Greenius. 2020. Available online: http://freegreenius.dlr.de (accessed on 1 March 2021).
20. Dersch, J.; Dieckmann, S.; Hennecke, K.; Pitz-Paal, R.; Taylor, M.; Ralon, P. LCOE reduction potential of parabolic trough and solar tower technology in G20 countries until 2030. *AIP Conf. Proc.* **2020**, *2303*, 120002. [CrossRef]
21. National Renewable Energy Laboratory (NREL). *Annual Technology Baseline (ATB) Cost and Performance Data for Electricity Generation Technologies [data set]*; National Renewable Energy Laboratory: Golden, CO, USA, 2021. [CrossRef]
22. Vignesh, R.; Feldman, D.; Desai, J.; Margolis, R.U.S. *Solar Photovoltaic System and Energy Storage Cost Benchmarks: Q1 2021*; NREL/TP-7A40-80694; National Renewable Energy Laboratory: Golden, CO, USA, 2021.
23. Electricity Prices. Available online: https://www.globalpetrolprices.com/electricity_prices/ (accessed on 10 May 2021).

Article

Development of a Simple Methodology Using Meteorological Data to Evaluate Concentrating Solar Power Production Capacity

Ailton M. Tavares [1,2], Ricardo Conceição [3], Francisco M. Lopes [4] and Hugo G. Silva [2,5,6,*]

[1] Instituto de Ciências da Terra (ICT), IIFA, Universidade de Évora, Rua Romão Ramalho 59, 7002-554 Évora, Portugal
[2] Laboratório Associado de Energia, Transporte e Aeronáutica (LAETA), Universidade de Évora, Rua Romão Ramalho 59, 7002-554 Évora, Portugal
[3] High Temperature Processes Unit, IMDEA Energy, Avda. Ramón de La Sagra 3, 28935 Móstoles, Madrid, Spain
[4] Instituto Dom Luiz (IDL), Faculdade de Ciências, Universidade de Lisboa, Campo Grande, 1749-016 Lisbon, Portugal
[5] Departamento de Física, ECT, Universidade de Évora, Rua Romão Ramalho 59, 7002-554 Évora, Portugal
[6] INEGI Alentejo, Universidade de Évora, Largo dos Colegiais 2, 7000-803 Évora, Portugal
* Correspondence: hgsilva@uevora.pt; Tel.: +351-967-480-736

Citation: Tavares, A.M.; Conceição, R.; Lopes, F.M.; Silva, H.G. Development of a Simple Methodology Using Meteorological Data to Evaluate Concentrating Solar Power Production Capacity. *Energies* **2022**, *15*, 7693. https://doi.org/10.3390/en15207693

Academic Editor: Juri Belikov

Received: 12 September 2022
Accepted: 3 October 2022
Published: 18 October 2022

Publisher's Note: MDPI stays neutral with regard to jurisdictional claims in published maps and institutional affiliations.

Abstract: Evaluation of the Concentrating Solar Power capacity factor is critical to support decision making on possible regional energy investments. For such evaluations, the System Advisor Model is used to perform capacity factor assessments. Among the required data, information concerning direct normal irradiance is fundamental. In this context, the Engerer model is used to estimate direct normal irradiance hourly values out of global horizontal irradiance ground measurements and other observed meteorological variables. Model parameters were calibrated for direct normal irradiance measurements in Évora (Southern Portugal), being then applied to a network of 90 stations, part of the Portuguese Meteorological Service. From the modelled direct normal irradiance, and for stations that comprise 20 years of data, typical meteorological years were determined. Finally, to identify locations of interest for possible installations of Concentrating Solar Power systems, annual direct normal irradiance availabilities and the respective capacity factor, for a predefined power plant using the System Advisor Model, were produced. Results show annual direct normal irradiance availabilities and capacity factors of up to ~2310 kWh/m^2 and ~36.2% in Castro Marim and in Faro, respectively. Moreover, this study supports energy policies that would promote Concentrating Solar Power investments in Southern Portugal (Alentejo and Algarve regions) and eastern centre Portugal (Beira Interior region), which have capacity factors above 30%.

Keywords: global horizontal irradiance; Engerer model; typical meteorological year; direct normal irradiance mapping; Concentrating Solar Power plant production capacity

1. Introduction

Electricity production from solar power plants is increasingly seen as one of the alternatives to conventional energy production systems. Photovoltaic (PV) systems' contributions to electric power generation have been growing exponentially [1]. Simultaneously, investments towards Concentrating Solar Power (CSP) plants have been increasing in regions where solar resources are suitable for such technology [1]. Although CSP has had a slower growth than PV, it has been widely used [2], with its global installed capacity expected to increase in the near future.

For installation of solar power systems, solar resource characterization is essential. Its assessment consists in solar irradiation availability (kWh/m^2) estimation at a given location considering a long time period [3]. Thus, a long-term database of solar irradiance

(W/m^2) is essential to establish: (i) a correct design for an efficient energy conversion; (ii) estimate electricity production; (iii) guarantee the bankability of CSP plants [1,4].

Acquisition of solar irradiance values at the surface can be performed with state-of-the-art instrumentation. For instance, Global Horizontal Irradiance (GHI) is generally measured with pyranometers. In contrast, measurements of Direct Normal Irradiance (DNI) require more equipment, including a sun tracker system where pyrheliometers are assembled. This makes the equipment more costly and regular maintenance necessary [5]. Such limitations constitute a problem for solar system implementation worldwide, not being restricted to the present study region (Portugal). Nonetheless, DNI can be estimated from measured GHI by remote sensed data or using separation models (semi-physical and empirical) for the respective solar components [6,7]. It is worth noting that satellite data accuracy can be affected by clouds and other atmospheric parameters, leading to frequent occurrence of missing data [8]. Since solar concentration systems are highly dependent on DNI, its accurate determination is fundamental [6,9], particularly at sites with significant solar resources in order to maximize solar harvesting. To solve DNI data scarcity, during the last 5 decades, research activity has grown regarding DNI availability estimation from GHI data. In parallel, there has been an increased number of meteorological stations capable of measuring GHI and DNI simultaneously, but most of these are recent (e.g., in Southern Portugal the stations have only roughly 7 years of operation). Therefore, DNI estimations lack statistical significance, which requires at least 15 years of information to have reliable availability estimates [10]. As opposed to these short time series, long-term time series do exist, although these mostly concern GHI.

In depth discussion of DNI estimation from GHI measurements can be found in [6,11]. In particular, in the work of Gueymard and Ruiz-Arias [6], a benchmark test comprising 140 models was performed, in which DNI values from GHI measurements were estimated. Amongst all presented models, the ones that produced the best scores were: the Engerer [12], Perez [13], and Hollands [14] models. Gueymard and Ruiz-Arias [6] tested and validated the Engerer model (Engerer2, henceforth), using high-quality 1 min data of GHI and DNI at 54 research-class stations distributed in seven continents, which were then grouped into in four distinct climate zones: arid, temperate, tropical, and high-albedo. In that study, the best results were found through Engerer2, since it showed consistent DNI estimations for almost all types of climates; it was called a "quasi-universal" 1 min separation model. According to Gueymard and Ruiz-Arias [6] and Starke [15], the success of Engerer2 is due to the fact that it was derived from 1 min data, hence allowing to predict fast transient phenomena, such as cloud enhancement. In the study by Kim et al. [1], Engerer2 was used as a deterministic prediction model to estimate DNI for the Korean peninsula at three ground stations (namely in Seoul, Buan, and Jeju), being then validated with observational data. Bright and Engerer [16] used Engerer2 to estimate DNI at different time resolutions (at 1, 5, 10, and 30 min), and also at 1 h and 1-day frequencies, which were under five main Köppen–Geiger climate classifications (polar, cold, temperate, arid, and equatorial). The authors used a deterministic clear-sky model to make the original model easy to use, where the clear-sky values were determined using the REST2 model [17]. Results showed good performances carried out by the model for most time resolutions (with the exception of the daily outputs), and for most climate types (except polar). Recently, Dazhi Yang [18] conducted a comprehensive and comparative evaluation of 10 recent separation models, namely, Engerer2 and Engerer4 [12], Starke1 and Starke2 [15], Starke3 [19], Abreu [20], Paulescu [21], Every1 and Every2 [22], and Yang4 [23]. Yang claims that the Yang4 model has the best overall performance, and therefore should replace Engerer2 as the new quasi-universal model [18]. Nevertheless, considering that Engerer2 has been validated in numerous studies, this model is chosen to be used as the basis model of the present work to estimate DNI from GHI, as it presents good scores along with a simple implementation methodology. It is also worth noting that Engerer2 is also considered due to its deterministic nature, and can be simply implemented worldwide.

The proposed methodology has the potential to be applied in high solar radiation regions of the globe. This study is centred on Portugal due to the exceptional conditions that the country offers for solar power harvesting. Moreover, it is expected to have an important role in solar energy projects in Europe. Mainland Portugal is seen as one of the top candidates to receive such investments [24]. An increase of installed solar power between ~8.1 and 9.9 GW is expected by 2030, corresponding to about 22–27% of the national annual electricity generation [8]. The present work provides a contribution to improve energy management of solar power systems, by developing reliable DNI availability maps for mainland Portugal. To achieve the proposed objective, a 20-year time series of meteorological data is used (including GHI, air temperature, relative air humidity, wind speed, atmospheric pressure, and precipitation), being measured through 90 ground stations, scattered across the country, as part of the *Instituto Português do Mar e da Atmosfera* (IPMA—the Portuguese national meteorological service) network. From the available data, DNI estimations based on Engerer2 [12] are performed for all stations. For the modelled DNI time series, data post-processing is performed, using multivariable regression coefficients to determine new and more accurate datasets. Engerer2 parameters and regression coefficients are then calibrated using DNI data measurements in Évora (reference station) along with other meteorological datasets. The post-processing model is based on the Multivariate Regression Model (MRM) implemented by Lopes et al. [25], which allows to adjust DNI predictions, provided by the European Centre for Medium-Range Weather Forecasts (ECMWF), to ground observations. Additionally, calculation of typical meteorological years (TMY) is performed with the aim of providing representative data of a given region considering a longer period, which is a key step for elaboration of projects of such dimensions [26–28]. To achieve a better assessment regarding CSP plant performance, investors need to have reliable and realistic climate and soiling [29] data to guarantee project viability. Therefore, TMY calculation is very useful, as it carries weather information for an entire year, consisting of representative months of regional climate over a long period of time [27,28]. The method described by Hall et al. [28], examines nine meteorological parameters such as daily means, maximum and minimum values of air and dew point temperatures, daily means and maximum values of wind speed, and daily values of GHI. The method is applied to compute the TMY for mainland Portugal, and, consequently, the mapping of annual DNI availabilities. Finally, the analysis is then focused on capacity factor (CF) mapping performed across the country with the goal of identifying the most suitable locations for the installation of CSP plants. CF outputs can be used as a comparative measure of the amount of energy produced by a power plant during one year to the energy that would be produced if the power plant was operated at nominal power during the same period [30]. The CF has the advantage of revealing the production capacity of a given power plant, regardless of its size. Moreover, to compute CF values, a freely available power plant model is used, namely the System Advisor Model (SAM), from the National Renewable Energy Laboratory, [31]. With the SAM, the user is able to set up a predefined power station, with a very detailed configuration that allows the simulated power to be close to actual operational plants, such as the Andasol3 power plant [32].

The presented work aims to fill the existing gap between solar assessment research and decision making on possible CSP plants by developing a simple and self-contained methodology that can translate common meteorological time series into a figure of merit of CSP plants that can be decisive for their implementation; CF was selected as figure of merit. To the authors' best knowledge, this is the first time that such a methodology has been developed, and it has enabled the earliest estimation of CF values for mainland Portugal. This procedure has led to the identification of three regions that possess high potential for CSP implementations: Alentejo and Algarve (Southern Portugal), and Beira Interior (Eastern Central Portugal). Furthermore, the most important aspect of this methodology is that it can be applied in a simple way to other regions in the world that possess high potential for CSP. Therefore, this work is considered as a contribution to the implementation of state-of-the-art power plants not only for Portugal, but for the world.

Considering the aforementioned aspects, this manuscript is organized as follows: in Section 2, the work methodology is presented; Section 3, focuses on the results and respective discussion; in Section 4, some implications at a decision making level are drawn for Portugal; lastly, Section 5 summarizes the main conclusions; additional information is added in and Appendices A and B.

2. Methodology

2.1. Measurements

A 20-year period of GHI time series is considered in this work (from 2000 to 2019), being continuously measured over 90 weather stations from the IPMA network, scattered across mainland Portugal (Figure 1), providing coverage of the entire country. Regarding the 90 stations shown in Figure 1, 17 are considered main stations (blue squares), while the remaining stations are used as secondary stations (red circles) and the reference station from *Instituto de Ciências da Terra* (ICT—Institute for Earth Sciences) (green diamond), located in Évora city, is also presented. Main stations are those at which hourly observations are made, or at least three times per day, while at secondary stations, at least one observation is performed per day [33]. More details concerning all stations used can be found in Table A1 (Appendix A).

Measurements of GHI were performed in 1 h intervals using state-of-the-art instrumentation, with Kipp&Zonen instrumentation, The Netherlands (www.kippzonen.com/, accessed on 2 October 2022), namely the CM11 and Hukseflux LP02 pyranometers, with the CM11 model being used in the majority of IPMA's stations. Both pyranometers are secondary standard instruments according to the International Organization of Standardization (ISO) 9060:1990 [34]. However, there is no detailed information regarding their calibration and maintenance, due to the large number of stations and their spatial distribution. Thus, in some cases, measurements may be affected by lack of calibration, malfunction, or, more commonly, soiling effect [35,36]. GHI and Diffuse Horizontal Irradiance (DHI) measurements at the reference station in the ICT observatory (University of Évora) were performed with pyranometers (model CMP11), secondary class instrument [34]. DHI was measured using a pyranometer with a shading ball that blocks the beam radiation. For DNI measurements, pyrheliometers were used, particularly with the model CHP1, a World Meteorological Organization (WMO) first class instrument with an estimated uncertainty on a daily basis of <1% [37]. The instruments used were mounted on a SOLYS2 Sun-Tracker (Kipp&Zonen) [25].

Regarding other meteorological variables measured at IPMA's stations, the air temperature and relative air humidity were recorded at a height of 1.5 m, from which the corresponding hourly averages are used for the analysis; while wind speed is recorded at a height of 10 m, in which the averages of the last 10 min of each hour were used [38]. Air temperature and relative humidity measurements were made with the Vaisala HMP45 sensor, whereas wind speed was measured with the Vaisala WAA 15A anemometer [39].

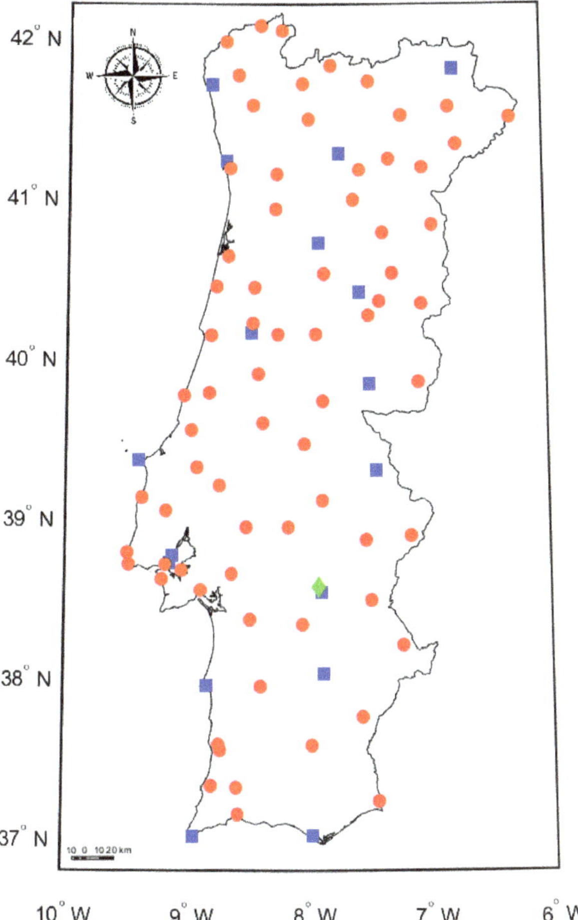

Figure 1. Geographical location of the 90 IPMA ground stations used for GHI measurements in mainland Portugal, considering a 20-year period (from 2000 to 2019). Blue squares represent the main stations, red circles the secondary stations, and the green diamond is the reference station from Instituto de Ciências da Terra (ICT), located in Évora city.

2.2. Data Pre-Processing

To use reliable data, a prior quality control procedure was performed for all the available measuring stations from IPMA's network, as well as the ICT station. All data points that do not follow the proposed filters (filters for DNI and DHI were only applied to data from the ICT reference station, in which these variables were available), as described in [6], were rejected:

1. $Z < 85°$;
2. $GHI > 0$, $DHI > 0$, $DNI \geq 0$;
3. $DNI < 1100 + 0.03 \times Elev$;
4. $DNI < E_{0n}$;
5. $DHI < 0.95 \times E_{0n} \times \cos^{1.2} Z + 50$;
6. $GHI < 1.50 \times E_{0n} \times \cos^{1.2} Z + 100$;
7. $DHI/GHI < 1.05$ for $GHI > 50$ and $Z < 75°$;
8. $DHI/GHI < 1.10$ for $GHI > 50$ and $Z > 75°$.

In these conditions, "Elev" is the elevation of each station (Table A1), and "E_{0n}" is the irradiance at the top of the atmosphere. Irradiance values below a solar zenith angle (Z) of $85°$ are considered in this analysis, thus eliminating situations of low irradiance that are not relevant for solar applications, as well as situations where the model has reduced precision (condition 1), mainly due to the high zenith angles: $\cos(Z) \approx 0$. Only positive values of GHI, DHI, and DNI are used (condition 2). DNI values below a certain reference value accounting for station elevation (condition 3) are considered [6]. GHI, DHI, and DNI values below the threshold regarding maximum limits of the baseline surface radiation network (BSRN) [40] (conditions 4, 5, and 6) are also used. Both "Z" and "E_{0n}" were calculated based on the formula proposed by [41], i.e., Equations (2.12) and (2.77), respectively. Following the application of proposed filters on the observed ICT data, the remaining data (which correspond to about 99.32% of the total data) were used to calibrate the corresponding Engerer model parameters, as discussed further on.

2.3. Engerer Model Application

The Engerer2 [12] is a deterministic prediction model, in which the diffuse fraction (K) can be calculated using five predictors, namely: GHI clearness index (k_t), apparent solar time (AST), zenith angle (Z), variability index (k_e), and the deviation (Δk_{tc}) between the observed value of k_t at the surface, and the one obtained under a clear sky (k_{tc}). The following relation between predictors allows one to obtain K:

$$K = \frac{DHI}{GHI}, \tag{1}$$

$$K = C + \frac{1 - C}{1 + \exp(\beta_0 + \beta_1 \times k_t + \beta_2 \times AST + \beta_3 \times Z + \beta_4 \times \Delta k_{tc})} + \beta_5 \times k_e, \tag{2}$$

where C is the value of lower asymptote and β_n ($n = 0, \ldots, 5$) is the nth Engerer coefficient, whereas Δk_{tc} is given by:

$$\Delta k_{tc} = k_t - k_{tc}, \tag{3}$$

In which k_{tc} is calculated as follows:

$$k_{tc} = \frac{GHI_{cs}}{E_{0n} \times \cos(Z)}. \tag{4}$$

Here, GHI_{cs} is the global horizontal irradiance under clear-sky conditions, and in this study it is defined deterministically as [16]:

$$GHI_{cs} = DNI_{cs} \times \cos(Z) + c \times DNI_{cs}, \tag{5}$$

where DNI_{cs} is the direct normal irradiance under clear-sky conditions, defined as [16]:

$$DNI_{cs} = A \times \exp\left[\frac{-k}{\cos(Z)}\right], \tag{6}$$

in which the coefficients A, k, and c depend on days of the year (d). These coefficients are calculated as follows:

$$\begin{aligned} A &= 1160 + 75\sin\left[\frac{360(d-275)}{365}\right], \\ k &= 0.174 + 0.035\sin\left[\frac{360(d-100)}{365}\right], \\ c &= 0.095 + 0.04\sin\left[\frac{360(d-275)}{365}\right]. \end{aligned} \tag{7}$$

The term k_e is the portion of K attributed in the case of cloud enhancement, defined as:

$$k_e = \max\left(0;\ 1 - \frac{GHI_{cs}}{GHI}\right). \tag{8}$$

Since Engerer2 is made for 1 min temporal resolution, and the fact that GHI data from IPMA's network are measured at 1 h resolution, this work makes use of the Engerer2 model configured for hourly irradiance values of GHI. As previously mentioned, Bright and Engerer [16] evaluated the performance of the Engerer2 model over various time scales, revealing good results for hourly data.

2.4. Data Post-Processing

To reduce the difference between the modelled and observed DNI, post-processing is performed for the modeled data. In this context, a multivariate regression model (MRM) is used to adjust hourly modelled DNI to the observations. The MRM model [25] uses an interactive stepwise function, with an adjusted r-squared criterion, to improve DNI predictions. The method allows to perform all possible combinations between predictors, until the best fitting option is found. In this work, a third-degree polynomial was used to calibrate regression coefficients for three years of measured DNI. The predictors used are the following: modelled DNI (DNI_i^{MOD}), GHI (GHI_i^{OBS}), clear-sky GHI (GHI_i^{CS}) and DNI (DNI_i^{CS}), air temperature (Ta_i^{OBS}), air relative humidity (RH_i^{OBS}), wind speed (WS_i^{OBS}), atmospheric pressure (P_i^{OBS}), and precipitation ($Prec_i^{OBS}$). Thus, a function is used to adjust the modelled DNI (DNI_i^{ADJ}) at hour i according to:

$$DNI_i^{ADJ} = F\left(DNI_i^{MOD}, GHI_i^{OBS}, GHI_i^{CS}, DNI_i^{CS}, Ta_i^{OBS}, RH_i^{OBS}, WS_i^{OBS}, P_i^{OBS}, Prec_i^{OBS}\right) \tag{9}$$

where F is adjusted to reduce the error between the DNI_i^{ADJ} and the observed DNI (DNI_i^{OBS}) values. The function F is then applied to all stations under study.

Finally, the post-processing DNI outputs are filtered for: (i) values greater than 1042 W/m^2 (i.e., the maximum value observed at Évora station), which are replaced by the corresponding clear-sky DNI value at that moment; (ii) negative values, which are replaced by values of modelled DNIs prior to the post-processing. It is worth noting that the parameters calibrated with the observed DNI data in Évora will be used to post-process the modelled DNI in all stations of the IPMA's network.

2.5. Data Gap Filling

Due to equipment malfunction, data gaps in time series are common, particularly for the case of solar radiation measurements. Therefore, gap filling methods are often used, such as the one followed here, in reference to works [3,42]. For data gaps of one or two hours of missing GHI values in a day, missing values are filled by linear interpolation from the non-missing values. For missing periods of three and more hours in a day, a geographical interpolation is performed. The algorithm starts to select the four closest neighbouring stations (k) of the station to which data gaps will be filled (j), and determines the long-term linear regression coefficients (slope and interception point) between this station and the neighbouring ones, as follows:

$$GHI_j = m_{jk} \times GHI_k + b_{jk}, \ k = 1 \dots 4, \tag{10}$$

With this set of four relations, hourly data gaps in station j are filled by the median of the four hourly values of the measured values on the k stations, GHI_k, on the corresponding hours of the missing data in GHI_j. To obtain reliable data from the linear regression coefficients, outliers are rejected from the correlation by applying the following restrictions: $GHI_j > m_{jk} \times GHI_k + b_{jk} + 2\sigma_{jk}^+$ and $GHI_j < m_{jk} \times GHI_k + b_{jk} - 2\sigma_{jk}^-$. The parameters σ_{jk}^+ and σ_{jk}^- are the standard deviations calculated from differences for every hour l, as given by:

$$\delta_{jk}(l) = GHI_j(l) - \left[m_{jk} \times GHI_k(l) + b_{jk}\right], \tag{11}$$

where δ_{jk}^+ means positive differences and δ_{jk}^- negative differences, through the following relation [3]:

$$\sigma_{jk}^{\pm} = \sqrt{\frac{\Sigma_l \left| \delta_{jk}^{\pm}(l) \right|^2}{N-1}}, \tag{12}$$

where N is the number of hours with mutually available data for stations j and k.

After removing the outliers, new values of m_{jk} and b_{jk} are then calculated for the neighbouring stations, while the GHI_j is calculated through Equation (10). For instance, for the station located at Évora, the four closest neighbouring stations are Viana do Alentejo (~26.6 km), Reguengos (~36.6 km), Estremoz (~48.7 km), and Avis (~50.9 km), where the values of m were approximately 1.13, 1.10, 0.98, and 1.12 W/m^2, and the values of b were 3.27, 6.27, 9.00, and 11.69 W/m^2 for the respective stations. The fact that $m \sim 1$ means that the data in the different stations have a similar trend, while $b < 12$ W/m^2 means that a small deviation exists. This demonstrates the suitability of the adopted interpolation method. Similar results are observed for the other stations of the IPMA's network.

It is noted that regarding the TMY calculation, gap filling was only performed for months that had gaps smaller than five days. Therefore, months with more than five days of missing data were not considered for analysis.

2.6. Typical Meteorological Year Calculation

According to the method proposed in Hall [28], a TMY can be calculated by concatenating twelve typical meteorological months (TMM). The selection of the TMM of each calendar month is calculated based on Finkelstein–Schafer (FS) daily values [43]. Selection is made by comparing cumulative frequency distribution functions (CDFs) of each parameter: mean, maximum and minimum daily value of air temperature, relative humidity, maximum and mean daily values of the wind speed, and daily global horizontal solar irradiation availability, for every month of every year to their long-term distribution. Comprehensive details for this method can be found in [28]. Following these aspects, characteristic annual GHI and DNI availabilities (kWh/m^2/year) are calculated from the summation of hourly irradiance values of TMY regarding each station.

2.7. The Power Plant Model

The power plant simulator, i.e., the SAM, is a freely available financial and performance model designed to facilitate decision making in renewable energy system projects. The model not only allows to make performance predictions, but also energy cost estimates that these systems require [31]. The SAM is used here to simulate a power plant and to estimate its production capacity from meteorological data belonging to IPMA's network. Since there are no DNI measurements in this network, DNI was estimated as previously explained. TMY is elaborated from that meteorological data and SAM simulations are performed using the TMY calculated for each station under study.

In this work, a case study is considered focusing an operational CSP facility with similar configurations to Andasol3, based on the model previously developed on [30,44]. The Andasol3 is a commercial power plant that uses a linear focus parabolic trough system that runs with thermal oil as heat transfer fluid and has 7.5 h indirect molten salt storage capacity at full load. Its solar field aperture area is 510,120 m^2, with a nominal capacity of 50 MW$_e$ and an expected power generation of 175 GWh/year. Further information on the configuration of the power plant can be found in [32].

3. Results and Discussion

3.1. DNI Validation and Calibration

To validate and calibrate the parameters from the Engerer2 and MRM models, four years of DNI measurements, ranging from 2016 to 2019, were used together with ground measurements performed at the ICT station. The Engerer2 parameters are determined from a non-linear regression model, where the model is fitted, with the corresponding parameters determined from an iterative procedure. The proposed validation consists in determining the parameters considering three years of measured GHI, which are then used to estimate DNI for the fourth year and compared with DNI observations for that year. The MRM parameters are also calibrated with three years of modelled DNI, clear-sky GHI and DNI, as well as observed GHI, Ta, RH, WS, P, and Prec. It is important to underline the fact that the modelled DNI is part of the predictors, allowing to improve the MRM. The post-processing method reduces the overestimation of the model when GHI values are strongly composed by DHI (that should correspond to very low DNI values of almost zero) and when GHI observations provide values above 1000 W/m^2, in which cases the Engerer2 model has difficulties in estimating DNI accurately. The MRM parameters are used to post-process the estimated DNI for the fourth year, and thus to compare with measured DNI for the same year. For instance, DNIs from 2016, 2017, and 2019 are used to calibrate the model (adjusting the model and MRM parameters to observations), while DNI from 2018 is used to validate the model estimations against the observations of that year. A similar procedure will be carried out for the remaining years.

To compare model estimations against observations, a statistical analysis was performed using conventional metrics, such as the root mean square error (RMSE), the mean bias error (MBE), and the mean absolute error (MAE), as follows:

$$RMSE = \sqrt{MSE} = \sqrt{\frac{1}{n} \times \sum_{j=1}^{n} \left[I(j) - \hat{I}(j) \right]^2}, \tag{13}$$

$$MBE = \frac{1}{n} \times \sum_{j=1}^{n} \left[I(j) - \hat{I}(j) \right], \tag{14}$$

$$MAE = \frac{1}{n} \times \sum_{j=1}^{n} \left| I(j) - \hat{I}(j) \right|, \tag{15}$$

where $I(j)$ and $\hat{I}(j)$ are the observed and modelled value at hour j, respectively, while n is the number of assessed pairs.

In Table 1, a statistical summary between model outputs and observations is provided. The acronyms MOD1 and MOD2 represent the analysis performed without and with the MRM post-processing method, respectively. Regarding the correlation coefficient (r) and the coefficient of determination (R^2), very small differences are observed when using MOD2 in comparison to MOD1, reaching an improvement of 1 and 2%, respectively, in 2018. MOD2 shows a decrease of 18% in 2016 to 25% in 2018 in relation to MOD1, with respect to RMSE values. Regarding the MBE values, there is a decrease of more than 60 and 97% in 2016 and 2019, respectively, when MOD2 is used instead of MOD1. There is an improvement of over 20% in 2018 when using MOD2 instead of MOD1, regarding MAE values. These error metric values show that MOD2 presents best scores in comparison to MOD1, which makes it useful for the estimation of DNI.

Table 1. Statistical and descriptive analysis summary for the model towards the measurements using standard error metrics for four consecutive years of hourly data: correlation coefficient (r), coefficient of determination (R^2), root mean square error (RMSE), mean bias error (MBE), and mean absolute error (MAE). MOD1 and MOD2 represent the analysis without and with the MRM model, respectively.

	2016		2017		2018		2019	
Error metrics	MOD1	MOD2	MOD1	MOD2	MOD1	MOD2	MOD1	MOD2
r (—)	0.97	0.98	0.97	0.98	0.97	0.98	0.97	0.98
R^2 (—)	0.94	0.95	0.95	0.96	0.94	0.96	0.95	0.96
RMSE (W/m^2)	92.93	76.33	83.45	66.02	93.73	70.45	87.39	66.06
MBE (W/m^2)	−39.36	−15.41	−31.63	−3.45	−36.06	−2.58	−36.69	0.96
MAE (W/m^2)	61.15	50.24	55.26	47.07	59.88	47.85	55.62	44.71

In Figure 2, a comparative analysis between modelled DNI (with MOD2) and observations is shown. Overall, results show good agreement between modelled and observed values. Within the range between 800 and 1000 W/m^2 of observed DNI, the model tends to underestimate observations. This range of values represents the clear-sky period and clean atmosphere, in which the model has its limitations, since a deterministic clear-sky model is being used instead of one that considers the actual aerosol load. In such circumstances, the observed GHI will be higher than the modelled GHI_{cs}, making k_e different from zero and increasing K, which results in lower DNI. It is also observed that for DNI values lower than ~30 W/m^2, the model overestimates observations, reaching values of ~730 W/m^2, i.e., about 24 times higher. This behaviour is commonly found under cloudy conditions, where measured DNI is significantly lower than the model estimates, due to, for instance, the passing of clouds. Under these conditions DNI is reduced to almost zero, while the decrease is not so significant for GHI, because of the increase in DHI (Figure 2d shows a point circled in red as an example). As a result, if GHI measurements under these conditions are used to estimate DNI, such estimations will have the tendency to be higher than the observed ones. Further explanation is provided along with case studies in what follows.

The typical daily behaviour for observed DNI and GHI, along with the modelled DNI for 2017 (considered here as a case study), is shown in Figure 3. Four different types of conditions are represented: (a) clear sky, (b) cloudy sky, (c) cloud passage, and (d) several cloud passages. In Figure 3a, it is possible to validate what has been previously reported, i.e., concerning the model behaviour in clear-sky conditions. An underestimation concerning the observations is found, especially near solar noon. In fact, the work conducted in [42] detailed similar behaviour, in which the use of climatologic aerosol values in numerical weather prediction models instead of actual aerosol observations plays an important role. Consequently, for days with a clear and very clean atmosphere, observed DNI will be higher than the one resulting from the model. In the present study, this should stem from the fact that a determinist GHI is being considered for clear sky (GHI_{cs}), which does not consider actual aerosol loads. The use of deterministic GHI for clear sky is justified by the simplicity and self-containment of the proposed model, enabling an easy application to other regions. In Figure 3b, it can be observed that under cloudy conditions, the model tends to overestimate observations. Similar overestimations are frequently observed in DNI modelling, being extensively discussed in [42]. These values result from the fact that clouds tend to affect more significantly the DNI than the GHI (where DNI reduction is being partially compensated by the DHI increase). Thus, when GHI is used to estimate DNI, the estimation is higher than the actual measured values. This aspect can be observed in Figure 3b for a cloudy day, where at 11 UTC, the observed DNI is ~50 W/m^2, while the modelled DNI exceeds ~150 W/m^2. This means that the modelled value is ~3 times higher than the observed in such circumstances. In Figure 3c, the slight increase in the modelled DNI is due to a fast cloud passing that causes the observed DNI to decrease. For the GHI, this effect is not clearly observed since the increase in DHI offsets the GHI value. In Figure 3d, when there are many cloud passages, the model tends to overestimate the observations as depicted by the high values of GHI. Despite these limitations, the modelled DNI follows the observations relatively well.

Moreover, as previously stated, good correlations between DNI measurements and the model exist for all testing years. Figure 4 shows the probability distribution functions (PDFs) for observed and modelled DNIs, which depict a very similar behaviour. For DNI values in the range between ~920 and ~1000 W/m^2, the model tends to underestimate the observations. For DNI values in the range between ~800 and ~900 W/m^2, the model is prone to overestimating the observations, in part due to clouds that affect the observed DNI more than the modelled one, as previously explained. This effect causes the modelled diffuse fraction (K) to decrease while, consequently, the modelled DNI increases more significantly than the observed one. Despite these differences, results show how good the performance of the model is concerning estimating DNI from GHI, thus proving that the adopted method is a reliable source of quality data.

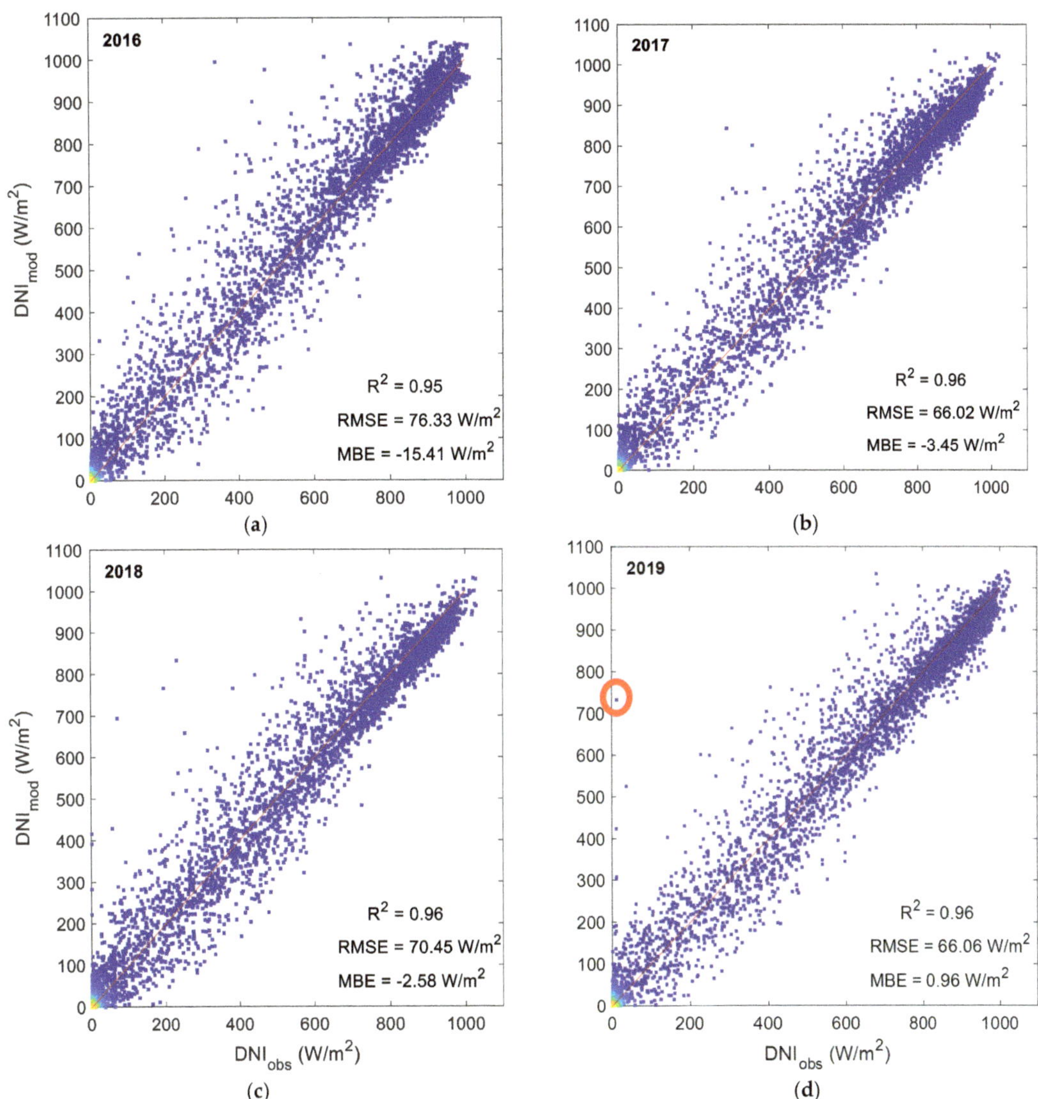

Figure 2. Modelled and observed hourly DNI for: (**a**) 2016, (**b**) 2017, (**c**) 2018, and (**d**) 2019.

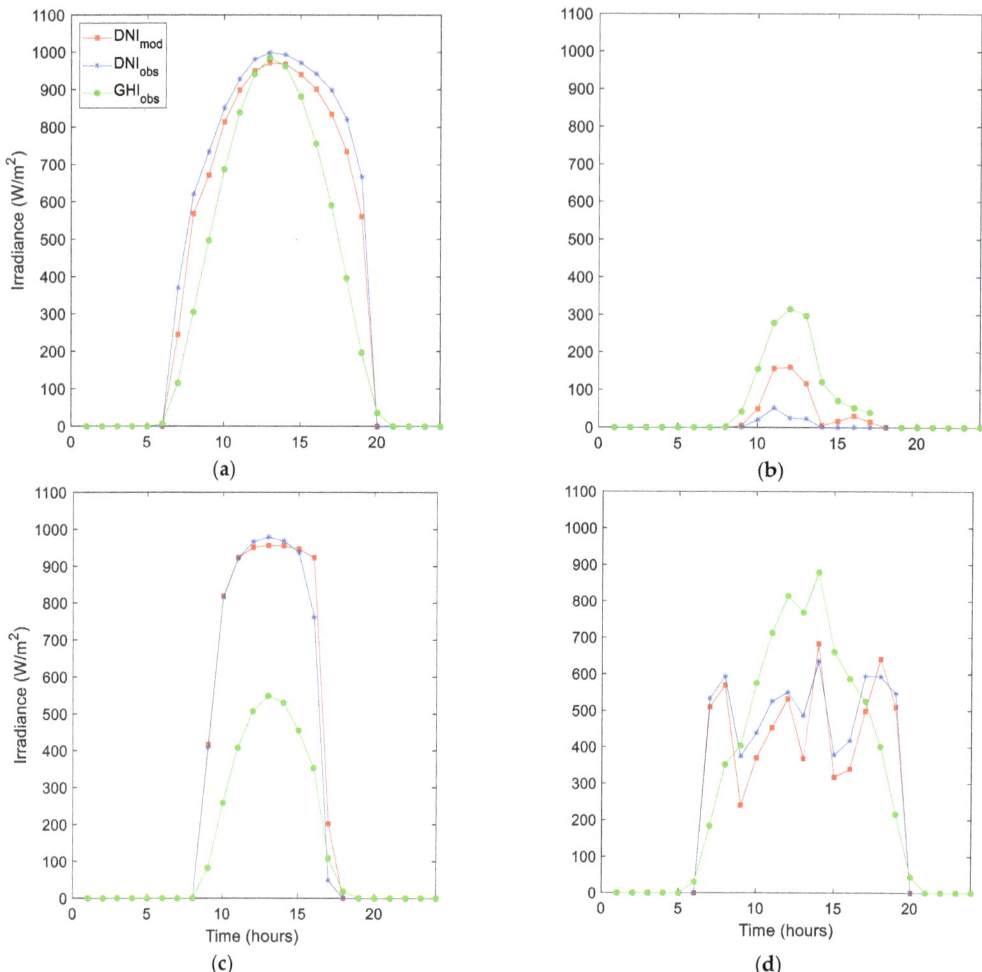

Figure 3. Representation of hourly observed and modelled DNI and observed GHI data: (**a**) clear sky; (**b**) cloudy sky; (**c**) cloud passing; (**d**) several clouds passing. Red square represents model DNI, green circle the observed GHI, and blue asterisk the observed DNI.

In Table 2, a statistical summary between measurements and model is provided. Results show that the modelled mean values are higher than the observed ones for all years. Regarding the standard deviation, modelled values are smaller than for observations. The DNI annual availability calculated through the model is greater than the observed one for most testing years, except 2019. The highest differences are ~2.93% for 2016, while a minimum of ~0.23% is found for 2019. Regarding the obtained MAE values, on average, the model overestimates or underestimates the observation by ~50.24, 47.07, 47.85, and 44.71 W/m^2, for 2016, 2017, 2018, and 2019, respectively. On the one hand, these values show the limitations of the adopted model, while on the other hand, in terms of annual availability, these differences are minimal. It is thus expected that the model can provide satisfactory values of direct irradiance availabilities at Évora, and, expectably, at other regions within Portugal.

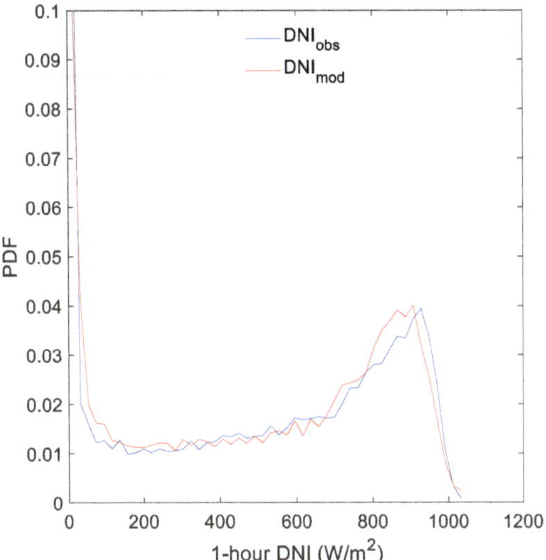

Figure 4. Probability density functions (PDFs) for hourly observed (blue line) and modelled (red line) DNI considering four years (2016 to 2019).

Table 2. Statistical and descriptive analysis summary for hourly measurements and MOD2 using standard error metrics for four consecutive years (2016–2019) of hourly data: mean, median, standard deviation (σ), DNI availability (E_b), and relative difference of DNI availability (ΔE_b).

	2016		2017		2018		2019	
Error metrics	Obs.	MOD2	Obs.	MOD2	Obs.	MOD2	Obs.	MOD2
Mean (W/m^2)	512.91	528.32	549.79	553.24	462.58	465.16	531.38	530.42
Median (W/m^2)	585.44	608.17	640.62	651.76	490.28	492.01	608.71	618.10
σ (W/m^2)	340.88	340.48	336.46	324.92	341.51	332.86	345.34	337.18
E_b (kWh/m^2/year)	2074.36	2135.19	2220.84	2234.65	1862.85	1871.18	2149.16	2144.13
ΔE_b (%)	2.93		0.62		0.45		0.23	

3.2. DNI Estimation

As shown in previous results, 2016 is considered an outlier. Therefore, 3 years (2017, 2018, and 2019) of simultaneous DNI and GHI measurements for Évora are used to perform the calibration of the Engerer2 model parameters. The model parameters, presented in Table 3, show the obtained values in this analysis, as well as the ones obtained from the literature (used for comparison reasons). Despite the differences between obtained and referenced values, which result from the different climatic zones, the parameters lie within reasonable agreement, being thus considered as a validation for the adjusted parameters. It is worth noting that, according to Bright and Engerer [16], no meaningful conclusions can be drawn based on the nature of the parameters alone, as there are some uncertainties related to the (i) clear-sky model and (ii) k_e, which also does not accurately describe situations like cloud enhancement, but rather the relationship of the clear-sky GHI and the measured GHI.

The calibrated parameters can be used to estimate DNI time series in all stations from IPMA's network, since mainland Portugal's climate can be divided into hot dry summers (Csa), cool dry summers (Csb), and a rainy winter throughout the country, according to the Köppen classification [45].

Table 3. Model parameters that result from three years (2017 to 2019) of data measured at the ICT station, in Évora city.

Parameters	Predictor	Present Study	Literature
C	-	−0.0861	−0.0097
β_0	-	−3.7884	−5.0317
β_1	k_t	6.8001	8.5084
β_2	AST	0.0050	0.0132
β_3	Z	−0.0003	0.0074
β_4	Δk_{tc}	−1.9639	−3.0329
β_5	k_e	0.0543	0.5640

Finally, post-processing of DNI time series is performed using the MRM parameters calibrated with 3 years (2017, 2018, and 2019) of modelled DNI, GHI, and DNI under clear-sky conditions, and observed *GHI*, T_a, *RH*, *WS*, *P*, and *Prec*, according to Equation (9). Table A2 in Appendix B shows all significant and possible relations between predictors and predictand (terms), the weight (estimates) that each relation has in the respective adjustment, the standard error (SE) obtained, the t-statistics (tStat), and the *p*-value (*p* Value), for DNI hourly adjustment. The coefficients x_1, x_2, x_3, …, x_8 represent, respectively, modelled *DNI* (x_1), observed *GHI* (x_2), *GHI* (x_3), and *DNI* under clear-sky conditions (x_4), and the observed meteorological variables: T_a (x_5), *RH* (x_6), *WS* (x_7), and *P* (x_8). Precipitation revealed negligible weight as predictor.

3.3. TMY Calculation

For the TMY calculation, days with more than 2 h of gaps, and months with more than 5 days of gaps, were rejected. As previously mentioned (Section 2.4), hourly data from the four nearest stations are used to fill the missing periods using the median of data at the same hour for those stations. It is worth noting that the stations of Viana do Castelo, Leiria/Barosa, and Amareleja 2, were removed from the analysis, since these did not have 5 full years of data, thus not having statistical significance. To minimize data loss, the density of the measuring network is such that stations near the ones removed from the analysis can be used as representative of that specific region.

In Figure 5b, the comparison between the PDFs for three years (from 2017 to 2019) of measured (in purple) and modelled DNI (in red and green), and the one calculated from TMY data (in blue), is shown. The PDFs of modelled DNI at the ICT station (in red) and at the IPMA station at Évora (in green) are shown to provide an easy interpretation of the results. For comparison reasons, the PDFs of the GHI observed at the ICT and IPMA stations are plotted along with the GHI found with the TMY (Figure 5a). Between DNI values of ~750 to 950 W/m², the PDF for observed data is higher than that the one from the model (calculated from TMY). It is noticeable that there was the same behaviour in the PDF plotted with the IPMA's DNI data calculated for 2017 to 2019 against the PDF obtained from the DNI observed at the ICT station during the same period. This means that values in this range are found more frequently for observations than the model data from IPMA. This is somehow the inverse behaviour to that found in Figure 4. For DNI values above 950 W/m², the PDFs of the model from IPMA data are higher than the PDF from observation. Again, there is an opposite behaviour to that seen in Figure 4, which leads to the conclusion that these values are directly related to the values of GHI observed at the IPMA station in Évora, as can be seen in Figure 5a.

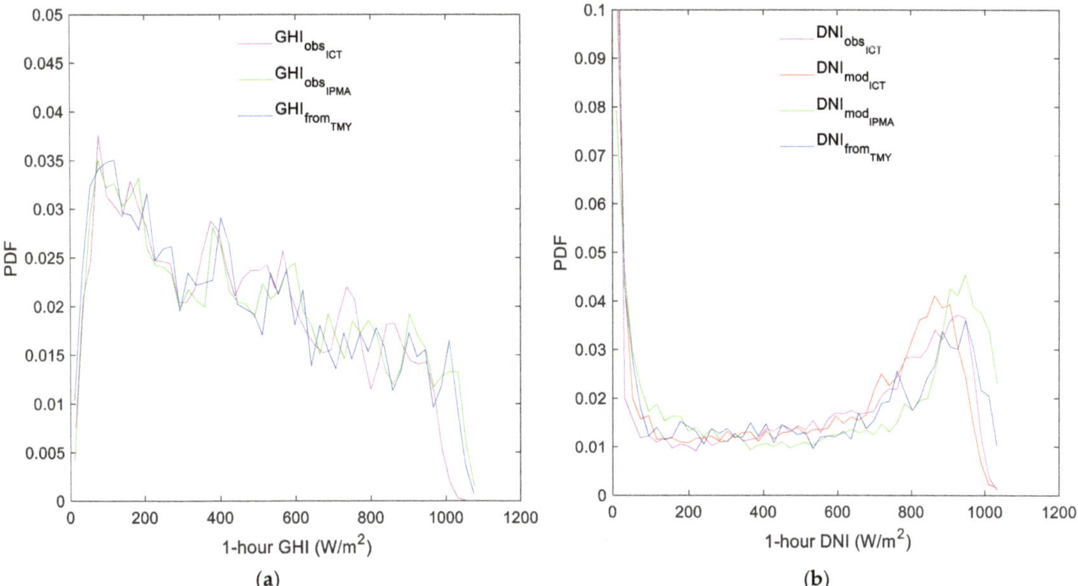

Figure 5. Probability Density Functions (PDFs) for 3 years (2017–2019) and for TMY: (**a**) GHI observed at ICT, IPMA stations and calculated from TMY, using IPMA data; (**b**) DNI observed and modelled at ICT station, DNI modelled at IPMA stations and calculated from TMY, using IPMA data.

3.4. IPMA's Main Stations: DNI Availabilities and CF Estimations

Considering IPMA's main stations, the results of DNI availability (estimated from TMY) are presented in Table 4, where stations are ordered by location, i.e., from the southernmost station to the northernmost one. Additional details regarding all IPMA network stations will be discussed in the next subsection (Section 3.5). By analysing the data presented in Table 4, it is possible to notice that the highest annual DNI and GHI availabilities are verified at Faro and Sagres stations (coastal Southern Portugal) with 2280.66 and 1988.94 kWh/m^2/year, respectively. In contrast, the smallest DNI and GHI availability values were obtained at Penhas Douradas (interior central Eastern Portugal) and Cabo Carvoeiro (coastal central Western Portugal), with 1521.87 and 1625.36 kWh/m^2/year, respectively.

At Faro, DNI availabilities of ~2280.66 kWh/m^2/year are obtained, while observed GHI availabilities are found to be ~1963.54 kWh/m^2/year. At Sines, the modelled DNI availability is ~2024.39 kWh/m^2/year, where GHI availability values are ~1854.89 kWh/m^2/year. In Cavaco et al. [8], DNI average availability was calculated for Sines from 4 years of measured data, having obtained a value of ~2081 kWh/m^2/year. This represents a relative difference of ~2.72%. At Évora, an availability value of ~2030.14 kWh/m^2/year was obtained for the modelled DNI and ~1808.25 kWh/m^2/year for observed GHI. Considering an average DNI availability for the 4 years of measured data used in the model validation, a value of ~2076.80 kWh/m^2/year is found. The difference in annual availability of measured and modelled DNI is ~2.25%, which shows good agreement between the two. Stations at Lisboa/Geofísico and Lisboa/Gago Coutinho (~5.6 km apart) show DNI availabilities of ~1888.81 and 1985.18 kWh/m^2/year, respectively, while respective GHI availabilities are found to be ~1755.44 and 1788.31 kWh/m^2/year. An average availability of DNI and GHI from 5 years of ground-based measurements at Lisboa/LNEG was found to be ~1918 and 1715 kWh/m^2/year, respectively, as shown in Cavaco et al. [8]. The Lisboa/LNEG station is located ~4.4 km from Lisboa/Gago Coutinho station, where differences in DNI and GHI availability between these stations are about ~3.50 and 4.27%, respectively. The high DNI availability value at the Lisboa/Gago Coutinho station is directly related to the high GHI availability value obtained at this station, comparing with the values measured at Lisboa/LNEG. It can be concluded that for the main IPMA stations located in the southern region of Portugal, the modelled DNI availabilities are in good agreement with those presented in Cavaco et al. [8] from observed data. Considering DNI availability obtained in Cabo Carvoeiro, located in interior central western Portugal, a value of ~1521.87 kWh/m^2/year is found for the model. The value obtained by the model is relatively low, considering that the availability of GHI measured is ~1625.36 kWh/m^2/year. It should be mentioned that GHI measurements at this station started in August 2002, with an interruption during February 2006. In this case, measurements were restarted in November 2007, where measured values were shown to be relatively higher. Furthermore, Santa Cruz (a secondary station) is ~26 km from Cabo Carvoeiro, where DNI and GHI availabilities of ~1889.36 and ~1703.69 kWh/m^2/year can be found, respectively. The disagreement between values for Cabo Carvoeiro and Santa Cruz may point to a poor DNI estimation at Cabo Carvoeiro, which may result from a poor source of quality data for GHI, thus contributing to a lower performance of the model output. At Castelo Branco, the DNI and GHI availabilities are ~2078.07 and ~1769.59 kWh/m^2/year, respectively. Castelo Branco is a station located in the interior in the central eastern region of Portugal. Penhas Douradas station has DNI and GHI availability values of ~1467.85 and 1672.30 kWh/m^2/year, respectively. The Aldeia Souto (secondary) station is located about 16 km east of Penhas Douradas, where it is possible to observe DNI and GHI availabilities of ~1825 and 1648 kWh/m^2/year, respectively. The Covilhã (secondary) station is located about 18 km south of Penhas Douradas, where availabilities of ~2091 and 1739 kWh/m^2/year for DNI and GHI are found, respectively. This seems to point to the fact that DNI estimations for Penhas Douradas station are anomalously low for that region. Still, this station is part of the Serra da Estrela natural park, and can be affected by local weather effects like the formation of clouds. At Porto, the DNI availability obtained with the model is ~1624.47 kWh/m^2/year and the GHI availability is ~1603.49 kWh/m^2/year. At Porto/S. gens, a secondary station that is at a distance of about 6 km from Porto station, the DNI and GHI availability is found to be ~1630.00 and ~1543.47 kWh/m^2/year. Both values are consistent and reveal lower levels of solar radiation in this region (i.e., in the northern part of the country and on the coast). Finally, at Bragança, the DNI and GHI availabilities are found to be ~1847.01 and ~1648.14 kWh/m^2/year, respectively.

Table 4. Estimated DNI availability, GHI availability, and capacity factor (CF), from IPMA data (IPMA main stations).

Station Names	E_b (kWh/m^2/year)	E_g (kWh/m^2/year)	CF (%)
Sagres	2245.30	1988.94	35.2
Faro	2280.66	1963.54	36.2
Sines	2024.39	1854.89	31.9
Beja	2145.45	1878.65	31.8
Évora	2030.14	1808.25	30.3
Lisboa/Geofísico	1888.81	1755.44	29.1
Lisboa/Gago Coutinho	1985.18	1788.31	30.1
Portalegre	1875.37	1698.47	27.4
Cabo Carvoeiro	1521.87	1625.36	22.8
Castelo Branco	2078.07	1769.59	31.2
Coimbra	1809.43	1645.21	25.7
Penhas Douradas	1467.85	1672.30	19.4
Viseu	1786.90	1622.90	26.1
Porto	1624.47	1603.49	24.2
Vila Real	1603.01	1561.46	23.6
Bragança	1847.01	1648.14	28.3

In Table 4, CF values (estimated from the SAM) for the main IPMA stations are presented. CF values higher than 30% are considered as benchmark values for the installation of CSP plants in a region of interest. Such a benchmark comes naturally from the fact that stations with higher CF have higher production capabilities, making them more suitable for bankable CSP projects. In this regard, the Faro station, which is a station located in a coastal area in the southeastern region of the country, depicts a CF of ~36.2%. This is the highest value for Portugal, making it a region of particular interest for the implementation of CSP power plants. Moreover, Sines is located in the southwestern coastal region, with a CF value of ~31.9%. Regarding the CF value at Évora, ~30.3% is obtained. CF values of ~29.1 and 30.1% are obtained at Lisboa/Geofísico and Lisboa/Gago Coutinho, respectively. Both these stations are in the southwestern region of Portugal. CF values of ~22.8% are found at Cabo Carvoeiro. This low value results from the low DNI availabilities estimated for this region, as previously mentioned. As a matter of fact, the CF value from Santa Cruz station (~28%), being located about 26 km away from Cabo Carvoeiro, tends to indicate that the value for Cabo Carvoeiro is anomalously low. At Castelo Branco, a CF of ~31.2% is obtained, while in Penhas Douradas station the CF is found to be ~19.4%, which is directly related to the low value of DNI availability, as previously discussed. Stations like Aldeia Souto and Covilhã, located ~16 and 18 km away from Penhas Douradas, have CF values of ~26.9 and 30.6%, respectively. The CF values calculated at Porto and Porto/S. gens were ~24.2 and 23.7%, respectively, while Bragança (located in the northeastern region) shows a CF of ~28.3%.

3.5. PMA Network: Assessment of Solar Availability

In Table A1 of Appendix A, a summary can be found for the DNI and GHI availabilities, and for the CF, found for all the stations within the IPMA network. Regarding DNI and GHI availabilities, it is possible to find DNI values between ~1263 kWh/m^2/year at Montalegre in the interior northwestern, and ~2310 kWh/m^2/year at Castro Marim in the interior southeastern. Regarding GHI availabilities, the values range between ~1383 kWh/m^2/year at Montalegre in the interior northwestern area, and ~1989 kWh/m^2/year at Sagres in the coastal southwestern area, the southernmost station of the IPMA network. In fact, there is a clear tendency for the increase of DNI and GHI from north to south and from west to east, as further discussed.

In Figure 6a, a GHI availability map is shown, being calculated from the TMY with the observed values. Considering observed GHI availability, the southern region of Portugal depicts higher values. These range from ~1725 kWh/m^2/year in Viana do Alentejo to ~1989 kWh/m^2/year in Sagres. In the eastern region, values range between ~1650 and 1780 kWh/m^2/year. In the north and west, values range between ~1383 and 1700 kWh/m^2/year. Similar ranges of GHI availability values can be found in ref. [46]. For the case of DNI availability maps with values estimated from IPMA data, presented in Figure 6b, in the southern region with the most sun exposure, values range between ~1800 and 2310 kWh/m^2/year. More to the eastern side, in the Beira Interior region, values range between ~1400 and 2100 kWh/m^2/year. Regarding the northern area, values range between ~1200 and 1847 kWh/m^2/year. In the western area, values of annual DNI availability ranging between ~1400 and ~1900 kWh/m^2/year can be found. These large ranges of values for the estimated DNI availability within the same region are a consequence of observed GHI spatial variation, which may be affected by local meteorological conditions of the areas where the stations are located or by errors that may result from measuring inaccuracies. It is worth noting the DNI availability values above 2000 kWh/m^2/year in the interior central eastern region of Portugal (Beira Interior region). Beira Interior has very particular climatic nuances, with cold to very cold winters and moderate to very hot summers [47].

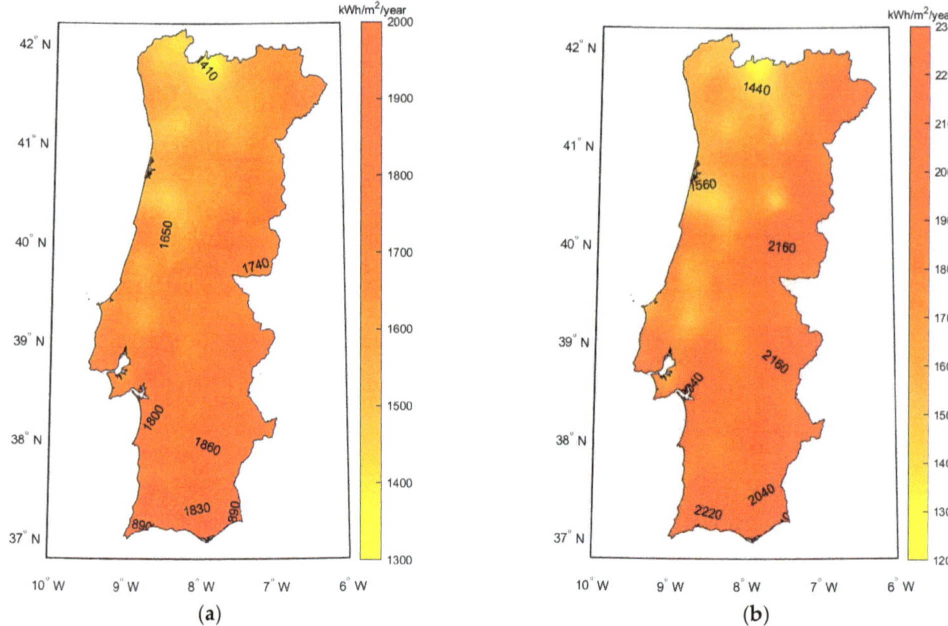

Figure 6. Availability maps (kWh/m²/year) calculated from the TMY (%) estimated from IPMA stations: (**a**) GHI, (**b**) DNI.

3.6. IPMA Network: Production Capacity of CSP Plants

In this work, a case study is considered focusing an operational CSP facility with a similar configuration to Andasol3. Figure 7 shows the Andasol3 solar power plant. Finally, regarding the obtained CF, illustrated in Figure 8, values range between ~17.8%, at Montalegre in the northwest, and 36.2%, at Faro. There is a general increase from north to the south and from west to east in Portugal, following the previously shown DNI and GHI availability values, as expected. Chaves station is 29 km from Montalegre and a CF value of 23.2% was found. The low CF value in Montalegre is directly related to the local GHI measurement. It is also worth mentioning that the CF value in Penhas Douradas, 19.4%, is low compared to other stations in the Beira Interior region (e.g., Covilhã, 30.6%). Anomalies have already been identified in the GHI measurements at Penhas Douradas and this is reflected in the DNI estimation and, consequently, in the CF value. The regions of Beira Interior (that includes the Castelo Branco and Covilhã stations), Alentejo (a region on the South of Portugal that encompasses Portalegre, Sines, Évora and Beja stations), and Algarve (southernmost region that comprises Sagres and Faro), which had already been highlighted by its DNI availability values, shows high CF levels, reaching values higher than 30%.

Figure 7. Andasol3 power plant located in Aldeire (Granada, Spain). All rights reserved to ©
Google Earth.

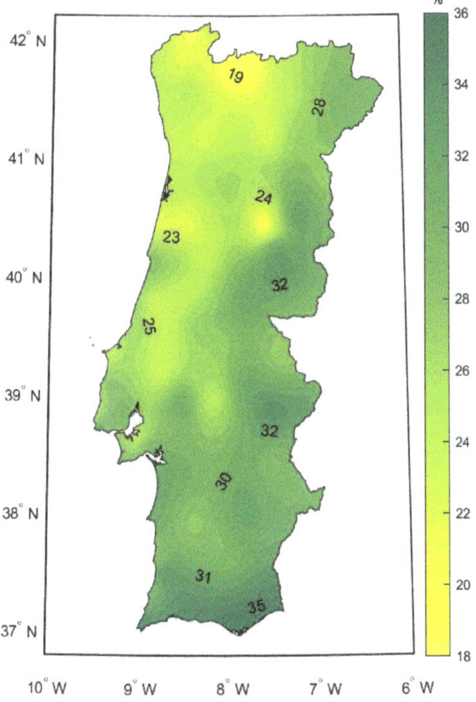

Figure 8. Capacity factor (CF) map for continental Portugal calculated from the TMY meteorological
data obtained through the IPMA stations.

4. Implications for Decision Making

It is worth noting that Portugal has a national plan for energy and climate (*Roteiro para a Neutralidade Carbono*—RNC2050, as proposed through the European Commission [48]) towards carbon neutrality until 2050. The plan sets an ambitious target that aims at a strong implementation of renewable energies in the energy sector to reach an energy mix with almost exclusively renewables. According to this document, CSP should act as an anchor for the irregularity of wind and PV, the latter only being produced during the day. This is because cheap storage is achievable for CSP technologies, allowing the storage to be sized in such way as to produce energy for almost 24 h a day. Additionally, CSP can be hybridized with PV such that PV produces energy during the day with no storage and CSP stores energy during the day for night use. Furthermore, in the Portuguese energy sector, one of the largest contributions of renewables comes from hydroelectric plants, 23% of the total 59% in 2021 [49], which is lower during summer months and will tend to decrease if the tendency for a reduction in precipitation is confirmed in the future. Thus, the implementation of CSP technology in Portugal, as shown in this study, can be complementary to other renewable sources, providing its contribution to reach important objectives of the RNC2050. The present work provides concrete indications for regions of great interest for the installation of CSP systems, such as Beira Interior, Alentejo, and Algarve. This contribution is based on the availability values of DNI, and CF, presented in this analysis. It is planned that a more in-depth study, to be carried out in the future, will enable the introduction of other factors, such as proximity to the electric grid, road access, and soiling, that influence the feasibility of the project.

In such a context, according to the report published by National Energy Grids (*Redes Energéticas Nacionais*—REN), in 2021 the power consumption of Portugal was ~49.5 TWh, from which 59% was produced by renewable energy [49]. This means that, to reach the goals of RNC2050, the remaining 41% of energy must be produced by renewable sources before 2050. Thus, to have an idea of how CSP could contribute to that objective, let us assume a hypothetical scenario that all this energy is produced from CSP plants installed in the most suitable region of Portugal; according to CF values, that is Faro; this means that approximately 128 plants like Andasol3 would have to be deployed in this region, occupying an area of ~65.3 km^2. In fact, the actual number of CSP plants in Spain is roughly 51, this means that the hypothetical scenario would have 2.51 times more CSP power plants than the number of plants already operating in Spain now. This reveals that such a scenario for a 100% renewable energy mix, may be difficult to reach, but it is not definitely unreachable, especially if new technologies like direct molten-salt CSP plants [30,50] reach the market development with higher efficiencies at lower level energy costs.

5. Conclusions

In this work, an evaluation concerning the production capacity of CSP plants is carried out by estimating CF values. The CF assessment is performed using the SAM software, which requires as input meteorological variables. The main one being DNI, which was modelled from a GHI, measured in a national meteorological network, using the Engerer2 model and MRM, both evaluated and calibrated for a benchmark station located at Évora (Southern Portugal). Additionally, DNI availability maps based on the regional TMY were also produced at different ground stations. The TMY were further used in simulations of an Andasol3-like power plant. From such simulations, CF values were obtained. In summary, the main conclusions of this work can be stated as follows:

- DNI values modelled from GHI with the use of Engerer2 show very good agreement with the observations;
- GHI and DNI annual availabilities estimated for the IPMA network show very good accordance with previous values found in the literature;
- Annual DNI availabilities and CFs were found to be as high as ~2310 kWh/m^2 and ~36.2% in Castro Marim and in Faro cities in Algarve, respectively.

- DNI availability and CF mapping showed the existence of three preferential regions for CSP installation: two in Southern Portugal—Alentejo and Algarve regions; and one in eastern central Portugal—Beira Interior region);

A more detailed regional study is planned, which should include the information of other factors, such as proximity to the electric grid, road access, and soiling. Finally, it is important to highlight that the method used in this work is not restricted to Portugal, and can be applied, as a tool to support decision making at similar regions worldwide that have a high potential for CSP.

Author Contributions: The concept of this work was formed by A.M.T., R.C., F.M.L. and H.G.S. The applied methodology was implemented by A.M.T., R.C., F.M.L. and H.G.S. The use of the SAM software was carried out by A.M.T. and F.M.L. Analysis, validation, investigation, resources, data curation were made by A.M.T. The original draft preparation was written and editing by A.M.T. and H.G.S. All authors have read and agreed to the published version of the manuscript.

Funding: This work is supported by national funding awarded by FCT—Foundation for Science and Technology, I.P., projects ICT—UIDB/04683/2020 and LAETA—UIDB/50022/2020. A.M. Tavares acknowledges the PhD grants *Programa de Bolsas CAMÕES I.P./MILLENNIUM*. F.M. Lopes acknowledges FCiênciasID for the postdoctoral research grants (PTDC/CTA-MET/28946/2017 and H2020-IBA-SPACE-CHE2-2019—reference 958927). R. Conceição acknowledges the European Union's Horizon 2020 research and innovation programme under grant agreement N° 823802 (SFERA-III) and wishes to thank to "Comunidad de Madrid" and European Structural Funds for their financial support to ACES2030-CM project (S2018/EMT-4319). H.G. Silva acknowledges the Physics Department of the University of Évora for supporting his work.

Institutional Review Board Statement: Not applicable.

Informed Consent Statement: Not applicable.

Acknowledgments: A.M. Tavares ICT to provide good working conditions and for making available the solar irradiance data. F.M. Lopes acknowledges IDL and IPMA. Finally, the authors are grateful to: Thomas Fasquelle for the development of SAM's Andasol 3 script; IPMA, particularly to Jorge Neto, for providing the meteorological data; P. Canhoto and E. Abreu, for the maintenance of solar radiation measurements at the ICT station in Évora.

Conflicts of Interest: The authors declare no conflict of interest.

Nomenclature

β_n	Coefficients of Engerer model
ΔE_b	Relative difference of DNI availability [%]
Δk_{tc}	Deviation between the observed value of k_t at surface and the one obtained under clear sky [dimensionless]
σ	standard deviation
AST	Apparent solar time [h]
C	Lower asymptote [dimensionless]
CF	Capacity factor [%]
DHI	Diffuse Horizontal Irradiance [W/m^2]
DNI	Direct Normal Irradiance [W/m^2]
DNI_{cs}	Direct Normal Irradiance at clear-sky [W/m^2]
GHI	Global Horizontal Irradiance [W/m^2]
GHI_{cs}	Global Horizontal Irradiance at clear-sky [W/m^2]
DNI_i^{ADJ}	adjusted modelled DNI [W/m^2]
E_b	DNI availability [kWh/m^2/year]
E_g	GHI availability [kWh/m^2/year]
Elev	Elevation of each station [m]
E_{0n}	Irradiance at the top of the atmosphere
K	diffuse fraction [dimensionless]

k_e	portion of K attributed in case of cloud enhancement [dimensionless]
k_t	Clearness index [dimensionless]
k_{tc}	Clearness index at clear-sky [dimensionless]
MAE	Mean Absolute Error [W/m^2]
MBE	Mean Bias Error [W/m^2]
P	Atmospheric pressure [hPa]
$Prec$	Precipitation [mm]
r	correlation coefficient [dimensionless]
R^2	coefficient of determination [dimensionless]
RH	Relative Humidity [%]
RMSE	Root Mean Square Error [W/m^2]
T_a	Ambient Temperature [°C]
WS	Wind Speed [m/s]
Z	Zenith angle [°]
CDF	Cumulative Frequency Distribution Function
CSP	Concentrating Solar Power
FS	Finkelstein-Schafer statistic
IPMA	Instituto Português do Mar e da Atmosfera
MRM	Multivariate Regression Model
SAM	System Advisor model
TMM	Typical Meteorological Month
TMY	Typical Meteorological Year

Appendix A

Table A1. List of all IPMA stations (latitude, longitude, elevation, years of available data, annual DNI and GHI availabilities, capacity factor, and annual power generation).

Station Names	Latitude (°N)	Longitude (°W)	Elevation (m)	Period of Data (Years)	E_b (kWh/m^2/year)	E_g (kWh/m^2/year)	CF (%)	Annual Power Generation (GWh/year)
Cabo Carvoeiro/Farol	39.36	−9.41	32.00	9.01	1521.87	1625.36	22.8	99.79
Sagres/Quartel da Marinha	37.01	−8.95	22.85	10.02	2245.30	1988.94	35.2	154.35
Lisboa/Geofísico	38.72	−9.15	77.00	13.30	1888.81	1755.44	29.1	127.50
Sines/Monte Chãos	37.95	−8.84	103.00	19.72	2024.39	1854.89	31.9	139.87
Viana do Castelo/Meadela	41.71	−8.80	16.00	4.79	-	-	-	-
Porto/Pedras Rubras	41.23	−8.68	69.00	17.63	1624.47	1603.49	24.2	105.95
Coimbra/Aeródromo	40.16	−8.47	171.00	17.01	1809.43	1645.21	25.7	112.64
Faro/Aeroporto	37.02	−7.97	5.00	16.96	2280.66	1963.54	36.2	158.71
Évora/Aeródromo	38.54	−7.89	247.56	13.21	2030.14	1808.25	30.3	132.53
Viseu/Aeródromo	40.71	−7.90	644.37	13.97	1786.90	1622.90	26.1	114.49
Beja	38.03	−7.87	246.00	17.02	2145.45	1878.65	31.8	139.48
Vila Real/Aeródromo	41.27	−7.72	561.00	19.92	1603.01	1561.46	23.6	103.28
Penhas Douradas/Observatório	40.41	−7.56	1380.00	14.80	1467.85	1672.30	19.4	84.98
Castelo Branco	39.84	−7.48	386.00	11.83	2078.07	1769.59	31.2	136.84
Portalegre	39.29	−7.42	597.00	17.07	1875.37	1698.47	27.4	120.20
Bragança	41.80	−6.74	690.00	17.57	1847.01	1648.14	28.3	123.87
Lisboa/Gago Coutinho	38.77	−9.13	103.88	17.14	1985.18	1788.31	30.1	131.73
Odemira/S.Teotónio	37.55	−8.73	120.54	15.92	2051.54	1825.73	32.4	141.95
Vila Nova de Cerveira/Aeródromo	41.97	−8.68	34.00	15.30	1618.96	1532.58	23	100.88
Monção/Valinha	42.07	−8.38	80.00	15.14	1555.61	1466.69	20.6	90.43
Lamas de Mouro	42.04	−8.20	880.00	15.12	1553.14	1464.82	22.5	98.48
Montalegre	41.82	−7.79	1005.00	15.78	1262.83	1383.36	17.8	77.85
Ponte de Lima/Escola Agrícola	41.76	−8.57	40.00	17.20	1538.06	1518.47	22.1	97.01
Chaves/Aeródromo	41.73	−7.47	360.00	17.97	1627.84	1568.18	23.2	101.46
Cabril/S. Lourenço	41.71	−8.03	585.00	6.70	1460.30	1458.28	19.9	87.27
Braga/Merelim	41.58	−8.45	68.35	16.30	1725.64	1571.64	25.1	109.81
Cabeceiras de Basto	41.49	−7.98	350.00	16.99	1549.57	1521.93	22.3	97.81
Mirandela	41.51	−7.19	250.00	13.33	1704.56	1597.46	25.3	110.66
Macedo de Cavaleiros/Izeda-Morais	41.57	−6.79	702.00	13.51	1991.53	1679.19	29	126.92
Miranda do Douro	41.50	−6.27	693.00	12.72	1992.55	1709.17	29.2	127.86

Table A1. *Cont.*

Station Names	Latitude (°N)	Longitude (°W)	Elevation (m)	Period of Data (Years)	E_b (kWh/m^2/year)	E_g (kWh/m^2/year)	CF (%)	Annual Power Generation (GWh/year)
Mogadouro	41.34	−6.73	644.00	17.08	1973.36	1685.47	29.3	128.30
Carrazêda de Ansiães	41.24	−7.30	715.00	16.97	1665.29	1574.04	26	113.84
Porto/S.Gens	41.18	−8.64	89.19	6.81	1630.00	1543.47	23.7	103.68
Moncorvo	41.19	−7.02	600.00	17.04	1889.45	1662.77	27.8	121.68
Pinhão	41.17	−7.55	130.00	9.92	1508.15	1535.83	22	96.52
Luzim	41.15	−8.25	287.17	9.39	1510.18	1492.18	22.3	97.71
Moimenta da Beira	40.99	−7.60	715.00	16.80	1776.39	1633.02	25.7	112.72
Trancoso/Bandarra	40.78	−7.35	840.00	15.89	1931.57	1692.99	28.6	125.06
Arouca	40.93	−8.26	270.00	5.88	1775.57	1612.02	25.5	111.64
Figueira de Castelo Rodrigo/V.Torpim	40.83	−6.94	635.00	16.23	1843.46	1653.39	27.5	120.40
Guarda	40.53	−7.28	1020.00	16.13	2111.23	1734.55	31.1	136.01
Nelas	40.52	−7.86	425.00	16.07	1745.60	1580.33	25.3	110.88
Pampilhosa da Serra	40.15	−7.93	835.59	13.55	1918.11	1664.43	27.5	120.49
Covilhã	40.26	−7.48	482.00	14.41	2090.86	1738.84	30.6	133.96
Aldeia Souto/Quinta Lageosa	40.35	−7.39	468.00	7.93	1824.67	1647.52	26.9	117.68
Lousã/Aeródromo	40.14	−8.24	193.77	11.85	1608.49	1555.31	23.4	102.40
Aveiro/Universidade	40.64	−8.66	5.00	17.79	1607.61	1586.50	23.7	103.92
Dunas de Mira	40.45	−8.76	14.00	7.57	1510.74	1535.10	21.3	93.42
Anadia/Estação Vitivinícola da Bairrada	40.44	−8.44	45.00	17.48	1399.34	1487.32	21.4	93.54
Coimbra/Bencanta	40.21	−8.46	35.00	10.22	1691.03	1593.23	24.7	108.26
Figueira da Foz/Vila Verde	40.14	−8.81	4.00	16.50	1963.77	1717.17	28.8	126.00
Ansião	39.90	−8.41	396.24	15.01	1721.54	1612.64	24	105.13
Leiria/Aeródromo	39.78	−8.82	45.00	11.38	1573.39	1567.91	23.2	101.81
Leiria/Barosa	39.75	−8.83	24.00	4.14	-	-	-	-
São Pedro de Moel	39.76	−9.03	40.00	9.15	1917.07	1675.57	27.3	119.62
Tomar/Vale Donas	39.59	−8.37	75.42	15.94	1860.74	1710.12	27.4	119.81
Alcobaça/Estação Fruticultura Vieira Natividade	39.55	−8.97	38.00	16.66	1781.07	1669.28	26.5	116.08
Rio Maior/ETAR	39.31	−8.92	52.83	14.55	1692.12	1648.63	25.1	110.04
Santarém	39.20	−8.74	71.91	17.04	1522.67	1554.56	22.9	100.17
Torres Vedras/Dois Portos	39.04	−9.18	110.00	16.05	1924.50	1726.04	28.3	123.76
Coruche/Estação de Regadio (INIA)	38.94	−8.51	18.75	16.15	2115.38	1825.18	31.2	136.68
Santa Cruz/Aeródromo	39.13	−9.38	40.71	8.81	1889.36	1703.69	28	122.52
Cabo da Roca	38.78	−9.50	141.23	6.68	1893.25	1695.90	26.4	115.59
Lisboa/Tapada da Ajuda	38.71	−9.18	69.96	8.90	1861.28	1722.95	27.9	122.40
Cabo Raso/Farol	38.71	−9.49	7.88	9.63	2218.63	1847.57	34.1	149.16
Barreiro/Lavradio	38.67	−9.05	6.00	16.04	1604.73	1649.61	23.4	102.37
Pegões	38.65	−8.64	64.00	6.80	2044.33	1810.87	28.9	126.75
Setúbal/Estação de Fruticultura	38.55	−8.89	35.00	16.06	1963.72	1769.06	29.9	130.77
Almada/Praia da Rainha	38.62	−9.21	5.51	16.63	1885.80	1743.98	28.6	125.25
Alcácer do Sal—Barrosinha	38.36	−8.48	29.00	16.49	1999.84	1801.29	30.5	133.69
Alvalade	37.95	−8.39	46.97	15.37	1838.25	1740.06	27.7	121.47
Zambujeira	37.58	−8.74	67.00	9.79	1901.81	1769.89	29.9	131.05
Aljezur	37.33	−8.80	11.95	13.79	1931.27	1787.92	30.3	132.60
Foía	37.31	−8.60	895.30	8.67	2213.31	1809.38	32.8	143.55
Sabugal/Martim Rei	40.34	−7.04	858.00	14.39	2067.15	1749.23	30.6	133.82
Zebreira	39.85	−7.07	374.00	15.48	1963.11	1742.58	29.8	130.43
Proença-a-Nova/Moitas	39.73	−7.87	379.00	14.73	2069.83	1751.58	31.1	136.19
Alvega	39.46	−8.03	51.05	15.87	1836.80	1693.62	27.6	120.75
Avis/Benavila	39.11	−7.88	152.25	16.30	1986.54	1753.16	30.6	133.97
Mora	38.94	−8.16	129.45	6.41	1730.08	1670.45	24.3	106.43
Elvas/Est. Melhoramento Plantas	38.89	−7.14	209.97	16.71	2127.34	1815.76	32.7	143.33
Estremoz/Techocas	38.86	−7.51	366.00	16.31	2198.18	1845.04	33.6	147.34
Reguengos/S.Pedro do Corval	38.48	−7.47	265.17	9.39	1905.44	1744.70	29	126.94
Viana do Alentejo	38.33	−8.05	202.00	8.66	1890.40	1724.96	29.9	130.93
Amareleja	38.21	−7.21	192.00	9.45	1996.22	1798.38	29.8	130.74
Amareleja2	38.20	−7.23	180.00	4.58	-	-	-	-
Mértola/Vale Formoso	37.76	−7.55	190.00	15.54	2044.28	1802.03	30.9	135.48
Castro Verde/Neves Corvo	37.58	−7.97	225.00	17.13	1882.23	1779.48	29.4	128.73
Castro Marim/Reserva Nacional do Sapal	37.23	−7.43	4.83	16.57	2310.08	1912.79	35.6	155.96
Portimão/Aeródromo	37.15	−8.58	2.00	16.19	2196.05	1882.81	34.1	149.24

Appendix B

Table A2. Hourly multivariate regression metrics obtained with the stepwise function considering a third-degree polynomial adjustment of DNI (MRM) for four years (from 2017 to 2019).

	DNI Adjusted			
Terms	**Estimates**	**SE**	**tStat**	***p* Value**
(Intercept)	0	0	-	-
x_1	-4.28	2.72	-1.57	0.11
x_2	25.26	6.68	3.78	1.55×10^{-4}
x_3	6.11	5.01	1.22	0.22
x_4	1.37	0.66	2.07	0.04
x_5	-16.78	40.65	-0.41	0.68
x_6	-10.35	7.46	-1.39	0.17
x_7	-5.80	6.18	-0.94	0.35
x_8	0.13	0.15	0.81	0.42
$x_1 \times x_2$	-1.16×10^{-3}	5.53×10^{-3}	-0.21	0.83
$x_1 \times x_3$	6.62×10^{-3}	2.11×10^{-3}	3.13	1.74×10^{-3}
$x_1 \times x_4$	1.08×10^{-3}	6.29×10^{-4}	1.72	0.09
$x_1 \times x_5$	0.08	0.10	0.79	0.43
$x_1 \times x_6$	-5.37×10^{-3}	4.36×10^{-3}	-1.23	0.22
$x_1 \times x_7$	-0.02	7.71×10^{-3}	-2.51	0.01
$x_1 \times x_8$	4.16×10^{-3}	2.65×10^{-3}	1.57	0.12
$x_2 \times x_3$	-0.01	9.91×10^{-3}	-1.43	0.15
$x_2 \times x_4$	-9.85×10^{-3}	2.15×10^{-3}	-4.59	4.47×10^{-6}
$x_2 \times x_5$	-0.65691	0.134572	-4.88	1.07×10^{-6}
$x_2 \times x_6$	0.03	0.01	2.41	0.01
$x_2 \times x_7$	0.04	0.01	3.28	1.05×10^{-3}
$x_2 \times x_8$	-0.02	6.43×10^{-3}	-3.49	4.88×10^{-4}
$x_3 \times x_4$	7.81×10^{-3}	2.50×10^{-3}	3.13	1.78×10^{-3}
$x_3 \times x_5$	-0.01	0.02	-0.52	0.61
$x_3 \times x_6$	-0.12	0.03	-4.46	8.20×10^{-6}
$x_3 \times x_7$	-0.01	0.01	-0.87	0.38
$x_3 \times x_8$	-6.17×10^{-3}	4.75×10^{-3}	-1.30	0.19
$x_4 \times x_5$	0.01	0.01	0.96	0.34
$x_4 \times x_6$	0.01	6.33×10^{-3}	2.38	0.02
$x_4 \times x_7$	$4.52 \times 10{-03}$	0.01	0.41	0.68
$x_5 \times x_6$	0.18	0.06	3.26	1.10×10^{-3}
$x_5 \times x_7$	0.50	0.23	2.20	0.03
$x_5 \times x_8$	4.00×10^{-3}	0.04	0.10	0.92
$x_6 \times x_8$	-2.91×10^{-3}	6.98×10^{-3}	-0.42	0.68
x_1^2	-5.81×10^{-3}	3.11×10^{-3}	-1.87	0.06
x_2^2	4.90×10^{-3}	3.23×10^{-3}	1.52	0.13
x_3^2	-1.84×10^{-3}	6.33×10^{-3}	-0.29	0.77
x_4^2	-5.21×10^{-3}	1.08×10^{-3}	-4.81	1.55×10^{-6}
x_6^2	0.15	0.03	4.70	2.66×10^{-6}
x_7^2	0.87	0.64	1.35	0.18
$x_1 \times x_2 \times x_3$	7.32×10^{-5}	8.25×10^{-6}	8.87	8.44×10^{-19}
$x_1 \times x_2 \times x_4$	-9.92×10^{-6}	1.97×10^{-6}	-5.03	5.00×10^{-7}
$x_1 \times x_2 \times x_5$	1.39×10^{-4}	3.91×10^{-5}	3.57	3.63×10^{-4}
$x_1 \times x_2 \times x_6$	1.49×10^{-5}	1.10×10^{-5}	1.36	0.17
$x_1 \times x_2 \times x_7$	-8.68×10^{-5}	4.48×10^{-5}	-1.94	0.05
$x_1 \times x_2 \times x_8$	-8.43×10^{-6}	4.68×10^{-6}	-1.80	0.07
$x_1 \times x_3 \times x_4$	-7.57×10^{-7}	2.60×10^{-6}	-0.29	0.77
$x_1 \times x_3 \times x_5$	-1.64×10^{-5}	3.42×10^{-5}	-0.48	0.63
$x_1 \times x_3 \times x_6$	1.77×10^{-5}	1.03×10^{-5}	1.73	0.08
$x_1 \times x_3 \times x_7$	8.08×10^{-5}	4.91×10^{-5}	1.64	0.10

Table A2. *Cont.*

	DNI Adjusted			
Terms	Estimates	SE	tStat	*p* Value
$x_1 \times x_4 \times x_5$	-3.81×10^{-5}	1.19×10^{-5}	-3.20	0.00
$x_1 \times x_4 \times x_6$	-4.69×10^{-6}	3.70×10^{-6}	-1.27	0.20
$x_1 \times x_4 \times x_7$	-2.97×10^{-5}	1.53×10^{-5}	-1.94	0.05
$x_1 \times x_5 \times x_6$	1.63×10^{-4}	5.95×10^{-5}	2.74	0.01
$x_1 \times x_5 \times x_8$	-6.10×10^{-5}	1.02×10^{-4}	-0.60	0.55
$x_2 \times x_3 \times x_4$	-7.39×10^{-6}	4.34×10^{-6}	-1.70	0.09
$x_2 \times x_3 \times x_5$	0.000125	5.71×10^{-5}	2.19	0.03
$x_2 \times x_3 \times x_6$	5.08×10^{-5}	1.48×10^{-5}	3.42	0.00
$x_2 \times x_3 \times x_8$	1.16×10^{-5}	9.08×10^{-6}	1.27	0.20
$x_2 \times x_4 \times x_6$	-2.23×10^{-5}	9.92×10^{-6}	-2.25	0.02
$x_2 \times x_5 \times x_6$	-2.89×10^{-4}	1.69×10^{-4}	-1.71	0.09
$x_2 \times x_5 \times x_7$	2.92×10^{-4}	5.27×10^{-4}	0.55	0.58
$x_2 \times x_5 \times x_8$	6.50×10^{-4}	1.31×10^{-4}	4.95	7.59×10^{-7}
$x_3 \times x_4 \times x_5$	-3.13×10^{-5}	2.83×10^{-5}	-1.10	0.27
$x_3 \times x_4 \times x_6$	1.12×10^{-5}	1.10×10^{-5}	1.02	0.31
$x_3 \times x_5 \times x_6$	4.54×10^{-4}	1.49×10^{-4}	3.03	0.00
$x_3 \times x_5 \times x_7$	-1.57×10^{-4}	6.05×10^{-4}	-0.26	0.80
$x_3 \times x_6 \times x_8$	9.31×10^{-5}	2.44×10^{-5}	3.82	0.00
$x_4 \times x_5 \times x_6$	-4.07×10^{-4}	8.32×10^{-5}	-4.90	9.80×10^{-7}
$x_4 \times x_5 \times x_7$	-1.38×10^{-3}	4.91×10^{-4}	-2.82	4.81×10^{-3}
$x_1^2 \times x_2$	2.47×10^{-5}	2.21×10^{-6}	11.19	6.57×10^{-29}
$x_1^2 \times x_3$	-1.24×10^{-5}	2.28×10^{-6}	-5.43	5.64×10^{-8}
$x_1^2 \times x_4$	8.43×10^{-7}	4.61×10^{-7}	1.83	0.07
$x_1^2 \times x_5$	-4.44×10^{-5}	9.31×10^{-6}	-4.77	1.89×10^{-6}
$x_1^2 \times x_6$	-6.60×10^{-6}	2.75×10^{-6}	-2.40	0.02
$x_1^2 \times x_7$	2.02×10^{-5}	1.34×10^{-5}	1.50	0.13
$x_1^2 \times x_8$	8.80×10^{-6}	2.99×10^{-6}	2.94	3.25×10^{-3}
$x_1 \times x_2^2$	-5.43×10^{-5}	4.34×10^{-6}	-12.50	1.31×10^{-35}
$x_1 \times x_3^2$	-2.90×10^{-5}	4.64×10^{-6}	-6.25	4.25×10^{-10}
$x_1 \times x_4^2$	1.12×10^{-6}	5.84×10^{-7}	1.92	0.05
$x_1 \times x_6^2$	3.51×10^{-5}	1.92×10^{-5}	1.83	0.07
$x_2^2 \times x_3$	-2.15×10^{-5}	3.10×10^{-6}	-6.95	3.92×10^{-12}
$x_2^2 \times x_4$	1.20×10^{-5}	3.46×10^{-6}	3.46	5.42×10^{-4}
$x_2^2 \times x_5$	-1.55×10^{-4}	4.62×10^{-5}	-3.36	7.86×10^{-4}
$x_2^2 \times x_6$	-5.26×10^{-5}	1.19×10^{-5}	-4.43	9.60×10^{-6}
$x_2 \times x_3^2$	6.95×10^{-6}	1.78×10^{-6}	3.90	9.58×10^{-5}
$x_2 \times x_4^2$	7.50×10^{-6}	1.94×10^{-6}	3.86	1.14×10^{-4}
$x_2 \times x_6^2$	-0.00018	5.10×10^{-5}	-3.44	5.89×10^{-4}
$x_3^2 \times x_4$	8.26×10^{-6}	2.04×10^{-6}	4.06	4.95×10^{-5}
$x_3^2 \times x_5$	-1.78×10^{-5}	2.31×10^{-5}	-0.77	0.44
$x_3^2 \times x_6$	-1.39×10^{-5}	6.21×10^{-6}	-2.24	0.03
$x_3^2 \times x_8$	-5.75×10^{-7}	6.09×10^{-6}	-0.09	0.92
$x_3 \times x_4^2$	-1.03×10^{-5}	2.17×10^{-6}	-4.76	1.99×10^{-6}
$x_3 \times x_6^2$	1.40×10^{-4}	4.45×10^{-5}	3.14	1.69×10^{-3}
$x_4^2 \times x_5$	3.18×10^{-5}	1.75×10^{-5}	1.82	0.07
$x_4^2 \times x_6$	2.43×10^{-6}	5.02×10^{-6}	0.49	0.63
$x_4 \times x_6^2$	-7.78×10^{-5}	2.58×10^{-5}	-3.01	2.59×10^{-3}
$x_4 \times x_7^2$	1.37×10^{-3}	6.21×10^{-4}	2.21	0.03
$x_5 \times x_6^2$	-3.16×10^{-4}	3.72×10^{-4}	-0.85	0.39
x_1^3	-3.94×10^{-6}	2.91×10^{-7}	-13.53	2.04×10^{-41}
x_2^3	1.63×10^{-5}	1.92×10^{-6}	8.51	2.02×10^{-17}
x_3^3	-2.07×10^{-6}	6.28×10^{-7}	-3.30	9.78×10^{-4}
x_4^3	4.02×10^{-6}	6.93×10^{-7}	5.80	6.80×10^{-9}
x_6^3	-0.00059	0.000122	-4.87	1.15×10^{-6}
x_7^3	-0.09	0.03	-2.81	5.00×10^{-3}

References

1. Kim, C.K.; Kim, H.G.; Kang, Y.H.; Yun, C.Y.; Kim, S.Y. Probabilistic Prediction of Direct Normal Irradiance Derived from Global Horizontal Irradiance over the Korean Peninsula by Using Monte-Carlo Simulation. *Sol. Energy* **2019**, *180*, 63–74. [CrossRef]
2. Conceicao, R.; Alami, A.; Romero, M. Soiling Effect in Solar Energy Conversion Systems: A Review. *Renew. Sustain. Energy Rev.* **2022**, *162*, 112434. [CrossRef]
3. Silva, H.G.; Abreu, E.F.M.; Lopes, F.M.; Cavaco, A.; Canhoto, P.; Neto, J.; Collares-Pereira, M. Solar Irradiation Data Processing Using Estimator MatriceS (SIMS) Validated for Portugal (Southern Europe). *Renew. Energy* **2020**, *147*, 515–528. [CrossRef]
4. Salazar, G.; Gueymard, C.; Galdino, J.B.; de Castro Vilela, O.; Fraidenraich, N. Solar Irradiance Time Series Derived from High-Quality Measurements, Satellite-Based Models, and Reanalyses at a near-Equatorial Site in Brazil. *Renew. Sustain. Energy Rev.* **2020**, *117*, 109478. [CrossRef]
5. Aler, R.; Galván, I.M.; Ruiz-Arias, J.A.; Gueymard, C.A. Improving the Separation of Direct and Diffuse Solar Radiation Components Using Machine Learning by Gradient Boosting. *Sol. Energy* **2017**, *150*, 558–569. [CrossRef]
6. Gueymard, C.A.; Ruiz-Arias, J.A. Extensive Worldwide Validation and Climate Sensitivity Analysis of Direct Irradiance Predictions from 1-Min Global Irradiance. *Sol. Energy* **2016**, *128*, 1–30. [CrossRef]
7. Padovan, A.; Del Col, D.; Sabatelli, V.; Marano, D. DNI Estimation Procedures for the Assessment of Solar Radiation Availability in Concentrating Systems. *Energy Procedia* **2014**, *57*, 1140–1149. [CrossRef]
8. Cavaco, A.; Canhoto, P.; Collares Pereira, M. Procedures for Solar Radiation Data Gathering and Processing and Their Application to DNI Assessment in Southern Portugal. *Renew. Energy* **2021**, *163*, 2208–2219. [CrossRef]
9. Schroedter-Homscheidt, M.; Benedetti, A.; Killius, N. Energy Meteorology Verification of ECMWF and ECMWF/MACC's Global and Direct Irradiance Forecasts with Respect to Solar Electricity Production Forecasts. *Meteorol. Z.* **2016**, *26*, 1–19. [CrossRef]
10. Stoffel, T.; Renné, D.; Myers, D.; Wilcox, S.; Sengupta, M.; George, R.; Turchi, C. *Concentrating Solar Power: Best Practices Handbook for the Collection and Use of Solar Resource Data*; National Renewable Energy Lab. (NREL): Golden, CO, USA, 2010. Available online: https://www.nrel.gov/docs/fy10osti/47465.pdf (accessed on 5 December 2020).
11. Gueymard, C.A. Parameterized transmittance model for direct beam and circumsolar spectral irradiance. *Sol. Energy* **2001**, *71*, 325–346. [CrossRef]
12. Engerer, N.A. ScienceDirect Minute Resolution Estimates of the Diffuse Fraction of Global Irradiance for Southeastern Australia. *Sol. Energy* **2015**, *116*, 215–237. [CrossRef]
13. Chain, C.; George, R.A.Y.; Vignola, F. A New Operational Model For Satellite-Derived Irradiances: Description and Validation. *Sol. Energy* **2003**, *73*, 307–317.
14. Hollands, K.G.T. An improved model for diffuse radiation: Correction for atmospheric back-scattering. *Sol. Energy* **1987**, 233–236. [CrossRef]
15. Starke, A.R.; Lemos, L.F.L.; Boland, J.; Cardemil, J.M.; Colle, S. Resolution of the Cloud Enhancement Problem for One-Minute Diffuse Radiation Prediction. *Renew. Energy* **2018**, *125*, 472–484. [CrossRef]
16. Bright, J.M.; Engerer, N.A. Engerer2: Global Re-Parameterisation, Update, and Validation of an Irradiance Separation Model at Different Temporal Resolutions. *J. Renew. Sustain. Energy* **2019**, *11*, 033701. [CrossRef]
17. Gueymard, C.A. REST2: High-Performance Solar Radiation Model for Cloudless-Sky Irradiance, Illuminance, and Photosynthetically Active Radiation—Validation with a Benchmark Dataset. *Sol. Energy* **2008**, *82*, 272–285. [CrossRef]
18. Yang, D. Estimating 1-Min Beam and Diffuse Irradiance from the Global Irradiance: A Review and an Extensive Worldwide Comparison of Latest Separation Models at 126 Stations. *Renew. Sustain. Energy Rev.* **2022**, *159*, 112195. [CrossRef]
19. Starke, A.R.; Lemos, L.F.L.; Barni, C.M.; Machado, R.D.; Cardemil, J.M.; Boland, J.; Colle, S. Assessing One-Minute Diffuse Fraction Models Based on Worldwide Climate Features. *Renew. Energy* **2021**, *177*, 700–714. [CrossRef]
20. Abreu, E.F.M.; Canhoto, P.; Costa, M.J. Prediction of Diffuse Horizontal Irradiance Using a New Climate Zone Model. *Renew. Sustain. Energy Rev.* **2019**, *110*, 28–42. [CrossRef]
21. Paulescu, E.; Blaga, R. A Simple and Reliable Empirical Model with Two Predictors for Estimating 1-Minute Diffuse Fraction. *Sol. Energy* **2019**, *180*, 75–84. [CrossRef]
22. Every, J.P.; Li, L.; Dorrel, D.G. Köppen-Geiger Climate Classification Adjustment of the BRL Diffuse Irradiation Model for Australian Locations. *Renew. Energy* **2020**, *147*, 2453–2469. [CrossRef]
23. Yang, D. Temporal-Resolution Cascade Model for Separation of 1-Min Beam and Diffuse Irradiance. *J. Renew. Sustain. Energy* **2021**, *13*, 056101. [CrossRef]
24. Comissão Europeia. PLANO NACIONAL ENERGIA E CLIMA 2021–2030 (PNEC 2030). 2019. Available online: https://ec.europa.eu/energy/sites/ener/files/documents/pt_final_necp_main_pt.pdf (accessed on 5 December 2020).
25. Lopes, F.M.; Conceição, R.; Silva, H.G.; Salgado, R.; Collares-Pereira, M. Improved ECMWF Forecasts of Direct Normal Irradiance: A Tool for Better Operational Strategies in Concentrating Solar Power Plants. *Renew. Energy* **2021**, *163*, 755–771. [CrossRef]
26. Kambezidis, H.D.; Psiloglou, B.E.; Kaskaoutis, D.G.; Karagiannis, D.; Petrinoli, K.; Gavriil, A.; Kavadias, K. Generation of Typical Meteorological Years for 33 Locations in Greece: Adaptation to the Needs of Various Applications. *Theor. Appl. Climatol.* **2020**, *141*, 1313–1330. [CrossRef]
27. Nielsen, K.P.; Vignola, F.; Ramírez, L.; Blanc, P.; Meyer, R.; Blanco, M. Excerpts from the Report: "BeyondTMY—Meteorological Data Sets for CSP/STE Performance Simulations". *AIP Conf. Proc.* **2017**, *1850*, 140017. [CrossRef]

28. Hall, I.J.; Prairie, R.R.; Anderson, H.E.; Boes, E.C. Generation of a Typical Meteorological Year. In Proceedings of the 1978 Annual Meeting of the American Section of the International Solar Energy Society, Denver, CO, USA, 28–31 August 1978; pp. 669–671.

29. Conceição, R.; Lopes, F.M.; Tavares, A.; Lopes, D. Soiling Effect in Second-Surface CSP Mirror and Improved Cleaning Strategies. *Renew. Energy* **2020**, *158*, 103–113. [CrossRef]

30. Lopes, T.; Fasquelle, T.; Silva, H.G. Pressure Drops, Heat Transfer Coefficient, Costs and Power Block Design for Direct Storage Parabolic Trough Power Plants Running Molten Salts. *Renew. Energy* **2021**, *163*, 530–543. [CrossRef]

31. Blair, N.; Dobos, A.P.; Freeman, J.; Neises, T.; Wagner, M.; Ferguson, T.; Gilman, P.; Janzou, S. *System Advisor Model, Sam 2014.1. 14: General Description*; NREL Rep. No. TP-6A20-61019; Natl. Renew. Energy Lab.: Golden, CO, USA, 2014; Volume 13. [CrossRef]

32. National Renewable Energy Laboratory, "Andasol 3 CSP Project," 2021. Available online: https://solarpaces.nrel.gov/project/andasol-3 (accessed on 16 June 2022).

33. Instituto Português do Mar e da Atmosfera. Glossário Climatológico/Meteorológico. Available online: https://www.ipma.pt/pt/educativa/glossario/meteorologico/index.jsp?page=glossario_ef.xml&print=true (accessed on 2 August 2021).

34. *ISO 9060:1990*; Solar Energy—Specification and Classifications of Instruments for Measuring Hemispherical Solar and Direct Solar Radiation. International Organization for Standardization: Geneva, Switzerland, 1990. Available online: www.iso.org/standard/16629.html (accessed on 5 December 2020).

35. Alami, A.; Conceiç, R.; Gonçalves, H.; Ghennioui, A. CSP Performance and Yield Analysis Including Soiling Measurements for Morocco and Portugal. *Renew. Energy* **2020**, *162*, 1777–1792. [CrossRef]

36. Lopes, D.; Conceição, R.; Gonçalves, H.; Aranzabe, E.; Pérez, G.; Collares-pereira, M. Anti-Soiling Coating Performance Assessment on the Reduction of Soiling Effect in Second-Surface Solar Mirror. *Sol. Energy* **2019**, *194*, 478–484. [CrossRef]

37. *ISO 9059:1990*; Solar Energy—Calibration of Field Pyrheliometers by Comparison to a Reference Pyrheliometer. International Organization for Standardization: Geneva, Switzerland, 2014. Available online: www.iso.org/standard/16628.html (accessed on 5 December 2020).

38. Instituto Português do Mar e da Atmosfera. Rede de Estações Meteorológicas. Available online: https://www.ipma.pt/pt/otempo/obs.superficie/ (accessed on 2 August 2021).

39. Instituto Português do Mar e da Atmosfera. Parques Meteorológicos e Equipamentos. Available online: https://www.ipma.pt/pt/educativa/observar.tempo/index.jsp?page=ema.index.xml (accessed on 2 August 2021).

40. Long, C.N.; Dutton, E.G. BSRN Global Network Recommended QC Tests, V2.0. Available online: https://bsrn.awi.de/fileadmin/user_upload/bsrn.awi.de/Publications/BSRN_recommended_QC_tests_V2.pdf (accessed on 5 December 2020).

41. Kalogirou, S.A. Environmental Characteristics. In *Solar Energy Engineering: Processes and Systems*; Elsevier: Amsterdam, The Netherlands, 2014; pp. 51–92.

42. Lopes, F.M.; Silva, H.G.; Salgado, R.; Cavaco, A.; Canhoto, P.; Collares-Pereira, M. Short-Term Forecasts of GHI and DNI for Solar Energy Systems Operation: Assessment of the ECMWF Integrated Forecasting System in Southern Portugal. *Sol. Energy* **2018**, *170*, 14–30. [CrossRef]

43. Finkelstein, J.M.; Schafer, R.E. Improved Goodness-of-Fit Tests. *Biometrika* **1971**, *58*, 641–645. [CrossRef]

44. Lopes, F.M.; Conceição, R.; Fasquelle, T.; Silva, H.G.; Salgado, R.; Canhoto, P.; Collares-Pereira, M. Predicted Direct Solar Radiation (ECMWF) for Optimized Operational Strategies of Linear Focus Parabolic-Trough Systems. *Renew. Energy* **2020**, *151*, 378–391. [CrossRef]

45. Instituto Português do Mar e da Atmosfera. Clima de Portugal Continental. Available online: https://www.ipma.pt/pt/educativa/tempo.clima/ (accessed on 3 August 2021).

46. Cavaco, A.; Silva, H.; Canhoto, P.; Neves, S.; Neto, J.; Pereira, M.C. Global Solar Radiation in Portugal and its variability, monthly and yearly. In WES 2016—Workshop on Earth Sciences, Institute of Earth Sciences. 2016, pp. 1–4. Available online: https://dspace.uevora.pt/rdpc/bitstream/10174/19395/1/Afonso_Cavaco_et_al_WES_2016_paper_28.pdf (accessed on 3 August 2021).

47. Cunha, L. A Beira Interior—Portugal: Caracterização Física. Rota da Lã Translana Percursos e Marcas um Territ. Front. Beira Inter. (Portugal), Comarc. Tajo-Salor-Almonte 2008. pp. 47–53. Available online: https://www.researchgate.net/publication/324089073_A_beira_Interior_-_Portugal_caracterizacao_fisica (accessed on 3 August 2021).

48. Governo Português. Roteiro para a Neutralidade Carbónica 2050. 2019. Available online: https://www.portugal.gov.pt/download-ficheiros/ficheiro.aspx?v=%3D%3DBAAAAB%2BLCAAAAAAABACzMDexAAAut9emBAAAAA%3D%3D.Governo (accessed on 3 August 2021).

49. Redes Energéticas Nacionais. DADOS TÉCNICOS. 2021. Available online: https://datahub.ren.pt/media/hkkdskwq/dados-t%C3%A9cnicos-2021.pdf (accessed on 3 August 2021).

50. Lopes, T.; Fasquelle, T.; Silva, H.G.; Schmitz, K. HPS2—Demonstration of Molten-Salt in Parabolic Trough Plants—Simulation Results from System Advisor Model. *AIP Conf. Proc.* **2020**, *2303*, 110003. [CrossRef]

Article

Efficient Output Photovoltaic Power Prediction Based on MPPT Fuzzy Logic Technique and Solar Spatio-Temporal Forecasting Approach in a Tropical Insular Region

Fateh Mehazzem *, Maina André and Rudy Calif

Univ Antilles, LaRGE Laboratoire de Recherche en Géosciences et Energies (EA 4539), F-97100 Pointe-à-Pitre, France
* Correspondence: fateh.mehazzem@univ-antilles.fr

Abstract: Photovoltaic (PV) energy source generation is becoming more and more common with a higher penetration level in the smart grid because of PV energy's falling production costs. PV energy is intermittent and uncertain due to its dependence on irradiance. To overcome these drawbacks, and to guarantee better smart grid energy management, we need to deal with PV power prediction. The work presented in this paper concerns the study of the performance of the fuzzy MPPT approach to extract a maximum of power from solar panels, associated with PV power estimation based on short time scale irradiance forecasting. It is particularly applied to a case study of a tropical insular region, considering extreme climatic variability. To validate our study with real solar data, measured and predicted irradiance profiles are used to feed the PV system, based on solar forecasting in a tropical insular context. For that, a spatio-temporal autoregressive model (STVAR) is applied. The measurements are collected at three sites located on Guadeloupe island. The high variability of the tropical irradiance profile allows us to test the robustness and stability of the used MPPT algorithms. Solar forecasting associated with the fuzzy MPPT technique allows us to estimate in advance the produced PV power, which is essential for optimal energy management in the case of smart energy production systems. Simulation of the proposed solution is validated under Matlab/Simulink software. The results clearly demonstrate that the proposed solution provides good PV power prediction and better optimization performance: a fast, dynamic response and stable static power output, even when irradiation is rapidly changing.

Keywords: boost converter; MPPT fuzzy; output PV power estimation; PV system; solar forecasting; STVAR model; tropical insular region

Citation: Mehazzem, F.; André, M.; Calif, R. Efficient Output Photovoltaic Power Prediction Based on MPPT Fuzzy Logic Technique and Solar Spatio-Temporal Forecasting Approach in a Tropical Insular Region. *Energies* **2022**, *15*, 8671. https://doi.org/10.3390/en15228671

Academic Editors: Juri Belikov and Jaroslaw Krzywanski

Received: 26 August 2022
Accepted: 2 November 2022
Published: 18 November 2022

Publisher's Note: MDPI stays neutral with regard to jurisdictional claims in published maps and institutional affiliations.

1. Introduction

Solar energy prevision helps to reduce grid imbalance charges and avoid penalties. Thus, the forecast for the operation of solar power plants gives visibility as to the resources that will be available in the coming minutes, hours, and days. Integrating forecasting into the production process helps operators to limit the uncertainty associated with photovoltaic energy production. As a result, this allows a massive and secure injection of solar energy into the grid.

In a tropical insular context, for which the costs of supplying electricity are higher than those of the rest of the world, the challenges induced by a massive development of electrical renewable energies for supply–demand balance and network stability are indeed particularly important [1]. Wind and solar energy are the most promising renewable energy resources because of their potential and availability. In the last decades, PV systems have become more and more common in electricity generation due to their: price reduction, efficiency, and increased life. In addition, they provide clean and sustainable energy. The overall output power of different PV systems can vary under dynamic irradiance fluctuations [2].

Nowadays, the rapid growth of the penetration of renewable energy sources into distribution grids brings new challenges to overcome: stability, cost system operation, and congestion management. An effective weather forecasting based on a great spatial resolution, and the integration of artificial intelligence, contribute to obtaining an improved state estimation of the power system and avoid it reaching a dangerous state [3,4].

The privileged location of the tropical insular regions, which have particularly favorable climatic conditions with more than 1400 h of insolation per year, is a major advantage for photovoltaic solar production. On the other hand, these regions are characterized by high variability of observed irradiance [5]. This high variability is the consequence of a solar microclimate's diversity due to its complex topography in a small space (mountainous/low orography, land/sea contrasts) and frequent cloud formation in the tropical zone [6–9]. In [5,9–11], the types of days and the fluctuating nature of solar radiation from Guadeloupe island (the island of our measurements sites) are highlighted, exhibiting regimes of variability. Moreover, these islands are not electrically interconnected, which makes their network fragile, and it requires a threshold for the power instantly consumed from intermittent renewable energies (solar and wind) to be set.

In the literature, several metrics have been defined for the variability quantification of irradiance. A few metrics can be quoted such as the variability index (VI) [12], which is a ratio including measured irradiance, clear sky irradiance signal, the number of consecutive observations, and the averaging interval of the time series. In [12], the author showed a classification of sites and periods of time when variability is significant. In [13], the intra-hour variability score (IHVS) [13] captures several key variability characteristics into a single metric for irradiance, or PV output power, data. It is shown that this metric facilitates the comparison of the variability of PV systems of different sizes, with different configurations, across different climates. In [14], the daily aggregate ramp rate (DARR) represents the sum of all absolute increment values during a given hour or day. This metric made it possible to characterize variability due to geographical dispersion by simulating a step-by-step increase in the plant size at the same location. In [15], the variability score (VS) is based on increments in temporally averaged data, with averaging time scale, expressed in a percentage. In [16], the nominal variability is defined as the ramp rate's variance of clear sky index. This metric can clearly distinguish two extremum cases of insolation conditions, namely, perfectly clear conditions (i.e., no variability) and heavily overcast conditions. In [17], the variability score VS, defined as the maximum value of ramp rate magnitude (RR0) times ramp rate probability, has been performed on the same measurements site under our study to categorize typical days as a function of their variability [11]. The study showed that 58% of ramp rates are higher than 50 $W \cdot m^{-2}$ concerning the intermittent cloudy sky days at 5 min time scales, and the type of days presenting the higher magnitude of ramp rate and higher VS values. Thirty percent of such days occurred in the year 2012. Moreover, the results revealed that VS increased over Δt.

In this challenging context, the forecasting technique needs to have robustness and predictive efficiency allowing high variability to be forecast at short time scales. The high variability of irradiance leads to increased grid variability and uncertainty for grid-connected solar photovoltaics (PV) as penetration increases. The development of an efficient forecasting model contributes to better management by power system operators and/or PV plant owners.

In the recent literature, the spatial and temporal characteristics of irradiance are considered by introducing the information observed at several neighboring locations. The spatio-temporal techniques have already shown an improvement in the forecast's quality [18–20] for several environmental processes: in wind velocity fields such as in [21] where neural and geostatistical techniques were developed to obtain the wind speed maps, or in [22] using a VAR (vector autoregressive) model to forecast wind speed and atmospheric concentration of carbon monoxide by integrating spatio-temporal parameters from spatially sparse sites. In [23], the authors developed Taylor kriging technics also applied to wind speed forecasting and improving the predictive performance compared to the ARIMA (autoregressive integrated moving average) model, and other techniques for modeling wind speed such as space–time covariance models were described and applied in [24]. In the solar forecasting area, several spatio-temporal tools have been developed such as kriging approaches: cross-validation with spatial kriging integrating parameters of cloud motion in [25] resulting in a detailed map of solar radiation, or space–time kriging to forecast irradiance at arbitrary spatial locations, such as in [26] with a variance–covariance structure explored and defined, in [27] integrating shrinkage parameters, or in [28] to forecast very short-term irradiance at unobserved locations. Moreover, other approaches have been developed in the literature such as lasso-based variable selection [29], spatio-temporal ARMA (autoregressive moving average) such as (STARMA) models ([30,31], or spatio-temporal vector autoregressive (STVAR) models [6,27] providing future irradiance based on a linear combination of past observations at the most relevant neighboring locations.

PV power production is considered a nonlinear probabilistic process because it is affected by the variation in:

- many weather parameters such as solar irradiance, temperature, wind, etc.;
- the PV cells nonlinear model;
- the DC-DC converter model;
- outages of the grid and possible islanding conditions of the PV parks;
- other operation conditions.

The output power of the PV system changes with time and depends on the variability of meteorological factors. Accurate PV power forecasting contributes to improving the stability, security, and reliability of grid operation, as well as scheduling and grid energy management [32]. It also affects power quality, voltage surges, reverse power flows, etc. PV power forecasting can be performed for different forecast time horizons, from very short time scales (minutes) to long time scales (days). The choice of forecast time horizon depends mainly on applications [33]. PV power forecasting for the short-term horizon is used, for example, in PV power ramp rate estimation [34]. Fast irradiance fluctuations affect the quality and reliability of PV power. PV power forecasting for the long-term horizon is used, for example, in energy management optimization in microgrids [35].

PV power forecasting methods can be classified into two categories: direct and indirect methods [36]. Direct methods consist of applying physical, statistical, artificial intelligence, or hybrid models to PV measurements. They propose several techniques and models. Among them are: Physical models, obtained from the modeling of solar radiation dynamics in the atmosphere using physics laws, which is more efficient in long-term prediction [37]. Statistical approaches that use historical data samples and give better performance in short-term prediction [38]. Artificial intelligence approaches, also known as machine learning approaches. They use different architectures of artificial neural network (ANN) and fuzzy techniques. They are based on basic operations of training and testing. Their advantage is that they deal with complex nonlinear features effectively, and their main drawback is the requirement for a large quantity of data during the training stage [39]. Deep learning structures are more popular nowadays [40]. Hybrid approaches, which are a combination of two or more approaches that give more accuracy in PV power forecasting in different cases [41].

Indirect methods consist of fragmenting the power prediction problem into two stages: forecasting weather parameters and introducing them in commercial PV simulation software to forecast the PV power generation. Solar forecasting was conducted for different time scale horizons and using different forecasting methods cited above.

Concerning PV power short-term horizon forecasting applications, other approaches are proposed in the literature to estimate the expected power of a PV system from a ramp rate characterization. An analysis of variability is therefore investigated, such as in [42], exhibiting the frequency of a given fluctuation from output PV power for a certain day by an analytic model, and [43] who demonstrated the smoothing of rapid ramps observed in a location point by large PV plants, which was strengthened in [16] by the result of the aggregation of multiple PV plants. In [14], a quantitative metric called the daily aggregate ramp rate (DARR) was proposed to quantify, categorize, and compare daily variability from power output, across multiple sites. In [16], models were developed exhibiting a capacity to extrapolate the resulting variability on an ensemble of power plants, and, in [34], a method was proposed to estimate the largest expected PV power ramp rates (RR).

The works in the literature clearly show the purposes of power forecasting, namely, the development of strategies to mitigate the impact of solar variability on electric power grids, the optimization of the design and size of PV power plants, and, consequently, a maximization of their effectiveness and minimization of costs by using the curve fitted to the largest RR ratios, irradiance, and cloud shadow velocities dataset. In [44], the authors propose the implementation of an optimized solar PV utilization using a low-cost hardware platform to improve efficiency, and for better saving of the grid energy. The proposed solution is based on an online intelligent algorithm.

As a reminder, in a photovoltaic system, the maximum power point (MPP) corresponds to the point of tangency between the characteristic points of the photovoltaic generator for a certain value of solar radiation [45]. The role of the MPPT is to precisely detect at every moment this point of maximum energy efficiency. The performance of an MPPT controller depends on how quickly it reaches the point of maximum power, how it oscillates around that point, and how robust it is in the face of abrupt atmospheric changes. Several power point maximum (PPM) tracking algorithms have been cited in the literature over the last few years [46–48]. These algorithms can be mainly classified as conventional, metaheuristic, artificial intelligence, nonlinear, and hybrid algorithms.

Conventional MPPT techniques such as Perturb and Observe (P&O) [49], Incremental Conductance (IC) [50], and Hill Climbing (HC) [51] give good results under uniform irradiance conditions. However, this performance is not guaranteed during transitional regimes. Other works are based on metaheuristic approaches such as particle swarm optimization (PSO) [52] and genetic algorithms (GA) [53]. Their performance depends on selected design parameters and initial conditions. MPPT techniques based on artificial intelligence such as fuzzy logic (FL) [54] and neural networks (NN) [55] have become more and more popular. They have the advantage of working even with imprecise input values and they do not need a high precision mathematical model. In addition, they can deal with nonlinearities. In recent years, many MPPT techniques based on nonlinear control have been proposed. Among them are sliding mode [56] and backstepping [57]. These approaches are known for their robustness and stability using Lyapunov functions. Hybrid MPPT techniques are a combination of two or more of the MPPT techniques cited before [58].

Fuzzy logic was developed by Zadeh in 1965 from his theory of fuzzy subsets [59]. It allows us to imitate human reasoning by using the different collected data in linguistic forms. One of the main interests of the use of fuzzy logic consists of being able to pass, simply through the intermediary of linguistic rules, the expertise that one may have of the process towards the controller. It is thus possible to transform the expert's knowledge into simple rules that the controller can implement. The ease of implementing solutions for complex problems is then associated with robustness for uncertainties and the possibility of integrating the expert's knowledge.

In MPP optimization problems of PV systems, it is recognized that classical methods [49–51] are widely used due to their simple structure and the fact that they require fewer measured parameters, but their speed of convergence is variable. In addition, the system will constantly oscillate around the MPP. These oscillations can be minimized by reducing the value of the increment step. However, a small increment value can slow down the search for the MPP. This drawback allowed us to find in permanence a compromise between precision and speed. In addition, it is an unsuitable method for rapidly changing weather conditions. Conversely, fuzzy MPPT controllers are well adapted to this type of problem and behave well in various atmospheric conditions. They converge quickly, although they can work with imprecise inputs, and do not require a well-known mathematical model [54,60,61]. Moreover, fuzzy control is known for its robustness and its adaptive nature, giving it high performance with respect to parameter variation and disturbances of the system.

In these works, a structure with several steps is defined to feed our PV production system, firstly performing irradiance forecasting from a STVAR model. The model was developed in [4] and showed good predictive performance for time scales from 5 min to 1 h. This spatio-temporal autoregressive model (STVAR) was applied for a very short time scale, 1 min ahead for the first time. From these forecasting model results, a predicted irradiance profile was built. This profile is composed of selected values, spread over one-year prediction data. It shows an important magnitude of irradiance fluctuations. A robust fuzzy MPPT optimization algorithm was applied for a PV system with a capacity of 3 kW peak to overcome this constraint and to estimate the output produced PV power. The fuzzy logic approach has the advantage of improving the transitional regime and reducing fluctuations in the static state whatever changes in the climate there are.

The main novelty of this article is the description and validation of an efficient indirect PV power forecasting method based on short time scale irradiance spatio-temporal forecasting and the MPPT approach. Thus, this work consisted of connecting the contributions of these two fields of research (statistical forecasting irradiance and the MPPT approach). This investigation highlights the results of their combination. The proposed method was applied in a challenging environment, a tropical insular context known by its high solar variability, using in situ real measured solar data. The proposed PV power forecasting method was validated and its performance was very close to that of irradiance forecasting. Efficiency, accuracy, and robustness were studied by simulations for the four seasonal days representing the fast variability of the tropical climate over one year. These results are advantageous to the design and component sizing of PV power plants.

This work is organized as follows. Section 2 (Proposed System Configuration) describes the global structure of the proposed PV system, and its different components: solar PV generator, DC-DC boost converter, and load. In the Section 3 (Solar Forecasting and Control of the Proposed System), the first part describes the used solar forecasting model. After sampling processing, measured and predicted irradiance profiles are given. The second part presents the MPPT fuzzy optimization algorithm. Section 4 (Results and Discussion) shows simulation results obtained for the proposed solution. Finally, the different contributions given in this paper are summarized in the Section 5 (Conclusions).

2. Proposed System Configuration

2.1. Global Structure Design

A block diagram of the proposed control system is shown in Figure 1. The global structure of the studied PV plant is mainly composed of a PV generator (GPV), DC-DC boost converter, and load.

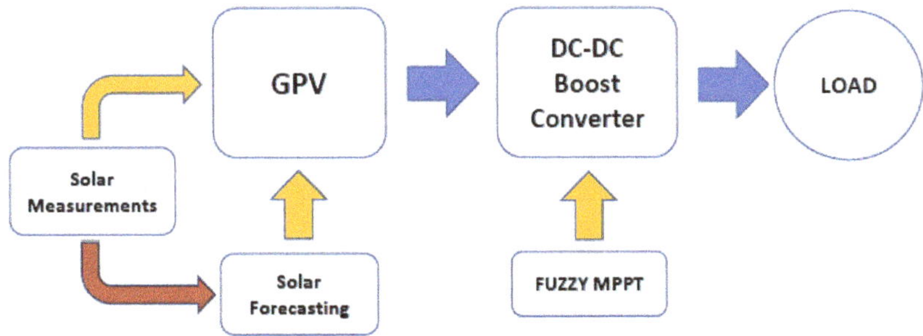

Figure 1. Global structure diagram for PV system.

The description of the proposed structure is as follows:

From solar measurements collected at three sites on Guadeloupe island for a year, and using the STVAR prediction model, we obtained prediction results for a short time scale.

Solar measurements and forecasting data are used to feed our PV production system. For the configuration of our PV system, we propose the use of a DC-DC boost converter, controlled by an MPPT fuzzy LOGIC algorithm to extract the maximum power from the photovoltaic panels.

The objective of this combination is to have an indirect estimate of the power produced by our PV system.

The following sections describe each part of the proposed structure illustrated in Figure 1.

2.2. Solar PV Generator

For more simplicity, photovoltaic cells can be represented using the single-diode model as shown in Figure 2.

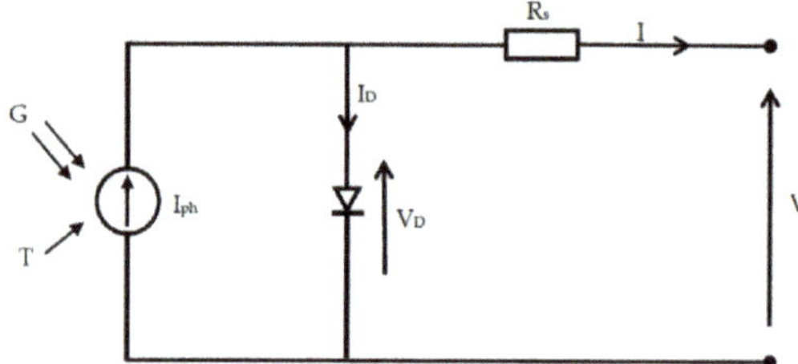

Figure 2. Single-diode model of the PV cell without a shunt resistor.

The series resistance is the internal resistance of the cell, whereby the shunt resistance represents the leakage current of the junction, being generally very large (equivalent to an open circuit), it can be neglected on the equivalent diagram.

The PV cell electrical characteristics are given by the following equations

$$I = I_{ph} - I_D \tag{1}$$

$$I_{ph} = I_{ph}(T_1) \times [1 + K_0 \times (T - T_1)] \tag{2}$$

$$I_{ph}(T_1) = I_{cc}(T_1) \times \left(\frac{G}{G_0}\right) \tag{3}$$

$$K_0 = \frac{I_{cc}(T_2) - I_{cc}(T_1)}{T_2 - T_1} \tag{4}$$

$$I_D = I_s(e^{\left(\frac{V_D}{V_T}\right)} - 1) \tag{5}$$

$$V_T = \frac{nKT}{q} \tag{6}$$

$$V_D = V + R_s I \tag{7}$$

$$R_s = -\frac{dV}{dI_{V_{co}}} - \frac{1}{X_V} \tag{8}$$

$$X_V = \frac{I_s(T_1)}{V_T(T_1)} \cdot e^{\frac{V_{co}(T_1)}{V_T(T_1)}} \tag{9}$$

$$I_s = I_s(T_1) * \left(\frac{T}{T_1}\right)^{3/n} e^{\left(\frac{-q \cdot V_g}{n \cdot K}\right)\left(\frac{1}{T} - \frac{1}{T_1}\right)} \tag{10}$$

$$I_s(T_1) = \frac{I_{cc}(T_1)}{e^{\left(\frac{V_{co}(T_1)}{V_T(T_1)}\right)} - 1} \tag{11}$$

where I, I_{ph}, I_D, I_{cc}, and I_S are, respectively, PV cell output current, photocurrent, diode current, short-circuit current, and saturation current in A. Reference temperature ($T_1 = 25\,°C = 298\,°K$), T: absolute surface temperature, G_0: reference irradiance ($G_0 = 1000\,W/m^2$), K_0: current variation coefficient, q: elementary electron charge, K: Boltzmann's constant, n: diode ideality factor. V_T: thermodynamic potential, V_D: voltage across the diode, V: voltage at the cell terminals, R_s: resistance series, V_{co}: the open circuit voltage.

For solar cell, the equation between current and voltage then becomes

$$I = I_{ph} - I_s\left(e^{\left(\frac{q(V+R_s I)}{nKT}\right)} - 1\right) \tag{12}$$

The PV generator represents a 3 kW PV module composed of 10 PV modules of 300 Wp (type SunPower SPR-300E-WHT-D) connected in series, and provide 547 V as the output voltage at the maximum power point. The PV panel parameters are given in Appendix A.

2.3. DC-DC Boost Converter

To reduce the number of PV panels used, we chose a boost converter to increase the output voltage of a DC voltage. It ensures the impedance matching between the PV panels and the load. It is composed of the inductor (Lo), input capacitor (Ci), output capacitor (Co), MOSFET transistor switch, and power diode (D). 1 and 2 are terminals for load connection. Figure 3 shows the circuit diagram of the boost converter.

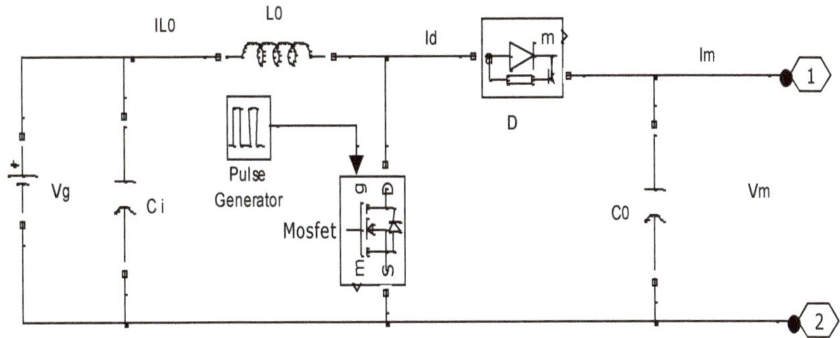

Figure 3. Boost converter circuit.

Equations of the DC-DC converter are given by

$$C_i = \frac{\partial V_g}{\partial t} = I_g - I_{L0} \tag{13}$$

$$L_0\frac{\partial I_{L0}}{\partial t} = V_g - (1-D)V_m \tag{14}$$

$$L_0\frac{\partial I_{L0}}{\partial t} = V_g - (1-D)V_m \tag{15}$$

where I_m is the output current of the boost converter (A), I_g is the output current of the PV panels (A), V_m is the output voltage of the boost converter (V), V_g is the output voltage of the PV panels, I_{LO} is the inductor current (A).

The switching state is also governed by a signal of period T and duty cycle D.

3. Solar Forecasting and Control of the Proposed System

3.1. Forecasting Model and Preprocessing of Datasets

3.1.1. Study Context and Experimental Datasets

Experimental measurements were collected on Guadeloupe island; therefore, an insular context located in a tropical climate.

This context favors very frequent cloud formations driven by the diversity of local topography on a small space (orographic clouds) and by synoptic systems due to land/sea contrast [7,8]. This meteorological context leads to the diversity and complexity of solar microclimates. Consequently, the irradiance observed is characterized by high spatial and temporal variability, particularly at short time scales [7,8,62–64]. In this complex context, solar radiation forecasting is challenging, particularly at very short time scales.

The global horizontal irradiance (GHI) was measured at 1 Hz with a pyranometer CM22 from Kipp and Zonen pyranometer (type SP Lite) whose response time is less than a second. The sensor accuracy given by the manufacturer is 3%. For this study we used an averaged database at 1 min of the global horizontal irradiance (GHI) measured for a period starting in January 2012 and ending in December 2012. The measurements were collected at site Petit-canal Gros-cap, along the cliffs (16°38 N, 61°49 W). The forecasting time series at Petit-canal was obtained by the STVAR model using past irradiance observations at the three measurement sites (see Figure 4). Figure 5 displays a five-day irradiance sequence showing high temporal variability. To perform the statistical forecasting model, the temporal trend of the irradiance time series must be removed. Classically, in solar areas, clear sky irradiance is used. This parameter removes the temporal trend of irradiance mainly due to the solar geometry effects by a ratio between the measured irradiance and theoretical clear sky irradiance for the considered 1 min/location.

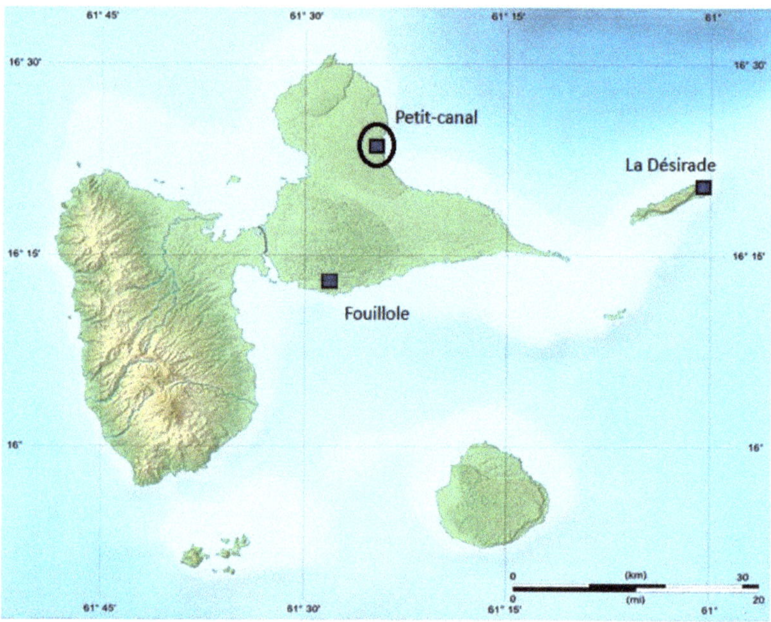

Figure 4. Guadeloupe archipelago and geographical location of three measurement sites: Petit-canal (location of predicted GHI time series selected for our study (black circle)), Fouillole, La Désirade.

3.1.2. Definition of the Forecasting Model: Spatio-Temporal Autoregressive Model—STVAR

The proposed model STVAR (spatio-temporal vector autoregressive) consists of forecasting GHI at a considered location (reference site) by a weighted linear combination of past observations at this location and neighboring locations (spatial predictors) at p lags, plus an error or innovations term, whereas the temporal VAR models uses a linear combination of past observations of n variables representing different parameters but located at one site. A spatio-temporal VAR (p) process is defined as follows:

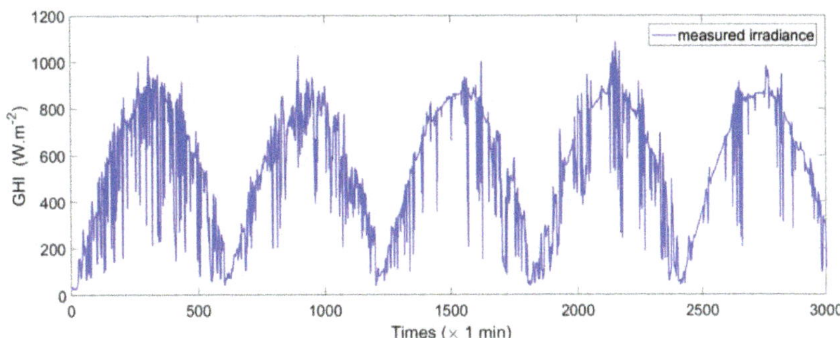

Figure 5. Sequence of five days at 1 min time scale times series of measured ground irradiance GHI.

Consider spatio-temporal variable $Z(s_i, t)$ representing the observations of a stochastic process indexed in $R^2 \times R$. Each observation is made at location s_i and at time t. The considered predictive VAR (vector autoregressive) model (see [22]) is given by

$$Z_t = \sum_{i=1}^{p} R_i(Z_{t-i}) + \varepsilon_t \qquad (16)$$

where p is the order corresponding to the time lag; $Z_t = (Z(s_1, t), Z(s_2, t), \ldots, Z(s_N, t))'$ are the spatio-temporal data, ε_t is white noise, and R_i is an $N \times N$ unknown parameter matrix. The N rows of these matrices correspond to the N locations at which time series are observed.

Estimation of the parameters in Equation (16) is obtained with least squares. For more details on the methodology (see Chap. 14 [64]).

If all lagged time series values associated with all measurement sites had to be integrated into the modeling process as input data, the computational process would not be optimized. For more efficiency, a process of the selection of optimal temporal input data (p lags optimal of past observations) and spatial input data (more relevant neighboring sites variable) was developed. This procedure is based on the calculation of the information criterion BIC (Bayesian information criterion) for sequential integration of each spatio-temporal explanatory variable. The relevant spatio-temporal input data are selected by minimization of BIC [6,22]. This procedure is built according to the method proposed in [6,22]. For more details on the modeling process see [6,9].

This process of spatio-temporal input selection was performed on a training dataset. The first month of the dataset was used as a model training dataset and the rest of the year as a test dataset.

STVAR model showed a good predictive performance for a time scale from 5 min to 1 h [6,9]. In this work, this model was performed at 1 min time scale regarding our study context.

3.2. Fuzzy MPPT Algorithm

The studied system is made up of nonlinear elements that need complex and heavy models (photovoltaic panels, electronic power system, etc.). However, most nonlinear control approaches require the availability of a mathematical model of the system. The assured performance is directly related to the accuracy of the model used. That is why we chose to use a fuzzy controller.

To obtain MPP from PV panels, the MPPT algorithm was developed using the fuzzy logic technique [54]. The principle of the MPPT algorithm is to find the maximum point of the curve $P(V)$. For that, we have to bring the error E between dP/dV and its optimal value to zero. When E is positive, the value of P increases, and, conversely, when E is negative, the value of *P* decreases. For that, a fuzzy logic (FL) controller is used. The classical fuzzy controller diagram is shown in the Figure 6:

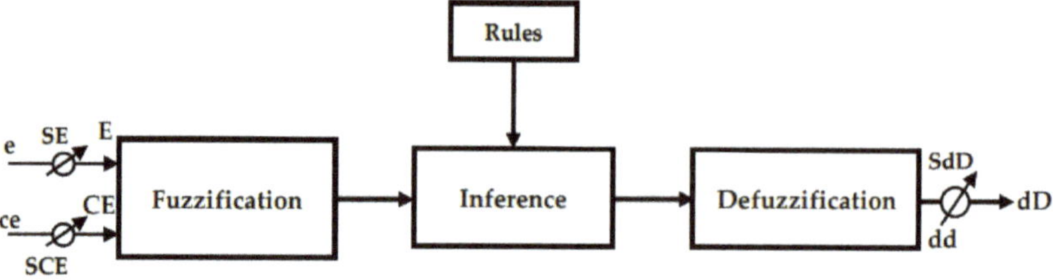

Figure 6. Fuzzy controller general structure.

E: error between *dP/dV* and its optimal value, *CE*: error variation of *E*, *dD*: duty cycle increment. SE, SCE: input gains, SdD: output gain.

1. **Fuzzification and defuzzification** transforms real input variables into fuzzy ones and the reverse. *E* and *CE* are inputs of the FL controller, and *dD* is the output. It represents the variable step of the duty cycle to control the DC-DC converter. The Figure 7 shows the membership functions for *E*, *CE*, and *dD*.

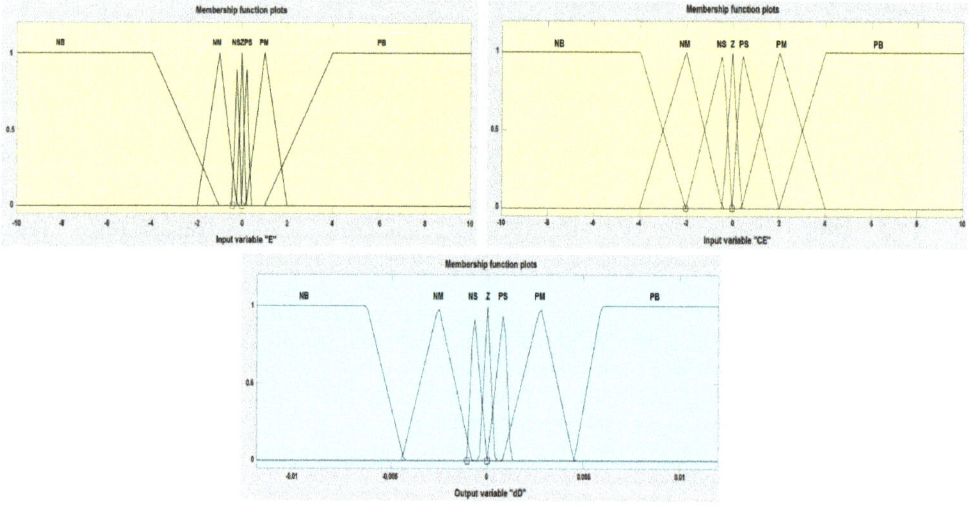

Figure 7. Membership functions of *E*, *CE*, and *dD*.

Among the different forms of membership functions (trapezoid, Gaussian, and triangular, etc.), the symmetrical triangular form is considered the most appropriate for its simplicity. The use of the nonuniform distributions of fuzzy sets is more suitable for controlling the nonlinear behavior of the PV system. In this study, the symmetrical triangular shape was selected and the boundaries were considered as: [−10, 10] for (*E* and *CE*), and [−0.012, 0.012] for *dD*, respectively.

2. **Inference** is performed using the Mamdani method. Seven linguistic variables are expressed as (NB: negative big), (NM: negative medium), (NS: negative small), (Z: zero), (PS: positive small), (PM: positive medium), (PB: positive big). Table 1 shows the applied rules that ensure the relationship between the inputs and the output of the FL controller. The symmetric rule base is usually used for constant growth systems.

Table 1. Fuzzy rules.

E \ ΔE	NB	NM	NS	Z	PS	PM	PB
NB	NB	NB	NB	NB	NM	NS	Z
NM	NB	NB	NB	NM	NS	Z	PS
NS	NB	NB	NM	NS	Z	PS	PM
Z	NB	NM	NS	Z	PS	PM	PB
PS	NM	NS	Z	PS	PM	PB	PB
PM	NS	Z	PS	PM	PB	PB	PB
PB	Z	PS	PM	PB	PB	PB	PB

The flowchart of the fuzzy controller is shown in the Figure 8.

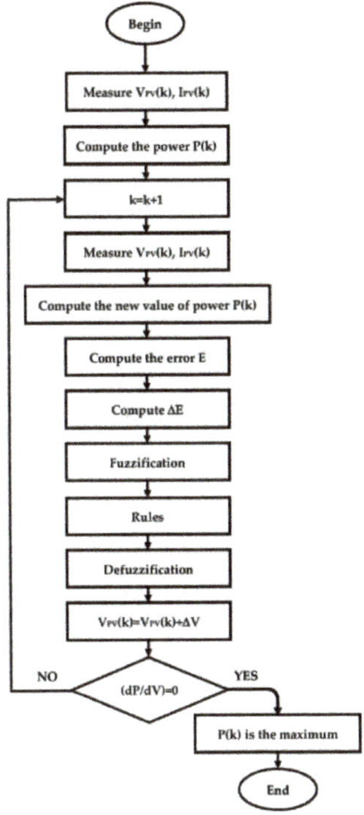

Figure 8. Fuzzy MPPT technique flowchart.

4. Results and Discussion

4.1. Solar Forecasting Results: Datasets Used for the Simulation

Irradiance in a tropical insular context presents particularly high intraday variability with days that can be clustered as a function of irradiance conditions [10,63,65].

The proposed structure in Figure 1 contains two dynamic systems, operating with two different time scales. The MPPT PV system is very fast (sampling time equal 10^{-4} s) compared with the solar forecasting system (sampling time equal 1 min). For that, the data selection operation is necessary.

A sequence of 11 values was selected with a high magnitude of fluctuations to assess the robustness of the system process. Hence, values of the sequence were chosen according to the high amplitude of fluctuations between two points; that is to say between time t and t + 1 (1 min). To represent a sequence representative of the high variabilities occurring over the year, a sequence was chosen on a day belonging to a month of a season. One season includes 3 months. Four seasons are considered in an intertropical climate: one dry and sunny season, one rainy and cyclonic season, and two intermediate seasons [8,62,65]. Therefore, four different observations and forecasting days were selected to extract the fluctuation with high amplitude (Figures 9–12). The built sequence is represented in Figure 13. One value was artificially (not consecutive value) chosen to simulate a plateau (see Figure 13, index point of time t_7 and t_8 are not consecutive).

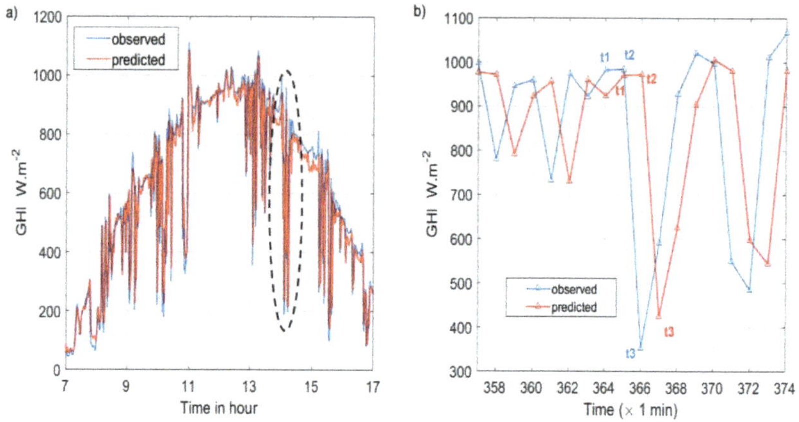

Figure 9. (**a**) Observed signal and predicted signal by STVAR model for a day (5 March). The black dotted-line is the selected part for the zoom. (**b**) Zoom of the part in a black dotted-line ellipse in (**a**).

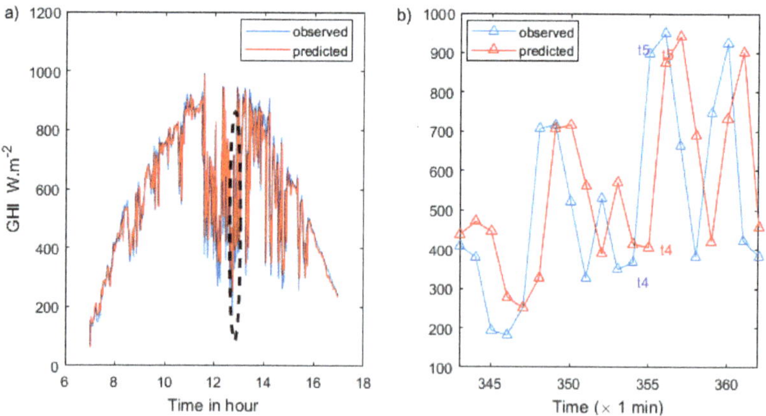

Figure 10. (**a**) Observed signal and predicted signal by STVAR model for a day (11 May). The black dotted-line is the selected part for the zoom. (**b**) Zoom of the part in a black dotted-line ellipse in (**a**).

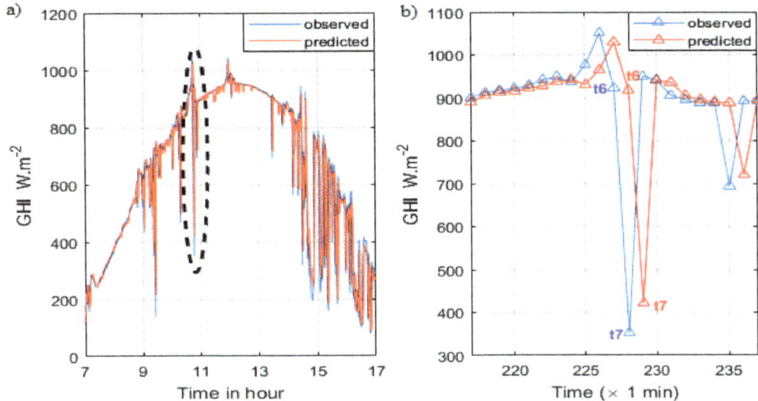

Figure 11. (**a**) Observed signal and predicted signal by STVAR model for a day (29 August). The black dotted-line is the selected part for the zoom. (**b**) Zoom of the part in a black dotted-line ellipse in (**a**).

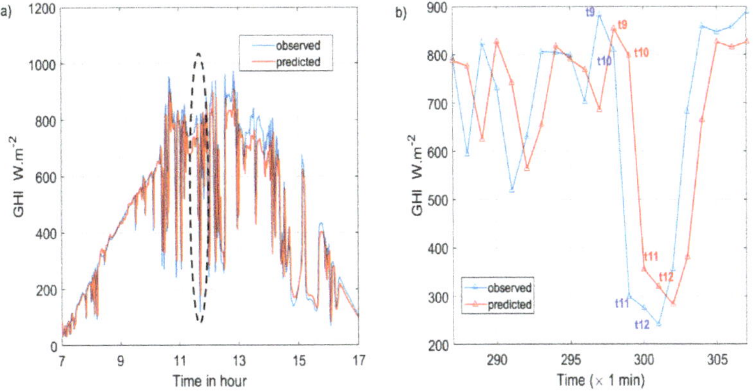

Figure 12. (**a**) Observed signal and predicted signal by STVAR model for a day (16 December). The black dotted-line is the selected part for the zoom. (**b**) Zoom of the part in a black dotted-line ellipse in (**a**).

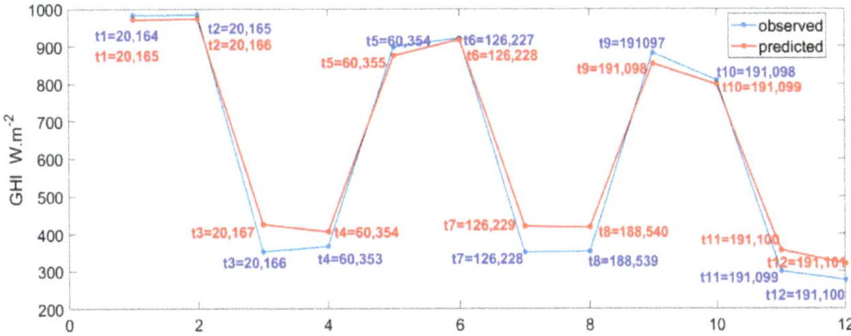

Figure 13. Sampling of observed and predicted fluctuations irradiance sequence.

There is a time lag between the observations sequence and forecast sequence, which is a particularity of the autoregressive models. This particularity can be observed in

Figures 9b, 10b, 11b and 12b. To build the sequence represented by Figure 13 we removed this time lag.

The values corresponding to time t_i with i from 1 to 3 in the sequence in Figure 9 are indicated.

The values corresponding to time t_i with i from 4 to 5 in the sequence in Figure 10 are indicated.

The values corresponding to time t_i with i varying from 6 to 7 in the sequence in Figure 11 are indicated.

Table 2 gives the solar forecasting performance obtained for 1 min ahead with the STVAR model concerning the four selected days. Model performance is usually assessed by the following statistical metrics: relative Root Mean Squared Error (rRMSE), relative Mean Absolute Error (rMAE), and relative Mean Bias Error (rMBE). Relative error metrics were normalized to the mean observed irradiance data for the considered period.

$$\text{rRMSE} = \frac{\sqrt{\frac{1}{N}\sum_{i=1}^{N}(\hat{G}-G)^2}}{mean(G)} \times 100\% \tag{17}$$

where $\hat{G}$ is the forecasted values and G the observed values.

$$\text{rMAE} = \frac{\frac{1}{N}\sum_{i=1}^{N}|\hat{G}-G|}{mean(G)} \times 100\% \tag{18}$$

$$\text{rMBE} = \frac{\frac{1}{N}\sum_{i=1}^{N}\hat{G}-G}{mean(G)} \times 100\% \tag{19}$$

Table 2. Solar forecasting performance for seasonal days.

	Day1 (5 March)	**Day2 (11 May)**	**Day3 (29 August)**	**Day4 (16 December)**
rMAE (%)	11.32	10.82	6.86	13.13
rMBE (%)	−1.65	0.19	−0.03	−2.99
rRMSE (%)	19.99	19.01	14.36	21.80

The values corresponding to time t_i with i from 9 to 12 in the sequence in Figure 12 are indicated.

4.2. MPPT Fuzzy Logic Controller Performance

Diagram blocks of the fuzzy logic controller are presented in Figure 14:

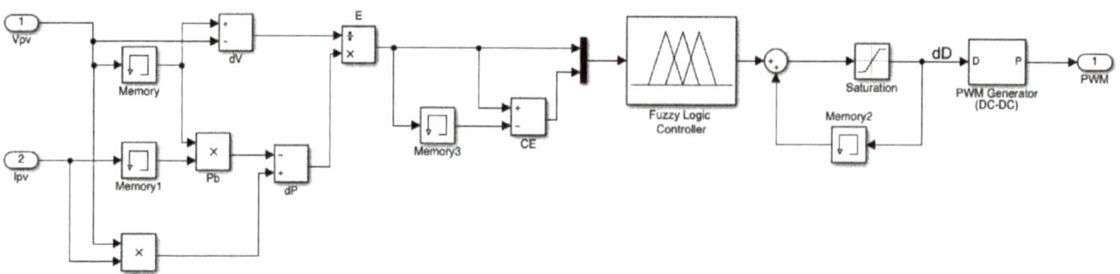

Figure 14. Diagram blocks of MPPT fuzzy logic controller.

Simulation results are given to check the PV MPPT fuzzy logic performance. The comparison in terms of robustness, rapidity, accuracy, and stability of the proposed controller with the conventional P&O one was tested for a real measured data irradiance profile presented in Figure 15, including sudden changes, which is presented in the solar forecasting section. The temperature was taken as constant and equal at 25 °C.

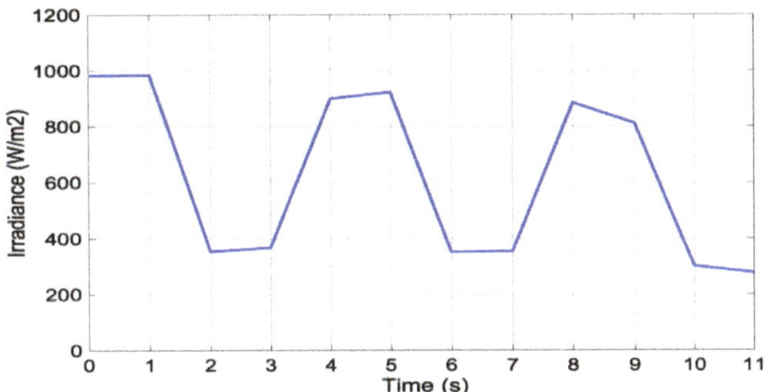

Figure 15. Measured irradiance profile.

Figure 16 shows the dynamic performance of PV voltage, PV current, and output PV power for the two controllers (fuzzy and P&O) under a variable irradiance profile.

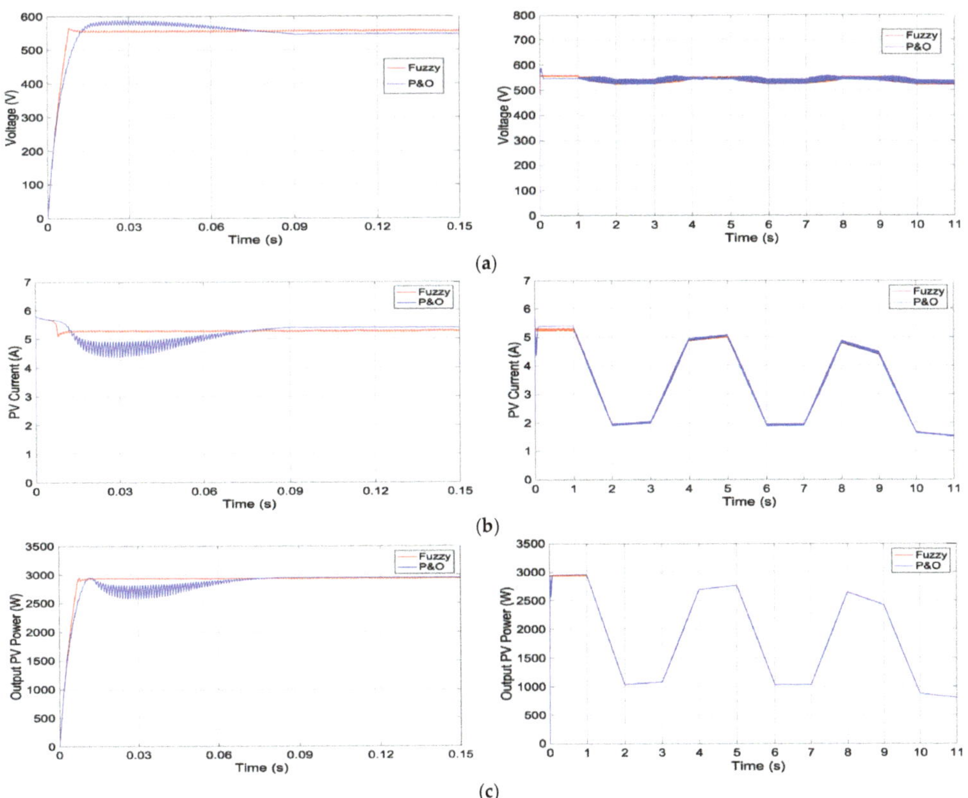

Figure 16. Comparative MPPT PV system performance (fuzzy vs. P&O) for two time scales. (**a**) Output voltage, (**b**) PV current, (**c**) output PV power.

It can be seen that, despite the multiple changes in solar data, the two controllers can follow the change in irradiance to reach the MPP.

The comparison between the two MPPT techniques can be made for performance during the transient and steady state. For that, responses are presented for two time scales. Regarding responses with a short simulation time of 0.15 s, we notice:

- The dynamic response of the P&O controller exhibits unwanted ripples, which are a dangerous drawback for PV systems. In addition, this dynamic response is a pseudo-oscillatory time response with an overshoot equal to 11.87%.
- Compared to P&O, the fuzzy controller converges and reaches the steady state faster with a response time ten times shorter (0.008 s against 0.08 s). We can clearly observe the superiority of the fuzzy controller response in terms of stability and accuracy with respect to the P&O one, which is less stable with ripples and oscillation, and less accurate with a slight permanent steady-state error.

Simulation results demonstrate very clearly that the proposed scheme (fuzzy MPPT controller) exhibits a fast and accurate dynamic response and stable steady-state output power against a rapid change in the irradiance profile.

4.3. Output PV Power Prediction

The PV power generation plant of 3 kW controlled by the MPPT fuzzy logic technique was fed with the measured and predicted solar data results given in Section 4.1. The solar measurements were collected from the Petit-canal site in Guadeloupe. Predicted solar data were obtained using the STVAR model for a very short time scale of 1 min ahead. Four seasonal days, representing the fast variability of a tropical climate over 10 h of sunshine every day, were selected. The produced output power of the PV system was calculated using measured and predicted data in the cases of the four seasonal days.

In order to evaluate the accuracy and efficiency of the PV power prediction, $rRMSE$, $rMAE$, and $rMBE$ were calculated for the four seasonal days. The results are given in Table 3.

Table 3. PV power prediction performance for seasonal days.

	Day 1	Day 2	Day 3	Day 4
rMAE (%)	11.58	11.09	6.99	13.70
rMBE (%)	−1.60	0.19	−0.03	−3.14
rRMSE (%)	20.43	19.46	14.65	22.60

From the simulation results given in Figure 17 and Table 3 for the same four days in Section 4.1 belonging to different seasons, we notice that:

- Output PV power variability calculated from predicted data in the cases of the four days is close to that calculated from the measured one, which demonstrates the efficiency of the proposed method.
- The influence of intraday variability on the predictive performance of the model is also observed. The highest errors are obtained for day 4 corresponding to the day presenting the highest intraday variability ($rRMSE = 22.60\%$, $rMAE = 13.70\%$), whereas the lowest errors are obtained for day 3 presenting the lowest intraday variability and with less pronounced variation ($rRMSE = 14.65\%$, $rMAE = 6.99\%$). Consequently, the results are an illustration of the predictive performance model as a function of the irradiance conditions as seen in the results in the literature [7,9,66].
- According to the $rMBE$ results, the model tends to underestimate the observed values (negative values of $rMBE$).

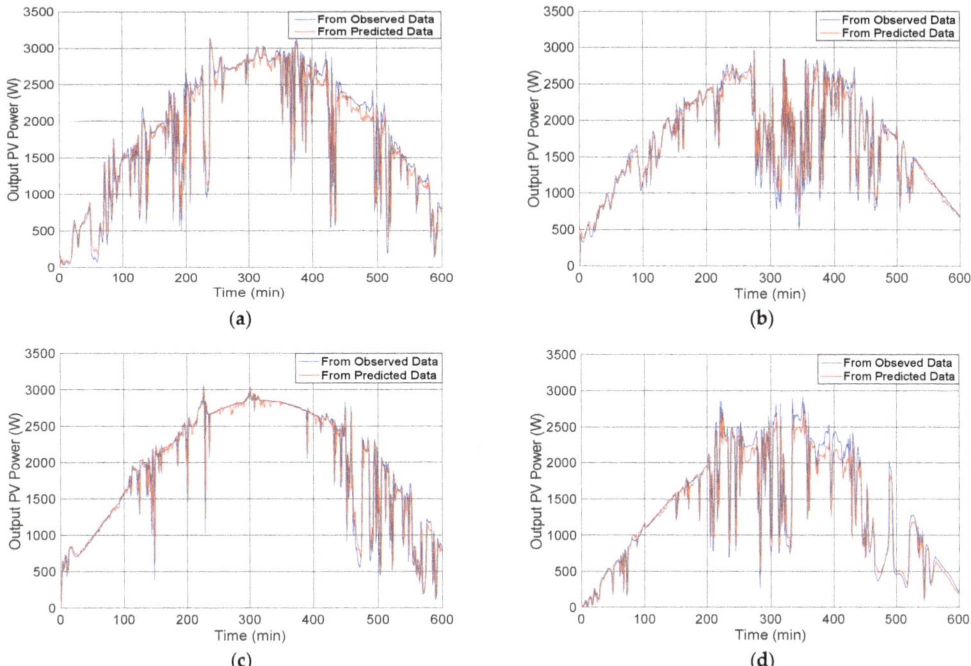

Figure 17. Output PV power prediction over 4 season days (1 min ahead). (**a**) Day1 (5 March); (**b**) Day2 (11 May); (**c**) Day3 (29 August); (**d**) Day4 (16 December).

Consequently, by comparison of the solar irradiance prediction previously mentioned in Figures 9–12 and Table 2 with the indirect output power prediction given in Figure 17 and Table 3, the output PV power model has the same predictive performance trend as the irradiance forecasting model (STVAR).

$$\text{skill} - \text{score} = (1 - \frac{RMSE_{model}}{RMSE_{persistence}}) \times 100 \tag{20}$$

Skill-score is an error metric that compares the model performance with a reference model described by Equation (20). According to [67], the skill parameter (skill-score) proposed is a comparison with the proposed model and the persistence model. Persistence forecasts, as the name implies, are defined as having the output PV power that persists for the next time step. It is a naive model often used as a reference forecasting model. Table 4 presents the results of the skill-score for the same 4 days as in Table 3. A positive skill-score significates that the proposed model has a better predictive performance than the persistence model. A negative skill-score shows that the proposed model has a lower predictive performance than the persistence model.

Table 4. PV power prediction skill-score performance for seasonal days.

	Day1 (5 March)	Day2 (11 May)	Day3 (29 August)	Day4 (16 December)
rRMSE (%) Persistence model	21.33	20.24	15.50	22.87
Skill-score (%)	4.22	3.85	6.71	1.18

Consequently, Table 4 exhibits a better predictive performance of our forecasting technic than the persistence model whatever the seasonal day. This metric error is also able to compare if our approach could be useful to other researchers.

5. Conclusions

The existence of a sudden and rapid change in the meteorological conditions, especially in a tropical context, led us to propose through the present work a robust and reliable solution based on real measured and predicted irradiance data. The proposed work deals with an indirect PV power production prediction of a PV power generation plant for a short time scale. The PV power generation plant is composed of 10 solar panels (3 kW) and a DC-DC boost converter. The PV power prediction approach is based on a combination of solar forecasting results and a fuzzy MPPT technique. Solar measurements data were collected on Guadeloupe Island.

Solar forecasting using the STVAR model, applied for the first time at 1 min time scales, allowed us to obtain a representative irradiance profile to test the efficiency of the proposed MPPT PV solution. The latter is designed to face two challenges: maximum power extraction from the PV generator, and output PV power prediction. The fuzzy MPPT technique ensures a better optimization performance presenting good robustness with respect to extreme climatic conditions. The output PV power prediction performance depends on the used solar prediction method, the MPPT PV technique, and irradiance profile variability. The same predictive performance trend of the output PV power model as the irradiance forecasting model (STVAR) is an expected result due to the structure of the PV system. Indeed, the structure combines the irradiance forecasting model and the forecasting PV output model. The strength and limits of the irradiance forecasting model (STVAR model) are consequently imposed on the final results (output PV forecasting); hence, the importance of an appropriate choice of forecasting irradiance model. The investigation of a new global structure diagram for a PV system integrating two approaches (statistical forecasting irradiance and an MPPT approach) showed the strengths and limitations of this system. In a subsequent investigation, we will test another forecasting model that has already shown good predictive performance in the literature, at this very short time scale (1 min). Operationally, such an approach would mesh with operational industry-targeted forecast services.

Author Contributions: Conceptualization, F.M.; Data curation, M.A.; Formal analysis, R.C.; Investigation, M.A.; Methodology, F.M.; Project administration, F.M.; Software, F.M.; Supervision, R.C.; Validation, F.M.; Visualization, R.C.; Writing—original draft, F.M. and M.A. All authors have read and agreed to the published version of the manuscript.

Funding: This research received no external funding.

Conflicts of Interest: The authors declare no conflict of interest.

Appendix A

We chose model SunPower SPR-300E-WHT-D. This panel is composed of 96 cells series-connected ($n_s = 96$).

- These data represent the typical performance of the panel SPR-300E, which is measured with output, and no additional equipment effect is included such as the diodes and the cables. The data are based on the measures under the standard conditions SRC (Standard Reporting Conditions, knowledge also: STC or Standard Test Conditions), which set:
- an Irradiance of 1 kW/m^2 (1 sun) to a spectrum AM 1.5;
- a temperature of the cell of 25 °C.

Table A1. Photovoltaic panel parameters.

Maximal power	P_m	300 W
Voltage for maximal power	V_{pm}	54.7 V
Current for maximal power	I_{pm}	5.49 A
Current of short circuit	I_{sc}	5.87 A
Voltage of open circuit	V_{oc}	64 V
Temperature coefficient of I_{sc}	T_{sc}	$(0.061738)\ \%/°C$
Temperature coefficient of V_{oc}	T_{oc}	$(-0.2727)\ \%/°C$

References

1. Gioutsos, D.M.; Blok, K.; van Velzen, L.; Moorman, S. Cost-optimal electricity systems with increasing renewable energy penetration for islands across the globe. *Appl. Energy* **2018**, *226*, 437–449. [CrossRef]
2. Silsirivanich, N. Fluctuation Characteristics effect analysis of Solar Irradiance Data by Wavelet transform. *Energy Procedia* **2017**, *138*, 301–306. [CrossRef]
3. Zafeiropoulou, M.; Mentis, I.; Sijakovic, N.; Terzic, A.; Fotis, G.; Maris, T.I.; Vita, V.; Zoulias, E.; Ristic, V.; Ekonomou, L. Forecasting Transmission and Distribution System Flexibility Needs for Severe Weather Condition Resilience and Outage Management. *Appl. Sci.* **2022**, *12*, 7334. [CrossRef]
4. Sijakovic, N.; Terzic, A.; Fotis, G.; Mentis, I.; Zafeiropoulou, M.; Maris, T.I.; Zoulias, E.; Elias, C.; Ristic, V.; Vita, V. Active System Management Approach for Flexibility Services to the Greek Transmission and Distribution System. *Energies* **2022**, *15*, 6134. [CrossRef]
5. Calif, R.; Schmitt, F.G.; Huang, Y.; Soubdhan, T. Intermittency study of high frequency global solar radiation sequences under a tropical climate. *Sol. Energy* **2013**, *98*, 349–365. [CrossRef]
6. André, M.; Dabo-Niang, S.; Soubdhan, T.; Ould-Baba, H. Predictive spatio-temporal model for spatially sparse global solar radiation data. *Energy* **2016**, *111*, 599–608. [CrossRef]
7. Lauret, P.; Voyant, C.; Soubdhan, T.; David, M.; Poggi, P. A benchmarking of machine learning techniques for solar radiation forecasting in an insular context. *Sol. Energy* **2015**, *112*, 446–457. [CrossRef]
8. Cécé, R.; Bernard, D.; Brioude, J.; Zahibo, N. Microscale anthropogenic pollution modelling in a small tropical island during weak trade winds: Lagrangian particle dispersion simulations using real nested LES meteorological fields. *Atmos. Environ.* **2016**, *139*, 98–112. [CrossRef]
9. André, M.; Perez, R.; Soubdhan, T.; Schlemmer, J.; Calif, R.; Monjoly, S. Preliminary assessment of two spatio-temporal forecasting technics for hourly satellite-derived irradiance in a complex meteorological context. *Sol. Energy* **2019**, *177*, 703–712. [CrossRef]
10. Soubdhan, T.; Emilion, R.; Calif, R. Classification of daily solar radiation distributions using a mixture of Dirichlet distributions. *Sol. Energy* **2009**, *83*, 1056–1063. [CrossRef]
11. Monjoly, S.; André, M.; Calif, R.; Soubdhan, T. Forecast Horizon and Solar Variability Influences on the Performances of Multiscale Hybrid Forecast Model. *Energies* **2019**, *12*, 2264. [CrossRef]
12. Stein, J.S.; Hansen, C.W.; Reno, M.J. The variability index: A new and novel metric for quantifying irradiance and PV output variability. In Proceedings of the World Renewable Energy Forum, Denver, CO, USA, 13–17 May 2012; pp. 13–17.
13. Lenox, C.; Nelson, L. Variability comparison of large-scale photovoltaic systems across diverse geographic climates. In Proceedings of the 25th European Photovoltaic Solar Energy Conference, Valencia, Spain, 6–10 September 2010.
14. Van Haaren, R.; Morjaria, M.; Fthenakis, V. Empirical assessment of short-term variability from utility-scale solar PV plants. *Prog. Photovolt. Res. Appl.* **2012**, *22*, 548–559. [CrossRef]
15. Lave, M.; Reno, M.J.; Broderick, R.J. Characterizing local high-frequency solar variability and its impact to distribution studies. *Sol. Energy* **2015**, *118*, 327–337. [CrossRef]
16. Perez, R.; David, M.; Hoff, T.E.; Jamaly, M.; Kivalov, S.; Kleissl, J.; Lauret, P.; Perez, M. Spatial and Temporal Variability of Solar Energy. *Found. TrendsR Renew. Energy* **2016**, *1*, 1–44. [CrossRef]
17. Kleissl, J. *Solar Energy Forecasting and Resource Assessment*; Academic Press: Cambridge, MA, USA, 2013.
18. Finkenstadt, B.; Held, L.; Isham, V. *Statistical Methods for Spatio-Temporal Systems*; CRC Press: Boca Raton, FL, USA, 2006.
19. Cressie, N.; Johannesson, G. Fixed rank Kriging for very large data sets. *J. R. Stat. Soc. Ser. B* **2008**, *70*, 209–226. [CrossRef]
20. Cressie, N.; Wikle, C.K. *Statistics for Spatio-Temporal Data*; John Wiley & Sons: Hoboken, NJ, USA, 2011.
21. Cellura, M.; Cirrincione, G.; Marvuglia, A.; Miraoui, A. Wind speed spatial estimation for energy planning in Sicily: A neural kriging application. *Renew. Energy* **2008**, *33*, 1251–1266. [CrossRef]
22. De Luna, X.; Genton, M.G. Predictive spatio-temporal models for spatially sparse enviromental data. *Stat. Sin.* **2005**, *15*, 547–568.
23. Liu, H.; Shi, J.; Erdem, E. Prediction of wind speed time series using modified Taylor Kriging method. The 3rd International Conference on Sustainable Energy and Environmental Protection, SEEP. *Energy* **2010**, *35*, 4870–4879. [CrossRef]
24. Porcu, E.; Mateu, J.; Saura, F. New classes of covariance and spectral density functions for spatio-temporal modelling. *Stoch. Environ. Res. Risk Assess.* **2008**, *22*, 65–79. [CrossRef]

25. Inoue, T.; Sasaki, T.; Washio, T. Spatio-Temporal Kriging of Solar Radiation Incorporating Direction and Speed of Cloud Movement. 2012. Available online: https://www.jstage.jst.go.jp/article/pjsai/JSAI2012/0/JSAI2012_1K2IOS1b3/_article/-char/ja/ (accessed on 25 August 2022).

26. Yang, D.; Gu, C.; Dong, Z.; Jirutitijaroen, P.; Chen, N.; Walsh, W.M. Solar irradiance forecasting using spatial-temporal covariance structures and time-forward kriging. *Renew. Energy* **2013**, *60*, 235–245. [CrossRef]

27. Yang, D.; Dong, Z.; Reindl, T.; Jirutitijaroen, P.; Walsh, W.M. Solar irradiance forecasting using spatio-temporal empirical kriging and vector autoregressive models with parameter shrinkage. *Sol. Energy* **2014**, *103*, 550–562. [CrossRef]

28. Aryaputera, A.W.; Yang, D.; Zhao, L.; Walsh, W.M. Very short-term irradiance forecasting at unobserved locations using spatio-temporal kriging. *Sol. Energy* **2015**, *122*, 1266–1278. [CrossRef]

29. Yang, D.; Ye, Z.; Lim, L.H.I.; Dong, Z. Very short term irradiance forecasting using the lasso. *Sol. Energy* **2015**, *114*, 314–326. [CrossRef]

30. Glasbey, A.; Allcroft, C.D. A spatiotemporal autoregressive moving average model for solar radiation. *J. R. Stat. Soc. Ser. C* **2008**, *57*, 343–355. [CrossRef]

31. Dambreville, R.; Blanc, P.; Chanussot, J.; Boldo, D. Very short-term forecasting of the Global Horizontal Irradiance using a spatio-temporal autoregressive model. *Renew. Energy* **2014**, *72*, 291–300. [CrossRef]

32. Ahmed, R.; Sreeram, V.; Mishra, Y.; Arif, M.D. A review and evaluation of the state-of-the-art in PV solar power forecasting: Techniques and optimization. *Renew. Sustain. Energy Rev.* **2020**, *124*, 109792. [CrossRef]

33. Rajagukguk, R.A.; Ramadhan, R.A.A.; Lee, H.-J. A Review on Deep Learning Models for Forecasting Time Series Data of Solar Irradiance and Photovoltaic Power. *Energies* **2020**, *13*, 6623. [CrossRef]

34. Lappalainen, K.; Wang, G.C.; Kleissl, J. Estimation of the largest expected photovoltaic power ramp rates. *Appl. Energy* **2020**, *278*, 115636. [CrossRef]

35. Husein, M.; Chung, I.-Y. Day-Ahead Solar Irradiance Forecasting for Microgrids Using a Long Short-Term Memory Recurrent Neural Network: A Deep Learning Approach. *Energies* **2019**, *12*, 1856. [CrossRef]

36. Das, U.K.; Tey, K.S.; Seyedmahmoudian, M.; Mekhilef, S.; Idris, M.Y.I.; van Deventer, W.; Horan, B.; Stojcevski, A. Forecasting of photovoltaic power generation and model optimization: A review. *Renew. Sustain. Energy Rev.* **2018**, *81*, 912–928. [CrossRef]

37. Sobri, S.; Koohi-Kamali, S.; Rahim, N.A. Solar photovoltaic generation forecasting methods: A review. *Energy Convers. Manag.* **2018**, *156*, 459–497. [CrossRef]

38. Raza, M.Q.; Nadarajah, M.; Ekanayake, C. On recent advances in PV output power forecast. *Sol. Energy* **2016**, *136*, 125–144. [CrossRef]

39. Radicioni, M.; Lucaferri, V.; De Lia, F.; Laudani, A.; Lo Presti, R.; Lozito, G.M.; Riganti Fulginei, F.; Schioppo, R.; Tucci, M. Power Forecasting of a Photovoltaic Plant Located in ENEA Casaccia Research Center. *Energies* **2021**, *14*, 707. [CrossRef]

40. Yu, D.; Choi, W.; Kim, M.; Liu, L. Forecasting Day-Ahead Hourly Photovoltaic Power Generation Using Convolutional Self-Attention Based Long Short-Term Memory. *Energies* **2020**, *13*, 4017. [CrossRef]

41. Niccolai, A.; Dolara, A.; Ogliari, E. Hybrid PV Power Forecasting Methods: A Comparison of Different Approaches. *Energies* **2021**, *14*, 451. [CrossRef]

42. Marcos, J.; Marroyo, L.; Lorenzo, E.; Alvira, D.; Izco, E. Power output fluctuations in large scale PV plants: One year observations with one second resolution and a derived analytic model. *Prog. Photovolt. Res. Appl.* **2011**, *19*, 218–227. [CrossRef]

43. Mills, A.; Ahlstrom, M.; Brower, M.; Ellis, A.; George, R.; Hoff, T.; Kroposki, B.; Lenox, C.; Nicholas, M.; Stein, J.; et al. *Understanding Variability and Uncertainty of Photovoltaics for Integration with the Electric Power System*; Lawrence Berkeley National Lab.(LBNL): Berkeley, CA, USA, 2009; Volume 9. [CrossRef]

44. Nasir, M.; Khan, H.A.; Khan, I.; Hassan, N.U.; Zaffar, N.A.; Mehmood, A.; Sauter, T.; Muyeen, S.M. Grid Load Reduction through Optimized PV Power Utilization in Intermittent Grids Using a Low-Cost Hardware Platform. *Energies* **2019**, *12*, 1764. [CrossRef]

45. Andrean, V.; Chang, P.C.; Lian, K.L. A Review and New Problems Discovery of Four Simple Decentralized Maximum Power Point Tracking Algorithms—Perturb and Observe, Incremental Conductance, Golden Section Search, and Newton's Quadratic Interpolation. *Energies* **2018**, *11*, 2966. [CrossRef]

46. Ali, A.; Almutairi, K.; Malik, M.Z.; Irshad, K.; Tirth, V.; Algarni, S.; Zahir, M.H.; Islam, S.; Shafiullah, M.; Shukla, N.K. Review of Online and Soft Computing Maximum Power Point Tracking Techniques under Non-Uniform Solar Irradiation Conditions. *Energies* **2020**, *13*, 3256. [CrossRef]

47. Karami, N.; Moubayed, N.; Outbib, R. General review and classification of different MPPT Techniques. *Renew. Sustain. Energy Rev.* **2017**, *68*, 1–18. [CrossRef]

48. Motahhir, S.; El Hammoumi, A.; El Ghzizal, A. The most used MPPT algorithms: Review and the suitable low-cost embedded board for each algorithm. *J. Clean. Prod.* **2020**, *246*, 118983. [CrossRef]

49. Zhu, Y.; Kim, M.K.; Wen, H. Simulation and Analysis of Perturbation and Observation-Based Self-Adaptable Step Size Maximum Power Point Tracking Strategy with Low Power Loss for Photovoltaics. *Energies* **2019**, *12*, 92. [CrossRef]

50. Li, C.; Chen, Y.; Zhou, D.; Liu, J.; Zeng, J. A High-Performance Adaptive Incremental Conductance MPPT Algorithm for Photovoltaic Systems. *Energies* **2016**, *9*, 288. [CrossRef]

51. Chalh, A.; El Hammoumi, A.; Motahhir, S.; El Ghzizal, A.; Subramaniam, U.; Derouich, A. Trusted Simulation Using Proteus Model for a PV System: Test Case of an Improved HC MPPT Algorithm. *Energies* **2020**, *13*, 1943. [CrossRef]

52. Alshareef, M.; Lin, Z.; Ma, M.; Cao, W. Accelerated Particle Swarm Optimization for Photovoltaic Maximum Power Point Tracking under Partial Shading Conditions. *Energies* **2019**, *12*, 623. [CrossRef]
53. Hadji, S.; Gaubert, J.-P.; Krim, F. Real-Time Genetic Algorithms-Based MPPT: Study and Comparison (Theoretical an Experimental) with Conventional Methods. *Energies* **2018**, *11*, 459. [CrossRef]
54. Robles Algarín, C.; Taborda Giraldo, J.; Rodríguez Álvarez, O. Fuzzy Logic Based MPPT Controller for a PV System. *Energies* **2017**, *10*, 2036. [CrossRef]
55. Chang, S.; Wang, Q.; Hu, H.; Ding, Z.; Guo, H. An NNwC MPPT-Based Energy Supply Solution for Sensor Nodes in Buildings and Its Feasibility Study. *Energies* **2019**, *12*, 101. [CrossRef]
56. Ahmed, S.; Muhammad Adil, H.M.; Ahmad, I.; Azeem, M.K.; e Huma, Z.; Abbas Khan, S. Supertwisting Sliding Mode Algorithm Based Nonlinear MPPT Control for a Solar PV System with Artificial Neural Networks Based Reference Generation. *Energies* **2020**, *13*, 3695. [CrossRef]
57. Ali, K.; Khan, L.; Khan, Q.; Ullah, S.; Ahmad, S.; Mumtaz, S.; Karam, F.W.; Naghmash. Robust Integral Backstepping Based Nonlinear MPPT Control for a PV System. *Energies* **2019**, *12*, 3180. [CrossRef]
58. Kececioglu, O.F.; Gani, A.; Sekkeli, M. Design and Hardware Implementation Based on Hybrid Structure for MPPT of PV System Using an Interval Type-2 TSK Fuzzy Logic Controller. *Energies* **2020**, *13*, 1842. [CrossRef]
59. Zadeh, L. Fuzzy sets. *Inf. Control.* **1965**, *8*, 338–353. [CrossRef]
60. Tchoketch Kebir, G.F.; Larbes, C.; Ilinca, A.; Obeidi, T.; Tchoketch Kebir, S. Study of the Intelligent Behavior of a Maximum Photovoltaic Energy Tracking Fuzzy Controller. *Energies* **2018**, *11*, 3263. [CrossRef]
61. Hassan, T.-U.; Abbassi, R.; Jerbi, H.; Mehmood, K.; Tahir, M.F.; Cheema, K.M.; Elavarasan, R.M.; Ali, F.; Khan, I.A. A Novel Algorithm for MPPT of an Isolated PV System Using Push Pull Converter with Fuzzy Logic Controller. *Energies* **2020**, *13*, 4007. [CrossRef]
62. Badosa, J.; Haeffelin, M.; Chepfer, H. Scales of spatial and temporal variation of solar irradiance on Reunion tropical island. *Sol. Energy* **2013**, *88*, 42–56. [CrossRef]
63. Voyant, C.; Soubdhan, T.; Lauret, P.; David, M.; Muselli, M. Statistical parameters as a means to a priori assess the accuracy of solar forecasting models. *Energy* **2015**, *90*, 671–679. [CrossRef]
64. Pena, D.; Tiao, G.C.; Tsay, R.S. *A Course in Time Series Analysis*; John Wiley & Sons: Hoboken, NJ, USA, 2001.
65. Soubdhan, T.; Abadi, M.; Emilion, R. Time dependent classification of solar radiation sequences using best information criterion. *Energy Procedia* **2014**, *57*, 1309–1316. [CrossRef]
66. Monjoly, S.; André, M.; Calif, R.; Soubdhan, T. Hourly forecasting of global solar radiation based on multiscale decomposition methods: A hybrid approach. *Energy* **2017**, *119*, 288–298. [CrossRef]
67. Coimbra, C.F.M.; Kleissl, J.; Marquez, R. Overview of solar forecasting method and a metric for accuracy evaluation. In *Solar Energy Forecasting and Resource Assessment*; Kleissl, J., Ed.; Elsevier: Amsterdam, The Netherlands, 2013; pp. 171–194.

Article

Negative Impact Mitigation on the Power Supply System of a Fans Group with Frequency-Variable Drive

Yerbol Yerbayev [1], Ivan Artyukhov [2],*, Artem Zemtsov [3], Denis Artyukhov [2], Svetlana Molot [2], Dinara Japarova [1] and Viktor Zakharov [4]

[1] Higher School of Electrical Engineering and Automation, Zhangir Khan West Kazakhstan Agrarian Technical University, Uralsk 090009, Kazakhstan
[2] Department of Power and Electrical Engineering, Yuri Gagarin State Technical University of Saratov, 410054 Saratov, Russia
[3] Department of Power Supply for Industrial Enterprises, Samara State Technical University, Syzran Branch, 446001 Syzran, Russia
[4] Department of Energy, Automation and Computer Technology, West Kazakhstan Innovative and Technological University, Uralsk 090009, Kazakhstan
* Correspondence: ivart54@mail.ru

Abstract: The technological installations' characteristics are possible to improve by equipping fans with a frequency-controlled electric drive. However, it can lead to an electromagnetic compatibility problem in the electrical supply system. This problem becomes worse if a large number of fans are included in the technological installation and the electric drives are powered from a substation connected to a limited power source. As an example, in this article we investigate the power supply system of a gas cooling unit with variable-frequency electric drives for fans. The electric drives' operating mode dependences characterizing the non-sinusoidal voltages and currents of the power source are obtained with the help of simulation modeling in the MATLAB environment with the Simulink expansion package. The typical substation circuit usage for the power supply of a group of fans with a frequency-controlled drive does not meet the requirements of IEEE Standard 519-2014. We can solve the problem of electromagnetic compatibility by changing the substation topology and organizing DC busbars and replacing frequency converters with inverters. We proposed forming DC busbars using 12-pulse rectifiers powered by transformers with two secondary windings with different connection schemes. The simulation results confirmed that the proposed substation topology provides the voltage and current harmonics level on the substation power busbars in accordance with the IEEE Standard 519-2014 requirements over the entire frequency range of the fans' motor control.

Keywords: gas cooling plant; power supply system; frequency variable drive; power quality; harmonic compensation

Citation: Yerbayev, Y.; Artyukhov, I.; Zemtsov, A.; Artyukhov, D.; Molot, S.; Japarova, D.; Zakharov, V. Negative Impact Mitigation on the Power Supply System of a Fans Group with Frequency-Variable Drive. *Energies* **2022**, *15*, 8858. https://doi.org/10.3390/en15238858

Academic Editor: Juri Belikov

Received: 23 October 2022
Accepted: 21 November 2022
Published: 23 November 2022

Publisher's Note: MDPI stays neutral with regard to jurisdictional claims in published maps and institutional affiliations.

1. Introduction

There is a wide range of technological installations including groups of fans. For example, cooling towers cool recycled water at thermal power plants, oil refineries, and petrochemical and other industrial enterprises [1]. A large number of fans contain gas cooling plants (GCPs), providing the required temperature regime of the main gas transport. The main structural element of the GCP is the air-cooling unit (ACU). Electrically driven fans pump air through their heat-exchange sections [2]. GACs, where fans are placed under heat-exchange sections and are driven by multipole asynchronous motors, have become widespread. Figure 1 shows the GCP, which consists of 6 GACs, type 2AVG-75. These GACs have two VASO-16-14-24 electric motors in their composition with fans on the shaft. The rated power of the motor is 37 kW; rotation speed of the fans wheels is about 250 rpm.

A GCP can include 24 or more electrically driven fans. The total power consumed by a single gas cooling plant alone amounts to hundreds of kilowatts. Therefore, a considerable amount of electricity is used to cool the gas. Power consumption for gas cooling after compression can reach 50–70% of total power consumption for gas transportation depending on climatic conditions [3]. Thus, gas cooling cost reduction is an actual problem, which can be solved in the following interrelated directions:

- Improving the aerodynamic characteristics of fans;
- Improving the design of heat exchange sections;
- Improving the algorithms and technical means of fan motor control.

GCP is a complex multidimensional system that implements the energy transfer functions in a distributed heat exchanger. The dependence of the gas temperature at the GCP outlet in the general case is:

$$T_{out} = \Psi\left(T_{in},\ G_{gas},\ \Theta_{air},\ G_{air},\ R_{\Theta}\right), \tag{1}$$

where

T_{in}, G_{gas} —temperature and mass flow rate of gas at the GCP inlet, respectively;
Θ_{air}, G_{air}—temperature and mass flow rate of air through the GCP heat exchange sections;
R_{Θ}—thermal resistance, characterizing the pollution level of the heat exchange surface.

The gas flow rate and temperature after compression, as well as the pollution level of the heat exchanger surfaces, are slow-moving disturbing influences. Outdoor temperature variations (daily and seasonal) are rapidly changing disturbing factors. It follows from (1) that the required gas temperature T_{out} at the GCP outlet can be provided by changing the cooling air mass flow rate G_{air}, which is determined by the sum of the cooling air mass flow rates created by simultaneously operating M fans:

$$G_{air} = \sum_{k=1}^{M} G_{air.k}. \tag{2}$$

In this case, the fan performance depends on the rotational speed n_k of the fan impeller, the blade angles α_k, and the air temperature Θ_{air}.

The simplest GCP control algorithm involves controlling the mass flow rate (2) by changing the number M of fans running simultaneously. A typical power supply scheme for such a control algorithm is shown in Figure 2. Two 6(10)/0.4 kV transformers are installed at the substation to ensure the required reliability of power supply. The voltages from the secondary windings of the transformers are fed through circuit breakers QF1, QF2 to the corresponding busbar sections, which can be connected through a sectional switch QF3.

(a)   (b)

Figure 1. Gas cooling plant (**a**) and VASO-16-14-24 electric motor with a fan on shaft (**b**).

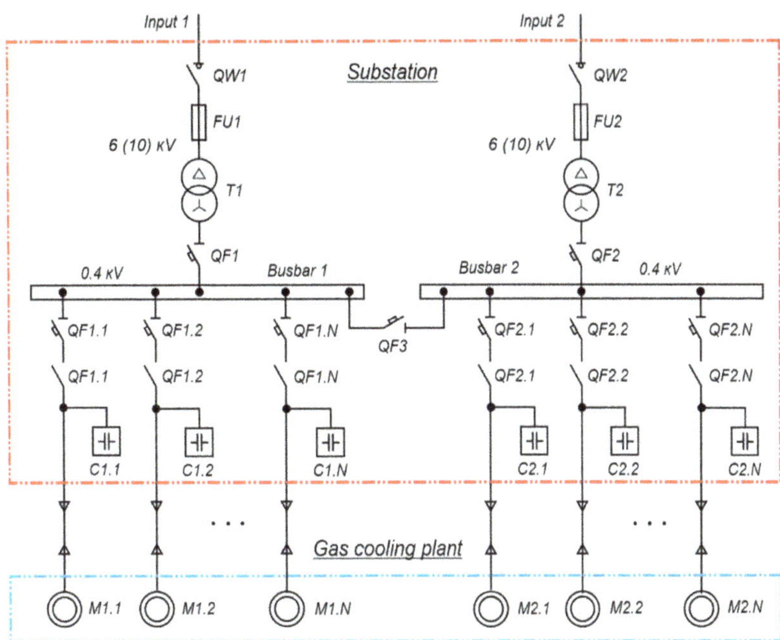

Figure 2. Typical power supply scheme for gas cooling plant.

Electric motors M1.1...M1.N and M2.1...M2.N are powered via cables laid on trestle bridges and are connected to 0.4 kV section buses via circuit breakers QF1.1...QF1.N, QF2.1...QF2.N and contactors K1.1...K1.N, K2.1...K2.N. Capacitors C1.1...C1.N and C2.1...C2.N are connected in parallel to the stator windings to compensate the reactive power of electric motors. Centralized reactive power compensation on busbar sections is also possible.

The discrete motor control algorithm has significant disadvantages. Due to the design features of the GAC, when the fan is running, part of the air blows back through the adjacent fan that is not running if the flow is sufficiently strong to ensure rotating in the opposite direction. Air recirculation has a major impact on the energy efficiency of the gas cooling process, increasing electrical energy losses and reducing overall plant efficiency. In addition, the subsequent direct start of the fan motor rotating in antiphase causes electrical and mechanical shock loads, much higher than the nominal allowed for the motor-fan system. Serious loads on the fan components create a short-term power outage, as a result of which the fan shuts down and restarts after the power supply is restored.

A special feature of the GCP (Figure 1) is the use of multipole asynchronous motors with a power factor of 0.65 to 0.68 in nominal mode. The starting current ratio of such motors is 4.5 to 5. The high inertia of the rotating masses leads to a slow start. A large amount of electrical energy is used to change the mechatronic system kinetic energy. The simultaneous activation of several fans, e.g., after a power outage, is impossible because the power source is overloaded and the protection is triggered.

Methods and technical solutions to reduce starting currents are well known for induction motors with a small number of pole pairs [4,5]. However, they do not have a positive effect for multipole induction motors. Experiments with the soft starter showed that the engine would first start accelerating and then "hang" at an intermediate speed. In spite of the long time, the engine did not go to the set mode. It has been suggested this happens due to the low value of the multipole motor power factor at the starting mode.

Figure 3 shows graphs of changes in apparent power S and its active P and reactive Q components during start-up of an induction motor with a fan on the shaft [6]. Rated motor power is 37 kW, number of pole pairs is 12. Diameter of fan wheel is 5 m.

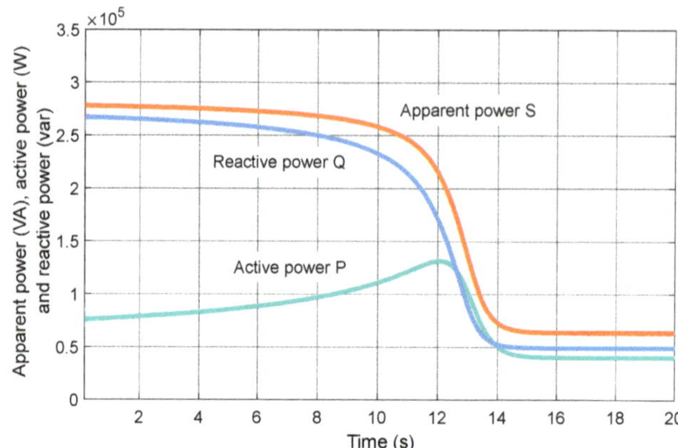

Figure 3. Changes of the active, reactive, and apparent power in the starting motor process.

Reactive power is several times higher than active power at the stages of induction motor magnetic field formation and rotor speed increases. Therefore, the solution to the problem is to use a device that performs controlled compensation of motor reactive power at the starting mode [7].

GCPs, where temperature mode is changed by fans' frequency control, provide more efficient use of electric power [8–10]. Hereinafter, we will use the term frequency-controlled gas cooling plant (FCGCP).

In general, fan impellers can rotate at different frequencies. However, experimental results have shown the turbulence of the cooling "wind field" must be removed in order to increase the efficiency of heat transfer processes in GCP sections. This requires all GCP fans to rotate at the same frequency, which is generated by the control system. The principle of gas temperature control is explained in the graphs in Figure 4 [11].

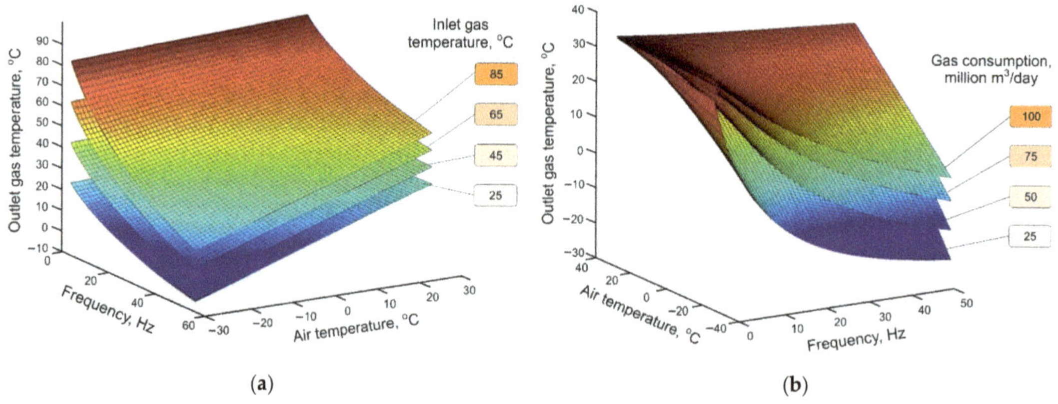

Figure 4. Dependencies of GCP outlet gas temperature on motor control frequency and ambient temperature: (**a**) at different GCP inlet gas temperatures; (**b**) at different mass flow rates of the cooled gas.

Let us assume the gas temperature T_{out} at the GCP outlet corresponded to the set value. Then, if the process parameters (T_{in}, G_{gas}) remain unchanged, the ambient air temperature Θ_{air} increases. This leads to an increase in the gas temperature T_{out} at the GCP outlet. The graphs in Figure 4 show that in order to restore the required gas temperature value, it is necessary to increase the frequency f of the voltage applied to the stator windings of electric motors. Thus, to stabilize the temperature T_{out} at the GCP output, the increments of the disturbing influences ΔT_{in}, ΔG_{gas}, $\Delta\Theta_{air}$ and the regulating influence Δf must have opposite signs.

The power supply scheme of GCP with a frequency-controlled electric drive of fans is shown in Figure 5. Frequency converters (FCs) FC1.1...FC1.N, FC2.1...FC2.N are connected to 0.4 kV busbar sections through switching devices QF1.1...QF1.N, QF2.1...QF2.N and line chokes L1.1...L1.N, L2.1...L2.N to control GCP fan motors.

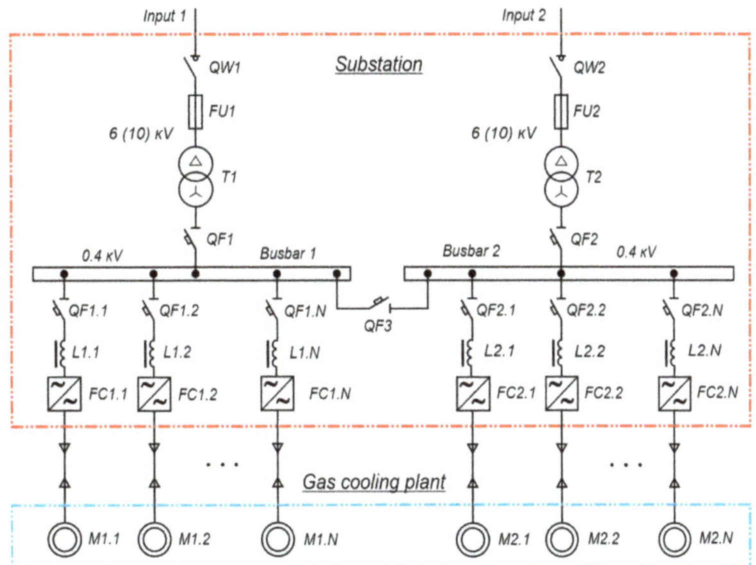

Figure 5. Power supply scheme for gas cooling plant with frequency variable drive of fans.

Connecting a large number of FCs to the substation busbars causes a powerful consumption of non-sinusoidal current. This is explained by the fact that modern frequency converters for motor control are made according to the scheme: rectifier—smoothing filter—autonomous voltage inverter on IGBT-modules. Capacitors are used as a smoothing filter, so the FC input current has a pulsed character with the dominance of the 5th and 7th harmonics [12,13].

Thus, there is a problem of electromagnetic compatibility (EMC) of the FCGCP substation for power supply with power sources and other electrical consumers of the compressor station. The problem of EMC gets worse if the compressor station power supply is coming from an autonomous power source, particularly from a gas turbine power plant [14].

The power source overloads occur when the motors start, causing voltage dips while using the discrete fan control method. With frequency variable fans, the GCP drive consumes a non-sinusoidal current whose magnitude and harmonic composition depend on the operating mode of the plant. The power quality indicators, characterizing the voltage form, go beyond the limits (set by the regulations, for example, Standard 519-2014 [15]) at a certain parameter ratio of power sources and GCP electric drives.

Given the relevance of the problem, the aim of this research is to develop a technical solution, the use of which can reduce the negative impact of the substation with a load in the form of a group of frequency-variable drives on the power source. To reach this aim, it is necessary to investigate the situation in detail with the application of typical solutions. The investigation tool is an adequate model, which for different operating modes of electric drives allows us to obtain data on the spectral composition of voltages and currents at the substation power buses. Next, based on the results obtained and the world experience in the field of EMC in power supply systems with a frequency-variable drive, it is necessary to propose a technical solution that allows us to reduce the harmonics of voltages and currents to values that are defined by regulatory documents. In this case, the required result should be achieved with minimal financial costs at the stages of implementation and operation, while maintaining other indicators, including reliability.

The problem of harmonic compensation in power supply systems with non-linear electric consumers, including variable-frequency drives, has been studied in a large number of papers. One way of solving the EMC problem involves the use of passive filters. Solution options are described in detail in [16–18]. For the studied object, the passive filters can be connected to the 0.4 kV buses of the substation. However, there is a risk of resonance phenomena.

Active harmonic filters (AHFs) allow for solving EMC problems more effectively [19–22]. When reconstructing a GCP that is in operation, this option may be appropriate, since the installation of the AHF does not require complex construction and installation work. The disadvantage of this solution is the high cost of AHF.

When FCGCP is designed for a new facility, it makes sense to use FCs that use active rectifiers on fully controlled power switches [23,24]. However, this solution also significantly increases the cost of the substation's electrical equipment.

Multipulse rectifiers are a good solution for creating converters with a reduced distorting effect on the power supply system [25–29]. The operating principle of such rectifiers assumes the presence of multi-winding transformers. For this reason, multipulse rectifier circuits are appropriate when it is necessary to change the value of the supply voltage. The FCGCP power supply system requires a reduction of the 6(10) kV line voltage to the value necessary to operate the electric drives. Therefore, the multipulse rectification option can achieve a positive effect.

This paper is structured as follows. Section 2 outlines the approach to modeling the power supply system for a group of frequency variable drives and presents the results of a study on the influence of a typical FCGP on the power source. In Section 3, a new substation topology for FCGP power supply is proposed and the simulation results confirming the positive effect of this solution are provided. Section 4 discusses the results obtained and provides future directions for the work. Finally, the conclusion is provided in Section 5.

2. Impact of the Typical FCGCP on the Power Supply System

2.1. Basic Theoretical Provisions

The mechanism of line voltage distortion can be explained by the scheme in Figure 6. One phase of the power source is represented as an equivalent bipolar with EMF $e(t)$, inductance L_0 and resistance R_0. A load as a GCP power supply substation is connected to the power source, which consumes current $i(t)$ from the grid.

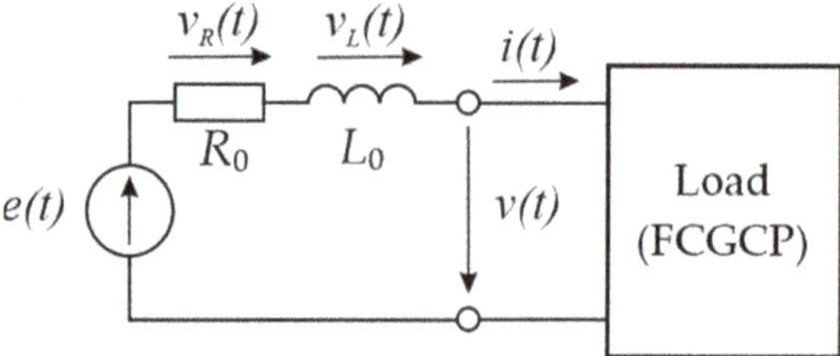

Figure 6. Calculation diagram for one phase of the FCGCP power supply system.

The following equation can be written for the instantaneous voltage value $v(t)$:

$$v(t) = e(t) - v_R(t) - v_L(t),\tag{3}$$

where $v_R(t) = R_0 \cdot i(t)$, $v_L(t) = L_0 \cdot \frac{di(t)}{dy}$–voltage drop on the resistance and inductance.

If the EMF of the power source $e(t)$ and the load current $i(t)$ are sinusoidal, as in the case of discrete GCP fan control, then the voltage $v(t)$ is also sinusoidal. The effective value of this voltage differs from the effective value of EMF by the voltage drop across the elements R_0 and L_0 connected in series.

When the GCP is equipped with a frequency-controlled fan drive, a non-sinusoidal current $i(t)$ is consumed from the power source. Current flowing through the elements R_0 and L_0 creates non-sinusoidal voltage drops on them. Therefore, even in the case of a sinusoidal EMF according to expression (3), the voltage $v(t)$ will be non-sinusoidal. The distortion of the voltage curve $v(t)$ depends on the relationship between the power supply and the load [30].

Equation (3) in complex form for the n-th harmonic of the line voltage is:

$$\underline{V}_n = \underline{E}_n - (R_0 + jn\omega_1 L_0) \cdot \underline{I}_n = \underline{E}_n - \underline{Z}_n \cdot \underline{I}_n \; ; \; n = 1, 2, 3, \ldots\tag{4}$$

where $\underline{E}_n$, $\underline{V}_n$, $\underline{I}_n$—the harmonic complexes of EMF, voltage, and current with the numbers n; $\underline{Z}_n$—the resistance complex of the power source at the n-th harmonic.

If the EMF $e(t)$ is sinusoidal, then Equation (4) for all harmonics with numbers $n \geq 2$ takes the form:

$$\underline{V}_n = -(R_0 + jn\omega_1 L_0) \cdot \underline{I}_n = -\underline{Z}_n \cdot \underline{I}_n .\tag{5}$$

Thus, the effective value of the n-th harmonic of the power source voltage for this situation is determined by the equation:

$$V_n = Z_n \cdot I_n.\tag{6}$$

The GCP with variable-frequency-drive fans is a dynamically changing system; its power source effect depends on many factors. Therefore, it is necessary to determine the conditions under which the influence of the FCGCP on the power supply system becomes critical, and make a decision how to solve the problem. This requires information about the harmonic composition of the current consumed by the GCP from the power source in various operation modes.

It is advisable to choose a solution to the problem of EMC in the power supply system of FCGCP based on the IEEE Standard 519-2014 [15]. This document does not regulate only non-sinusoidal voltage values. Compared to other standards, it regulates the current distortion limits of powerful consumers.

According to IEEE 519-2014, total harmonic distortion (THD) of voltage on power systems rated 1 kV through 69 kV is limited to 5%, with each individual harmonic limited to 3%.

Voltage THD is calculated by the formula:

$$\mathrm{THD} = \frac{\sqrt{U_2^2 + U_3^2 + U_4^2 + U_5^2 + \ldots}}{U_1} \cdot 100\%.\tag{7}$$

Current distortion limits are ranked in relation to I_{SC}/I_L ratio on point of common coupling (PCC), where I_{SC}–maximum short-circuit current at PCC, I_L–maximum demand load current (fundamental frequency component) at the PCC under normal load operating conditions. Current distortion limits for power supply systems rated 120 V through 69 kV are provided in Table 1. All values should be in percent of the maximum demand current, I_L. Therefore, even harmonics are limited to 25% of the odd harmonic limits above.

Table 1. Maximum harmonic current in percent of I_L.

I_{SC}/I_L	Individual Harmonic Order (Odd Harmonics)					
	$3 \le n < 11$	$11 \le n < 17$	$17 \le n < 23$	$23 \le n < 35$	$35 \le n < 50$	TDD
<20	4.0	2.0	1.5	0.6	0.3	5.0
20 < 50	7.0	3.5	2.5	1.0	0.5	8.0
50 < 100	10.0	4.5	4.0	1.5	0.7	12.0
100 < 1000	12.0	5.5	5.0	2.0	1.0	15.0
>1000	15.0	7.0	6.0	2.5	1.4	20.0

The IEEE 519-2014 standard for current also regulates the total demand distortion (TDD), which is calculated by the formula:

$$TDD = \frac{\sqrt{I_2^2 + I_3^2 + I_4^2 + I_5^2 + \dots}}{I_L} \cdot 100\%. \tag{8}$$

The row in Table 1, which shows the values of individual harmonics when $I_{SC}/I_L < 20$, is of major interest for the studied object. The situation $I_{SC}/I_L = 20$ corresponds to the practically important case, when one of the inputs of the substation for FCGCP power supply is connected to a transformer 110/6(10) power of 10 MVA. In these cases, EMC problems in the power supply system become visible. When the FCGCP is powered by off-line sources, the I_{SC}/I_L ratio becomes even smaller.

2.2. Modeling the FCGCP Power Supply System

We need a model that reflects the electric drive features as a power-source load to study power-quality issues in the power supply system of a compressor station when GCP is equipped with a frequency-controlled electric drive of fans.

The calculation diagram for one section of the substation busbars, supplying power to the FCGCP, is shown in Figure 7.

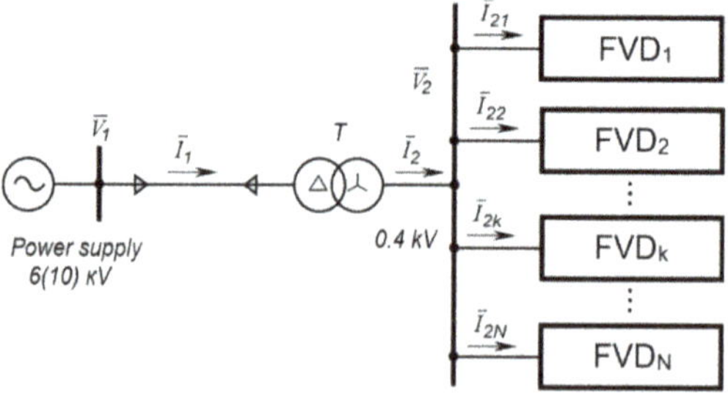

Figure 7. Calculation scheme for one busbar section.

For example, N electric drives are connected to the 0.4 kV busbar section, and each of them consumes a vector of non-sinusoidal currents $\bar{I}_{2k}$. The secondary current of transformer T will be equal to the sum of these currents:

$$\bar{I}_2 = \sum_{k=1}^{N} \bar{I}_{2k}. \tag{9}$$

The voltage on the secondary side of the transformer is characterized by the vector $\overline{V}_2$. Its components, according to (3)–(5), have a non-sinusoidal shape. A vector of non-sinusoidal currents $\overline{I}_1$ is drawn from the power supply; as a result, the components of the vector $\overline{V}_1$ also have a non-sinusoidal shape.

Each of the N drives can be represented by an equivalent scheme, which is shown in Figure 8. In this scheme: L_{sk}, R_{sk}—inductance and resistance of the line choke, respectively, which can be installed at the input of the k-th FC to ensure EMC standards; C_k, L_{dk}, R_{dk}—capacitance, inductance and resistance of the smoothing choke at the rectifier output, respectively; $R_{inv\cdot k}$ is the equivalent input resistance of the inverter on the k-th FC.

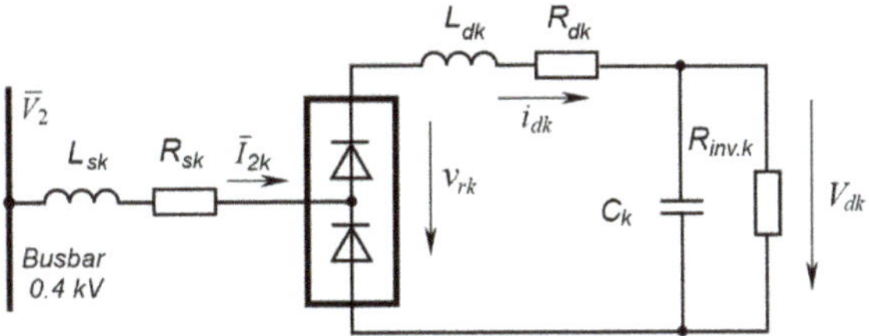

Figure 8. Calculation scheme of the variable-frequency drive.

The inverter presentation as the equivalent input resistance $R_{inv.k}$ is valid because the purpose of the electrical complex modeling is to analyze the influence of canonical harmonics in the spectrum of current consumed by the FC on the electricity quality on the substation buses. The spectrum of harmonics caused by the IGBT module switching is shifted to the high-frequency region. The analysis of this spectrum requires a model that takes into account the parasitic inductances and capacitances of the elements.

The inverter consumes power, which can be found using the equation:

$$P_k = \eta^{-1} P_{0.k} \cdot (f / f_0)^3 \tag{10}$$

where $P_{0.k}$, η are the rated power and efficiency of the electric motor connected to the inverter, respectively; f_0, f—nominal and current value of the motor control frequency.

The power that is dissipated on the equivalent resistance:

$$P_k = \frac{V_{dk}^2}{R_{inv.k}}. \tag{11}$$

From (8) and (9) we obtain the value of the equivalent resistance of the k-th inverter:

$$R_{inv.k} = \frac{\eta \cdot V_{dk}^2}{P_{0.k} \cdot (f / f_0)^3}. \tag{12}$$

Figure 9 shows a simulation model scheme to study electromagnetic processes in one section of a typical GCP power supply system, consisting of 12 gas air-coolers with a frequency-controlled fan drive of 37 kW each. The power of the 10/0.4 kV transformer was 1000 kVA, short-circuit voltage—5.5%.

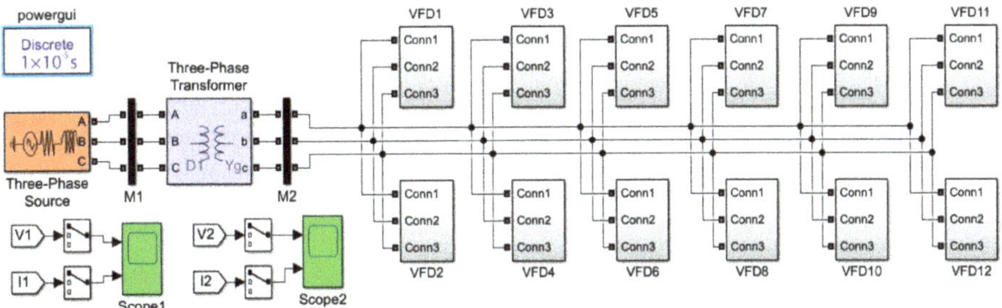

Figure 9. Schematic of the simulation model of a typical power supply system.

A Three-Phase Source unit represents the substation power supply. The Three-Phase Transformer unit simulates a transformer 10/0.4 kV. Simulation models' parameters are shown in Tables A1 and A2.

The schematic of the VFD subsystem is shown in Figure 10. The subsystem consists of a rectifier (Diode Bridge block), line reactors at the rectifier input (Line Reactors block), and a smoothing filter (LD, C blocks). Resistor R, the value of which is calculated by Formula (12), represents the inverter. The motor efficiency was assumed to be 0.915.

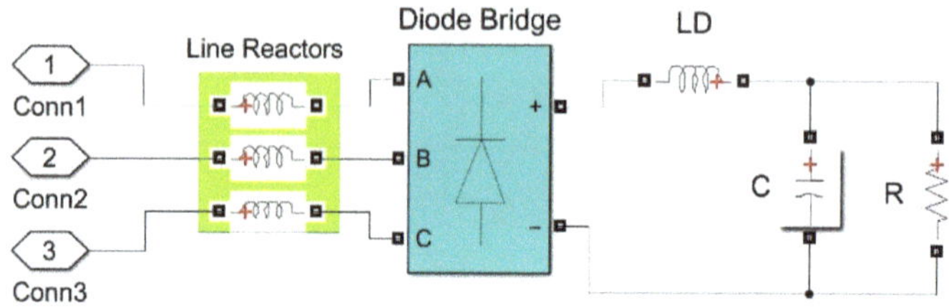

Figure 10. VFD subsystem scheme.

The line chokes inductances were 24 μH. The inductance and capacitance of the smoothing filter were 120 μH and 1000 μF, respectively.

Virtual oscilloscopes Scope 1 and Scope 2 display voltage and current curves. Blocks M1 and M2 (Three-Phase VI Measurements) generate the corresponding signals.

Simulations were performed for a three-phase power source with a voltage of 10 kV and a frequency of 50 Hz, the parameters of which were varied to obtain results for a range of values $I_{SC}/I_L \leq 20$. The control frequency of the fan motors varied from 15 to 50 Hz. The lower limit of the control frequency was adopted based on the conditions to avoid vibration and motors overheating. For each set of parameters f, I_{SC}/I_L virtual oscillograms of phase voltage in PCC and input current in the substation were obtained. Using the built-in MATLAB tool FFT Analysis, the harmonic composition of these curves was investigated.

Figures 11–14 show simulation results for I_{SC}/I_L = 3.2; 5; 8 and 20.

The graphs in Figure 11 characterize the non-sinusoidal voltage values in the PCC. They show that harmonics with numbers 5 and 7 appear and increase in the voltage curve spectrum as the fan motor control frequency increases. When I_{SC}/I_L = 8; 20 THD is in the acceptable range. If I_{SC}/I_L = 3.2; 5, then at certain frequencies of motor control THD exceeds the value of 5%. Voltage harmonics with numbers 7 and above at all I_{SC}/I_L values are less than 3%. The 5th harmonic reaches this value at f = 47 Hz for I_{SC}/I_L = 5 and at f = 42 Hz for I_{SC}/I_L = 3.2.

The graphs in Figure 12 show that the greatest contribution to the distortion of the current curve shape is made by the 5th and 7th harmonics. The relative values of these harmonics actually exceed the permissible value of 4% in the entire range of motor control frequency variation at all I_{SC}/I_L ratios.

The maximum voltage waveform distortion in PCC is observed at motor control frequency f = 50 Hz and I_{SC}/I_L = 3.2. The substation voltage and input current curves for this set of parameters are shown in Figure 13. The voltage spectrum is shown in Figure 14.

Analysis of the simulation results shows that a radical way to reduce the negative impact of the FCGCP on the power supply system is to minimize the 5th and 7th harmonics in the consumed current. The following section offers a solution to this problem.

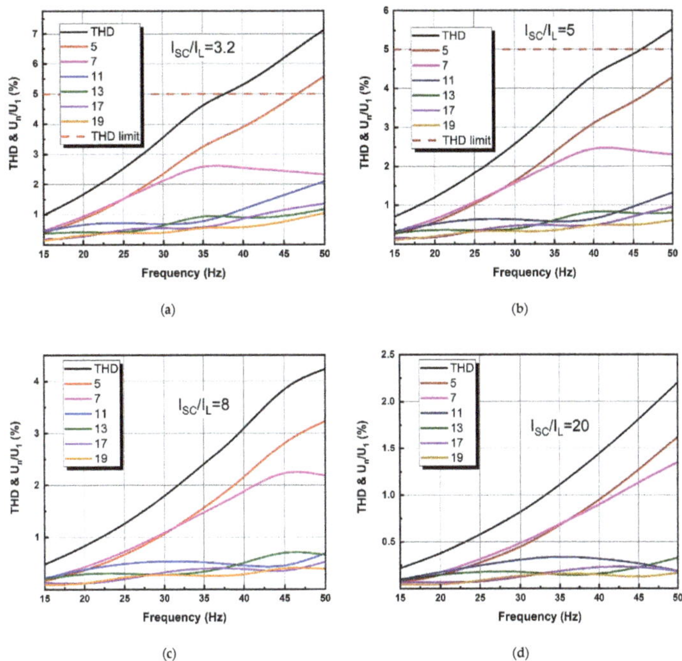

Figure 11. Dependence of voltage THD and voltage harmonics in PCC on motor control frequency in the scheme in Figure 5.

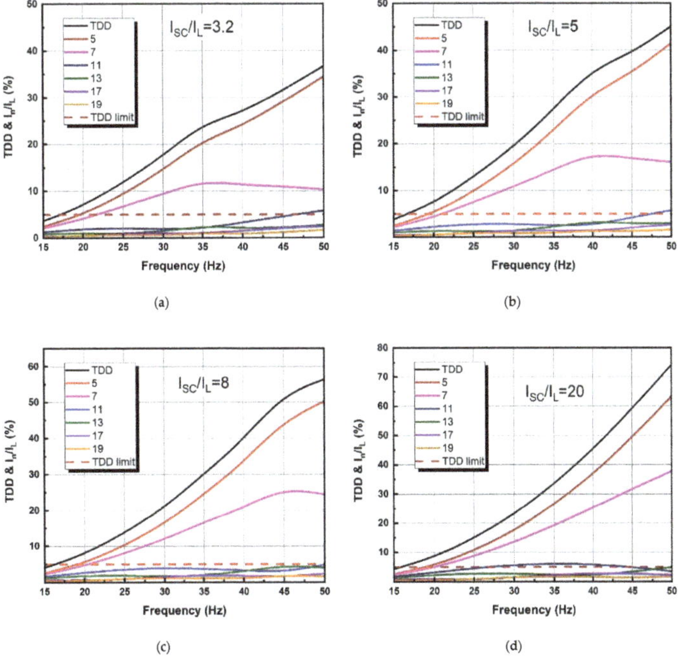

Figure 12. Dependence of current TDD and current harmonics on motor control frequency in the scheme in Figure 5.

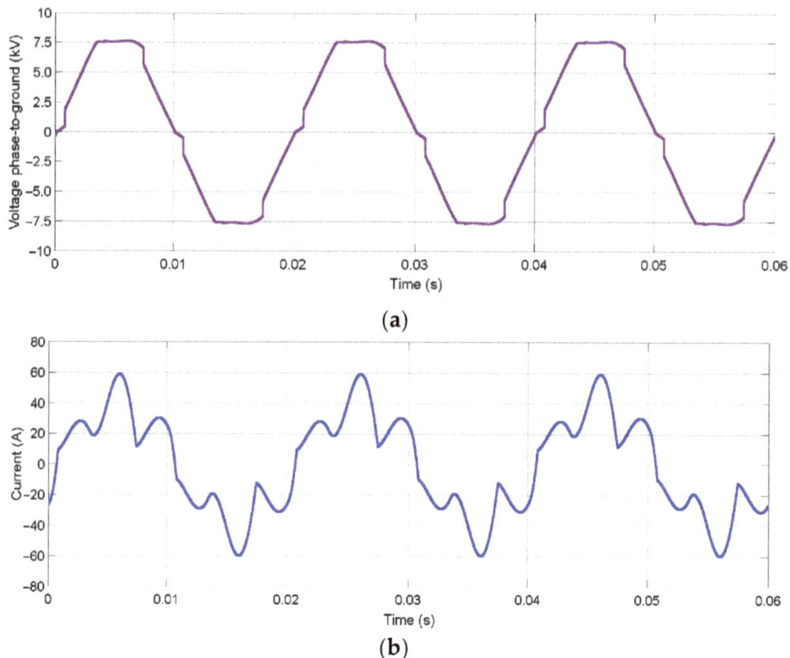

Figure 13. Voltage (**a**) and current (**b**) curves in the scheme in Figure 5 at motor control frequency 50 Hz and $I_{SC}/I_L = 3.2$.

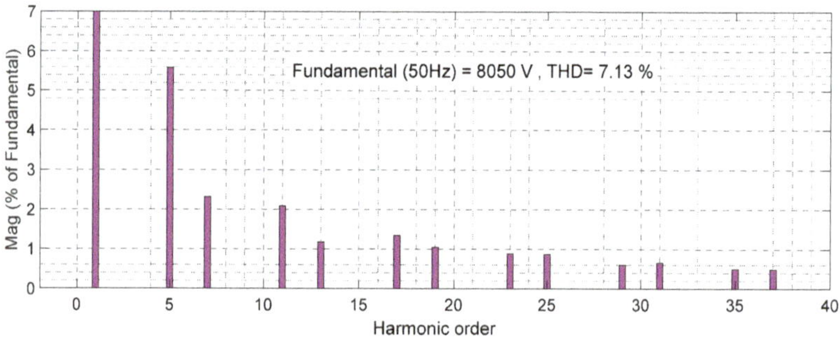

Figure 14. The voltage spectrum in the scheme in Figure 5 at motor control frequency 50 Hz and $I_{SC}/I_L = 3.2$.

3. Novel Substation Topology for Power Supply to the FCGCP

3.1. Proposed Technical Solution

Different options can be considered depending on whether the GCP is an existing facility undergoing renovation or is in the design phase. In any case, the comparison of variants should be made on the condition that the substations have the same spectrum of currents consumed for all FCGCP operation modes.

An effective solution to the EMC problem can be provided by a fundamental change in the substation architecture for FCGCP power supply by organizing DC buses and changing the frequency converters to inverters. As shown in Figure 15, this section consists of transformer and rectifier units [27], which include three-winding transformers T1, T2, and rectifiers AC/DC1.1, AC/DC1.2 and AC/DC2.1, AC/DC2.2. The transformers have one secondary winding in a star connection and one in a delta connection. The outputs of rectifiers are connected to DC busbar through chokes Ld1 and Ld2, which not only smooth the output voltage ripples, but also serve to eliminate equalizing currents. These currents arise because the output voltage ripples of rectifiers AC/DC1.1 and AC/DC1.2 of one module (AC/DC2.1 and AC/DC2.2 of the other module) are shifted relative to each other by $\pi/6$.

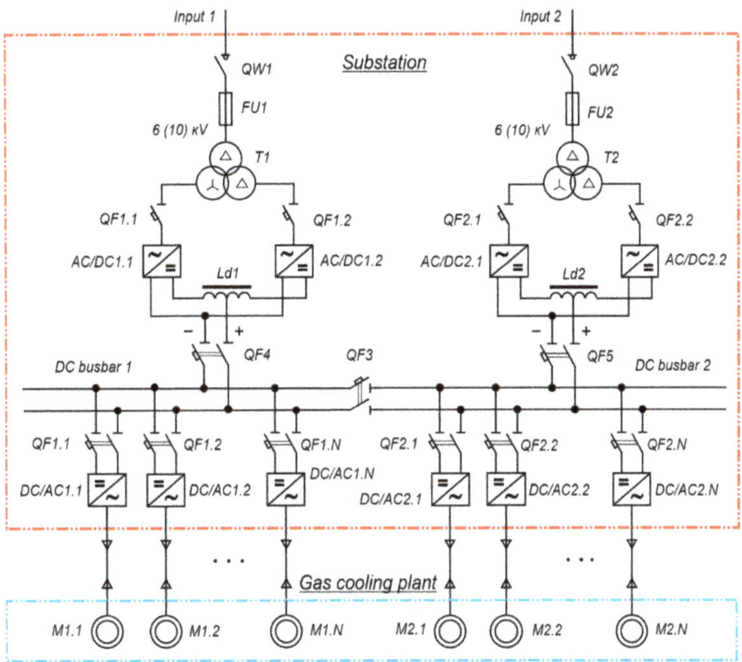

Figure 15. Power supply scheme with three-winding transformers and DC busbars.

DC/AC1.1...DC/AC1.N and DC/AC2.1...DC/AC2.N inverters receive power from the DC link and control the electric motors of M1.1...M1.N, M2.1...M2.N fans. It is necessary to note an important feature of the power supply scheme with the united DC link. Here are the DC-operated switching devices. These are QF1.1...QF1.N and QF2.1...QF2.N circuit breakers for connecting AC/DC1.1...AC/DC1.N and AC/DC2.1...AC/DC2.N inverters to the DC bus. The QF4 and QF5 switchgear, which connect the transformer and rectifier units to the DC bus, as well as the QF3 section switch, also operate on direct current.

Since the secondary winding voltages of transformers T1, T2 are shifted in phase with respect to each other by an angle equal to $\pi/6$, when the secondary winding currents are equal, the 5th and 7th harmonics of magnetic flux are compensated. As a result, these harmonics are also compensated in the primary winding currents of the transformers. The spectrum of currents begins with the 11th harmonic.

3.2. Simulation of the Proposed Electrical Complex

A simulation model in MATLAB with Simulink extension package was developed to study the electrical complex. A simulation model scheme for one busbar section is shown in Figure 16.

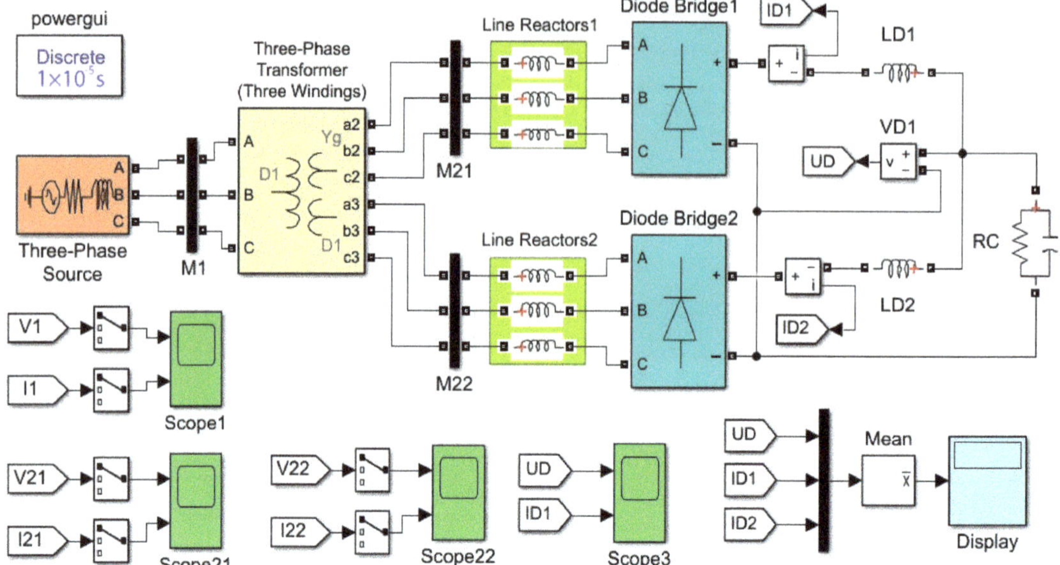

Figure 16. Simulation model scheme of a single busbar section of an electrical complex with an integrated DC link.

A Three-Phase Source unit represents the substation power supply. The Three-Phase Transformer (Three Windings) block is used to simulate a transformer with two secondary windings. Simulation models' parameters are shown in Tables A1 and A3.

Diode Bridge units model rectifiers. Line Reactors represent the line chokes at the input of the rectifiers. An RC unit as a parallel-connected resistor and capacitor represents the inverters connected to the DC bus. The value of the equivalent resistance R depends on the number of simultaneously working inverters and their load. The formula for its calculation can be obtained using the power balance.

Each of the inverters on the DC bus consumes power, which can be found by Formula (10). Therefore, the total power consumed by a group of N inverters will be determined by the formula:

$$P_{d.in} = \sum_{k=1}^{N} P_k = \sum_{k=1}^{N} \eta^{-1} P_{0.k} (f/f_0)^3. \tag{13}$$

If all N drives have the same power, then taking into account (8) and (9) we acquire an expression for calculating the equivalent load resistance R on the DC bus, which is necessary to know in the simulation:

$$R = \frac{\eta \cdot V_{dk}^2}{N \cdot P_{0.k} \cdot (f/f_0)^3}. \tag{14}$$

The Formula (14) shows that the equivalent load resistance on the DC bus decreases when the number of simultaneously connected inverters increases, and increases when the control frequency of the electric motors decreases.

The capacitor capacity, written in the setup window of the RC block of the simulation model in Figure 16, depends on the number of simultaneously working inverters. Therefore, if any inverter is connected to the DC bus, the corresponding filter capacitor is connected.

To calculate the equivalent resistance according to Formula (14), the following conditions are taken: number of variable-frequency drives $N = 12$, rated power of each drive $P_0 = 37$ kW.

A three-phase bridge rectifier forms the DC bus voltage. Therefore, this formula can be used to calculate it:

$$V_d = \frac{3\sqrt{2}}{\pi} V_2 \tag{15}$$

where V_2 is the effective value of voltage on the secondary windings of the transformer.

A 1000 μF filter capacitor is connected in parallel to each inverter, and therefore the total capacitor capacity in the model circuit is $12 \times 1000 = 12000$ μF. The inductance of smoothing reactors LD1 and LD2 is assumed to be 100 μH.

The results of the simulation are shown in Figures 17–20. The control frequency range of the fan motors and the power source parameters were the same as in the circuit analysis in Figure 5.

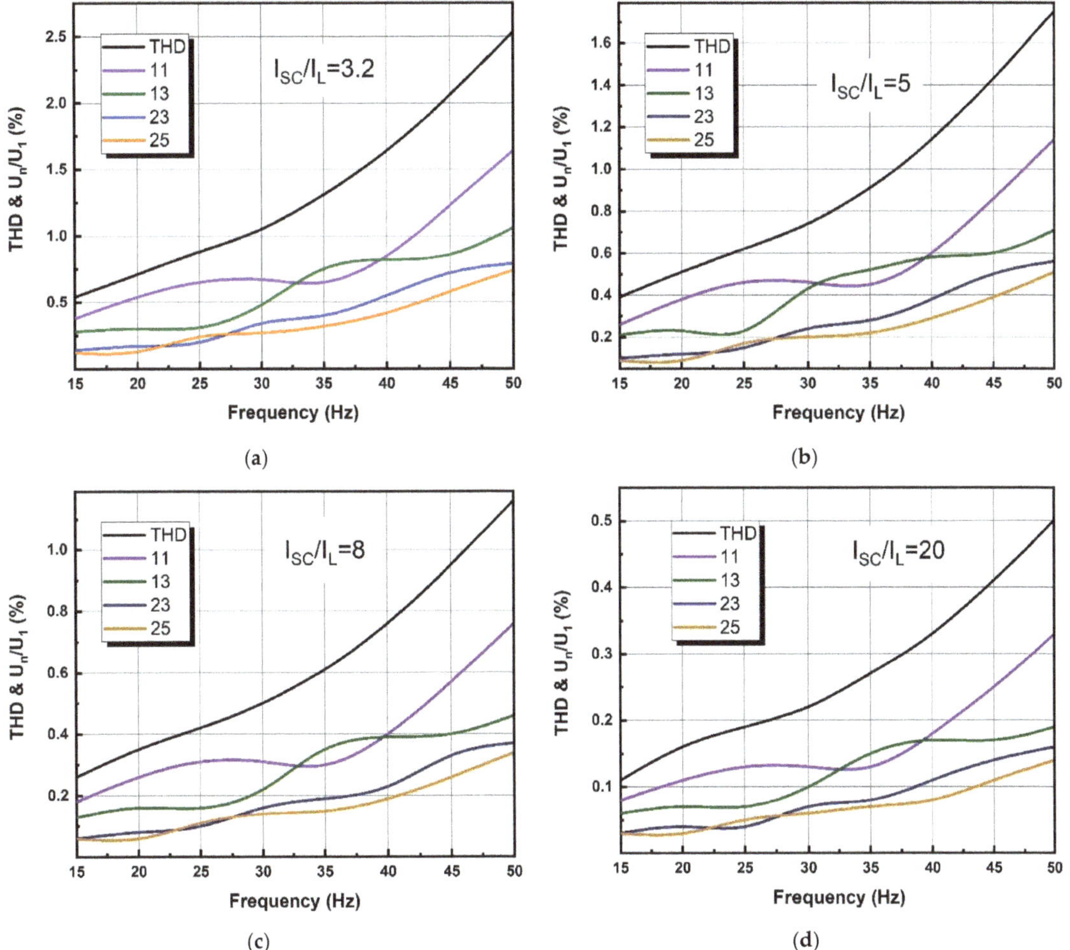

Figure 17. Dependence of THD and voltage harmonics in PCC on motor control frequency in the scheme in Figure 15.

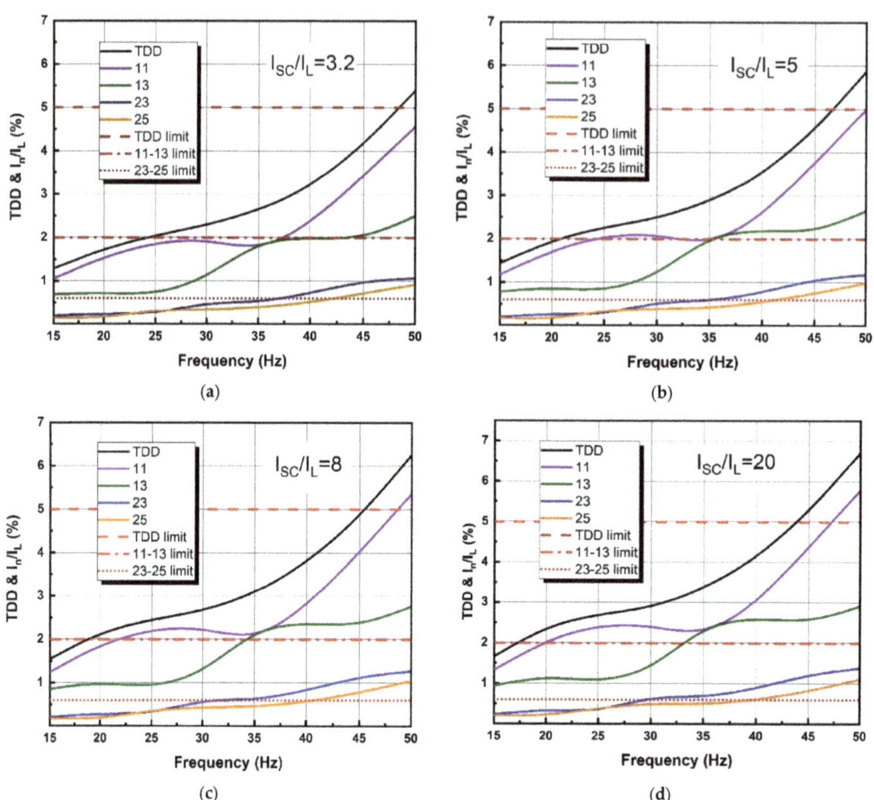

Figure 18. Dependence of TDD and current harmonics on motor control frequency in the scheme in Figure 15.

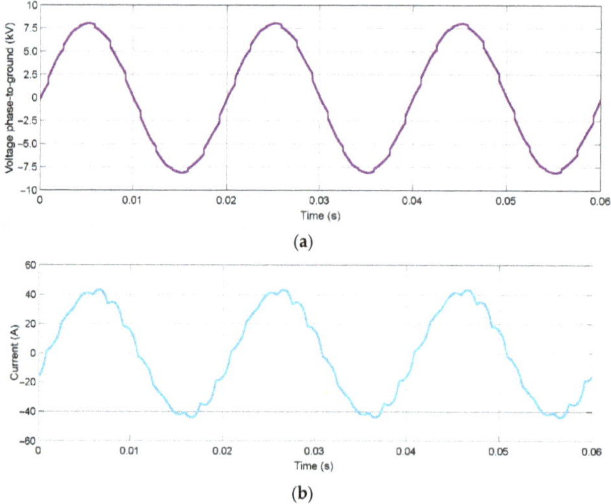

Figure 19. Voltage (**a**) and current (**b**) curves in the scheme in Figure 15 at motor control frequency 50 Hz and $I_{SC}/I_L = 3$.

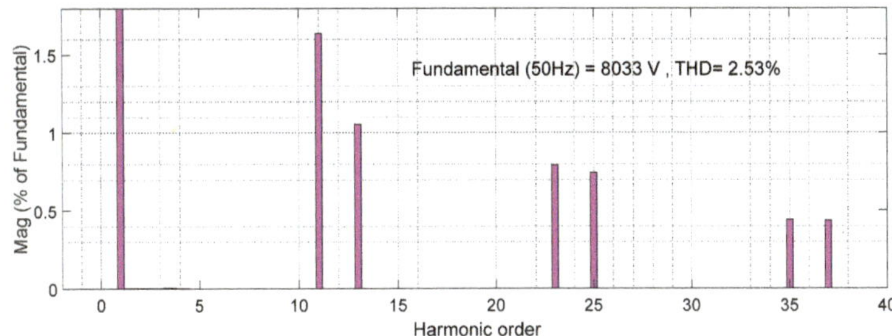

Figure 20. The voltage spectrum in the scheme in Figure 15 at motor control frequency 50 Hz and $I_{SC}/I_L = 3.2$.

The graphs in Figure 17 show that for all values of the I_{SC}/I_L ratio and motor control frequencies, the voltages THD and individual harmonics are in the range established by IEEE Standard 519-2014.

The current TDD (Figure 18) slightly exceeds the allowable value of 5% while controlling motors with frequencies of 45 Hz and above. Exceeding the 11th and 13th harmonics limit of 2% is seen at control frequencies of 35 Hz and above. In this case, the harmonic number 11 requires attention, and therefore, when increasing the control frequency to 50 Hz, this harmonic value ranges from 4.57 to 5.78%. The current harmonics with numbers 23 and 25 have no significant effect on the distortion of the voltage curve in the PCC.

Substation voltage and input current curves at motor control frequency f = 50 Hz and $I_{SC}/I_L = 3.2$ ratio are shown in Figure 19. The spectral content of the voltage is shown in Figure 20.

4. Discussion

The results of modeling are the basis for making a decision on the option to ensure EMC in the power supply system of a group of fans with a frequency-controlled drive. As mentioned above, the option is determined by the current situation.

If GCP fans are equipped with a frequency-variable drive on a running plant, installation of active filters or the use of frequency converters with active rectifiers is an effective solution. Both options require large capital expenditures, but they can start working quickly enough.

The option of changing the architecture of the substation to supply power to the FCGCP by organizing DC buses, replacing FC with inverters and using transformers with two secondary windings is much cheaper. The substation cost increase is only due to the increased cost of transformers and the need to purchase and install rectifiers with equalizing reactors. Some increase in the cost of the substation will occur due to the use of DC switching equipment. The disadvantages of this option include a large amount of construction and installation work, and the replacement of transformers' and rectifiers' installation. Therefore, this option has certain advantages when designing new gas transmission facilities.

Further direction of research involves modeling and study of transients during start-up of electric motors and changes in their operation mode.

5. Conclusions

The negative impact reduction on the power supply system of a group of fans with a variable-frequency drive is an urgent task. Its solution makes it possible to ensure the required quality of electricity. The article studies the effect of a frequency-controlled gas cooling plant on a power-limited source and proposes a solution to reduce it. The authors propose changing the architecture of the substation for FCGCP power supply by organizing DC buses, replacing the FC with inverters, and using transformers with two secondary windings with different connection schemes. The advantage of this option is the following: While supplying the substation from an autonomous power source, at which the output voltage frequency is subjected to changes, the proposed option in any situation provides full compensation of harmonics with numbers 5, 7, 17, and 19 in the current consumption curve. These positive properties of the proposed solution can be applied most effectively in the design and construction of new gas transmission facilities.

Author Contributions: Conceptualization, I.A.; methodology, I.A.; software, D.A. and I.A.; validation, I.A. and D.A.; formal analysis, S.M.; investigation, A.Z.; resources, Y.Y. and V.Z.; data curation, V.Z. and D.J.; writing—original draft preparation, I.A. and D.A.; writing—review and editing, I.A.; visualization, D.A. and I.A.; supervision, Y.Y. and S.M.; project administration, I.A. All authors have read and agreed to the published version of the manuscript.

Funding: This research received no external funding.

Data Availability Statement: Not applicable.

Conflicts of Interest: The authors declare no conflict of interest.

Abbreviations

The following abbreviations are used in this manuscript:

Abbreviations

ACU	Air-cooling unit
AHF	Active harmonic filter
EMC	Electromagnetic compatibility
EMF	Electromotive force
FC	Frequency converter
FCGCP	Frequency-controlled gas cooling plant
FFT	Fast Fourier transformation
GCP	Gas cooling plant
IGBT	Insulated Gate Bipolar Transistors
PCC	Point of common coupling
TDD	Total demand distortion
THD	Total harmonic distortion
VFD	Variable-frequency drive

Nomenclature

P	Active power
Q	Reactive power
S	Apparent power
T_{in}	Temperature of gas at the GCP inlet
G_{gas}	Mass flow rate of gas at the GCP inlet
Θ_{air}	Temperature of air through the GCP heat exchange sections
G_{air}	Mass flow rate of air through the GCP heat exchange sections
R_{Θ}	Thermal resistance
T_{out}	Gas temperature at the GCP outlet
$e(t)$	Electromotive force
$v(t)$	Voltage on the load
$v_R(t)$	Voltage on the resistance
$v_L(t)$	Voltage on the inductance
$i(t)$	Current consumed from the grid
$\overline{V}_1$	Grid voltages vector
$\overline{I}_1$	Grid currents vector
$\overline{V}_2$	Busbar 0.4 kV voltage vector
$\overline{I}_2$	Secondary currents vector
f_0	Nominal value of the motor control frequency
f	Current value of the motor control frequency
$P_{0.k}$	Rated power of the electric motor
P_k	Electric motor power at control frequency f
η	Efficiency of the electric motor

V_2 RMS voltage on the transformer secondary winding
V_d DC busbar voltage
$P_{d.in}$ Power consumed by a group of N inverters
$R_{inv.k}$ Value of the equivalent resistance of the k-th inverter
L_{sk} Inductance of the line choke
R_{sk} Resistance of the line choke
C_k Capacitance of the smoothing choke at the rectifier output
L_{dk} Inductance of the smoothing choke at the rectifier output
R_{dk} Resistance of the smoothing choke at the rectifier output

Appendix A

Table A1 shows the simulation model parameters for the energy source in Figures 9 and 16. Tables A2 and A3 show the simulation model parameters for the transformers in Figures 9 and 16, respectively.

Table A1. Simulation model parameters for the Three-Phase Source block.

Parameter	Value	Unit
Phase-to-phase rated voltage	10	kV
Frequency	50	Hz
Source resistance	0.2 … 1.33	Ω
Source inductance	$(3 \dots 18.75) \times 10^{-3}$	H

Table A2. Simulation model parameters for the Three-Phase Transformer block in Figure 9.

Parameter	Value	Unit
Rated power	1000	kVA
Frequency	50	Hz
Winding 1 parameters		
Phase-to-phase rated voltage	10	kV
Winding resistance	1.05	Ω
Leakage inductance	17.5×10^{-3}	H
Winding 2 parameters		
Phase-to-phase rated voltage	0.4	kV
Winding resistance	0.004	Ω
Leakage inductance	28×10^{-6}	H
Magnetization parameters		
Resistance	16×10^3	Ω
Inductance	22.5	H

Table A3. Simulation model parameters for the Three-Phase Transformer block in Figure 16.

Parameter	Value	Unit
Rated power	1000	kVA
Frequency	50	Hz
Winding 1 parameters		
Phase-to-phase rated voltage	10	kV
Winding resistance	1.05	Ω

Table A3. *Cont.*

Parameter	Value	Unit
Leakage inductance	17.5×10^{-3}	H
Winding 2 parameters		
Phase-to-phase rated voltage	0.4	kV
Winding resistance	0.012	Ω
Leakage inductance	84×10^{-6}	H
Winding 3 parameters		
Phase-to-phase rated voltage	0.4	kV
Winding resistance	0.004	Ω
Leakage inductance	28×10^{-6}	H
Magnetization parameters		
Resistance	16×10^{3}	Ω
Inductance	22.5	H

References

1. Kroger, D.G. Air-Cooled Heat Exchangers and Cooling Towers: Thermal-flower Performance Evaluation and Design. Ph.D. Thesis, University of Stellenbosch, Stellenbosch, South Africa, 2004.
2. Mokhatab, S.; Poe William, A.; Speight, J.G. *Handbook of Natural Gas Transmission and Processing–Principles and Practices*, 3rd ed.; Elsevier: Amsterdam, The Netherlands, 2015.
3. Artyukhov, I.I.; Arshakyan, I.I.; Trimbach, A.A. Analysis of circuits of frequency control of gas-cooling device's fans. In Proceedings of the International Conference on Actual Problems of Electron Devices Engineering, Saratov, Russia, 24–25 September 2008; pp. 381–389. [CrossRef]
4. Larabee, J.; Pellegrino, B.; Flick, B. Induction Motor Starting Methods and Issues. In Proceedings of the Record of Conference Papers Industry Applications Society 52nd Annual Petroleum and Chemical Industry Conference, Athens, Greece, 12–14 September 2005; pp. 217–222.
5. Habyarimana, M.; Dorrell, D.G. Methods to reduce the starting current of an induction motor. In Proceedings of the International Conference on Power, Control, Signals and Instrumentation Engineering (ICPCSI), Chennai, India, 21–22 September 2017; pp. 34–38. [CrossRef]
6. Artyukhov, I.I.; Stepanov, S.F.; Arshakyan, I.I.; Korotkov, A.V.; Pogodin, N.V. Dynamic compensation of reactive power at power supply system for air cooler of gases. *Promyshlennaya Energ.* **2004**, *6*, 47–51.
7. Artyukhov, I.I.; Stepanov, S.F.; Mirgorodskaya, E.E.; Mityashin, N.P.; Zemtsov, A.I. Transient Processes with Starting of a Multi-Pole Asynchronous Motor with a Fan on the Shaft. In Proceedings of the 2021 17th Conference on Electrical Machines, Drives and Power Systems (ELMA 2021), Sofia, Bulgaria, 1–4 July 2021; pp. 1–4. [CrossRef]
8. Abakumov, A.M.; Kuznetsov, P.K.; Stepashkin, I.P. Adaptive automatic control system for air-cooled gas apparatus. *J. Phys. Conf. Ser.* **2018**, *1111*, 012031. [CrossRef]
9. Yanvarev, I.; Vanyashov, A.; Krupnikov, A. Thermal Management Technologies Development for the Gas Transport on the Gas-main Pipeline. *Procedia Eng.* **2015**, *113*, 237–243. [CrossRef]
10. Artyukhov, I.; Abakumov, A.; Zemtsov, A.; Yerbayev, Y.; Zakharov, V. Energy Efficiency Analysis of Control Algorithms for Fan Electric Drives in Gas Air-Cooling Plants. In *Lecture Notes in Civil Engineering, Proceedings of ICEPP 2021, Kazan, Russia, 26 February 2021*; Springer: Cham, Switzerland, 2022; Volume 190, pp. 46–55. [CrossRef]
11. Artyukhov, I.I.; Stepanov, S.F.; Pylskaya, E.K.; Zemtsov, A.I. Dynamics of Compressed Gas Temperature Stabilization System with Variable-Frequency Drive of Fans in a Gas Air-Cooling Unit. In Proceedings of the International Conference on Electrotechnical Complexes and Systems (ICOECS), Ufa, Russia, 27–30 October 2020; pp. 1–5. [CrossRef]
12. Novák, J.; Šimánek, J.; Černý, O.; Doleček, R. EMC of frequency controlled electric drives. *Radioengineering* **2008**, *17*, 101–105.
13. Zare, F.; Soltani, H.; Kumar, D.; Davari, P.; Delpino, H.A.M.; Blaabjerg, F. Harmonic Emissions of Three-Phase Diode Rectifiers in Distribution Networks. *IEEE Access* **2017**, *5*, 2819–2833. [CrossRef]
14. Artyukhov, I.; Bochkareva, I.; Molot, S.; Kalganova, S.; Stepanov, S.; Trigorly, S. Power Quality in Industrial Isolated Generation Power Systems with Powerful Nonlinear Consumers. In Proceedings of the 9th International Scientific Symposium EL-EKTROENERGETIKA 2017, Stará Lesná, Slovak Republic, 12–14 September 2017; pp. 44–49.
15. *Standard 519-2014*; IEEE Recommended Practice and Requirements for Harmonic Control in Electric Power Systems. IEEE: New York, NY, USA, 2014.
16. Jaafari, K.A.A.; Poshtan, M.; Beig, A.R. Passive wide spectrum filter for variable speed drives in oil and gas industry. In Proceedings of the 11th International Conference on Electrical Power Quality and Utilisation, Lisbon, Portugal, 17–19 October 2011; pp. 1–6.

17. Park, B.; Lee, J.; Yoo, H.; Jang, G. Harmonic Mitigation Using Passive Harmonic Filters: Case Study in a Steel Mill Power System. *Energies* **2021**, *14*, 2278. [CrossRef]
18. Lumbreras, D.; Gálvez, E.; Collado, A.; Zaragoza, J. Trends in Power Quality, Harmonic Mitigation and Standards for Light and Heavy Industries: A Review. *Energies* **2020**, *13*, 5792. [CrossRef]
19. Špelko, A.; Blažič, B.; Papič, I.; Herman, L. Active Filter Reference Calculations Based on Customers' Current Harmonic Emissions. *Energies* **2021**, *14*, 220. [CrossRef]
20. Vitoi, L.; Brandao, D.; Tedeschi, E. Active Power Filter Pre-Selection Tool to Enhance the Power Quality in Oil and Gas Platforms. *Energies* **2021**, *14*, 1024. [CrossRef]
21. Raman, R.; Sadhu, P.K.; Kumar, R.; Rangarajan, S.S.; Subramaniam, U.; Collins, E.R.; Senjyu, T. Feasible Evaluation and Implementation of Shunt Active Filter for Harmonic Mitigation in Induction Heating System. *Electronics* **2022**, *11*, 3464. [CrossRef]
22. Shankar, V.A.; Kumar, N.S. Implementation of Shunt Active Filter for Harmonic Compensation in a 3 Phase 3 Wire Distri-bution Network. *Energy Procedia* **2017**, *117*, 172–179. [CrossRef]
23. Hudson, R.; Hong, S.; Hoft, R. Modeling and simulation of a digitally controlled active rectifier for power conditioning. In Proceedings of the of Sixth Annual Applied Power Electronics Conference and Exhibition, Dallas, TX, USA, 10–15 March 1991; pp. 423–429. [CrossRef]
24. Chimonyo, K.B.; Kumar, K.S.; Kumar, B.K.; Ravi, K. Design and Analysis of Electrical Drives Using Active Front End Converter. In Proceedings of the 2018 Second International Conference on Inventive Communication and Computational Technologies (ICICCT), Coimbatore, India, 20–21 April 2018; pp. 115–119. [CrossRef]
25. Vilberger, M.E.; Kulekina, A.V.; Bakholdin, P.A. The twelfth-pulse rectifier for traction substations of electric transport. *IOP Conf. Ser. Earth Environ. Sci.* **2017**, *87*, 032052. [CrossRef]
26. Qian, G.; Wang, Q.; He, S.; Dai, W.; Wei, N.; Zhou, N. Harmonic Modeling and Analysis for Parallel 12-Pulse Rectifier under Unbalanced Voltage Condition in Frequency-Domain. *Energies* **2022**, *15*, 3946. [CrossRef]
27. Corti, F.; Shehata, A.H.; Laudani, A.; Cardelli, E. Design and Comparison of the Performance of 12-Pulse Rectifiers for Aerospace Applications. *Energies* **2021**, *14*, 6312. [CrossRef]
28. Setlak, L.; Kowalik, R. Examination of Multi-Pulse Rectifiers of PES Systems Used on Airplanes Compliant with the Concept of Electrified Aircraft. *Appl. Sci.* **2019**, *9*, 1520. [CrossRef]
29. Surya, H. The Effect of 12 Pulse Converter Input Transformer Configuration on Harmonics of Input Current. In *IOP Conference Series: Materials Science and Engineering*; IOP Publishing, Bristol, UK, 2021; Volume 1158. [CrossRef]
30. Shklyarskiy, Y.; Dobush, I.; Carrizosa, M.J.; Dobush, V.; Skamyin, A. Method for Evaluation of the Utility's and Consumers' Contribution to the Current and Voltage Distortions at the PCC. *Energies* **2021**, *14*, 8416. [CrossRef]

Review

Small-Scale Hybrid and Polygeneration Renewable Energy Systems: Energy Generation and Storage Technologies, Applications, and Analysis Methodology

Maksymilian Homa, Anna Pałac, Maciej Żołądek * and Rafał Figaj

Department of Sustainable Energy Development, Faculty of Energy and Fuels,
AGH University of Science and Technology, 30-059 Krakow, Poland
* Correspondence: mzoladek@agh.edu.pl

Abstract: The energy sector is nowadays facing new challenges, mainly in the form of a massive shifting towards renewable energy sources as an alternative to fossil fuels and a diffusion of the distributed generation paradigm, which involves the application of small-scale energy generation systems. In this scenario, systems adopting one or more renewable energy sources and capable of producing several forms of energy along with some useful substances, such as fresh water and hydrogen, are a particularly interesting solution. A hybrid polygeneration system based on renewable energy sources can overcome operation problems regarding energy systems where only one energy source is used (solar, wind, biomass) and allows one to use an all-in-one integrated systems in order to match the different loads of a utility. From the point of view of scientific literature, medium- and large-scale systems are the most investigated; nevertheless, more and more attention has also started to be given to small-scale layouts and applications. The growing diffusion of distributed generation applications along with the interest in multipurpose energy systems based on renewables and capable of matching different energy demands create the necessity of developing an overview on the topic of small-scale hybrid and polygeneration systems. Therefore, this paper provides a comprehensive review of the technology, operation, performance, and economical aspects of hybrid and polygeneration renewable energy systems in small-scale applications. In particular, the review presents the technologies used for energy generation from renewables and the ones that may be adopted for energy storage. A significant focus is also given to the adoption of renewable energy sources in hybrid and polygeneration systems, designs/modeling approaches and tools, and main methodologies of assessment. The review shows that investigations on the proposed topic have significant potential for expansion from the point of view of system configuration, hybridization, and applications.

Keywords: solar; wind; biomass; energy storage; energy system; renewables; polygeneration; hybrid; systems analysis

Citation: Homa, M.; Pałac, A.; Żołądek, M.; Figaj, R. Small-Scale Hybrid and Polygeneration Renewable Energy Systems: Energy Generation and Storage Technologies, Applications, and Analysis Methodology. *Energies* **2022**, *15*, 9152. https://doi.org/10.3390/en15239152

Academic Editor: Juri Belikov

Received: 31 October 2022
Accepted: 29 November 2022
Published: 2 December 2022

Publisher's Note: MDPI stays neutral with regard to jurisdictional claims in published maps and institutional affiliations.

1. Introduction

Among the greatest challenges of the modern world, the goal of reducing greenhouse gas emissions produced by anthropogenic activity stands out the most. One of the tasks set for all countries in the world is to achieve zero-carbon energy generation by 2050 [1]. For many years, international and national organizations have been persuading the private and public sector to switch to renewable energy sources (RES), increase the efficiency of energy generation from conventional sources, or phase them out altogether. Unfortunately, this is a very difficult goal to achieve.

Getting energy from unconventional sources is not only a modern idea. Large hydroelectric power plants have been popular around the world since the beginning of the last century [2], and the conversion of wind energy into mechanical power, for example, in windmills, has been known for hundreds of years [3]. Today, solar power plants and wind farms are commonplace, and the first part of the 21st century has seen a big focus on distributed energy [4]. Nowadays, many homes are equipped with small photovoltaic power plants or wind turbines, and this fact is primarily influenced by development and ongoing RES application and research. One of the main directions of development is to increase the efficiency of energy generation or search for new solutions. In particular, this can involve the conversion of renewable energy itself [5,6] or individual components of the installation [7–9]. Despite its great popularity, energy generation from renewable sources is characterized by several disadvantages compared to conventional sources [10]. The most significant of these are the instability of operation [11] and the economic unprofitability of individual solutions [12].

The efficiency of most ways of producing energy from RES is closely related to weather conditions. As a result, the exact duration of their operation cannot be accurately predicted, as occurs in the case of photovoltaic plants [13,14], where during operation, a peak in energy production is observed during the sunny hours, while there is almost null or no generation in the evenings or at night. In addition, it is necessary to take into account the period of the year and the weather of each day. These facts would translate into significant limits and disadvantages if the energy mixes of entire countries were based mainly on PV. We live in a time when even a temporary blackout would cause huge disservices to systems and property damage or could lead to dangerous situations that put lives and health at risk. However, there are various ways to solve this problem. The first is to introduce a technology characterized by stable operation into the power system based on RES. This could be, for example, a system based on biomass combustion [15] or nuclear power [16]. The second way can be to secure against a possible drop in production by creating a system [17] based on different RES technologies, whose projected production hours during the day are complementary. This way can also be further improved by using energy storage technologies [18,19]. Despite such possibilities, adapting such a system to the national power grid is not an easy operation from the technical and grid management points of view. However, thanks to recent technological advances and research, they are finding very high potential in the application of distributed energy.

A system that produces energy from at least two different renewable sources is called a hybrid renewable energy system (HRES) [20–22]. It can be based on the simultaneous generation of useful energy from solar, wind, hydro, or geothermal energy sources. A biomass energy conversion system can also be used for this purpose. Obviously, energy taken from storage can be considered a source of renewable energy as well, once the storage component has been charged using RES. As mentioned earlier, HRES has a number of advantages including the most important one, which is the stabilization of the operation of the entire system. An example is a photovoltaic (PV) system operating simultaneously with a wind turbine [23,24].

When using renewable energy sources in distributed power generation, it is very important to match the system to the user's needs. To show that, two parameters should be highlighted: the self-consumption rate [25] and the ratio of energy demand coverage [26]. As for the first parameter, it should be understood as the ratio of energy directly used by the user to the total energy produced from RES. The values of this indicator depend on several parameters, including the location of the installation, the energy demand of the user, the output of the system itself, or weather conditions. An example would be a photovoltaic system installed on a typical single-family house. For example, a value of this indicator at 25% means that of all the energy produced by the PV, only 25% was used directly by the user, while the rest was wasted or supplied to the electric grid. The second parameter can be calculated as the ratio of energy extracted from RES to the total demand for this type of energy. These values are usually given for a longer period of time. For example, if user has an annual demand for electricity equal to 5000 kWh and a photovoltaic installation from which is possible to self-consume 2000 kWh directly, this means that energy demand coverage in this case will be 40%.

An ideal HRES system should be characterized by 100% self-consumption and 100% energy demand coverage. This would result in zero losses of generated energy and a negligible risk of running out of energy, only due to the failure of system components. Unfortunately, this ideal scenario is practically impossible to achieve due to the technical constraints of the devices, variability of user demand, and economic limits of the potential investments. The focus, however, should be on analyses showing the possibilities of better and better systems, which are as close as possible to the ideal ones.

When generating one type of energy, there is always some loss in the system. This can be seen in the case of electricity generation with a biomass or coal boiler Rankine cycle system. In this case, energy is lost through a pipe, mainly due to the dissipation of heat by the condenser. In the early 19th century, it was realized that after leaving the turbine, water still has a lot of useful energy that can be recovered [27]. It can be used, for example, to heat domestic hot water. A plant that simultaneously generates electricity and useful heat is called a combined heat and power (CHP) plant, and the process itself is called cogeneration [28–30]. Simultaneous generation of two types of energy is very economical and increases the efficiency of the entire plant compared to a separated production of energies. Chill generators in the form of sorption chillers can also be included in CHP systems [31,32]. In this case, the process that allows the simultaneous generation of three types of energy is called trigeneration. Furthermore, when the generation of energy is also coupled with the production of useful by-products, we are referring to polygeneration [33–35]. Passing from cogeneration to polygeneration, it is possible to observe that the grade of energy utilization significantly increases, and this may also have significant effects on the effectiveness of the energy systems adopted.

Each of the aforementioned methods of increasing the efficiency of plant operation can be used in distributed power generation. Maximization of energy yield and achievement of the highest possible efficiency rates are essential aspects of small hybrid RES-based polygeneration systems. Currently, great attention is paid to the optimization of such systems. It is necessary to select energy sources appropriately, and if necessary, to select energy storage technologies. The most important features of such a system should be maximum autonomy and efficiency, and thus cost-effectiveness. It is the second feature that is usually the biggest problem to be solved.

The present review aims to provide a comprehensive overview of recent advances regarding small-scale hybrid renewable energy systems. In particular, the review is focused on the technical aspects of the different energy generation and storage technologies, their use in HRES, and approaches and tools used in the design of integrated HRES. The review is organized as follows. Firstly, the available technologies for the production of energy are presented, and after that, the possibilities of energy storage are introduced. The review continues with a description of the different configurations of HRES technologies and concludes with the presentation of methods to investigate HRES. In the end, some general conclusions regarding small-scale HRES are given in order to summarize the review.

2. Available Renewable Energy Technologies

Several technologies are used to generate energy in HRES depending on the type of energy to be generated—electrical or thermal—or on the availability of renewable energy sources to be used by the system. The presented technologies are the ones that are conventionally used in HRES to produce energy from the specific renewable energy source and are commercially available on the market. The technologies are:

- photovoltaic modules;
- solar collectors;
- wind turbines;
- water turbines;
- biomass units;
- heat pumps.

2.1. Photovoltaic Systems

When considering the world's energy production in 2021 from renewable energy sources, photovoltaic installations rank third in terms of installed capacity [36]. Recent reports show that this value has increased sevenfold between 2010 and 2020 indicating very dynamic growth and suggesting ever-increasing installed capacity in the future. The very high popularity and affordability of this technology are contributing to its continuous development and thus increasing its efficiency. Photovoltaic installations are very often used as a key energy source in small HRES [37]. The basic solution is to use standard first-generation modules and place them on the available surfaces of the utility being powered. When planning even the most basic photovoltaic installation, it is important to keep in mind several factors that affect its operation. This directly includes the geographic location of the facility being powered, and thus the angle of the modules from the floor [13]. The efficiency of an installation located in Scandinavian countries will be different from one in the vicinity of the equator. The difference is primarily due to the different values of average horizontal irradiation in these locations, but also the average annual air temperature. The higher it is, the lower the efficiency of PV cells [38]. This problem is being addressed by the latest research on module cooling systems to enhance their performance. Praveenkumar et al. [39] showed that it is possible to integrate a PV module with a heat pipe. The research indicated that this decreased the module's surface temperature by an average of 6.72 °C on a sunny day. The temperature reduction resulted in an increase in cell efficiency of about 2.98% over the comparison module. A different approach was followed by Sornek et al. [40] in designing a system to cool a PV module by spraying its surface with water. Their research showed that this procedure allowed an increase in the maximum instantaneous power of the modules by about 10% compared to the comparative system.

A PV system can be placed not only on the roof or ground—a very popular approach in Asia [41] is floating PV systems. This solution not only saves space for the installation by using a body of water but also increases energy production by cooling the PV panels [42]. Studies [43] show that a floating installation has a lower average surface temperature of the modules by 2–4% compared to an installation placed on the ground. In addition, the authors pointed out that this method of installation reduces the intensity of water evaporation in the water tank used. In recent years, an increasing emphasis on research on photovoltaic cells of the second, third, or fourth generation can be observed [44]. However, from the point of view of commercial applications, they are characterized in most cases by lower efficiency than first-generation solutions except for multijunction cells, the cost of which prevents their cost-effective use for most small-scale polygeneration systems.

Attention should also be paid to technology using focused solar radiation. Using mirrors or lenses, it is possible to increase the efficiency of electricity production from a photovoltaic cell. This technology is called concentrated photovoltaics (CPV). However, a review of this technology [45] indicates that this solution is not valid for receiving only electricity, due to the occurrence of excessive temperatures. It is advisable to additionally dissipate effectively or use the heat produced.

When considering the operation of a statistical PV system, the largest energy yield losses are associated with the shading or fouling of individual PV cells [46]. This is due to the characteristics of connecting modules in series with each other in a classical solution. This results in small values of current with voltage increase, which positively affects the safety of the installation and limits the ohmic losses on the solar cables. In order to make the entire photovoltaic circuit operate at uniform current parameters, a DC/DC voltage converter is present at each PV installation. It is designed to track the maximum power point of the circuit and adjust the installation voltage to achieve it. This component is called a maximum power point tracker (MPPT). With this configuration, partial shading of the installation or even of a few cells of one module negatively affects the performance of the entire installation. To cope with this problem, more and more research is being dedicated to the proper configuration of MPPT control. There are many options for tracking the point of maximum power [47,48], including: traditional—based on direct measurements of voltage and current; mathematical modeling—showing the locations of the point of maximum power; and intelligent algorithms—allowing the creation of a neural network that is taught to look for the right parameters to obtain the best results. Photovoltaic installations have plenty of advantages that demonstrate their versatility of use in small-scale polygeneration systems. These include, first and foremost, the relative stability of operation under conditions of adequate insolation and low price.

2.2. Solar Collectors

The abovementioned photovoltaic modules are not the only way to use solar energy. Thermal solar collectors are also used for this purpose. The most common are classic flat-plate collectors, which have a layer that absorbs solar energy and a piping in which the working medium is heated. The heat is then transferred usually to a domestic water circuit. In addition to flat-plate collectors, tubular vacuum variants, which use heat pipes that allow the heating medium to move, are also popular. The adoption of a vacuum allows the absorber surface to be well insulated, resulting in less heat loss to the environment.

It is possible to integrate a solar collector with a photovoltaic module, creating a photovoltaic–thermal (PV/T) hybrid system [49], which allows not only the supply of electricity but also heat. The hybrid cogeneration approach to energy production increases the efficiency of the entire system. However, this solution requires suitable climatic conditions for profitable operation. Tracking the performance of existing installations [50] shows that, depending on the latitude, the thermal or photovoltaic part of the hybrid module has better performance. In addition, using such a system to heat water provides temperatures ranging from 40 °C to 60 °C, which typically limits the possibilities of its use in relatively low-temperature applications [51].

In the case of photovoltaic cells, the concentration of solar radiation is associated with large losses due to excessive temperature. For solar power systems, however, the possibility of achieving relatively high temperatures is key. Several types of collectors using concentrated solar radiation can be distinguished [52]: compound parabolic collectors [53], trough parabolic collectors [54], parabolic dishes [55,56], Fresnel lenses [57], and power towers [58]. Each of these differs in a number of factors, but comparisons should mainly be made on the basis of operating temperature, from which the appropriate solution can be selected to meet needs. Collectors operating on concentrated solar radiation are not often used with small-scale systems, so can be observed as large-scale solar farms.

In order to improve the performance of a collector system, there is a lot of research involving the proper adjustment of the water circulation system. This can be carried out in a direct way, i.e., the heating medium in the collector is water, which when heated flows directly into the building pipelines. The second option is a system in which a separate heating medium is present, flowing in a separate circuit near the collector and transferring energy to a second circuit with domestic water. The movement of the medium is typically assisted by circulating pumps and is rarely achieved by natural convection, taking advantage of the difference in fluid density at different temperatures. Optimization of the system can also involve selecting the most efficient heating medium [59,60] or locating heat storage. In most solar collector systems, tanks can be observed inside the building; however, especially on a small scale, external tanks sometimes integrated into the collector itself are popular. In countries with high temperatures, this can allow additional reheating of the water in such a tank [61].

Collector systems can serve as the main source of hot water in a hybrid system, but they are increasingly encountered at the same time as auxiliaries, which are mainly intended to reheat the circulating medium to improve the performance of the entire system [62]. These are not the most popular methods of obtaining thermal energy; however, under the right environmental circumstances, they can be a very good source of energy for small-scale polygeneration systems or as the main source of heat in large-scale systems [63].

2.3. Wind Turbines

Wind energy is the second-largest RES in the world [64]. Wind turbines are mainly used to generate energy from this source, and when considering this technology, the first classification points to turbines with vertical (VAWT) and horizontal (HAWT) axes of rotation. The first group can be divided into two basic types [65]. The first is the thrust type [66], which includes the Savonius rotor [5] and Sistan rotor [67], where blades are usually bowl-shaped. The rotation of turbines of this type is caused directly by the force of wind pressure. They are characterized by the lowest rotational speed and therefore the lowest power yield. The second type consists of turbines based on lift force, which requires the use of blades in the shape of airfoils, and this group can include the Darrieus turbine [68] or H-rotor [69]. They are characterized by a higher rotational speed, which is able to match or even exceed turbines with a horizontal rotation axis. Unfortunately, the big problem of this solution is the relatively high wind speed required to start operation. Nevertheless, it is possible to combine both types to reduce its value [70]. Due to their design, VAWTs always generate negative torque, which reduces their efficiency. One way to deal with negative torques is to use augmentation methods, which allow changing the direction, to a certain extent, of the wind flowing through the turbine [71].

The scientific literature indicates that it is possible to improve the efficiency of turbines with a vertical axis of rotation using elements that increase wind speed [65]. These could be a deflector plane or a diffuser using the Venturi effect. Also very promising are the results of studies of the usage of methods such as air currents near buildings [72] or suitably shaped terrain [73]. These methods allow wind turbines to be placed in urban centers or next to individual homes, demonstrating the possibilities of their usage in distributed energy. The possibilities are supported by the fact that VAWTs take up less space than HAWTs.

As concerns VAWT design details, Qiang Gao et al. [74] showed a novel approach to controlling the blade arrangement in a Darrieus turbine, which can also positively affect its operation. In the case of lift-based types, it is very important to choose the right shape of airfoils for the blades [75], which is also a vital aspect of designing turbines with a horizontal axis of rotation [74].

Because of their higher efficiency, HAWTs are the most popular among commercial applications. A significant number of wind farms are based on this technology. In addition to dimensions, they can differ in the number of rotor blades. The best choice turns out to be a three-blade turbine. It allows very high efficiency while minimizing noise and manufacturing costs. Considering a different number of blades, the performance of the turbines is close [76], with the single-blade rotor, which by shifting the center of gravity falls into a strong vibration, the most distant. As with VAWTs, the operation of this type of turbine can be improved with the use of diffusers, but one of the most prosperous ideas is to equip the rotors with systems designed to adjust the rotor blade arrangement according to wind parameters [77]. Changing the angle of attack is also important for safety reasons, such as in situations where the turbine must be stopped.

Both groups of wind turbines (VAWTs and HAWTs) can be installed on land or water. The second way may allow them to work better due to the absence of any obstacles and high wind speeds. For this purpose, turbines with a horizontal axis of rotation are best suited, as they can withstand higher overloads. There are studies [78] testifying to the possibility of creating floating wind farms. This would allow the use of waters too deep to place a foundation and could also minimize costs.

For all types of turbines, it is very important to choose the right material to create their blades. It must be light enough, but also strong enough to achieve the best performance. Most often composite materials (carbon and glass fibers) are used. However, this is one of the biggest problems for wind turbines because these materials are difficult to recycle. Initial research shows the possibility of using used rotor blades to make composite plates, which can be used to make structural components for bridges or buildings. Researchers [79] have shown that proper processing allows even higher strength than analogous plates created for this purpose.

2.4. Water Turbines

The use of the potential or kinetic energy of water to produce energy is the main RES worldwide [36]. From the point of view of technologies, the types of hydropower plants adopted are dammed [80] and pumped storage [81]. The first type involves restricting the flow of a river by damming the water, creating an artificial basin, and at the same time increasing the possibility of utilizing the high head to produce electricity. For this purpose, Francis [82] or Pelton [83] water turbines are used, depending on the available head of the water. The second type uses mainly the potential energy of water.

Hydropower plants may be used as a type of energy storage. In fact, when there is a peak of power production on the electrical grid coupled with a relatively low price of electrical energy, i.e., excess energy is produced and the energy is cheap, water from the lower reservoir can be pumped into the upper one, while when there is a peak of demand of energy in the grid and the price is relatively high, the water can be released into the lower reservoir, giving back the energy stored.

These types of turbines are rarely used for small hydropower plants [84]. For this purpose, a large proportion of other means of energy production, consisting of flow or tidal power plants, can be used. The first of these are located in rivers, where they use the kinetic energy of water to produce mechanical energy and then electricity. Historically, river mill wheels were used for these purposes. As of today, the best universal system cannot be directly selected. Depending on the case, it could be, for example, an Archimedes screw [85] or a Kaplan turbine [86]. Properly fitting a small hydropower plant into a given landscape is a very difficult task. One should keep in mind the relatively negative impact of such an installation on the environment. The construction of such a power plant can cause problems in fish migration, disturb the biological balance of the river or pollute the water itself. However, studies indicate that if a number of guidelines are followed and particular attention is paid to the development of the plant, these problems can be minimized [87,88].

2.5. Biomass Technologies

In the era of transitioning away from conventional energy sources, there is an increased search for zero-carbon alternatives. One possible method of clean energy generation is the combustion of biomass. Biomass can be defined as any organic substance of biological origin (vegetable or animal) available in the world. When using biomass processing technology, the most chosen biomasses are municipal waste, agricultural waste, vegetables, and energy crops [89], as well as wood in unprocessed form or pellet form. In addition, these biomass types can be converted into biogas or biofuels, which can have much better energy conversion parameters compared to the raw/origin material [90].

In addition to the type of fuel, different ways of burning biomass can be distinguished, depending on the scale of customer needs. For high-capacity units, fluidized beds [91] and grate boilers [92] are commonly used devices. The former has the highest efficiency, which derives from the characteristics of the combustion itself. Fuel is pulverized in the combustion chamber to form a slurry, with sand or ash particles adopted as inert material. For residential customers, the most popular way to obtain heat from biomass is by a fireplace. The installation can be adapted to transfer energy around the building, as well as to heat domestic water by burning wood in the fireplace [93,94].

Commercial biomass boilers are based on the use of wood or pellets, while in the case of using straw, batch boilers are often used. These boilers can be characterized by very high output (up to 1 MW), but there are also smaller units for a single household.

Multistage biomass combustion shows promising results [95]. This consists of drying, pyrolysis, oxidation, and reduction stages. By converting solid fuel into a gaseous state, higher efficiencies are achieved throughout the process. Recent studies indicate that despite the inclusion of biomass as a renewable energy source, emphasis should be placed on the development of cogeneration technologies that allow obtaining not only heat but also electricity or cooling from combustion [96].

Biomass combustion is one of the most popular methods of heat generation in distributed renewable energy applications. Households without access to district heating or the natural gas infrastructure most often use units that burn wood or pellets. It is worth noting that suitably adapted systems can be used to generate electricity. The heat generated by a biomass combustion unit is transferred to a substance that changes the state of matter or, in general, to the working fluid of a power cycle. It then drives a steam or gas turbine, which, connected by a shaft to a generator, produces electricity. However, this way of biomass use can be more efficient when it is coupled with a combined heat and power (CHP) approach. In such a CHP system, components that take off excess heat allow one to supply thermal energy to the user. Such a solution works well in large central CHP plants, but also in distributed power generation [97].

2.6. Heat Pumps

Currently, there are relatively few methods of heating buildings that are not connected to a district heating network. Focusing on the use of renewable energy sources, in addition to the solar or biomass conversion systems already discussed, heat pumps are becoming increasingly popular. The use of these devices, particularly in the distributed energy sector, is considered a critical method for decarbonizing heat production worldwide [98].

Their operation is based on extracting energy from a low-temperature source to a higher-temperature source. The whole process must be assisted by supplying external energy, such as electricity. Heat transfer is carried out by means of a working medium consisting typically of an organic fluid, which must have appropriate operating thermodynamic parameters depending on the temperature of the lower source [99]. The most important of these is the boiling point. The medium is supposed to receive heat in the evaporator, from a medium with a relatively low temperature.

The lower heat source can be a variety of different media. Often, it is the air surrounding the building [100]. Pump systems developed for this purpose have fans that allow heat transfer from the air to the working medium. The great popularity of this solution is directly due to its relative versatility. An air heat pump can be installed on a building in a variety of climatic conditions. The most stable operation of these devices is observed for ambient temperatures in the range of $-3\ ^{\circ}\text{C}$–$10\ ^{\circ}\text{C}$ [101]; however, after appropriate adjustment of the operation of the units, optimal operation is observed even at temperatures down to $-30\ ^{\circ}\text{C}$ [102].

The lower source of the heat pump can also be the ground. Direct energy extraction from globally available deposits is often uneconomical and requires specific conditions [103]. Most often, small geothermal deposits have too low a temperature to heat domestic water and supply heating directly or generate electricity. This energy stored in the ground can easily be used in heat pump systems. The main advantage of these solutions is the relatively stable temperature of the lower source, which translates into better heating efficiency. Ground-source heat pump applications can be divided into two types: vertical or horizontal heat exchangers. The first [104] are based on drilling a borehole reaching typically from 30 to 100 m deep or more. This procedure is designed to ensure a lower heat source with the highest possible constant temperature. With depth, the temperature of the ground tends to be more constant over the year; however, at the same time the investment costs associated not only with digging the borehole itself but also with the circulating pump with the required head increase. As regards the horizontal ground heat exchangers, they involve laying a ground collector with a large area at a depth of 50–120 cm [105], making them more susceptible to ground temperature variation at shallow depths.

In addition to extracting energy from the ground, heat pumps can also extract thermal energy from surface water [106] or groundwater [107] on a similar basis to ground pumps.

Also worth mentioning are the possibilities of integrating heat pumps with solar [108] or photovoltaic [109] systems. In the first case, solar collectors can act as a lower heat source, reducing the temperature difference between the evaporator and condenser, and increasing the efficiency ratio of the system. A photovoltaic system very often goes hand in hand with heat pumps, since the electricity produced by PV can easily be used to power a circulating pump.

The use of electric compressor heat pumps is not the only way to transfer heat energy from a medium with a lower temperature to a higher one. It is also possible to distinguish systems that base their functionality on the process of absorption [110] or adsorption [111]. Unlike compressor pumps, these require the application of energy in the form of heat to the system. The main part of the absorption-based system is the circulation of usually a solution of water and lithium bromide (LiBr) salt. It is also possible to use other substances with similar sorption characteristics. During the operation of the system, when heat is supplied to the mixture, the absorbent material separates from the water and becomes the energy carrier. After giving up the heat, it reunites with water, and the process repeats itself. If it comes to adsorption processes, a mixture of two substances is not observed in it. Water in this process is deposited on the adsorbent material, often silica gel [112]. In this activity, the system gives up heat, and conversely, in the case of heating the sorbent, water is released. This type of heat pump requires an external heat source, most often a gas burner or hot water in traditional applications or thermal energy from renewables. For the latter, it is possible to use a number of heat sources, such as solar collectors [113] or a biomass burning unit [114].

The described heat pump systems are assumed to have the process of heating domestic water or indoor air. However, it is possible to change the operation mode of the device from heating to cooling, shifting the role of the condenser to one of an evaporator, and vice versa, by means of a valve system. This allows you to receive energy from the utility. Apart from reversible vapor compressor heat pumps, a common application consists of the use of the previously mentioned sorption heat pumps as a chiller. This is a very common approach in hybrid renewable polygeneration systems in order to produce cooling through the use of heat.

In order to give a good idea of the approaches taken by the authors of the various works to improve the performance of the cited technologies, Table 1 was created. It describes how the researchers dealt with the problems of the various RES technologies and shows what results they obtained.

Table 1 shows that there are many ways to improve the operation of RES technologies. However, it is worth mentioning that not all the methods shown above carry only positive aspects. An example is the idea of concentrated PV cells [21]. This idea, despite the theoretical increase in efficiency, also translates into a significant increase in the surface temperature of the cell, which results in a decrease in efficiency. Similarly, the idea of creating hybrid PV/T [26] modules can be problematic. The issue with this solution is the requirement of specific weather conditions to take full advantage of this system, which directly translates into its cost-effectiveness. An analogy can be made with the creation of diffusers for wind turbines [41], which can increase their efficiency, but depending on the situation, may or may not be useful. The problem here may be the space utilized by this component. Instead of a diffuser, the turbine itself could have larger dimensions, which would translate into better performance.

The ideas described above show that in most cases, there is not a universal way to improve the performance of RES technology, since several ways require specific conditions. This shows how much emphasis should be given to the proper selection of components of a classic or hybrid installation.

2.7. RES Technologies in Literature

Selected examples of systems based on renewable energy sources from the literature were collected and are summarized in Table 2. In order to denote the climate zone of the specified location, a symbol has been added next to the location of a given solution. The choice of symbols was guided by Köppen's classification [115]. In the present paper, it was decided to use the first level of classification, which was described on the basis of average annual precipitation, average monthly precipitation, and average monthly temperature.

The adopted classification includes the tropical humid (A), dry (B), mild midlatitude (C), severe midlatitude (D), polar (E), and highland (H) climate zones. It is important to note that this classification is also adopted for other tables in the paper.

Table 1. Summary of methods on improving the performance of the listed technologies.

Technology	Problem	Solution	Result	Reference
PV	PV module surface temperature too high	Integration with a system that uses excess heat	Decreased cell temperature, increased cell efficiency	[39]
		Integration with a system spraying surface with water	Decreased cell temperature, increased cell efficiency	[40]
	PV module efficiency or annual irradiation too low	Focusing solar irradiation with lens or mirrors	Increased efficiency, very high surface temperature	[45]
	High influence of shading on efficiency of PV module	New approaches in MPPT optimization	Lowered sensitivity of PV modules to changing external conditions	[48]
Solar thermal	Possible higher efficiency of whole system	Creation of hybrid PV/T module	Increased system efficiency, relatively low temperature of heated water	[49]
	Possible higher collector efficiency	Focusing solar irradiation with lens or mirrors	Increased temperature and efficiency of collector	[52]
VAWT	Negative torques of WT while operating	Integration with an augmentation system	Increased WT efficiency but waste of space which could be used for bigger rotor	[71]
	Possible higher WT efficiency	Integration with wind accelerating construction (diffuser, building)		[65]
	Relatively high needed starting wind velocity	Introduction of system controlling arrangement of blades	Increased WT efficiency	[77]
HAWT	Requiring large tracts of land, high sensitivity of wind conditions to land shapes, hard to place on deep waters	Building floating wind farms	Increased efficiency, space saving	[78]
WT	Hard to recycle components	Usage WT elements for other purposes	Positive impact on environment	[79]
Water turbine	Difficult to implement as a small system	Possible places (near rivers) in-depth analysis, creation of guideline for such investments	Possibility of creating modern small hydropower plants in a greater number of locations	[87]
Biomass	Possible higher efficiency of the system	Introduction of multistage combustion	Increased efficiency of system	[95]
		Introduction of CHP components		[97]
Heat pumps	Possibility of lower source temperatures being too low	Introduction of different working media	Improved system reliability	[102]

Table 2. Selected research on small-scale RES systems.

Technology	Power Installed	Energy Demand	Location	Energy Storage	Evaluation Method	Economic Review		Experiment	Reference
						Indicator	Value		
PV	5.6 kW	5.7 MWh/year	Lublin, Poland (D)	No	PVSYST	ROI	8.9 y	No	[116]
	48 kW	383 MWh/year	Lublin, Poland (D)	No	PVSYST	ROI	5.7 y	No	[116]
	36 kW	-	Sverdlovsk, Russia (D)	No	PVSYST	-	-	No	[117]
	270 W	159.42 kWh/m^2/year	Kumasi, Ghana (A)	Yes	TRNSYS	LCOE		Yes	[118]
	13.5 kW	93.8 kWh/month	Beijing, China (C)	Yes	Energy balance equations	LCR	0.95	No	[119]
WT	3 kW	3600 kWh (heat demand)	Edinburgh, Scotland (C)	Yes	DesignBuilder, TRNSYS	Comprehensive score	0.407–0.594	No	[120]
	5 kW	3600 kWh (heat demand)	Edinburgh, Scotland (C)	Yes	DesignBuilder, TRNSYS	Comprehensive score	0.473–0.638	No	[120]
	10 kW	3600 kWh (heat demand)	Edinburgh, Scotland (C)	Yes	DesignBuilder, TRNSYS	Comprehensive score	0.455–0.524	No	[120]
	1 kW	-	Forlì, Italy (C)	No	Mathematical analysis—Weibull	LCOE	0.61 €/kWh	Yes	[121]
	3 kW	2330 kWh	Ankara, Turkey (B)	No	MATLAB—Weibull	-	-	No	[122]
Hydrokinetic Turbines	5 kW	10,715 kWh	Baton Rouge, USA (C)	No	CFD	ROI	4–5 years	No	[123]
	5 kW	2620 kWh	Itacoatiara, Brazil (A)	No	CFD	ROI	6–7 years	No	[123]
	3 kW	10,715 kWh	Baton Rouge, USA (C)	No	CFD	ROI	7–8 years	No	[123]
	3 kW	2620 kWh	Itacoatiara, Brazil (A)	No	CFD	ROI	15–16 years	No	[123]

When considering photovoltaic (PV) technologies, studies showing the operation of such systems in various locations around the world were collected. In this regard, it is a fairly universal technology. However, it should be noted that in locations characterized by low average irradiation, it will be less effective. Cieślak [116] pointed out that in the case of creating a basic system without an energy storage system, it is more cost-effective to install one that does not cover all the energy demand. In this case, an off-grid system would not be reasonable. The authors in [117] proposed a system with a solar tracer. They pointed out that such a solution is more efficient than a classic system facing in one direction. In addition, Agyekum et al. [118] analyzed the effect of high temperature on the energy loss of the system. The paper was based on a comparison of the economics and efficiency of PV and PVT systems in Ghana. The results indicated that a stand-alone PV system had higher cost-effectiveness than a stand-alone hybrid system. It is worth mentioning that this situation becomes reversed when energy storage is added to the system. Li et al. [119] proposed an off-grid PV system with energy storage. The authors compared different system configurations and identified the most cost-effective approach. It consisted of 66% building demand energy storage and 1.4 energy penetration of the PV system.

When analyzing stand-alone PV systems, the authors of publications are most often supported by software based on meteorological data and basic PV models (PVsyst) [116,117]. Such methodology allows one to analyze the performance of the systems depending on the geographic location and to analyze the efficiency of the system operation accounting for several design factors. In the case of a more complicated polygeneration system [118], one should turn to more complex software (such as TRNSYS).

Most of the works available in the literature include an economic analysis. The most commonly used indicators are return on investment (ROI) [116] or levelized cost of electricity (LCOE) [118]. These approaches allow an assessment of the profitability of proposed solutions.

Wind turbines (WTs) seemingly can perform well in a wider range of locations around the world. However, it should be noted that stand-alone wind turbines are not the most cost-effective solution in most cases. A suggestion from numerous studies is to create a hybrid system incorporating other systems in addition to the turbine. The authors of [120] indicated that a cost-effective approach would be to add an energy storage system to WT installation. The purpose of this system was to power a heat pump to cover the entire heat demand of the building. The authors undertook a comparison of such systems with different WT capacities and battery capacities at different locations in Scotland. Based on the comprehensive score method, they compared the investigated options. The results presented indicated that such systems could not qualify as the most profitable. However, the authors suggested that their use in Scotland would be possible with the assumption of subsidies from the government. The authors of [121] came to interesting conclusions. The study was based on a comparison between the actual production of a wind turbine and an estimated value based on weather data. The study showed the difference between the two situations and allowed for their economic analysis. The authors noted that the studied installation was not characterized by high profitability, and they concluded the analysis under the assumption of better wind conditions allowed satisfactory results. In general, the work showed that stand-alone wind systems are usually not economically efficient and that they could play a greater role in hybrid systems.

Location is very important for wind power plants. Bilir et al. [122] undertook an analysis of the feasibility of an area in the vicinity of Ankara, Turkey for WT siting. The study indicated that the area does not have sufficient wind for large-capacity wind turbines, so the authors chose an alternative solution of three smaller WTs. The results showed that such an installation can produce enough energy to cover the electricity needs of small housing. It is worth noting that no energy storage system was assumed in this case, and thus most of the energy yield would be lost or transferred to the grid.

Distributed hydropower is also worth mentioning. Puertas-Frías et al. [123] indicates that it is possible to use medium- and high-flow rivers for this purpose. Two locations were selected for analysis: Baton Rouge, USA on the Mississippi River, and Itacoatiara, Brazil on the Amazon River. The first step in the study was to select a suitable hydroturbine for the task of power production. A number of different profiles were considered to create a rotor with a horizontal axis of rotation. The selected parameters then made it possible to calculate their efficiency and also the cost-effectiveness of this solution. The systems studied were intended to be on-grid, and this fact showed that the economic viability of the investment is directly affected by a country's policy regarding support for renewable energy sources. Without possible additional subsidies, the installation would not be profitable. ROI was used for the economic analysis, and the whole simulation was carried out in CFD software.

It is worth noting that in the modern literature, it is difficult to find works based on the analysis of a small system entirely based on biomass. Common knowledge shows that while heating a household with biomass is profitable, in terms of electricity generation, it should go hand in hand with the production of other types of energy [124].

As devices characterized by relatively high electricity consumption, heat pumps should be used in tandem with systems that obtain energy from renewable sources. The use of grid energy in most cases is not among the most economical treatments [125].

2.8. Summary

Considering all the RES technologies mentioned above, it is hard to point out unconditionally the best one. Each has many advantages as well as disadvantages. However, it is possible to notice certain trends when it comes to the popularity of using a given system in scientific studies and also in practice. Biomass-based technologies can be considered the most popular. Although the data presented in Section 2.7 show that systems based exclusively on these technologies are no longer being developed to any great extent, it should be remembered that they are among one of the most popular sources of energy in the thermal sector in distributed and central [126] power generation. This is supported primarily by the fact that these technologies have existed for decades while the massive use of other RES is relatively new. These systems, in distributed power generation, are characterized by relatively low investment and operating costs. However, it should be noted that solar systems, which are becoming increasingly popular despite their relatively high investment costs, are a cheaper solution with a longer operating period [127]. It is worth noting that these days, there is a very strong emphasis on replacing fossil fuels. One of the solutions is the creation of biogas plants and biomass-based combined heat and power plants, thus covering the electricity and heat needs of small towns and villages [128]. The idea is to simultaneously produce energy and use the excess organic matter available in such agglomerations.

Solar energy systems seem to take second place in terms of popularity. This is supported primarily by a certain versatility of this solution, the low complexity of basic installations, or the very large government support worldwide [129]. However, they have major drawbacks related to the stability of operation, which is also characteristic of wind energy-based technologies [130]. For this reason, the authors of several studies suggest the development of hybrid systems or the use of energy storage to improve the efficiency of the entire system when using this technology.

Hydropower technologies account for a significant share of the RES mix in the world. However, these are large run-of-river, dam or pumped storage power plants. The use of water power in small systems is not a very difficult undertaking in terms of technology. However, it requires specific conditions, in particular the presence of flowing watercourses and the possibility to use natural or artificial basins. This fact translates into their lower accessibility and low popularity.

Heat pumps are considered the future of the thermal sector. However, it is worth noting that for their operation they require the application of electricity, the price of which has been rising rapidly in recent years (2020–2022) [98]. This fact translates into an obligatory parallel use of renewable energy sources with heat pump systems in order for such an investment to be profitable.

The technologies presented in Section 2 were brought together to show the feasibility of their use in small-scale hybrid polygeneration RES-based systems. It should be noted that the authors here were not looking for novel approaches that may work only in a narrow range of solutions, but were guided by research showing improvements in their properties in the abovementioned systems. It is very important, especially from an economic perspective, to use well-studied, reliable ways to improve efficiency in such systems. However, this does not mean that there is no room for innovative approaches. An example is the previously described approach [39,40] showing ways to extract heat from PV cells. Such or similar technologies, however, need to be properly optimized and improved to find their place in small-scale hybrid polygeneration RES-based systems.

3. Storage Technologies

To respond the climate change and minimize its impact, energy storage technologies have become a priority in many countries around the world [131]. This has led to a major increase in the number of technologies using renewable energy sources [132].

The disadvantage of this direction of development is the presence of devices characterized by intermittent power generation, which leads to a decrease in system reliability and feasibility. This situation requires the use of compensation elements to avoid potential power shortages or store surpluses [133]. An energy storage unit is exactly this type of compensation element. However, storage technologies are met with some skepticism due to the high initial cost of the system and the associated transformation losses [134]. Recently, a significant price drop was observed for some energy storage technologies, e.g., lithium ion batteries [135], but the price of a novel manufacturing methods may offset any cost savings until economies of scale take over. This is the difficulty in bringing new technology to market; however, suitable energy storage has a significant role in improving the energy efficiency of renewable-based systems, which leads to savings [136].

Energy storage methods can be divided according to a number of criteria. The most important of them are presented in the following subchapters. Due to the variety of values describing the characteristic sizes of each energy storage, the most important ones are listed in Table 3.

Table 3. Technical characteristics of the systems.

Storage Type	Energy Efficiency	Volumetric Energy Density	Lifetime	Storage Period	Cost	Small-Scale Applicability
[-]	[%]	[kWh/m^3]	[year or cycles]	[time]	Euro/kWh	[-]
Thermal Energy Storage						
Sensible storage	50–90	93	10–30 years	days/months	0.1–10	+
Latent storage	75–90	50–3210	10–20 years	hours/months	10–50	?
Thermochemical storage	75–100	200–500	15–30 years	hours/days	8–100	+
Electrical Energy Storage						
Supercapacitors	90–95	1.5–15 Wh/kg	1 million	seconds/minutes	300–2000	+
SMES	90–95	0.5–10	>1 million	minutes/hours	13 k–76 k	-
Electrochemical Energy Storage						
Lead–acid batteries	74–95	50–80	203–1500	days/months	100–830	+
Lithium batteries	90–97	200–500	3500–20,000	days/months	500–2000	+
Nickel batteries	71	60–150	350–2000	days	450–1800	+
Sodium sulfur batteries	75–85	156–255	2500–8250	days/months	280–700	+/-

Table 3. *Cont.*

Storage Type	Energy Efficiency	Volumetric Energy Density	Lifetime	Storage Period	Cost	Small-Scale Applicability
Redox flow batteries	60–80	16–60	7000–15,000	days/ months	110–1000	?
Chemical Energy Storage						
Methane	49–79	1200	-			
Hydrogen	54–84	400	-	days/ months	2–15	+
Mechanical Energy Storage						
Solid media (flywheel)	85–95%	80–200	10,000–100,000	hours	650–2625	+
Liquid media (PHES)	65–87%	0.01–0.12	-	days/ months	500–1700	-
Gaseous media (CAES)	30–50%	0.04–10	20–40 years	days	2–140	-
References	[137–142]					

"-" no, "+" yes, "?" experiential applications.

3.1. Thermal Energy Storage

Thermal energy storage (TES) allows excess heat energy to be stored and used after hours, days or months. It can be stored on a scale from a single process to a region [143]. Depending on the temperature range of the storage medium, we can distinguish between low-temperature storage (up to 120 °C), medium-temperature storage (120–500 °C), and high-temperature storage (>500 °C) [137]. TES storage systems are commonly integrated with concentrated solar power (CSP) plants: 80% of power plants rely on this type of energy storage, increasing efficiency by smoothening out fluctuations in energy demand throughout the day [144]. CSPs use all three basic types of TES: sensible, latent and thermochemical. The classification of main thermal energy storage technologies is presented in Figure 1.

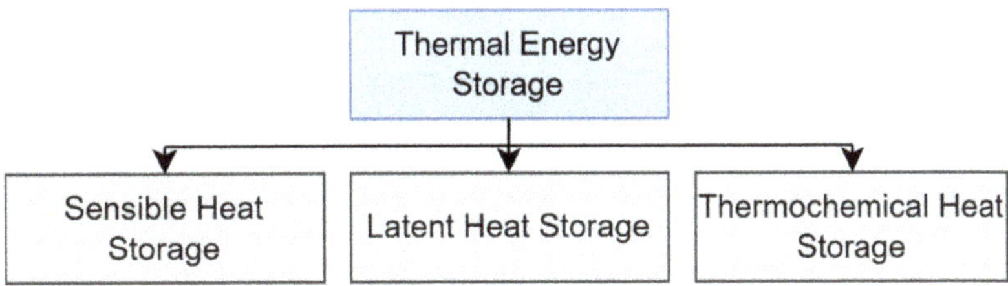

Figure 1. The main classification of thermal energy storage technologies.

3.1.1. Sensible Heat Storage (SHS)

In this method, thermal energy is stored in a material, leading to a change in its temperature. The change in the temperature of a body and its heat capacity is used to carry out the charging and discharging processes [145]. The amount of heat stored depends on the temperature difference between the initial and final states of the substance, its mass, and specific heat. Water is most commonly used as the storage medium [146], although a water–glycol mixture, concrete, and rock are also used [146]. The technology is most often integrated with heat pumps, solar thermal systems [147] and heating systems as buffer storage [148]. Compared to latent or thermochemical thermal, sensible thermal storage has a lower energy density when we consider a limited temperature range. However, sensitive thermal storage technology is standardized and has a much lower price than other types of storage [149].

Commonly used components for SHS are packed-bed storage tanks (PBSS), which store thermal energy by heating and cooling the solids with a heat transfer fluid (usual air) flowing through the beds [150]. The most widely cited advantages of these storage media are a broad operating temperature range, reduced corrosion, and low cost of storage material [151]. Detailed studies for different solids and their parameters to evaluate the thermal performance of a sensible heat storage bed with air as the heat transfer fluid have been discussed by A. Elouali et al. [152].

For long-term storage, underground heat storage is used [153]. This method has been considered by Recep Yumrutaş et al. [154]. They presented a mathematical analysis to determine the long-term performance of solar-assisted home heating systems using a heat pump and underground thermal energy storage (UTES). The temperature of the TES reservoir increases with the years, resulting in a reduction in the annual requirement for heat pump operation. After the fifth year of operation, the annual periodic operating conditions have been reached. A TES reservoir is a viable energy storage solution, as long as it is located sufficiently deep underground. Researchers noted that the efficiency of mentioned systems increases with the volume of the reservoir.

An interesting new technique with promising results in the test phase is the combination of sensible and latent heat storage [155]. In [156], the authors proposed an installation including the storage of thermal energy in quartz, with part of the energy stored in phase-change material (PCM). A simulation-based analysis showed that the implementation of energy recycling from the compressor and heat storage led to an increase in system efficiency from 34.4% to 60.6%.

SHS can be also a valid solution for extreme conditions. G. Hailu et al. [157] presented an installation for a building designed for net zero energy. The house additionally uses high-performance building techniques, e.g., Arctic walls combined with photovoltaic panels, collectors and sand-bed storage. The work shows that thermal storage systems made of readily available materials (e.g., sand) can be a cost-effective way of storing energy.

3.1.2. Latent Heat Storage (LHS)

Latent heat storage systems adopt the phenomenon of energy storage during phase change. Materials that enable LHS storage are known as phase-change materials (PCMs). The primary field of application is residential and industrial heating and air conditioning [158].

Storage of large amounts of energy with small temperature differences is the main advantage of storage systems using PCMs. Because adding energy to the system does not increase the temperature difference with the environment, exergy losses are lower than with SHS [159]. However, this system is technically difficult to implement due to leakage that occurs during the phase transition with conventional phase-change materials [61]. PCM must be characterized by high specific heat, thermal conductivity, and density. The material should also have a melting point in a suitable temperature range [160]. All of the properties mentioned are associated with an increase in the price of the system.

There is a group of materials suitable for phase-change heat storage [161]. The criteria they must meet and examples of substances are given in Table 4. The following table is based on information from [162,163].

The evaporation process, despite its high enthalpy, involves a volume change that is difficult to control [164]. For this reason, it is rarely used. Materials such as salt hydrate and paraffin have found practical use in LHS systems [165]. Despite its high storage density and good thermal properties, salt hydrate is characterized by inconsistent melting. This results in a decrease in capacity with the number of cycles [166]. A solution to this problem has been investigated by Tyagi et al. [167]. In their work, they observed that the intensity of phase separation increased at lower mass flow rates. They showed that measurements on smaller samples help to verify that the PCM composition is not degraded due to incongruent melting. Paraffins, on the other hand, allow their melting temperature to be controlled by changing the length of the alkane chain [168]. They present high melt compatibility and cyclic stability, and they are environmentally safe and noncorrosive. Problems with their use are their flammability and high price [169].

Table 4. Latent heat storage materials.

Substance Examples					
Class	Material	Melting Point [°C]	Density [kg/m^3]	Thermal Conductivity [W/m·K]	Latent Heat of Fusion [kJ/kg]
Organic	Paraffin wax	64	916 (solid, 33.6 °C) 769 (liquid, 65 °C)	0.346 (solid, 33.6 °C) 0.167 (liquid, 63.5 °C)	173.6
Fatty acids	Palmitic acid	64	850 (at 65 °C)	0.162 (at 68.4 °C)	185.4
Salt hydrate	$CaCl_2 \cdot 6H_2O$	29	1562 (at 32 °C) 1802 (at 24 °C)	0.540 (at 38.7 °C) 1.008 (at 23 °C)	190.8
Metallics	Bi-in eutectic	72	-	-	25.0
Desirable characteristics					
Thermal Properties	Physical Properties		Kinetic Properties	Chemical Properties	Economics
Melting temperature in desired operating range	Small vapor pressure at operating temperatures		Little or no super-cooling during freezing	Chemical stability	Abundant
High latent heat of fusion per unit volume	Small volume variation on phase change		High nucleation rate to avoid supercooling	Complete reversible freezing/melting cycle	Large-scale availabilities
High specific heat	High density		Adequate rate of crystallization	Compatibility with container materials	Effective cost
High thermal conductivity of both phases				No toxic, flammable, or explosive material	

An interesting study was conducted by Johansen et al. [170]. They presented the results of a study of a solar system combined with a sodium acetate trihydrate (SAT) heat storage tank used to heat up a water tank. During a 6-month test period, the SAT was heated 53 times by the solar collectors above 80 °C. During the test period, it supplied 135 kWh of heat to the water buffer tank. The research shows the core concepts of supercooling and the discharge of heat from the PCM. Additionally, in [171], Englmairab et al. introduced a numerical simulation of the previously mentioned system [170]. Validation results of the component models showed a high degree of similarity to the measured data. Numerical simulation was used to optimize components of a system, which showed very promising results. A full charge of a single 200 L PCM unit and a 2.8 m^3 water tank enabled the supply of heat for 18 days in January. As a result, the building's heating needs can be covered by almost 100% renewable energy sources.

At present, the search for phase-change materials focuses on finding materials with high enthalpy at different temperatures [172]. For conventional solid–liquid phase-change materials, solutions are being explored to eliminate phase-change leakage. To solve this problem, Yao Menga et al. proposed the use of a composite [173], while Ali Usman et al. have seen an opportunity for the development of solid–solid phase-transition materials [174].

3.1.3. Thermochemical Heat Storage (THS)

This type uses reaction energy from reversible chemical processes or physical surface reactions. The absence of thermal losses is a consequence of storing energy in the form of reaction energy rather than heat [175]. Despite their particularly high energy density and technically possible long energy storage times [176], these systems are rarely used for economic reasons [177]. THS systems can be divided into three categories: chemically reversable processes, adsorption storage systems and absorption storage systems [178]. For chemically reversable processes, the thermodynamic equilibrium temperature is decisive. Discharge of the system occurs at a temperature lower than the equilibrium temperature. Charging, on the other hand, occurs at a lower temperature.

K. Kant et al. [179] presented a study of long-term energy storage via an absorption process for heating buildings. The heat storage was based on an aqueous solution of LiBr (lithium bromide). The authors designed and built a prototype, which they tested under static and dynamic operating conditions compatible with domestic solar and heating systems. The tests showed that the absorption process could allow the house to be heated. However, the absorber design used by the authors prevented heat recovery in the charging–discharge processes. They thus suggested using a two-piece absorber or a pool absorber.

The authors of [180] pointed out that despite the potential high energy storage density and low losses, poor heat and mass transport significantly hinders the commercialization and implementation of THS systems. Above all, the development of this technology requires fundamental innovation in the development and discovery of new materials.

3.1.4. Thermal Energy Storage: Summary

The main difference between the three mentioned types of TES is the range of operating temperatures. Most SHS storage systems have the broadest operating temperature range, as the temperature increases through the whole charging process, in some cases reaching 500 °C and above. However, high temperatures cause energy loss due to limitation on insulation and heat creep of materials. When only a limited range of operating temperatures is possible, this type of TES has on average lower energy density compared to other types of TES. On the other hand, LHS and THS storages avoid the downsides of high temperatures by containing the energy in state change and chemical processes subsequently. LHS has the potential of a very long-lasting energy storage, as long as it is kept at constant temperature in order to avoid spontaneous phase change. From a practical point of view, the melting phase is the most commonly used. The evaporation process, despite its high enthalpy, involves a volume change that is difficult to control.

From an economical point of view, SHS systems use the cheapest and most easily available energy storage medium and are also safe. Media used in LHS must meet specific criteria and come with their drawbacks. Similarly, THS systems require media that can improve their heat and mass transport, which would make them more economically viable.

In terms of applicability, SHS systems are widely used, starting with household buffers with a few kWh capacity through seasonal storage systems up to district heating with capacity in the range of GWh. On the contrary, SHS and THS systems are generally integrated in niche applications, such as thermal management in dishwashers or air-conditioning of a room.

3.2. Electrical Energy Storage

The major advantage of direct electrical energy storage is the absence of energy conversion, which translates directly into reduced losses. However, due to its high costs and very low energy density, this solution is currently not widely used [149]. Classification of technologies for direct electrical storage is presented in Figure 2.

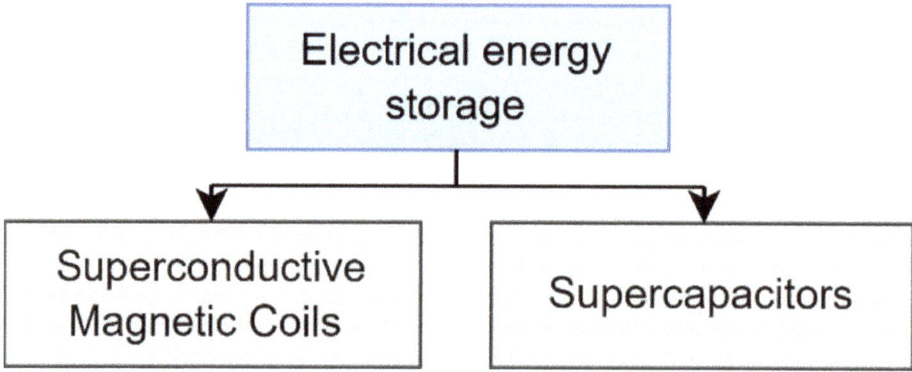

Figure 2. The main classification of electrical energy storage technologies.

3.2.1. Supercapacitors

A supercapacitor is a capacitor of high capacity, frequently much higher than other types of capacitors used in electronic devices, but with lower voltage limits. In terms of energy storage, they are transitional devices between electrolytic capacitors and rechargeable batteries [181], storing even 100 times more energy per volume than standard capacitors [182]. Additionally, supercapacitors can charge and discharge with higher power than any type of battery, and can repeat this cycle many more times. However, they are not suitable for long-term energy storage due to their tendency to self-discharge.

Due to their nature, supercapacitors are a frequent choice in applications that require fast energy absorption or delivery. This capability is required to match demand and control the ramping up or down of the energy supplied by renewable technologies. Shifting cloud cover, rapid change in wind direction, or changing power loads results in large real-time variation in power surplus or deficit. Supercapacitors can absorb these surpluses or supply energy during energy deficits, in turn smoothening the power output of renewable energy sources [183].

However, in small-scale applications, minute-by-minute variation in energy generation has a much smaller amplitude and is less impactful on energy installation. This way, supercapacitors are currently applied only in large-scale renewable energy plants [184].

3.2.2. Superconducting Electromagnetic Energy Storage (SMES) Systems

Superconducting magnetic energy storage (SMES) systems use cryogenically cooled superconducting coil. Direct current flowing through the coil creates a magnetic field, which is used for the purposes of storing energy [185]. Typically, an SMES system is composed of three parts: a superconducting coil, power conditioning system, and cryogenically cooled isolated refrigerator. All of them are stationary components, making them extremely stable. Once the superconducting coil is charged, the current will not decay and the magnetic energy can be stored almost indefinitely. The stored energy can be retrieved by discharging the coil.

Originally, SMES was developed for large-scale load leveling; however, with its rapid-discharge capabilities, it has been deployed on electric power systems for pulsed-power and system-stability applications [186]. Another advantage of SMES is the ability to almost instantly shift between charging and discharging the storage. Furthermore, similarly to supercapacitors, SMES systems can be used in conjunction with renewable energy sources in order to smoothen its energy output. The most important advantage of SMES is that the time delay during charge and discharge is quite short.

Padimiti et al. [187] researched the applicability of SMES technology as a power quality device and for damping power system oscillations. The conclusion was that the system is capable of rapid discharge of large amounts of energy, which is a necessary factor in the improvement of the dynamic performance of the power system.

The potential application of SMES systems on a small scale is limited by their large economic costs of implementation coupled with the high prices of superconductors. Additionally, keeping cryogenic conditions in an SMES system is a power-consuming process justifiable only on a large scale.

3.2.3. Electrical Energy Storage: Summary

Storage systems based on supercapacitors and SMES have much in common. They can absorb a large amount of electrical energy in a very short time and discharge it equally fast. Additionally, their charging process can be repeated multiple times. Both systems are applied in situations where a quick switch between absorbing and discharging of electrical energy is required, i.e., renewable power plants.

However, supercapacitors are not suitable for holding charge over long periods of time, while SMES systems can contain their charge almost indefinitely as long as they are kept in suitable conditions. Compared to supercapacitors, SMES systems require much larger and more advanced infrastructure, which greatly increases their cost of implementation.

3.3. Electrochemical Energy Storage Systems

Electrochemical energy storage has high efficiency, a fast response rate, and a relatively low price [149]. These systems use electrodes connected by an ion-conducting electrolyte phase, and chemical reactions are used to transfer the electrical charge. The battery parameter most often taken for comparison is the capacity, which represents the ability of the battery to store an electrical charge. The possible technologies of electrochemical energy storage are classified in Figure 3.

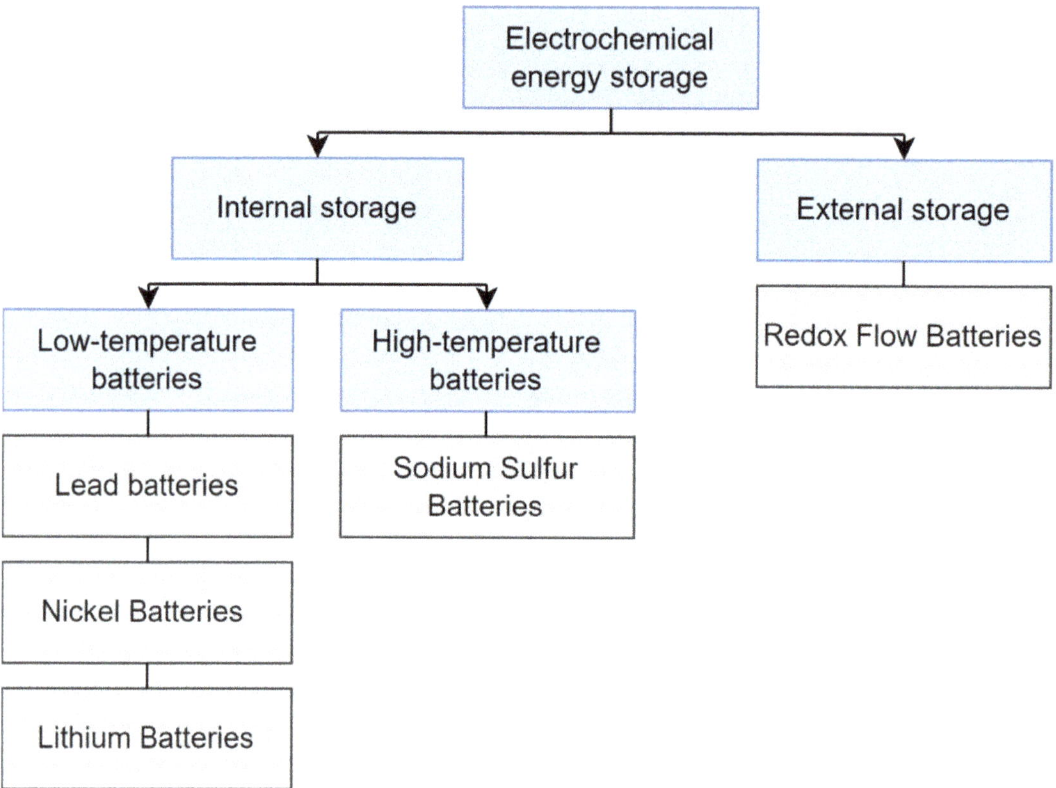

Figure 3. General classification of the electrochemical energy storage system.

3.3.1. Lead–Acid Batteries

A lead–acid battery consists of a negative electrode made of porous lead that facilitates the formation and dissolution of lead. The positive electrode consists of lead oxide. Both are immersed in an electrolytic solution of sulfuric acid and water. The technology of the production of lead–acid batteries is very mature, which lowers their price, making them a popular and economical choice. Due to their long lifetime, they are frequently used in renewable energy systems. On the other hand, they have relatively low energy density, only moderate efficiency, and high maintenance requirements.

The efficiency of this type of battery is assumed to be between 90% and 97% and depends on its application [187]. These batteries are particularly well suited as a backup power source when used in conjunction with a solar installation that is not continuously used. An example of such a system is the solar boat design proposed by Ahmad Nasirudina et al. [188]. However, even in this field, lead–acid batteries are being replaced by lithium ion batteries, which will be the storage systems of the future for electric yachts, as predicted in [189].

The ability of a lead–acid battery to operate over a wide temperature range is an undoubted advantage. However, changes in its parameters must always be monitored, as the state of charge of the battery depends on the season. In summer, the state of charge is close to 100% and its value decreases in winter. The minimum charge level is controlled by an automatic system that disconnects the battery at the minimum discharge voltage [190]. Otherwise, the battery may be completely discharged, which negatively impacts its condition [149]. At higher discharge currents in lead–acid batteries, there is an undesirable polarization effect. This effect requires special methods for determining the state of charge [191].

Another advantage of lead–acid batteries is a well-developed method of disposal, which significantly reduces their negative environmental impact [192], and systems using them have the potential to become emission-free. Amutha et al. [193] analyzed the performance of seven different off-grid systems using lead–acid batteries for different types of loads. The study considered seven combinations of components, such as a wind turbine, solar system, hydroelectric power plant, and diesel generator. The result of the simulation using the HOMER software showed that the battery improved the performance of the renewable energy system and enabled zero-carbon solutions.

Although these batteries are already applied in many fields, they are gradually being replaced by lithium ion batteries, which often prove to be a cheaper and longer-lasting solution. Abraham Alem Kebedeet et al. [194] compared the efficiency of a grid-connected photovoltaic system (PVGCS) integrated with a lithium ion battery and a lead–acid battery, respectively. In order to analyze the efficiency of both systems, the connection of the battery to the PVGCS was modeled using HOMER-Pro software. The analyses showed that the lithium ion battery had better discharge characteristics, providing a longer service life. This techno-economic study based on realistic load profiles and resource data showed that the cost of energy for a system with a lithium ion battery was EUR 0.02/kWh lower compared to a lead–acid battery. The study showed that lithium ion batteries are the preferred choice for a PVGCS system over lead–acid batteries.

3.3.2. Lithium Batteries

Lithium ion battery technology is distinguished by the possibility to use commercially available storages with relatively high energy concentration and to adopt them in a wide range of applications [195]. Due to the process of gradual expansion of their use from the portable electronics sector to the power grid sector, they are in a phase of continuous development [196].

The positive electrode of these batteries is made up of lithiated metal oxide, the anode consists of layered graphite, and the electrolyte is lithium salts dissolved in organic carbonates [197]. The way in which the reactions occur allows three basic classifications of lithium technologies to be defined: lithium ion, lithium sulfur and lithium air [198].

Currently, the simple association with lithium ion batteries is the automotive industry. Electric cars can also function as energy storage. Cars equipped with the vehicle-to-home (V2H) function can be charged and then return the energy when it is needed. This requires special two-way chargers for electric cars, by means of which energy storage in the car can be the answer to the summer overproduction of electricity. An example of such a system can be found in the work of García-Vázquez et al. [199]. The authors investigated a hybrid renewable energy system including V2H energy storage. However, under the restrictions described by the authors, the addition of the V2H only slightly reduced the necessary capacity of energy storage. The feasibility of using EVs as a support storage system is greatly dependent on the driver's profile and the daily trip range of the vehicle.

The energy storage in the car, despite its large capacity, is not able to replace traditional stationary energy storage. The problem is that the greatest production of electricity from a photovoltaic installation often occurs during people's working hours. Nevertheless, using an electric car for energy storage increases the self-consumption of energy produced by PV, speeding up the return on investment for both PV and the electric car [200].

Ghassan Zubi et al. [201] investigated the deployment of lithium ion battery potential considering technical and economic aspects. It was found that despite the relatively high initial cost, the lithium ion battery is competitive in the cost of energy storage and the ongoing cost reduction will promote the accelerated use of lithium ion batteries in this application. The researchers pointed out that a more favorable environment for reducing the barriers to deployment would be created through appropriately selected incentive strategies and schemes.

Large-scale lithium batteries may find application in microgeneration systems. Darcovich et al. [202] evaluated the performance of large-scale batteries working with a microgeneration system. In addition, they compared two novel cathode materials—Li-NCA and LiMnO—and conventional $LiMn_2O_4$. The use of the new materials resulted in a 30% increase in battery life compared to conventional materials. This occurred because of the higher capacity of the batteries, which caused them to operate through a smaller part of their capacity range, resulting in a lower net load on the battery materials.

3.3.3. Nickel Batteries

There are four main groups of rechargeable batteries that are based on nickel: nickel–cadmium battery (NiCd), nickel–iron battery (NiFe), nickel–metal hydride (NiMH), and nickel–hydrogen battery (NiH_2).

NiCd and NiFe batteries use nickel oxide hydroxide as one of the electrodes. NiCd batteries use metallic cadmium as the other electrode; however, as it is toxic, it was banned for most uses by European Union directives in 2002 and 2011 [203]. On the other hand, NiFe batteries are still in use as they are very robust and tolerant of abuse such as overcharge, overdischarge, and short-circuiting [204]. They have a very long life span even in harsh conditions and they are often used as a backup in situations when they are continuously charged. Despite those advantages, nickel–iron batteries are considered unsuitable for electrical storage systems due to their low efficiency. Additional disadvantageous factors are self-discharge effect and easily corrodible iron anode [205].

NiMH batteries replaced the NiCd cell, as its negative electrodes are made of a hydrogen-absorbing alloy instead of cadmium. Additionally, the capacity of the NiMH batteries is triple that of NiCd batteries of the same size. NiMH batteries also offer higher energy densities, though still lower than ones offered by lithium ion batteries [206].

Nirmal-Kumar et al. performed a comparison of NiCd, NiMH, lead–acid and lithium ion batteries in small-scale energy storage systems [207]. They used HOMEr software to simulate the behavior of the considered batteries in a system with a photovoltaic energy source. The conclusion was that the NiMH batteries had the best electrical performance, slightly better than NiCd. On the other hand, lead–acid batteries had the lowest initial cost, making them a very popular choice. However, the authors predicted the rising popularity of lithium ion batteries in small-scale storage systems, as their price is dropping with advancing technology. Also, the results of the simulations show that they have the lowest annual operating cost.

3.3.4. Sodium–Sulfur Batteries

In sodium–sulfur batteries, molten sodium acts as a negative electrode and molten sulfur is used as a negative electrode. They are separated by a solid electrolyte made of ceramic sodium alumina [208]. Sodium ions travelling between the cathode and anode are the driving force of the battery. However, the travelling ion can disrupt the crystalline structure of the battery, over time leading to fracture and reduced battery life [209]. This is one of the fundamental problems blocking the commercialization of these batteries. Nevertheless, they are still often considered an alternative to lithium ion batteries due to their safety, material availability and lower toxicity [208]. Interesting results in improvement of sodium–sulfur cells were obtained by et al. Jiaru He [210]. They created an electrolyte that prevents the dissolution of sulfur and thus solves transport problems. This provided a longer battery life, demonstrating stable performance for more than 300 charge and discharge cycles. Currently, research is also focused on developing suitable anode materials [211].

Sodium–sulfur (NaS) batteries are classified as high-temperature batteries. However, the high operating temperature results in a low self-discharge rate of the battery, but with long periods of inactivity, this causes the battery to discharge rapidly. These batteries also require heating during long idle phases. Due to their high operating temperature oscillating around 300 °C, they are used in large-scale installations. They have been introduced on a utility scale in Japan, where they support the electricity transmission and distribution system [212]. A large-scale energy storage unit using NaS technology has been built at the BASF site located in Antwerp [213]. The unit has power of 950 kW and reaches a capacity of 5.8 MWh.

It is possible that room-temperature sodium–sulfur (RT NaS) batteries could be used for small-scale solutions in the future. These are intended to address the safety and corrosion problems of their high-temperature predecessors [214]. However, they are in an ongoing phase of development. In this context, Peng Chen et al. [215] presented the direction of development of these energy storages. In their review, they paid particular attention to aspects of separator and cathode modification, ways to protect the metal anode, and electrolyte optimization.

3.3.5. Redox Flow Batteries

The redox battery is the most common type of flow battery. They adopt a reduction and oxidation reaction to trigger the movement of electrons, which are discharged through an external circuit. Due to the reversibility of the redox reaction after discharge, it is possible to change the oxidation state of the reactants again and reuse them. Redox batteries can also be quickly recharged by replacing used electrolytes with new ones [216]. On the market, they are used for peak load balancing. An example of this type of application is a Japanese power utility, Kansai Electric Power Corporation, using a vanadium redox flow battery (VRB) for this purpose [217]. The vanadium battery is currently the most advanced type of flow battery. This is due to its stability, which allows it to perform many discharge and charge cycles without undesirable reactions occurring. However, the low energy density of stored energy and the price of vanadium encourage the development of other types of redox batteries [218].

Tao Zou et al. [219] constructed a vanadium redox flow battery energy storage system with necessary perpetual components to study the operation conditions of the redox flow batteries. They concluded that in order to maintain the stability of the stack and improve the service life of the system, it should operate under an appropriate current. The efficiency of their system can reach 61.0% while working at the 74 A current level.

On a larger scale, Jefimowski et al. presented the concept of VRB batteries acting as stationary energy storage for optimal energy and cost performance of microgrids. In the studied case, they achieved a reduction in daily energy consumption of 1.77 MWh and a reduction in peak power of 581 kW [220].

Further possible applications of the redox flow batteries depend on ongoing research, which aims to improve the material properties of the electrodes and electrolyte [219].

3.3.6. Electrochemical Energy Storage: Summary

Electrochemical storage systems have been developing for a very long time and they are one of the most common types of electrical energy storage. Lithium batteries especially have become increasingly popular in recent years, as their energy density is greater than both acid and nickel batteries. However, lithium batteries are on average more expensive. Additionally, they are less safe than nickel batteries and prone to burning if short-circuited. Nickel batteries also last longer in harsh conditions and are less prone to damage through abusive treatment such as overdischarging and overcharging.

It is important to consider the environmental aspect of batteries, as the materials used in acid, nickel, and lithium batteries have a negative environmental impact in terms of both obtaining the materials and utilization of the batteries. Sodium–sulfur batteries were developed as an alternative, as the materials are widely available, and the battery itself is safer and less toxic.

3.4. Chemical Energy Storage

The chemicals used in this type of storage system have a higher energy density and a longer discharge time than battery technologies [221]. Thus, the diversity of their use arises. They can be used as raw materials for the chemical industry, for direct electricity generation, and in the transport sector as a substitute fuel instead of fossil fuel. Synthetic fuels produced from renewable energy can complement or supplement batteries in the transport sector [222]. The concept of converting electricity into a chemical energy carrier is called power to gas (PtG). The essence of PtG technology is to produce a gas (hydrogen or methane) [223]. The types of chemical energy storage are shown in Figure 4.

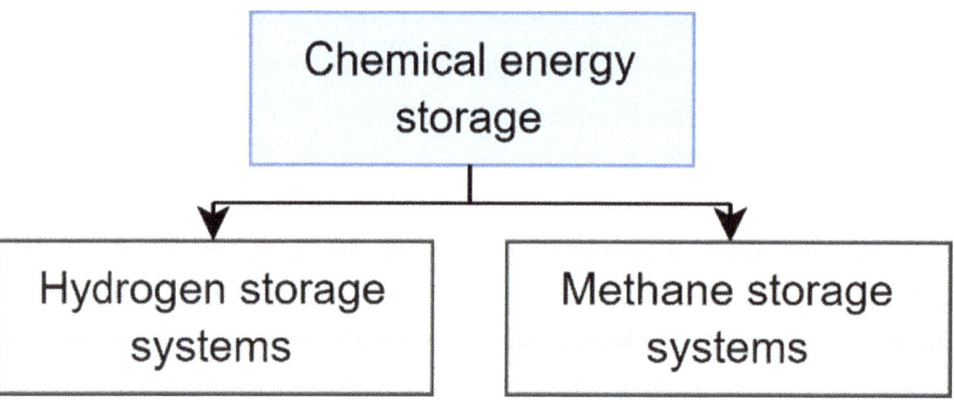

Figure 4. The main classification of chemical energy storage systems.

3.4.1. Methane

Methane is an alternative to hydrogen energy storage. Production occurs through the methanation process of CO or CO_2. In the context of PtG applications, methanation of CO_2 is more common [224]. Methane is an interesting concept from the point of view of energy transformation because of its similarity to natural gas, which means that it is compatible with existing infrastructure and combustion systems [225]. With a goal to demonstrate the readiness for integration into existing energy networks, the EU has undertaken a four-year project called STORE&GO [226]. The project demonstrated three separate large-scale test sites showing wide climatic variation: southern Italy (200 kW), Switzerland (700 kW), and northern Germany (1 MW). Applied methanation technologies were millistructured catalytic methanation using CO_2 from the atmosphere, biological methanation at wastewater treatment plants, and methanation through isothermal catalytic honeycomb wall reactors, respectively. Obtained methane was stored in existing local infrastructure, and transported by different location-dependent methods from long-distance transport grids to regional LNG distribution networks via cryogenic trucks. The project was completed in 2020 and the results proved that methanation methods work well under realistic conditions, with the process achieving 76% overall efficiency in Switzerland. Therefore, PtG competes well with both power-to-hydrogen and all-electric applications.

An integration of PtG in small-scale energy systems is still under development. Ligang Wang et al. [227] presented a solid oxide electrolyzer (SOEC) as a promising small-scale power-to-methane system. The work focused on increasing the efficiency of the system with additional heat integration. Similarly, Biswas et al. [225] also tried to compensate high electrical energy demand of electrolysis with the utilization of waste heat from exothermic methanation reactions. Additionally, they researched the capacity of a SOEC to co-electrolyze both steam and carbon dioxide as opposed to only water.

However, the downsides of the presented type of methane synthesis method in the context of small-scale applications are the specific high-temperature working conditions and complex infrastructure [228]. Newly developed materials and advancements in methane synthesis methods are required to implement methane PtG storage into small-scale systems.

3.4.2. Hydrogen

Hydrogen appears in many studies as the fuel of the future. However, one problem is the efficiency of hydrogen extraction. Due to the molecular structure of fossil fuels, they are the most commonly used for hydrogen production. Among these, we can distinguish oil, natural gas, and coal [229]. From the point of view of the processes, steam methane reforming (SMR) currently accounts for the largest share of global hydrogen production due to economic aspects. Conversely, only 4% of the hydrogen in global production is obtained by electrolysis from water. The electrolysis process can also serve as a form of energy storage. In the electrolysis process, electrical energy can be converted into hydrogen, and the hydrogen stored can then be re-electrified. The efficiency of this process is low—around 50% [230].

Due to the low energy density of hydrogen in standard conditions, it is difficult to store. For storage, it is compressed to high pressure or cooled to the condensation temperature. Both methods, therefore, require additional energy inputs and suitable structures for safe and permanent storage.

Low-pressure hydrogen storage is available through a chemical reaction involving a hydrogen-absorbing alloy, which results in the formation of a metal hydride. Metal hydrides offer a higher volumetric energy density than liquid hydrogen [231]. Metal alloys (based on, e.g., magnesium, aluminum) adsorb hydrogen in their structure, causing the gas molecules to be closely packed [232]. This solution is not without drawbacks, such as weight, price, and slow filling process [233]. The available studies mainly focus on improving heat transfer efficiency. An example is the study by Larpruenrudee et al. [234], in which the authors designed and optimized a semicylindrical coil for hydrogen storage. The results show that by using the proposed system, the hydrogen absorption time can be reduced by 59% compared to a spiral heat exchanger.

On the other hand, the review in [235] of hydride materials outlined that of all complex hydrides, alanate-based systems are the most frequently investigated. In particular, $NaAlH_4$ was highlighted to be suitable for mobile applications. However, for sectors requiring higher capacities, such as transport, two borohydrides have been distinguished: $NaBH_4$ and $LiBH_4$.

Puranen et al. [236] analyzed an off-grid system including energy storage in the form of batteries (short term) and hydrogen (long term). They used data from an existing installation in Finland connecting a 21 kW_p photovoltaic (PV) field and a heat pump-based heating system with a single household. The simulation results showed that neither the battery nor the hydrogen energy storage system alone was sufficient to maintain year-round off-grid operation under northern climate conditions.

3.4.3. Chemical Energy Storage: Summary

Both hydrogen and methane can be obtained, among other means, by utilizing electrical energy. This gaseous carrier can then be stored and transported. This is where the two gases differ, as hydrogen has much smaller and lighter molecules, which increases the potential of unwanted leakage. Additionally, hydrogen has around a third the energy density of methane for the same storage pressure and temperature conditions. However, hydrogen, unlike methane burns, without carbon emissions, which may increase the number of possible application locations. In terms of applicability, both hydrogen and methane present the advantage of being used in present gas installations, meaning they can be easily integrated into the existing power networks.

3.5. Mechanical Energy Storage (MES)

Mechanical energy storage systems take advantage of kinetic or gravitational forces to store produced energy. This energy is stored in storage media, and those systems can be divided based on state. The challenging part in the categorization of those systems is that in some cases, storage media changes state in the process and frequently exchanges heat. This is especially true for some types of gaseous and liquid media storage. Those systems are also classified as thermomechanical systems (TMS). The main types of MES systems are presented in the scheme reported in Figure 5.

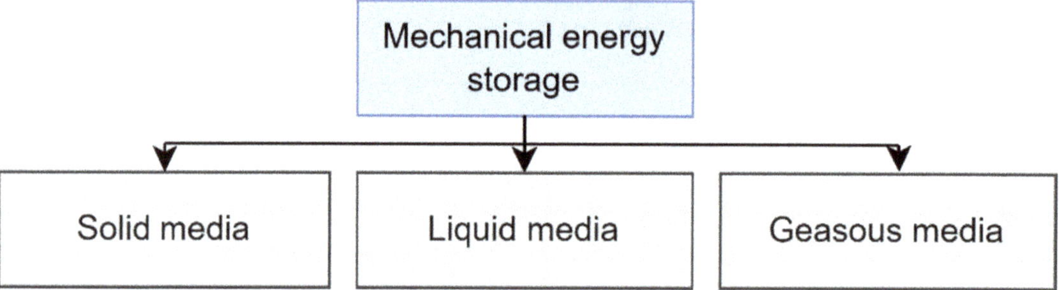

Figure 5. Main classification of mechanical energy storage systems.

3.5.1. Solid Media

Flywheel is one of the oldest known energy storage systems, as this principle was first applied in the potter's wheel [236]. In its simplest form, a flywheel consists of a suspended, rotating mass that stores the energy in the form of kinetic energy of the rotational motion [237]. Moment of inertia the angular velocity of a flywheel can be modified to adjust the storage system to the given application.

A flywheel can act as an electrical energy storage system when coupled with a reversible electric motor. This is a common case, in which surplus electrical energy powers the motor that is connected to the shaft of the flywheel, in turn increasing its angular velocity. The process can be reversed, as the electric motor can act as a generator that gradually turns the flywheel's kinetic energy into electrical power. Such configuration of a flywheel is commonly found in commercial energy storage systems, one of which was analyzed by Okou et al. [238] in their analysis of small-scale flywheel energy storage technology. In their work, they compared a lead–acid battery and a flywheel as a means of energy storage in rural areas of Uganda. It was presented that over a 10-year period, the application of flywheel storage systems lowered the energy costs by 15%, mainly because of the high maintenance costs of the batteries. The authors also highlighted that the use of an electromechanical flywheel storage system would mitigate the environmental problems associated with lead–acid battery disposal.

Other types of storage systems using solid media are gravity energy storage (GES) and buoyance energy storage (BES). Both of them rely on the relative positioning of a static load in a potential energy field [221]. The energy storage capacity of the system is proportional to the weight and the distance it can travel between its maximum and minimum elevation.

In the category of mechanical energy storage, pumped hydroenergy systems (PHES) and flywheels are overwhelmingly more popular and commercially implemented storage systems than others. However, GES and BES systems are still being developed as they do not lose any of their stored energy over time, in contrast to flywheels, whose energy is drained by friction.

Ruoso et al. [239] created a simulation model of a gravitational potential energy storage system (GES). The modeled system was compact, and it consisted of a 12 m height shaft with a diameter of 4 m. A 5 m tall piston was being moved along the shaft, increasing its potential energy during the charging period and releasing it on demand. This system has a capacity of 11 kWh, an efficiency of about 90%, and a lifetime of 50 years. The developed model presented a promising solution for small-scale storage applications.

3.5.2. Gaseous Media

Compressed air energy storage (CAES) systems use surplus energy to compress air or other gases and inject it into reservoir. It commonly utilizes a depleted underground natural gas reservoir. The compressed air can expand and power an electrical generator during peak periods when the energy is needed most [240]. During the process of gas compression, excess heat is generated, and in the basic CAES system it is dumped into the atmosphere, thus requiring a second injection of heat before the re-expansion of the gas [241]. However, advanced adiabatic CAES stores the excess heat produced during the compression of the gas and reuses it to heat the gas at the expansion.

An interesting investigation regarding CAES has been performed: Cheayb et al. [242] performed a validated simulation of a small-scale CAES with additional trigeneration in their experimental setup. Although the validation provided results very close to the experiment, they also revealed the very low efficiency of a small-scale CAES system. In their setup, the amount of stored energy was equal to 12.1 kWh, and they obtained only 0.5 kWh during discharge. They concluded that further research and mechanical improvements to the system are required to make it a feasible option as a small-scale storage system.

3.5.3. Liquid Media

Liquid air energy storage (LAES) or cryogenic energy storage (CES) works similarly to a CAES system, but the major difference is that the air is liquefied and it is stored in a reservoir. During charging of storage, electricity is used to cool air until it liquefies and then stores the liquid air in a reservoir. When the system is discharged, previously liquefied air is brought back to a gaseous state, heated by the exposure to ambient air or waste heat. Once decompressed, the expanding gas is used to power a turbine and generate electricity. LAES systems are a mature solution that uses off the shelf components that have a lifetime of over 30 years, thus resulting in low technology risk. LAES systems have similar performance characteristics to pumped hydro and can utilize low-grade waste heat. The sizes of such systems extend from around 5 MW to 100+ MW and are very well suited to long duration applications. It is worth mentioning that due to this exchange of heat of the storage media, LAES systems are also categorized as thermomechanical systems (TMS), although the majority of LAES systems currently operational are large-scale installations. Researchers at the Birmingham Centre for Energy Storage [243] designed a micronetwork LAES. The novel part of their approach consisted of the utilization of excess heat produced by LAES, especially when working in the low-pressure range. This way, the overall efficiency of the system was optimized. The resulting hybrid LAES system had a maximum efficiency of around 76% and allowed one to reduce the annual energy consumption by 12.1 MWh compared to the stand-alone LAES. Researchers claim that the new findings suggest that small-scale LAES systems have great potential for applications in local decentralized micro energy networks.

Table 5. Literature review of energy systems with storage.

Project Scale	System Type	Storage Type	Application	Evaluation Tool	Experiment	Economic Review	Location	Ref.
10 [kWp]	PVGCS	Lithium ion	One House	Mathematical/HOMER/Matlab	-	+	Bahir Dar, Ethiopia (A)	[194]
10 [kWp]	PVGCS	Lead–acid	One House	Mathematical/HOMER/Matlab	-	+	Bahir Dar, Ethiopia (A)	
21 [kWh]	PV/WT/Hydro	Lead–acid	Household	HOMER	-	+	Kadayam, India (B)	[193]
4 [kW]	PV/grid/TSC	SHS					Palmer, Alaska (D)	[157]
10 [kW]	HP,/SC	TTES	One House	Mathematical	-	-	Gaziantep, Turkey (B)	[154]
11.4 [kWh]	SC	LHS (PCMs:Sodium acetate trihydrate)	One House	Laboratory test	+	-	Lyngby, Denmark (C)	[170]
-	SC	LHS (PCMs:Sodium acetate trihydrate)	One House	TRNSYS 17	+	+	Lyngby, Denmark (C)	[171]
20 [kWh]	PV	Hydrogen/Fuel cell	One House	Simulation	-	-	Southern Finland (D)	[235]
20 [kW]	PV/WT	V2H/Lithium Ion	One House	HOMER	-	+	Los Barrios, Spain (B)	[199]
6 [kW]	SE/P2a	Lithium ion	One House	OpenFOAM	-	-	Ottawa, Canada (D)	[202]
11 [kW]	GES	GES (piston type)	One House	Simulation	-	+	-	[239]
-	CAES	gas—liquid air	Household	Simulation	-	+	-	[243]

PV—photovoltaics; WT—wind turbine; SC—solar thermal collector; PVGCS—grid-connected photovoltaic system; TTES—tank thermal energy storage; LHS—latent heat storage, PCM—phase-change material; V2H—vehicle-to-home, SHS—sensible energy storage; TSC—tube solar collector, SE—Stirling engine, P2a—pump air to air; "+"—yes; "-"—no.

3.5.4. Mechanical Energy Storage: Summary

From the mechanical energy storage systems, pumped hydro is the most used, especially in large-scale applications. However, pumped hydro typically requires natural basins in suitable geological conditions, while LAES systems present similar performance over a broader spectrum of possible locations. Currently, the performance of CAES systems lags behind LAES and pumped hydro and requires further development to become economically competitive.

An additional advantage of potential energy storage systems such as GES, BES, and pumped hydro is that they do not lose any of their charge over time, in contrast to the flywheel, which loses its charge over time relatively quickly.

Solid media storage systems mostly consist of simple mechanical components, which makes them economically appealing; however, as those are moving mechanical components, they require frequent maintenance, which must be considered.

3.6. Storage Technologies: Summary

Table 5 collects multiple researches concerning the topic of energy storage systems in small-scale renewable energy applications. They are focused on small-scale solutions, with the energy input in the range of 10–20 kWh. Each of them presents a different approach to storing energy, and they are briefly discussed below.

The presented examples of small-scale storage systems show that there is no system without its drawbacks, but under certain conditions, they become a beneficial part of a renewable energy system. Incorporation of the energy storage system permits utilization of the surplus energy produced by most renewable energy sources, and this helps to further reduce emissions and decrease the cost of electrical energy. The type of energy storage system is chosen based on multiple factors, with the most significant being storage period, energy conversion, and charge/discharge rate. With a variety of possible solutions, there is a suitable energy storage system for most of the applications.

4. Use of RES Technologies in Polygeneration Hybrid Systems

The described renewable energy and storage technologies are very well suited for use in polygeneration systems. Many studies are being conducted to properly optimize such systems and appropriately adapt the available technologies to consumer needs. Table 6 shows the collected work demonstrating the feasibility of using various RES in small-scale polygeneration systems.

Table 6. Literature review of polygeneration hybrid systems.

Technology	Power Installed	Energy Demand	Location	Energy Storage	Load	Evaluation Method	Economic Assessment		Ref.
							Indicator	Value	
PV	50 kW	El: 100 kWh/day, Th: 25 kWh/day	India (B)	Hydrogen, lead–acid and LIB battery	DH, DHW, EL, H_2	HOMER	-	-	[244]
PVT	5701 kwh/yr	El: 4932 kWh/yr, Th: 3651 kWh/yr	Spain, Almería (B)	TES	DH, C, DHW, EL, FW	TRNSYS	-	-	[245]
CPVT, Bio	1164 MWh/yr	-	Italy, Naples (C)	TES	DH, C, DHW, EL, FW	TRNSYS	NPV, PI	-	[246]
WT, PV, CHP	Varied	El: 1.682 MWh/yr, Th: 2.028 MWh/yr	Italy (C)	TES (heat, cool), battery	DH, C, DHW, EL	Analytical	SPB, internal rate of return, LCOE	Varied	[247]
WT, PV, CHP	70 kW	El: 30 MWh/yr, Th: 79.5 MWh/yr	Italy, Pantelleria (C)	TES (heat, cool), battery	DH, C, DHW, EL, FW	TRNSYS	SPB	5.67–12.2 years	[114]
Solar, Bio	4349 MWh/year	-	Italy, Naples (C)	TES	DH, C, DHW, FW	TRNSYS	SPB	3.71 years	[248]
Solar, bio	280 kW	-	India, Chennai (B)	NO	C, EL, FW	Analytical	NO	-	[249]
Bio, WT, PV	60 kW	El: 50 MWh/yr, Th 103.5 MWh/yr	Poland, Gdansk (D)	TES	DH, C, EL	TRNSYS	SPB	10 years	[250]
Bio, PV	62 kW EL, 87.7 kW Th	Varied	Spain, various locations (B)	TES	DH, C, EL	TRNSYS	Polygeneration indicators	Varied	[251]
Bio, PV	150 kW, 20 kW	Varied	Spain (B)	TES, lead acid batteries	DH, C, EL	MATLAB, TRNSYS	NPC	Varied	[252]
Heat pump, PVT	300 m^2 PVT field	El: 46.3 MWh	Italy, Naples (C)	TES, lead–acid batteries	DH, C, DHW, EL	TRNSYS	SPB	Varied	[253]

DH—domestic heat, DHW—domestic hot water, C—cooling, EL—electrical energy, H_2—hydrogen, FW—fresh water.

When considering Table 6, the first thing to note is the choice of the location of the systems investigated in the literature. Popular choices of authors are countries with relatively high average annual temperatures, such as India, Spain and Italy. Very few works propose hybrid polygeneration systems in rather cold climates. Countries in northern Europe are a case in point [250]. This fact points directly to problems related to the economic viability of such hybrid system projects. Among all the possibilities, one of the most popular choices for the main source is solar energy. This translates into a relatively low yield of such plants in locations with low values of average annual irradiation. A solar-based installation could very easily become uneconomical compared to conventional solutions under such weather conditions [254]. The single articles whose authors undertook the creation of a polygeneration system in relatively cold climates were mostly based on biomass-burning CHP systems [255].

Photovoltaic technologies are developing rapidly. Recent discoveries include perovskites [256] or other thin-film technologies [257], which at the current stage of research do not have the highest efficiency ratings [258]. The highest efficiencies are enjoyed by multijunction cells [259], but despite this fact, the most common approaches consist of standard, monocrystalline/polycrystalline, photovoltaic modules. As an example, Murthy et al. [244] suggested creating a photovoltaic field to provide electricity and heat in an Indian village. It is worth noting that the system involved the generation of electricity from PV, in parallel with the generation of hydrogen, which was then used to produce electricity and heat using fuel cells when needed. The authors undertook a performance comparison of the proposed system using different energy storage methods. The article is based on the creation of an installation characterized by the minimum amount of energy returned to the country's power grid, as well as maximum coverage of user demand. The authors pointed out that in this case the best solution is to use lithium ion batteries and hydrogen storage in parallel. The work gives a good indication of how important these technologies are in polygeneration systems. The operation of photovoltaic systems is closely tied to the day and night cycle and seasons, which translates into large energy shortages during periods of low sunshine and large energy surpluses in the opposite situations.

A very popular solution, especially in Mediterranean countries, are hybrid PVT systems. The literature indicates that with the help of such panels, a well-chosen RES-based system can be used to produce electricity, domestic heat and cooling, as well as to heat and desalinate water [245]. The authors proposed using the excess electricity generated to desalinate seawater using the reverse osmosis (RO) phenomenon [260]. Thanks to this procedure, the loss of generated energy is minimized, and thus the economy of the solution increases. RO and other methods of seawater desalination find their use in a large part of modern small polygeneration system concepts [113,114,246,249]. Systems that generate different types of energy while producing fresh water can be a solution to a number of problems of the modern world dealing with fresh water availability. A very popular approach is the adoption of a system based on biomass combustion, where a boiler can be used as the main or additional source of energy. Optimization of a system using mainly solar energy with the presence of an additional biomass boiler was carried out by Calise et al. [246] The paper included a description of the proposed polygeneration system based on CPVT to produce electricity, cooling and utility heat, as well as heat and desalinate water. The authors presented the possibility of optimizing such a system by adjusting the size of the collector field, the number of multieffect distillation (MED) units, the flow rate of chilled water and the heat storage capacity. Economic and exergy analyses were carried out. The first indicated that the solution involving large number of collector systems was highly profitable for the economical point of view, but this fact was contradicted by the conclusions of the second analysis, which revealed a very low utilization rate of the exergy supplied to the system under bigger collector areas. This work indicates that the design of similar systems should be guided by various efficiency indicators of the whole process.

Wind turbines as single energy sources are not very common in polygeneration systems. Due to their unstable operation, even in hybrid systems, the best efficacy of power generation is achieved by systems equipped with energy storage. The authors of [114] showed the possibility of creating a complex polygeneration system incorporating a wooden-chip fueled CHP boiler, a wind turbine, and a photovoltaic system. As for the system reported in [245], it has a seawater desalination unit based on the process of reverse osmosis. The authors also included an additional LPG boiler to support the system in case of energy shortages. However, dynamic simulation indicated that it must only be turned on in a minimal amount of time during the year. The proposed polygeneration system showed high operating efficiency and a variable simple payback period (SPB) between 5.67 and 12.2 years depending on the selected reference scenario. The system is an example of well-chosen components of a hybrid plant, which allowed one to achieve stability of operation and cost-effectiveness of the proposed solution. The importance of thermal and electrical energy storage should be highlighted here, as well as the seawater desalination system, which allowed the use of excess electrical energy, minimizing losses. A very similar solution was described in [247] showing similar use of renewable energy sources (WT, PV, biomass). It is worth noting that the authors carried out system optimizations using different approaches. Two strategies were based on covering the total electricity or heating demand and last one consisted of following modified base load. These optimization approaches made it possible to conduct an economic analysis in each case and identify the best operation strategy of the system.

Apart from systems using hybrid PVT solutions, it should be mentioned that standard solar thermal systems can also be suitable in individual cases. In [248], it is shown that a sufficiently large collector system, working together with a biomass boiler, may be characterized by high cost-effectiveness. It is worth noting that the simulation showed that during the winter period, almost all the heating demand is covered by the boiler, while the summer operation of the system, including the generation of cooling, is ensured practically entirely by solar energy. The authors calculated a SPB of 3.71 years, indicating the system is feasible. However, it should be borne in mind that the proposed system was located in a place with a relatively high average annual insolation. This is an additional fact showing the location-dependent versatility of solar systems. The system proposed in [250] is interesting since it assumes the use of a biomass boiler, PV and WT. in a polygeneration system located in the northern part of Poland. This is a region characterized by relatively low values of average annual solar radiation with respect to southern Europe locations. The author optimized the system by considering four different scenarios for biomass source. Due to the location of the system, the main source of thermal energy and electricity was a biomass boiler coupled with a simple steam Rankine cycle. However, it is worth noting that energy from the wind turbine and the photovoltaic system accounted for 11.1% and 35.1% of the total electricity produced, respectively. This shows that despite the relatively scarce solar energy availability for PV, the region has good prerequisites for wind turbine operation. When designing similar installations, one should keep in mind to maximize the potential of a given location in the world when it comes to energy generation from renewables.

A large part of the systems presented above included some form of energy storage. The most common was a basic hot-water tank, considered as an indispensable element of a standard heating system. Typically, the design and optimization of thermal storages consists only in volume changes, while other forms of improvement, as use of advanced thermal storages (thermochemical, phase-change materials, etc.) are rarely used. Furthermore, individual authors undertake the use of cold storage technology, but this is still not among the popular solutions.

As for electrical energy storage, the use of lithium ion or lead–acid batteries can be seen most often in individual papers. This shows that the authors of the works follow proven and well-known solutions when designing polygeneration systems, and new technologies, which are described extensively in Chapter 3, are often overlooked. The reason consists in the fact that such technologies themselves are not very common and sufficiently researched for the application in energy systems, or it would not be cost-effective to use them on a small scale. However, it is worth noting that more and more installations are observed using hydrogen generation and storage. Fuel cells, once considered a novel approach, are becoming more common, even in commercial applications [261].

Currently in the literature, one can find numerous different proposals for small hybrid polygeneration systems. These are primarily systems based on biomass CHP boilers, as they are characterized by stable operation and are not highly location-dependent. It is worth noting that the use of other technologies, such as PV, PVT or wind turbine installations, carries many limitations related to the geolocation of the entire system. The creation of a hybrid installation is typically characterized by higher efficiency than a single system acquiring energy from one RES. The best locations, where RES-based installations are showing the highest efficiency, are coastal regions characterized by high windiness and relatively high average annual temperature. In this framework, it is worth mentioning that the economy of the systems can be increased with the use of possible means of receiving or storing the excess of generated energy. Recent studies presented in this section indicate that these aspects are intrinsic of modern polygeneration systems.

5. Approaches and Tools in the Design of Integrated Energy Systems

The aim of an integrated energy system is to maximize efficiency and minimize losses [262]. Therefore, sizing the individual components of such a system involves increasing the number of variables and parameters that need to be considered in the design.

Supporting the design process with energy simulations allows one to eliminate the number of errors in the output system and to assess the impact of many variables on its performance. In addition, the occurrence of diverse challenges that have a significant impact on the energy sector, such as combating climate change, economic recession or ensuring energy security [263], results in a significant expansion of aspects to be considered during energy planning. Energy system models are of particular relevance in planning the energy transition and studying its impacts on the energy sector and applications [264].

5.1. Approaches in Modeling of Hybrid Renewable Energy Systems

Energy systems operate in a variable mode by adjusting their operating point to a given energy demand. Renewable energy sources contribute to the computational complexity of such systems due to intermittent energy production and uncertainty in resource availability [265]. The need to understand and predict the performance of the various components of the energy system drives the development of various models. Many approaches to system modeling can be found in the literature, depending on the objective. They are usually classified into three groups—computational, mathematical, and physical—as shown in Figure 6, prepared based on [266], showing the classification for energy models depending on the modeling approach. All these approaches are characterized by varying capabilities in describing the occurring phenomena.

In agent-based models (ABM), the system is modeled as a collection of autonomous decision-making units called agents. The agents' decisions and evaluations are made based on a set of rules [267]. This approach is most commonly used to forecast electricity prices and also to simulate quantitative trends in customer behavior [268]. However, this is not the only field of application for this approach. Klein et al. [269] presented an agent-based modeling process used in energy systems analysis and energy scenario studies. They present the ability to integrate different algorithms and modeling approaches as a particular advantage.

Knowledge-based models consist of domain-specific knowledge bases and an inference engine that derives new knowledge based on specific rules and a user interface [270]. The approach discussed above was used by Abbey et al. [271] for power planning of a two-stage energy storage system applied to wind systems. Their research showed that the proposed approach can serve as a design procedure for the controller and for the sizing of a short-term storage device.

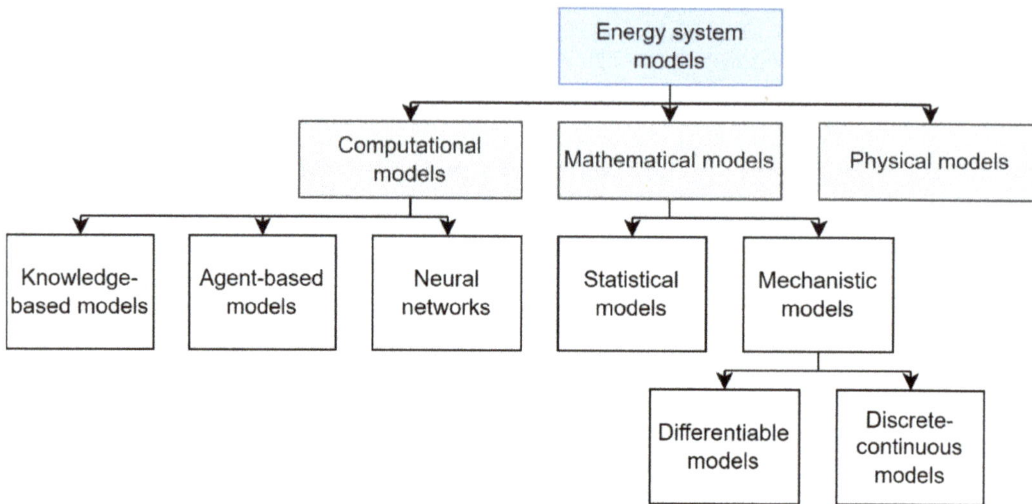

Figure 6. Classification of energy system models according to modeling approach.

Neural networks, unlike the knowledge-based models discussed above, do not have explicit rules. They consist of collections of cells that process input and output information. Their training is done by adjusting the connection weights between these nodes [272]. This technique has been used for maximum power point tracking (MPPT) of photovoltaic generators [273], biomass energy prediction [274] or wind energy resource assessment with a forecasting procedure [275]. In a review carried out by Thiaw et al. [276], the MPPT controller design process for photovoltaic generators was presented to improve their efficiency. They also presented a possibility of assessing the available and recoverable wind energy potential of a site, by finding a suitable wind distribution based on a neural model.

With mathematical models, a distinction is usually made between statistical (empirical) and mechanistic (theoretical) methods. The mechanistic approach is based on a mathematical description of a phenomenon or process. The nature of the phenomenon may be chemical, biological or mechanical. It can be used to predict process loads when analyzing components for renewable energy systems, e.g., wind turbine hubs [277].

Modeling and simulation techniques can encompass testing, configuration activities and the development and adaptation processes of the system under study. They also provide a useful tool to support the decision-making process. A simulation is a result of testing a model, which is an abstract representation of the real system [278]. Several simulations can be run with a single model to test alternative solutions. A suitable model should represent a less complex version of the real system. For analysis and evaluation of energy systems, the model should include a description of its properties or performance [279]. Tasking this information will describe the system design processes, its operational patterns and changes in behavior and performance.

Optimization is a certain modeling approach in which there is a calculation of decision variables that minimize or maximize the objective function under constraints [280]. These decision variables are usually the design characteristics of the energy system. In contrast to simulation, optimization mimics the evolution of a situation or system over time [281]. In the literature, we can find many optimization techniques applicable to the dimensioning of hybrid renewable energy systems. These approaches vary according to the optimization objectives. The optimization is required for sizing, combinations, determining operation strategies and scheduling of sources. A number of commercial software have proved useful in sizing and optimizing HRES [282]. Examples of these are described in the following subsections. Figure 7 shows the parameters that make up the corresponding optimization process for hybrid renewable systems. The figure was developed based on [283].

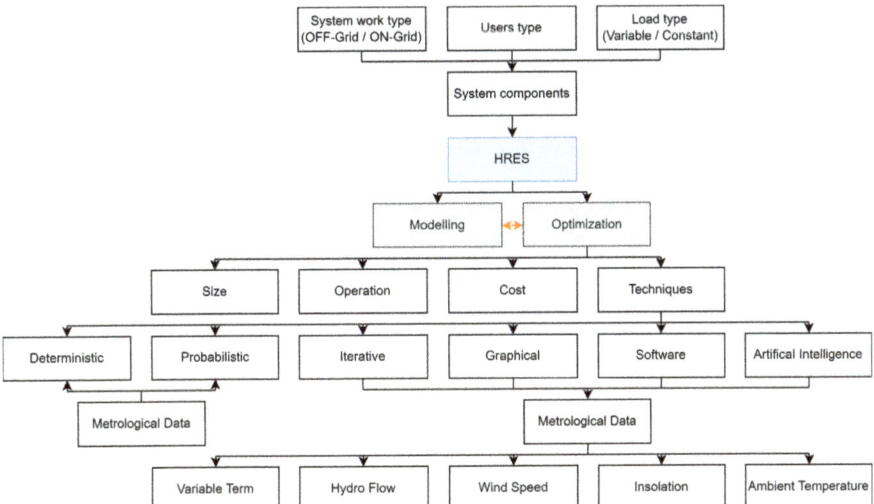

Figure 7. The general optimization process for hybrid renewable systems (classifications developed based on [283]).

With the iterative method, results are obtained by solving a series of calculation steps starting from a certain starting value. The simulation is carried out until the desired criteria are met [284]. In general, the iterative approach allows the model to be built step by step, and at the same time, any defects in the system under consideration can be caught at early stages [285]. It can also be used for large systems with a large number of components [286]. An example of such is the work written by Geleta et al. [287], where an iterative method was used to select the number of wind turbines and photovoltaic panels for an autonomous system intended to satisfy the desired specific load for a given area. The deterministic method provides accurate weather forecast values for a specific location. On the other hand, the probabilistic method suggests the probability of certain weather events [288]. Both methods work on the basis of weather data implemented from historical data.

5.2. Tools

In addition to the previously presented models used to analyze the behavior of the various components of hybrid systems, simulation and optimization software also plays an important role. Their areas of application can include system design, control strategies, and both multiobjective and economic optimization [289]. This section describes the most commonly used tools in a simulation of a hybrid energy system. Each of them has different possibilities and applications [290,291]. There is software such as TRNSYS and HOMER that can simulate entire complex energy systems, measuring their performance, applicability and economics. With their flexibility, those tools can be used to accomplish vast range of tasks, which are extended with a significant number of additional plugins and incorporated software.

TRNSYS is a highly flexible and graphically based software used to simulate the behavior of transient systems [292]. It has an open and modular structure that allows the incorporation of other previously independent tools. TRNSYS is primarily used to simulate entire energy systems i.e., local communities, off-grid systems or island power systems. It can simulate all thermal and renewable energy generation systems, with the exception of nuclear, wave, tidal, and hydropower. In TRNSYS, the user can define a time step of a simulation in the range from 0.01 s up to typically 1 h with a time span of multiple years.

HOMER-Pro is software designed for microgrid optimization, allowing the calculation of both stand-alone and grid-connected systems. Systems include power sources such as photovoltaics, wind turbines, biomass generators, combustion engines, microturbines and hydro together with storage technologies and loads. Additionally, all economic factors can be included in the process. The simulation can cover a period of one year with a minimum time year of 1 min [293]. It also allows users to create their own dispatch strategy by integrating the software with an algorithm created in MATLAB.

IHOGA/MHOGA are two versions of the Hybrid Optimization by Genetic Algorithms (HOGA) software. IHOGA is dedicated to systems up to 5 MW, while MHOGA covers everything above [294] Both versions can simulate systems including photovoltaic generators, wind and hydroelectric turbines, auxiliary generators, energy storage as well as components of hydrogen (electrolyzer, hydrogen tank and fuel cell). It includes optimized control strategies that can be used for both off-grid systems and grid-connected systems. The software includes multiperiod simulation, multiobjective optimization, sensitivity analysis and probability analysis with time steps of up to 1 min.

The analysis of a hybrid energy system can be additionally supported by tools used to simulate specific parts of the system or physical phenomena that influence the system. When photovoltaics are considered, PVSOL software can be used to perform detailed shading analysis with a 3D visualization including surrounding objects [295]. It can simulate systems from small single-roof scale to large solar parks in combination with appliances, battery systems and electric vehicles.

Where a wind turbine is a part of the hybrid energy system, QBLADE is a tool that can be utilized. It can run highly detailed simulations of any wind turbine design, with advanced physics models more than 20 times as fast as real time [296]. It is primarily used for aero-servo-hydro-elastic design, prototyping, simulation, and certification of wind turbines. Some of its many features are an aerodynamic model of a wind turbine, structural multibody simulation, wind and wave generation, a hydrodynamic model for offshore applications and controller integration.

Further analysis of the complex flow of air or water around turbines can be performed in OpenFOAM. It has an extensive range of features, including fluid flows, chemical reactions, heat transfer or solid mechanics [297]. Additional software that can be used to support physical simulations of the parts of hybrid energy systems is ANSYS. It is a software package that integrates a wide range of tools that can be used to simulate almost any real phenomenon [298].

Lastly, MATLAB software can support the tools mentioned by providing integration of user-defined functions, algorithms, or neural networks [299]. It can generate data describing the behavior of required subcomponents of larger energy systems. Additionally, it is useful in the post-processing of simulation data, providing additional user-defined analysis and visualization.

5.3. Application Examples

For efficient utilization of renewable energy sources, the system's feasibility needs to be studied before implementation. This approach was applied by Hiendro et al. [300] that presented a technological and economic feasibility analysis of a hybrid system composed of photovoltaic, wind turbine and battery. In order to obtain optimal size determination and cost minimization, HOMER software was used. Researchers found that the most important components in the proposed system are the battery and the wind turbine. HOMER-Pro was also used by Kebedeet et al. [194] in their work to compare the efficiency of a grid-connected photovoltaic system (PVGCS) integrated with a lithium ion battery or a lead–acid battery. Equivalent Circuit Model (ECM) was utilized in order to investigate the performance of batteries. Batteries were included in HOMER-Pro techno-economic analysis considering realistic commercial load profiles. The software allowed the researchers to conclude that in a typical application scenario, lithium ion batteries are recommended as they are more profitable and reduce the overall cost of energy produced by the hybrid system.

In other research, Amutha et al. [193] used the HOMER software to estimate the different types of load requirements in an off-grid hybrid energy system, for the installation of optimal energy sources and battery energy storage. The possible extension of a grid in order to connect to the system was evaluated. Obtained results indicated that the extension of the grid is not economically viable with high initial and maintenance costs.

The use of TRNSYS software can be found in the work of Figaj et al. [114]. The research described a hybrid polygeneration system with wind turbine, photovoltaic field, biomass-fired Rankine cycle, thermal and electrical energy storage, absorption chiller and reverse osmosis water desalination unit. Backup energy in the form of an LPG generator was also included. The system was arranged to satisfy the needs of 10 households on Pantelleria Island, Italy, in terms of electrical energy, heating, cooling and freshwater. The minimum payback period for the investigated installation was 7 years for the reference scenario considering the comparison of the proposed and reference system both connected to the electric grid.

IHOGA was applied in multiple works of Carroquino et al. [301]. One of them is the comparison of lead–acid and lithium ion batteries in the standalone photovoltaic system in Spain. Ten simulations were performed in five different locations, half with a full PV system configuration and the other half with a hybrid system. Optimization processes were carried out in iHOGA software, which simulated the economic and technical operation of the possible solutions of the system. Results regarding the economic comparison show that in only three cases out of ten the optimum was obtained with lithium ion batteries. In the remaining seven cases, lead–acid batteries were better from the point of view of economic performance.

PVSOL was also used in the work of Sharma et al. [302] to design and simulate grid-connected solar PV systems for campus hostels in India. The system designed in PVSOL has a power of 234 kW and covers an area of approximately 1560 m^2. Its economic analysis was performed with the same software to determine the feasibility of the system. It was concluded that the expected return on investment is 11 years.

Abdullah et al. [303] used ANSYS for numerical analysis of a hybrid PV thermal air collector. Prior to simulation, a 3D geometric model of the PVT collector was created using ANSYS Design Modeler considering real-world components, which were later used in validating the experiment. A mesh of the mentioned model was created in ANSYS Meshing, which was then imported into the simulation set-up in ANSYS Fluent. The fluid flow and heat transfer characteristics of the PVT collector were determined using computational fluid dynamics (CFD) simulation in mentioned Fluent module. The validation experiment showed that the difference between the simulation and real-world data was not exceeding 2%. By means of simulation, a revised design of the PVT collector was proposed, with the aim to increase its efficiency.

Using Simulink included in MATLAB, El-Hady et al. [304] modeled a photovoltaic and wind turbine hybrid energy system that can supply a load of around 10 kW. The system was tested under changing conditions of wind speed and insolation. Using MATLAB, researchers prepared different load cases, including step changes in load level in order to observe the response of the system.

Moreover, Matlab Simulink was used to simulate systems in the field of electric vehicles. Srujana et al. [305] investigated the components of a battery-electric vehicle framework, and performed a simulation in Simulink to obtain correct sizing. Additionally, a study by Fotouhi et al. [306] presented the development of an estimation model of energy consumption in an EV fleet management system. A usage model of a single EV was created in Simulink and utilized in the analysis of the energy consumption and performance of the entire fleet.

The software listed above is some of the most commonly used in the available literature to investigate hybrid and polygeneration systems. However, there is also other software that can be used to assess the performance of energy systems. There are decision-making programs enabling analysis of energy systems and projects such as RETScreen [307] and EnergyPro [308]. Moreover, EnergyPLAN [309] software is used to simulate the operation of large-scale or national energy systems rather than small-scale ones. Another program, EnergyPlus [310] is also used for building energy simulations that may be coupled with simulation of energy systems. Depending on the type of system under investigation, one can also conduct detailed analyses of individual system components. An example of such a tool is Zemax, which finds its application in optical engineering [311]. Garcia et al. [312] used the software to model parabolic ring array concentrators and calculate the solar flux. Similarly, Tracepro is software for the design and analysis of lighting and optical systems [313]. It was used in [314] to calculate the optical performance of a novel umbrella heliostat.

6. Review Indicators

The approaches and tools described in the previous chapter and used in the modeling and analysis of hybrid and polygeneration systems are intrinsically connected to the calculation of some indicators allowing one to evaluate the performance of chosen design concepts. In fact, the investigation of combinations of technologies adopted in small-scale hybrid and polygeneration systems is performed typically in terms of energy and economy. The energy-related performance indexes are the ones characteristic of technical assessment of energy systems, such as efficiency, energy yield, etc. However, the main aspect that is considered in the investigation of hybrid and polygeneration systems is the economic performance and feasibility. Indeed, such systems may be highly challenging in terms of cost-effectiveness due to their complexity and novelty with respect to conventional technologies. In order to address this challenge, in the scientific literature, there are available several approaches for the investigation of economic performance.

The first approach is to calculate the levelized cost of energy (LCOE). This method is intended to show how much, taking into account investment and maintenance costs, it will cost to generate electricity with the help of the proposed solution. The lower the value of the index, the more profitable the investment. This method is used primarily when comparing different system concepts in the same location with the same energy demand of the consumer. An example is in [118] where a comparison of the performance of a classic PV module and a hybrid PVT was presented. This method was also used by Sigarchian et al. [249], where different strategies were shown when optimizing a polygeneration system, and the LCOE value showed which approach was the best in terms of cost-effectiveness. The calculation of *LCOE* is presented in Equation (1) [122]:

$$LCOE = \frac{I_0 + \sum_{t=1}^{n} C_0^{OandM} (1 + e + r_{OandM})^t / (1 + i)^t}{\sum_{t=0}^{n} E_{el,y} / (1 + i)^t} \tag{1}$$

where I_0 corresponds to the initial investment cost, n is the time period considered in the analysis, t is the time in years, C_0^{OandM} is operation and maintenance costs of considered system, e is yearly inflation rate, r_{OandM} is operation and maintenance escalation rate, this value directly shows any increases in maintenance costs during the operation of the system, i is the discount rate, and $E_{el,y}$ shows the total amount of electricity produced by the system in a year.

As presented above, in order to calculate the LCOE, it is necessary to gather information about the costs of the project, but also to predict the operating costs of the system itself. This indicator relates only to the electricity generated, and thus is very rarely observed for polygeneration installations, where the generation of other types of energy or by-products would have to be evaluated by other calculations.

Another approach is to calculate the net present value (NPV) of an investment, and consequently also the profit index (PI). The higher the PI value, the better the economic efficiency of the investment. This method is more versatile and better suited for verifying the profitability of hybrid polygeneration systems. NPV is the direct difference between the net investment earnings over a given period and the investment cost. The method of calculating PI is presented in Equation (2):

$$PI = \frac{NPV}{I_0} \tag{2}$$

where I_0 is the initial investment cost. This method is relatively versatile, since a range of cash flows can be taken into account when calculating NPV. This can be shown from [246], where cash flows are presented in the form of savings on CO_2 tariffs and energy sales to the country's electricity grid and operating costs.

However, the calculation of NPV refers to investments, and therefore to possible systems whose operation is characterized by direct profit. For systems in which the sale of produced energy or other products is not the most important, it is possible to use the net present cost (NPC) indicator [315]. The calculation of this indicator is similar to NPV; however, mainly investment costs and the costs that go along with the operation of a given system are taken into account. This can be illustrated by the example in [252]. The authors took into account investment costs, costs of possible component replacements, operations and maintenance costs, as well as fuel costs. In this case, there were sales of electricity. When considering NPC values, the lowest value of this indicator will be characterized by the most cost-effective solutions.

One of the most popular methods of assessing the economic viability of an investment is the calculation of simple payback (SPB) period. Many authors of works on RES-based hybrid polygeneration systems use this method for economic analysis [113,114,250,253]. This indicator shows after how many years the total return on investment costs of a given system will be observed, which directly allows comparing different system concepts. The calculation of SPB is shown in Equation (3):

$$SPB = \frac{I_0}{I_n} \tag{3}$$

where I_0 is the initial investment cost, and I_n describes the net annual revenue. When using this method, it should be kept in mind that its simplicity can directly affect the accuracy of the results. During the calculation, the evolution of the value of money over time, and therefore the possible variation in revenue over the year, is usually not taken into account. Despite these limitations, the method is well suited for comparing the economic viability of different solutions, primarily different variants of systems in the same location.

In addition to economic approaches, it is also worth mentioning other methods of comparing the hybrid systems. One of them can be the levelized cover ratio (LCR), which directly shows how much of the user's energy demand is covered by the proposed system. This method is used to optimize the system, but its highest indications do not necessarily coincide with the best economical solution. The LCR calculation must involve an additional economic evaluation of each of the compared solutions [251]. To calculate the LCR, it is possible to use Equation (4):

$$LCR = \frac{E_{user} - E_{ex}}{E_{user}} \tag{4}$$

where E_{user} is the energy demand of the system user and E_{ex} is the value of energy that must be supplied from an external source, which can be an additional generator or the country's power grid. For polygeneration systems in particular, polygeneration indicators are a popular method for evaluating a given concept. They allow us to illustrate what positive aspects come from polygeneration in relation to separate generation. This method was used by Picallo-Perez et al. [251], calculating the percentage of energy savings (PES). This parameter can also be referred to the calculation of exergy efficiency. The authors presented a thorough description of the entire calculation of these parameters in their work.

7. Conclusions

Hybrid and polygeneration renewable energy systems have received particular attention in the recent period from researchers, technicians, policymakers and stakeholders, making them a possible means of achieving the goals of present-day energy systems and distributed generation. In this context, the adoption of hybrid and polygeneration schemes is possible on different scales, from big plants to small-scale applications at the level of a single household. The adoption of renewable energy systems based on the use of multiple energy sources may lead to significant advantages, such as the reduction of negative effects of energy generation fluctuation, improved energy independence of the user, and significant reduction of energy bills. These advantages can be even more if the adopted energy system is not only limited to a monogeneration regime, but instead allows one to produce more forms of energy along with other useful substances or final products. Investigations of such systems are relatively common in the scientific literature for medium or large systems, while for small-scale systems, as pointed put by this review, there is significant potential for improvement of technical and scientific knowledge regarding the combination of energy sources, adoption of different technologies and applications.

The development the investigations on hybrid and polygeneration systems powered by renewables is related to several aspects of such systems, as the ones involving the available technologies for energy generation and storage, use of renewable technologies and the configurations of systems, and the methodology and tools adopted in the assessment of energy and economic performance. In the present review, the authors focused on the latter standpoints to outline the state of the art regarding the research of hybrid polygeneration renewable energy systems. The main outcomes from this review are summarized as follows.

- From the point of view of energy generation technologies, the availability of commercially ready devices is vast and mature, since it ranges between solar collector and photovoltaic panels, wind and water turbines, biomass boilers and heat pumps. Typically, investigations on a small scale rely on proven universal technologies since the goal is to assess the performance of a specific system, rather than the investigation of the performance of a novel and advanced component. As regards systems based on only one technology, wind turbines and photovoltaic panels are the most common solution, while systems based exclusively on biomass are scarcely investigated.
- All storage technologies in small-scale storage systems have their drawbacks, but under appropriate conditions, they are an essential part of a renewable energy system. Despite a wide availability of thermal and electrical energy storage technologies, the systems are mainly based on common solutions, such as lead–acid or lithium ion batteries or liquid storage tanks. Moreover, hydrogen systems are also a possibility for storage of electrical energy in several applications available in literature. In general, the type of energy storage system is chosen on the basis of several factors, such as level of autonomy, efficiency, and energy charge and discharge rate. With a variety of possible solutions, there is a suitable energy storage system for most of the applications.
- The systems investigated in the literature mainly focus on the combination of solar, wind and biomass, for which the investigations are conducted to properly optimize the configuration and appropriately adapt the available technologies to user needs. Due to the massive adoption of solar energy in systems, locations with relatively high average annual temperatures and solar availability are considered, while only few works propose hybrid polygeneration systems in cold climates. There are also examples based on biomass boilers, since they are characterized by stable operation and are not affected by the weather conditions.
- The tools and methodologies for investigation of the performance of hybrid polygeneration systems based on renewables involved in almost all the cases the adoption of an analytical/numerical approach. This is obviously connected to the relatively high cost of development of pilot and experimental setups for small-scale applications. Therefore, the investigations are performed by means of tools such as TRNSYS and HOMER in order to select the individual components and design the configuration

of whole systems. In this process, a significant number of variables and parameters are considered, in the form of weather conditions, technical specification of devices, control strategies and economic scenarios adopted for the analysis. In general, the main goal used in such tools is to assess the operational characteristics of the system, perform energy and economic analyses, and optimize the system by means of selected criteria.

In the context of the present review, it possible to identify some pathways in the investigation of hybrid polygeneration renewable energy systems in small-scale applications, as listed below:

-   investigation of pairing of renewable energy sources other than the conventional solar–wind one, and the integration of three or more renewable energy sources, as for solar–wind–biomass systems;
-   analysis of the possibilities of integration of advanced energy storages, as phase-change materials, methane (methanation), compressed air storage, and thermochemical heat storage;
-   investigation of suitable applications/users for hybrid and polygeneration systems that are different from the ones typically selected as case studies, such as villages and small communities.

Author Contributions: Conceptualization, M.Ż. and R.F.; methodology, M.Ż. and R.F.; formal analysis, M.Ż. and R.F.; investigation, M.H. and A.P.; data curation, M.H., A.P. and M.Ż.; writing—original draft preparation, M.H., A.P., M.Ż. and R.F.; writing—review and editing, M.Ż. and R.F.; visualization, M.H. and A.P.; supervision, R.F. All authors have read and agreed to the published version of the manuscript.

Funding: Part of this research was funded by the Polish Ministry of Higher Education on the basis of decision 0086/DIA/2019/48.

Data Availability Statement: Not applicable.

Acknowledgments: This work was carried out under Subvention and Subvention for Young Scientists, Faculty of Energy and Fuels, AGH University of Science and Technology, Krakow, Poland. The authors acknowledge the use of the infrastructure of the Center of Energy, AGH UST in Krakow. The research was supported by the Foundation for Polish Science (FNP).

Conflicts of Interest: The authors declare no conflict of interest.

References

1. Net Zero by 2050—Analysis—IEA. Available online: https://www.iea.org/reports/net-zero-by-2050 (accessed on 28 October 2022).
2. Kałuża, T.; Hämmerling, M.; Zawadzki, P.; Czekała, W.; Kasperek, R.; Sojka, M.; Mokwa, M.; Ptak, M.; Szkudlarek, A.; Czechlowski, M.; et al. The Hydropower Sector in Poland: Historical Development and Current Status. *Renew. Sustain. Energy Rev.* **2022**, *158*, 112150. [CrossRef]
3. Kaldellis, J.K.; Zafirakis, D. The Wind Energy (r)Evolution: A Short Review of a Long History. *Renew. Energy* **2011**, *36*, 1887–1901. [CrossRef]
4. Executive Summary—Unlocking the Potential of Distributed Energy Resources—Analysis—IEA. Available online: https://www.iea.org/reports/unlocking-the-potential-of-distributed-energy-resources/executive-summary (accessed on 28 October 2022).
5. Norouztabar, R.; Ajarostaghi, S.S.M.; Mousavi, S.S.; Nejat, P.; Koloor, S.S.R.; Eldessouki, M. On the Performance of a Modified Triple Stack Blade Savonius Wind Turbine as a Function of Geometrical Parameters. *Sustainability* **2022**, *14*, 9816. [CrossRef]
6. Yang, J.; Luo, X.; Zhou, Y.; Li, Y.; Qiu, Q.; Xie, T. Recent Advances in Inverted Perovskite Solar Cells: Designing and Fabrication. *Int. J. Mol. Sci.* **2022**, *23*, 11792. [CrossRef]
7. Feinauer, M.; Uhlmann, N.; Ziebert, C.; Blank, T. Simulation, Set-Up, and Thermal Characterization of a Water-Cooled Li-Ion Battery System. *Batteries* **2022**, *8*, 177. [CrossRef]
8. Minai, F.; Márquez, G.; Jiménez, A.; Gonzalez Lamar, D.; Faiz Minai, A.; Ahmad Khan, A.; Kumar Pachauri, R.; Malik, H.; Pedro García Márquez, F.; Arcos Jiménez, A. Performance Evaluation of Solar PV-Based Z-Source Cascaded Multilevel Inverter with Optimized Switching Scheme. *Electronics* **2022**, *11*, 3706. [CrossRef]
9. Wincukiewicz, A.; Wierzyńska, E.; Bohdan, A.; Tokarczyk, M.; Korona, K.P.; Skompska, M.; Kamińska, M. Enhanced Performance of Camphorsulfonic Acid-Doped Perovskite Solar Cells. *Molecules* **2022**, *27*, 7850. [CrossRef]

10. Ang, T.Z.; Salem, M.; Kamarol, M.; Das, H.S.; Nazari, M.A.; Prabaharan, N. A Comprehensive Study of Renewable Energy Sources: Classifications, Challenges and Suggestions. *Energy Strategy Rev.* **2022**, *43*, 100939. [CrossRef]
11. Benavides, D.; Arévalo, P.; Tostado-Véliz, M.; Vera, D.; Escamez, A.; Aguado, J.A.; Jurado, F. An Experimental Study of Power Smoothing Methods to Reduce Renewable Sources Fluctuations Using Supercapacitors and Lithium-Ion Batteries. *Batteries* **2022**, *8*, 228. [CrossRef]
12. Pasetti, M.; Amini, S.; Bahramara, S.; Golpîra, H.; Francois, B.; Soares, J. Techno-Economic Analysis of Renewable-Energy-Based Micro-Grids Considering Incentive Policies. *Energies* **2022**, *15*, 8285.
13. Bekun, F.V.; Kambule, N.; Adepoju, O.; Karahüseyin, T.; Abbaso, S. Performance Loss Rates of a 1 MWp PV Plant with Various Tilt Angle, Orientation and Installed Environment in the Capital of Cyprus. *Sustainability* **2022**, *14*, 9084.
14. Tao, S.; Li, C.; Zhang, L.; Tang, Y. Operational Risk Assessment of Grid-Connected PV System Considering Weather Variability and Component Availability. *Energy Procedia* **2018**, *145*, 252–258. [CrossRef]
15. Oyekale, J.; Petrollese, M.; Cocco, D.; Cau, G. Annualized Exergoenvironmental Comparison of Solar-Only and Hybrid Solar-Biomass Heat Interactions with an Organic Rankine Cycle Power Plant. *Energy Convers. Manag. X* **2022**, *15*, 100229. [CrossRef]
16. Son, I.W.; Jeong, Y.; Son, S.; Park, J.H.; Lee, J.I. Techno-Economic Evaluation of Solar-Nuclear Hybrid System for Isolated Grid. *Appl. Energy* **2022**, *306*, 118046. [CrossRef]
17. Kang, D.; Jung, T.Y. Renewable Energy Options for a Rural Village in North Korea. *Sustainability* **2020**, *12*, 2452. [CrossRef]
18. Silva, F.M.Q.; el Kattel, M.B.; Pires, I.A.; Maia, T.A.C. Development of a Supervisory System Using Open-Source for a Power Micro-Grid Composed of a Photovoltaic (PV) Plant Connected to a Battery Energy Storage System and Loads. *Energies* **2022**, *15*, 8324. [CrossRef]
19. Dzhonova-Atanasova, D.; Georgiev, A.; Nakov, S.; Panyovska, S.; Petrova, T.; Maiti, S. Compact Thermal Storage with Phase Change Material for Low-Temperature Waste Heat Recovery—Advances and Perspectives. *Energies* **2022**, *15*, 8269. [CrossRef]
20. Islam, M.R.; Akter, H.; Howlader, H.O.R.; Senjyu, T. Optimal Sizing and Techno-Economic Analysis of Grid-Independent Hybrid Energy System for Sustained Rural Electrification in Developing Countries: A Case Study in Bangladesh. *Energies* **2022**, *15*, 6381. [CrossRef]
21. Nazir, S.; Farah, S.; Paend Bakht, M.; Salam, Z.; Gul, M.; Anjum, W.; Kamaruddin, M.A.; Khan, N.; Lawan Bukar, A. The Potential Role of Hybrid Renewable Energy System for Grid Intermittency Problem: A Techno-Economic Optimisation and Comparative Analysis. *Sustainability* **2022**, *14*, 14045.
22. Liang, T.; Webley, P.A.; Chen, Y.C.; She, X.; Li, Y.; Ding, Y. The Optimal Design and Operation of a Hybrid Renewable Micro-Grid with the Decoupled Liquid Air Energy Storage. *J. Clean. Prod.* **2022**, *334*, 130189. [CrossRef]
23. Deltenre, Q.; de Troyer, T.; Runacres, M.C. Performance Assessment of Hybrid PV-Wind Systems on High-Rise Rooftops in the Brussels-Capital Region. *Energy Build.* **2020**, *224*, 110137. [CrossRef]
24. Harrison-Atlas, D.; Murphy, C.; Schleifer, A.; Grue, N. Temporal Complementarity and Value of Wind-PV Hybrid Systems across the United States. *Renew. Energy* **2022**, *201*, 111–123. [CrossRef]
25. Tercan, S.M.; Demirci, A.; Gokalp, E.; Cali, U. Maximizing Self-Consumption Rates and Power Quality towards Two-Stage Evaluation for Solar Energy and Shared Energy Storage Empowered Microgrids. *J. Energy Storage* **2022**, *51*, 104561. [CrossRef]
26. Orioli, A.; di Gangi, A. Five-Years-Long Effects of the Italian Policies for Photovoltaics on the Energy Demand Coverage of Grid-Connected PV Systems Installed in Urban Contexts. *Energy* **2016**, *113*, 444–460. [CrossRef]
27. Knowles, J. Overview of Small and Micro Combined Heat and Power (CHP) Systems. In *Small and Micro Combined Heat and Power (CHP) Systems*; Woodhead Publishing: Soston, UK, 2011; pp. 3–16.
28. Im, Y. Assessment of the Impact of Renewable Energy Expansion on the Technological Competitiveness of the Cogeneration Model. *Energies* **2022**, *15*, 6844. [CrossRef]
29. Bagherian, M.A.; Mehranzamir, K. A Comprehensive Review on Renewable Energy Integration for Combined Heat and Power Production. *Energy Convers. Manag.* **2020**, *224*, 113454. [CrossRef]
30. Tahir, M.F.; Haoyong, C.; Guangze, H. Evaluating Individual Heating Alternatives in Integrated Energy System by Employing Energy and Exergy Analysis. *Energy* **2022**, *249*, 123753. [CrossRef]
31. Pieper, H.; Kirs, T.; Krupenski, I.; Ledvanov, A.; Lepiksaar, K.; Volkova, A. Efficient Use of Heat from CHP Distributed by District Heating System in District Cooling Networks. *Energy Rep.* **2021**, *7*, 47–54. [CrossRef]
32. Maidment, G.G.; Prosser, G. The Use of CHP and Absorption Cooling in Cold Storage. *Appl. Therm. Eng.* **2000**, *20*, 1059–1073. [CrossRef]
33. Gimelli, A.; Muccillo, M. Development of a 1 KW Micro-Polygeneration System Fueled by Natural Gas for Single-Family Users. *Energies* **2021**, *14*, 8372. [CrossRef]
34. Ramadhani, F.; Hussain, M.A.; Mokhlis, H.; Erixno, O. Solid Oxide Fuel Cell-Based Polygeneration Systems in Residential Applications: A Review of Technology, Energy Planning and Guidelines for Optimizing the Design. *Processes* **2022**, *10*, 2126. [CrossRef]
35. Rong, A.; Lahdelma, R. Role of Polygeneration in Sustainable Energy System Development Challenges and Opportunities from Optimization Viewpoints. *Renew. Sustain. Energy Rev.* **2016**, *53*, 363–372. [CrossRef]
36. *Renewables—Global Energy Review 2021—Analysis*; International Energy Agency (IEA): Paris, France, 2021.
37. Żołądek, M.; Filipowicz, M.; Sornek, K.; Figaj, R.D. Energy Performance of the Photovoltaic System in Urban Area—Case Study. *IOP Conf. Ser. Earth Environ. Sci.* **2019**, *214*, 012123. [CrossRef]

38. Aslam, A.; Ahmed, N.; Qureshi, S.A.; Assadi, M.; Ahmed, N. Advances in Solar PV Systems; A Comprehensive Review of PV Performance, Influencing Factors, and Mitigation Techniques. *Energies* **2022**, *15*, 7595. [CrossRef]

39. Praveenkumar, S.; Gulakhmadov, A.; Agyekum, E.B.; Alwan, N.T.; Velkin, V.I.; Sharipov, P.; Safaraliev, M.; Chen, X. Experimental Study on Performance Enhancement of a Photovoltaic Module Incorporated with CPU Heat Pipe—A 5E Analysis. *Sensors* **2022**, *22*, 6367. [CrossRef]

40. Sornek, K.; Goryl, W.; Figaj, R.; Dąbrowska, G.; Brezdeń, J. Development and Tests of the Water Cooling System Dedicated to Photovoltaic Panels. *Energies* **2022**, *15*, 5884. [CrossRef]

41. Inside Asia's Floating Solar Panels Boom. Available online: https://energytracker.asia/inside-asias-floating-solar-panels-boom/ (accessed on 28 October 2022).

42. Cazzaniga, R.; Cicu, M.; Rosa-Clot, M.; Rosa-Clot, P.; Tina, G.M.; Ventura, C. Floating Photovoltaic Plants: Performance Analysis and Design Solutions. *Renew. Sustain. Energy Rev.* **2018**, *81*, 1730–1741. [CrossRef]

43. Nisar, H.; Janjua, A.K.; Hafeez, H.; Shakir, S.; Shahzad, N.; Waqas, A. Thermal and Electrical Performance of Solar Floating PV System Compared to On-Ground PV System-an Experimental Investigation. *Sol. Energy* **2022**, *241*, 231–247. [CrossRef]

44. Al-Ezzi, A.S.; Ansari, M.N.M. Photovoltaic Solar Cells: A Review. *Appl. Syst. Innov.* **2022**, *5*, 67. [CrossRef]

45. Powell, K.M.; Rashid, K.; Ellingwood, K.; Tuttle, J.; Iverson, B.D. Hybrid Concentrated Solar Thermal Power Systems: A Review. *Renew. Sustain. Energy Rev.* **2017**, *80*, 215–237. [CrossRef]

46. El-Mahallawi, I.; Elshazly, E.; Ramadan, M.; Nasser, R.; Yasser, M.; El-Badry, S.; Elthakaby, M.; Oladinrin, O.T.; Rana, M.Q. Solar PV Panels-Self-Cleaning Coating Material for Egyptian Climatic Conditions. *Sustainability* **2022**, *14*, 11001. [CrossRef]

47. Ahmed, M.; Harbi, I.; Kennel, R.; Rodríguez, J.; Abdelrahem, M. Evaluation of the Main Control Strategies for Grid-Connected PV Systems. *Sustainability* **2022**, *14*, 11142. [CrossRef]

48. Vankadara, S.K.; Chatterjee, S.; Balachandran, P.K.; Mihet-Popa, L. Marine Predator Algorithm (MPA)-Based MPPT Technique for Solar PV Systems under Partial Shading Conditions. *Energies* **2022**, *15*, 6172. [CrossRef]

49. Giwa, A.; Yusuf, A.; Dindi, A.; Balogun, H.A. Polygeneration in Desalination by Photovoltaic Thermal Systems: A Comprehensive Review. *Renew. Sustain. Energy Rev.* **2020**, *130*, 109946. [CrossRef]

50. Yao, W.; Kong, X.; Han, X.; Wang, Y.; Cao, J.; Gao, W. Research on the Efficiency Evaluation of Heat Pipe PV/T Systems and Its Applicability in Different Regions of China. *Energy Convers. Manag.* **2022**, *269*, 116136. [CrossRef]

51. Figaj, R.; Żołądek, M.; Gory, W. Dynamic Simulation and Energy Economic Analysis of a Household Hybrid Ground-Solar-Wind System Using TRNSYS Software. *Energies* **2020**, *13*, 3523. [CrossRef]

52. Ahmed, S.F.; Khalid, M.; Vaka, M.; Walvekar, R.; Numan, A.; Rasheed, A.K.; Mubarak, N.M. Recent Progress in Solar Water Heaters and Solar Collectors: A Comprehensive Review. *Therm. Sci. Eng. Prog.* **2021**, *25*, 100981. [CrossRef]

53. Salek, F.; Eshghi, H.; Zamen, M.; Ahmadi, M.H. Energy and Exergy Analysis of an Atmospheric Water Generator Integrated with the Compound Parabolic Collector with Storage Tank in Various Climates. *Energy Rep.* **2022**, *8*, 2401–2412. [CrossRef]

54. Gao, X.; Wei, S.; Xia, C.; Li, Y. Flexible Operation of Concentrating Solar Power Plant with Thermal Energy Storage Based on a Coordinated Control Strategy. *Energies* **2022**, *15*, 4929. [CrossRef]

55. Figaj, R.; Żołądek, M. Operation and Performance Assessment of a Hybrid Solar Heating and Cooling System for Different Configurations and Climatic Conditions. *Energies* **2021**, *14*, 1142. [CrossRef]

56. Figaj, R.; Żołądek, M. Experimental and Numerical Analysis of Hybrid Solar Heating and Cooling System for a Residential User. *Renew. Energy* **2021**, *172*, 955–967. [CrossRef]

57. Zhang, K.; Su, Y.; Wang, H.; Wang, Q.; Wang, K.; Niu, Y.; Song, J. Highly Concentrated Solar Flux of Large Fresnel Lens Using CCD Camera-Based Method. *Sustainability* **2022**, *14*, 11062. [CrossRef]

58. Adnan, M.; Zaman, M.; Ullah, A.; Gungor, A.; Rizwan, M.; Raza Naqvi, S. Thermo-Economic Analysis of Integrated Gasification Combined Cycle Co-Generation System Hybridized with Concentrated Solar Power Tower. *Renew. Energy* **2022**, *198*, 654–666. [CrossRef]

59. Jamshed, W.; Şirin, C.; Selimefendigil, F.; Shamshuddin, M.D.; Altowairqi, Y.; Eid, M.R. Thermal Characterization of Coolant Maxwell Type Nanofluid Flowing in Parabolic Trough Solar Collector (PTSC) Used Inside Solar Powered Ship Application. *Coatings* **2021**, *11*, 1552. [CrossRef]

60. Wohld, J.; Beck, J.; Inman, K.; Palmer, M.; Cummings, M.; Fulmer, R.; Vafaei, S. Hybrid Nanofluid Thermal Conductivity and Optimization: Original Approach and Background. *Nanomaterials* **2022**, *12*, 2847. [CrossRef]

61. George, A. *Lane Solar Heat Storage: Volume I: Latent Heat Material*, 1st ed.; CRC Press: Boca Raton, FL, USA, 2017.

62. Mouaky, A.; Rachek, A. Thermodynamic and Thermo-Economic Assessment of a Hybrid Solar/Biomass Polygeneration System under the Semi-Arid Climate Conditions. *Renew. Energy* **2020**, *156*, 14–30. [CrossRef]

63. Kasaeian, A.; Bellos, E.; Shamaeizadeh, A.; Tzivanidis, C. Solar-Driven Polygeneration Systems: Recent Progress and Outlook. *Appl. Energy* **2020**, *264*, 114764. [CrossRef]

64. Bošnjaković, M.; Katinić, M.; Santa, R.; Marić, D. Wind Turbine Technology Trends. *Appl. Sci.* **2022**, *12*, 8653. [CrossRef]

65. Das Karmakar, S.; Chattopadhyay, H. A Review of Augmentation Methods to Enhance the Performance of Vertical Axis Wind Turbine. *Sustain. Energy Technol. Assess.* **2022**, *53*, 102469. [CrossRef]

66. Kim, S.; Cheong, C. Development of Low-Noise Drag-Type Vertical Wind Turbines. *Renew. Energy* **2015**, *79*, 199–208. [CrossRef]

67. Müller, G.; Chavushoglu, M.; Kerri, M.; Tsuzaki, T. A Resistance Type Vertical Axis Wind Turbine for Building Integration. *Renew. Energy* **2017**, *111*, 803–814. [CrossRef]

68. Dabachi, M.A.; Rouway, M.; Rahmouni, A.; Bouksour, O.; Sbai, S.J.; Laaouidi, H.; Tarfaoui, M.; Aamir, A.; Lagdani, O. Numerical Investigation of the Structural Behavior of an Innovative Offshore Floating Darrieus-Type Wind Turbines with Three-Stage Rotors. *J. Compos. Sci.* **2022**, *6*, 167. [CrossRef]

69. Aihara, A.; Mendoza, V.; Goude, A.; Bernhoff, H. Comparison of Three-Dimensional Numerical Methods for Modeling of Strut Effect on the Performance of a Vertical Axis Wind Turbine. *Energies* **2022**, *15*, 2361. [CrossRef]

70. Riegler, H. HAWT versus VAWT: Small VAWTs Find a Clear Niche. *Refocus* **2003**, *4*, 44–46.

71. Zhu, C.; Wang, T.; Zhong, W. Combined Effect of Rotational Augmentation and Dynamic Stall on a Horizontal Axis Wind Turbine. *Energies* **2019**, *12*, 1434. [CrossRef]

72. Gil-García, I.C.; García-Cascales, M.S.; Molina-García, A. Urban Wind: An Alternative for Sustainable Cities. *Energies* **2022**, *15*, 4759. [CrossRef]

73. Giorgi, D.; Li, R.; Wang, Y.; Lin, H.; Du, H.; Wang, C.; Chen, X.; Huang, M. A Mesoscale CFD Simulation Study of Basic Wind Pressure in Complex Terrain—A Case Study of Taizhou City. *Appl. Sci.* **2022**, *12*, 10481.

74. Noronha, N.P.; Krishna, M. Aerodynamic Performance Comparison of Airfoils Suggested for Small Horizontal Axis Wind Turbines. *Mater. Today Proc.* **2021**, *46*, 2450–2455. [CrossRef]

75. Zhang, J.H.; Lien, F.S.; Yee, E. Investigations of Vertical-Axis Wind-Turbine Group Synergy Using an Actuator Line Model. *Energies* **2022**, *15*, 6211. [CrossRef]

76. Adeyeye, K.A.; Ijumba, N.; Colton, J. The Effect of the Number of Blades on the Efficiency of A Wind Turbine. *IOP Conf. Ser. Earth Environ. Sci.* **2021**, *801*, 12020. [CrossRef]

77. Ranjbar, M.H.; Rafiei, B.; Nasrazadani, S.A.; Gharali, K.; Soltani, M.; Al-Haq, A.; Nathwani, J. Power Enhancement of a Vertical Axis Wind Turbine Equipped with an Improved Duct. *Energies* **2021**, *14*, 5780. [CrossRef]

78. Basbas, H.; Liu, Y.C.; Laghrouche, S.; Hilairet, M.; Plestan, F. Review on Floating Offshore Wind Turbine Models for Nonlinear Control Design. *Energies* **2022**, *15*, 5477. [CrossRef]

79. Smoleń, J.; Olesik, P.; Jała, J.; Adamcio, A.; Kurtyka, K.; Godzierz, M.; Kozera, R.; Kozioł, M.; Boczkowska, A. The Use of Carbon Fibers Recovered by Pyrolysis from End-of-Life Wind Turbine Blades in Epoxy-Based Composite Panels. *Polymers* **2022**, *14*, 2925. [CrossRef] [PubMed]

80. Hamidifar, H.; Akbari, F.; Rowiński, P.M. Assessment of Environmental Water Requirement Allocation in Anthropogenic Rivers with a Hydropower Dam Using Hydrologically Based Methods—Case Study. *Water* **2022**, *14*, 893. [CrossRef]

81. Esquivel-Sancho, L.M.; Muñoz-Arias, M.; Phillips-Brenes, H.; Pereira-Arroyo, R. A Reversible Hydropump–Turbine System. *Appl. Sci.* **2022**, *12*, 9086. [CrossRef]

82. Joy, J.; Raisee, M.; Cervantes, M.J. Hydraulic Performance of a Francis Turbine with a Variable Draft Tube Guide Vane System to Mitigate Pressure Pulsations. *Energies* **2022**, *15*, 6542. [CrossRef]

83. Qasim, M.A.; Velkin, V.I.; Shcheklein, S.E.; Hanfesh, A.O.; Farge, T.Z.; Essa, F.A. A Numerical Analysis of Fluid Flow and Torque for Hydropower Pelton Turbine Performance Using Computational Fluid Dynamics. *Inventions* **2022**, *7*, 22. [CrossRef]

84. Okot, D.K. Review of Small Hydropower Technology. *Renew. Sustain. Energy Rev.* **2013**, *26*, 515–520. [CrossRef]

85. Yoosefdoost, A.; Lubitz, W.D.; Dario, O.; Mejia, L. Design Guideline for Hydropower Plants Using One or Multiple Archimedes Screws. *Processes* **2021**, *9*, 2128. [CrossRef]

86. Sinagra, M.; Picone, C.; Picone, P.; Aricò, C.; Tucciarelli, T.; Ramos, H.M. Low-Head Hydropower for Energy Recovery in Wastewater Systems. *Water* **2022**, *14*, 1649. [CrossRef]

87. Rotilio, M.; Marchionni, C.; de Berardinis, P. The Small-Scale Hydropower Plants in Sites of Environmental Value: An Italian Case Study. *Sustainability* **2017**, *9*, 2211. [CrossRef]

88. Bravo-Córdoba, F.J.; Torrens, J.; Fuentes-Pérez, J.F.; García-Vega, A.; Sanz-Ronda, F.J. Fishway Attraction Efficiency during Upstream and Down-Stream Migration: Field Tests in a Small Hydropower Plant with Run-of-the-River Configuration. *Biol. Life Sci. Forum* **2022**, *13*, 40.

89. Sornek, K.; Żołądek, M.; Goryl, W.; Figaj, R. The Operation of a Micro-Scale Cogeneration System Prototype—A Comprehensive Experimental and Numerical Analysis. *Fuel* **2021**, *295*, 120563. [CrossRef]

90. Żołądek, M.; Kafetzis, A.; Figaj, R.; Panopoulos, K. Energy-Economic Assessment of Islanded Microgrid with Wind Turbine, Photovoltaic Field, Wood Gasifier, Battery, and Hydrogen Energy Storage. *Sustainability* **2022**, *14*, 12470. [CrossRef]

91. Zhang, W.; Wei, C.; Liu, X.; Zhang, Z. Frost Resistance and Mechanism of Circulating Fluidized Bed Fly Ash-Blast Furnace Slag-Red Mud-Clinker Based Cementitious Materials. *Materials* **2022**, *15*, 6311. [CrossRef] [PubMed]

92. Vershinina, K.; Nyashina, G.; Strizhak, P. Combustion, Pyrolysis, and Gasification of Waste-Derived Fuel Slurries, Low-Grade Liquids, and High-Moisture Waste: Review. *Appl. Sci.* **2022**, *12*, 1039. [CrossRef]

93. Pedretti, F.; European Hophornbeam, E.; Ilari, A.; Fabrizi, S.; Foppa Pedretti, E. European Hophornbeam Biomass for Energy Application: Influence of Different Production Processes and Heating Devices on Environmental Sustainability. *Resources* **2022**, *11*, 11.

94. Żołądek, M.; Figaj, R.; Sornek, K. Energy Analysis of a Micro-Scale Biomass Cogeneration System. *Energy Convers. Manag.* **2021**, *236*, 114079. [CrossRef]

95. Wagle, A.; Angove, M.J.; Mahara, A.; Wagle, A.; Mainali, B.; Martins, M.; Goldbeck, R.; Paudel, S.R. Multi-Stage Pre-Treatment of Lignocellulosic Biomass for Multi-Product Biorefinery: A Review. *Sustain. Energy Technol. Assess.* **2022**, *49*, 101702. [CrossRef]

96. Bagherian, M.A.; Mehranzamir, K.; Rezania, S.; Abdul-Malek, Z.; Pour, A.B.; Alizadeh, S.M. Analyzing Utilization of Biomass in Combined Heat and Power and Combined Cooling, Heating, and Power Systems. *Processes* **2021**, *9*, 1002. [CrossRef]
97. Wołowicz, M.; Kolasiński, P.; Badyda, K. Modern Small and Microcogeneration Systems—A Review. *Energies* **2021**, *14*, 785. [CrossRef]
98. Heat Pumps—Analysis—IEA. Available online: https://www.iea.org/reports/heat-pumps (accessed on 18 November 2022).
99. Xiao, S.; Nefodov, D.; McLinden, M.O.; Richter, M.; Urbaneck, T. Working Fluid Selection for Heat Pumps in Solar District Heating Systems. *Sol. Energy* **2022**, *236*, 499–511. [CrossRef]
100. Neubert, D.; Glück, C.; Schnitzius, J.; Marko, A.; Wapler, J.; Bongs, C.; Felsmann, C. Analysis of the Operation Characteristics of a Hybrid Heat Pump in an Existing Multifamily House Based on Field Test Data and Simulation. *Energies* **2022**, *15*, 5611. [CrossRef]
101. Bahman, A.M.; Parikhani, T.; Ziviani, D. Multi-Objective Optimization of a Cold-Climate Two-Stage Economized Heat Pump for Residential Heating Applications. *J. Build. Eng.* **2022**, *46*, 103799. [CrossRef]
102. Gschwend, A.; Menzi, T.; Caskey, S.; Groll, E.A.; Bertsch, S.S. Energy Consumption of Cold Climate Heat Pumps in Different Climates—Comparison of Single-Stage and Two-Stage Systems. *Int. J. Refrig.* **2016**, *62*, 193–206. [CrossRef]
103. Vicidomini, M.; D'Agostino, D.D. Geothermal Source Exploitation for Energy Saving and Environmental Energy Production. *Energies* **2022**, *15*, 6420. [CrossRef]
104. Luo, J.; Rohn, J.; Bertermann, D.; Salhein, K.; Kobus, C.J.; Zohdy, M. Control of Heat Transfer in a Vertical Ground Heat Exchanger for a Geothermal Heat Pump System. *Energies* **2022**, *15*, 5300.
105. Gabániová, L.; Kudelas, D.; Prčík, M. Modelling Ground Collectors and Determination of the Influence of Technical Parameters, Installation and Geometry on the Soil. *Energies* **2021**, *14*, 7153. [CrossRef]
106. Kwon, Y.; Bae, S.; Nam, Y. Development of Design Method for River Water Source Heat Pump System Using an Optimization Algorithm. *Energies* **2022**, *15*, 4019. [CrossRef]
107. Al-Nimr, M.A.; Al-Waked, R.F.; Al-Zu'bi, O.I. Enhancing the Performance of Heat Pumps by Immersing the External Unit in Underground Water Storage Tanks. *J. Build. Eng.* **2021**, *40*, 102732. [CrossRef]
108. Lin, Y.; Bu, Z.; Yang, W.; Zhang, H.; Francis, V.; Li, C.-Q. A Review on the Research and Development of Solar-Assisted Heat Pump for Buildings in China. *Buildings* **2022**, *12*, 1435. [CrossRef]
109. Niekurzak, M.; Lewicki, W.; Drożdż, W.; Miązek, P. Measures for Assessing the Effectiveness of Investments for Electricity and Heat Generation from the Hybrid Cooperation of a Photovoltaic Installation with a Heat Pump on the Example of a Household. *Energies* **2022**, *15*, 6089. [CrossRef]
110. Hernández-Magallanes, J.A.; Domínguez-Inzunza, L.A.; Lugo-Loredo, S.; Sanal, K.C.; Cerdán-Pasarán, A.; Tututi-Avila, S.; Morales, L.I. Energy and Exergy Analysis of a Modified Absorption Heat Pump (MAHP) to Produce Electrical Energy and Revaluated Heat. *Processes* **2022**, *10*, 1567. [CrossRef]
111. Pinheiro, J.M.; Salústio, S.; Rocha, J.; Valente, A.A.; Silva, C.M. Adsorption Heat Pumps for Heating Applications. *Renew. Sustain. Energy Rev.* **2020**, *119*, 109528. [CrossRef]
112. Dias, J.M.S.; Costa, V.A.F. Modelling and Analysis of a Complete Adsorption Heat Pump System. *Appl. Ther. Eng.* **2022**, *213*, 118782. [CrossRef]
113. Calise, F.; d'Accadia, M.D.; Figaj, R.D. A novel solar-assisted heat-pump driven by photovoltaic/thermal collectors: Dynamic simulation and thermoeconomic optimization. *Energy* **2016**, *95*, 346–366. [CrossRef]
114. Figaj, R.; Żołądek, M.; Homa, M.; Pałac, A. A Novel Hybrid Polygeneration System Based on Biomass, Wind and Solar Energy for Micro-Scale Isolated Communities. *Energies* **2022**, *15*, 6331. [CrossRef]
115. Köppen, W. *Grundriß Der Klimakunde*; De Gruyter: Berlin, Germany, 1931.
116. Cieślak, K.J. Multivariant Analysis of Photovoltaic Performance with Consideration of Self-Consumption. *Energies* **2022**, *15*, 6732. [CrossRef]
117. Agyekum, E.B.; Mehmood, U.; Kamel, S.; Shouran, M.; Elgamli, E.; Adebayo, T.S. Technical Performance Prediction and Employment Potential of Solar PV Systems in Cold Countries. *Sustainability* **2022**, *14*, 3546. [CrossRef]
118. Abdul-Ganiyu, S.; Quansah, D.A.; Ramde, E.W.; Seidu, R.; Adaramola, M.S. Techno-Economic Analysis of Solar Photovoltaic (PV) and Solar Photovoltaic Thermal (PVT) Systems Using Exergy Analysis. *Sustain. Energy Technol. Assess.* **2021**, *47*, 101520. [CrossRef]
119. Li, S.; Zhang, T.; Liu, X.; Xue, Z.; Liu, X. Performance Investigation of a Grid-Connected System Integrated Photovoltaic, Battery Storage and Electric Vehicles: A Case Study for Gymnasium Building. *Energy Build.* **2022**, *270*, 112255. [CrossRef]
120. Ji, Q.; He, H.; Kennedy, S.; Wang, J.; Peng, Z.; Xu, Z.; Zhang, Y. Design and Evaluation of a Wind Turbine-Driven Heat Pump System for Domestic Heating in Scotland. *Sustain. Energy Technol. Assess.* **2022**, *52*, 101987. [CrossRef]
121. Pellegrini, M.; Guzzini, A.; Saccani, C. Experimental Measurements of the Performance of a Micro-Wind Turbine Located in an Urban Area. *Energy Rep.* **2021**, *7*, 3922–3934. [CrossRef]
122. Bilir, L.; Imir, M.; Devrim, Y.; Albostan, A. An Investigation on Wind Energy Potential and Small Scale Wind Turbine Performance at İncek Region—Ankara, Turkey. *Energy Convers. Manag.* **2015**, *103*, 910–923. [CrossRef]
123. Puertas-Frías, C.M.; Willson, C.S.; García-Salaberri, P.A. Design and Economic Analysis of a Hydrokinetic Turbine for Household Applications. *Renew. Energy* **2022**, *199*, 587–598. [CrossRef]
124. Pastore, L.M.; lo Basso, G.; de Santoli, L. Can the Renewable Energy Share Increase in Electricity and Gas Grids Takes out the Competitiveness of Gas-Driven CHP Plants for Distributed Generation? *Energy* **2022**, *256*, 124659. [CrossRef]

125. Zhou, W.; Wang, B.; Wang, M.; Chen, Y. Performance Analysis of the Coupled Heating System of the Air-Source Heat Pump, the Energy Accumulator and the Water-Source Heat Pump. *Energies* **2022**, *15*, 7305. [CrossRef]

126. Bioenergy—Analysis—IEA. Available online: https://www.iea.org/reports/bioenergy (accessed on 18 November 2022).

127. Are Renewable Heating Options Cost-Competitive with Fossil Fuels in the Residential Sector?—Analysis—IEA. Available online: https://www.iea.org/articles/are-renewable-heating-options-cost-competitive-with-fossil-fuels-in-the-residential-sector (accessed on 18 November 2022).

128. Saleem, M. Possibility of Utilizing Agriculture Biomass as a Renewable and Sustainable Future Energy Source. *Heliyon* **2022**, *8*, 08905. [CrossRef]

129. Lee, C.C.; Hussain, J.; Chen, Y. The Optimal Behavior of Renewable Energy Resources and Government's Energy Consumption Subsidy Design from the Perspective of Green Technology Implementation. *Renew. Energy* **2022**, *195*, 670–680. [CrossRef]

130. Harb, A.; Tadeo, F. Introduction to the Special Section on the Stability Impacts of Renewable Energy Sources Integrated with Power Grid (VSI-Srpg). *Comput. Electr. Eng.* **2022**, *98*, 107681. [CrossRef]

131. Shamoon, A.; Haleem, A.; Bahl, S.; Javaid, M.; Bala Garg, S. Role of Energy Technologies in Response to Climate Change. *Mater. Today Proc.* **2022**, *62*, 63–69. [CrossRef]

132. Eder, L.V.; Provornaya, I.V.; Filimonova, I.V.; Kozhevin, V.D.; Komarova, A.V. World Energy Market in the Conditions of Low Oil Prices, the Role of Renewable Energy Sources. *Energy Procedia* **2018**, *153*, 112–117. [CrossRef]

133. Zohuri, B. *Hybrid Energy Systems: Driving Reliable Renewable Sources of Energy Storage*; Springer: Berlin/Heidelberg, Germany, 2017; pp. 1–287.

134. Kaldellis, J.K. *Stand-Alone and Hybrid Wind Energy Systems: Technology, Energy Storage and Applications*; Elsevier: Amsterdam, The Netherlands, 2010; p. 554.

135. Study Reveals Plunge in Lithium-Ion Battery Costs | MIT News | Massachusetts Institute of Technology. Available online: https://news.mit.edu/2021/lithium-ion-battery-costs-0323 (accessed on 28 October 2022).

136. Sánchez, A.; Zhang, Q.; Martín, M.; Vega, P. Towards a New Renewable Power System Using Energy Storage: An Economic and Social Analysis. *Energy Convers. Manag.* **2022**, *252*, 115056. [CrossRef]

137. IRENA—International Renewable Energy Agency. Available online: https://www.irena.org/ (accessed on 28 October 2022).

138. Olympios, A.V.; McTigue, J.D.; Farres-Antunez, P.; Tafone, A.; Romagnoli, A.; Li, Y.; Ding, Y.; Steinmann, W.D.; Wang, L.; Chen, H.; et al. Progress and prospects of thermo-mechanical energy storage—A critical review. *Prog. Energy* **2021**, *3*, 022001. [CrossRef]

139. Li, J.; Zhang, M.; Yang, Q.; Zhang, Z.; Yuan, W.; Gee, A.M.; Bae, S.; Jeon, S.U.; Park, J.-W. The State of the Art of the Development of SMES for Bridging Instantaneous Voltage Dips in Japan. *IEEE Trans. Appl. Supercond.* **2018**, *22*, 507–511.

140. The European Association for Storage of Energy. Available online: https://ease-storage.eu/ (accessed on 28 October 2022).

141. Ibrahim, H.; Ilinca, A.; Perron, J. Energy Storage Systems—Characteristics and Comparisons. *Renew. Sustain. Energy Rev.* **2008**, *12*, 1221–1250. [CrossRef]

142. Chen, H.; Cong, T.N.; Yang, W.; Tan, C.; Li, Y.; Ding, Y. Progress in Electrical Energy Storage System: A Critical Review. *Prog. Nat. Sci.* **2009**, *19*, 291–312. [CrossRef]

143. Ter-Gazarian, A. *Energy Storage for Power Systems*; Peter Peregrinus Ltd.: London, UK, 1994.

144. Shukla, A.; Sharma, A.; Biwole, P.A. *Latent Heat-Based Thermal Energy Storage Systems: Materials, Applications, and the Energy Market*; Apple Academic Press Inc.: Cambridge, MA, USA, 2022; ISBN 9781774639641.

145. Aggarwal, A.; Goyal, N.; Kumar, A. Thermal Characteristics of Sensible Heat Storage Materials Applicable for Concentrated Solar Power Systems. *Mater. Today Proc.* **2021**, *47*, 5812–5817. [CrossRef]

146. Liu, M.; Steven Tay, N.H.; Bell, S.; Belusko, M.; Jacob, R.; Will, G.; Saman, W.; Bruno, F. Review on Concentrating Solar Power Plants and New Developments in High Temperature Thermal Energy Storage Technologies. *Renew. Sustain. Energy Rev.* **2016**, *53*, 1411–1432. [CrossRef]

147. Bhale, P.v.; Rathod, M.K.; Sahoo, L. Thermal Analysis of a Solar Concentrating System Integrated with Sensible and Latent Heat Storage. *Energy Procedia* **2015**, *75*, 2157–2162. [CrossRef]

148. Tamme, R.; Laing, D.; Steinmann, W.-D.; Bauer, T. Thermal Energy Storage. In *Encyclopedia of Sustainability Science and Technology*; Springer: New York, NY, USA, 2012; pp. 10551–10577.

149. Budt, M.; Wolf, D. *Handbook of Energy Storage*; Sterner, M., Stadler, I., Eds.; Springer: New York, NY, USA, 2019; ISBN 978-3-662-55503-3.

150. Singh, H.; Saini, R.P.; Saini, J.S. A Review on Packed Bed Solar Energy Storage Systems. *Renew. Sustain. Energy Rev.* **2010**, *14*, 1059–1069. [CrossRef]

151. Li, G. Sensible Heat Thermal Storage Energy and Exergy Performance Evaluations. *Renew. Sustain. Energy Rev.* **2016**, *53*, 897–923. [CrossRef]

152. Elouali, A.; Kousksou, T.; el Rhafiki, T.; Hamdaoui, S.; Mahdaoui, M.; Allouhi, A.; Zeraouli, Y. Physical Models for Packed Bed: Sensible Heat Storage Systems. *J. Energy Storage* **2019**, *23*, 69–78. [CrossRef]

153. Givoni, B. Underground Longterm Storage of Solar Energy—An Overview. *Sol. Energy* **1977**, *19*, 617–623. [CrossRef]

154. Yumrutaş, R.; Ünsal, M. Energy Analysis and Modeling of a Solar Assisted House Heating System with a Heat Pump and an Underground Energy Storage Tank. *Sol. Energy* **2012**, *86*, 983–993. [CrossRef]

155. Suresh, C.; Saini, R.P. Experimental Study on Combined Sensible-Latent Heat Storage System for Different Volume Fractions of PCM. *Sol. Energy* **2020**, *212*, 282–296. [CrossRef]

156. Ryu, J.Y.; Alford, A.; Lewis, G.; Ding, Y.; Li, Y.; Ahmad, A.; Kim, H.; Park, S.H.; Park, J.P.; Branch, S.; et al. A Novel Liquid Air Energy Storage System Using a Combination of Sensible and Latent Heat Storage. *Appl. Ther. Eng.* **2022**, *203*, 117890. [CrossRef]

157. Hailu, G.; Hayes, P.; Masteller, M. Long-Term Monitoring of Sensible Thermal Storage in an Extremely Cold Region. *Energies* **2019**, *12*, 1821. [CrossRef]

158. Sarbu, I.; Sebarchievici, C. A Comprehensive Review of Thermal Energy Storage. *Sustainability* **2018**, *10*, 191. [CrossRef]

159. Faraj, K.; Khaled, M.; Faraj, J.; Hachem, F.; Castelain, C. Phase Change Material Thermal Energy Storage Systems for Cooling Applications in Buildings: A Review. *Renew. Sustain. Energy Rev.* **2020**, *119*, 109579. [CrossRef]

160. Pasupathy, A.; Velraj, R.; Seeniraj, R.v. Phase Change Material-Based Building Architecture for Thermal Management in Residential and Commercial Establishments. *Renew. Sustain. Energy Rev.* **2008**, *12*, 39–64. [CrossRef]

161. Sharma, A.; Tyagi, V.v.; Chen, C.R.; Buddhi, D. Review on Thermal Energy Storage with Phase Change Materials and Applications. *Renew. Sustain. Energy Rev.* **2009**, *13*, 318–345. [CrossRef]

162. Hasanuzzaman, M.; Abd Rahim, N. *Energy for Sustainable Development: Demand, Supply, Conversion and Management*; Academic Press: Cambridge, MA, USA, 2019; ISBN 9780128146460.

163. Sarbu, I.; Dorca, A. Review on Heat Transfer Analysis in Thermal Energy Storage Using Latent Heat Storage Systems and Phase Change Materials. *Int. J. Energy Res.* **2019**, *43*, 29–64. [CrossRef]

164. Raoux, S.; Wuttig, M. *Phase Change Materials: Science and Applications*; Springer: New York, NY, USA, 2009; ISBN 0387848746.

165. Delgado, J.M.P.Q.; Martinho, J.C.; Vaz Sá, A.; Guimarães, A.S.; Abrantes, V. *Thermal Energy Storage with Phase Change Materials*; Springer Briefs in Applied Sciences and Technology; Springer International Publishing: Cham, Switzerland, 2019; ISBN 978-3-319-97498-9.

166. Li, W.; Klemeš, J.J.; Wang, Q.; Zeng, M. Salt Hydrate—Based Gas-Solid Thermochemical Energy Storage: Current Progress, Challenges, and Perspectives. *Renew. Sustain. Energy Rev.* **2022**, *154*, 111846. [CrossRef]

167. Tyagi, V.v.; Pandey, A.K.; Giridhar, G.; Bandyopadhyay, B.; Park, S.R.; Tyagi, S.K. Comparative Study Based on Exergy Analysis of Solar Air Heater Collector Using Thermal Energy Storage. *Int. J. Energy Res.* **2012**, *36*, 724–736. [CrossRef]

168. Flory, P.J.; Vrij, A. Melting Points of Linear-Chain Homologs. *The Normal Paraffin Hydrocarbons. J. Am. Chem. Soc.* **1963**, *85*, 3548–3553. [CrossRef]

169. Vakhshouri, A.R. Paraffin as Phase Change Material. In *Paraffin—An Overview*; Intech Open: London, UK, 2019.

170. Johansen, J.B.; Englmair, G.; Dannemand, M.; Kong, W.; Fan, J.; Dragsted, J.; Perers, B.; Furbo, S. Laboratory Testing of Solar Combi System with Compact Long Term PCM Heat Storage. *Energy Procedia* **2016**, *91*, 330–337. [CrossRef]

171. Englmair, G.; Moser, C.; Schranzhofer, H.; Fan, J.; Furbo, S. A Solar Combi-System Utilizing Stable Supercooling of Sodium Acetate Trihydrate for Heat Storage: Numerical Performance Investigation. *Appl. Energy* **2019**, *242*, 1108–1120. [CrossRef]

172. Mehling, H. Enthalpy and Temperature of the Phase Change Solid–Liquid—An Analysis of Data of Compounds Employing Entropy. *Sol. Energy* **2013**, *95*, 290–299. [CrossRef]

173. Meng, Y.; Zhao, Y.; Zhang, Y.; Tang, B. Induced Dipole Force Driven PEG/PPEGMA Form-Stable Phase Change Energy Storage Materials with High Latent Heat. *Chem. Eng. J.* **2020**, *390*, 124618. [CrossRef]

174. Usman, A.; Xiong, F.; Aftab, W.; Qin, M.; Zou, R. Emerging Solid-to-Solid Phase-Change Materials for Thermal-Energy Harvesting, Storage, and Utilization. *Adv. Mater.* **2022**, *34*, 2202457. [CrossRef] [PubMed]

175. Abedin, A.H. A Critical Review of Thermochemical Energy Storage Systems. *Open Renew. Energy J.* **2011**, *4*, 42–46. [CrossRef]

176. Zhang, H.; Baeyens, J.; Cáceres, G.; Degrève, J.; Lv, Y. Thermal Energy Storage: Recent Developments and Practical Aspects. *Prog. Energy Combust. Sci.* **2016**, *53*, 1–40. [CrossRef]

177. Dinçer, I.; Rosen, M. *Thermal Energy Storage Systems and Applications*; John Wiley & Sons: Hoboken, NJ, USA, 2021; ISBN 1119956625.

178. Ding, Y.; Riffat, S.B. Thermochemical Energy Storage Technologies for Building Applications: A State-of-the-Art Review. *Int. J. Low-Carbon Technol.* **2013**, *8*, 106–116. [CrossRef]

179. Kant, K.; Pitchumani, R. Advances and Opportunities in Thermochemical Heat Storage Systems for Buildings Applications. *Appl. Energy* **2022**, *321*, 119299. [CrossRef]

180. N'Tsoukpoe, K.E.; le Pierrès, N.; Luo, L. Experimentation of a LiBr–H_2O Absorption Process for Long-Term Solar Thermal Storage: Prototype Design and First Results. *Energy* **2013**, *53*, 179–198. [CrossRef]

181. Qi, Z.; Koenig, G.M., Jr. Review Article: Flow Battery Systems with Solid Electroactive Materials. *J. Vac. Sci. Technol. B Nanotechnol.* **2017**, *35*, 040801. [CrossRef]

182. Häggström, F.; Delsing, J. IoT Energy Storage—A Forecast. *Energy Harvest. Syst.* **2018**, *5*, 43–51. [CrossRef]

183. Oh, K. Supercapacitors Strengthen Renewable Energy Utilization. Available online: https://www.eaton.com/content/dam/eaton/products/electronic-components/resources/brochure/eaton-supercapacitors-strengthen-renewable-energy-utilization.pdf (accessed on 18 November 2022).

184. Supercapacitors for Renewable Energy Applications | Electronics360. Available online: https://electronics360.globalspec.com/article/14903/supercapacitors-for-renewable-energy-applications (accessed on 28 October 2022).

185. CR5—Superconducting Magnetic Energy Storage: Status and Perspective | Superconductivity News Forum. Available online: https://snf.ieeecsc.org/abstracts/cr5-superconducting-magnetic-energy-storage-status-and-perspective (accessed on 28 October 2022).

186. Superconducting Magnetic Energy Storage: 2021 Guide | Linquip. Available online: https://www.linquip.com/blog/superconducting-magnetic-energy-storage/ (accessed on 28 October 2022).

187. Nguyen, T.M.P. Lead Acid Batteries in Extreme Conditions: Accelerated Charge, Maintaining the Charge with Imposed Low Current, Polarity Inversions Introducing Non-Conventional Charge Methods. Ph.D. Thesis, Université Montpellier II, Montpellier, France, 2009.

188. Nasirudin, A.; Chao, R.M.; Utama, I.K.A.P. Solar Powered Boat Design Optimization. *Procedia Eng.* **2017**, *194*, 260–267. [CrossRef]

189. Malla, U. Design and Sizing of Battery System for Electric Yacht and Ferry. *Int. J. Interact. Des. Manuf. (IJIDeM)* **2019**, *14*, 137–142. [CrossRef]

190. Pangaribowo, T.; Utomo, W.M.; Bakar, A.A.; Khaerudini, D.S. Battery Charging and Discharging Control of a Hybrid Energy System Using Microcontroller. *Indones. J. Electr. Eng. Comput. Sci.* **2020**, *17*, 575–582. [CrossRef]

191. Sun, R.L.; Hu, P.Q.; Wang, R.; Qi, L.Y. A New Method for Charging and Repairing Lead-Acid Batteries. *IOP Conf. Ser. Earth Environ. Sci.* **2020**, *461*, 012031. [CrossRef]

192. Sutanto, J. *Environmental Study of Lead Acid Batteries Technologies*; GRIN Publishing: Munich, Germany, 2011; ISBN 3656033846.

193. Amutha, W.M.; Rajini, V. Cost Benefit and Technical Analysis of Rural Electrification Alternatives in Southern India Using HOMER. *Renew. Sustain. Energy Rev.* **2016**, *62*, 236–246. [CrossRef]

194. Kebede, A.A.; Coosemans, T.; Messagie, M.; Jemal, T.; Behabtu, H.A.; van Mierlo, J.; Berecibar, M. Techno-Economic Analysis of Lithium-Ion and Lead-Acid Batteries in Stationary Energy Storage Application. *J. Energy Storage* **2021**, *40*, 102748. [CrossRef]

195. Pistoia, G. *Lithium-Ion Batteries: Advances and Applications*; Elsevier: Amsterdam, The Netherlands, 2013; ISBN 0444595139.

196. Liu, Y.; Zhang, R.; Wang, J.; Wang, Y. Current and Future Lithium-Ion Battery Manufacturing. *iScience* **2021**, *24*, 102332. [CrossRef]

197. Asenbauer, J.; Eisenmann, T.; Kuenzel, M.; Kazzazi, A.; Chen, Z.; Bresser, D. The Success Story of Graphite as a Lithium-Ion Anode Material—Fundamentals, Remaining Challenges, and Recent Developments Including Silicon (Oxide) Composites. *Sustain. Energy Fuels* **2020**, *4*, 5387–5416. [CrossRef]

198. Bini, M.; Capsoni, D.; Ferrari, S.; Quartarone, E.; Mustarelli, P. Rechargeable Lithium Batteries: Key Scientific and Technological Challenges. In *Rechargeable Lithium Batteries*; Woodhead Publishing: Cambridge, UK, 2015; pp. 1–17.

199. García-Vázquez, C.A.; Espinoza-Ortega, H.; Llorens-Iborra, F.; Fernández-Ramírez, L.M. Feasibility Analysis of a Hybrid Renewable Energy System with Vehicle-to-Home Operations for a House in off-Grid and Grid-Connected Applications. *Sustain. Cities Soc.* **2022**, *86*, 104124. [CrossRef]

200. Małek, A.; Caban, J.; Wojciechowski, Ł. Charging Electric Cars as a Way to Increase the Use of Energy Produced from RES. *Open Eng.* **2020**, *10*, 98–104. [CrossRef]

201. Zubi, G.; Adhikari, R.S.; Sánchez, N.E.; Acuña-Bravo, W. Lithium-Ion Battery-Packs for Solar Home Systems: Layout, Cost and Implementation Perspectives. *J. Energy Storage* **2020**, *32*, 101985. [CrossRef]

202. Darcovich, K.; Henquin, E.R.; Kenney, B.; Davidson, I.J.; Saldanha, N.; Beausoleil-Morrison, I. Higher-Capacity Lithium Ion Battery Chemistries for Improved Residential Energy Storage with Micro-Cogeneration. *Appl. Energy* **2013**, *111*, 853–861. [CrossRef]

203. Han, A.; Han, E.; Han Hv, E. *Directive 2011/65/eu of the European Parliament and of the Council of 8 June 2011 on the Restriction of the Use of Certain Hazardous Substances in Electrical and Electronic Equipment (Recast) (Text with EEA Relevance)*; Harting: Espelkamp, Germany, 2011.

204. Linden, D.; Reddy Editor, T.B.; York, N.; San, C.; Lisbon, F.; Madrid, L.; City, M.; New, M.; San, D.; Seoul, J. *Handbook of Batteries*; McGraw-Hill Education Ltd.: New York, NY, USA, 2002; ISBN 0071359788.

205. Dell, R.M.; Rand, D.A.J. Energy Storage—A Key Technology for Global Energy Sustainability. *J. Power Sources* **2001**, *100*, 2–17. [CrossRef]

206. Lithium-Ion Battery—Clean Energy Institute. Available online: https://www.cei.washington.edu/education/science-of-solar/battery-technology/ (accessed on 28 October 2022).

207. Nair, N.K.C.; Garimella, N. Battery Energy Storage Systems: Assessment for Small-Scale Renewable Energy Integration. *Energy Build.* **2010**, *42*, 2124–2130. [CrossRef]

208. Wen, Z.; Hu, Y.; Wu, X.; Han, J.; Gu, Z. Main Challenges for High Performance NAS Battery: Materials and Interfaces. *Adv. Funct. Mater.* **2013**, *23*, 1005–1018. [CrossRef]

209. Huang, J.J.; Weinstock, D.; Hirsh, H.; Bouck, R.; Zhang, M.; Gorobtsov, O.Y.; Okamura, M.; Harder, R.; Cha, W.; Ruff, J.P.C.; et al. Disorder Dynamics in Battery Nanoparticles During Phase Transitions Revealed by Operando Single-Particle Diffraction. *Adv. Energy Mater.* **2022**, *12*, 2201644. [CrossRef]

210. He, J.; Bhargav, A.; Shin, W.; Manthiram, A. Stable Dendrite-Free Sodium-Sulfur Batteries Enabled by a Localized High-Concentration Electrolyte. *J. Am. Chem. Soc.* **2021**, *143*, 20241–20248. [CrossRef] [PubMed]

211. Lee, B.; Paek, E.; Mitlin, D.; Lee, S.W. Sodium Metal Anodes: Emerging Solutions to Dendrite Growth. *Chem. Rev.* **2019**, *119*, 5416–5460. [CrossRef] [PubMed]

212. Oshima, T.; Kajita, M.; Okuno, A. Development of Sodium-Sulfur Batteries. *Int. J. Appl. Ceram. Technol.* **2004**, *1*, 269–276. [CrossRef]

213. BASF and NGK to Partner on Developing the Next Generation of Sodium-Sulfur Batteries. Available online: https://www.basf.com/global/en/who-we-are/sustainability/whats-new/sustainability-news/2019/nas-battery-systems.html (accessed on 18 November 2022).

214. Kumar, D.; Rajouria, S.K.; Kuhar, S.B.; Kanchan, D.K. Progress and Prospects of Sodium-Sulfur Batteries: A Review. *Solid State Ion* **2017**, *312*, 8–16. [CrossRef]

215. Chen, P.; Wang, C.; Wang, T. Review and Prospects for Room-Temperature Sodium-Sulfur Batteries. *Mater. Res. Lett.* **2022**, *10*, 691–719. [CrossRef]
216. Zhang, H.; Li, X.; Zhang, J. *Redox Flow Batteries: Fundamentals and Applications*; CRC Press: Boca Raton, FL, USA, 2017; ISBN 1351648721.
217. Kepco—The Kansai Electric Power Co., Inc. Available online: https://www.kepco.co.jp/english/ (accessed on 28 October 2022).
218. Huang, Z.; Mu, A.; Wu, L.; Yang, B.; Qian, Y.; Wang, J. Comprehensive Analysis of Critical Issues in All-Vanadium Redox Flow Battery. *ACS Sustain. Chem. Eng.* **2022**, *10*, 7786–7810. [CrossRef]
219. Zou, T.; Luo, L.; Liao, Y.; Wang, P.; Zhang, J.; Yu, L. Study on Operating Conditions of Household Vanadium Redox Flow Battery Energy Storage System. *J. Energy Storage* **2022**, *46*, 103859. [CrossRef]
220. Jefimowski, W.; Szeląg, A.; Steczek, M.; Nikitenko, A. Vanadium Redox Flow Battery Parameters Optimization in a Transportation Microgrid: A Case Study. *Energy* **2020**, *195*, 116943. [CrossRef]
221. Brun, K.; Allison, T.; Dennis, R. *Thermal, Mechanical, and Hybrid Chemical Energy Storage Systems*; Academic Press: Cambridge, MA, USA, 2021; ISBN 9780128198926.
222. Bindra, H.; Revankar, S. *Storage and Hybridization of Nuclear Energy: Techno-Economic Integration of Renewable and Nuclear Energy*; Academic Press: Cambridge, MA, USA, 2018; ISBN 0128139757.
223. Kooijinga, J. Large Scale Electricity Storage Using Power-to-Gas in a 100% Renewable Power System Comparison Study P2G Storage Using H_2, NH_3 or CH_4. Doctoral Dissertation, University of Southampton, Southampton, UK, 2019.
224. Carriveau, R.; Ting, D.S.K. *Methane and Hydrogen for Energy Storage*; The Institution of Engineering and Technology: London, UK, 2016; pp. 1–173.
225. Biswas, S.; Kulkarni, A.P.; Giddey, S.; Bhattacharya, S. A Review on Synthesis of Methane as a Pathway for Renewable Energy Storage with a Focus on Solid Oxide Electrolytic Cell-Based Processes. *Front. Energy Res.* **2020**, *8*, 229. [CrossRef]
226. STORE&GO ǀ STORE&GO. Available online: https://www.storeandgo.info/ (accessed on 28 October 2022).
227. Wang, L.; Chen, M.; Küngas, R.; Lin, T.E.; Diethelm, S.; Maréchal, F.; van Herle, J. Power-to-Fuels via Solid-Oxide Electrolyzer: Operating Window and Techno-Economics. *Renew. Sustain. Energy Rev.* **2019**, *110*, 174–187. [CrossRef]
228. Varbanov, P.S.; Su, R.; Lam, H.L.; Liu, X.; Klemeš, J.J.; Wang, L.; Mian, A.; de Sousa, C.R.; Diethelm, S.; van Herle, B.J.; et al. Integrated System Design of a Small-Scale Power-to-Methane Demonstrator. *Chem. Eng. Trans.* **2017**, *61*, 1339–1344.
229. Mazloomi, K.; Gomes, C. Hydrogen as an Energy Carrier: Prospects and Challenges. *Renew. Sustain. Energy Rev.* **2012**, *16*, 3024–3033. [CrossRef]
230. International Energy Agency, Global Hydrogen Review 2021. Available online: https://www.iea.org/reports/global-hydrogen-review-2021 (accessed on 18 November 2022).
231. el Kharbachi, A.; Dematteis, E.M.; Shinzato, K.; Stevenson, S.C.; Bannenberg, L.J.; Heere, M.; Zlotea, C.; Szilágyi, P.; Bonnet, J.P.; Grochala, W.; et al. Metal Hydrides and Related Materials. Energy Carriers for Novel Hydrogen and Electrochemical Storage. *J. Phys. Chem. C* **2020**, *124*, 7599–7607.
232. Liu, Y.; Huang, Y.; Xiao, Z.; Reng, X. Study of Adsorption of Hydrogen on Al, Cu, Mg, Ti Surfaces in Al Alloy Melt via First Principles Calculation. *Metals* **2017**, *7*, 21. [CrossRef]
233. Ni, M. An Overview of Hydrogen Storage Technologies. *Energy Explor. Exploit.* **2006**, *24*, 197–209. [CrossRef]
234. Larpruenrudee, P.; Bennett, N.S.; Gu, Y.T.; Fitch, R.; Islam, M.S. Design Optimization of a Magnesium-Based Metal Hydride Hydrogen Energy Storage System. *Sci. Rep.* **2022**, *12*, 1–16. [CrossRef]
235. Mechanical Electricity Storage Technology ǀ Energy Storage Association. Available online: https://energystorage.org/why-energy-storage/technologies/mechanical-energy-storage/ (accessed on 28 October 2022).
236. Puranen, P.; Kosonen, A.; Ahola, J. Technical Feasibility Evaluation of a Solar PV Based Off-Grid Domestic Energy System with Battery and Hydrogen Energy Storage in Northern Climates. *Sol. Energy* **2021**, *213*, 246–259. [CrossRef]
237. Kokkotis, P.; Psomopoulos, C.S.; Ioannidis, G.C.; Kaminaris, S.D.; Kokkotis, P.I.; Ch Ioannidis, G.; Kaminaris, S.D. Small scale energy storage systems. A short review in their potential environmental impact. *Fresenius Environ. Bull.* **2017**, *26*, 5658–5665.
238. The Potential Impact of Small-Scale Flywheel Energy Storage Technology on Uganda's Energy Sector. Available online: http://www.scielo.org.za/scielo.php?script=sci_arttext&pid=S1021-447X2009000100002 (accessed on 28 October 2022).
239. Ruoso, A.C.; Caetano, N.R.; Rocha, L.A.O. Storage Gravitational Energy for Small Scale Industrial and Residential Applications. *Inventions* **2019**, *4*, 64. [CrossRef]
240. Mucci, S.; Bischi, A.; Briola, S.; Baccioli, A. Small-Scale Adiabatic Compressed Air Energy Storage: Control Strategy Analysis via Dynamic Modelling. *Energy Convers. Manag.* **2021**, *243*, 114358. [CrossRef]
241. Lund, H.; Salgi, G. The Role of Compressed Air Energy Storage (CAES) in Future Sustainable Energy Systems. *Energy Convers. Manag.* **2009**, *50*, 1172–1179. [CrossRef]
242. Cheayb, M.; Marin Gallego, M.; Tazerout, M.; Poncet, S. Modelling and Experimental Validation of a Small-Scale Trigenerative Compressed Air Energy Storage System. *Appl. Energy* **2019**, *239*, 1371–1384. [CrossRef]
243. She, X.; Zhang, T.; Peng, X.; Wang, L.; Tong, L.; Luo, Y.; Zhang, X.; Ding, Y. Liquid Air Energy Storage for Decentralized Micro Energy Networks with Combined Cooling, Heating, Hot Water and Power Supply. *J. Therm. Sci.* **2020**, *30*, 1–17. [CrossRef]
244. Murthy, S.S.; Dutta, P.; Sharma, R.; Rao, B.S. Parametric Studies on a Stand-Alone Polygeneration Microgrid with Battery Storage. *Therm. Sci. Eng. Prog.* **2020**, *19*, 100608. [CrossRef]

245. Gesteira, L.G.; Uche, J. A Novel Polygeneration System Based on a Solar-Assisted Desiccant Cooling System for Residential Buildings: An Energy and Environmental Analysis. *Sustainability* **2022**, *14*, 3449. [CrossRef]

246. Calise, F.; d'Accadia, M.D.; Piacentino, A.; Vicidomini, M. Thermoeconomic Optimization of a Renewable Polygeneration System Serving a Small Isolated Community. *Energies* **2015**, *8*, 995–1024. [CrossRef]

247. Sigarchian, S.G.; Malmquist, A.; Martin, V. Design Optimization of a Small-Scale Polygeneration Energy System in Different Climate Zones in Iran. *Energies* **2018**, *11*, 1115. [CrossRef]

248. Calise, F.; d'Accadia, M.D.; Vicidomini, M. Optimization and Dynamic Analysis of a Novel Polygeneration System Producing Heat, Cool and Fresh Water. *Renew. Energy* **2019**, *143*, 1331–1347. [CrossRef]

249. Govindasamy, P.K.; Rajagopal, S.; Coronas, A. Integrated Polygeneration System for Coastal Areas. *Therm. Sci. Eng. Prog.* **2020**, *20*, 100739. [CrossRef]

250. Figaj, R. Performance Assessment of a Renewable Micro-Scale Trigeneration System Based on Biomass Steam Cycle, Wind Turbine, Photovoltaic Field. *Renew. Energy* **2021**, *177*, 193–208. [CrossRef]

251. Picallo-Perez, A.; Sala-Lizarraga, J.M. Design and Operation of a Polygeneration System in Spanish Climate Buildings under an Exergetic Perspective. *Energies* **2021**, *14*, 7636. [CrossRef]

252. Wegener, M.; Malmquist, A.; Isalgue, A.; Martin, A.; Arranz, P.; Camara, O.; Velo, E. A Techno-Economic Optimization Model of a Biomass-Based CCHP/Heat Pump System under Evolving Climate Conditions. *Energy Convers. Manag.* **2020**, *223*, 113256. [CrossRef]

253. Calise, F.; Figaj, R.D.; Vanoli, L. A Novel Polygeneration System Integrating Photovoltaic/Thermal Collectors, Solar Assisted Heat Pump, Adsorption Chiller and Electrical Energy Storage: Dynamic and Energy-Economic Analysis. *Energy Convers. Manag.* **2017**, *149*, 798–814. [CrossRef]

254. de León, C.M.; Ríos, C.; Brey, J.J. Cost of Green Hydrogen: Limitations of Production from a Stand-Alone Photovoltaic System. *Int. J. Hydrogen Energy* **2022**. [CrossRef]

255. Grunow, P. Decentral Hydrogen. *Energies* **2022**, *15*, 2820. [CrossRef]

256. Jabeen, N.; Zaidi, A.; Hussain, A.; Hassan, N.U.; Ali, J.; Ahmed, F.; Khan, M.U.; Iqbal, N.; Elnasr, T.A.S.; Helal, M.H. Single- and Multilayered Perovskite Thin Films for Photovoltaic Applications. *Nanomaterials* **2022**, *12*, 3208. [CrossRef]

257. Antohe, V.-A. Advances in Nanomaterials for Photovoltaic Applications. *Nanomaterials* **2022**, *12*, 3702. [CrossRef]

258. Seroka, N.S.; Taziwa, R.; Khotseng, L. Solar Energy Materials-Evolution and Niche Applications: A Literature Review. *Materials* **2022**, *15*, 5338. [CrossRef]

259. Das, N.; Wongsodihardjo, H.; Islam, S. Modeling of Multi-Junction Photovoltaic Cell Using MATLAB/Simulink to Improve the Conversion Efficiency. *Renew. Energy* **2015**, *74*, 917–924. [CrossRef]

260. Shafaghat, A.H.; Eslami, M.; Baneshi, M. Techno-Enviro-Economic Study of a Reverse Osmosis Desalination System Equipped with Photovoltaic-Thermal Collectors. *Appl. Ther. Eng.* **2023**, *218*, 119289. [CrossRef]

261. Baroutaji, A.; Arjunan, A.; Robinson, J.; Wilberforce, T.; Abdelkareem, M.A.; Olabi, A.G. PEMFC Poly-Generation Systems: Developments, Merits, and Challenges. *Sustainability* **2021**, *13*, 11696. [CrossRef]

262. Mittal, S.; Ruth, M.; Pratt, A.; Lunacek, M.; Krishnamurthy, D.; Jones, W. *A System-of-Systems Approach for Integrated Energy Systems Modeling and Simulation*; National Renewable Energy Lab.: Golden, CO, USA, 2015.

263. Huang, X. Ecologically Unequal Exchange, Recessions, and Climate Change: A Longitudinal Study. *Soc. Sci. Res.* **2018**, *73*, 1–12. [CrossRef]

264. Chang, M.; Thellufsen, J.Z.; Zakeri, B.; Pickering, B.; Pfenninger, S.; Lund, H.; Østergaard, P.A. Trends in Tools and Approaches for Modelling the Energy Transition. *Appl. Energy* **2021**, *290*, 116731. [CrossRef]

265. Henriot, A. Economics of Intermittent Renewable Energy Sources: Four Essays on Large-Scale Integration into European Power Systems. Ph.D. Thesis, University of Paris, Paris, France, 2014.

266. Subramanian, A.S.R.; Gundersen, T.; Adams, T.A. Modeling and Simulation of Energy Systems: A Review. *Processes* **2018**, *6*, 238. [CrossRef]

267. Bonabeau, E. Agent-Based Modeling: Methods and Techniques for Simulating Human Systems. *Proc. Natl. Acad. Sci. USA* **2002**, *99*, 7280–7287. [CrossRef]

268. Rai, V.; Henry, A.D. Agent-Based Modelling of Consumer Energy Choices. *Nat. Clim. Chang.* **2016**, *6*, 556–562. [CrossRef]

269. Klein, M.; Frey, U.J.; Reeg, M. Models Within Models—Agent-Based Modelling and Simulation in Energy Systems Analysis. *J. Artif. Soc. Soc. Simul.* **2019**, *22*, 6. [CrossRef]

270. Dhaliwal, J.S.; Benbasat, I. The Use and Effects of Knowledge-Based System Explanations: Theoretical Foundations and a Framework for Empirical Evaluation. *Inf. Syst. Res.* **1996**, *7*, 342–362. [CrossRef]

271. Abbey, C.; Strunz, K.; Joós, G. A Knowledge-Based Approach for Control of Two-Level Energy Storage for Wind Energy Systems. *IEEE Trans. Energy Convers.* **2009**, *24*, 539–547. [CrossRef]

272. Adams, T.A., II. Modeling and Simulation of Energy Systems. *Processes* **2019**, *7*, 523. [CrossRef]

273. Jyothy, L.P.N.; Sindhu, M.R. An Artificial Neural Network Based MPPT Algorithm for Solar PV System. In Proceedings of the 4th International Conference on Electrical Energy Systems, ICEES 2018, Chennai, India, 7–9 February 2018; pp. 375–380.

274. Obafemi, O.; Stephen, A.; Ajayi, O.; Nkosinathi, M. A Survey of Artificial Neural Network-Based Prediction Models for Thermal Properties of Biomass. *Procedia Manuf.* **2019**, *33*, 184–191. [CrossRef]

275. Manero, J.; Béjar, J.; Cortés, U. Wind Energy Forecasting with Neural Networks: A Literature Review. *Comput. Sist.* **2018**, *22*, 1085–1098. [CrossRef]

276. Thiaw, L.; Sow, G.; Fall, S. Application of Neural Networks Technique in Renewable Energy Systems. In Proceedings of the 2014 First International Conference on Systems Informatics, Modelling and Simulation, Sheffield, UK, 29 April–1 May 2014.

277. Checchi, A.; Bissacco, G.; Hansen, H.N. A Mechanistic Model for the Prediction of Cutting Forces in the Face-Milling of Ductile Spheroidal Cast Iron Components for Wind Industry Application. *Procedia CIRP* **2018**, *77*, 231–234. [CrossRef]

278. Council, N.R. *Statistics, Testing, and Defense Acquisition: New Approaches and Methodological Improvements*; Statistics, Testing, and Defense Acquisition 1AD; National Academies Press: Washington, DC, USA, 1998.

279. Hall, L.M.H.; Buckley, A.R. A Review of Energy Systems Models in the UK: Prevalent Usage and Categorisation. *Appl. Energy* **2016**, *169*, 607–628. [CrossRef]

280. Lund, H.; Arler, F.; Østergaard, P.A.; Hvelplund, F.; Connolly, D.; Mathiesen, B.V.; Karnøe, P. Simulation versus Optimisation: Theoretical Positions in Energy System Modelling. *Energies* **2017**, *10*, 840. [CrossRef]

281. Yoro, K.O.; Daramola, M.O.; Sekoai, P.T.; Wilson, U.N.; Eterigho-Ikelegbe, O. Update on Current Approaches, Challenges, and Prospects of Modeling and Simulation in Renewable and Sustainable Energy Systems. *Renew. Sustain. Energy Rev.* **2021**, *150*, 111506. [CrossRef]

282. Memon, S.A.; Patel, R.N. An Overview of Optimization Techniques Used for Sizing of Hybrid Renewable Energy Systems. *Renew. Energy Focus* **2021**, *39*, 1–26. [CrossRef]

283. Khan, A.A.; Minai, A.F.; Pachauri, R.K.; Malik, H. Optimal Sizing, Control, and Management Strategies for Hybrid Renewable Energy Systems: A Comprehensive Review. *Energies* **2022**, *15*, 6249. [CrossRef]

284. Modeling and Simulation. Available online: http://home.ubalt.edu/ntsbarsh/simulation/sim.htm (accessed on 28 October 2022).

285. All about the Iterative Design Process | Smartsheet. Available online: https://www.smartsheet.com/iterative-process-guide (accessed on 28 October 2022).

286. Seyyedi, S.M.; Ajam, H.; Farahat, S. A New Iterative Approach to the Optimization of Thermal Energy Systems: Application to the Regenerative Brayton Cycle. *Proc. Inst. Mech. Eng. Part A J. Power Energy* **2010**, *224*, 313–327. [CrossRef]

287. Geleta, D.K.; Manshahia, M.S. Optimization of Hybrid Wind and Solar Renewable Energy System by Iteration Method. *Adv. Intell. Syst. Comput.* **2019**, *866*, 98–107.

288. Jaseena, K.U.; Kovoor, B.C. Deterministic Weather Forecasting Models Based on Intelligent Predictors: A Survey. *J. King Saud Univ.-Comput. Inf. Sci.* **2022**, *34*, 3393–3412. [CrossRef]

289. Shah, Y.T. *Hybrid Power Generation, Storage, and Grids*; CRC Press: Boca Raton, FL, USA, 2021; ISBN 9780367678401.

290. Connolly, D.; Lund, H.; Mathiesen, B.V.; Leahy, M. A Review of Computer Tools for Analysing the Integration of Renewable Energy into Various Energy Systems. *Appl. Energy* **2010**, *87*, 1059–1082. [CrossRef]

291. Chen, Y.; Guo, M.; Chen, Z.; Chen, Z.; Ji, Y. Physical Energy and Data-Driven Models in Building Energy Prediction: A Review. *Energy Rep.* **2022**, *8*, 2656–2671. [CrossRef]

292. Klein, S.A. *TRNSYS 18: A Transient System Simulation Program*; Solar Energy Laboratory, University of Wisconsin: Madison, WI, USA, 2017.

293. HOMER Pro—Microgrid Software for Designing Optimized Hybrid Microgrids. Available online: https://www.homerenergy.com/products/pro/index.html (accessed on 28 October 2022).

294. IHOGA/MHOGA—Simulation and Optimization of Stand-Alone and Grid-Connected Hybrid Renewable Systems. Available online: https://ihoga.unizar.es/en/ (accessed on 28 October 2022).

295. PV*SOL—Plan and Design Better Pv Systems with Professional Solar Software | PV*SOL and PV*SOL Premium. Available online: https://pvsol.software/en/ (accessed on 28 October 2022).

296. QBlade—Next Generation Wind Turbine Simulation. Available online: https://qblade.org/ (accessed on 28 October 2022).

297. OpenFOAM. Available online: https://www.openfoam.com/ (accessed on 28 October 2022).

298. Ansys | Engineering Simulation Software. Available online: https://www.ansys.com/ (accessed on 28 October 2022).

299. MathWorks—Makers of MATLAB and Simulink—MATLAB & Simulink. Available online: https://www.mathworks.com/ (accessed on 28 October 2022).

300. Hiendro, A.; Kurnianto, R.; Rajagukguk, M.; Simanjuntak, Y.M. Junaidi Techno-Economic Analysis of Photovoltaic/Wind Hybrid System for Onshore/Remote Area in Indonesia. *Energy* **2013**, *59*, 652–657. [CrossRef]

301. Carroquino, J.; Escriche-Martínez, C.; Valiño, L.; Dufo-López, R. Comparison of Economic Performance of Lead-Acid and Li-Ion Batteries in Standalone Photovoltaic Energy Systems. *Appl. Sci.* **2021**, *11*, 3587. [CrossRef]

302. Sharma, R.; Gidwani, L. Grid Connected Solar PV System Design and Calculation by Using PV SOL Premium Simulation Tool for Campus Hostels of RTU Kota. In Proceedings of the IEEE International Conference on Circuit, Power and Computing Technologies, ICCPCT 2017, Kollam, India, 20–21 April 2017.

303. Abdullah, A.L.; Misha, S.; Tamaldin, N.; Rosli, M.A.M.; Sachit, F.A. Theoretical Study and Indoor Experimental Validation of Performance of the New Photovoltaic Thermal Solar Collector (PVT) Based Water System. *Case Stud. Therm. Eng.* **2020**, *18*, 100595. [CrossRef]

304. El-Hady, A.A.M.; Arafa, S.H.; Nashed, M.N.F.; Ramadan, S.G. Modeling and Simulation for Hybrid of PV-Wind System. *Int. J. Eng. Res.* **2015**, *4*, 178–183. [CrossRef]

305. Srujana, A.; Srilatha, M.A.; Suresh, C.M.S. Electric Vehicle Battery Modelling and Simulation Using MATLAB-Simulink. *Turk. J. Comput. Math. Educ.* **2021**, *12*, 4604–4609.

306. Fotouhi, A.; Auger, D.J.; Cleaver, T.; Shateri, N.; Propp, K.; Longo, S. Influence of Battery Capacity on Performance of an Electric Vehicle Fleet. In Proceedings of the 2016 IEEE International Conference on Renewable Energy Research and Applications, ICRERA 2016, Birmingham, UK, 20–23 November 2016; pp. 928–933.

307. RETScreen. Available online: https://www.nrcan.gc.ca/maps-tools-and-publications/tools/modelling-tools/retscreen/7465 (accessed on 21 November 2022).

308. EnergyPRO—Software for Modelling and Analysis of Complex Energy ProjectsEMD International. Available online: https://www.emd-international.com/energypro/ (accessed on 21 November 2022).

309. Schmeling, L.; Klement, P.; Erfurth, T.; Kästner, J.; Hanke, B.; von Maydell, K.; Agert, C. Review of different software solutions for the holistic simulation of distributed hybrid energy systems for the commercial energy supply. In Proceedings of the 33rd European Photovoltaic Solar Energy Conference and Exhibition, Amsterdam, The Netherlands, 27–28 September 2017.

310. EnergyPlus. Available online: https://energyplus.net/ (accessed on 21 November 2022).

311. Zemax. Available online: https://www.zemax.com/ (accessed on 21 November 2022).

312. Garcia, D.; Liang, D.; Tibúrcio, B.D.; Almeida, J.; Vistas, C.R. A Three-Dimensional Ring-Array Concentrator Solar Furnace. *Sol. Energy* **2019**, *193*, 915–928. [CrossRef]

313. Optical and Illumination Simulation, Design & Analysis Tool. TracePro. Available online: https://lambdares.com/tracepro (accessed on 21 November 2022).

314. Yang, M.; Zhang, Y.; Wang, Q.; Zhu, Y.; Taylor, R.A. A Coupled Structural-Optical Analysis of a Novel Umbrella Heliostat. *Sol. Energy* **2022**, *231*, 880–888. [CrossRef]

315. Net Present Cost. Available online: https://www.homerenergy.com/products/pro/docs/latest/net_present_cost.html (accessed on 29 October 2022).

Article

Load Frequency Model Predictive Control of a Large-Scale Multi-Source Power System

Tayma Afaneh, Omar Mohamed * and Wejdan Abu Elhaija

Department of Electrical Engineering, King Abdullah II School of Engineering, Princess Sumaya University for Technology (PSUT), Amman 11941, Jordan
* Correspondence: o.mohamed@psut.edu.jo

Abstract: With increased interests in affordable energy resources, a cleaner environment, and sustainability, more objectives and operational obligations have been introduced to recent power plant control systems. This paper presents a verified load frequency model predictive control (MPC) that aims to satisfy the load demand of three practical generation technologies, which are wind energy systems, clean coal supercritical (SC) power plants, and dual-fuel gas turbines (GTs). Simplified state-space models for the two thermal units were constructed by concepts of subspace identification, whereas the individual wind turbine integration was implicated by the Hammerstein–Wiener (HW) model and then augmented from the output to simulate the effect of a wind farm, assuming similar power harvesting from all turbines in the farm. A practical strategy of control was then suggested, which was as follows: with a changing load demand, the available harvested wind energy must be fully admitted to the network to cover part of the load demand with the free energy, and the resultant load signal will then be instructed to the MPCs designed for the coal and gas units for the coordination of generation. The load signal, after being penetrated by wind, has more transients and faster changes, and needs a more sophisticated control in order to follow the load demand of the flexible coal and gas units. Furthermore, as the level of wind penetration increases, the power system frequency excursions are higher. The simulation results show an acceptable performance for linear MPCs embedded to the GT and coal units, with around a 90 MW share of wind without exceeding the safe restrictions of the plants and allowable reasonable frequency excursions. The complete simulation framework can be used to facilitate wind energy penetration in such power systems and train the operators and future engineers with subsequent power system frequency simulation studies.

Keywords: supercritical power plant; model predictive control; wind farm; gas turbines; load frequency control; optimization

Citation: Afaneh, T.; Mohamed, O.; Abu Elhaija, W. Load Frequency Model Predictive Control of a Large-Scale Multi-Source Power System. *Energies* **2022**, *15*, 9210. https://doi.org/10.3390/en15239210

Academic Editors: Juri Belikov and Abu-Siada Ahmed

Received: 28 October 2022
Accepted: 25 November 2022
Published: 5 December 2022

Publisher's Note: MDPI stays neutral with regard to jurisdictional claims in published maps and institutional affiliations.

1. Introduction

1.1. Aims

The adequacy, security, and stability of power systems are major issues in energy utilities and power engineering research, and have recently received much interest for enhancing power system operations after the integration of considerable levels of renewable energy (RE). The dominant RE technologies are well known to be wind and solar. Despite the fact that solar and wind power capacities are the fastest growing RE technologies, with rates of around 40% and 20%, respectively [1], solar energy is recognized to be easier to model and predict than wind. On the other hand, with the research efforts conducted on clean fossil technologies with the aims of improving energy efficiency and reducing CO_2 emissions, the power system has been integrated with hybrid power generation devices and auxiliaries that have different operational characteristics and, thus, have introduced more challenging objectives to load frequency control (LFC) systems. The usual way to deal with the load frequency control of a two-area power system or multiple generation systems is shown in Figure 1, which is a standard figure that is commonly found in many textbooks for basic control actions carried out for the aim of generation control [2–4].

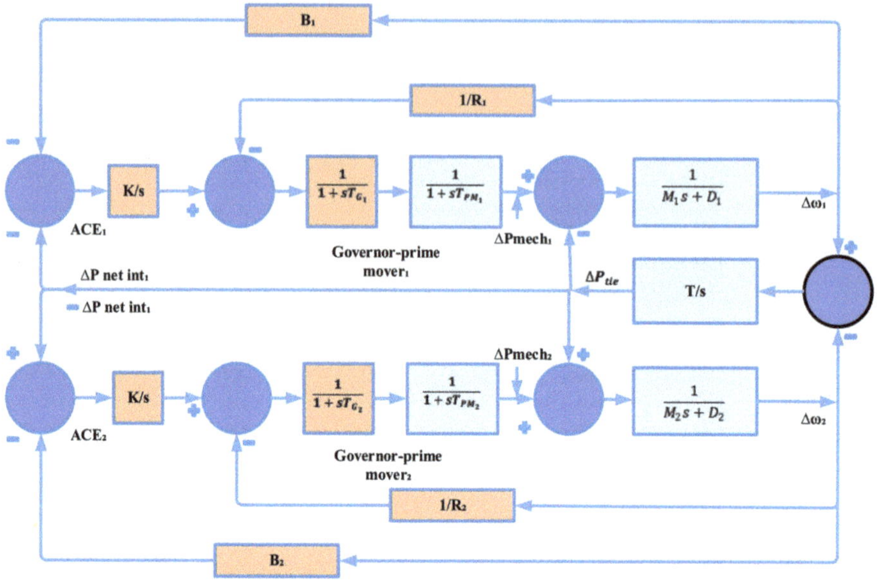

Figure 1. Conventional load frequency control model for two units.

However, it must be emphasized that more sophisticated models and control are needed when the issue is further complicated by the presence of mixed technologies of wind turbines, modern clean coal technologies, and dual-fuel gas turbines. The system is made more realistic using the system identification of state-space models, which can then be embedded with an appropriate MPC strategy in order to modernize the conventional load frequency control.

The one that we suggest in this paper is based on verified state-space models of a real 600 MW supercritical (SC) cleaner power plant, a 250 MW Siemens Dual-Fuel gas turbine GT, and a Hammerstein–Wiener model for a wind farm that is assumed to be composed of 190 wind turbines (TWT-1.65). The data set for every energy resource mentioned was gathered from previous research work of the corresponding author [5–7]. Furthermore, every model has an individual MISO, structured to enhance the accuracy and produce more realistic results. The proposed upgraded system in this paper is shown in Figure 2. The new proposed system with the aforementioned resources would offer better environmental insight while keeping the system secure and stable through a promising MPC strategy. The Hammerstein–Wiener model in the figure was used to predict the harvested wind power to satisfy part of the load demand signal, and the remaining load instructed the MPCs that control the clean coal unit and gas unit to cover the remaining load with minimum fuel. With this insight, the significance of the paper can be stated as follows:

- Facilitating a cleaner and environmentally friendly power system from verified models and simulations of a hybrid power generation system.
- Realizing the importance of mixed generation technology in terms of system strength and environmental effect.
- Investigating the effect of wind penetration on traditional power systems fed by coal power plant and coal/gas stations.
- Determination of the most practically feasible level of wind energy penetration without causing extensive negative effects on the frequency, which is necessary for all stability classes (dynamic and steady-state stabilities).

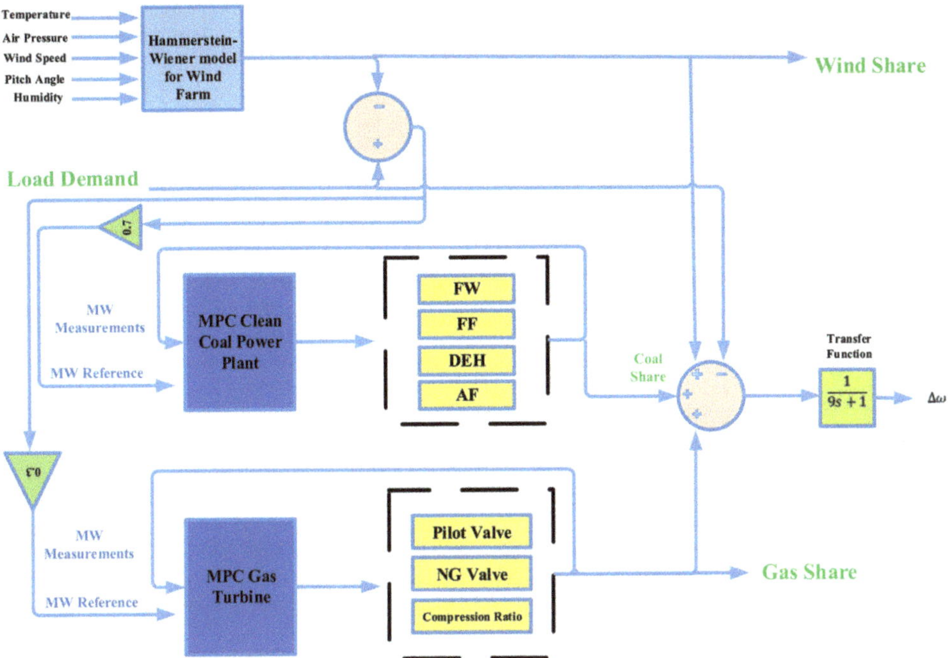

Figure 2. The proposed MPC-based LFC system for the multi-source power system.

1.2. Literature Review and Paper Contributions

The review of literature has focused on modelling and control of multi-source power systems, which are naturally hybrid from the generation side. The review should then cover different features and methodologies of modelling and power systems, with emphasis on system dynamics and control. A wide range of publications were studied in order to have a clear picture of the practical importance of the research area.

Mohamed et al. [8] have reviewed new trends for supercritical and ultra-supercritical power plants with two main categories of models: physical models, and empirical models of SC and USC power plants, which have been simulated using APROS®, APD®, FORTRAN, SimuEngine, GSE, MATLAB®, and Thermolib. Because of the distinct advantages of each method, both methods are unquestionably eligible for modelling SC and USC units. The control system strategies have been also reviewed and the dominant control approach was MPC.

Rehman [9] has presented a study that establishes a reliable coordination between renewable resources in Hybrid Power Systems (HPSs). The most popular scheme has been found to be the Wind-PV systems with and without storage. Wind-PV total capacity has been varied from 1.6 to 120.0 kW, with an average size of 21.0 kW and an average COE of 0.458 US$/kWh. Simulation has been carried out by HOMER and HOMER Pro software for the design and optimization of HPS, in addition to HOGA and MATLAB. The favoured targets of hybrid power systems have been compiled during the survey. However, the system is relatively small and may not be suitable for large-scale loads due to the low energy density of renewables.

Li et al. [10] has implemented a model of a hydrogen storage wind and gas complementary power generation system for energy management. A dynamic mathematical model of a micro gas turbine has been combined with an aerodynamic model of a wind turbine. The alkaline electrolyser model has developed the economic and environmental cost of the system as an objective function, and establishes the capacity optimization model of the wind–gas complementary power generation system. The results have been compared with variable loads in a specific area during the northern winter and it has been concluded that the wind–gas complementary system can boost energy use while lowering wind curtailment.

Tsoutsanis [11] has implemented a generic model from MATLAB/SIMULINK for hybrid power plants, consisting of 14 wind turbines and a gas turbine. The author has investigated the performance and challenges of hybrid power plants, has studied different scenarios, and has compared the dynamic response of a hybrid power plant and a double gas turbine power plant for 10 h, which have shown improvements in terms of emissions and fuel consumption.

Barelli et al. [12] have presented an MPC strategy and applied it to a hybrid system that is structured by a solid oxide fuel cell (SOFC) and a recuperative GT for following a typical daily load demand in microgrids. The simulation has been carried out in MATLAB® and a Simulink environment, and has shown a promising system performance with high average efficiency (approximately 54.5%) and load changes with quick response .

Di Gaeta et al. [13] have shown a dynamic model of a 100 kW commercial micro gas turbine (MGT) fed with a mixture of standard and alternative fuels and the MGT emulated dynamics have been based on first order differential equations with different operating conditions. Customized genetic algorithms (GAs) and non-linear east squares (NLS) direct search methods have been used. The simulations have been carried out via the Dormand–Prince integration solver with a time step of 0.25 s on a MATLAB/SIMULINK environment.

Bizon [14] has proposed proton exchange membrane fuel dell (FC) HPSs to study optimization, switching function, and balance of DC power flow. The model has been built and simulated using MATLAB/SIMULINK. The study has shown the classification of seven FC HPSs with the batteries operated in charge-sustaining mode in order to notice the decrease in size, maintenance, and lifetime. The best design in each category has been chosen to focus on in the experiment, obtaining a 99.56% conversion of hydrogen.

Bensaber et al. [15] have proposed HPSs using PV, wind turbines, and batteries to meet load demand and control the voltage and frequency. The challenges of the proposed HPS in terms of controlling and managing the power flow have been discussed.

Avagianos et al. [16] have published a review of the solid-fuel flexible thermal power plants, discussing the effect of the penetration of non-dispatchable power plants upon wind and solar, for example, and analysing flexible operation requirements that influence their economic viability: fast start-up process, low minimum load, and ramping.

Zhang et al. [17] have proposed a fuzzy model predictive controller (MPC) for wide-range load tracking. The researchers have assembled an extended state observer for the plant behaviour and the MPC with unknown disturbances. The research has been carried out to simulate the extended state observer-based fuzzy model predictive control (ESOFMC) on a 1000 MW ultra-supercritical (USC) boiler–turbine unit. The final results represent the performance of load following over a wide range. However, despite the detailed consideration of the USC controller, no renewable penetrations have been practically considered.

Annaluru [18] has studied the flexibility of the Indian power grid and its challenges in terms of Renewable Energy (RE) with different variable generation (VG) penetration, such as solar PV and wind. The author has studied the Indian grid's net load curve, with emphasis on three potential penetration levels for RE scenarios. Improving the plant flexibility has been presented with three alternative solutions.

Qatamin et al. [7] have proposed System Identification (SI) techniques for prediction of wind turbine output power using a generalized state-space model. The method uses subspace methods mainly based on a numerical algorithm for subspace state space system sdentification (N4SID) and prediction error method (PEM), with multi input single output (MISO) to investigate the differences in computational capabilities. It has been noticed that there is a considerable relation between model order and simulation accuracy, which motivates the study of optimal order that achieves the maximum possible accuracy. A comparison between PEM and the N4SID has been carried out in terms of the accuracy, complexity, and speed of simulation. However, the research lacks extension of LFC studies.

Ersdal et al. [19] have proposed a model predictive controller for automatic generator control (AGC) of the Nordic power system, based on a simplified system model. The study considers the limitations of different parts of the power system, such as tie-line power flow, capacity, and rate of change in generation, which lead to more flexibility and coordination between multiple inputs. The Kalman filter has been integrated with the MPC algorithm for load frequency control with state estimation. The research finally has given a comparison between MPC and the conventional (LFC/AGC) with proportional-integral (PI) controllers.

Shakibjoo et al. [20] have proposed a new load frequency control (LFC)/type-2 fuzzy control (T2FLC) with multi-areas, thermal units, wind farms, demand response (DR), and a battery energy storage system (BESS). The proposed linear model has been implemented using MATLAB software with four simulation scenarios of the 10-machine New England 39-bus test system (NETS-39b) to obtain an approximately 20% improvement.

From the literature analysis, it has been deduced that the load frequency control still needs further study that should be preferably closer to the practical sense, which can be applicable—when normalized—for conventional centralized power systems or embedded generation. The paper contributions can be then clarified as follows:

- A cleaner power system framework has been presented from verified simulators of three different generating units, which has been built with the aim of satisfying system stability and environmental requirements. These units are a clean coal supercritical unit, a dual-fuel gas turbine unit, and a wind farm.
- A Hammerstein–Wiener Model has been built to predict the wind penetration level according to weather data inputs. The typical example has given many levels of approximately 90 MW penetration peak. On the other hand, the other two flexible units are emulated with identified state-space models. The parameters of all models have been identified and verified with real sets of data, unlike the previously published literature of load frequency control studies that usually adopt standard transfer functions and form a block diagram without enough attention to parameter calibration to fit a real set of data.
- A simplified and practically feasible operation and control strategy has been suggested for heavily loaded power systems: the wind farm output should be admitted to the network from economic and environmental view-points, the remaining load signal, heavily affected by the randomness of the wind, is instructed to the MPC controllers of the flexible generators (coal and gas), and the wind farm is relieved from duty of load frequency control because this obligation is completely carried out by the flexible units and their associated controls, which in turn, reduces the storage technology requirements of the wind turbines where that storage would be needed only in case of curtailed or low load.

The paper has been organized as follows: Section 2 presents an overview on state-space and HW identification algorithms; Section 3 shows the general mechanism of MPC through its mathematical description with its application to the coal and gas units; Section 4 depicts the simulations of the verified models and of the control performance analysis; and Section 5 concludes the research work and states the future suggestions.

2. An Overview of State-Space and Hammerstein-Wiener Systems' Identification

2.1. Theory and Basic Mechanism for State-Space Subspace Identification

In system identification, an informative set of data must be available in advance to setup the optimized parameters of a new model or calibration of existing model parameters. Once the identification phase has been finished, the identified model should be checked via a different data set in order to ensure the model validity and integrity [21]. In this paper, every set of data has been split to have 50% of the samples for identification and the other 50% of data samples for verification. The approach of identification that has been adopted in this paper is the subspace N4SID algorithm. The general following explanation is applicable to any subspace algorithm, which have been published in many references [22–24]. Using the following linear, discrete time-invariant state-space model (SSM):

$$x_{k+1} = Ax_k + Bu_k + w_k \tag{1}$$

$$y_k = Cx_k + Du_k + v_k \tag{2}$$

The assumed pair (C, A) is observable and the pair (A, B) is controllable. Where $x_k \in R^n$ is the state vector, $u_k \in R^r$ is the input vector, $y_k \in R^m$ is the output vector, and w_k, v_k are zero mean white Gaussian noise. The most used tool to be highlighted is the singular value decomposition (SVD), which is used to produce the required state-space matrices. However, some concepts and preliminaries must be first explained, as in the following subsections.

2.1.1. Inputs and Outputs

Arranged inputs and outputs data as a Hankel matrix, shown in extended data vectors and extended data matrices:

The extended data vectors, given a number L refers to known output vectors and a number $L + g$ refers to known input vectors, are given as:

$$y_{k|L} \stackrel{\text{def}}{=} \begin{bmatrix} y_k \\ y_{k+1} \\ \vdots \\ y_{k+L-1} \end{bmatrix} \in R^{Lm} \tag{3}$$

$$u_{k|L+g} \stackrel{\text{def}}{=} \begin{bmatrix} u_k \\ u_{k+1} \\ \vdots \\ u_{k+L+g-2} \\ u_{k+L+g-1} \end{bmatrix} \in R^{(L+g)m} \tag{4}$$

The extended data matrices define the following output data matrix with L block rows and K columns:

$$Y_{k|L} \stackrel{\text{def}}{=} \begin{bmatrix} y_k & y_{k+1} & y_{k+2} & \cdots & y_{k+K-1} \\ y_{k+1} & y_{k+2} & y_{k+3} & \cdots & y_{k+K} \\ \vdots & \vdots & \vdots & \ddots & \vdots \\ y_{k+L-1} & y_{k+L} & y_{k+L+1} & \cdots & y_{k+L+K-2} \end{bmatrix} \in R^{Lm \times K} \tag{5}$$

This is the known data matrix of output variables, and the matrix:

$$U_{k|L+g} \stackrel{\text{def}}{=} \begin{bmatrix} u_k & u_{k+1} & u_{k+2} & \cdots & u_{k+K-1} \\ u_{k+1} & u_{k+2} & u_{k+3} & \cdots & u_{k+K} \\ \vdots & \vdots & \vdots & \ddots & \vdots \\ u_{k+L+g-2} & u_{k+L+g-1} & u_{k+L+g} & \cdots & u_{k+L+K+g-3} \\ u_{k+L+g-1} & u_{k+L+g} & u_{k+L+g+1} & \cdots & u_{k+L+K+g-2} \end{bmatrix} \in R^{(L+g)r \times K} \tag{6}$$

is the known data matrix of input variables.

The reversed extended observability matrix (O_i) are defined as the following definitions according to SSM, Equations (1) and (2). And the number of block rows is denoted by subscript i.

$$O_i \overset{\text{def}}{=} \begin{bmatrix} C \\ CA \\ \vdots \\ CA^{i-1} \end{bmatrix} \in R^{im \times n} \tag{7}$$

and a reversed extended controllability matrix (C_i^d) for pair (A, B):

$$C_i^d \overset{\text{def}}{=} \begin{bmatrix} A^{i-1}B & A^{i-2}B & \cdots & B \end{bmatrix} \in R^{n \times ir} \tag{8}$$

It is important to mention that there are many applications for observability and controllability matrices, such as monitoring lithium-ion battery [25]. However, the application here is somehow different as it focuses on building the matrices necessary for off-line identification of linearized discrete time models of the flexible generating units, and the decomposition of those matrices in the final stage of identification.

The triangular Toeplitz matrix (H_i^d): The columns result from multiplying between the extended observability matrix and the transposed extended controllability matrix.

This matrix contains the four matrices (C, A, B, D).

The triangular Toeplitz matrix (H_i^d).

$$H_i^d \overset{\text{def}}{=} \begin{bmatrix} D & 0 & \cdots & 0 \\ CB & D & \cdots & 0 \\ \vdots & \vdots & \ddots & \vdots \\ CA^{i-2}B & CA^{i-3}B & \cdots & D \end{bmatrix} \in \mathbb{R}^{im \times (i+g-1)} \tag{9}$$

The number of block rows is denoted by subscript i. The block columns are denoted by $i + g - 1$.

The lower block stochastic triangular Toeplitz matrix $\mathcal{H}_i^s$ for the pair (C, A).

Define the Toeplitz matrix (H_i^s)

$$H_i^s \overset{\text{def}}{=} \begin{bmatrix} 0 & 0 & \cdots & 0 \\ C & 0 & \cdots & 0 \\ \vdots & \vdots & \ddots & \vdots \\ CA^{i-2} & CA^{i-3} & \cdots & 0 \end{bmatrix} \in \mathbb{R}^{im \times il} \tag{10}$$

Finally, X_i is the state vector is denoted as:

$$X_i = \begin{bmatrix} x_i & x_{i+1} & \cdots & x_{i+j-1} \end{bmatrix} \tag{11}$$

The main target here is to recover the hidden state x_k and if this hidden state is recovered, by using singular value decomposition (SVD), the X_i and the system order n can be recovered and known.

The basic realization theory of subspace identification as a part of the deterministic case is explained briefly in the next subsection.

2.1.2. Realization Theory

Realization theory is discussed by defining the impulse response model: A linear state-space model $x_{x+1} = Ax_k + Bu_k$ and $y_k = Cx_k + Du_k$ with the initial state x_0. The impulse response model can be given as:

$$y_k = CA^k x_0 + \sum_{i=1}^{k} CA^{k-i} Bu_{i-1} + Du_k \tag{12}$$

So the matrix at time instant $k - i + 1$ is as follows:

$$H_{k-i+1} = CA^{k-i}B \quad \in \mathbb{R}^{ny \times nu},\tag{13}$$

The impulse response matrix at time is $k - i + 1$. Then, the output, y_k, at time k is defined in terms of impulse response matrices as following: $H_1 = CB$, $H_2 = CAB$, and so on $\ldots$, $H_k = CA^{k-1}B$.

The impulse responses $H_i \ \forall \ i = 1, \ldots, L+J.$, can be used to develop the Hankel matrices that contain necessary information of the system matrices.

The Hankel matrices

$$H = \begin{bmatrix} H_1 & H_2 & H_3 & \cdots & H_J \\ H_2 & H_3 & H_4 & \cdots & H_{J+1} \\ H_3 & H_4 & H_5 & \ddots & H_{J+2} \\ \vdots & \vdots & \ddots & \ddots & \vdots \\ H_{L+1} & H_{L+2} & \cdots & \cdots & H_{L+J} \end{bmatrix} \in R^{ny(L+1) \times nu.J}\tag{14}$$

the submatrices H_A, H_B, H_C can be then extracted [22].

$$H_{1|L} = H_n = \begin{bmatrix} H_1 & H_2 & H_3 & \cdots & H_J \\ H_2 & H_3 & H_4 & \cdots & H_{J+1} \\ H_3 & H_4 & H_5 & \ddots & H_{J+2} \\ \vdots & \vdots & \ddots & \ddots & \vdots \\ H_{L+1} & H_{L+2} & \cdots & \cdots & H_{L+J-1} \end{bmatrix} \in R^{ny.L \ \times nu.J}\tag{15}$$

$$H_{2|L} = H_A = \begin{bmatrix} H_2 & H_3 & H_3 & \cdots & H_{J+1} \\ H_3 & H_4 & H_4 & \cdots & H_{J+2} \\ H_4 & H_5 & H_5 & \ddots & H_{J+3} \\ \vdots & \vdots & \ddots & \ddots & \vdots \\ H_{L+1} & H_{L+2} & \cdots & \cdots & H_{L+J} \end{bmatrix} \in R^{ny.L \ \times nu.J}\tag{16}$$

$$H_B = \begin{bmatrix} H_1 \\ H_2 \\ H_3 \\ \vdots \\ H_L \end{bmatrix} \in R^{ny.L \times nu}\tag{17}$$

$$H_C = \begin{bmatrix} H_1 & H_2 & H_3 & \cdots & H_J \end{bmatrix} \in R^{ny \times nu.J}\tag{18}$$

These matrices are closely related to the observability and the controllability matrices (O_L, C_J), respectively, which are given to facilitate of realization theory, where O_L is the observability matrix defined by:

$$O_L = \begin{bmatrix} C \\ CA \\ CA^2 \\ \vdots \\ CA^{L-1} \end{bmatrix} \in R^{ny.L \ \times nx}\tag{19}$$

and C_J is the controllability matrix defined by:

$$C_J = \begin{bmatrix} B & AB & \cdots & A^{J-1}B \end{bmatrix} \in R^{nx \times nu.J}\tag{20}$$

a suitable factorization of the Hankel matrix, $H_n = H_{1|L}$, may be used to determine O_L, C_J. Then the system matrices to satisfy the following three matrix relations are chosen.

$$H_{1|L} = H_A = O_L A C_J, \; H_B = O_L B, \; H_C = C C_J \tag{21}$$

Conclude the system matrices as follows [22]:

$$A = \left(O_L^T O_L\right)^{-1} O_L^T H_A C_J^T \left(C_J C_J^T\right)^{-1} \tag{22}$$

$$B = \left(O_L^T O_L\right)^{-1} O_L^T H_B \tag{23}$$

$$C = H_C C_J^T \left(C_J C_J^T\right)^{-1} \tag{24}$$

The realization of the parameter of matrices is obtained from SVD.

In linear algebra, the singular value decomposition (SVD) is defined as follows [26], in an analysis of real or complex matrices. Now we can apply SVD of the finite block to Hankel matrix H_n.

$$H_{1|L} = H_n = O_L C_J = U S V^T = U S_1 S_2 V^T \tag{25}$$

Results are directly obtained from the SVD when applying factorization of the Hankel matrix into the product of the observability and controllability matrices. For example, the internally balanced:

$$O_L = U S_1, \; C_J = S_2 V^T \tag{26}$$

Then the system matrices are estimated as follows and defined as internally balanced:

$$\begin{aligned} A &= S_1^{-T} U^T H_A V S_2^{-T} \\ B &= S_1^{-T} U^T H_B \\ C &= H_C V S_2^{-T} \end{aligned} \tag{27}$$

The different subspace identification algorithms are discussed originally in [23]. Since we have used the subspace method as tool identification, any subspace algorithm can be used to fulfil the modeling aims in the paper. N4SID has been selected to identify the models of SCPP and GT. The next section discusses the basic theory of MPC.

2.2. The Hammerstein-Wiener Model

The Hammerstein–Wiener (HW) model is introduced here as nonlinear component in multisource load frequency control schemes. The HW model in this paper is used to predict the harvested wind power from the individual wind turbine given. The inputs of temperature, air pressure, wind speed, pitch angle, and humidity, are admitted to an input nonlinearity block, to a linear transfer function, then to the output nonlinearity block to form the well-known Hammerstein–Wiener model. The rule of this component is to accurately predict the wind share, although there is no reason to prevent it from being used for control system design. However, the wind subsystem is left as open loop in the proposed model for the aforementioned reasons in the paper contributions. Therefore, several benefits of the Hammerstein–Wiener models are mentioned, such as achieving higher accuracy by simplifying the control process by allowing for the checking of each parameter separately. As stated earlier in the paper, the Hammerstein–Wiener model has been built to ensure the integrity of the input parameters of the wind model, but this shall not affect the linear MPC performance of the thermal units.

The Hammerstein–Wiener model is mostly used when the system output depends nonlinearly on its inputs. The input–output relationship, in this case, could be decomposed into two interconnected elements. The linear transfer function is used to represent the system's dynamics, and the nonlinear functions are used to capture nonlinearities for the inputs and outputs of the linear system [27]. The standard block diagram of the Hammerstein–Wiener model component is shown in Figure 3.

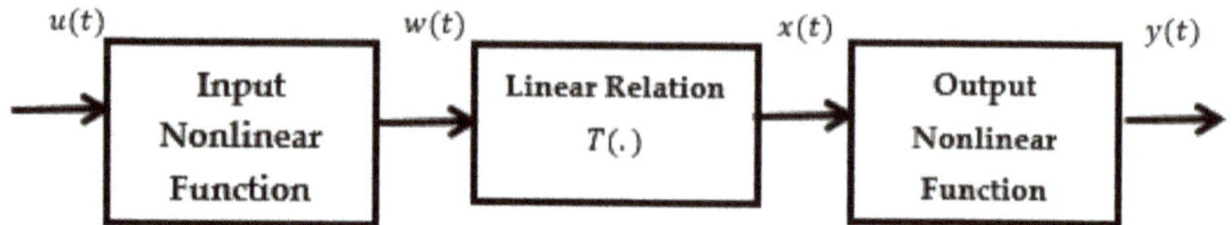

Figure 3. Generalized structure of HW model.

Where:

- $\omega(t) = f(u(t))$ is a nonlinear function for transforming input data $u(t)$.
- $x(t) = T(.).w(t)$ and $y(t)$ have the same dimensions, $x(t)$ has a linear relation with $w(t)$; where $w(t)$ and $x(t)$ are the input and output of the linear block, respectively, and are defined as internal variables.
- $y(t) = h(x(t))$ is a nonlinear function that maps the output of the linear block $x(t)$.

Various scalar nonlinearity estimators are provided by MATLAB® through the system identification toolbox. Hammerstein–Wiener models have nonlinearity estimators for both input nonlinearity f and output nonlinearity h. The nonlinearity estimators could be configured as a saturation, dead zone, unit gain, wavelet network, sigmoid network, piecewise linear function, or other options. In this paper, the piecewise functions for input and output nonlinearities are configured. Finally, an iterative search algorithm has been used with appropriate settings to estimate the HW model of the wind turbine. The individual turbine output is then augmented through a linear gain to formulate the effect of a huge wind farm of 190 turbines.

3. The General Mechanism of Model Predictive Control and Its Application to the Flexible Units

The model predictive control (MPC) is considered as one of the well-recognized control technologies for industrial process applications, especially in thermal power generation plants, used to calculate the predicted plant output after simulation and to control the output of SCPP. There are two reasons to utilize these advanced control techniques. Firstly, there are several variety control topologies in the MPC model. Secondly, there is the ability to use wide operational constraints in the MPC model to ensure the safety and reliability of the power plant [28].

The action of the MPC can be summarized as follows:

1. Predict the output trends.
2. Perform optimization to reduce the errors.
3. Calculate the control shifts or changes in inputs, then execute only the first sample of control inputs.
4. Repeat these steps every sampling time.

In the section "Control Strategy Testing with MPC", a validated state-space mathematical model was used for the 600-MW cleaner coal power plant and 250-MW gas turbine unit, which represents the production of a supercritical coal-fired power plant and gas power plant. The linear state-space model was identified as the control strategy. The discrete-time state-space model could be used for prediction as follows [29]:

3.1. Prediction of State and Output Variables

The future control trajectory (predicted niput) is represented by:
Assume the initial condition u equal zero.

$$\Delta u(k_i), \ \Delta u(k_i + 1), \ldots, \ \Delta u(k_i + N_C - 1). \tag{28}$$

The future state variables are represented by:

$$x(k_i + 1|\,k_i), x(k_i + 2|\,k_i), \ldots,\; x(k_i + m|\,k_i), \ldots, x\big(k_i + N_p\big|\,k_i\big) \tag{29}$$

where $k_i > 0$ is the sampling instant, N_C is the control horizon, and N_p is the prediction horizon. Also referring to the length of the optimization, N_C should be less than N_p. Prediction horizon is the length of time for the predicted future output, whereas the control horizon represents the length of time for the optimized trajectory of the manipulated inputs.

The future state variables (means future output), based on the state-space model (A, B, C), are represented as:

$$x(k_i + 1|\,k_i) = Ax(k_i) + B\Delta u(k_i)$$
$$x(k_i + 2|\,k_i) = Ax(k_i + 1|\,k_i) + B\Delta u(k_i + 1)$$
$$= A^2 x(k_i) + AB\Delta u(k_i) + B\Delta u(k_i + 1)$$
$$\vdots$$
$$x\big(k_i + N_p\big|\,k_i\big) = A^{N_p} x(k_i) + A^{N_p-1} B\Delta u(k_i) + A^{N_p-2} B\Delta u(k_i + 1)$$
$$+ \ldots + A^{N_p - N_C} B\Delta u(k_i + N_C - 1).$$

Predict output variables (power) from predicted state variables directly after assume $D = 0$:

$$y(k_i + 1|\,k_i) = CAx(k_i) + CB\Delta u(k_i)$$
$$y(k_i + 2|\,k_i) = CA^2 x(k_i) + CAB\Delta u(k_i) + CB\Delta u(k_i + 1)$$
$$\vdots$$
$$y\big(k_i + N_p\big|\,k_i\big) = CA^{N_p} x(k_i) + CA^{N_p-1} B\Delta u(k_i) + CA^{N_p-2} B\Delta u(k_i + 1)$$
$$+ \ldots + CA^{N_p - N_C} B\Delta u(k_i + N_C - 1).$$

Then arrange input and output in vectors:

$$Y = \big[y(k_i + 1|\,k_i)\; y(k_i + 2|\,k_i) y(k_i + 3|\,k_i) \ldots y\big(k_i + N_p\big|\,k_i\big)\big]^T \tag{30}$$

$$U = [\Delta u(k_i)\; \Delta u(k_i + 1)\; \Delta u(k_i + 2)\; \ldots\; \Delta u(k_i + N_C - 1)]^T \tag{31}$$

Collect the two vectors Y and U together in a compact matrix form as:

$$Y = Fx(k_i) + \varphi \Delta U \tag{32}$$

where:

$$F = \begin{bmatrix} CA \\ CA^2 \\ CA^3 \\ \vdots \\ CA^{N_p} \end{bmatrix}; \varphi = \begin{pmatrix} CB & 0 & 0 & \cdots & 0 \\ CAB & CB & 0 & \cdots & 0 \\ CA^2 & CAB & CB & \cdots & 0 \\ \vdots & \vdots & \vdots & \cdots & \vdots \\ CA^{N_p-1}B & CA^{N_p-2}B & CA^{N_p-3}B & \cdots & CA^{N_p-N_C}B \end{pmatrix} \tag{33}$$

3.2. Optimization of Control Signals

The very basic case is to find the first derivative of the cost function with respect to ΔU and obtain the vector sequence ΔU as the future control law.

- Set data vector that contains the set-point:

$$R_s^T = [1\,1\,\ldots 1]r(k_i) \tag{34}$$

Define cost function J as following:

$$J = (R_s - Y)^T (R_s - Y) + \Delta U^T \overline{R} \Delta U \tag{35}$$

where $\overline{R}$ is a diagonal matrix and equal $r_w\, I_{N_c \times N_c}$; r_w is used as a tuning parameter for the MPC, and they mainly affect the performance of the controller and computation time demands, the aim would be solely to make the error as small as possible.

- Find the optimal ΔU that will minimize J, write J as following:

$$J = (R_s - Fx(k_i))^T (R_s - Fx(k_i)) - 2\Delta U^T \varphi^T (R_s - Fx(k_i)) + \Delta U^T \left(\varphi^T \varphi + \overline{R}\right)\Delta U$$

- Take the first derivative of the J and to obtain the minimum J put condition $\frac{\partial J}{\partial \Delta U}$ equal to zero:

$$\frac{\partial J}{\partial \Delta U} = -2\varphi^T (R_s - Fx(k_i)) + 2\left(\varphi^T \varphi + \overline{R}\right)\Delta U$$

- The optimal solution for the control signal:

$$\Delta U = \left(\varphi^T \varphi + \overline{R}\right)^{-1} \varphi^T (R_s - Fx(k_i)) \tag{36}$$

Applying the aforementioned concepts to the hybrid system, the advancement in control strategy with the MPC leads the system for the two generating units to be as shown in Figure 4.

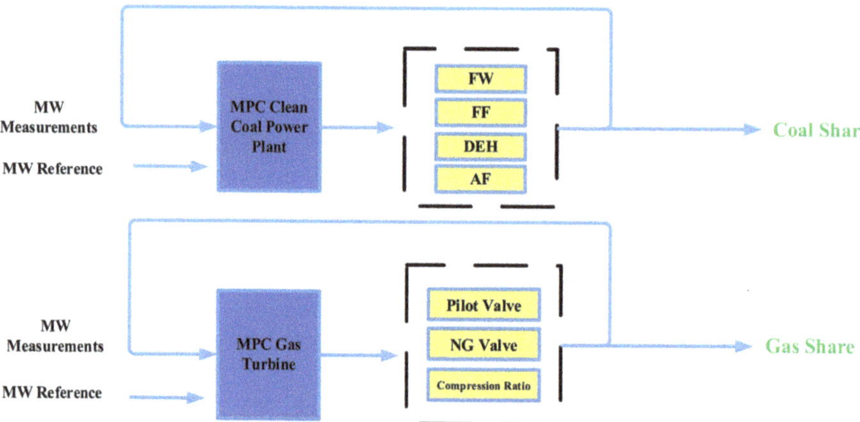

Figure 4. The advancement control strategy with MPC.

The inputs and outputs are adopted with regard to the available operational data and previously published research that indicates those inputs are the most influential variables on the plant performance and characteristics. There are many parameters that affect the performance of the MPC. One of the most effective parameters is the prediction horizon (H_p); in this model, the prediction horizon is reduced to 8 samples for the coal power plant and 6 samples for the gas turbine (GT), which achieves the best performance results. The load demand signal has been assumed to have a sudden increase followed by a sudden decrease in load demand. The choice of these parameters values has to be decided on the basis of many simulation scenarios [30].

The main objective of the advancement control strategy is to integrate 600-MW of clean coal supercritical (SC) power plant and 250-MW of dual-fuel gas turbine (GT) units into the grid to develop a feasible controller for the coal and gas unit for the load following capability, with integration of considerable wind energy penetration. The global model will be used to design predictive controllers for the clean coal supercritical (SC) power plant and dual-fuel gas turbine (GT) units, which allow different levels of wind energy penetration and cover the remaining amount of load that varies over a 12 h time window.

In summary of this section, the MPC algorithm computes the plant inputs so that the plant output follows the adopted reference signal. It utilizes an internal plant model to forecast the future production, uses optimization so that the plant's output tracks the instructed reference, and calculates many steps in the optimum sequence. However, the MPC applies the first sample and ignores the others. Then, the prediction horizon is shifted forward one-time step, and the process of computation should be then repeated. The following section discusses the simulation results for testing the applied MPC strategy.

4. Simulation Results

4.1. Modeling Identification Results

For successful control system implementation, verified models must be available at earlier stage. This sub-section briefly reports the models' features for the three resources under study, which are the wind, the coal, and the gas power generators, respectively. Because the wind resources embed higher variability than coal and gas, it is preferable to be simulated by nonlinear system identification, whereas the coal and gas tend to be more deterministic and could be simply described by linear system identification. Another reason that supports this argument is that the classical MPC needs linear models to act as an internal model for prediction of the MPC application of the coal and gas units. In this paper, the wind generator should be integrated into the grid such that it freely injects the harvested power to the network to cover the highest possible portion of the load with free energy. Therefore, the LFC should be the duty of the coal and gas units to attain the required balance. The designed MPCs have been applied to the coal and gas units for the purpose of LFC [31,32].

Regarding the wind model, because the wind is left uncontrollable, any model structure, no matter how overcomplicated, could be chosen in order to capture the practical dynamics with reasonable accuracy. The Hammerstein-Wiener (HW) model structure has been selected for the prediction of the harvested power of the wind farm. The HW model for the wind turbine generator system has been tuned and the parameters obtained are mentioned in Table 1.

Table 1. The HW model description for the wind generator.

Input Nonlinearity Function: Type and Features	The Transfer Function: No, of Poles and Zeros	The Output Nonlinearity Function: Type and Feature
Piecewise function with 30 break-points for all inputs.	Data No. of poles = 13 No. of zeros = 2 For all Functions	Piecewise function with 30 break-points for all inputs.

The implemented models of the flexible generators are two multi-input single-output (MISO) discrete state-space mathematical models. The first one is for SCPP and has four inputs, U1, U2, U3, and U4: the air flow (AF), fuel flow (FF), feed water flow (FW), and digital electro-hydraulic (DEH) governor reference, respectively. It has one output, y, the SCPP output power. In addition, the SSM for GT has three inputs, Ug1, Ug2, and Ug3, for the natural gas valve, the pilot valve, and the compression ratio, respectively, and one output y, which is the GT output power. The parameters of the matrices A, B, C, and D have been identified to be as follows:

The parameters of the matrices for clean coal SCPP are as follows:

$$
A = \begin{bmatrix}
0 & 1 & 0 & 0 & 0 & 0 & 0 & 0 \\
0 & 0 & 1 & 0 & 0 & 0 & 0 & 0 \\
0 & 0 & 0 & 1 & 0 & 0 & 0 & 0 \\
0 & 0 & 0 & 0 & 1 & 0 & 0 & 0 \\
0 & 0 & 0 & 0 & 0 & 1 & 0 & 0 \\
0 & 0 & 0 & 0 & 0 & 0 & 1 & 0 \\
0 & 0 & 0 & 0 & 0 & 0 & 0 & 1 \\
-0.2513 & 1.182 & -2.696 & 4.288 & -5.5 & 5.681 & -4.813 & 3.108
\end{bmatrix}
$$

$$B = \begin{bmatrix} 0.2286 & 0.324 & -0.02447 & -0.08589 \\ -0.06089 & 0.06483 & 0.09362 & 0.006508 \\ -0.01787 & -0.3207 & 0.1021 & 0.02841 \\ 0.02413 & 0.3035 & -0.05365 & -0.02026 \\ -0.03907 & 0.1484 & -0.1002 & 0.004572 \\ -0.01455 & -0.2376 & 0.03089 & 0.01863 \\ 0.00596 & 0.2219 & 0.07073 & -0.01692 \\ -0.04258 & 0.2435 & -0.01026 & 0.000967 \end{bmatrix}$$

$$C = \begin{bmatrix} 1 & 0 & 0 & 0 & 0 & 0 & 0 & 0 \end{bmatrix}$$

$$D = \begin{bmatrix} 0 & 0 & 0 & 0 \end{bmatrix}$$

Also, the parameters of the matrices for GT are as follows:

$$A = \begin{bmatrix} -0.01395 & -0.8446 & 0.4796 & 0.07193 \\ 0.6643 & 0.4801 & 0.1731 & -0.0450 \\ -0.06218 & 0.2890 & 0.4074 & -0.8663 \\ 0.05671 & 0.07252 & 0.1488 & 0.4045 \end{bmatrix}$$

$$B = \begin{bmatrix} 0.2365 & 0.3046 & -0.1499 \\ 0.2378 & 0.1942 & -0.7624 \\ -0.169 & -0.03077 & -0.4752 \\ 0.05271 & 0.04153 & -0.1551 \end{bmatrix}$$

$$C = \begin{bmatrix} -10.8 & -35.95 & -25.41 & 15.54 \end{bmatrix}$$

$$D = \begin{bmatrix} 0 & 0 & 0 \end{bmatrix}$$

Figure 5 shows the performance of an identified model over 24 h of operation of the individual wind turbine generator system. The HW model of the wind energy conversion system follows the main dynamical trends of the real system, and the performance of the model has been quantified to be 80.88% accurate. The system has been augmented to be a typical wind farm by multiplying this output by the typical number of turbines in wind farms.

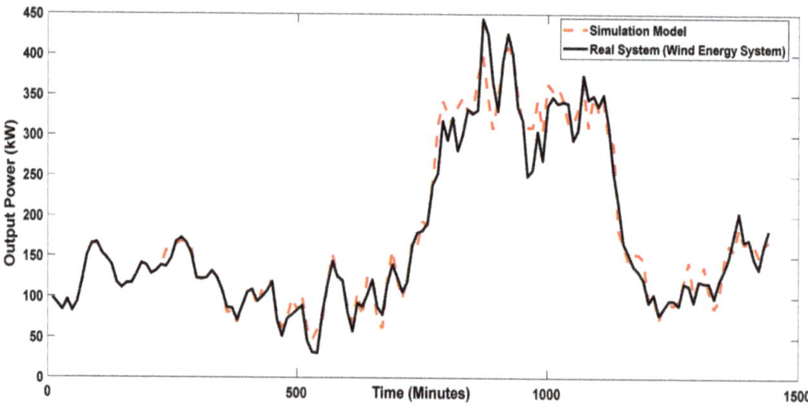

Figure 5. Measured and simulated wind turbine generator output.

The linear models of the SCPP and GT have been identified using subspace identification that has been described in Section 2.1. The simulations have been depicted in Figures 6 and 7. Fourth order models for both processes have reasonably represented the actual behavior of the SCPP and GT, with an accuracy for each model of 92.19% and 82.31%, respectively.

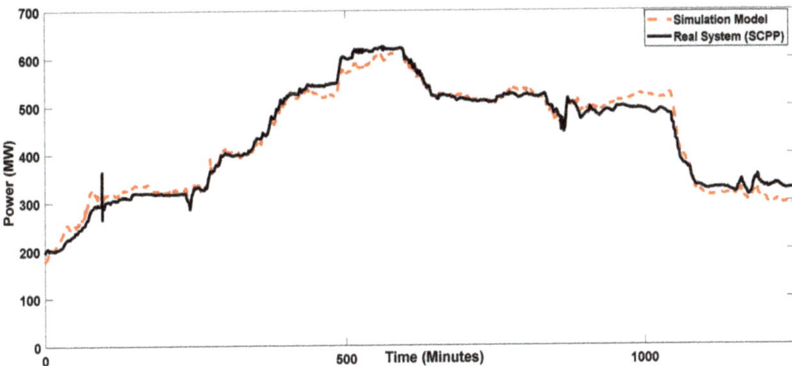

Figure 6. Measured and simulated SCPP output.

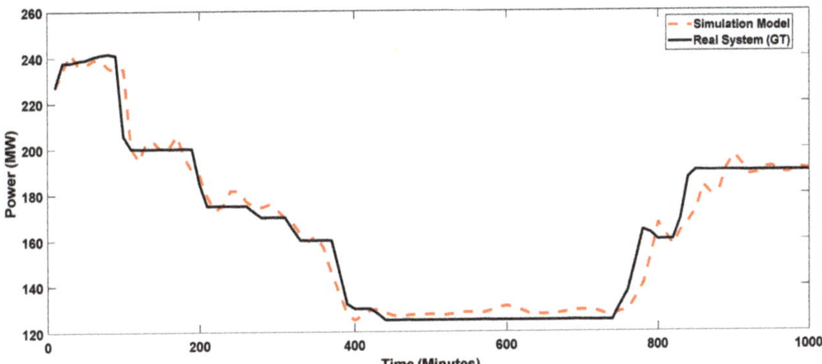

Figure 7. Measured and simulated GT output.

Every data set has been split to be 50% for identification and 50% for verification. The model quality has been ensured for the three sets of data as shown above, in which the first half of the samples are used for identification and the second half are used for verification. However, one must make sure that the overall sample time is unique for the whole system, as the gathered data samples differ from one system to another, which results in unifying the sample time of 5 s for every sample in the LFC simulator.

4.2. Controller Decisions for a Practical Wind Penetration

The proposed control strategy for the 600-MW clean coal supercritical (SC) power plant and 250-MW dual-fuel gas turbine (GT) units has been simulated to enhance the efficiency and reduce the negative effects on the frequency. This necessary for all stability classes (dynamic and steady-state stability) which allow different levels of wind energy penetration and cover the remaining amount of load.

The reason for developing a linear MPC strategy for both a coal and gas power plant, is to regulate them with focus on operational, safety and emissions constraints. In addition, the aim was to design a feasible controller for the coal and gas unit for the load following capability, with focus on prediction to obtain better results. Linear MPC provides a reasonable computational demand compared to the other nonlinear MPC. For this reason, the temperature, pressure, and humidity are not involved as model outputs. However, temperature, air pressure, wind speed, pitch angle and humidity are ensured by developing a Hammerstein–Wiener model, which is used as a safety detector for these five parameters.

The model has been developed and simulated in MATLAB Simulink on a PC environment with the assumed demand signal changes in a step from 640 MW to 600 MW (see Figure 8). Simulation results of Hammerstein–Wiener model for wind farm is reported and shows the maximum output of around 90 MW as follows in Figure 9. The advancement control strategy with MPC is simulated, and the results are reported in Figures below.

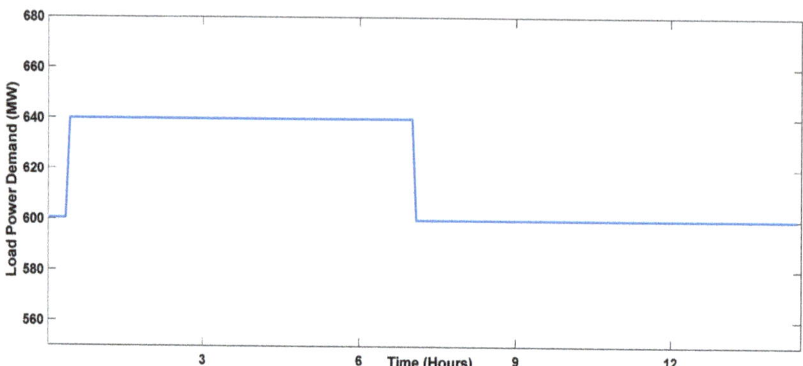

Figure 8. Load power demand.

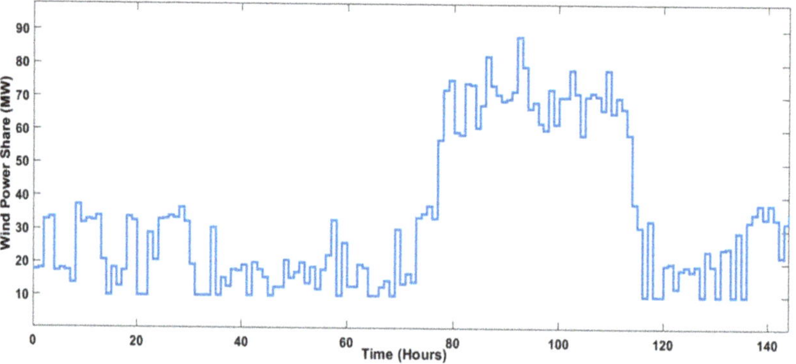

Figure 9. Simulated output power of the wind farm for practical inputs.

The Hammerstein–Wiener model has been built to control five parameters as an input to get single output with different penetration. This output is used as a reference for coal and gas units. Figures 10 and 11, respectively, show that the manipulated inputs of the plant, the air flow (AF), fuel flow (FF), feed water flow (FW), and digital electro-hydraulic (DEH) governor signal are maintained within their constraints, are restricted by the 600 MW SCPP. Figure 11 shows the output power at the maximum of approximately 440 MW.

As stated earlier in the paper, the Hammerstein–Wiener model has been built to ensure the integrity of the input parameters of the wind model, but this shall not affect the linear MPC performance of the thermal units for two reasons. Firstly, the linear MPC is preferable in practice due to its simplicity and lower computational burdens. Secondly, because the performance is tested on linear models for the thermal plants, for practical implementations, any glitches can be compensated by the testing of the parameters and calibration. After a wide range of experiments and trials, the parameters of the Hammerstein–Wiener models have been determined and selected with the most satisfactory results.

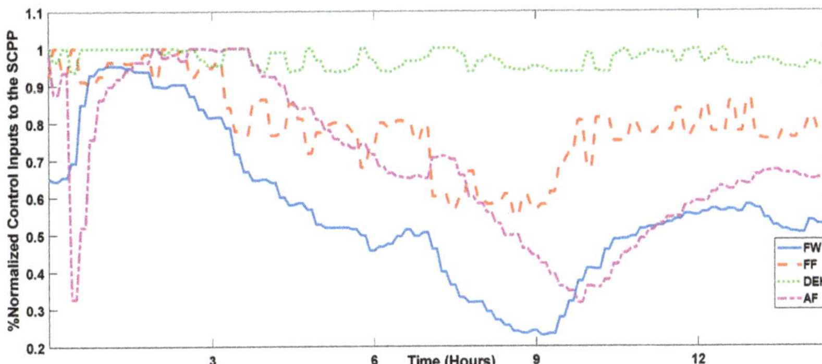

Figure 10. Normalized clean coal power plant control inputs to the SCPP.

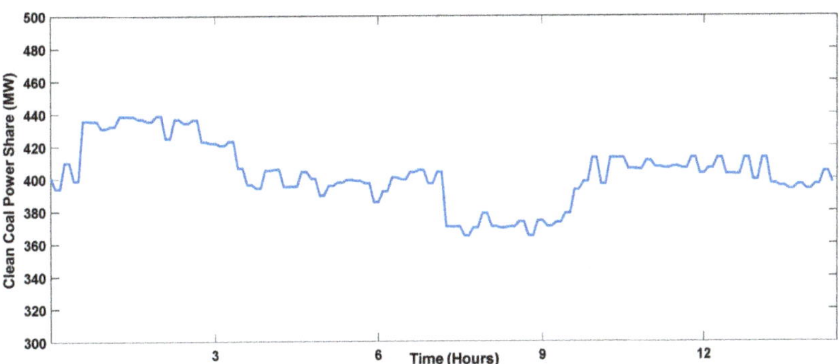

Figure 11. Output power share of the clean coal SC power plant.

Figures 12 and 13, respectively, show the manipulated inputs and the output shared by the GT plant, where the inputs are the pilot valve, natural gas valve (NGCTRL), and the compression ratio (COMPPO), are maintained within their constraints, those restricted by the 250-MW dual-fuel gas turbine (GT).

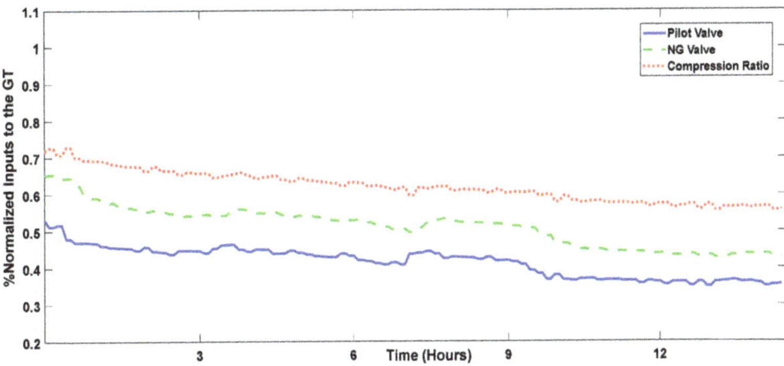

Figure 12. Normalized inputs to the GT.

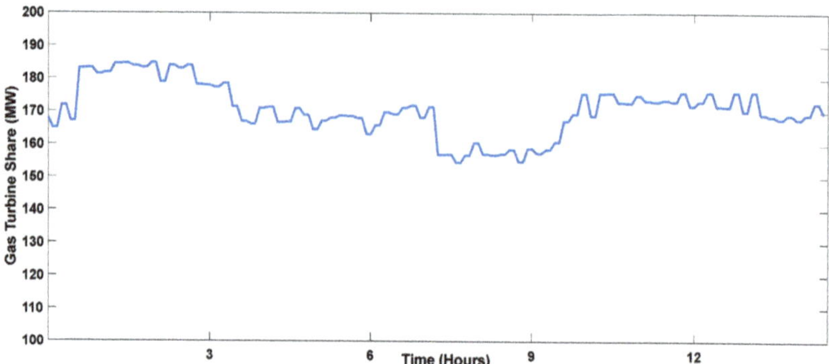

Figure 13. Output power share of the gas turbine.

The next three figures show the different scenarios of total generation. The total generation for wind farm and clean thermal generation units is shown in Figure 14, the total generation compared with load following capability shown in Figure 15, and the total imbalance represented in Figure 16.

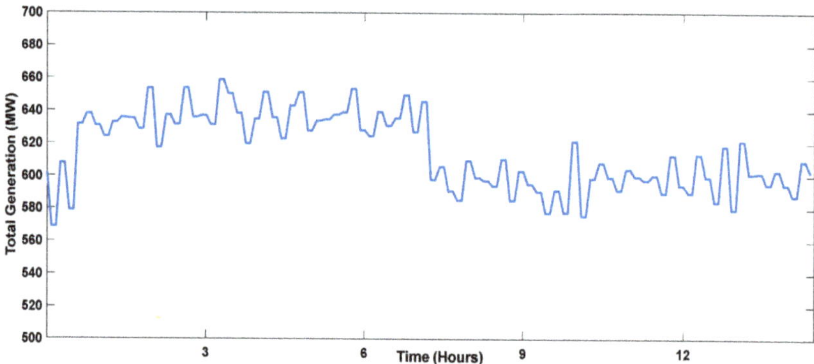

Figure 14. Total Generation.

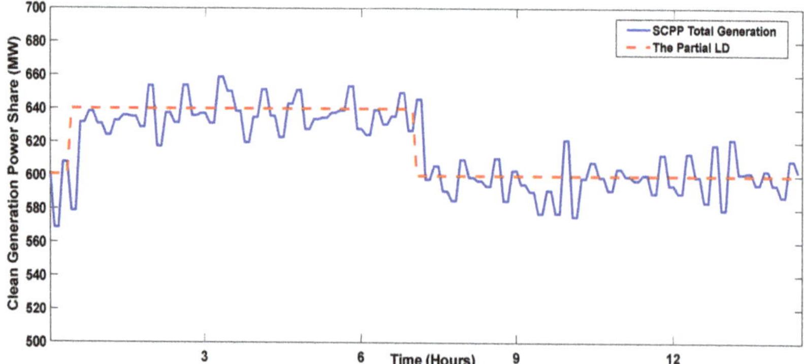

Figure 15. Total Generation and Load Demand.

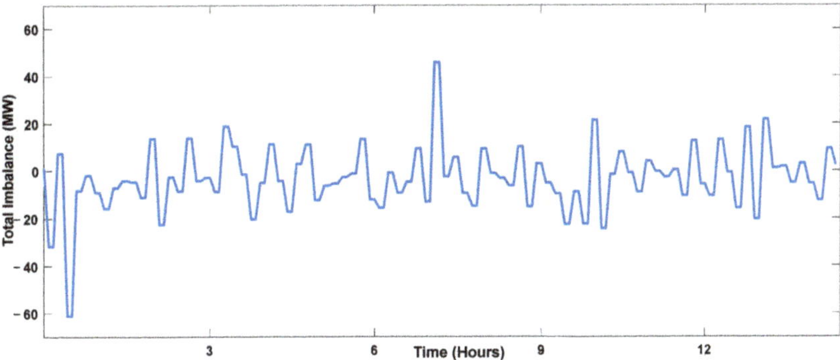

Figure 16. Total Imbalance.

In order to enhance the efficiency and reduce the negative effects on the frequency, Figure 17 shows the actual frequency, approximately equal 50 Hz.

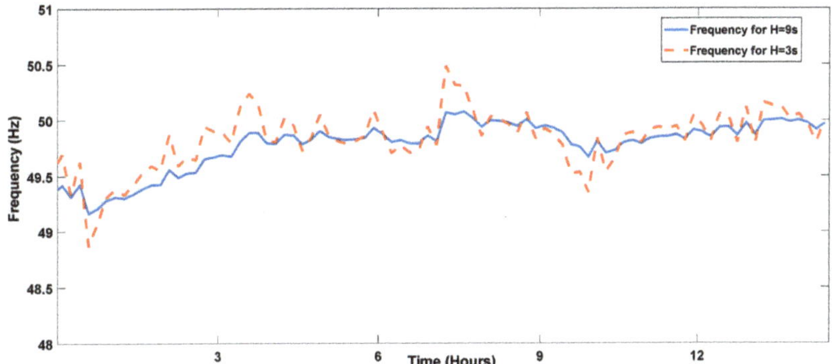

Figure 17. Actual Frequency.

It can be deduced that the feasible penetration of the wind energy results in considerable reductions in the fuel consumption of coal in the SCPP and the combined signals of fuel oil and natural gas injected to the GT. The water consumption is also reduced in the SCPP, with those indictors ensuring the economic viability of wind energy integration. In addition, there are enhancements to the environmental conditions, especially when it is combined with cleaner thermal units such as those proposed in this paper; these are the modern SCPP cleaner coal power plant and the GT, as a part of the combined-cycle energy-efficient unit. With a maximum penetration of approximately 90 MW, the total production follows the total demand signal with acceptable imbalance of around 6% in the load transition interval and 3% in the load steady interval. This can be investigated in an alternative way, meaning the acceptable imbalance cannot be standardized because it depends on some factors, such as the system strength or inertia, which is different from one grid to another. Here, we have considered three second s inertia and nine second inertia in order to show the effect of the domination of rotating machines in the power system in general, whether for generating or motoring/condensing actions. The frequency excursions for the worst-case inertia has been typically acceptable, and hence the model predictive load frequency control is a promising technology for enhancing the power system behavior in response to sudden load changes or load small disturbances.

5. Conclusions

This paper presents a feasible control strategy for clean thermal generation units, which are wind energy, clean coal SC power plant, and dual-fuel gas turbine (GT). This study has been based on a Hammerstein–Wiener Model for wind plants, the theory of subspace system identification, and MPC strategy for both coal and gas power plants. The practical advantages of the proposed strategy, in terms of stable and efficient operation, have been confirmed using validated simulations by analyzing the results of this study with emphasis on efficiency, frequency, and output shared power, regulated by the MPC algorithm. In addition, this study can be considered a general guideline for preliminary research into a further improvement in power system frequency responses to different levels of penetration from the wind plant [33].

In general, linear MPC is known to be preferred in the industry over non-linear, because of appropriate results and reasonable computational burdens. However, there is a subject that can be resolved in future research related to the computational burdens of the nonlinear MPC, due to its theoretical complication in comparison with linear MPC, to facilitate the application of nonlinear MPC in practical power plants. This can be achieved by suggesting more advanced optimization techniques. However, to use linear MPC in the future, the MPC algorithm can be modified with the extension of measured and unmeasured disturbances, and the linear specific model can be augmented by noise terms representing the practical noises that happen in power plants.

As a future recommendation, adaptive model predictive control could be a feasible suggestion for further improvements. The explicit MPC is not recommended to be applied to such studies as it exhibits longer computation time, which leads to an unsatisfactory on-line performance. Therefore, an adaptive MPC or multiple MPC are highly recommended, instead of the explicit MPC, to be a future point of research for further enhancements. In terms of modeling, it is suggested that the state-space linearized models are replaced with more accurate paradigms, such as nonlinear models, which are either derived from system physics, nonlinear system identification, or machine learning modeling. Further closer models to reality can be created through the inclusion of the tie lines between units, which form the multisource interconnected power system.

Author Contributions: Conceptualization, O.M.; methodology, T.A., O.M., and W.A.E.; software, T.A., and O.M.; validation, O.M., and W.A.E.; formal analysis, T.A., and O.M.; investigation, O.M., and W.A.E.; resources, O.M.; data curation, O.M.; writing—original draft preparation, T.A.; writing—review and editing, O.M., and W.A.E.; visualization, T.A.; supervision, O.M., and W.A.E.; project administration, O.M., and W.A.E. All authors have read and agreed to the published version of the manuscript.

Funding: This research received no external funding.

Data Availability Statement: Data available on request due to the preference of requesting it from the corresponding author rather than via direct availability.

Conflicts of Interest: The authors declare no conflict of interest.

References

1. Staffell, I.; Jansen, M.; Chase, A.; Cotton, E.; Lewis, C. *Energy Revolution: Global Outlook*; Drax: Selby, UK, 2018.
2. Wood, A.J.; Wollenberg, B.F.; Sheblé, G.B. *Power Generation, Operation, and Control*; John Wiley & Sons: Hoboken, NJ, USA, 2013.
3. Saadat, H. *Power System Analysis TMH*; McGraw Hill: New York, NY, USA, 2002.
4. Weedy, B.M.; Cory, B.J.; Jenkins, N.; Ekanayake, J.B.; Strbac, G. *Electric Power Systems*; John Wiley & Sons: Hoboken, NJ, USA, 2012.
5. Mohamed, O.R.I. Study of Energy Efficient Supercritical Coal-Fired Power Plant Dynamic Responses and Control Strategies. Ph.D. Thesis, University of Birmingham, Birmingham, UK, 2012.
6. Mohamed, O.; Wang, J.; Khalil, A.; Limhabrash, M. Predictive control strategy of a gas turbine for improvement of combined cycle power plant dynamic performance and efficiency. *SpringerPlus* **2016**, *5*, 980. [CrossRef] [PubMed]
7. Qatamin, R. Prediction of Power Output of Wind Turbines Using System Identification Techniques. Ph.D. Thesis, Princess Sumaya University for Technology, Amman, Jordan, 2020.

8. Mohamed, O.; Khalil, A.; Wang, J. Modeling and Control of Supercritical and Ultra-Supercritical Power Plants: A Review. *Energies* **2020**, *13*, 2935. [CrossRef]
9. Rehman, S. Hybrid power systems–Sizes, efficiencies, and economics. *Energy Explor. Exploit.* **2021**, *39*, 3–43. [CrossRef]
10. Li, Z.; Hou, S.; Cao, X.; Qin, Y.; Wang, P.; Che, S.; Sun, H. Modeling and analysis of hydrogen storage wind and gas complementary power generation system. *Energy Explor. Exploit.* **2021**, *39*, 1306–1323. [CrossRef]
11. Tsoutsanis, E. A Dynamic Performance Model for Hybrid Wind/Gas Power Plants. In *Modeling, Simulation and Optimization of Wind Farms and Hybrid Systems*; Maalawi, K.Y., Ed.; IntechOpen: London, UK, 2020.
12. Barelli, L.; Bidini, G.; Ottaviano, A. Integration of SOFC/GT hybrid systems in Micro-Grids. *Energy* **2017**, *118*, 716–728. [CrossRef]
13. Di Gaeta, A.; Reale, F.; Chiariello, F.; Massoli, P. A dynamic model of a 100 kW micro gas turbine fuelled with natural gas and hydrogen blends and its application in a hybrid energy grid. *Energy* **2017**, *129*, 299–320. [CrossRef]
14. Bizon, N. Real-time optimization strategies of Fuel Cell Hybrid Power Systems based on Load-following control: A new strategy, and a comparative study of topologies and fuel economy obtained. *Appl. Energy* **2019**, *241*, 444–460. [CrossRef]
15. Amar Bensaber, A.; Benghanem, M.; Guerouad, A.; Amar Bensaber, M. Power flow control and management of a Hybrid Power System. *Przegląd Elektrotechniczny* **2019**, *95*, 188–192. [CrossRef]
16. Avagianos, I.; Rakopoulos, D.; Karellas, S.; Kakaras, E. Review of process modeling of solid-fuel thermal power plants for flexible and off-design operation. *Energies* **2020**, *13*, 6587. [CrossRef]
17. Zhang, F.; Zhang, Y.; Wu, X.; Shen, J.; Lee, K.Y. Control of ultra-supercritical once-through boiler-turbine unit using MPC and ESO approaches. In Proceedings of the 2017 IEEE Conference on Control Technology and Applications (CCTA), Kohala Coast, HI, USA, 27–30 August 2017; pp. 994–999.
18. Annaluru, R. *Managing the Power Grid Ramping Challenges Critical to Success of India's Renewable Energy Targets*; IDEAS: Bloomington, MN, USA, 2017.
19. Ersdal, A.M.; Imsland, L.; Uhlen, K. Model predictive load-frequency control. *IEEE Trans. Power Syst.* **2015**, *31*, 777–785. [CrossRef]
20. Shakibjoo, A.D.; Moradzadeh, M.; Din, S.U.; Mohammadzadeh, A.; Mosavi, A.H.; Vandevelde, L. Optimized Type-2 Fuzzy Frequency Control for Multi-area Power Systems. *IEEE Access* **2021**, *10*, 6989–7002. [CrossRef]
21. Ljung, L. Theory for the user. In *System Identification*; Prentice-Hall: Hoboken, NJ, USA, 1987.
22. Di Ruscio, D. *Subspace System Identifcation: Theory and Applications. Lecture Notes*; Telemark University College: Porsgrunn, Norway, 2009.
23. Favoreel, W.; De Moor, B.; Van Overschee, P. Subspace state space system identification for industrial processes. *J. Process Control.* **2000**, *10*, 149–155. [CrossRef]
24. De Moor, B.; Van Overschee, P. *Subspace Identification for Linear Systems: Theory Implementation Application*; Kluwer Academic Publishers: New York, NY, USA, 1996.
25. Meng, J.; Boukhnifer, M.; Diallo, D. Lithium-Ion Battery Monitoring and Observability Analysis with Extended Equivalent Circuit Model. In Proceedings of the 2020 28th Mediterranean Conference on Control and Automation (MED), Saint-Raphaël, France, 16–19 June 2020; pp. 764–769.
26. Meyer, C.D. *Matrix Analysis and Applied Linear Algebra*; Siam: Philadelphia, PA, USA, 2000; Volume 71.
27. Ljung, L. *System Identification Toolbox™ User's Guide*; MathWorks Incorporated: Natick, MA, USA, 2021.
28. Wang, L. *Model Predictive Control System Design and Implementation Using MATLAB®*; Springer Science & Business Media: Berlin/Heidelberg, Germany, 2009.
29. Maciejowski, J.M. *Predictive Control: With Constraints*; Pearson Education Limited: London, UK, 2002.
30. Znad, O.A.; Mohamed, O.; Elhaija, W.A. Speeding-up Startup Process of a Clean Coal Supercritical Power Generation Station via Classical Model Predictive Control. *Process Integr. Optim. Sustain.* **2022**, *6*, 751–764. [CrossRef]
31. Mohamed, O.; Wang, J.; Guo, S.; Wei, J.; Al-Duri, B.; Lv, J.; Gao, Q. Mathematical modelling for coal fired supercritical power plants and model parameter identification using genetic algorithms. In *Electrical Engineering and Applied Computing*; Springer: Dordrecht, The Netherlands, 2011; pp. 1–13.
32. Ljung, L. Wiley Encyclopedia of Electrical and Electronics Engineering. In *System Identification*; Prentice-Hall: Hoboken, NJ, USA, 1999.
33. Tielens, P.; Henneaux, P.; Cole, S. *Penetration of Renewables and Reduction of Synchronous Inertia in the European Power System: Analysis and Solutions*; European Commission, Directorate-General for Energy, Publications Office: Brussels, Belgium, 2020.

Article

Impact of Reverse Power Flow on Distributed Transformers in a Solar-Photovoltaic-Integrated Low-Voltage Network

Issah Babatunde Majeed [1,*] and Nnamdi I. Nwulu [2]

[1] Electrical and Electronic Department, University of Johannesburg, Auckland Park 2006, South Africa
[2] Center for Cyber-Physical Food, Energy and Water Systems, University of Johannesburg, Auckland Park 2006, South Africa
* Correspondence: issahmajeed@gmail.com; Tel.: +233-546238256

Abstract: Modern low-voltage distribution systems necessitate solar photovoltaic (PV) penetration. One of the primary concerns with this grid-connected PV system is overloading due to reverse power flow, which degrades the life of distribution transformers. This study investigates transformer overload issues due to reverse power flow in a low-voltage network with high PV penetration. A simulation model of a real urban electricity company in Ghana is investigated against various PV penetration levels by load flows with ETAP software. The impact of reverse power flow on the radial network transformer loadings is examined for high PV penetrations. Using the least squares method, simulation results are modelled in Excel software. Transformer backflow limitations are determined by correlating operating loads with PV penetration. At high PV penetration, the models predict reverse power flow into the transformer. Interpolations from the correlation models show transformer backflow operating limits of 78.04 kVA and 24.77% at the threshold of reverse power flow. These limits correspond to a maximum PV penetration limit of 88.30%. In low-voltage networks with high PV penetration; therefore, planners should consider transformer overload limits caused by reverse power flow, which degrades transformer life. This helps select control schemes near substation transformers to limit reverse power flow.

Keywords: solar photovoltaic; simulation data; reverse power flow; low-voltage network; substation transformer; penetration levels; grid integration

Citation: Majeed, I.B.; Nwulu, N.I. Impact of Reverse Power Flow on Distributed Transformers in a Solar-Photovoltaic-Integrated Low-Voltage Network. *Energies* **2022**, *15*, 9238. https://doi.org/10.3390/en15239238

Academic Editor: Juri Belikov

Received: 10 October 2022
Accepted: 1 December 2022
Published: 6 December 2022

Publisher's Note: MDPI stays neutral with regard to jurisdictional claims in published maps and institutional affiliations.

1. Introduction

The expanding worldwide need for energy necessitates the leverage of renewable energy technologies (RETs). Renewable energy technologies have the potential to become the dominant form of future energy technology, given the ease with which they can be deployed and the low cost at which they can be generated. They are also capable of mitigating global warming and enhancing energy sustainability [1]. One of these RETs, the photovoltaic (PV) solar system, is being widely used as a renewable energy source worldwide [2–4]. As a result, PV grid integration has advanced as a renewable energy technology that promises energy sufficiency and long-term sustainability. The benefits of PV grid integration include voltage support, improved power quality, loss reduction, postponement of new or upgraded transmission and distribution infrastructure, and increased utility system resilience [5].

Despite these advantages, there are some concerns and constraints that limit the use of PV in the grid. Some of these challenges include protective measures, problems with reverse power flow, and hosting capacity [6,7]. Additionally, variability and intermittency pose reliability challenges for PV grid integration [8]. These adverse challenges depend on the level of PV penetration into the grid and the network load demand response [9]. For instance, low-level PV penetration is known to benefit the distribution network by increasing bus voltages, reducing losses, and extending the transformer's usable life [10,11]. On the other hand, high-level penetration challenges include power losses and problems with protection equipment [12,13]. Similarly, in high PV penetration networks, the development of reverse power flow (RPF), which can cause transformer overload, has been reported to increase network load, overvoltage, and losses [14–16].

The reverse power flow phenomenon occurs when the PV power generation in a grid-connected network exceeds the local load demand [17]. This is an indication that RPF is more likely to occur in network regions with lower peak loads. Likewise, the overgeneration of PV solar production may lead to the appearance of RPFs in low-voltage networks [7,18]. Reverse power flow in a low-voltage (LV) network can cause instability, such as in the line sections and distribution transformers [19,20]. The overloading of the distribution transformer is one consequence of a low-load, high-PV penetration network; higher voltages are also seen at low-voltage (LV) and medium-voltage (MV) levels. [21,22].

In [22], the authors address the effect of thermal loads on transformer technical life, as a result of an increase in PV penetration in an LV network.. Results revealed that significant reversed power flow can increase insulation degradation and shorten the technical life of a transformer. Similarly, reference [23] has demonstrated that, under reverse power flow, an increase in winding temperature in the transformer results in an increase in winding losses. The effect is insulation degradation that leads to a shortened lifetime and earlier breakdown of the transformer, a result which is also supported by [24]. This conclusion is supported by [25] who showed that increased excitation voltages due to RPF raise the transformer magnetizing current. Therefore, core losses are increased by increased winding temperature, which reduces the life of the transformer.

Reference [26] assesses how PV penetration levels impact utility transformers by altering the thermal stress to which the components are exposed. The authors demonstrated that overload periods cause losses and have a significant impact on the lifespan of transformers. They concluded that various simulation scenarios must be run to estimate the maximum PV penetration depth that the utility transformer can withstand.

To avoid transformer loss of life due to overload from solar PV production, control schemes can be implemented where the excess production is used to charge energy storage systems. Such is the case in [27] where the authors propose a smart charging scheme to coordinate electric vehicles (EVs) and battery energy storage systems (BESS) in the presence of PV generation. The scheme is designed to prevent the overloading of transformers above their nameplate capacities. Such a coordination scheme is shown to improve transformer life expectancy.

Reference [28] addresses the transformer overload issue caused by RPF, by including transformer constraints in their optimisation problem formulation to provide more accurate solar PV allocation. Notably, battery energy storage systems (BESS) are utilised to demonstrate how transformer overloads may be minimised in the presence of high solar PV penetrations into the grid. Similarly, Ref. [29] employed the BESS scheme in their study to prevent the progression of RPF in a real distribution network due to high PV penetration. They demonstrated how excess energy resulting from high PV penetration is used to charge energy storage systems (ESS) to upgrade the grid and enable network expansion. They concluded that storage capacity systems should be built next to LV-MV transformers to prevent overvoltage conditions and transformer overloads when RPF is present.

In their study of solar roof potential assessments, Ref. [16] used a deterministic approach to evaluate the maximum and minimum PV penetration levels in a medium voltage feeder in Ulm, Germany. To evaluate the minimum penetration level, RPF is used as a constraint for overloading the MV/LV transformers in the investigated feeders. The limitation of this study is that the analysis is performed at the secondary MV/LV substation. Reference [30] proposes smart transformers to control reverse power flow in the LV network. This analysis is performed and verified using the control-hardware-in-loop (CHIL) real-time simulation methodology. As a result, the proposed RPF limitation controller reduces the power output from the solar PV to avoid RPF in the MV grid.

Reference [25] demonstrates how RPF affects the performance of distribution transformers. Using analytical techniques, the authors showed that distribution transformer losses increase significantly during reverse power flow, resulting in a 25% reduction in the transformer's lifespan. Similarly, In the presence of varying levels of grid-connected PV penetration, the main goal of [31] is to investigate the impact of ageing in overhead distribution transformers caused by reverse power flow and the impact on utilities' permitted revenue. The methodology included a Monte Carlo simulation and depreciation rates, which depend on transformer ageing. According to the findings, many transformers are replaced after 40 per cent PV penetration because their degradation accelerates and their operational lifespan shortens, reducing utility revenue. In other research, the impact of high penetration PV on the ageing of distribution transformers in a PV grid network has been thoroughly established [26,32–34].

Most of the investigated cases of high PV penetration mainly focus on feeder representative metrics due to reverse power flow [2,30,35]. For instance, in [16] an overall system analysis due to PV penetration is carried out on the MV network. Certain studies also focus on distribution transformer constraints and strategies used to determine maximum PV penetration in the distribution system [28,36]. For instance, in [28] transformer constraints are applied to ensure that they are not loaded beyond their power rating due to reverse power flow caused by high PV impacts. However, these constraints are transformer power rating-tied and do not address the issue of critical operating conditions for reverse power to flow. To address this gap, we provide the needed insight into the backflow limits analysis of the operating conditions of the transformer. These limits are reached before the reverse flow can cause substantial overload on the transformer.

In addition, there are few studies on the assessments and impacts of PV penetration on the national grid in Ghana [37–40]. These investigations are carried out on grid-connected or isolated LV feeders on 33 kV/161 kV sub-transmission networks as case and feasibility studies. None of these publications demonstrate the relationship between transformer operating loads and PV penetration due to power flow dynamics in the distribution system. The study will fill the aforementioned gaps by deploying the methodologies on a solar PV-tied real urban ECG LV network. The motivation for the present study is to provide utilities with insights into substation transformer operating limits beyond which reverse power flows as a result of high solar PV penetration. The main objective of this study is to predict the reverse power flow and transformer backflow limits in a radial LV network under high solar PV penetration

Using the ETAP software, the study models and analyses the distribution network to quantify the effects of reverse power flow on transformer loading, which results in losses. In addition, the analysis establishes the maximum penetration depth at the margin of RPF in the substation transformer. This is a utility-based study aimed at assessing safety thresholds for substation transformers due to excessive PV installations in LV radial networks. The study provides a one-case scenario of high constant solar production in the presence of average real static loads in a radial LV network. The findings provided in this study would serve as a recommendation for utilities to set safe margins to safeguard the flow of reverse power into the substation transformer. This general framework is required for establishing pre-defined threshold parameters to protect LV networks from RPF caused by excessive solar PV installations.

The major contributions that will address the outlined gaps in the literature are as follows:

Statistical models are developed for threshold analysis to predict:

- Transformer backflow limits due to high solar PV impacts;
- The maximum depth of penetration of solar PV at the margin of RPF in the substation transformer.

The rest of the paper is organized as follows: The methodology is covered in Section 2, which includes a case study network, as well as modelling and simulation.

Section 3 contains the results and discussion, including the determination of transformer backflow limits, and Section 4 contains the conclusion.

2. System Modelling and Power Flow Analysis

A mathematical analysis to consider the steady state output analysis of the inverter grid current injection is considered in this section.

2.1. Equivalent Circuit of Inverter Grid Interphase

As seen in Figure 1, an equivalent inverter grid circuit is required to send active and reactive current into the grid [41]. At the point of connection into the grid, a current filter is used to mitigate harmonic current. Given the inverter voltage, $\mathbf{V}_{inv}$, the vector sum of the grid voltage, $\mathbf{V}_{grid}$, and the voltage across the filter, $\mathbf{V}_L$, can be obtained from Kirchoff's voltage law.

$$\mathbf{V}_{inv} = \mathbf{V}_{grid} + \mathbf{V}_L \tag{1}$$

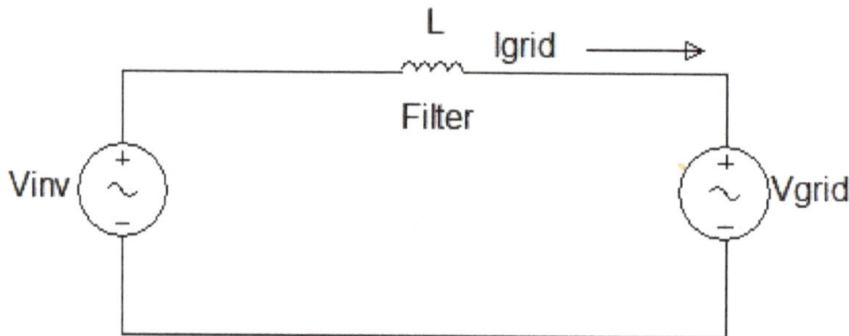

Figure 1. Equivalent circuit of a PV grid interphase.

For active power transfer, at the point of connection, the condition in (2) must be satisfied.

$$\left|\mathbf{V}_{inv}\right| = \left|\mathbf{V}_{grid}\right| \tag{2}$$

In the power flow of the PV grid system, the output power of the PV power varies at random with light intensity [42]. Assuming constant inverter output, it follows from [43] that the active power injected into the grid, P_{active}, is characterised by (3).

$$P_{active} = \frac{\left|\mathbf{V}_{inv}\right|\left|\mathbf{V}_{grid}\right|\sin \delta}{\omega L} \tag{3}$$

where δ is the angle between the inverter output voltage and the grid voltage, L is the inductance of the coupling inductor, and ω is the grid frequency. By varying the phase angle δ, the inverter's active power flow into the grid can be controlled [41]. Additionally, grid-connected inverters are current sources and can deliver voltage output by synchronising with the grid voltage and frequency [44]. This condition is satisfied by (2). For maximum power output, the inverter is designed to operate at a power factor of unity, with no reactive power injection into the grid [44,45]. From Figure 1, the steady-state current injection into the grid is obtained from the expression in (4):

$$I_{inv} = \frac{\left|\mathbf{V}_{inv}\right| - \left|\mathbf{V}_{grid}\right|}{\omega L} \tag{4}$$

2.2. Solar PV Grid Load-Point Analysis

It is expected that the amount of solar energy produced in a grid system will fluctuate during the day. Likewise, the load profiles of the grid system are dynamic. Therefore, the instantaneous real power dynamics, defined at any given time, t, at a local load point is given in Equation (5):

$$P_{\text{net power } i}(t) = P_{\text{load } i}(t) - P_{\text{PV } i}(t) \; \forall \, t \tag{5}$$

where, $P_{\text{net power } i}(t)$ and $P_{\text{load } i}(t)$ are the net active power at the ith bus and load active power at the ith bus, respectively. $P_{\text{pv } i}(t)$ is the active power of solar PV at the *I* th bus. For multiple transformers, the total net active power, $P_T(t)$, in the following relation:

$$P_T(t) = \sum_{i=1}^{N} P_{\text{net power } i}(t) \;\; \forall \, t \tag{6}$$

It follows from (5) and (6), that when

$$\sum_{i=1}^{N} P_{\text{load } i}(t) > \sum_{i=1}^{M} P_{\text{PV } i}(t) \;\; \forall \, t \tag{7}$$

conventional current flows, leading to a positive value of $P_T(t)$, represent a power flow from the transformer to the loads (7). However, in the case of (8), a negative value of $P_T(t)$ represents a reverse flow of power into the substation transformer.

$$\sum_{i=1}^{N} P_{\text{load } i}(t) < \sum_{i=1}^{M} P_{\text{PV } i}(t) \;\; \forall \, t \tag{8}$$

In particular, when $P_T(t) = 0$, a critical point is reached, beyond which there is a reverse flow of power. This is the point where the transformer backflow limits can be determined. A major challenge is that RPF is not concurrent across the different parts of the network [46]. This means that different zones within the same grid may not experience RPF within the same period. However, the net flow of power is what is seen by the transformer. Neglecting energy losses on line components and inverters, the global net power contributions to the grid are represented by (9):

$$P_T(t) = \sum_{i=1}^{N} P_{\text{load } i}(t) - \sum_{i=1}^{M} P_{\text{PV } i}(t) \;\; \forall \, t \tag{9}$$

$$P_{\min} \le \sum_{i=1}^{N} P_{\text{load } i}(t) - \sum_{i=1}^{M} P_{\text{PV } i}(t) \le P_{\max} \forall \, t \tag{10}$$

Equation (10) places power flow limits on the transformer, defined as the transformer loading constraints, $P_{\min}$ and $P_{\max}$. These limits protect the transformer from under-loading and overloading conditions, respectively, thereby preventing reverse power flow due to excess energy from solar PV, which can cause overloading in the transformer [29]. It should be noted that these limits are pre-defined, depending on the loading patterns in the network. However, the transformer loading constraint, $P_{\max}$, differs from the backflow limit due to reverse power flow, $P_T(t)$. So, while $P_T(t)$ sets the limit at the margin of reverse power flow, $P_{\max}$ sets a limit beyond which the transformer is overloaded by reverse power flow due to increased PV penetration.

There is no absolute definition for PV penetration level. Various definitions are advanced from both the distribution system point of view and the bulk system point of view [47]. In this study, the depth of penetration, D, is defined for the system loadings connected to the substation transformer. Thus,

$$D = \frac{\text{Aggregated PV rating on feeder in kVA}}{\text{Transformer full load kVA rating}} \tag{11}$$

In (11), the depth of penetration is seen to be a function of the transformer's net active power, expressed as

$$D = f(P_T(t) > 0) \forall \, t \tag{12}$$

Equation (12) indicates that as long as the net active power flow is always greater than zero, no RPM will be established in the substation transformer for increasing levels of PV penetration. It follows from (12) that the maximum depth of penetration, D_{max}, can be obtained at the margin of RPF in the substation transformer, as follows:

$$D_{max} = \lim_{P_T(t) \to 0} D \quad \forall t \tag{13}$$

It can be deduced from (13) that the maximum depth of penetration is obtained at the threshold of the RPF when the theoretical total net active power flow into the substation transformer is zero.

3. Materials and Methods

This is a utility-based study aimed at determining safety thresholds for substation transformers due to excessive PV installations in LV radial networks. In this simulation-based research design, a test model was developed with field data to represent a real LV network. Next, a solar PV inverter system was designed as the distributed generator in the LV network, which is powered by a single substation transformer. This study used the power flow calculation tool in ETAP software to model and simulate the LV network using field data [48]. In the base case, the network was simulated to determine the overload operating conditions of the substation transformer. In the second scenario, the network was reset to the normal loadings with the transformer operating without overload. Solar PV units were deployed to the grid cumulatively, based on a dispersion rule and transformer loading constraints. Simulation data were obtained at different levels of PV penetration to build correlation models, using the Excel data analysis tool. These correlation models were used to extrapolate transformer backflow limits. Figure 2 presents an overview of the design process of the study.

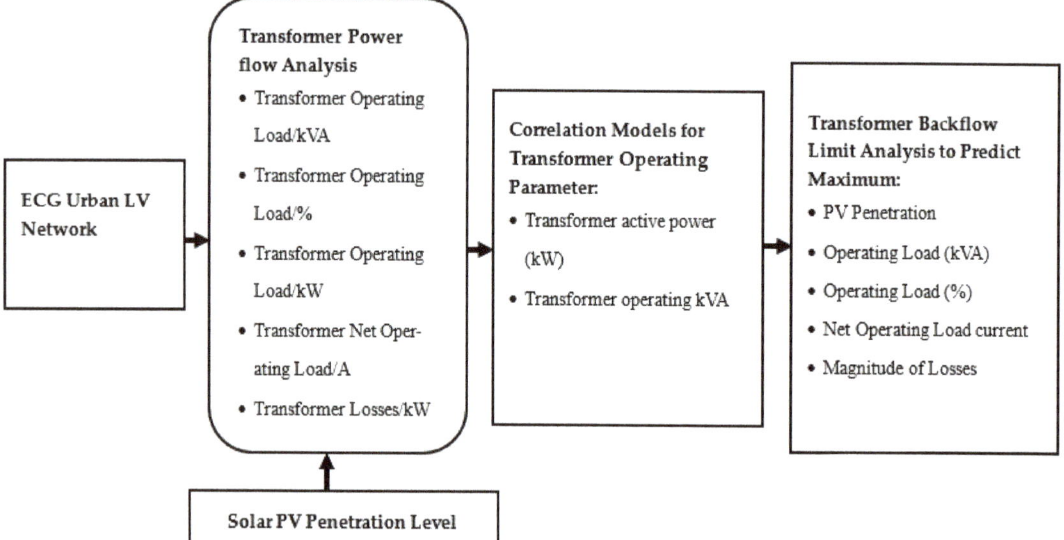

Figure 2. Design framework for distribution transformer assessment due to the impact of reverse power flow in a PV solar-photovoltaic-integrated low-voltage network.

3.1. Case Study Network

This research considers a real urban ECG LV residential network in the western region of Ghana as a case study grid. The investigated network has a radial configuration and is connected to the medium voltage (MV) grid through a three-phase four-wire system with a 315 kVA, 11/0.415 kV, Delta-Y connected transformer (known as C96). The primary feeder lines of 50 mm^2 aluminium bare conductor size are LV lines. For a pole height of about 8 m, the maximum span is about 50 m. Spur lines of the same material are sized 25 mm^2. The network has one feeder and numerous laterals, most of which are single phases. It is characterised by poor bus voltages, registering as low as 208 V at the load points, typically because of load growth issues and an increase in load demand by existing customers. The usual practice used to improve these voltages requires the zoning of the LV network, stringing new lines to the customer vicinity, and injecting new transformers to upgrade the network. The above challenges justify the proposed introduction of solar PV to improve the voltage profile on the network.

To determine the peak load on the 11 kV C10 feeder, we used data from the load monitoring exercise carried out by ECG in 2019 on the C96 distribution transformer. The peak current obtained from the load current monitoring was 223 A. The total maximum operating load on the transformer is obtained as follows:

$$\text{Total kVA on C96 transformer} = \frac{\sqrt{3}\,I_1 V_1}{1000} = \frac{\sqrt{3} \times 223 \times 415}{1000} \tag{14}$$

Therefore, transformer full load kVA rating is 160.29 kVA, which is used to determine the depth of penetration for the solar PV rating according to (11).

3.2. Modelling of System Components

The existing LV network was modelled to exhibit the characteristics of the real network. Power flow components such as LV lines, loads, and transformers are modelled in the Etap environment using field data collected during load monitoring.

3.2.1. The 11 kV Source Feeder

A network usually begins from a source. This source is usually represented by the bus of a distribution substation. In a typical case, a substation transformer steps down 33 kV into an 11 kV source. From this source, the main feeder is run through the network. The source impedance is the equivalent impedance of the transformer, transmission lines, and generators supplying the 11 kV bus. The equivalent source used in this research is defined by the base power, source equivalent impedance, short-circuit power, etc.

3.2.2. Network Loads

Individual loads are attached to the bus bars and modelled as lumped loads consisting of three and single-phase loads with a total system load of 158.95 kVA. These residential loads were modelled as constant impedance, since they consist mostly of heating devices and lighting bulbs. Based on the results from the load monitoring exercise on ECG networks, loads on service poles were lumped as load points and shared unequally across the phases. These loads are modelled as lumped three-phase static loads with rated average loads of 1.5 kW.

The required data used for the modelling and simulation of the loads and 11 kV feeder source are presented in Table 1.

Table 1. Data of a typical residential customer and an 11 kV source representing the primary substation used in the simulation.

Parameter		Value
Load	Section Id	Lump 76
	Load type	Constant kVA = 80% constant Z = 20%
	Nominal voltage	230 V/415 V
	Typical Connected load	1.5 kW
	Configuration	Delta
	Power factor	0.85
	Customer type	Residential
	Load factor	0.9
	Load distribution	lumped load, unbalanced
11 kV Feeder Source	Nominal voltage	11.5 kV
	Operating voltage	11 kV
	Base Power	100 MVA
	Short-circuit rating	31.8 MVA (three-phase)
	Source Configuration	Wye

3.2.3. Substation Transformer

The transformer is modelled to exhibit the characteristics of the field substation transformer. Load monitoring was carried out on transformer legs to obtain the average maximum system loadings. In the Etap software, the two-winding transformer is modelled as 11 kV/0.415 kV with a rating of 315 kVA. Other modelling parameters are shown in Table 2.

Table 2. Data on overhead line showing conductor type and nominal data on windings of distribution transformer for the LV network.

Parameter		Value
Overhead Line	Equivalent Impedance	Positive sequence Zero sequence, 21 ohms
	Conductor Type	LV, 50 mm^2 ACSR
	Maximum Span	LV, 50 m
	Nominal Ampacity	LV, 209 A
	Maximum Temperature	75 °C
Distribution Transformer Windings	Section Id	C96
	Frequency	50 Hz
	Type	Three-phase core
	Nominal Rating	315 kVA
	Primary Voltage	11 kV$_{line}$
	Secondary Voltage	0.415 kV$_{line}$
	Sequence Impedance	$Z_1 = 4\%$, $Z_0 = 4\%$ $X_1/R_1 = 1.5\%$, $X_0/R_0 = 1.5\%$
	Configuration	Primary, delta Secondary, star
	Phase Shift	Dyn11

3.2.4. Overhead Lines

In the model, the overhead lines are the primary feeder lines. The phase conductors are modelled as ACSR (aluminium conductor steel reinforced)-type with an ampacity of 209 A at a maximum operating temperature of 75 degrees Celsius. These LV lines have a maximum span of 50 m. Other modelling parameters are shown in Table 2. Table 3 gives the convergence criteria for the simulation.

Table 3. Data for load flow simulation convergence criteria.

Convergence Criteria Parameter	Value
Calculation Method	Adaptive Newton–Raphson
Convergence Parameters	0.0001 tolerance 99 iterations
Calculation Options	Assume line transposition Include line charging

3.2.5. Solar PV Dispersion Criteria

A three-phase solar PV inverter system was designed as an integral part of a solar PV system. The inverter was sized for constant output power and unity power factor using the LV network system loadings and ETAP software simulation data. The inverter size was based on network system loadings and simulation data provided by ETAP software. In the ETAP solar PV interface, the selection of Photowatt power plants for the grid-connected large-scale system was appropriate for the design. In this study, harmonics were not considered in the inverter design. The simulation data used for the solar PV design are shown in Tables 4 and 5.

Table 4. Solar PV parameters used for modelling PV panel obtained from ETAP software.

PV Panel	
Manufacturer	Photowatt
Model	PW6-110
Type	Multi-crystalline
Size	110
Number of cells	36
Maximum Vdc	1000
Power factor	1
Watt/Panel	110.3
Number in series	20
Number in Parallel	10
Irradiance/W/m^2	1000
Ta (Ambient temperature)/degree Celsius	30
Tc (Cell temperature)/degree Celsius	5
MPP (Maximum power point)/kW	21.69
Amps, dc	64.2

Table 5. Data for three-phase inverter unit for solar PV system obtained from ETAP software.

DC Rating	
kW	22.06
V	343.6
FLA (Full load Ampere)	64.2
%Efficiency	90.34
AC rating	
kVA	19.93
kV	0.415
FLA	27.73
%PF	100
Imax	150%

To maximise the depth of penetration based on network constraints, solar PV units have been installed in distribution networks using a variety of strategies [49–52]. In contrast with the customer-based randomised PV allocation, the utility-based PV allocation is well defined. This study adopts the latter approach, which is done systematically for future planning and to regulate customer allocation for injection into the network. In this study, the selection criteria for the next solar PV unit allocation involves prioritising the busbar with the worst voltage profile. Where the worst voltage bus is defined as any bus with under-voltage conditions below 0.95 per unit.

The following were the steps taken to allocate distributed solar PVs at the busbars on the modelled LV network.

3.2.6. Algorithm for Solar PV Dispersion in LV Network

1. Start;
2. Run a load flow calculation for the LV network without a solar PV unit;
3. Identify and place a solar PV unit at the load bus with the worst voltage profile;
4. Run a load flow calculation for the network;
5. Repeat steps 2 and 3 until transformer loadings (kW) register negative values;
6. End.

The PV penetration involves adding new PV generators at locations to the base case model by increasing the size of the potential PV generators until reverse current flows in the transformer. The simulation results obtained were used to set up correlation models using the least square method in Excel software [53]. The models for operating loads and PV penetration were analysed to determine transformer backflow limits.

4. Results and Discussions

Several studies [25,28,46] have investigated power backflow limits for grid upgrades in distribution networks. What is not so clear in the literature is the transformer-based backflow limits due to high-level solar PV grid penetration. The simulation results obtained in this study explain the relationship between transformer operating loads and solar PV depth of penetration due to power flow dynamics in the distribution system. The established casework also determined the maximum depth of penetration and the operational locations of the PV systems in the grid. These findings are significant because, as long as the distribution of loads and PV installations are known, restricting transformer backflows to pre-defined limits could prevent over-voltages in the network feeder [54].

Additionally, reverse power flow may violate voltage and line capacity margins as a result of excessive PV deployments in LV networks. This could be avoided by establishing pre-defined transformer backflow limits, above which surplus photovoltaic energy is exported back to energy storage devices [28].

The modelled network in Figure 3 shows the positions of the solar PV units dispersed in the LV network. They are incrementally placed at successive load points where the worst voltages are recorded after each iteration. Table 6 is a summary of the results of the simulations on the LV network.

The transformer kVA loading base case is 169 kVA without PV penetration for the period (Table 6). An initial PV penetration of 12% represents a 19.93 kVA inverter output. Additional inverter constant operating conditions are 147.3 kVA, 121.6 kW, and 2.75 kVA (losses) and 7.73 A. At this initial condition of no PV penetration, the worst bus voltage after the initial simulation is registered at location A6/10, as shown in Table 6. Therefore, the next solar PV injection is situated at A6/10, and the simulation is run to determine the next operating conditions of the distribution transformer. It is anticipated that any excess generation is fed back into the grid upstream of the transformer when the injected PV production exceeds the load requirement, as suggested in (8).

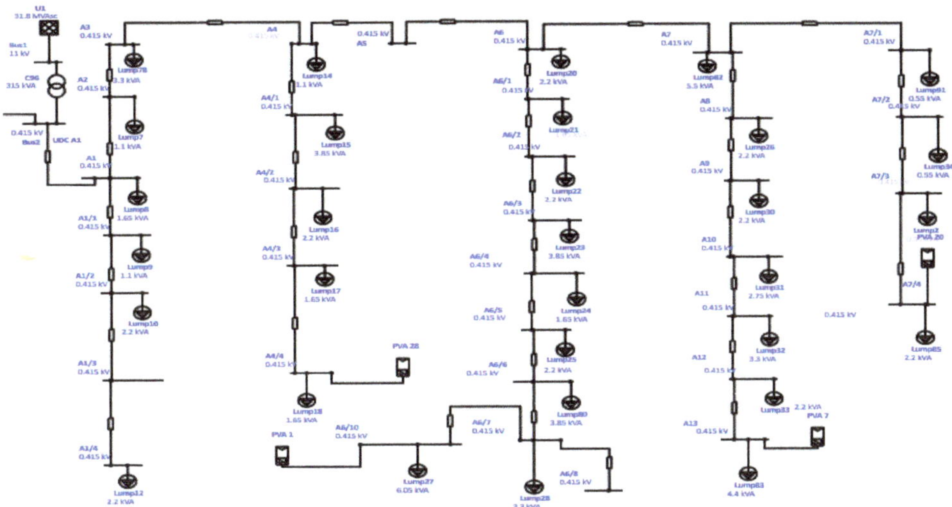

Figure 3. ETAP one-line diagram for a section of the PV-solar-integrated real urban ECG network.

Table 6. Summary of simulation results of PV-solar-integrated real urban ECG network.

PV Peak Inverter Output/KVA	PV Location with Worst Bus Voltage	PV Penetration/%	Transformer Operating Load/kVA	Transformer Operating Load/KW	Magnitude of Transformer Losses (kVA)	Transformer Operating Current/A
0	A6/10	0	169	144.5	3.61	8.87
19.93	A13	12	147.3	121.6	2.75	7.73
39.86	D7/4	25	128.6	101.1	2.08	6.75
59.79	D8/3	37	109.5	79.55	1.53	5.75
79.72	C12	50	92.73	59.04	1.08	4.87
99.65	A7/4	62	81.76	39.57	0.86	4.29
119.58	D6/4	75	78.31	21.08	0.72	4.11
139.51	C6/3	87	79.14	2.13	0.81	4.15
159.44	A1/4	99	81.3	−16.42	0.86	4.27
179.37	-	112	87.5	−35.51	0.94	4.59

4.1. Analysis of Transformer Operating Loads

Simulation results from the ECG network were used to create statistical models and graphs to show how PV penetration and transformer operating parameters are correlated. The analysis of these models is presented.

4.1.1. Transformer kVA Loading

As the PV penetration increases, the transformer operating load (kVA) decreases. In this case, capacity is freed at the transformer due to solar production in the network. Further PV penetration results in reversed power flow, leading to current reversal into the substation transformer. Figure 4 is the characteristic graph showing the transformer decreasing operating kVA loading with an increase in PV penetration. The turning point indicates current reversal into the substation transformer. The increased RPF increases the transformer operating load beyond full load conditions [23]. Reference [16] obtained similar results for impact studies in the MV/LV substation transformers.

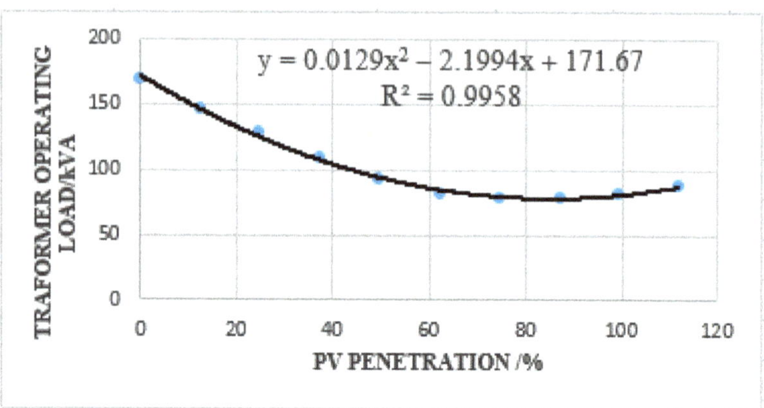

Figure 4. Graphical analysis of transformer kVA loading.

4.1.2. Transformer kW Loading

Figure 5 characterises the transformer's active power flow per penetration, resulting from increased PV injection into the network. This shows a negative linear relationship between the active power operating load and the PV depth of penetration. The graph involves a sign change at the zero-crossing, beyond which reverse power flows. A similar result is obtained in [55] for the line real power in the IEEE 9-bus system under high PV impact. The maximum depth of penetration is determined at the zero-crossing point.

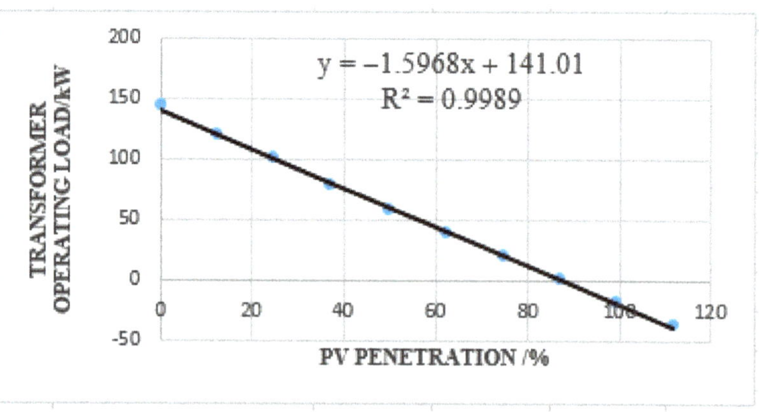

Figure 5. Graphical analysis of transformer kW loading.

4.1.3. Transformer Percentage Loading

Transformer loadings are required to meet the requirement for loads and load growth in a distribution system. With the increased solar PV penetration, capacity is initially freed up to reduce the percentage loading on the substation transformer. In Figure 6, it is shown that beyond the maximum penetration level of 88.3%, the percentage loading increases.

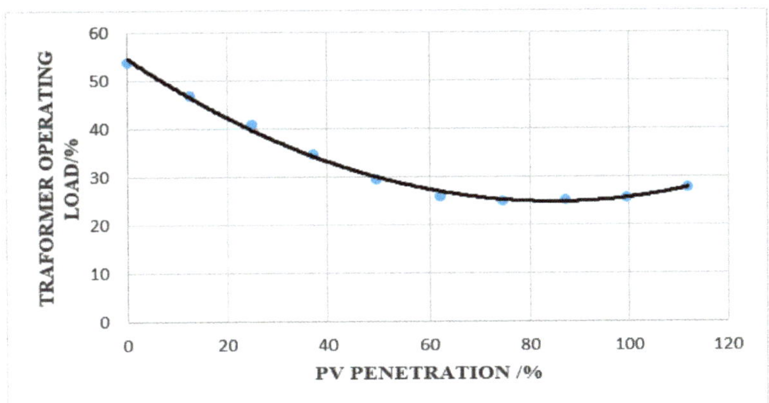

Figure 6. Graphical analysis of transformer percentage loading.

Reference [1] obtained similar results on the effect of high-impact PV in an LV network transformer. It is also shown that the overload conditions increase the transformer load losses beyond the maximum depth of PV penetration.

4.1.4. Transformer Load Losses

The impact of PV penetration on system losses is illustrated in Figure 6. The magnitude of the transformer load losses decreases when the PV penetration is increased. When the PV generation increases, capacity is freed on the grid, resulting in reduced losses in the transformer. At one particular point and beyond, the PV generation exceeds the local demand and RPF occurs, which could overload the transformer. Overloaded transformers incur more load losses as a result of increased active and reactive currents. In Figure 7, it is observed that the magnitude of the losses follows a U-shaped curve such that increasing the penetration level decreases energy losses; however, beyond 88.30% penetration the losses start to increase. Reference [55] obtained similar results for overall system losses for the IEEE 9-bus system. Using photovoltaic micro-installations in a low-voltage network, the authors in [56] obtained similar results for the system losses while considering variable weather conditions.

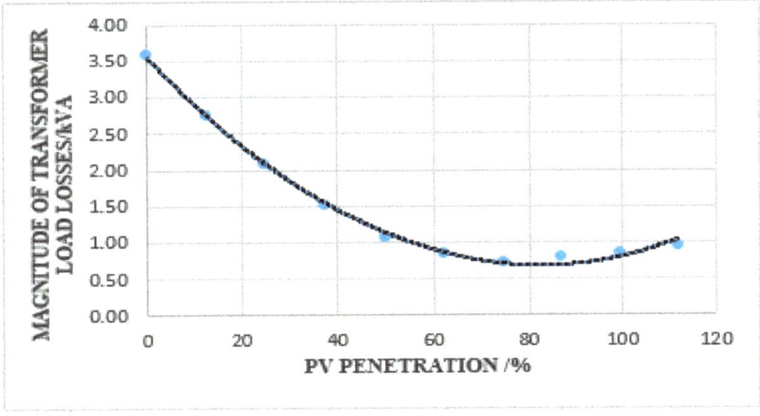

Figure 7. Graphical analysis of total transformer losses.

4.1.5. Transformer Load Current

The impact of PV penetration on transformer loading parameters must be considered while planning the network [57]. In general, the capacity of the substation transformer to accommodate power flow with the injection of solar PV is limited by the thermal current rating of the transformer [58]. It is observed in Figure 8 that the PV penetration initially reduces the load current, freeing up capacity on the transformer. As the penetration increases, resulting in RPF, the load current starts to increase. As the case shows, successive penetrations of the PV system would increase the load current beyond the full load current rating to damage the transformer.

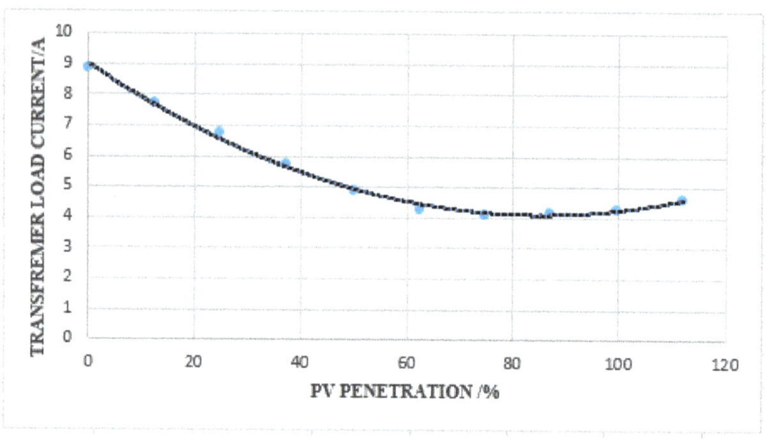

Figure 8. Graphical analysis of transformer load current.

4.1.6. Overloading of Distribution Transformers

In a continuous operation, distribution transformers should not be loaded beyond their power ratings. Overloading a transformer can reduce its lifespan [26]. In some cases, however, harmonics can cause a transformer to generate heat. This means that, though the transformer is not overloaded, harmonic loads in the network can cause high currents to overheat the transformer [59]. In the case of solar PV penetration into the LV network, reverse power flows into the substation transformer, overloading it beyond its rated power. Therefore, increased penetration must be limited to prevent cases of transformer overload due to reverse power flow. These limitations are different from the backflow limits due to reverse power flow in a PV-connected grid system considered in this study.

4.2. Transformer Backflow and Overload Limits

In this study, loading criteria are set for the substation transformer based on solar PV penetration. Base case limit criteria are, however, necessary for comparing the load limits of the transformer without PV injection and with PV injection. The base case requires that the transformer be loaded at its rated value to determine the maximum operating parameters for safe operations. In Table 7, the operating parameters for transformer overload at zero penetration are shown. A 100% operating load corresponds to 315.1 kVA and 265 kW operating apparent power and active power, respectively. Beyond these threshold values, the transformer is overloaded.

Table 7. Summary of transformer overload and backflow limits.

Transformer Loading Limits	PV Penetration/%	Transformer Operating Load/kVA	Transformer Operating Load/%	Transformer Operating Load/kW	Transformer NetOperating Load/A	Transformer Load Losses/kW
Without PV Injection	0	315.1	100	265	16.54	6.99
Injection With PV	88.30	78.04	24.77	0	4.28	0.55

Sustained and increased reverse power flow can result in transformer overload beyond its rated value [57]. With solar PV injection, each distribution circuit should have a maximum capacity for accommodating distributed production. In the present study, this implies that when the installed generation on a circuit has reached its maximum, at a point just before RPF, no further applications can be accepted for a solar PV unit, regardless of size. This is the basic idea behind this research. Hence, the statistical models obtained from the simulation results are used to interpolate results for the critical backflow limits of the distribution transformer. At a critical state of current flow, the net power flow in the transformer is zero, referring to (6). The model for the transformer's active power, P_T^{kW}, obtained from Figure 5, is presented as follows:

$$P_T^{kW} = -1.60D + 141.01 \tag{15}$$

Equation (15) predicts the active power flow at each solar PV depth of penetration. Hence, the maximum depth of penetration at the zero-crossing point, beyond which RPF exists, can be estimated. When this threshold is exceeded, RPF begins to develop and the active power component of the transformer loadings adopts negative values. In this scenario, the transformer net current decreases to a critical minimum value of 4.28 A and begins to rise in the reverse direction towards the medium voltage substation within the reverse power flow mode (Figure 8). The model in (16) obtained from Figure 4 predicts the transformer operating kVA at various PV depths of penetrations:

$$P_T^{kVA} = 0.0129D^2 - 2.1994D + 171.67 \tag{16}$$

Using the models in (15) and (16), the maximum depth of penetration is estimated as 88.30%, which corresponds to transformer backflow limits of 78.04 kVA and 24.77%, respectively (Table 7). At the maximum depth of penetration, the predicted load current, beyond which reverse power flow can be found, is 4.28 A. It is observed that the loading limits with and without PV penetration are significantly wide apart. This is because the backflow limits are supposed to be the minimum operating conditions of the transformer just before reverse power flows. With increased PV penetration, these operating conditions approach the overload conditions of the transformer obtained in the base case scenario. This can be a result of sustained and increasing reverse power flow beyond the backflow limits.

From the above analysis, it follows that the advanced knowledge of these backflow limits necessitates control schemes to prevent damage to protection systems. In addition, the technique is necessary for system planning purposes, especially for the adoption of battery energy storage systems [28].

The implementation of the techniques in this study faces several challenges. We assumed lumped-distributed loads without regard to the load distributions in the phases. Loads were also modelled as static loads, which are not true representations of the constantly changing load patterns in a real network. The study does not consider the temporal and spatial variations in solar PV output across the network.

Furthermore, the PV production is dispersed across the grid, which minimises current violations more than if the PV production were concentrated in a single location. Additionally, the results obtained are tailored to the grid layout and load profiles for a typical urban network. Other regions with differing load patterns have to be explored on an individual basis. However, this should not have an impact on the case study's applicability.

5. Conclusions

One of the concerns of utility planners is the loss or degradation of transformer life caused by an overload due to increased PV penetration. Studies show that reverse power flow due to increased PV penetration creates overload conditions in substation transformers. To mitigate this, researchers suggest utilising various control energy storage schemes to store excess PV production during peak load periods.

However, the methodology adopted in this study is to find transformer operating thresholds for reverse power flow, which can result in overload conditions as a result of excessive PV penetration. Using simulation results from a radial test LV network, a statistical approach is used to create correlation models between solar PV penetration depth and transformer operating loads.

The correlation models predict transformer backflow limits due to high solar PV grid penetration as follows: Based on the transformer loading threshold, a maximum depth of penetration of 88.30% is obtained. At this penetration limit, interpolations using correlation models indicate transformer backflow operating load limits of 78.04 kVA and 24.77% at an operating current of 4.28 A. These limitations are contrasted with the transformer overload criteria, determined without PV penetration to demonstrate how a sustained and increasing reverse power flow, which exceeds the backflow limits, can cause transformer overload.

The simulation studies' results provide useful information not only on the impact of RPF on distribution transformer loadings but also on the depth of penetration in a solar PV-integrated LV network. The study determines a set of safe margins to safeguard the flow of reverse power into the substation transformer. Further studies to investigate the impacts of PV penetration should consider new mitigation techniques aimed at protecting substation transformers from overload conditions with high PV penetration. The backflow limitations must be established for the entire network heuristically by performing typical-year simulations.

Author Contributions: Conceptualization, I.B.M.; Formal analysis, N.I.N.; Methodology, I.B.M.; Software, I.B.M.; Supervision, N.I.N.; Writing—original draft, I.B.M.; Writing—review and editing, N.I.N. All authors have read and agreed to the published version of the manuscript.

Funding: This research received no external funding.

Data Availability Statement: Not applicable.

Acknowledgments: The authors would like to acknowledge the technical support provided by the Electricity Company of Ghana (ECG) in providing needed data for this research.

Conflicts of Interest: The authors declare no conflict of interest.

References

1. Kenneth, A.P.; Folly, K. Voltage Rise Issue with High Penetration of Grid Connected PV. *IFAC Proc. Vol.* **2014**, *19*, 4959–4966. [CrossRef]
2. Dondariya, C.; Porwal, D.; Awasthi, A.; Shukla, A.K.; Sudhakar, K.; Murali, M.M.; Bhimte, A. Performance Simulation of Grid-Connected Rooftop Solar PV System for Small Households: A Case Study of Ujjain, India. *Energy Rep.* **2018**, *4*, 546–553. [CrossRef]
3. Panigrahi, R.; Mishra, S.K.; Srivastava, S.C.; Srivastava, A.K.; Schulz, N.N. Grid Integration of Small-Scale Photovoltaic Systems in Secondary Distribution Network—A Review. *IEEE Trans. Ind. Appl.* **2020**, *56*, 3178–3195. [CrossRef]
4. Chathurangi, D.; Jayatunga, U.; Perera, S.; Agalgaonkar, A.P.; Siyambalapitiya, T. A Nomographic Tool to Assess Solar PV Hosting Capacity Constrained by Voltage Rise in Low-Voltage Distribution Networks. *Int. J. Electr. Power Energy Syst.* **2022**, *134*, 107409. [CrossRef]
5. Kadir, A.F.A.; Khatib, T.; Elmenreich, W. Integrating Photovoltaic Systems in Power System: Power Quality Impacts and Optimal Planning Challenges. *Int. J. Photoenergy* **2014**, *2014*, 321826.
6. Ismael, S.M.; Aleem, S.H.E.A.; Abdelaziz, A.Y.; Zobaa, A.F. Probabilistic Hosting Capacity Enhancement in Non-Sinusoidal Power Distribution Systems Using a Hybrid PSOGSA Optimization Algorithm. *Energies* **2019**, *12*, 1018. [CrossRef]
7. Holguin, J.P.; Rodriguez, D.C.; Ramos, G. Reverse Power Flow (RPF) Detection and Impact on Protection Coordination of Distribution Systems. *IEEE Trans. Ind. Appl.* **2020**, *56*, 2393–2401. [CrossRef]
8. Saad, S.N.M.; van der Weijde, A.H. Evaluating the Potential of Hosting Capacity Enhancement Using Integrated Grid Planning Modeling Methods. *Energies* **2019**, *12*, 3610. [CrossRef]
9. Afonaa-Mensah, S.; Wang, Q.; Uzoejinwa, B.B. Investigation of Daytime Peak Loads to Improve the Power Generation Costs of Solar-Integrated Power Systems. *Int. J. Photoenergy* **2019**, *2019*, 5986874. [CrossRef]
10. Cohen, M.A.; Callaway, D.S. Effects of Distributed PV Generation on California's Distribution System, Part 1: Engineering Simulations. *Sol. Energy* **2016**, *128*, 126–138. [CrossRef]
11. Agah, S.M.M.; Abyaneh, H.A. Quantification of the Distribution Transformer Life Extension Value of Distributed Generation. *IEEE Trans. Power Deliv.* **2011**, *26*, 1820–1828. [CrossRef]
12. Von Appen, J.; Braun, M.; Stetz, T.; Diwold, K.; Geibel, D. Time in the sun: The challenge of high PV penetration in the German electric grid. *IEEE Power Energy Mag.* **2013**, *11*, 55–64. [CrossRef]
13. Jahangiri, P.; Aliprantis, D.C. Distributed Volt/VAr Control by PV Inverters. *IEEE Trans. Power Syst.* **2013**, *28*, 3429–3439. [CrossRef]
14. Ameur, A.; Berrada, A.; Loudiyi, K.; Aggour, M. Analysis of Renewable Energy Integration into the Transmission Network. *Electr. J.* **2019**, *32*, 106676. [CrossRef]
15. Mulenga, E.; Bollen, M.H.J.; Etherden, N. Distribution Networks Measured Background Voltage Variations, Probability Distributions Characterization and Solar PV Hosting Capacity Estimations. *Electr. Power Syst. Res.* **2021**, *192*, 106979. [CrossRef]
16. Ebe, F.; Idlbi, B.; Morris, J.; Heilscher, G.; Meier, F. Evaluation of PV Hosting Capacities of Distribution Grids with Utilisation of Solar Roof Potential Analyses. *CIRED—Open Access Proceedings Journal* **2017**, *2017*, 2265–2269. [CrossRef]
17. Sharma, V.; Aziz, S.M.; Haque, M.H.; Kauschke, T. Effects of High Solar Photovoltaic Penetration on Distribution Feeders and the Economic Impact. *Renew. Sustain. Energy Rev.* **2020**, *131*, 110021. [CrossRef]
18. Guo, Y.; Wu, Q.; Gao, H.; Shen, F. Distributed Voltage Regulation of Smart Distribution Networks: Consensus-Based Information Synchronization and Distributed Model Predictive Control Scheme. *Int. J. Electr. Power Energy Syst.* **2019**, *111*, 58–65. [CrossRef]
19. Daher, N.A.; Saliba, L.; Mougharbel, I.; Kanaan, H.Y.; Saad, M. Hybrid Algorithm for Monitoring Reverse Power Flow & Ause G E\Distributed Renewable Energy Sources. In Proceedings of the 2020 5th International Conference on Renewable Energies for Developing Countries (REDEC), Marrakech, Morocco, 29–30 June 2020; Volume 5, pp. 8–11.
20. Torres, I.C.; Negreiros, G.F.; Tiba, C. Theoretical and Experimental Study to Determine Voltage Violation, Reverse Electric Current and Losses in Prosumers Connected to Low-Voltage Power Grid. *Energies* **2019**, *12*, 4568. [CrossRef]
21. Aziz, T.; Ketjoy, N. PV Penetration Limits in Low Voltage Networks and Voltage Variations. *IEEE Access* **2017**, *5*, 16784–16792. [CrossRef]
22. Hajeforosh, S.F.; Khatun, A.; Bollen, M. Enhancing the Hosting Capacity of Distribution Transformers for Using Dynamic Component Rating. *Int. J. Electr. Power Energy Syst.* **2022**, *142*, 108130. [CrossRef]
23. Awadallah, M.A.; Xu, T.; Venkatesh, B.; Member, S.; Singh, B.N. On the Effects of Solar Panels on Distribution Transformers. *IEEE Trans. Power Deliv.* **2016**, *31*, 1176–1185. [CrossRef]
24. Wang, X.; Dsouza, K.; Tang, W.; Baran, M. Assessing the Impact of High Penetration PV on the Power Transformer Loss of Life on a Distribution System. In Proceedings of the 2021 IEEE PES Innovative Smart Grid Technologies Europe (ISGT Europe), Espoo, Finland, 18–21 October 2021; pp. 8–12. [CrossRef]

25. Upadhyay, P.; Kern, J.; Vadlamani, V. Distributed Energy Resources (Ders): Impact of Reverse Power Flow on Transformer. *CIGRE Sci. Eng.* **2020**, *19*, 99–107.
26. Manito, A.R.A.; Pinto, A.; Zilles, R. Evaluation of Utility Transformers' Lifespan with Different Levels of Grid-Connected Photovoltaic Systems Penetration. *Renew. Energy* **2016**, *96*, 700–714. [CrossRef]
27. Affonso, C.D.M.; Kezunovic, M. Technical and Economic Impact of Pv-Bess Charging Station on Transformer Life: A Case Study. *IEEE Trans. Smart Grid* **2019**, *10*, 4683–4692. [CrossRef]
28. Novoa, L.; Flores, R.; Brouwer, J. Optimal Renewable Generation and Battery Storage Sizing and Siting Considering Local Transformer Limits. *Appl. Energy* **2019**, *256*, 113926. [CrossRef]
29. Rafael, F.; Sevilla, S.; Knazkins, V.; Korba, P.; Kienzle, F. Limiting Transformer Overload on Distribution Systems with High Penetration Limiting Transformer Overload on Distribution Systems with High Penetration of PV Using Energy Storage Systems. *Int. J. Emerg. Technol. Adv. Eng.* **2016**, *6*, 294–303.
30. De Carne, G.; Buticchi, G.; Zou, Z.; Liserre, M. Reverse Power Flow Control in a ST-Fed Distribution Grid. *IEEE Trans. Smart Grid* **2018**, *9*, 3811–3819. [CrossRef]
31. Baroni, B.R.; Uturbey, W.; Costa, A.M.G.; Rocha, S.P. Impact of Photovoltaic Generation on the Allowed Revenue of the Utilities Considering the Lifespan of Transformers: A Brazilian Case Study. *Electr. Power Syst. Res.* **2021**, *192*, 106906. [CrossRef]
32. Uçar, B.; Bağrıyanık, M.; Kömürgöz, G. Influence of PV Penetration on Distribution Transformer Aging. *J. Clean Energy Technol.* **2017**, *5*, 131–134. [CrossRef]
33. El Batawy, S.A.; Morsi, W.G. On the Impact of High Penetration of Rooftop Solar Photovoltaics on the Aging of Distribution Transformers. *Can. J. Electr. Comput. Eng.* **2017**, *40*, 93–100. [CrossRef]
34. Pezeshki, H.; Wolfs, P.J.; Ledwich, G. Impact of High PV Penetration on Distribution Transformer Insulation Life. *IEEE Trans. Power Deliv.* **2014**, *29*, 1212–1220. [CrossRef]
35. Pollock, J.; Hill, D. Overcoming the Issues Associated with Operating a Distribution System in Reverse Power Flow. *IET Conf. Publ.* **2016**, *2016*, 2–7. [CrossRef]
36. Liu, H.; Guo, Y.; Ge, S.; Zhao, M. Impact of DG Configuration on Maximum Use of Load Supply Capability in Distribution Power Systems. *J. Appl. Math.* **2014**, *2014*, 136726. [CrossRef]
37. Anto, E.K.; Frimpong, E.A.; Okyere, P.Y. Impact Assessment of Increasing Solar PV Penetrations on Voltage and Total Harmonic Distortion of a Distribution Network. *J. Multidiscip. Eng. Sci. Stud.* **2015**, *1*, 116–127.
38. Obeng, M.; Gyam, S.; Sarfo, N.; Kabo-bah, A.T. Technical and Economic Feasibility of a 50 MW Grid-Connected Solar PV at UENR Nsoatre Campus. *J. Clean Prod.* **2020**, *247*, 119159. [CrossRef]
39. Kwofie, E.A.; Mensah, G.; Anto, E.K. Determination of the Optimal Power Factor at Which DG PV Should Be Operated. In Proceedings of the 2017 IEEE PES PowerAfrica, Accra, Ghana, 27–30 June 2017; pp. 391–395.
40. Kwofie, E.A.; Mensah, G.; Antwi, V.S. Post Commission Grid Impact Assessment of a 20 MWp Solar PV Grid Connected System on the ECG 33 KV Network in Winneba. In Proceedings of the IEEE PES/IAS PowerAfrica Conference: Power Economics and Energy Innovation in Africa, PowerAfrica 2019, Abuja, Nigeria, 20–23 August 2019; pp. 521–526.
41. Paghdar, S.; Sipai, U.; Ambasana, K.; Chauhan, P.J. Active and Reactive Power Control of Grid Connected Distributed Generation System. In Proceedings of the 2017 Second International Conference on Electrical, Computer and Communication Technologies (ICECCT), Coimbatore, India, 22–24 February 2017; pp. 1–7. [CrossRef]
42. Sadeghian, H.; Athari, M.H.; Wang, Z. Optimized Solar Photovoltaic Generation in a Real Local Distribution Network. In Proceedings of the 2017 IEEE Power and Energy Society Innovative Smart Grid Technologies Conference, ISGT 2017, Washington, DC, USA, 23–26 April 2017; pp. 1–5.
43. Alsafasfeh, Q.; Saraereh, O.A.; Khan, I.; Kim, S. Solar PV Grid Power Flow Analysis. *Sustainability* **2019**, *11*, 1744. [CrossRef]
44. Gnanadass, B.M.R. Solar Integrated Time Series Load Flow Analysis for Practical Distribution System. *J. Inst. Eng. Ser. B* **2021**, *102*, 829–841. [CrossRef]
45. Vinayagam, A.; Aziz, A.; PM, B.; Chandran, J.; Veerasamy, V.; Gargoom, A. Harmonics Assessment and Mitigation in a Photovoltaic Integrated Network. *Sustain. Energy Grids Netw.* **2019**, *20*, 100264. [CrossRef]
46. Thomson, M.; Infield, D.G. Impact of Widespread Photovoltaics Generation on Distribution Systems. *IET Renew. Power Gener.* **2007**, *1*, 33–40. [CrossRef]
47. Zain, M.; Ellabban, O.; Refaat, S.S.; Abu-rub, H.; Al-fagih, L. A Novel Methodology to Determine the Maximum PV Penetration in Distribution Networks. In Proceedings of the 2019 2nd International Conference on Smart Grid and Renewable Energy (SGRE), Doha, Qatar, 19–21 November 2019; pp. 1–6. [CrossRef]
48. Kirui, K.P.; Murage, D.K.; Kihato, P.K. Fuse-Fuse Protection Scheme ETAP Model for IEEE 13 Node Radial Test Distribution Feeder. *Eur. J. Eng. Res. Sci.* **2019**, *4*, 224–234. [CrossRef]
49. Setiawan, A.; Qashtalani, H.; Pranadi, A.D.; Ali, F.C.; Setiawan, E.A. Determination of Optimal PV Locations and Capacity in Radial Distribution System to Reduce Power Losses. *Energy Procedia* **2019**, *156*, 384–390. [CrossRef]
50. Broderick, R.; Williaams, J.; Munoz-Ramos, K. Clustering Methods and Feeder Selection for PV System Impact Analysis. Available online: https://www.epri.com/research/products/000000003002002562 (accessed on 28 June 2022).
51. Al-Saadi, H.; Zivanovic, R.; Al-Sarawi, S.F. Probabilistic Hosting Capacity for Active Distribution Networks. *IEEE Trans. Ind. Inform.* **2017**, *13*, 2519–2532. [CrossRef]

52. Hoke, A.; Butler, R.; Hambrick, J.; Kroposki, B. Steady-State Analysis of Maximum Photovoltaic Penetration Levels on Typical Distribution Feeders. *IEEE Trans. Sustain. Energy* **2013**, *4*, 350–357. [CrossRef]
53. Tsamaase, K.; Moyo, U.; Zibani, I.; Ngebani, I.; Mahindroo, P. Approximate Mathematical Model for Load Profiling and Demand Forecasting. *IOSR J. Electr. Electron. Eng. (IOSR-JEEE)* **2017**, *12*, 29–34. [CrossRef]
54. Li, X.; Borsche, T.; Andersson, G. PV Integration in Low-Voltage Feeders with Demand Response. In Proceedings of the 2015 IEEE Eindhoven PowerTech, Eindhoven, The Netherlands, 29 June–2 July 2015. [CrossRef]
55. Khan, M.A.; Arbab, N.; Huma, Z. Voltage Profile and Stability Analysis for High Penetration Solar Photovoltaics. *Int. J. Eng. Work.* **2018**, *5*, 109–114.
56. Korab, R.; Połomski, M.; Smołka, M. Evaluating the Risk of Exceeding the Normal Operating Conditions of a Low-Voltage Distribution Network Due to Photovoltaic Generation. *Energies* **2022**, *15*, 1969. [CrossRef]
57. Datta, U.; Kalam, A.; Shi, J. Smart Control of BESS in PV Integrated EV Charging Station for Reducing Transformer Overloading and Providing Battery-to-Grid Service. *J. Energy Storage* **2020**, *28*, 101224. [CrossRef]
58. Raouf Mohamed, A.A.; Best, R.J.; John Morrow, D.; Cupples, A.; Bailie, I. Impact of the Deployment of Solar Photovoltaic and Electrical Vehicle on the Low Voltage Unbalanced Networks and the Role of Battery Energy Storage Systems. *J. Energy Storage* **2021**, *42*, 102975. [CrossRef]
59. Yaghoobi, J.; Alduraibi, A.; Martin, D.; Zare, F.; Eghbal, D.; Memisevic, R. Impact of High-Frequency Harmonics (0–9 KHz) Generated by Grid-Connected Inverters on Distribution Transformers. *Int. J. Electr. Power Energy Syst.* **2020**, *122*, 106177. [CrossRef]

Article

A Model Predictive Control Based Optimal Task Allocation among Multiple Energy Storage Systems for Secondary Frequency Regulation Service Provision

Xiuli Wang [1], Xudong Li [1,*], Weidong Ni [2] and Fushuan Wen [3]

1   School of Electric Power, Civil Engineering and Architecture, Shanxi University, Taiyuan 030031, China
2   Guodian Nanjing Automation Co., Ltd., Nanjing 210032, China
3   School of Electrical Engineering, Zhejiang University, Hangzhou 310027, China
*   Correspondence: 202023504012@e-mail.sxu.edu.cn; Tel.: +86-150-3400-6366

Abstract: Power system stability has been suffering increasing threats with the ever-growing penetration of intermittent renewable generation such as wind power and solar power. Battery energy storage systems (BESSs) could mitigate frequency fluctuation of the power system because of their accurate regulation capability and rapid response. By dividing the Area Control Error (ACE) signal and the State of Charge (SOC) of battery energy storage systems into different intervals, the frequency control task of BESSs could be determined by considering the frequency control demand of the power grid in each interval and SOC self-recovery. The well-developed model predictive control can be employed to attain the optimal control variable sequence of BESSs in the following time, which can determine the output depths of BESSs and action timing at different intervals. The simulation platform MATLAB/Simulink is used to build two secondary frequency control scenarios of BESSs for providing frequency regulation service. The feasibility of the presented strategy is demonstrated by simulation results of a sample system.

Keywords: secondary frequency control; model predictive control; self-recovery of State of Charge (SOC); frequency regulation; area control error (ACE)

Citation: Wang, X.; Li, X.; Ni, W.; Wen, F. A Model Predictive Control Based Optimal Task Allocation among Multiple Energy Storage Systems for Secondary Frequency Regulation Service Provision. *Energies* **2023**, *16*, 1228. https://doi.org/10.3390/en16031228

Academic Editor: Juri Belikov

Received: 9 September 2022
Revised: 5 October 2022
Accepted: 10 October 2022
Published: 23 January 2023

1. Introduction

Extensive integration of renewable energy sources has become one of the key measures to address global resource shortage and climate change [1]. However, the widely employed renewable energy based generators, such as wind power and solar power, do not have the inertia and damping characteristics of traditional thermal power units. In addition, the intermittent and uncertain generation outputs impose challenges for power system security and stability, and frequency control/regulation is becoming more difficult [2,3]. Traditional frequency control no longer meets the ever-increasing requirement for frequency control. Hence, more efficient frequency control is highly demanding.

In recent years, electrochemical energy storage technology has developed rapidly, with material and production costs greatly reduced. Battery energy storage technology has been widely applied in power grid frequency control. The Battery Energy Storage System (BESS) has the advantages of large capacity, fast response, and bidirectional regulation; therefore, it has good prospects in power grid frequency control [4,5]. Energy storage-assisted thermal power unit frequency control is more suitable for the current power grid frequency control, considering the capacity and cost of energy storage. For example, Shanxi Jingyu Power Plant added a 9 MW lithium battery energy storage system to assist the frequency control of thermal power units [6]. This BESS improved the power plant's frequency quality and frequency control income. The Beijing Shijingshan Thermal Power Plant implemented a 2 WM demonstration project, which uses a BESS to improve frequency control performance in the context of high renewable energy integration [7]. The successful operation of the above projects demonstrate the feasibility of BESSs in secondary frequency regulation provision.

Much research work has been done on the employment of BESSs in providing secondary frequency control/regulation. In reference [8], frequency control signals were allocated according to the rated power of the BESS, without considering the SOC state of the battery, and it was easy to reach the capacity limit of the battery. Some recent literature considers the state of SOC, such as reference [9]: the coefficients of virtual droop control and virtual inertia control are adaptively controlled and the output power of the BESS is controlled according to SOC state so as to improve the economy of the whole unit operation. The authors in [10,11] discussed the energy storage battery's smooth output according to SOC state based on the fuzzy control theory, effectively reducing the system's demand for energy storage capacity. In order to avoid the phenomenon of over-limit in the actual working process of the energy storage system, the control strategy of switching state between charging, discharging mode, and Frequency regulation mode for battery energy storage was proposed in [12]. The literature [13] has proposed using the logistic regression function to construct the control law of energy storage adaptive frequency control and self-recovery state of charge, ensuring good SOC while adjusting the frequency. In [14], based on the Area Control Error (ACE) signal mode, the ACE high-frequency signal obtained through the low-pass filter is allocated to the BESS, reducing its frequency control pressure.

Most of the above studies focus on the coordination strategy between the power grid frequency control demand and SOC self-recovery without considering the loss caused by frequent unit operations. They also failed to acknowledge that BESS can accurately respond to high-frequency signals. For example, in [14], high-frequency components were allocated to energy storage, which reduced the operation times of the unit to a certain extent but could not change the magnitude of high-frequency components borne by the BESS in real time according to the SOC. Moreover, most of these strategies followed the traditional PI control method, which cannot achieve satisfactory frequency control effects in the current complex frequency control scene.

Considering the above problems, we propose a control strategy based on the Model Predictive Control (MPC) for energy storage systems to help thermal power units participate in secondary frequency control, which studies and formulates the output and self-recovery strategies of the energy storage battery. In this strategy, Empirical Mode Decomposition (EMD) partitions the frequency control signal ACE into high- and low-frequency components. According to the different output characteristics of the thermal power unit and the BESS, the high-frequency components are assigned to the BESS, and the unit bears the low-frequency components. In each ACE interval, the SOC recovery demands of the BESS and power grid frequency control are considered to dynamically change the high-frequency component size of the energy storage system and determine the optimal model predictive control output weighting matrix, the output target, and output depth of the traditional unit and energy storage system. The simulation results showed that our proposed method maintains good SOC, improves frequency control performance, and reduces unit wear.

2. Dynamic Model of Secondary Frequency Control with Energy Storage

2.1. Frequency Control Model of Two Area Interconnection Systems

We studied the frequency control dynamic model for two interconnected areas and established a power grid equivalent model for two interconnected area systems with BESS. The AGC control mode is set to Tie-Line Bias Control Mode. Figure 1 shows the energy storage system configured for area 2.

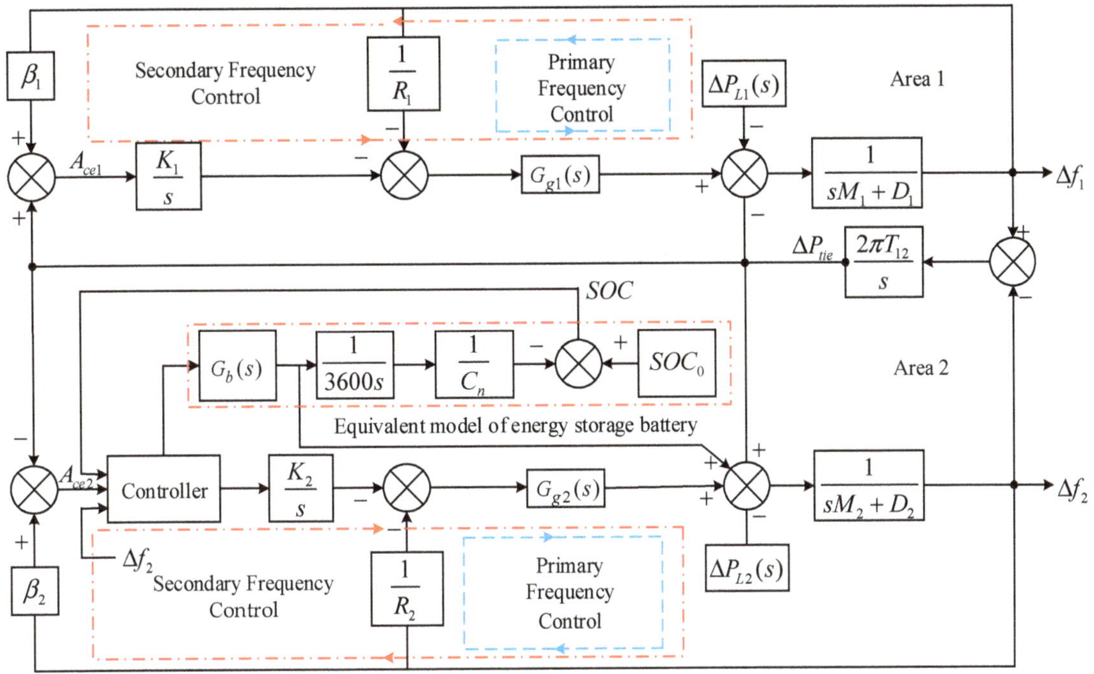

Figure 1. Frequency control model of the two-zone system with energy storage devices.

In Figure 1, $G_{gi}(s)$ is the simplified model of the generator set comprising the governor model $G_{gov}(s)$ and reheat turbine model $G_T(s)$ in series, and $i = 1, 2$ represents the two regions. ΔP_{Li}, ΔP_{ti}, ΔP_{ri}, ΔX_{gi}, ΔP_{tie}, and Δf_i are the load disturbance, turbine output power increment, reheater output power increment, governor position increment, tie-line power deviation, and frequency deviation, respectively. β_i and R_i are the power system deviation coefficient and unit adjustment coefficient, respectively. M_i and D_i are the regional inertia coefficient and regional damping coefficient, respectively. T_{12} is the synchronization coefficient of the tie line; A_{cei} is the regional control deviation; C_n is the rated capacity of energy storage; S_{oc0} and S_{oc} represent the initial SOC and real-time SOC for BESSs.

The transfer function of the governor model for traditional thermal power units:

$$G_{gov}(s) = \frac{1}{1 + sT_g} \tag{1}$$

where T_g represents the governor model's time constant for the thermal power unit.

The transfer function of the conventional reheat turbine model:

$$G_T(s) = \frac{1 + sF_{HP}T_{RH}}{(1 + sT_{CH})(1 + sT_{RH})} \tag{2}$$

where F_{HP}, T_{RH}, and T_{CH} are the reheater gain, reheater time constant, and turbine time constant, respectively.

The transfer function of the traditional thermal power unit model is:

$$G_{gi}(s) = G_{gov}(s)G_T(s) \tag{3}$$

ACE can be attained by

$$\begin{cases} A_{ce1} = \Delta P_{12} + \beta_1 \Delta f_1 \\ A_{ce2} = \Delta P_{21} + \beta_2 \Delta f_2 \end{cases} \tag{4}$$

where A_{ce1} and A_{ce2} are respectively the ACE signals of area 1 and area 2.

2.2. Equivalent Model of Battery Energy Storage Systems in Power Grid Frequency Control

Electrochemical energy storage usually comprises a battery body and a corresponding energy conversion system. The general model for electrochemical energy storage is greatly simplified compared to the entity [15]. The energy conversion system determines the main output characteristics of the BESS. The first-order inertia link usually replaces the delay characteristics of the energy storage system, and the battery gain replaces the relationship between power and frequency change. The product model of the first-order inertia link and the battery gain is used as the equivalent model for the BESS. This method is widely used in current research on BESSs participating in power grid frequency control. The author of [16] verifies that this model can be applied to frequency control research under appropriate circumstances.

The equivalent transfer function model of BESSs:

$$G_b(s) = \frac{K_B}{1 + sT_b} \tag{5}$$

where T_b and K_B are the time constant and battery gain of the BESS, respectively.

The SOC of BESS cannot be measured directly; therefore, it is difficult to determine high-precision monitoring. Currently, there are two indirect methods to monitor the SOC of a BESS. The first method is to measure the terminal voltage of the battery and then transfer the voltage to the SOC. The second method directly integrates the BESS output and obtains the SOC. We adopted the second method to measure the SOC of BESS.

3. Dynamic Model of Secondary Frequency Control with Energy Storage

3.1. Model Predictive Control

MPC has been used in the control field since the 1990s. Compared to the traditional PI controller, the MPC controller has superior performance and satisfies all operational constraints by minimizing the value of the objective function to determine the control variables [17]. Different operation objectives can be achieved by setting the appropriate weight factor value in the MPC controller's objective function. The MPC algorithm includes three parts: system prediction model, rolling optimization, and feedback correction, which have obvious advantages in dealing with constraints.

3.2. Application of MPC in Secondary Frequency Control

Firstly, we need to collect model data of AGC system in grid, model data of the BESS and load disturbance data before we use MPC. Secondly the state space expression of the system is established according to the model data of the BESS and load disturbance data. Finally, the optimal control variables of BESS are obtained through the prediction model, rolling optimization, and feedback correction, which can balance the SOC recovery demand of BESS and the demand of frequency control to grid.

3.2.1. System Prediction Model

The prediction model predicts the system's future state according to historical information on the system's state parameters and control variables, which determine the future control input information. Therefore, a model describing the system's dynamic behavior is needed as a prediction model. This model has no strict requirements on the system's expression form as long as the model can predict the system's future state value according to the input information.

The state equation of the two-area power frequency control system with energy storage is:

$$\left\{ \begin{array}{l} \dot{X} = AX + BU + RW \\ Y = CX \end{array} \right\} \tag{6}$$

where X, U, W, and Y represent the state variable, input variable, disturbance quantity, and output variable, respectively. A, B, R, and C represent corresponding matrices determined by the parameters of the system and controller. The specific elements are as follows:

$$X = \begin{bmatrix} R_{eg1} & \Delta P_{tie} & R_{eg2} & P_b & S_{oc} \end{bmatrix}^T$$

where R_{eg1} is the state variable of the regional generator set in area i, which is expressed as follows:

$$R_{eg1} = \begin{bmatrix} \Delta f_i & \Delta P_{ti} & \Delta P_{ri} & \Delta X_{gi} & \Delta P_{ci} \end{bmatrix}^T \; i = 1,2$$

$$U = u_b$$

$$W = \begin{bmatrix} \Delta P_{L1} & \Delta P_{L2} \end{bmatrix}^T$$

$$Y = \begin{bmatrix} A_{cei} & \Delta f_i & S_{oc} \end{bmatrix}^T$$

Equation (5) uses T_s as the sampling period for discretization, and the system's discrete state-space model is obtained, as shown in Equation (7):

$$\left\{ \begin{array}{l} X(k+1) = \overline{A}X(k) + \overline{B}U(k) + \overline{R}W(k) \\ Y(k) = CX(k) \end{array} \right. \tag{7}$$

where $\overline{A} = e^{AT_s}$, $\overline{B} = \int_0^{T_s} e^{At} B dt$, $\overline{R} = \int_0^{T_s} e^{At} R dt$.

3.2.2. Rolling Optimization

Rolling optimization determines future control effects through the optimum of a certain performance index. According to the discrete state space model (7), the frequency deviation of the system, ACE signal, and SOC change of the BESS starting from time K can be predicted, and a quadratic performance index function satisfying certain constraints can be constructed, which is expressed as:

$$\min J_k = \sum_{j=1}^{p} (Y(k+j|k) - Y_r(k+j))^T Q(Y(k+j|k) - Y_r(k+j)) + \\ \sum_{i=1}^{m} U^T(k+i-1|k) R U(k+i-1|k) \tag{8}$$

$$s.t. U_{\min} \leq U(k) \leq U_{\max} \tag{9}$$

$$Y_{\min} \leq Y(k) \leq Y_{\max} \tag{10}$$

In Equation (8), Q and R are the output weighting matrix and control weighting matrix of the model predictive control, respectively; $Y(k+j|k)$ represents the output state at time $k+j$ predicted by the system at time k; $Y_r(k+j)$ represents the output reference value of the system at $k+j$ time in the future, where $j \in (1,p)$, p is the prediction time domain. In this system, the frequency deviation and ACE signal's reference values are set to 0, and the SOC reference value of the BESS is 0.5. $U(k+i-1|k)$ is the prediction of system control variables at time k at time $k+i-1$ in the future, where $i \in (1,m)$. m is the control time domain. In Equation (9), $U_{\min}$ and $U_{\max}$ represent the lower and upper limits of system control variables and refer to the power constraint of the BESS. In Equation (10), $Y_{\min}$ and $Y_{\max}$ represent the lower and upper limits of the system output variables and refer to the SOC constraints of the BESS.

The future optimal control sequence $U(k+i-1)$ starting from time k is obtained by calculating the optimal value of the objective function at the specified time. The first control variable U_k in the obtained control sequence is applied to the BESS. The process above is repeated at the next sampling time. The optimization problem is refreshed and solved after the new measurement value is obtained, and the cycle is repeated for online optimization.

3.2.3. Feedback Correction

The above prediction model and rolling optimization process belong to open-loop control. When the model predictive control is used to predict the future output, the deviation between the predicted value and the actual value is caused by interference factors, so the control fails. To establish closed-loop control, the deviation between the predicted value and the actual value is introduced as feedback in MPC for feedback correction:

$$Y_P(k+1+j) = Y(k+1+j|k+1) + h(k+1) \tag{11}$$

where Y_P is the predicted value of the output after correction; h is the weighting coefficient matrix.

The idea of applying MPC in frequency control of BESS is as shown in Figure 2.

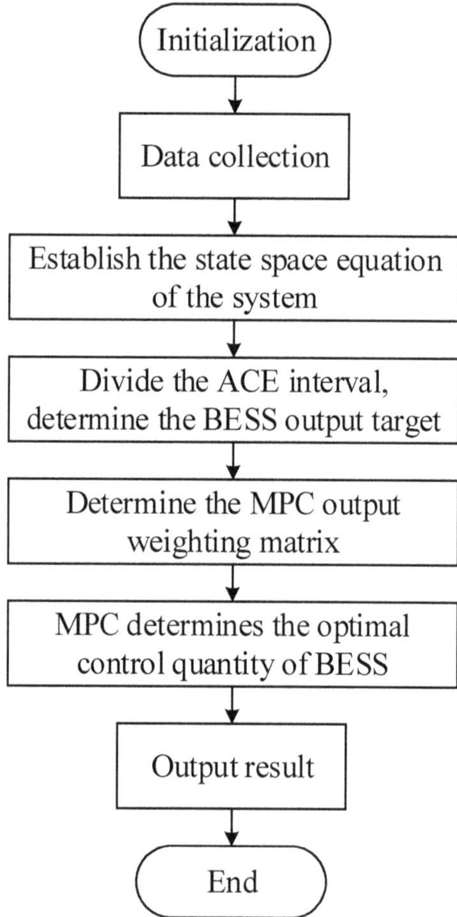

Figure 2. The flowchart of MPC application.

4. Control Strategy of Battery Energy Storage System

By formulating a reasonable output control strategy for BESSs, the frequency control effect of the system improves, and the SOC maintains a healthy state. The ACE signal's size is closely related to the output size of the BESS. According to the absolute value of ACE, it can be divided into several control intervals, including dead zone $(0 - ACE_{ce,d})$, normal state zone $(ACE_{ce,d} - ACE_{ce,n})$, sub-emergency state zone $(ACE_{ce,n} - ACE_{ce,e})$, and emergency state zone $(>ACE_{ce,e})$. Combined with the capacity limit and SOC of the battery, the output and self-recovery strategies of the BESS can be determined based on meeting the demands of frequency control while using the coordinated operation of thermal power units and BESSs.

4.1. Output Strategy of Battery Energy Storage System

4.1.1. Dead Zone

When the ACE signal is in the dead zone, the load disturbance to the system is minor, and the thermal power unit and BESS do not temporarily participate in the secondary frequency control. Moreover, in this case, we can induce the energy storage battery self-recovery condition. Keeping the SOC within the normal operating range provides more frequency control capacity in subsequent frequency control phases. In this state, the output weighting matrix of MPC is set to:

$$Q = diag(0, 0, 1) \tag{12}$$

4.1.2. Normal State Zone

The power grid is slightly disturbed when the ACE signal is in the normal state area and the system frequency control demand is minor. The battery energy storage system has sufficient capacity and can independently undertake the system frequency control demand. In this state, the advantages of fast response speed and accurate output of the BESS are exploited, and the priority output target of the BESS is set to eliminate frequency deviation. At the same time, when the SOC of the BESS is poor, the self-recovery action of energy storage begins under the power grid constraints. In this state, the output weighting matrix of MPC is set to:

$$Q = diag(1, 1, 0.2) \tag{13}$$

If the battery SOC enters the emergency recovery interval, it begins self-recovery, and the traditional unit undertakes the frequency control task.

4.1.3. Sub-Emergency State Zone

The grid system is greatly disturbed when the ACE signal reaches the sub-emergency state range. At this time, the frequency control demand on the system is significant, and the frequency control capacity of the BESS is relatively small and cannot undertake all the frequency control tasks alone. Therefore, the ACE signal is divided into high and low frequencies through EMD. The traditional unit assumes the ACE signal's low-frequency part, and the BESS assumes the ACE signal's high-frequency part. In addition, the recovery of the SOC of the BESS can be stopped in this state area to be in the safe operating range. In this state, the output weighting matrix of MPC is set as:

$$Q = diag(1, 1, 0) \tag{14}$$

4.1.4. Emergency State Zone

When the ACE signal is in the emergency state interval, the system frequency deviation is too large, and the BESS and the thermal power unit cannot meet the system frequency control demand. In this state, the BESS should be out of operation, and the power grid system should be used to cut the machine and dump the load, maintaining the safe and stable operation of the whole system and avoiding the rapid deterioration of the power grid frequency.

4.2. Self-Recovery Strategy of Battery Energy Storage System

When a BESS absorbs or releases only electrical energy for a long time, its SOC deteriorates gradually. Therefore, if the system is in the normal frequency control area, the BESS self-recovery strategy can use the residual capacity of the traditional unit's frequency control. Under the premise that the self-recovery operation of energy storage will not cause the system frequency deviation to exceed the normal regulation area, the BESS can properly reverse the output and improve the SOC. Therefore, the BESS can be fully powered in the sub-emergency regulation area. The self-recovery output limit of energy storage is shown in Figure 3.

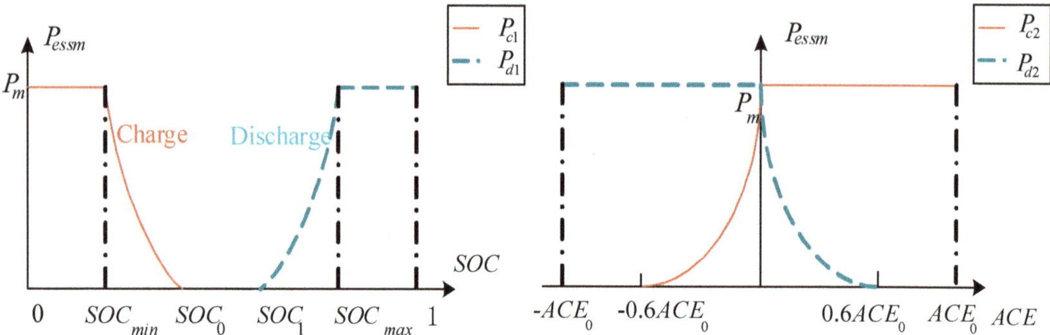

Figure 3. Diagram of output restrictions in self-recovery of energy storage.

The self-recovery power demand of energy storage and constraint curve of the power grid's frequency control system in Figure 3 obtain the control law of self-recovery action for the BESS, as shown in Equations (15) and (16).

$$
\begin{cases}
P_{c1} = \begin{cases} P_{\mathrm{m}} & SOC \leq SOC_{\min} \\ P_{\mathrm{m}} \cdot \left(20 \cdot x^2 - 17x + 3.6\right) & SOC_{\min} < SOC \leq SOC_0 \\ 0 & SOC \geq SOC_0 \end{cases} \\[2mm]
P_{d1} = \begin{cases} 0 & SOC \leq SOC_1 \\ P_{\mathrm{m}} \cdot \left(20 \cdot x^2 - 23x + 6.6\right) & SOC_1 < SOC \leq SOC_{\max} \\ P_{\mathrm{m}} & SOC > SOC_{\max} \end{cases}
\end{cases}
\tag{15}
$$

$$
\begin{cases}
P_{c2} = \begin{cases} 0 & ACE \leq -0.7ACE_0 \\ P_{\mathrm{m}} \cdot \left(2.22 \cdot y^2 + 3y + 1\right) & -0.7ACE_0 < ACE \leq 0 \\ P_{\mathrm{m}} & ACE > 0 \end{cases} \\[2mm]
P_{d2} = \begin{cases} P_{\mathrm{m}} & ACE \leq 0 \\ P_{\mathrm{m}} \cdot \left(2.22 \cdot y^2 - 3y + 1\right) & 0 < ACE \leq 0.7ACE_0 \\ 0 & ACE > 0.7ACE_0 \end{cases}
\end{cases}
\tag{16}
$$

where P_{c1}, P_{d1} are both (15) the self-recovery power of energy storage; P_{c2}, P_{d2} are both (16) the battery self-recovery output limit in the normal adjustment interval of ACE; P_{m} is the rated power of energy storage; ACE_0 is the normal adjustment range.

Equation (15) determines the self-recovery demand power of the BESS according to SOC. Equation (16) limits the self-recovery power of the BESS according to the power grid's ACE state. Considering the limitations of Equations (15) and (16), the minimum values of the two are selected as the actual self-recovery power of the BESS in every moment, the details are shown in Equation (17).

$$P_{ess} = \begin{cases} -\min\{|P_{c1}|, |P_{c2}|\}, SOC < SOC_0 \\ \min\{|P_{d1}|, |P_{d2}|\}, SOC > SOC_1 \end{cases} \tag{17}$$

where P_{ess} is the actual self-recovery power of BESS; P_{c1}, P_{d1} are both determined by Equation (15), and P_{c2}, P_{d2} are both determined by Equation (16).

According to Equation (17), the self-recovery diagram of energy storage can be obtained, as shown in Figure 4.

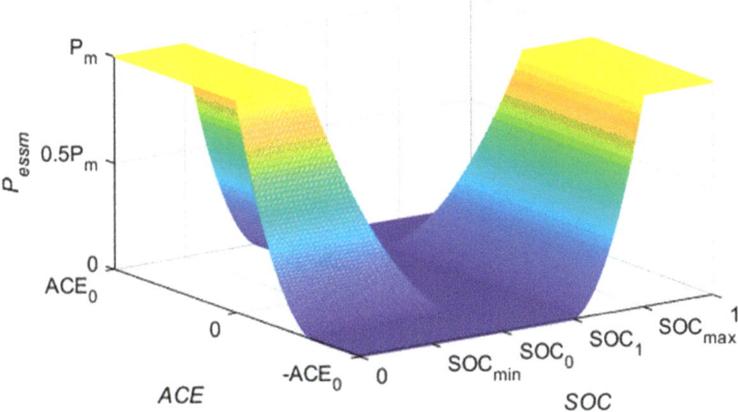

Figure 4. Energy storage battery recovery chart.

In Figure 4, when $SOC_0 < SOC < SOC_1$, the SOC state is better and does not require self-recovery. When $SOC_0 > SOC > SOC_{min}$ or $SOC_1 < SOC < SOC_{max}$, the SOC state is poor and needs self-recovery. The self-recovery power is determined by considering SOC and ACE signals comprehensively. When $SOC < SOC_{min}$ or $SOC > SOC_{max}$, the SOC state is worst and stay in a state of emergency. BESS will stop participating in frequency control to enter in self-recover at the maximum power.

By designing the self-recovery strategy of SOC, the BESS and the power grid system can sense each other's state in real-time. When the SOC deteriorates and the power grid is in the normal frequency control area, the energy storage self-recovery action can be conducted in real-time. The self-recovery power can be changed according to the ACE size and SOC so that the SOC of the BESS can maintain its optimal state, providing enough frequency control capacity for the later frequency control task.

4.3. Decomposition of ACE Signal

In the sub-emergency adjustment interval of ACE, high and low frequencies should be divided. The EMD method is used to decompose the ACE signal in real-time. Thus, the IMF1-IMF9 of each frequency band component can be obtained. The lower order modal component has a higher frequency, whereas the higher-order modal component has a lower frequency, as shown in Figure 5. Therefore, the first K order IMF components are assigned to the BESS, and the traditional unit bears the rest. The SOC and ACE size determines the value of K. When the system's ACE signal is positive, and the SOC is high (>0.65), or when the ACE signal is negative, and the SOC is low (SOC is <0.35), appropriately reducing the high-frequency component borne by the BESS is conducive to maintaining the SOC and reducing the deterioration speed. Similarly, when the ACE signal is positive, and the SOC is low, or when the ACE signal is negative, and the SOC is high, appropriately increasing the high-frequency component undertaken by the BESS improves the Frequency regulation effect and accelerates the SOC recovery to the normal interval. The value of K is shown in Equations (18) and (19).

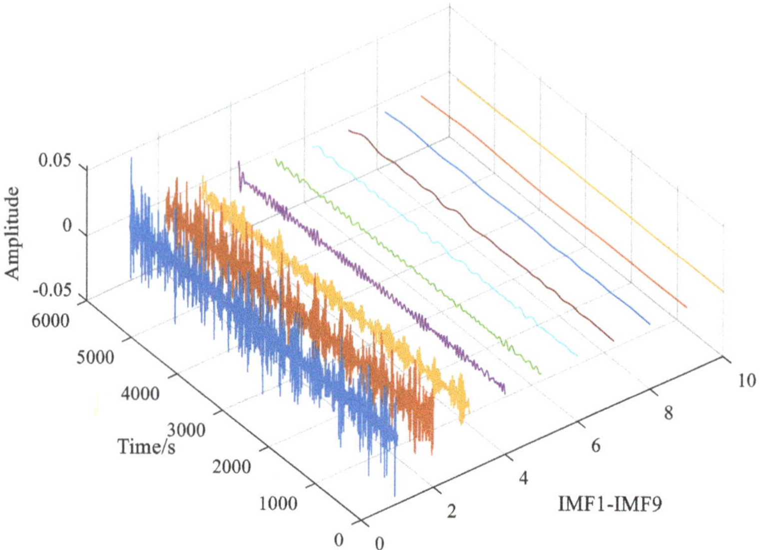

Figure 5. ACE signal decomposition curve based on EMD.

When the ACE signal is positive, the value of K is shown in Equation (18):

$$k = \begin{cases} 4 & 0 \leq SOC < SOC_{low} \\ 2 & SOC_{low} \leq SOC \leq SOC_{high} \\ 1 & SOC_{high} < SOC \leq 1 \end{cases} \tag{18}$$

When the ACE signal is negative, the value of K is shown in Equation (19):

$$k = \begin{cases} 1 & 0 \leq SOC < SOC_{low} \\ 2 & SOC_{low} \leq SOC \leq SOC_{high} \\ 4 & SOC_{high} < SOC \leq 1 \end{cases} \tag{19}$$

The ACE signal's high-frequency component has zero mean and small amplitude characteristics, etc. BESS assumes the ACE signal's high-frequency part with the advantage of rapid response and helps maintain the stability of internal SOC. It reduces the frequency control pressure of BESS and its frequency control capacity configuration requirements in frequency control power plants. The thermal power unit's assumption of the ACE signal's low-frequency component reduces the wear caused by frequent operation of the unit and improves the unit operation's economy. The integrated control policy flow is shown in Figure 6.

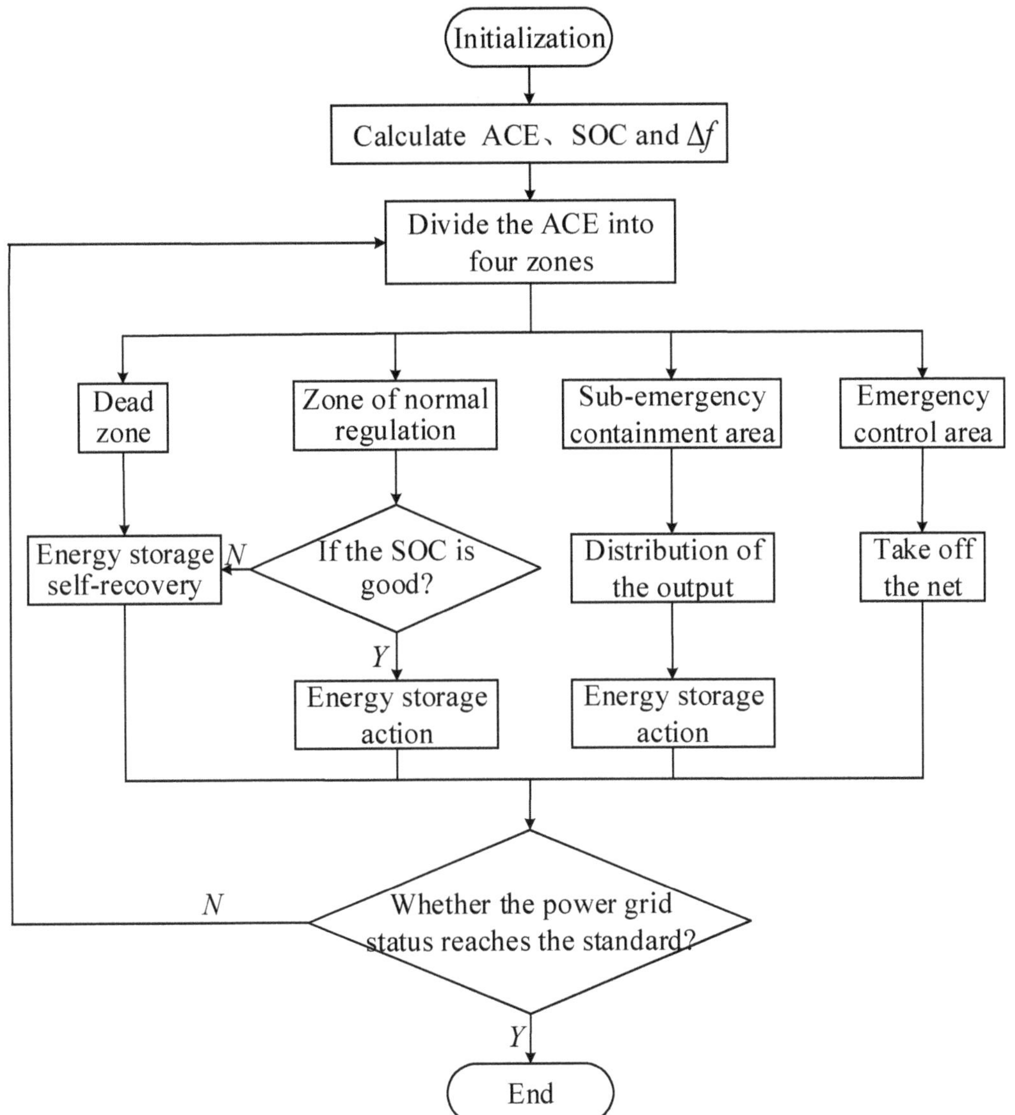

Figure 6. Flowchart of the control strategy.

5. The Simulation Verification

5.1. The Simulation Parameters

We adopted the control method of Tie-Line Bias Control for this paper. The frequency response model of two area interconnection systems with energy storage batteries was established by MATLAB/Simulink. Then, we compared the proposed method, the scheme not considering energy storage (Method 1), and the PI control method considering energy storage SOC self-recovery (Method 2) to verify the effectiveness of these strategies. Relevant parameters are shown in Tables 1 and 2.

Table 1. Parameters of simulation system model.

Parameters	Value	Parameters	Value
Power reference/MW	300	T_{CH_1}, T_{CH_2}	0.2, 0.3
Frequency reference/Hz	50	T_{RH_1}, T_{RH_2}	10, 8
Energy storage capacity/MW·h	4.5	F_{HP_1}, F_{HP_2}	0.25, 0.37
Energy storage power/MW	9	T_b	0.05
T_{g1}, T_{g2}	0.1, 0.08	B_1, B_2	35, 21
R_1, R_2	0.03, 0.05	D_1, D_2	2.75, 2
K_1, K_2	0.5, 0.5	M_1, M_2	10, 12

Table 2. Parameters of control rule.

Parameters	Value	Parameters	Value	Parameters	Value
SOC_{min}	0.2	SOC_{low}	0.35	SOC_0	0.4
SOC_{max}	0.8	SOC_{high}	0.65	SOC_1	0.6
$ACE_{ce,d}$	1×10^{-7}	$ACE_{ce,n}$	0.03	$ACE_{ce,e}$	0.05

5.2. Analysis of Simulation Results

The evaluation index of the secondary frequency control is defined as follows: Δf_m, Δf_{rms} is the maximum value and root mean square value of frequency deviation, ΔP_{tie_m}, ΔP_{tie_rms} is the maximum value and root mean square value of the tie-line power deviation; ACE_m and ACE_{rms} are the maximum value and root mean square value of regional control deviation; W_{gen} and W_{ess} are the amount of electricity contributed by thermal power units and energy storage batteries. S_{oc_rms} is the root mean square value of SOC.

5.2.1. Scene under Step Load Disturbance

At 0.03 s, a step disturbance of 0.03 pu was applied to area 2. The simulation time was set to 100 s, and the range of the prediction model was 0.1 s. Figure 7 compares the system response curves of the proposed strategy, Method 1, and Method 2 in area 2.

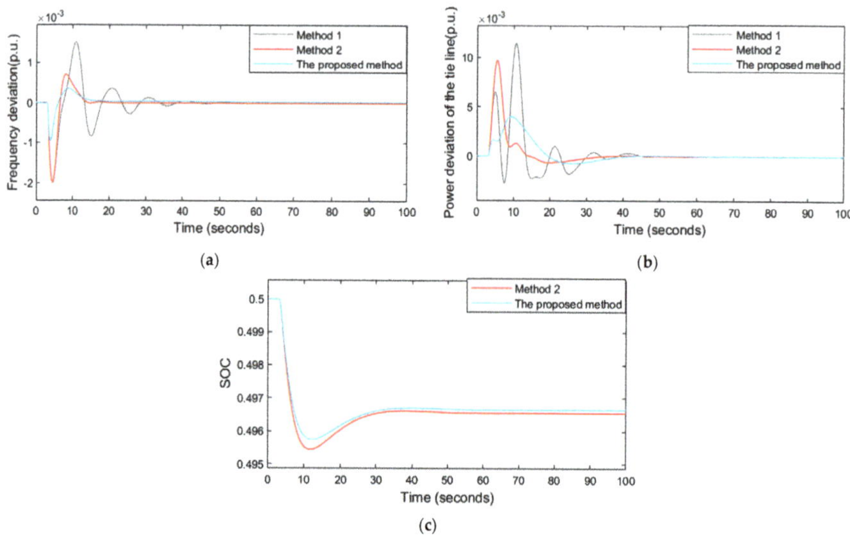

Figure 7. System response with step disturbance. (**a**) Description of the frequency deviation response curve. (**b**) Description of the tie-line power bias. (**c**) Description of change in SOC.

According to the experimental analysis results in Figure 6 and Table 3, we observed that the following.

(1) In Figure 7a,b, and Table 3, we compare the proposed method with Method 1(without BESS) and Method 2 (PI). From the results, we obtained the maximum values and root mean square values of system frequency deviation, the maximum values and root mean square values of tie-line power deviation, and the adjustment time, all of which decreased with our proposed method and Method 2. The overshoot of the proposed scheme is the smallest, indicating that the dynamic stability is superior to the other two methods.

(2) The system's response time is significantly faster in the MPC method than in Method 2, which fully uses the rapid response characteristics of the BESS. Figure 7c shows that SOC has a better maintenance effect.

Table 3. Frequency regulation evaluation index under step disturbance.

Evaluation Index	The Proposed Method	Method 1	Method 2
Δf_m/p.u.	9.560×10^{-4}	1.998×10^{-3}	1.690×10^{-3}
Δf_{rms}/p.u.	2.084×10^{-4}	5.722×10^{-4}	4.365×10^{-4}
ΔP_{tie_m}/p.u.	7.298×10^{-4}	2.776×10^{-2}	9.718×10^{-3}
ΔP_{tie_rms}/p.u.	1.016×10^{-3}	1.995×10^{-3}	1.483×10^{-3}
S_{oc_rms}	9.560×10^{-4}	1.998×10^{-3}	1.690×10^{-3}

5.2.2. Scene under Continuous Load Disturbance

Typical continuous load disturbance conditions of 100 min in a certain area are selected, as shown in Figure 8a; the simulation time is set to 6000 s, and the range of the prediction model is 0.1 s. Figure 8 compares the system response curves of the proposed method, Method 1, and Method 2 in area 2.

From the simulation analysis results in Figure 8 and Table 4, we observed that:

(1) The maximum value of frequency deviation and the root mean square value of frequency deviation for the proposed method are the minimum values compared to the other two methods, and the frequency control effect is significantly improved. In addition, the frequency decline rate of the proposed method is minor, and the recovery speed is faster.

(2) Figure 8b shows the final ACE signal allocation. The high-frequency component is assigned to the BESS and the traditional unit bears the low-frequency component of the ACE. According to Figure 8d,e and Table 4, the operation frequency of the unit is reduced and the operation loss is reduced. Moreover, the unit can contribute more energy, which is beneficial to frequency control. At the same time, the output of BESS is reduced, which reduces the frequency control capacity configuration requirements of BESS.

(3) In Figure 8f, the SOC fluctuation range of our strategy is between 0.3 and 0.7, the state-keeping effect is better, and the burden of BESS is reduced. In addition, it can be seen that when SOC is lower than 0.3, the SOC falling speed is significantly reduced and the rising speed is increased. It is proven that changing the component of high frequency for energy storage in real time is beneficial to maintain SOC state. The ACE signal's high-frequency component has the characteristics of zero mean value and small amplitude. Assuming the ACE signal's high-frequency component by the BESS has its advantages of fast response speed and maintaining SOC.

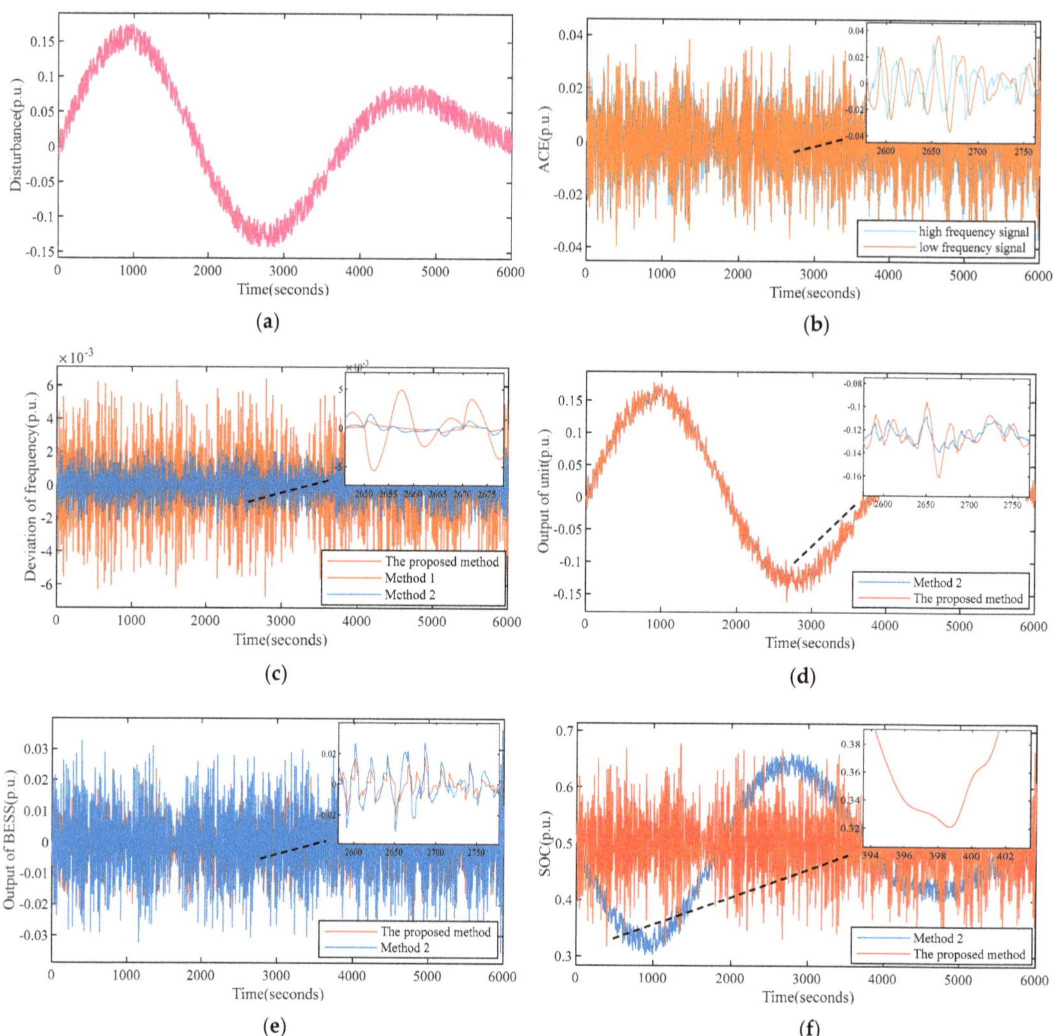

Figure 8. System response applying with continuous perturbation applied. (**a**) Continuous loading disturbance. (**b**) Low frequency signal and high frequency signal. (**c**) Deviation of frequency. (**d**) Output of unit. (**e**) Output of BESS. (**f**) The change of SOC.

Table 4. Frequency regulation evaluation index under continuous disturbance.

Evaluation Index	The Proposed Method	Method 1	Method 2
Δf_m/p.u.	1.100×10^{-3}	6.463×10^{-3}	2.316×10^{-3}
Δf_{rms}/p.u.	2.763×10^{-4}	1.933×10^{-3}	6.015×10^{-4}
W_{gen}	0.242	0.274	0.265
W_{ess}	0.021	–	0.0189
$S_{\mathrm{oc\_rms}}$	0.019	–	0.038

6. Conclusions

In this paper, EMD was used to decompose the ACE signal into high- and low-frequency signals. Considering the ACE signal and SOC, the MPC output weighting matrix and constraints were determined to change the high-frequency component of BESS in real-time and perform SOC self-recovery. The simulation results showed that the strategy fully used the advantages of BESSs and traditional thermal power units, reduced the frequent operation of thermal power units, and dynamically adjusted the distribution of frequency control demand signals. It improved the frequency control effect, maintained a good SOC, and simultaneously reduced the capacity configuration demand for BESS. In actual engineering applications, the performance index of frequency control, the benefit of the frequency control effect, and the interaction between them will be considered. Our future research work will include a more comprehensive study of this content.

Author Contributions: Conceptualization, X.L. and X.W.; methodology, X.L.; software, X.L.; validation, X.L., X.W. and W.N.; formal analysis, X.L. and F.W.; investigation, W.N.; resources, X.W.; data curation, X.L.; writing—original draft preparation, X.L.; writing—review and editing, X.L. and X.W.; visualization, W.N.; supervision, X.L.; project administration, F.W.; funding acquisition, X.W. All authors have read and agreed to the published version of the manuscript.

Funding: This research was funded by The Youth Science and Technology Fund of Gansu Province under grant 17JR5RA346, Science and Technology Project of State Grid Gansu Electric Power Company under grant SGGSJYOOPSJS1800.

Institutional Review Board Statement: Not applicable.

Informed Consent Statement: Not applicable.

Data Availability Statement: Not applicable.

Conflicts of Interest: The authors declare no conflict of interest.

References

1. Liu, Y.; Chen, L.; Han, X. Analysis on key issues of new energy substitution in China under energy transformation. *J. Proc. CSEE* **2022**, *42*, 515–524.
2. Zhang, J.; Li, Q.; Wang, X. Power supply planning method for large-scale new energy grid-connected system. *J. Proc. CSEE* **2020**, *40*, 3114–3124.
3. Li, J.; Feng, X.; Yan, G. Review on research of power system under high wind power permeability. *J. Power Syst. Prot. Control* **2018**, *46*, 163–170.
4. Yao, L.; Yang, B.; Cui, H. Challenges and progresses of energy storage technology and its application in power systems. *J. Mod. Power Syst. Clean Energy* **2016**, *4*, 519–528. [CrossRef]
5. Li, X.; Huang, J.; Yang, Y. Review on frequency control of large-scale energy storage power supply. *J. Power Syst. Prot. Control* **2016**, *44*, 145–153.
6. Sun, B.; Yang, S.; Liu, Z. Analysis and enlightenment of current status of frequency control demonstration and application of megawatt energy storage at home and abroad. *J. Autom. Electr. Power Syst.* **2017**, *41*, 8–16.
7. Xie, X.; Guo, Y.; Wang, B. Improving AGC performance of coal-fueled thermal generators using multi-mw scale BESS: A Practical Application. *J. IEEE Trans. Smart Grid* **2018**, *9*, 1769–1777. [CrossRef]
8. Cheng, Y.; Tabrizi, M.; Sahni, M. Dynamic available AGC based approach for enhancing utility scale energy storage performance. *IEEE Trans Smart Grid.* **2014**, *5*, 1070–1078. [CrossRef]
9. Ma, Z.; Li, X.; Tan, Z. The primary frequency regulation control method of energy storage frequency modulation dead zone is considered. *J. Trans. China Electrotech. Soc.* **2019**, *34*, 2102–2115.
10. Meng, G.; Chang, Q.; Sun, Y. Energy storage auxiliary frequency modulation control Strategy Considering ACE and SOC of Energy Storage. *J. IEEE Access* **2021**, *9*, 26271–26277.
11. Cui, H.; Yang, B.; Jiang, Y. Based on fuzzy control and SOC self-recovery energy storage participating in secondary frequency control control strategy. *J. Power Syst. Prot. Control* **2019**, *47*, 89–97.
12. Zhao, T.; Parisio, A.; Milanovicj, V. Distributed control of battery energy storage systems for improved frequency control. *J. IEEE Trans. Power Syst.* **2020**, *35*, 3729–3738. [CrossRef]
13. Li, R.; Li, X.; Tan, Z. Integrated control strategy considering energy storage battery participating in secondary frequency control. *J. Autom. Electr. Power Syst.* **2018**, *42*, 74–82.
14. Liu, Q.; He, S.; Lu, W. Model predictive control method for secondary frequency control assisted by battery energy storage. *J. Electr. Meas. Instrum.* **2020**, *57*, 119–125.

15. Yan, G.; Liu, Y.; Duan, S. Battery energy storage unit group participates in power distribution strategy of secondary frequency control in power system. *J. Autom. Electr. Power Syst.* **2020**, *44*, 26–34.
16. Lv, L.; Chen, S.; Zhang, X. Secondary frequency modulation control strategy for power system considering SOC consistency of large-scale battery storage. *J. Therm. Power Gener.* **2021**, *50*, 108–117.
17. Le, J.; Liao, X.; Zhang, Y. Review and prospect of distributed model predictive control methods for power system. *J. Autom. Electr. Power Syst.* **2020**, *44*, 179–191.

MDPI

St. Alban-Anlage 66

4052 Basel

Switzerland

www.mdpi.com

Energies Editorial Office

E-mail: energies@mdpi.com

www.mdpi.com/journal/energies